T0191605

Communications
in Computer and Information Science   1042

Commenced Publication in 2007
Founding and Former Series Editors:
Phoebe Chen, Alfredo Cuzzocrea, Xiaoyong Du, Orhun Kara, Ting Liu,
Krishna M. Sivalingam, Dominik Ślęzak, Takashi Washio, Xiaokang Yang,
and Junsong Yuan

More information about this series at http://www.springer.com/series/7899

Yuqing Sun · Tun Lu · Zhengtao Yu ·
Hongfei Fan · Liping Gao (Eds.)

Computer Supported Cooperative Work and Social Computing

14th CCF Conference, ChineseCSCW 2019
Kunming, China, August 16–18, 2019
Revised Selected Papers

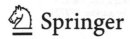

Springer

Editors
Yuqing Sun
Shandong University
Jinan, China

Zhengtao Yu
Kunming University
of Science and Technology
Kunming, China

Liping Gao
University of Shanghai
for Science and Technology
Shanghai, China

Tun Lu
Fudan University
Shanghai, China

Hongfei Fan
Tongji University
Shanghai, China

ISSN 1865-0929   ISSN 1865-0937 (electronic)
Communications in Computer and Information Science
ISBN 978-981-15-1376-3   ISBN 978-981-15-1377-0 (eBook)
https://doi.org/10.1007/978-981-15-1377-0

This Springer imprint is published by the registered company Springer Nature Singapore Pte Ltd.
The registered company address is: 152 Beach Road, #21-01/04 Gateway East, Singapore 189721, Singapore

Preface

Welcome to ChineseCSCW 2019, the 14th CCF Conference on Computer Supported Cooperative Work and Social Computing. ChineseCSCW 2019 was sponsored by the China Computer Federation (CCF), and co-organized by the Technical Committee on Cooperative Computing (TCCC) of CCF and the Kunming University of Science and Technology. The theme of the conference was "Crowd Intelligence Cooperation and Social Computing," which reflected the emerging trend of the combination of AI and human-centered computing.

ChineseCSCW (initially named CCSCW) is a highly reputable conference series on computer supported cooperative work (CSCW) and social computing in China with a long history. It aims at bridging Chinese and overseas CSCW researchers, practitioners, and educators, with a particular focus on innovative models, theories, techniques, algorithms, and methods, as well as domain-specific applications and systems, from both technical and social aspects in CSCW and social computing. The conference was initially held biennially since 1998, and has been held annually since 2014.

This year, the conference received 169 submissions, and after a rigorous double-blind peer review process, only 52 of them were eventually accepted as full papers to be orally presented, resulting in an acceptance rate of 31%. The program also included 10 short papers, which were presented as posters. In addition, the conference featured 4 keynote speeches, 8 technical seminars, and the second special panel discussion on "Social Computing and AI for Legal Affairs." We are grateful to the distinguished keynote speakers, Prof. Jianzhong Li from Harbin Institute of Technology, Prof. Jonathan Jianhua Zhu (ICA Fellow) from City University of Hong Kong, Prof. Weiming Shen (IEEE Fellow, CAE Fellow, and EIC Fellow) from National Research Council Canada, Prof. Huadong Ma from Beijing University of Posts and Telecommunications, Prof. Guoliang Li from Tsinghua University, and Prof. Haiyi Zhu from Carnegie Mellon University.

We hope that you enjoyed ChineseCSCW 2019.

August 2019
Ning Gu
Hongchun Shu

Organization

ChineseCSCW 2019 was organized by the China Computer Federation (CCF) and the Technical Committee of Cooperative Computing of CCF, and locally hosted by Kunming University of Science and Technology during August 16–18, 2019.

Steering Committee

Ning Gu	Fudan University, China
Bin Hu	Lanzhou University, China
Xiaoping Liu	Hefei University of Technology, China
Yong Tang	South China Normal University, China
Weiqing Tang	Chinese Academy of Sciences, China
Shaozi Li	Xiamen University, China
Yuqing Sun	Shandong University, China
Xiangwei Zheng	Shandong Normal University, China

General Chairs

Ning Gu	Fudan University, China
Hongchun Shu	Kunming University of Science and Technology, China

Program Committee Chairs

Yuqing Sun	Shandong University, China
Zhengtao Yu	Kunming University of Science and Technology, China
Tun Lu	Fudan University, China

Organization Committee Chairs

Xiaoping Liu	Hefei University of Technology, China
Xiangwei Zheng	Shandong Normal University, China
Cunli Mao	Kunming University of Science and Technology, China
Dongning Liu	Guangdong University of Technology, China

Publicity Chairs

Yong Tang	South China Normal University, China
Hongbin Wang	Kunming University of Science and Technology, China

Publication Chairs

Bin Hu	Lanzhou University, China
Hailong Sun	Beijing University of Aeronautics and Astronautics, China

Finance Chair

Shengxiang Gao	Kunming University of Science and Technology, China

Paper Award Chairs

Shaozi Li	Xiamen University, China
Yichuan Jiang	Southeast University, China

Program Committee

Tie Bao	Jilin University, China
Hongming Cai	Shanghai Jiao Tong University, China
Zhicheng Cai	Nanjing University of Science and Technology, China
Buqing Cao	Hunan University of Science and Technology, China
Donglin Cao	Xiamen University, China
Jian Cao	Shanghai Jiao Tong University, China
Chao Chen	Chongqing University, China
Jianhui Chen	Beijing University of Technology, China
Liangyin Chen	Sichuan University, China
Longbiao Chen	Xiamen University, China
Ningjiang Chen	Guangxi University, China
Qingkui Chen	University of Shanghai for Science and Technology, China
Qingzhang Chen	Zhejiang University of Technology, China
Yang Chen	Fudan University, China
Weineng Chen	South China University of Technology, China
Shiwei Cheng	Zhejiang University of Technology, China
Xiaohui Cheng	Guilin university of technology, China
Yuan Cheng	Wuhan University, China
Lizhen Cui	Shandong University, China
Weihui Dai	Fudan University, China
Xianghua Ding	Fudan University, China
Wanchun Dou	Nanjing University, China
Bowen Du	University of Warwick, UK
Hongfei Fan	Tongji University, China
Shanshan Feng	Shandong Normal University, China
Liping Gao	University of Shanghai for Science and Technology, China
Qiang Gao	Beihang University, China

Ning Gu	Fudan University, China
Kun Guo	Fuzhou University, China
Wei Guo	Shandong University, China
Yinzhang Guo	Taiyuan University of Science and Technology, China
Fazhi He	Wuhan University, China
Haiwu He	Chinese Academy of Sciences, China
Bin Hu	Lanzhou University, China
Xiping Hu	Lanzhou University, China
Wenting Hu	Jiangsu Open University, China
Yanmei Hu	Chengdu University of Technology, China
Changqin Huang	South China Normal University, China
Huan Huo	University of Shanghai for Science and Technology, China
Bo Jiang	Zhejiang Gongshang University, China
Bin Jiang	Hunan University, China
Weijin Jiang	Hunan University, China
Yichuan Jiang	Southeast University, China
Lanju Kong	Shandong University, China
Yi Lai	Xi'an University of Posts and Telecommunications, China
Fangpeng Lan	Taiyuan University of Technology, China
Dongsheng Li	IBM Research, China
Feng Li	Jiangsu University, China
Li Li	Southwest University, China
Jianguo Li	South China Normal University, China
Renfa Li	Hunan University, China
Shaozi Li	Xiamen University, China
Shiying Li	Hunan University, China
Taoshen Li	Guangxi University, China
Xiaoping Li	Southeast University, China
Yuanxi Li	Beijing Normal University, China
Lu Liang	Guangdong University of Technology, China
Bing Lin	Fujian Normal University, China
Hong Liu	Shandong Normal University, China
Dongning Liu	Guangdong University of Technology, China
Shijun Liu	Shandong University, China
Xiaoping Liu	Hefei University of Technology, China
Li Liu	Chongqing University, China
Tun Lu	Fudan University, China
DianJie Lu	Shandong Normal University, China
Huijuan Lu	China Jiliang University, China
Qiang Lu	Hefei University of Technology, China
Chen Lv	Shandong Normal University, China
Li Pan	Shandong University, China
Hui Ma	University of Electronic Science and Technology of China and Zhongshan Institute, China

KeJi Mao	Zhejiang University of Technology, China
Chunyu Miao	Zhejiang Normal University, China
Haiwei Pan	Harbin Engineering University, China
Youtian Qu	Zhejiang University of Media and Communications, China
Huawei Shen	Institute of Computing Technology, Chinese Academy of Sciences, China
Limin Shen	Yanshan University, China
Dejia Shi	Hunan University of Technology and Business, China
Yuliang Shi	ShanDa Dareway Company Limited, China
Xiaoxia Song	Datong University, China
Kehua Su	Wuhan University, China
Hailong Sun	Beihang University, China
Ruizhi Sun	China Agricultural University, China
Yuqing Sun	Shandong University, China
Yuling Sun	East China Normal University, China
Lina Tan	Hunan University of Technology and Business, China
Wen'an Tan	Nanjing University of Aeronautics and Astronautics, China
Shan Tang	Shanghai Polytechnic University, China
Yong Tang	South China Normal University, China
Yan Tang	Hohai University, China
Weiqing Tang	Beijing Zhongke Fulong Computer Technology Company Ltd., China
Yiming Tang	HeFei University of Technology, China
Yizheng Tao	China Academy of Engineering Physics, China
Shaohua Teng	Guangdong University of Technology, China
Dakuo Wang	IBM Research, USA
Lei Wang	Dalian University of Technology, China
Li Wang	Taiyuan University of Technology, China
Tong Wang	Harbin Engineering University, China
Jiangtao Wang	Peking University, China
Tianbo Wang	Beijing University of Aeronautics and Astronautics, China
Hongbo Wang	University of Science and Technology Beijing, China
Wanyuan Wang	Southeast University, China
Yijie Wang	National University of Defense Technology, China
Xiaodong Wang	National University of Defense Technology, China
Zhiwen Wang	Guangxi University of Science and Technology, China
Wei Wei	Xi'an University of Technology, China
Yiping Wen	Hunan University of Science and Technology, China
Jiyi Wu	Hangzhou Normal University, China
Chunhe Xia	Beihang University, China
Yong Xiang	Tsinghua University, China
Fangxion Xiao	Jinling Institute of Technology, China
Yu Xin	Harbin University of Science and Technology, China

Meng Xu	Shandong Technology and Business University, China
Xiaolan Xie	Guilin University of Technology, China
Zhiqiang Xie	Harbin University of Science and Technology, China
Heyang Xu	Henan University of Technology, China
Jianbo Xu	Hunan University of Science and Technology, China
Jiuyun Xu	China University of Petroleum, China
Bo Yang	University of Electronic Science and Technology of China, China
Chao Yang	Hunan University, China
Gang Yang	Northwestern Polytechnical University, China
Jing Yang	Lanzhou University, China
Xiaochun Yang	Northeastern University, China
Xiaojun Zhu	Taiyuan University of Technology, China
Dingyu Yang	Shanghai DianJi University, China
Lin Yang	Shanghai Computer Software Technology Development Center, China
Yang Yu	Sun Yat-sen University, China
Lei Yu	Information Engineering University, China
Xianchuan Yu	Beijing Normal University, China
Zhengtao Yu	Kunming University of Science and Technology, China
Zhiwen Yu	Northwestern Polytechnical University, China
An Zeng	Guangdong University of Technology, China
Changyou Zhang	Chinese Academy of Sciences, China
Dajun Zeng	Institute of Automation, Chinese Academy of Sciences, China
Guijuan Zhang	Shandong Normal University, China
Guangquan Zhang	Soochow University, China
Liang Zhang	Fudan University, China
Jifu Zhang	Taiyuan University of Science and Technology, China
Shaohua Zhang	Shanghai Software Technology Development Center, China
Wei Zhang	Guangdong University of Technology, China
Zili Zhang	Southwest University, China
Zhiqiang Zhang	Harbin Engineering University, China
Qiang Zhao	China Academy of Engineering Physics, China
Junlan Zhao	Inner Mongolia University of Finance and Economics, China
Yifeng Zhou	Southeast University, China
Xianjun Zhu	Jinling Institute of Technology, China
Xiangwei Zheng	Shandong Normal University, China
Ning Zhong	Beijing University of Technology, China
Qiaohong Zu	Wuhan University of Technology, China

Contents

Social Computing (Online Communities, Crowdsourcing, Recommendation, Sentiment Analysis, etc.)

Collaborative Models, Approaches, Algorithms, and Systems

Higher-Order Network Structure Embedding in Supply Chain Partner Link Prediction

Miao Xie[1], Tengjiang Wang[2], Qianyu Jiang[1], Li Pan[1], and Shijun Liu[1($\boxtimes$)]

[1] Shandong University, Jinan 250101, China
lsj@sdu.edu.cn
[2] Inspur General Software Co., Ltd., Jinan 250101, China

Abstract. Enterprise partner link prediction is a research direction of the recommendation system, which is used to predict the possibility of links between nodes in the enterprise network, and recommend potential high-quality partners for enterprises. This paper is based on the automobile enterprise network, and study how to recommend high-quality parts suppliers for auto manufacturers, then propose a supply chain corporate partner link prediction algorithm embedding in higher-order network structure. The coupled rating matrix and triad tensor model is constructed by mining the higher-order link patterns in the enterprise network, and considering the interaction between user demand and automobile manufacturers, which explicitly reflects the auto manufacturer's choice of its part suppliers. The model uses the Alternating Direction Multiplier Method (ADMM) to solve the problem, which effectively alleviates the data sparsity problem in the recommendation system. On real data crawled from automobile-related websites, experiments show that the algorithm can obtain more accurate link prediction effects than traditional algorithms.

Keywords: Link prediction · Tensor decomposition · Higher-order pattern · ADMM

1 Introduction

In recent years, with the development of network science, the research of link prediction become more and more closely related to the structure and evolution of networks [1]. This paper takes the automobile production industry as an example of enterprise network, and predicts the possibility of cooperation between enterprises, so as to recommend high quality suppliers for automobile manufacturers.

Enterprises play the role of producers in the network, providing automotive products for consumers, which can be subdivided into manufacturers role and suppliers role according to their different positions in the supply chain. There

© Springer Nature Singapore Pte Ltd. 2019
Y. Sun et al. (Eds.): ChineseCSCW 2019, CCIS 1042, pp. 3–17, 2019.
https://doi.org/10.1007/978-981-15-1377-0_1

are supply chain scenarios in the network: suppliers produce parts and supply them to automobile manufacturers, manufacturers produce cars to users, finally users purchase cars to consume products, which is the process of supply chain formation. Conversely, most users will rate cars after consumption. These evaluations reflect users' different needs for cars, which are fed back to manufacturers and then to suppliers. This process is in the form of demand chain (Fig. 1).

Fig. 1. Supply and demand chain in automobile enterprise network

Enterprise relationship network is developing dynamically. It will change with the establishment or breakdown of cooperation among enterprises. New links and new enterprise nodes often appear. At the same time, some old enterprise nodes and links will disappear. Traditional enterprise partner link prediction algorithms mainly depend on node attributes and network topology, usually only consider the low-order network structure in the network. However, in the case of large-scale network and sparse links, these algorithms are facing problems such as data imbalance, and it is difficult to get good prediction results. Binary structure can only show the supply relationship between two nodes. The higher-order pattern is a complex structure composed of more nodes in the network, such as Triad, Quad, Ego Network [2].

Higher-order connection patterns reveal higher-level interactions, reflecting nodes or regions with specific attributes in network structures, and are essential for understanding the structure and behavior of many complex networks [3]. We hope to use these higher-order network structures to study the interaction between enterprises and make more accurate link prediction. In fact, enterprises tend to get more supply links to gain stronger business competitiveness, thus forming a transitive ternary relationship [4].

As a basic higher-order connection pattern, the study of triads has been practiced in many fields [5]. Sociologists first use triads to study human social relationship tendencies and find that friends of friends often become friends [6]. In the enterprise network, triads contains the essential attributes and development trend of the network, which is of great significance to the study of the evolution of the network structure.

2 Related Work

The network connection pattern includes low-order binary link relation and complex higher-order connection pattern. Compared with the low-order connection structure obtained at the level of single node or edge, there is relatively little research work on the higher-order organization of network. Higher-order connection structure in enterprise network includes Triad, Quad, Ego network, structure hole, Community, etc. [7].

Triads are the core structure of higher-order networks. The researchers compared various factors that affect the evolution of the network, and finally came to the conclusion that triadic closure can be the basic principle of the formation and evolution of social networks [8]. In recent years, Tang's team at tsinghua university has systematically studied the closure of triads, including the mining of triads, the formation of triads, and the impact of closed triads on the strength of social network relationships [9].

Link prediction is an important research direction in relational networks, which can show the evolution process of complex networks. Early link prediction studies were based on markov chains. Zhu et al. [10] proposed that for large-scale data sets, a high-level model of Markov chain, could be used for prediction. Popescul and Ungar [11] established a regression model to study the citation relationship between literatures by using the information of literature nodes themselves, such as authors, published journals and literature contents. Lin [12] defined the similarity between nodes and directly conducted link prediction based on node attributes, achieving a better prediction effect.

In recent years, the prediction model is constructed by combining the higher-order network structure in the link prediction. Shi [13] studied the evolution process from triads to quads in enterprise networks, proposed algorithms for mining triads and quads in networks, and added quads as features to matrix decomposition for link prediction. Lou [14] proposed to use graph model to mine network triads for studying reciprocal relationships in networks. Huang [15] studied closed triads in dynamic social networks and proposed a probability factor model for link prediction.

3 Enterprise Network Structure

Automobile enterprise network includes many dimensions, and each dimension plays a unique role in the supply chain. In order to predict the supply relationship among enterprises, we statistic the different types of triad among enterprises, and further analyze the impact of triad connection structure on the development of the entire enterprise network structure.

3.1 Supply Chain in Enterprise Network

Figure 2 shows an example of how automotive products can be produced and sold in an automotive enterprise network. Customer John purchased a car named

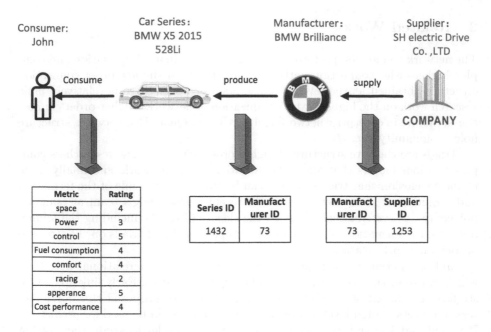

Fig. 2. An example of data in supply chain

BMW X5 2015 528Li and scored the performance of the car in different aspects. The score of each dimension is expressed in digits between 1 and 5. The higher the value, the better the evaluation. BMW Brilliance made this car as a manufacturer, and its suppliers were also found in corporate networks, such as Shanghai Electric Drive Co., Ltd. John has a high score on the car, which shows that the quality of this series produced by BMW is better from his perspective, which further indicates that BMW's supplier is high-quality partner.

Figure 2 contains the following entity dimensions:

Product Dimension: This dimension is the product model produced by the manufacturer in the supply chain. In our dataset, it refers to the car series, its parts are supplied by several suppliers. Users usually evaluate the model when buying it.

Metric Dimension: This dimension is the user's feedback on automobile performance after purchasing automobiles, which is divided into eight dimensions: space, power, control, fuel consumption, comfort, appearance, racing, cost performance. The higher the score on each dimension, the higher the satisfaction of the car, which reflects the reputation and satisfaction of the car in the user's mind.

Enterprise Dimension: The enterprise dimension has two different roles, manufacturer role and supplier role. A company may play two roles at the same time. When an enterprise supplies parts to it, it is the role of a manufacturer. When it supplies to another advanced manufacturer, it is the role of an supplier.

These dimensional entities are interrelated in the enterprise network and play different roles in the supply chain: supplier enterprises produce auto parts to manufacturers, manufacturers produce different auto products to supply to users, and users feedback the evaluation information to enterprises after purchasing cars. User ratings play an auxiliary role in enterprise relational link prediction, because users' evaluation of cars can reflect the quality of cars in different aspects. The quality of models is implicitly related to manufacturers' choice of suppliers, so we hope to extract users' evaluation information and the production relationship from the supply chain, so as to help get more accurate enterprise relationship.

3.2 Higher-Order Connection Patterns in Enterprise Network

Triad is a higher-order network structure, which is a closed structure formed by three nodes in the network and contains at least three edges. The links between nodes can be directed or undirected. In this paper, the triad in the enterprise network is composed of directed edges, which represent the supply relationship.

Using the triad mining algorithm [16] to count the number of triads in the enterprise network, we find several characteristics of the relationship between automobile enterprises: Firstly, there is no bidirectional supply relationship between enterprises, namely, there is no supply from A to B while B also supplies to A, so there are only eight different types of closed triads. Figure 3(c) utilizes three-digit binary coding to represent these eight different types of triads. A, B, C represent different enterprise nodes, A→B, B→C, and C→A are defined as forward edges, represented by 1, and backward edges are represented by 0. For example, the type 101 contains edges A→B, C→B and C→A.

Secondly, there is no circle in the enterprise network, accordingly, there is actually no triad types 111 and 000 in Fig. 3(a). This is because the upper and lower relationship between enterprises is relatively fixed, and circular supply will not occur. If node numbers in triads are ignored, the remaining six types of triads can be classified into one type, as shown in Fig. 3(b), which represents the most basic type of triads in enterprise network. This structure means that when there are two suppliers supplying to the same manufacturer, there is also a supply relationship between the two suppliers.

Through the above analysis, we found the relationship between two nodes, if the enterprise A supply to enterprise B, so there are only three types of traids made up of A and B, as shown in Fig. 3(c), respectively corresponding to the Fig. 3(a) of the type 101, 100 and 110. When the role of enterprise A and B is fixed, the composition of the traid can only be one of the three conditions.

With the dynamic development of the network, enterprises tend to form more closed triads [17]. Because triads have an important influence on the evolution of enterprise network structure and reveal some implicit relationships between enterprises, we utilize these three types of triads as features. The number of triads formed by any two enterprise nodes is obtained by statistics, and then a tensor data model is constructed to predict the linkage relationship between enterprises.

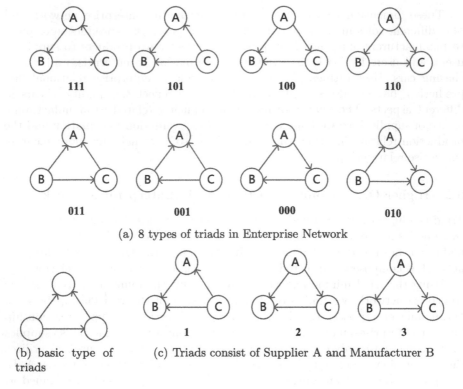

(a) 8 types of triads in Enterprise Network

(b) basic type of triads

(c) Triads consist of Supplier A and Manufacturer B

Fig. 3. Traids in enterprise network

4   Higher-Order Connection Structure Embedding in Link Prediction Algorithms

In this section, we will describe the details of our algorithm.

4.1   Data Model

Considering the influence of triads on link prediction in enterprise networks, we construct a tensor Y. In this tensor, the X-axis represents the manufacturer role of the enterprise, the Y-axis represents the supplier role of the enterprise, and the Z-axis represents different network structures in the enterprise network. We build this tensor because we hope to find out the potential information in the higher-order connection pattern by using the triad structure as the feature for training, so as to get more accurate prediction results.

However, although the scale of enterprise network is large, there are few existing link relations between enterprises, and the available data is sparse. It is not ideal to use only enterprise network structure for link prediction. To alleviate the problem of tensor sparsity, we utilize the abundant auxiliary information in the supply chain. After buying a car, users tend to rate its performance in different aspects, which partly reflects the quality of its parts. If the user gives a good evaluation, then it proves that the parts of the car are of high quality. Feedback of these evaluation information may affect the manufacturer's choice of supplier. The information in the supply chain consists of two parts: user rating information and production relationship information between manufacturers and car series. Combined with this auxiliary information, we can get better prediction effect and help manufacturers to choose more high-quality suppliers.

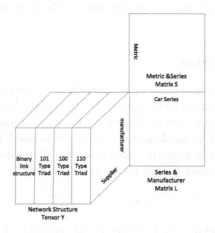

Fig. 4. Data model

Combining network structure and supply chain, we extract a forth-order tensor and two matrices to construct a coupled data model, as shown in Fig. 4. The dimensions of model include enterprise dimension, network structure dimension, series dimension and evaluation metric dimension.

$S \in R^{I \times J}$ is used to represent Metric × Series matrix, where I is the number of evaluation metric and J is the number of car series. The value range of element S_{ij} in matrix S is the integer between 1 and 5, representing the average score of series j on metric i. We obtained 169229 scoring information of 2159 car series and calculated the average score of each series on different metrics, which reflects the quality of parts in different aspects of the car series.

$L \in R^{J \times M}$ is used to represent Series × Manufacturer matrix, where J is the number of car series, M is the number of manufacturers. Each car series has a unique manufacturer. If $L_{jm} = 1$, represents series j produced by manufacturer m; otherwise, it is 0.

$Y \in R^{M \times M \times N}$ is used to represent Manufacturer × Supplier × Network Structure Feature, where M is the number of manufacturer and supplier, and N is the number of network structure feature. The element Y_{xyz} in the tensor Y represents the number of network structure z consisting of manufacturer x and supplier y. Because the enterprises in the network can be either manufacturers or suppliers, manufacturers and suppliers is of same size. In this paper, the number of network structure is 4, the first three layers correspond to three types of triads, and the last layer corresponds to the binary link structure. In the experiment, we need to normalize the first three layers of data in the tensor to maintain the consistency of data in the tensor, because the value of Y_{xy4} is a digit between 0 and 1, representing the possibility of establishing a supply relationship between x and y.

Some dimensions are shared in the data model, such as matrix S and matrix L sharing the car series dimension, matrix L and tensor Y sharing the manufacturer dimension. Our ultimate goal is to decompose the data model to predict the possibility of links.

4.2   Objective Function

Since the higher-order connection mode and the information in the supply chain are simultaneously integrated, the objective function needs to decompose a forth-order tensor and two matrices at the same time, and maintain the consistency of shared dimensions. The method in this paper is to use CP decomposition [18] to construct the joint decomposition loss function. The objective function of this multidimensional recommendation problem is as follows:

$$Z = Z_S + Z_L + Z_Y \tag{1}$$

Z_S is the least square error between the product of factor matrix $A \in R^{I \times K}, B_1 \in R^{J \times K}$ and Metric $\times$ Series matrix S, A and B_1 respectively represent metric factor matrix and car series factor matrix, which are obtained by decomposition matrix S, and K is the size of factor vectors, i.e. the rank of matrix. λ_1 is the regularization parameter. $\|A\|_F$ and $\|B\|_F$ correspond to the F-norm of matrix A, B_1. The objective function of Z_S is

$$Z_S = \frac{1}{2} \left\| S - AB_1^T \right\|_F^2 + \frac{\lambda_1}{2} \left(\|A\|_F^2 + \|B_1\|_F^2 \right) \tag{2}$$

Z_L is the least square error between the product of factor matrix $B_2 \in R^{J \times K}, F_1 \in R^{M \times K}$ and Series $\times$ Manufacturer matrix L. B_2 and F_1 respectively represent car series factor matrix and manufacturer factor matrix, which are obtained by decomposition matrix S, and λ_2 is the regularization parameter. The objective function of Z_L is

$$Z_L = \frac{1}{2} \left\| L - B_2 F_1^T \right\|_F^2 + \frac{\lambda_2}{2} \left(\|B_2\|_F^2 + \|F_1\|_F^2 \right) \tag{3}$$

Z_Y is the least square error between the product of factor matrix $F_2 \in R^{W \times K}, Q \in R^{W \times K}, H \in R^{V \times K}$ and Manufacturer $\times$ Supplier $\times$ Network Structure Feature tensor Y. F_2, Q and H respectively represent manufacturer factor matrix, supplier factor matrix and network structure feature factor matrix, which are obtained by CP decomposition of tensor Y, and λ_3 is the regularization parameter. Y_1 is a matrix obtained by expanding tensor Y according to mode-1 [19]. $\odot$ represents the Khatri-rao product [20]. The objective function of Z_Y is:

$$Z_Y = \frac{1}{2} \left\| Y_1 - F_2(Q \odot H)^T \right\|_F^2 + \frac{\lambda_3}{2} \left(\|F_2\|_F^2 + \|H\|_F^2 + \|Q\|_F^2 \right) \quad (4)$$

The series dimension is shared by matrix S and matrix L, and the manufacturer dimension is shared by tensor Y and matrix L. We expect factor matrix corresponding to these dimensions to be the same, so the global variable matrix is constructed to limit the factor matrix of shared dimension. For series dimension, we expect the decomposed factor matrix B_1 and B_2 are equal, so to build the global variable matrix $\overline{B}$ to limit B_1 and B_2. For manufacturers dimension, we expect the decomposed factor matrix F_1 and F_2 are equal, so to build the global variable matrix $\overline{F}$ to limit F_1 and F_2, as follows:

$$\begin{aligned} B_1 - \overline{B} = 0, B_2 - \overline{B} = 0 \\ F_1 - \overline{F} = 0, F_2 - \overline{F} = 0 \end{aligned} \quad (5)$$

Our objective function can be written as the following constrained optimization problem:

$$\min_{\psi} Z = Z_s + Z_L + Z_Y \\ s.t. \begin{cases} B_1 - \overline{B} = 0, B_2 - \overline{B} = 0 \\ F_1 - \overline{F} = 0, F_2 - \overline{F} = 0 \end{cases} \quad (6)$$

$\psi = \{A, B_1, B_2, F_1, F_2, Q, H\}$ represents the collection of factor matrix. For optimization problems under equality constraints, ADMM performs better in terms of convergence and accuracy than traditional ALS method. Therefore, we utilize the ADMM method to solve the problem.

4.3 Joint Decomposition Based on ADMM

For a constrained problem in the following form:

$$\min_{x,z} f(x) + g(z) \\ s.t. \ Ax + Bz = c \quad (7)$$

$F'(x)$ and $G'(z)$ are functions of x and z, $x \in R^n$ and $z \in R^m$ are variables, $A \in R^{p \times n}, B \in R^{p \times m}$ and $c \in R^p$ are known quantities. The above constrained optimization problem can be converted into the following form by using the augmented Lagrange method [21]:

$$L_\rho(x, z, \theta) = f(x) + g(z) \\ + \theta^T (Ax + Bz - c) + (\rho/2)\|Ax + Bz - c\|_F^2 \quad (8)$$

Θ is a Lagrange multiplier and $\rho > 0$ is a penalty parameter. Then, ADMM method is used to solve the above equation, and the following equation is iteratively calculated:

$$\begin{aligned} x^{k+1} &\leftarrow \arg\min_x \min L_\rho \left(x, z^k, \theta^k\right) \\ z^{k+1} &\leftarrow \arg\min_z \min L_\rho \left(x^{k+1}, z, \theta^k\right) \\ \theta^{k+1} &\leftarrow \theta^k + \rho \left(Ax^{k+1} + Bz^{k+1} - c\right) \end{aligned} \quad (9)$$

x^k, z^k and θ^k represent the results of the k-th iteration of x, z and θ, respectively.

Therefore, for Eq. (6), this constrained optimization problem can be transformed into an unconstrained optimization problem by means of augmented Lagrange method. The objective function is as follows:

$$
\begin{aligned}
&L_p\left(\Psi, \Theta_B^1, \Theta_B^2, \Theta_F^1, \Theta_F^2, \overline{B}, \overline{F}\right) \\
&= Z + \sum_{i=1}^2 \{ tr\left(\left[\Theta_F^i\right]^T \left(F_i - \overline{F}\right)\right) + \tfrac{\rho}{2} \left\|F_i - \overline{F}\right\|_F^2 \\
&\quad + tr\left(\left[\Theta_B^i\right]^T \left(B_i - \overline{B}\right)\right) + \tfrac{\rho}{2} \left\|B_i - \overline{B}\right\|_F^2 \}
\end{aligned}
\tag{10}
$$

For the factor submatrices in ψ, the partial derivatives of them with respect to the function L_p are calculated respectively, so that the partial derivatives are equal to 0. The updating rule of A is given here as Eq. (11), and the solving method of other factor matrices in ψ is similar to that of A.

$$
A = (SB_1)\left[B_1^T B_1 + \lambda_1 I_R\right]^{-1}
\tag{11}
$$

For Lagrange multiplier Θ_B^1, Θ_B^2, Θ_F^1, Θ_F^2, the gradient descent method is used to update, and the updated rules are as follows:

$$
\begin{aligned}
\Theta_F^i &= \Theta_F^i + \rho\left(F_i - \overline{F}\right) \\
\Theta_B^i &= \Theta_B^i + \rho\left(B_i - \overline{B}\right); \forall i \in \{1,2\}
\end{aligned}
\tag{12}
$$

For global variables $\overline{B}$ and $\overline{F}$, take partial derivatives of function L_p respectively, make partial derivatives equal to 0, and obtain the update rules of global variables. Due to the space limitation of the paper, taking B as an example, the updated Equation of $\overline{B}$ is:

$$
\overline{B} = \frac{1}{2\rho}\left(\Theta_B^1 + \Theta_B^2 + \rho B_1 + \rho B_2\right)
\tag{13}
$$

By setting $\left(\Theta_A^i\right)_0 = 0, i = 1, 2$, the following results can be proved:

$$
\sum_{i=1}^2 \left(\Theta_A^i\right)_k = 0, k = 1, 2, \ldots I_{\max}
\tag{14}
$$

Therefore, the updating rule of $\overline{B}$ and $\overline{F}$ can be simplified to Eq. (15):

$$
\begin{aligned}
\overline{B} &= \tfrac{1}{2}\left(B_1 + B_2\right) \\
\overline{F} &= \tfrac{1}{2}\left(F_1 + F_2\right)
\end{aligned}
\tag{15}
$$

Algorithm 1 shows how to minimize the objective function L_p by using the ADMM method. Sparse tensor Y, auxiliary information matrix S, matrix L, rank K of tensor, max iteration number MaxIter, initial iteration number iter, etc. are used as input conditions. ADMM algorithm is used to update each factor matrix and parameter matrix. When the algorithm reaches the maximum number of iterations, $Y = F_2 \circ Q \circ H$ can be used to restore tensor Y and complete its missing value.

Algorithm 1. Joint decomposition algorithm based on ADMM

Input: sparse tensor Y, matrix S, L, rank K of tensor, MaxIter, iter
Output: complete tensor Y
1: Initialize parameter ρ, λ_1, λ_2, λ_3, K, iter
2: Calculate matrix $\overline{B}$ and $\overline{F}$ by Equation (15)
3: **while** $iter < Maxiter$ **do**
4:    Update A $B_1, B_2, F_1, F_2, Q, H)$ based on Equation (11)
5:    Update $\overline{B}$ and $\overline{F}$ based on Equation (15)
6:    Update Θ_B^1, Θ_B^2, Θ_F^1, Θ_F^2 based on Equation (12)
7:    iter=iter $+ 1$
8:    **if** Achieving convergence criteria **then**
9:       **Break**
10:   **end if**
11: **end while**
12: **Return** $Y = F_2 \circ Q \circ H$

---

5 Experiment

5.1 Data

Auto enterprise network data set is crawled from relevant websites on the Internet. The data set contains users, car series, ratings, manufacturers and suppliers information. When cleaning the data, only users and series with previous purchase records are retained. The tensor and matrices used in the experiment are constructed from the relationships shown in Table 1, which lists their dimensions, number of contained tuples, and range of values. Among them, tensor Y and matrix L are extracted from http://auto.gasgoo.com, and matrix S is extracted from https://www.autohome.com.cn.

Table 1. Data set relational summary

Relational data	Dimensions	Tuples	Range
Rating matrix	Metric × Series	17272	1–5
Product matrix	Series × Manufacturer	2159	0 or 1
Supply matrix	Manufacturer × Supplier	43384	0 or 1

The goal of the decomposition of tensors is to predict the linkage relationship between enterprises, namely the data on the fourth layer matrix of tensors. The existing data on the fourth layer matrix is the real enterprise supply relationship, and these data are divided into training set and test set by the ratio of $2 : 1$. The parameters in the model include ρ, λ_1, λ_2, λ_3 and K, where ρ is the penalty parameter, λ_1, λ_2, λ_3 are regularization parameters and K is the rank of the tensor.

We use $\hat{T}$ to represent the test set, which contains n elements. For each element in the $\hat{T}$ set, y_i s the true value, and $\hat{y}_i$ is the predicted value. RMSE is the Root Mean Square Error between the predicted value and the actual value in the test set, which is the main reference standard of adjustment parameters. The smaller the RMSE value is, the better the prediction result will be. The calculation of rmse is as follows:

$$RMSE = \sqrt{\frac{\sum_{t=1}^{n}\left(\hat{y}_i - y_i\right)^2}{n}} \qquad (16)$$

For tuning the parameters in the model, first initialize them to 0.0001 to minimize their impact on model performance, making sure that the matrix and the tensor have the same contribution to the target function. $\lambda_1, \lambda_2, \lambda_3$ are regularization parameters for the rating matrix, product matrix, and network structure tensor, respectively. We fixed the other parameters, picked some different λ_1 values, calculate RMSE and draw polygons, as shown in Fig. 5.

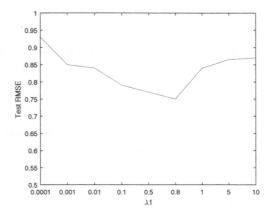

Fig. 5. Impact of λ_1 on RMSE

As can be seen from Fig. 5, RMSE first increases and then decreases with λ_1 value increasing. When λ_1 is too big, it will limit the influence of the rating matrix on the model. When λ_1 is too small, the scoring matrix is dominant, meaning the influence of the other factor matrices is ignored. $\lambda_1 = 0.8$ is our choice. For the other regularization parameters, we use the same parameter adjustment method, the final choice is $\lambda_2 = 0.5$, $\lambda_3 = 1.5$. For K, we choose a value in the range of 2–10, start from 2, increase by 1 each time, observe that when K is greater than 4, the performance change of the algorithm will be very small, so we choose $K = 4$. ρ is the penalty parameter of ADMM algorithm, fixed several other parameters on its adjustment, found that when $1 < \rho < 8$, the impact on algorithm performance is small, so choose $\rho = 5$.

5.2   Results

In order to verify the superiority of our algorithm, many link prediction algorithms are selected for comparison. The first is the traditional link prediction method, which directly decomsolves 0–1 type enterprise relationship matrix, including PMF [22], GPLVM [23], SVD++ [24], Social-MF [25] algorithm. In our data set, the scale of enterprise relationship matrix is 12492×12492, containing 43384 enterprise supply relationships, and the sparsity is as high as 99.97%. This algorithm only uses low-order binary relations, and does not consider the enterprise network structure such as the higher-order connection structure such as triad, nor does it consider the feedback information of users after buying cars. In large-scale enterprise networks, the available data information is limited.

Another approach is to only consider the information in the network structure, but not using the supply chain as a auxiliary information, directly use probability tensor decomposition (or CP decomposition method) to decompose the tensor Y. By predicting the missing value in the forth layer of tensor, the possibility of generating link relations between enterprises is obtained, and the objective function is shown as follows:

$$\min Z_Y = \frac{1}{2} \left\| Y_1 - F_2(Q \odot H)^T \right\|_F^2 + \frac{\lambda_2}{2} \left(\|F_2\|_F^2 + \|Q\|_F^2 + \|H\|_F^2 \right) \qquad (17)$$

The final predicted value is the probability of supply relationship between enterprises, which is between 0 and 1. Therefore, Area under Curve (AUC) is used to evaluate the performance of the algorithm. The higher the AUC value, the better the performance of algorithm. We have run each of the other comparison methods for 10 times. Table 2 records the average of 10 experiments, which shows that our algorithm performs better than other algorithms.

Table 2. AUC comparison of different algorithms

Algorithm	AUC
PMF	0.743
GPLVM	0.794
SVD++	0.812
Social-MF	0.808
Tensor decomposition	0.810
Our Algorithm	0.834

6   Conclusions

Based on the field of automobile enterprise network, this paper proposes a new algorithm embedded with higher-order connection structure to predict partner links in supply chain. We find that triads as the representative of higher-order connection structure is of great significance to enterprise relationship prediction. In addition, the user feedback information in the supply chain also has a positive impact on the relationship prediction between enterprises. Combining the two information can effectively improve the accuracy of prediction. Aiming at the problem of data sparsity in enterprise network, we construct a rating matrix-triad tensor coupled data model, transform the objective function into unconstrained optimization problem by augmented Lagrange method, and solve it by alternating direction multiplier method (ADMM), enhance the accuracy of the decomposition of tensor model, to ensure the convergence. Compared with other link prediction algorithms, our algorithm can get more accurate link prediction results.

Acknowledgements. This research work is supported by the National Key Research and Development Program (2018YFB1404501) and the Shandong Key Research and Development Program (2017CXGC0604, 2017CXGC0605, 2018GGX101019).

References

1. Sergey, N.D., Jose, F.M.: Evolution of networks. Adv. Phys. **51**(4), 1079–1187 (2002)
2. Linton, C.F.: Centered graphs and the structure of ego networks. Math. Soc. Sci. **3**(3), 291–304 (1982)
3. Tom, M., Anagha, J., Bruno, N., Yves, V.P.: Enrichment and aggregation of topological motifs are independent organizational principles of integrated interaction networks. Mol. BioSyst. **7**(10), 2769–2778 (2011)
4. Tore, O.: Triadic closure in two-mode networks: redefining the global and local clustering coefficients. Soc. Netw. **35**(2), 159–167 (2013)
5. Paul, W.H., Samuel, L.: A method for detecting structure in sociometric data. In: Social Networks, pp. 411–432. Elsevier (1977)
6. Dana, L.H.: Friendship networks and delinquency: the relative nature of peer delinquency. J. Quant. Criminol. **18**(2), 99–134 (2002)
7. Ranjay, G., Martin, G.: Where do interorganizational networks come from? Am. J. Sociol. **104**(5), 1439–1493 (1999)
8. Peter, K., Stefan, T.: Triadic closure dynamics drives scaling laws in social multiplex networks. New J. Phys. **15**(6), 063008 (2013)
9. Hong, H., Yuxiao, D., Jie, T., Hongxia, Y., Nitesh, V.C., Xiaoming, F.: Will triadic closure strengthen ties in social networks? ACM Trans. Knowl. Discov. Data (TKDD) **12**(3), 30 (2008)
10. Zhu, J., Hong, J., Hughes, J.G.: Using Markov chains for link prediction in adaptive web sites. In: Bustard, D., Liu, W., Sterritt, R. (eds.) Soft-Ware 2002. LNCS, vol. 2311, pp. 60–73. Springer, Heidelberg (2002). https://doi.org/10.1007/3-540-46019-5_5
11. Alexandrin, P., Lyle, H.U.: Statistical relational learning for link prediction. In: IJCAI Workshop on Learning Statistical Models from Relational Data, vol. 2003. Citeseer (2003)
12. Dekang, L.: An information-theoretic definition of similarity. In: ICML, vol. 98, pp. 296–304. Citeseer (1998)

13. Shi, X., Wang, L., Liu, S., Wang, Y., Pan, L., Wu, L.: Investigating microstructure patterns of enterprise network in perspective of ego network. In: Chen, L., Jensen, C.S., Shahabi, C., Yang, X., Lian, X. (eds.) APWeb-WAIM 2017. LNCS, vol. 10366, pp. 444–459. Springer, Cham (2017). https://doi.org/10.1007/978-3-319-63579-8_34
14. Tiancheng, L., Jie, T., John, H., Zhanpeng, F., Xiaowen, D.: Learning to predict reciprocity and triadic closure in social networks. ACM Trans. Knowl. Discov. Data (TKDD) **7**(2), 5 (2013)
15. Hong, H., Jie, T., Sen, W., Lu, L.: Mining triadic closure patterns in social networks. In: Proceedings of the 23rd International Conference on World Wide Web, pp. 499–504. ACM (2014)
16. Paul, W.H., Samuel, L.: Transitivity in structural models of small groups. Comp. Group Stud. **2**(2), 107–124 (1971)
17. Brian, R.F., Frank, S., John, P.: The theory of triadic influence. Emerg. Theor. Health Promot. Pract. Res. **2**, 451–510 (2009)
18. Miguel, A.V., Jeremy, E.C., Rodrigo, C.F., Jocelyn, C., Pierre, C.: Nonnegative tensor CP decomposition of hyperspectral data. IEEE Trans. Geosci. Remote Sens. **54**(5), 2577–2588 (2015)
19. Brett, W.B., Tamara, G.K.: Algorithm 862: MATLAB tensor classes for fast algorithm prototyping. ACM Trans. Math. Softw. (TOMS) **32**(4), 635–653 (2006)
20. Shuangzhe, L., Götz, T.: Hadamard, Khatri-Rao, Kronecker and other matrix products. Int. J. Inf. Syst. Sci. **4**(1), 160–177 (2008)
21. Zhouchen, L., Minming, C., Yi, M.: The augmented lagrange multiplier method for exact recovery of corrupted low-rank matrices. arXiv preprint arXiv:1009.5055 (2010)
22. Neil, D.L.: Gaussian process latent variable models for visualisation of high dimensional data. In: Advances in Neural Information Processing Systems, pp. 329–336 (2004)
23. Pierre, C.: Tensor decompositions. In: Mathematics in Signal Processing V, pp. 1–24 (2002)
24. Kumar, R., Verma, B.K., Rastogi, S.S.: Social popularity based SVD++ recommender system. Int. J. Comput. Appl. **87**(14) (2014)
25. Purushotham, S., Liu, Y., Kuo, C.C.J.: Collaborative topic regression with social matrix factorization for recommendation systems. arXiv preprint arXiv:1206.4684 (2012)

Solving the Signal Relay Problem of UAV in Disaster Relief via Group Role Assignment

Qian Jiang[1], Haibin Zhu[2], Ming Liao[1], Baoying Huang[1], Xiaozhao Fang[1], and Dongning Liu[1(✉)]

[1] School of Computer Science and Technology, Guangdong University of Technology, Guangzhou, China
liudn@gdut.edu.cn
[2] Department of Computer Science and Mathematics, Nipissing University, North Bay, Canada

Abstract. When an earthquake occurs, disaster relief is an urgent, complex and critical mission. Decision makers must possess critical role awareness to prioritize the numerous interrelated tasks arising from such an event. High on the list is communication network recovery within the disaster area. Unmanned aerial vehicles (UAVs) are often used in this regard. Some of them are used as repeaters to provide the required network coverage. Their timely, efficient deployment to specific locations is a challenging exercise for disaster relief planners. In response to such a need, this paper formalizes and solves the problem of UAV deployment for signal relay via group role assignment (GRA). The minimum spanning tree algorithm is applied to model a rapidly deployed optimal relay network. It can help establish the minimum number of relay points necessary to ensure communication stability. In this scenario, UAVs (agents) adopt roles as communication relays. The task of distributing UAVs to relay points can be solved quickly via the assignment process of GRA, which can solve the x-ILP problem with the help of the PuLP package of Python. Results from thousands of experimental simulations indicate that our solutions are effective, robust and practical. The process can be used to establish an optimal, efficient relay network using UAVs. Their rapid deployment can be a significant contribution in earthquake disaster relief.

Keywords: Disaster relief in earthquake · Signal relay · Unmanned aerial vehicle (UAV) · Group role assignment (GRA)

1 Introduction

China experiences many earthquake disasters. When such an event occurs, communication recovery within the disaster area plays a vital role in disaster relief. Unmanned aerial vehicles (UAVs) can be rapidly deployed to reestablish communications when physical infrastructure has been damaged or destroyed.

UAVs [1–3] can perform various roles in disaster relief such as patrolling, detecting and relaying. Decision makers face the challenge of quickly determining the numbers of and roles adopted by such equipment. In communication recovery, it is necessary to

© Springer Nature Singapore Pte Ltd. 2019
Y. Sun et al. (Eds.): ChineseCSCW 2019, CCIS 1042, pp. 18–29, 2019.
https://doi.org/10.1007/978-981-15-1377-0_2

have efficient tools to send the minimum number of UAVs to locations that provide stable coverage over the disaster area.

The key of deploying UAVs as communication relays is an awareness of the number and locations of relay points. Next comes the task of role and location assignment of drones from different base stations into a successfully coordinated network. There are many combination methods which can do this job. However, the specific scenario and scale of an earthquake often change. A comprehensive model is needed to deal with different but equally urgent incidents. It is critical that a solution distributes a minimum number of mobile communication devices into a robust, economical network in the shortest time possible. Consideration must also be given to other important tasks that a UAV fleet must perform disaster relief and re-construction.

This paper formalizes and solves the problem of establishing UAV signal relay networks via group role assignment (GRA). GRA [4, 5] is a complex process throughout the life cycle of RBC [6–8]. RBC has become a practical method for decision makers relative to task distribution.

Signal relay networks over a disaster area can be established by considering relay points as roles and UAVs as agents. This paper firstly provides an effective and comprehensive solution to the problem via GRA. In modeling the rapid reestablishment of communications over a disaster area, we initially use the minimum spanning tree algorithm to determine an optimal relay network that requires the minimum number of UAVs at locations that offer stable, full coverage of the affected area. Using the assignment process of GRA, relay locations are considered roles and UAVs are agents.

This paper has listed the contributions below:

(1) Clearly formalizing the problem of establishing signal relay networks using UAVs over a disaster area via GRA, which can provide a uniform method;
(2) An efficient, comprehensive method for deploying UAVs as repeaters to the right places within an acceptable time; and
(3) Creation of an economical yet stable communication network that uses the minimum number of UAVs as a timely, significant contribution to earthquake disaster relief.

The organization of the paper is listed: It firstly uses a real-world scenario to illustrate the problem in Sect. 2. Then, we formally define the problem via RBC and its E-CARGO model in Sect. 3. Section 4 proposes a solution to the problem via GRA, with simulation experiments. Section 5 discusses related work. The paper concludes with future work in Sect. 6.

2 A Real-World Scenario

Sichuan is a frequent earthquake area. For instance, in 2008, the Wenchuan earthquake of magnitude 8 occurred in Sichuan. Unfortunately, another 7.0-magnitude earthquake jolted Jiuzhaigou County on August 8th, 2017. According to the China Earthquake

Networks Center, the quake struck at a depth of 20 km. More than 90 emergency vehicles and 1,200 personnel were dispatched to participate in the rescue work.

The communication system was affected by the quake. This essential component of search and rescue required the deployment of several mobile communication vehicles to resume communication. Because the earthquake destroyed most of the roads in Jiuzhaigou, the vehicles were only capable of communication within a 2 km radius (see Fig. 1(a)).

To restore communication over the disaster area, the communication department decided to deploy UAVs carrying communication equipment. These UAVs departed from various bases in Jiuzhaigou simultaneously. We assume that each UAV hover stably in the air with a small radius so that we treat it as stationary state. Due to load limitations of the drones, the on-board communication equipment had to be within 3 km of the mobile communication vehicle (see Fig. 1(b)). The communication distance between drones was limited to 6 km.

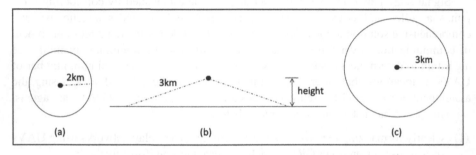

Fig. 1. The covered shape of the communication vehicle and the drone.

To ensure communication quality, the altitude of each drone is much smaller than its coverage radius; hence the administrators can consider that a drone's coverage shape is also a circle (see Fig. 1(c)). In general, for robust performance, the rescue center assumed that one UAV could only guarantee stable communication with one other link. However, it can be used as a back-up communication drone for other links if some of the intersected UAVs fail to work.

To guarantee the effectiveness of our solution, partial real-world elevation data of Jiuzhaigou was used to carry out all of the simulation experiments. Figure 2 is the contour map drawn from the partial elevation data of Jiuzhaigou. The square shape in Fig. 2 represents the base station, and the circle shape is the communication vehicle. To keep our visualization distribution result clearly displayed, it set the covered radius of a UAV as 3 km, and the number of bases is 8. The number of communication vehicles is set as 15.

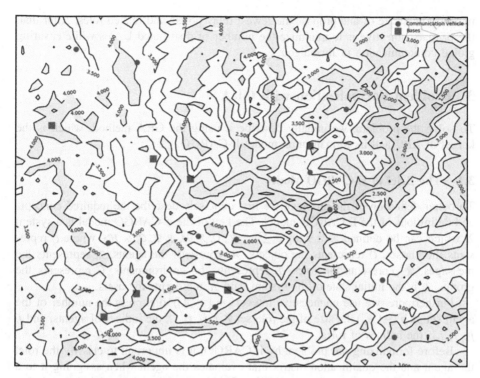

Fig. 2. The contour map of the partial real-world elevation data of Jiuzhaigou. Notes: A red square represents a base station, and a green circle represents a communication vehicle. The location of the communication vehicles in a range from 3 km to 98 km. (Color figure online)

As each drone has a maximum flight time, it needs to be replaced by other drones from the same base station when it nears its maximum flight time. Also, each base station needs to dedicate some drones to carry out patrolling and detecting tasks in the disaster area. Therefore, each station has a limited number of assigned drones for communication (see Table 1).

Table 1. The limited numbers of UAV

Base station	1	2	3	4	5	6	7	8
The limited numbers of UAV	10	13	16	9	11	14	12	15

This signal relay task is a great challenge in the deployment of UAVs in Jiuzhaigou. It is vital to quickly determine the number of signal relaying UAVs and send them to the right locations within an acceptable time, while having other UAVs available to do other relief jobs.

From the above scenario, the problem can be dealt with by following the initial steps of RBC and a related GRA problem. The first part is role negotiation, which needs to be aware of and propose the fewest relay points and related roles. After that, the primary objective is to minimize the number of distributed UAVs while ensuring global communication in Jiuzhaigou.

3   Problem Formalization with GRA

To solve the problem, we initially describe it using the GRA method, which is the kernel of RBC and its formalization model E-CARGO.

3.1   The Brief of E-CARGO Model

To better definition of the problem, we first concisely describe the required concepts and definitions of the E-CARGO model. The key of the E-CARGO model is a system Σ which can be defined as a 9-tuple $\Sigma ::= <C, O, \mathcal{A}, \mathcal{M}, \mathcal{R}, \mathcal{E}, \mathcal{G}, s_0, \mathcal{H}>$, where C represents the classes, O represents the objects, $\mathcal{A}$ represents the agents, $\mathcal{M}$ represents the messages, $\mathcal{R}$ represents the roles, $\mathcal{E}$ represents the environments, $\mathcal{G}$ represents the groups, s_0 is the initial state of the system, and $\mathcal{H}$ represents the users.

Furthermore, we use nonnegative integers m $(=| \mathcal{A} |)$ to express the number of the agent set $\mathcal{A}$, n $(= | \mathcal{R}|)$ the size of the role set $\mathcal{R}$, i, i_1, i_2, ... the indices of agents, and j, j_1, j_2, ... the indices of roles.

Before formalizing our problem by E-CARGO, it is important to classify the role, the agent and the qualification matrix that evaluates the agent when assigning it to a specific role. In our scenario, considering UAVs as agents is straightforward, but considering the relay points as roles poses quite challenging. For this reason, role negotiation was dealt with first. This was needed prior to using the GRA method.

3.2   Role Negotiation

The role negotiation process in this paper, called role awareness, aims to set up a relay network by specifying the specific properties of relay points, i.e., roles.

Our target is to achieve global communication in the disaster area through the establishment of a practical, optimal relay network formed by UAVs from different base stations. It is clear that drones (UAVs) from different bases can be regarded as agents. As for the relay network, we need to decompose it into the relay points so as to assign the UAVs specifically and evaluate the qualification of the drones from different base stations. There exists a mutually exclusive relationship and a collaborative relationship among these relay points, which are suited to become roles.

Based on the relay network's property, we decide to use the prim algorithm to set up the relay network (see Fig. 3). It is a kind of minimum spanning tree algorithm. We use it here because it has the property of the least global communication cost if the positions of communication vehicles are determined before the relay points are selected and the number of UAVs is sufficient for fully covering all of the communication vehicles. Using the Minimum Spanning Tree is practical and it makes the relay network easier to expand.

After getting the practically optimal relay network, the next task is to determine the location of relay points. To solve it, we propose a method (see Sect. 4) find the proper relay points from the optimal relay network.

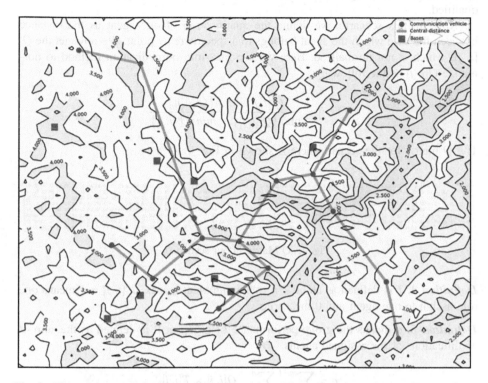

Fig. 3. The minimum spanning tree of relay points. Notes: A red square represents a base station, and a green circle represents a communication vehicle. (Color figure online)

3.3 Problem Formalizations

According to the analysis above, we define the relay points obtained by role awareness as roles and base stations as agents. To better describe RBC and its E-CARGO model, we use one small scale experiment as an example here. The role number we found by using role awareness based on our proposed method is 58, and the value for the base station is 8; The covered radius of the drone is 3 km, and the max communication distance between the UAVs is 6 km.

Definition 1. L is a lower bound of range vector for roles in environment e of group g.

Note: In our scenario, each role can only be assigned to one agent, so the L vector here can be formalized as: $L =$ [1 1 1 1 1 1 1 1 1 1 1 1 1 1 1 1 1 1 1 1 1 1 1 1 1 1 1 1 1 1 1 1 1 1 1 1 1 1 1 1 1 1 1 1 1 1 1 1 1 1 1 1 1 1 1 1 1 1].

Definition 2. L^a is an m ability limit vector, where $L^a[i]$ ($0 \leq i < m$) indicates the maximum number of agents for each role.

The L^a here is [10 13 16 9 11 14 12 15].

Definition 3. Q is an $m \times n$ qualification matrix, where $Q[i, j] \in [0, 1]$ expresses the qualification value of agent i for role j. $Q[i, j] = 0$ indicates not qualified and 1 means qualified.

Note: In this scenario, we define the qualification of one agent as the distance between the base which the agent belongs to a specific relay point. We now get the Q' $[i, j]$, which is not normalized. Here we use max-min normalization method to normalize $Q[i, j]$ as the following:

$$Q[i,j] = \frac{Q'[i,j] - min\{Q'[i][j]\}}{max\{Q'[i,j]\} - min\{Q'[i,j]\}}$$

The Q matrix for the scenario here is shown in Fig. 4.

Fig. 4. Q Matrix in our scenario.

Definition 4. T is defined as an $m \times n$ role assignment matrix. $T[i, j] = 1$ means one role assign to one agent and 0 means no.

Definition 5. σ means the group performance of group g. It is defined as the sum of qualifications of the assigned agents,

$$\sigma = \sum_{i=0}^{m-1} \sum_{j=0}^{n-1} Q[i,j] \times T[i,j].$$

Definition 6. The UAVs assignment problem can be formalized as follows:

$$\min \sigma = \sum_{i=0}^{m-1} \sum_{j=0}^{n-1} Q[i,j] \times T[i,j] \tag{1}$$

subject to

$$T[i,j] \in \{0, 1\}(0 \leq i < m, 0 \leq j < n) \tag{2}$$

$$\sum_{i=0}^{m-1} T[i,j] = L[j](0 \leq j < n) \tag{3}$$

$$\sum_{j=0}^{j-1} T[i,j] \leq \frac{L^a[i]}{2}(0 \leq i < m) \tag{4}$$

Where expression (1) is the objective function; expression (2) is a 0–1 constraint; and (3) indicates that each role must have the requested quantity of agents to solve its assignment. (4) is different from the traditional GRA problem, and it describes that the number of each agent has been limited to assign to the roles; each base needs to keep at least half of the limited number of roles for replacing those assigned drones with low battery.

For instance, Fig. 5 is a T with L = [1 1 1 1 1 1 1 1 1 1 1 1 1 1 1 1 1 1 1 1 1 1 1 1 1 1 1 1 1 1 1 1 1 1 1 1 1 1 1 1 1 1 1 1 1 1 1 1 1 1 1 1 1 1 1 1 1 1 1 1] and L^a = [10 13 16 9 11 14 12 15]. The total cost of the drones from 8 bases is 12.30, which is an optimal result via GRA.

Fig. 5. T Matrix in our scenario

Figure 6 shows the assignment result for our solutions.

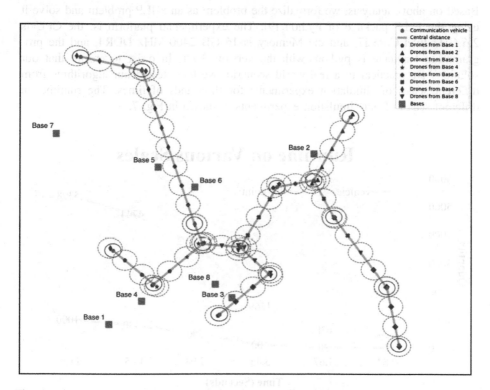

Fig. 6. The assignment result of our solution. Notes: Different kinds of shapes of the minimum spanning tree or near the minimum spanning tree represent the UAVs from different base stations, and a solid circle represents the communication coverage radius of a communication vehicle while a dotted circle is the communication coverage radius of a UAV.

4   Solutions and Experiments

4.1   Role Awareness

Here we propose a role awareness method to find the relay point as follows.

The Method of Finding the Relay Points
Step 1: Evaluate the minimum UAVs' number for each edge of the minimum spanning tree and find the central point of each edge.
Step 2: For the edges with the even minimum UAVs' number, parallelly put the UAVs from two sides of the central point with multiple distance of the UAV covered radius from the central point's location and finally check the boundary drones' position and adjust the boundary drones' positions based on the distance between the boundary drone and the closed edge point if necessary.
Step 3: For the edges with the odd minimum UAVs' number, put one UAV over the central point and then follow the same procedure as **Step 2**.

4.2   Experiments

Based on above analysis, we formalize the problem as an x-ILP problem and solve it using the PuLP package of Python [9]. The experimental platform is: the CPU is 2.6 GHz Intel Core i7, and the Memory is 16 GB 2400 MHz DDR4, and the programming language is python, with the version 3.7.1. In order to ensure that our solutions are practical in a real-world scenario, we have tested our algorithms from different scales of simulation experiments for thousands of times. The runtime on different scales of our simulation experiments is shown in Fig. 7.

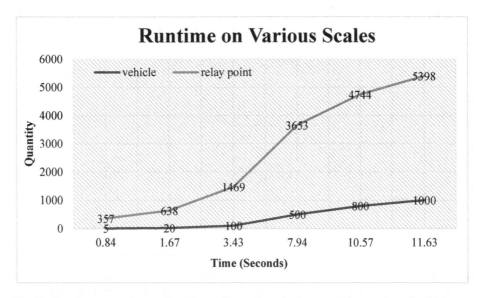

Fig. 7. Runtime on various scales. Notes: To analyze the impact of the number of vehicles on time, we set the covered radius of the UAVs as 3 km, the number of bases is 8, and the location of the communication vehicles range from 3 km to 1200 km for discretely generating them so as to cover the whole disaster area.

5 Related Work

The Unmanned Aerial Vehicle has become a popular tool for search and rescue because it is fast and not affected by obstacles like heavy traffic congestion or road damage. There are many types of research involving the use of drones as a rescue tool. Such uses are communication recovery, emergency response and rescue, etc. [1–3]. As a matter of fact, they are essentially assignment problems, including our research.

Many researches allocate their tasks based on data-related or agent-related methods [10–13]. Many of these researches analyze the properties of the tasks or the distribution of the data, and then select a specific algorithm which is suitable for the tasks.

Geng *et al.* [10] use a modified centralized algorithm based on particle swarm optimization to solve the task allocation problem in the search and rescue domain. They suppose that a centralized algorithm should perform better than distributed algorithms because it has all the available information at hand to solve the problem.

It is difficult to compare our solution with these algorithms due to differences in experimental scenarios and objectives. Our approach can seek a practically global optimal solution by defining a formalization model with an assignment algorithm to deal with the x-ILP problem in our scenario. Thousands of simulation experiments demonstrate that our solution is scale-independent; it can swiftly determine the practically optimal result even though the task's scale becomes large.

Allocating tasks based on the role instead of the agent may help build a model regardless of the distribution of the tasks. It helps formalize the problem and build a robust model. There are some role-based assignment researches [5, 14–16].

Group Role Assignment (GRA) [7, 15] is a vital methodology. It greatly improves the collaboration efficiency by seeking an optimal team execution based on agent evaluations.

By proposing the KM_B algorithm based on the RBC and its E-CARGO model, Zhu *et al.* [17] solve the M–M assignment problem. They first formalize the problem in a second order bipartite graph. Next, they solve it by improving the K–M algorithm with backtracking, i.e., KM_B, which is verified to be practical through simulative experiments.

The primary steps to use the Role-Based Collaboration include defining the roles, the agents, and the qualification matrix for the agents to be assigned with a specific role. In our scenario, the role is not clear. Consequently, we improve the RBC by appending the role awareness procedure so as to transform our problem into an x-ILP problem and use GRA to get the practically optimal solution for our research.

In summary, our research clarifies the relay points problem using RBC and its E-CARGO model. The key of Role-Based Collaboration is Group Role Assignment. It concentrates on constructing a practically optimal relay network while minimizing communication cost using a minimum number of UAVs.

6 Conclusion

A successful approach is proposed in this paper to solve the x-ILP problem with the help of PuLP package of Python. It designs a uniform solution concerning the signal relay problem of UAV in a disaster area. And thousands of random simulation experiments show that our solution can quickly find the relay points and assign the UAVs from different base stations to these points. It also demonstrates that our solution is practical, robust, and practically global optimal.

This paper formalized and solve the signal relay problem of the UAVs in disaster relief via group role assignment (GRA). In modeling, it first uses the minimum spanning tree algorithm to design the practically optimal relay network. And our proposed method in Sect. 3 can establish an awareness of and propose the relay points as, with respect to roles, which UAVs (as agents) should be assigned. Secondly, distributing the UAVs to these relay points can be solved quickly via the assignment process of GRA. Through these strategies, it can quickly recover global communication in the disaster area with the least communication cost regardless of the scale and the distribution.

There are some future works for this paper:

(1) More constraints may be required for further investigations like the mutually exclusive constraints between drones from different bases, and the multi-task assignments for each drone.
(2) Some UAV properties such as stability, endurance and maximum flight altitude raise concerns worthy of additional study.
(3) Elevation differences in a disaster area suggest that the third dimension needs to be considered when assigning drones.

Acknowledgement. We appreciate Mike Brewes for his assistance in proofreading this paper.

Thanks to National Natural Science Foundation of China for supporting part of this work under Grant No. 61402118, Natural Sciences and Engineering Research Council of Canada (NSERC) under Grants RGPIN2018-04818, S&T Project of Guangdong Province, grant No. 2016B010108007 and S&T Project of Guangzhou, grant No. 201604020145.

References

1. Liu, X., Ansari, N.: Resource allocation in UAV-assisted M2M communications for disaster rescue. IEEE Wirel. Commun. Lett. **8**(2), 580–583 (2018)
2. Wankmüller, C., Truden, C., Korzen, C., Hungerländer, P., Kolesnik, E., Reiner, G.: Base Station Allocation of Defibrillator Drones in Mountainous Regions. arXiv e-prints, arXiv: 1902.06685 (2019)
3. Maza, I., Caballero, F., Capitan, J., Martinez-De-Dios, J.R., Ollero, A.: Firemen monitoring with multiple UAVs for search and rescue missions. In: 8th IEEE International Workshop on Safety, Security, and Rescue Robotics, SSRR-2010, Philadelphia, PA, USA (2010)
4. Zhu, H., Zhou, M.C.: Role-based collaboration and its kernel mechanisms. IEEE Trans. Syst. Man Cybern. **36**(4), 578–589 (2018)
5. Liu, D., Huang, B., Zhu, H.: Solving the tree-structured task allocation problem via group multirole assignment. IEEE Trans. Autom. Sci. Eng. (2019). https://doi.org/10.1109/tase.2019.2908762
6. Liu, D., Yuan, Y., Zhu, H., Teng, S., Huang, C.: Balance preferences with performance in group role assignment. IEEE Trans. Cybern. **48**(6), 1800–1813 (2017)

7. Zhu, H., Zhou, M.C., Alkins, R.: Group role assignment via a Kuhn–Munkres algorithm-based solution. IEEE Trans. Syst. Man Cybern. **42**(3), 739–750 (2012)
8. Zhu, H.: Avoiding conflicts by group role assignment. IEEE Trans. Syst. Man Cybern. Syst. **46**(4), 535–547 (2015)
9. Mitchell, S., O'Sullivan, M., Dunning, I.: PuLP: A Linear Programming Toolkit for Python. Department of Engineering Science, The University of Auckland, Auckland (2011)
10. Geng, N., Meng, Q., Gong, D., Chung, P.W.H.: How good are distributed allocation algorithms for solving urban search and rescue problems? A comparative study with centralized algorithms. IEEE Trans. Autom. Sci. Eng. **16**(1), 478–485 (2019)
11. Aşık, O., Akın, H.L.: Effective multi-robot spatial task allocation using model approximations. In: Behnke, S., Sheh, R., Sarıel, S., Lee, D.D. (eds.) RoboCup 2016. LNCS (LNAI), vol. 9776, pp. 243–255. Springer, Cham (2017). https://doi.org/10.1007/978-3-319-68792-6_20
12. Singh, S., Singh, R.: Earthquake disaster based efficient resource utilization technique in IaaS cloud. Int. J. Adv. Res. Comput. Eng. Technol. **2**(6), 225–229 (2013)
13. Fortino, G., Russo, W., Savaglio, C., Shen, W., Zhou, M.: Agent-oriented cooperative smart objects: from IoT system design to implementation. IEEE Trans. Syst. Man Cybern.: Syst. **48**(11), 1939–1956 (2018)
14. Nair, R., Tambe, M.: Hybrid BDI-POMDP framework for multiagent teaming. J. Artif. Intell. Res. (2015). https://doi.org/10.1613/jair.1549
15. Burkard, R., DellAmico, M., Martello, S.: Assignment Problems (Revised Reprint). SIAM, Philadelphia (2012)
16. Dastani, M., Dignum, V., Dignum, F.: Role-assignment in open agent societies. In: Proceedings of the Second International Joint Conference on Autonomous Agents and Multiagent Systems, pp. 489–496. ACM, New York (2003)
17. Zhu, H., Liu, D., Zhang, S., Zhu, Y., Teng, L., Teng, S.: Solving the many to many assignment problem by improving the Kuhn-Munkres algorithm with backtracking. Theor. Comput. Sci. **618**(7), 30–41 (2016)

Service Discovery Method for Agile Mashup Development

Bo Jiang, Yezhi Chen, Ye Wang$^{(\boxtimes)}$, and Pengxiang Liu

Computer and Information Engineering, Zhejiang Gongshang University,
Hangzhou 310018, China
yewang@mail.zjgsu.edu.cn

Abstract. With the rapid expansion of services on the Internet, Mashup development has become a trend toward mainstream development. How to efficiently and quickly discover available services in Mashup development and make full use of existing services to meet the changing needs of users has become a new concern. Although there are a lot of work for service discovery, there are still some problems in the existing methods, such as limiting the service description to a single structured document, limiting the service search statement to keywords, and rarely mining the deeper semantics of the service text. information. In view of the above problems, this paper proposes the Service Discovery approach for Agile Mashup Development (SDAMD), which breaks the limitation of the single document and drives the user story in agile development as a service search. The original text, through the natural language processing technology, extracts the three elements of agile requirements, and then extracts the three service attributes of the agile service; then finds and recommends similar services by calculating the similarity between the service description and the search text. This article uses the real data of the services on the Programmable Web to verify the validity of SDAMD.

Keywords: Service computing · Service requirements · Agile development · Service matching · Functional semantics

1 Introduction

Today, the Internet is moving from data and web applications to data and service platform frameworks, encouraging users to create their own applications (Users that is developers). As a result, there is an increasing requirement for support for content reuse, collaborative sharing, and user content compilation principles [1]. Mashups have attracted much attention as a class of composite, Shared Web applications and development approaches. At present, the increasingly rich open service resources have provided sufficient basic support for Mashup [2] development. At the same time, users face many challenges in developing Mashup applications. First, the quality of open services is uneven [3]; Secondly, the large number of service resources makes it difficult to make decisions [4]. Furthermore, how to respond to changing, dynamic and fuzzy user requirements in a timely manner is also a problem. This makes service discovery methods and related technologies particularly important. Many of the

© Springer Nature Singapore Pte Ltd. 2019
Y. Sun et al. (Eds.): ChineseCSCW 2019, CCIS 1042, pp. 30–49, 2019.
https://doi.org/10.1007/978-981-15-1377-0_3

existing service discovery methods match only by keywords and cannot keep up with the speed of large-scale service growth. At the same time, the ability to express common techniques for describing and discovering services is poor for users.

In order to overcome the above obstacles, This paper proposes a service discovery method for agile [5] development–SDAMD (service discovery method for agile mashup development). This method introduces the user story concept of agile development to capture user requirements, and refines the fuzzy, dynamic and changeable user requirements in an agile way, and combines the refined user requirements with product development to develop in a circular and iterative manner. In each iteration process, three kinds of agile requirement elements are extracted from the user search text, that is, three elements of agile requirement. Then extract the similar three elements of agile service from the service description. Finally, the similarity calculation is carried out for the three elements of agile requirement and the three elements of agile service. Here, in order to realize the semantic similarity calculation of the extracted data, a new similarity calculation method is proposed in this paper. Finally, we use the real data of the API service on PWeb as an example to verify the effectiveness of the proposed method.

The rest of the paper is structured as follows: Sect. 2 discusses service discovery and text mining related work based on agile development. Section 3 details the SDAMD method, including service text preprocessing, service requirement element extraction, and similarity calculation. The fourth section compares the SDAMD method. Finally, the conclusions and outlook.

2 Related Work

Service discovery refers to the calculation of the matching algorithm based on the user's functional and non-functional requirements and constraints on the expected service [6]. Retrieving service sets that meet user requirements from the service library, efficient and accurate service discovery is an important prerequisite and foundation for service reuse. The general service discovery method is divided into two main steps: one is based on the extraction of key elements of the service description, and the other is based on the key elements for service matching calculation. Many research scholars at home and abroad have done a lot of related research.

2.1 Service Description Method

Service description is one of the two key points to solve the problem of service discovery. The service description is both a description of the available operations of the service and the expected implementation of the service by the requester in special cases. Service discovery methods can be divided into two categories based on the richness of description information in the service description document:

First, the grammatical level description language, which focuses on the description of the service interface syntax, a typical example is the tool proposed in the literature [7] using metaprogramming and other related technologies for text mining and service processing of WSDL documents.

Second, the semantic level description language compensates for the heterogeneity defects in the traditional grammatical level service description and enhances the semantics in the service function and behavior description. For example, the RESTful-style service based on natural language description [8], Literature [9] proposed a multi-hybrid clustering MHC method, which uses the vector space model (VSM) to cluster similar documents on the web document described by OWS-L ontology language.

Semantic-level services are becoming more popular than traditional-based grammar-level services. However, due to insufficient semantic information extraction, most of the service discovery methods are based on keyword matching for service matching based on WSDL document services, that is, biased towards traditional grammar-level services, and a few service discovery methods support semantic-level services [7, 10]. The structured nature of traditional SOAP service description documents leads to the lack of semantic information for contextual links in documents. This means that the traditional method of service discovery does not work at the semantic level. For example, "book hotel" and "order hotel" mean the same thing, but the grammar level service does not link the two well. Therefore, how to carry out deep text mining on the semantic level becomes a big challenge.

2.2 Service Matching Algorithm

The service matching algorithm refers to comparing the degree of matching between two descriptions by some algorithm. Service discovery is a process of mapping service search statement to service requirement description, thus transforming the process of service discovery into a process of matching service requirement description with service description. At present, domestic and foreign scholars have done a lot of research on service discovery, which can be roughly divided into two categories from the perspectives of grammar and semantics:

The grammar-based service discovery method is based on keyword discovery. This type of method is based on grammatical level description language, which is mainly based on simple classification and keyword matching. For example, [11] is to cluster the service registration information based on service description and UDDI, and then use the potential semantic index (LSI) to achieve the matching. In general, the existing grammar-level matching algorithms are relatively simple, but the precision rate is low, and it is difficult to ensure the compatibility of service composition.

Semantic-based service discovery method, using ontology to enhance semantic descriptive literature [12], literature [13] combined with subgraph matching, transforming service matching problem into bipartite graph expansion best matching problem, combined The topic vector model [14] mines the underlying topics of the document. In general, the existing semantic level matching algorithm is not conducive to the performance of the matching algorithm due to the complexity of natural language.

In addition to the deficiencies mentioned above, existing service discovery methods all pay too much attention to the algorithm implementation of discovery and ignore the application scenarios of service discovery. As a result, few discovery methods and development methods are combined: Existing service discovery methods only start from the service description of the rule or the service tag to match the corresponding service and are out of line with the user's needs. In software development, user requirements can be used to improve the understanding of the development team's goals, improve team project awareness, and thus improve team efficiency and product quality. Rapid response requirements in Mashup application development have become a major challenge.

Based on the above analysis, this paper proposes to use the user story template and the method of syntactic dependency to try to break through the constraints of traditional requirement extraction and service description information extraction, and improve the semantic quality of the acquired information to achieve the expectation of improving the calculation effect. For the service matching algorithm, this paper proposes a semantic similarity calculation method combined with WordNet, which makes the similarity calculation more semantic and achieves the expectation of semantic computing.

3   SDAMD Method

The SDAMD method uses user stories as an entry point to extract requirements. Analyze the syntactic dependencies [16] of the requirements document, and extract the requirement elements in the service requirement text using natural language processing techniques. At the same time, the service element is extracted from the service description file, defined as the service attribute, and then the similarity between the service requirement element and the service attribute is measured. Finally, the service to be recommended is sorted according to the similarity, and the most similar N service lists are returned to the previous service list. User, for their choice. The following sections focus on service requirement element extraction and service attribute extraction. Figure 1 shows the overall framework of the service recommendation method in this paper. The framework is divided into three parts:

(1) Extract the service requirement elements written with the user story as a template;
(2) Extraction of service attributes;
(3) Semantic matching calculation

The data set for the second part of the Fig. comes from the service registration platform PWeb.

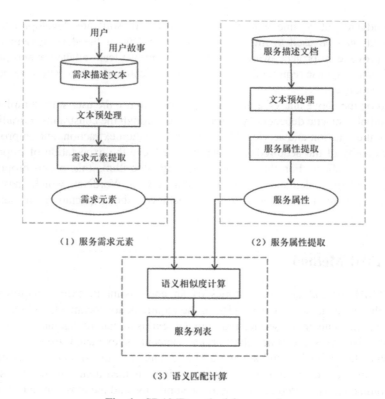

Fig. 1. SDAMD method framework

3.1   Service Requirement Element Extraction

Service Requirement Element. Service requirements are stored in the service requirements text and are the most direct and specific form of user requirements. Traditional service requirements texts are described in natural language [17, 18]. In this article, the user story [19] is used as a template to describe the specific service requirement elements. For example, As a music enthusiast, I want to search music in website so that I can listen music in website. Service requirement elements refer to the key elements that can reflect service functional requirements from the perspective of users in software engineering, including the role of finding services, requirements on services, and the limited scenarios of requirements, as shown below:

(1) Role: The role who looking for and using services; in the example, it refers to: "music enthusiast".
(2) Goal: requirements for finding a service; in the example, it refers to "search music in website."
(3) Motivation: The purpose of the user requesting the service. in the example, it refers to "listen music in website."

Definition 1. (service requirement element) sre = <ro,go,mo>, where ro represents the user to whom the service requirement is oriented, acting as an element of the object or subject in the sentence; go represents a set of service function requirements consisting of verbs, nouns and adjectives, which is a manifestation of service functions; mo represents the purpose and motivation of the service, and is also a set of requirement motivation consisting of verbs, nouns and adjectives.

Definition 2. (user) ro = <noun, adj>, noun is the specific role and identity information of the user. Adj denotes an adjective, which is an extension of the description of the role and identity information.

Definition 3. (Service function requirement set) go = <verb, do, io, adj>, which is called the service function requirement set because there may be multiple cases in which the requirements may be juxtaposed. The verb is the operation initiated by the role. do is the direct bearer of the role action. io represents an indirect object that provides a defined set of requirements. adj is a description expansion of the corresponding noun. Such as service function requirements {upload, music, website}.

Definition 4. (Requirement Motivation Set) mo = <verb,do,io,adj>, ibid., as requirement motivation {listen, music, website}.

Service Requirement Element Part of Speech Tagging. Polysemy is the universality of language. In the text described by natural language, there are some words that can act as nouns and can also act as verbs. But in context, the meaning of verbs and nouns is often quite different. For example, "book cheap hotel" and "English book" also have the word "book", but the meaning is different. Therefore, in the process of extracting the requirement elements, the gov and dep in the SD are respectively marked with the part of speech. For example, the gov in the dobj(gov, dep) acts as a verb, and the dep is a noun; the nn (gov, dep) in the gov and dep All are nouns.

Requirement Element Extraction. This paper extracts the service requirement elements based on the syntax dependency relationship. First, a parse analysis of the service requirement text is required through natural language processing techniques. This article uses the open source tool Stanford Parser developed by The Stanford NLP Group for parsing of text. Stanford Parser is the mainstream natural language processing tool, and we mainly use it to achieve the following functions: (1) word part recognition and labeling; (2) temporal restoration; (3) extraction of Stanford Dependency (SD) collection. Finally, extracting elements based on the SD collection, the specific extraction process is as follows:

Service User Extraction. We fully analyzed the SD defined in the Stanford Dependencies Manual, and summarized the following three scenarios, which can cover the possibility of all combinations of service users:

(1) pobj(As, dep):

Since the service user description part of the user story template is As a..., the corresponding term can be extracted through As, and the main noun in the Role can be extracted. For example, As a music enthusiast, you can get the dependency pobj(As, enthusiast), you can get enthusiast directly.

(2) nn(gov, dep):

gov is the main noun, and dep is the modified noun. All modified nouns can be extracted from the sentence. For example, As a music enthusiast, you can get the dependency nn (enthusiast, music), you can get enthusiast, music directly.

(3) amod(gov, dep):

gov is the main noun and dep is the adjective. All adjectives be extracted. For example, As a crazy music enthusiast, you can get enthusiast, crazy from dependency amod(enthusiast, crazy).

Service Function Information Set. In the process of extracting the service function information set, due to the complexity of natural language, some service function information needs to integrate several dependencies for combined reasoning and judgment. To cover all possible combinations of all service feature requirements, at least 8 scenarios are required:

(1) xcomp(want, dep):

The want is the word of the service function requirement description part (I want to…) in the user story template, and the first verb indicating the user-initiated action is found by want.

(2) dobj(gov, dep):

Gov is a verb, an action initiated by a user, through which the corresponding direct object is found. For example, "I want to search music in website", which contains the dependency dobj (search, music), can directly get search, music.

(3) iobj(gov, dep):

Gov is a verb, an action initiated by a user, through which the corresponding indirect object is found. For example, "gave customer a raise", which contains the dependency iobj(gave, customer), can get gave, customer directly.

(4) prep(a, b) & pobj(b, c):

If a is verb, we get the verb a and the preposition b through the dependency prep, a represents the action initiated by the user. At the same time, b is a preposition in prep, and the object can be found through the combination of relations. For example, "play with students", which contains dependencies prep(play, with) and pobj(with, students), can be converted indirectly to play students.

If a is noun, the noun a and the preposition b and b are obtained by pobj, and in the prep, the corresponding object can be found through the combination of relations. For example, "search music in website", which contains the dependencies prep (music, in) and pobj (in, website), can be converted indirectly to music and website.

(5) conj(gov, dep):

Both gov and dep are verbs that find all verbs that are in parallel with each other through the original verb. For example, "Upload and download", which contains the dependency conj (upload, download), can be directly converted to upload and download.

(6) pobj(gov, ep):

Find all indirect objects through this relationship. Such as in website, you can directly convert to get the website.

The last two dependencies (7) nn and (8) amod have been described in detail in the previous section and will not be described here.

Requirement Motivation Extraction. Since the service function requirement extraction and the requirement motivation extraction logic are almost the same, we only need to change the first extraction scenario, and the rest of the scenarios are exactly the same as those in the previous section service function requirement extraction, and therefore will not be described again.

(1) aux(dep, can):

Can is the word of the service function requirement description part in the user story template. The can find the first verb in the service scene that indicates the user initiated the action.

Data Cleaning and Part of Speech Eestoration. There are a small amount of noise in the extracted service requirement elements, such as ask, focus, and other words. We organize the words that need to be filtered into a stop word list. The service requirement element needs to filter some words that do not reflect functional semantics by using the stop word table to reduce the noise impact. Increasing the accuracy of extraction requires not only filtering vocabulary that has no practical meaning, but also requiring morphological restoration. Because of the complexity of natural language, temporal and singular-plural changes will also affect the next step of semantic similarity calculation.

By referring to the above dependency scenario, the specific service requirement element triplet is extracted as follows:

Algorithm 1. Service requirement element extraction algorithm

Input: a single user story based service description text text

Output: SRE record

1. *listSD=extract(text)*// extract dependency set listSD
2. FOR *sd* in *listSD*: // Requires element extraction based on dependencies
3.   IF(*sd* meets 3 scenarios in the service user extraction):
4.   *ro.append(sd)*
5.   ELSE IF(*sd* meets 8 scenarios for service function extraction):
6.   *go.append(sd)*
7.   ELSE IF(*sd* meets 8 scenarios extracted from requirement motivation):
8.   *mo.append(sd)*
9.   END IF
10. *SRE=create(ro, go, mo)* // Synthesize the set of requirement elements
11. FOR *sre* in *SRE*: // Data cleaning and part of speech reduction for the set of requirement elements
12.   *removeStopWord(sf)*
13.   *Lemmatization(sf)*
14. END FOR

3.2   Service Attribute Extraction

The service attribute is stored in the service description text, and the service description text is a function of the service, a record of the detailed information, and a storage carrier, which is written by the developer to assist the user in understanding the service. Existing service description texts are written in natural language in any format. For example: Customers can use the service to book hotels over the Internet.

A service attribute is a set of attributes that are extracted from the service description text and correspond to the service requirement element. It is a set of concerns that can reflect the service function. It includes the location service target of the service, the function information of the service, and the application scenario of the service. As follows:

(1) Service target (Agent): The object that the service is oriented to, in the example, refers to Customers, which is the main service object of the service.
(2) Activity: The function provided by the service. The service target needs to perform certain special operations and the desired effects after the operation. In the example, it is book hotels.
(3) Scenario: The usage scenario of the service, in the example, the Internet.

Definition 5. (service attribute) sa = <ag,ac,sc>,where ag represents the user facing the service requirement, acts as a component of the object or subject in the sentence; ac represents the service key function information set consisting of verbs and nouns, embodies the service function; sc represents the purpose, scene, etc. of the called service, such as ac refers to how to affect the user, the special scene of the user calling the service.

Definition 6. (Service feature set) ag = <verb, object>, in the form of a binary group. The verb is the operation initiated by ag. Object exists in the form of nouns, noun phrases, or noun phrases in a two-tuple, such as the service key function {upload, music}.

The service description text is described by natural language, has no format, and has no template. Therefore, there is no obvious regional division in each part, and it can only be distinguished by dependency. We cover all possible relationships through 14 dependencies. details as follows:

(1) nsubj(gov, dep):
Dep is a noun or noun phrase that acts as a subject in a clause. Gov and dep can be used as the verb part and the noun part of the basic service function information. For example, the sentence "Users can effectively control the zoom level of the image" contains the dependency nsubj(control, users), which gives control, users.
(2) nsubjpass(gov, dep):
Gov represents the main verb in a clause, and dep is a noun or a noun phrase that acts as a subject in a clause. The sentence "The size of the image can be specified." contains nsubjpass(specified, size), which gives you the specified, size.
(3) xsubj(gov, dep):
Gov represents the main verb in the clause, and dep is the subject of the verb. The sentence "Tom likes to eat fish" contains the dependency xsubj(eat, Tom), which gives you eat, Tom.

(4) agent(gov, dep):

Usually followed by the preposition "by", gov is the verb of the past tense, and dep is the supplementary noun. The sentence "The man has been killed by the police" contains the dependency agent (killed, police), which can be directly converted to {killed, police}.

(5) csubj(gov, dep):

Gov is the action initiated by the service target, and dep is the main component of the subject clause, through which the corresponding predicate verb can be found. For example, "What she said makes sense", which contains the dependency csubj (makes, said), can get makes, said.

(6) csubjpass(gov, dep):

Gov is a predicate verb (passive), dep is the main component of the subject clause, and can find the corresponding subject and predicate verb. Such as "That she lied was suspected by everyone", which contains the dependency csubjpass (suspected, lied), you can get suspected, lily.

(7) cop(gov, dep):

Where gov is the expression and dep is the verb. For example, "Bill is an honest man" cop (man, is), thereby obtaining the corresponding object.

The remaining 7 scenes: nn, dobj, iobj, prep&pobj, pobj, amod, conj. They have been elaborated in the last two sections, so I will not repeat them here.

Referring to the above scenario, the specific method for extracting service attributes is as follows:

Algorithm 2. Service attribute extraction algorithm

Input: a single user story based service description text text_S

Output: SS record

1. *listSD=extract(text_S)*// extract dependency set listSD
2. FOR *sd* in *listSD*:   // Service attribute extraction based on dependencies
3.   IF(*sd* meets the scenario in the service target extraction):
4.   *ag.append(sd)*
5.   ELSE IF(*sd* conforms to the scenario of service function extraction):
6.   *ac.append(sd)*
7.   ELSE IF(*sd* conforms to the scenario of service function extraction):
8.   *sc.append(sd)*
9.   END IF
10. *SS=create(ag, ac, sc)* // Combine the service attribute set
11. FOR *sre* in *SS*: // Data cleaning and part of speech restoration of service attribute set
12.   *removeStopWord(sf)*
13.   *Lemmatization(sf)*
14. END FOR

3.3   Service Similarity Measure

This section focuses on how to calculate the similarity between the user story q and the registration service s through the service requirement element and the service attribute. Here, the user story q analyzes the syntactic structure through the natural language processing technique, the morphological restoration, and generates the service requirement element set se according to the agile three elements; the registration service s also generates the service attribute set ss through the related operations such as the natural language processing technology.

The specific formula for calculating service similarity is as follows:

$$sim(q, s) = a \times usim(u_q, u_s) + basim(SF_q, SF_s) + dgsim(g_q, g_s) \tag{1}$$

The weight of the three requirement elements in Eq. (2), the parameter $\alpha + \beta + \gamma = 1$. In the experimental part we set $\alpha = 0.2$, $\beta = 0.6$, $\gamma = 0.2$ Specific parameter setting instructions can be seen in the next section of the experimental analysis. $usim(uq, us)$ represents the similarity between the Role element in the service requirement text q and the Agent element in the registration service s, $asim(SFq, SFs)$ represents the similarity calculated between the conjugate and the Action element in s, the combination is obtained by combining av and do of the Goal element in q with av and do in the Motivation element. $gsim(gq, gs)$ a is the similarity between the combination of io and adj in the Goal element and the Motivation element in q and the Scene element in s. The specific formula is as follows:

$$usim(u_q, u_s) = \begin{bmatrix} sim(w_{q_1} w_{s_1}) & \cdots & sim(w_{q_1} w_{s_j}) \\ \vdots & \ddots & \vdots \\ sim(w_{q_k} w_{s_1}) & \cdots & sim(w_{q_k} w_{s_j}) \end{bmatrix} \tag{2}$$

$$gsim(g_q, g_s) = \begin{bmatrix} sim(w_{q_1} w_{s_1}) & \cdots & sim(w_{q_1} w_{s_j}) \\ \vdots & \ddots & \vdots \\ sim(w_{q_k} w_{s_1}) & \cdots & sim(w_{q_k} w_{s_j}) \end{bmatrix} \tag{3}$$

$$sim(w_i, w_j) = \begin{cases} 1, & if \ w_i \ equals \ to \ w_j \\ WNSim(w_i, w_j), & otherwise \end{cases} \tag{4}$$

In the matrix (2), k means that uq is composed of k words, and j means that us consists of j words. Then iterate through the entire matrix, taking the maximum value of each row as the similarity between each word of the Role element in q and each word of the Agent element in s. And put these similarities into the set u_K, the specific form of u_K is $\{u_1, u_2, \cdots, u_K\}$. The matrix (3) can be obtained according to the same method.

Then use the method of averaging to obtain the similarity $usim(u_q, u_s)$ and the similarity $gsim(g_q, g_s)$ respectively.

$$usim(u_q, u_s) = \frac{1}{k} \sum_{i=1}^{k} u_i \tag{5}$$

$$gsim(g_q, g_s) = \frac{1}{k} \sum_{i=1}^{k} u_i \tag{6}$$

In formula (4), $WNSim(w_i, w_j)$ is the semantic similarity between two words calculated by WordNet.

$$asim(SF_q, SF_s) = \frac{\sum_{i=0}^{m} sim(SF_{qi}, SF_{s1 \sim sn})}{n} \tag{7}$$

Where SFq is a service requirement function information set corresponding to the user story q, and SFs is a service function information set corresponding to a single service in the data set. n is the number of parties having more service function information sets in q and s, and m is the number of parties having fewer service function information sets. $sim(SFqi, SFsi)$ represents the similarity of individual service service function information in q and s. Calculated as follows:

$$sim(SF_{qi}, SF_{si}) = w1 \times Sword(V1, V2) + w2 \frac{\sum_{i=1}^{m} \max(Sword(N_i, N_{1 \sim n}))}{n} \tag{8}$$

In the formula (8), is a verb contained in the function information, is a verb contained in the function information, is a noun contained in the function information, is a noun contained in the function information, In the formula, is the verb weight, and is the noun weight. Because the nouns are as important as the verbs in the service function information, set, in the experiment. and are semantic similarities between words, calculated by WordNet.

4 Experimental Analysis

We conducted a series of experiments to evaluate the algorithms extracted in this paper, including service service requirement elements, service attribute extraction algorithms and service discovery algorithms. All experiments were carried out on a platform developed based on the Python language. The operating environment was Win7 operating system, I5-2400 CPU, memory 4G, clocked at 3.2 GHz.

4.1 Experimental Data Collection and Analysis

PWeb is a well-known API registration and retrieval platform on the Internet. The platform contains a wide variety of APIs, covering various fields. As of now, the site has registered more than 10,000 API services. In the API page, detailed information about the API is listed, including description information, provider, API home page, category, service name, request and return format. It has been observed that the service description information is described in natural language and has a short length and is presented in short text. The number of words is about 100 words. The service description text is stored in the service requirements document and serves as a description of the service. Therefore, the API service description information provided on the Pweb can be used as the service requirement text, which meets the requirements of the data of this paper.

In the experimental data collection phase, by writing a crawl script, parsing the corresponding URL, simulating the browser operation, crawling all the information about the API services and services on the PWeb, and storing them in the database. During the crawl process, we designed some rules to filter non-compliant API services: (1) The service description text information is empty; (2) the service classification information is empty; (3) the abandoned service. Deposit the useful services into the database and wait for further analysis. At the same time, since the service registration of the platform is developed for all users, there are individual services that are repeatedly submitted. For duplicate service issues, we only keep one service. Through the above rules, 15928 API service entries were finally collected, including 398 categories.

Before the experiment, because the service description text was written by the user, and because of the randomness and non-standardity of natural language, There are the following problems: (1) there are special characters, such as: "â", "$", "&", etc.; (2) the description of the service requirements is too short; (3) the service description text has nothing to do with the service; The above problems will affect the accuracy of sentence extraction. Therefore, preprocessing is required before the service description elements are extracted. For question (1), we will filter or replace special characters, such as removing "â" and converting "&" into "and"; for question (2), we set a threshold, and the required text is automatically filtered after the number of words is less than the threshold. For the problem (3), eliminate this type of service. After text preprocessing, this experiment selected 1438 services, the fields are: Email (263), Video (258), Transportation (256), Photos (214), Travel (244), Music (203), which The number in parentheses represents the number of choices in the current field.

In order to evaluate the effects of the description element extraction algorithm and the service discovery algorithm, we constructed a manual standard set. Due to the huge workload, we assigned three agile development programmers to complete the establishment of the standard set. Three programmers created three different sets of standards. Finally, we used the precision to evaluate the effect of the algorithm. Since the three programmers had different understandings, after calculating the precision of each copy, the three averages were taken as the last evaluation data.

In this paper, we introduce the Jacard similarity coefficient (Jacard similarity coefficient) into the accuracy calculation. Jacard formula is as follows:

$$J(E, F) = \frac{|E \cap F|}{|E \cup F|} \tag{9}$$

Where E and F represent two different sets and the number of the two is different. The precision formula is as follows:

$$Precision = \frac{|S_A \cap S_M|}{|S_A|} \qquad (10)$$

The SA indicates that the service extracts the requirement element or the recommended service, and SM represents a manually extracted service requirement element or a recommended service.

4.2 Evaluation of Experimental Results

Extraction Algorithm Evaluation. According to the algorithm for extracting service attributes according to this paper, generate service attribute SS for each service s. Table 1 shows some of the service attribute extractions. Table 2 shows the service attribute extraction for agile Mashup development. In the 1438 service description texts, the number of user extractions is 1431, the extraction rate is 99.5%, the number of service function extractions is 1403, the extraction rate is 97.4%, and the number of service targets 1 is 1418, and the extraction rate is 98.6%. Due to the non-normative part of natural language, the description does not conform to the English grammar or the statement is incomplete, and the sentence component is missing, which can not be parsed by the extraction algorithm, so the extracted service attribute is empty, which affects the service attribute extraction rate.

Table 1. Service attribute extraction

Name	Category	Service attribute
PicMonkey	Photos	<null, null, {Picnik maker}>, <null, {upload image, edit image, save image}, null> <documention, <null, {contact provider}, {talk@picmonkey.com}>
Google Maps	Mapping	<null, null, <{embedding, map}> <{language localization, geocoding}, {utilize api}, {language developer, intranet}> <api http service, {access http service}, {connection, customer}>
Google-Adsense	Advertising	<null, {create content, generate content}, {developer, blogg}> <null, {choose, report, generate report, share revenue, share program}, {adsense, site}> <account, {create account, generate snippet, generate filter}, null>

Table 2. Service attribute extraction

Kind	Quantity	Extraction rate
Number of services	1438	100%
User extraction number	1431	99.5%
Service function extraction number	1403	97.4%
Service goal extraction number	1418	98.6%

According to the algorithm for extracting service requirement element extraction in this paper, a service requirement element SE is generated for each user story q, and Table 3 shows the extraction of some user stories. Table 4 shows the extraction of service requirement elements for agile Mashup development. In the 200 user stories, the number of user extractions is 200, the extraction rate is 100%, the number of service function extractions is 199, the extraction rate is 99.5%, and the number of service scene extractions is 148 (due to the fact that the 52 user stories themselves are not Describe the motivation of the requirement), the extraction rate is 100%. It can be seen that the extraction effect is superior when the template constraint is slightly applied. One of the user story grammars is not in conformity with the specification, resulting in an algorithm that cannot be parsed, reducing the extraction rate.

Table 3. Service requirement text (user story) extraction situation

User story	Number	Service Category	Service requirement element
As a user, I want to upload and edit photos online on server	S1	Photo	<user, upload photos; edit photos, server>
As a developer, I want to upload and search music in application	S2	Music	<developer, upload music; search music, application>
As a developer, I want to upload and share videos from application to store them forever	S3	Video	<developer, upload video; share video, application, store>
As a music enthusiast, I want to search music so that i can listen music in website	S4	music	<enthusiast, search music; listen music, website>
As a traveler, I want to search trip and hotel on website so that I can plan trip	S5	Travel	<traveler, search trip; search hotel; plan trip, website>

Table 4. Service requirement element extraction

Kind	Quantity	Extraction rate
Number of user stories	200	100%
User extraction number	200	100%
Service function requirement extraction number	195	99.5%
Requirement motivation extraction number	148	100%

Service Discovery Algorithm Parameter Selection. The service discovery algorithm proposed in this paper has three parameters α, β, γ, which correspond to users, function, motivation, respectively. $\alpha + \beta + \gamma = 1$. We developed service description texts for three different areas (Photo, Video, Music) as input and compared them to three standard sets. At the same time, we compare the precision of different numbers, such as Top5, indicating the precision when the number of services is 5. As can be seen from Fig. 2, the abscissa indicates the range of services 5–30, and the ordinate indicates the

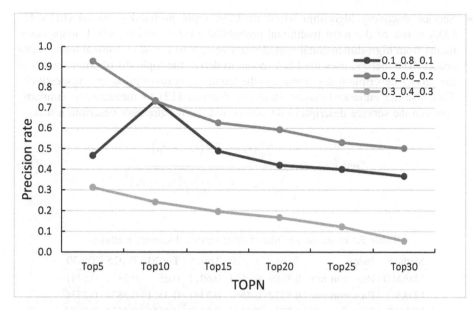

Fig. 2. Precision of different parameters.

percentage of Precison in the case of the corresponding TOPN. When the α, β, and γ values were 0.2, 0.6, and 0.2, the precision was substantially higher than that of the other groups. This indicates that when calculating the similarity between the user story q and the service description text, the user and the motivation are necessary, but the function should be emphasized, so the weight given to the function is relatively high. Since the experiment requires a lot of time and effort, after setting the α, β, and γ values to 0.2, 0.6, and 0.2, we select 5 user stories (see Table 3 for experiments).

Comparative experiment. This section compares the SDAMD method with the existing service discovery method VSMSD method [20] and the LDA topic vector method [10] to verify the validity of the SDAMD method. The following two algorithms will be introduced separately:

- Service Discovery Algorithms Based on Vector Space Model (VSMSD): Within the method, the user story q is preprocessed by the bag of words (Bag of Word). Expressed as a vector form $q = \{v1, v2, v3, v4, ... vi\}$, $i = $ the size of the corpus vocabulary. At the same time, the data in the data set service request text is preprocessed and the word frequency is statistically calculated, and finally expressed as a vector $s = \{v1, v2, v3, ..., vi\}$ (i ditto). Finally, the similarity between the vector q and the vector trumpet is calculated by similarity calculation formula. The formula is as follows:

$$\cos(\overrightarrow{q}, \overrightarrow{s}) = \frac{\overrightarrow{q} \cdot \overrightarrow{s}}{||\overrightarrow{q}||\,||\overrightarrow{s}||} = \frac{\sum_{i=1}^{v} \overrightarrow{q} \cdot \overrightarrow{s}}{\sqrt{\sum_{i=1}^{v} \overrightarrow{q}_i^2 \sum_{i=1}^{v} \overrightarrow{s}_i^2}} \tag{11}$$

- Service discovery algorithm based on LDA topic probability model (LDASD): LDA is one of the most traditional probabilistic topic models, which maps documents from high-dimensional spatial word vectors to low-dimensional topic vector spaces. The method uses the LDA model to derive the topic distribution vectors of the service description document and the requirement description text, respectively. Then use the enhanced cosine similarity formula (12) to measure the similarity between the service description document and the requirement description text.

$$Sim(a, u) = \frac{\sum_{i \in I} \left(r_{a,i} - \bar{r}_a\right)\left(r_{u,i} - \bar{r}_u\right)}{\sqrt{\sum_{i \in I} \left(r_{a,i} - \bar{r}_a\right)^2}\sqrt{\sum_{i \in I} \left(r_{u,i} - \bar{r}_u\right)^2}} \tag{12}$$

Table 5. Evaluation results of three service discovery methods

Method	Index	Top5	Top10	Top15	Top20	Top25	Top30
SDAMD	Precision rate	0.8889	0.7611	0.6911	0.65	0.58	0.5741
LDASD	Precision rate	0.5222	0.4667	0.3741	0.3511	0.3874	0.2556
VSMSD	Precision rate	0.8011	0.6444	0.5444	0.4556	0.4574	0.3667

We have developed five service user story statements, which are measured using the precision index for each statement q.

Figure 3 shows the precision of each service discovery method. We can see from the Fig. 3 that the method proposed in this paper is better than the VSMSD and LDASD methods. The SDAMD method has a precision of more than 0.88, which is far better than the LDASD method and has obvious advantages over the VSMSD. This means that the SDAMD method can query to match most of the associated services, while other methods match some of the unrelated services. The downward trend of SDAMD between top20 and top15 is obvious. The main reason for this phenomenon is that the extraction of service requirement elements is not complete. Especially in the following scenarios, Stanford Parse can't parse well: (1) service function information is represented by noun phrase, e.g. "photo uploading"; (2) sentence structure is incomplete or grammatical error.

It can be seen from Fig. 3 and Table 5 that when the three methods are applied to the five requirement statements in Table 3, the SDAMD algorithm proposed in this paper finds a higher precision than LDASD and VMSSD, and has the best discovery effect on service discovery.

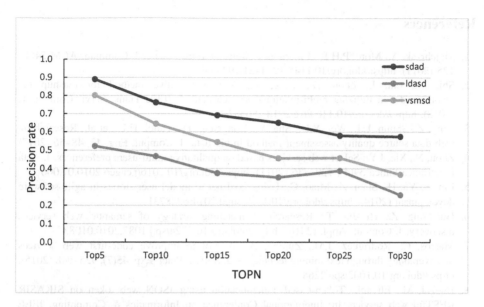

Fig. 3. Comparison of precision ratios of SDAMD, LDASD, and VSMSD methods.

5 Conclusions and Prospects

We propose a service discovery method SDAMD for agile mashup development, which uses natural language processing methods to analyze the grammatical structure of sentences. Given the corresponding agile requirements, i.e. user stories, extract service requirement elements, measure the similarity between user stories and services, and find the N services that are most similar. The data comes from the API service record on PWeb, and the experiment validates the effectiveness of our method. The experimental results show that our method is better than other traditional methods in terms of performance.

Next we will consider: (1) Due to the complexity of natural language, we can not only extract key elements, regardless of the order between words and words, which will have a certain impact on the accuracy of service discovery. Therefore, the next work will consider introducing Word Embedding and Attention, which will express each word as a word vector containing semantic information and word order, and calculate the service requirement document by word vector. The similarity between the search and the requirements search document. (2) At the same time, after training the word vector of a certain dimension, the naive Bayes is used to extract the key features of the service with semantic word vectors, which improves the accuracy of feature extraction and improves the accuracy of service discovery.

References

1. Majchrzak, A., More, P.H.B.: Emergency! web 2.0 to the rescue! J. Commun. ACM **54**(4), 125 (2011). https://doi.org/10.1145/1924421.1924449
2. Shi, M., Liu, J., Zhou, D., Tang, Y.: A topic-sensitive method for mashup tag recommendation utilizing multi-relational service data. J. IEEE Trans. Servi. Comput. 1, (2018). https://doi.org/10.1109/tsc.2018.2805826
3. Xing, Z., Shijun, L.I., Wei, Y.U., Sha, Y., Yonggang, D., Yahui, H.U., et al.: Research on web data source quality assessment method in big data. J. Comput. Eng. **43**, 48–56 (2017)
4. Zhou, N., Xie, J.Y.: Select web services based on qualitative multi-users preferences. J. Acta Electronica Sinica **39**(4), 729–736 (2011). https://doi.org/10.1016/j.cageo.2010.07.006
5. Lin, J., Yu, H., Shen, Z., Miao, C.: Using goal net to model user stories in agile software development (2014). https://doi.org/10.1109/snpd.2014.6888731
6. Hui-Ming, Z., Hui-Jia, T.: Research of matching strategy of semantic web services discovery. J. Comput. Appl. (2010). https://doi.org/10.3724/sp.j.1087.2010.01083
7. Mateos, C., Rodriguez, J.M., Zunino, A.: A tool to improve code-first web services discoverability through text mining techniques. J. Softw.: Pract. Exp. **45**(7), 925–948 (2015). https://doi.org/10.1002/spe.2268
8. Haekal, M., Eliyani.: Token-based authentication using JSON web token on SIKASIR RESTful web service. In: International Conference on Informatics & Computing. IEEE (2017). https://doi.org/10.1109/iac.2016.7905711
9. Yi-Song, L., Yu-Cheng, Y.: Semantic web service discovery based on text clustering and similarity of concepts. J. Comput. Sci. **11**, 46 (2013)
10. Zhang, N., Wang, J., He, K., Li, Z.: An approach of service discovery based on service goal clustering. In: IEEE International Conference on Services Computing IEEE (2016). https://doi.org/10.1109/scc.2016.22
11. Paliwal, A.V., Shafiq, B., Vaidya, J., Xiong, H., Adam, N.: Semantics-based automated service discovery. J. IEEE Trans. Serv. Comput. **5**(2), 260–275 (2012). https://doi.org/10.1109/TSC.2011.19
12. Roman, D., Kopecký, J., Vitvar, T., Domingue, J., Fensel, D.: WSMO-lite and hRESTS: lightweight semantic annotations for web services and RESTful APIs. J. Web Semant.: Sci. Serv. Agents World Wide Web **31**, 39–58 (2015). https://doi.org/10.1016/j.websem.2014.11.006
13. Deng, S.G., Yin, J.W., Li, Y., Wu, Z.: A method of semantic web service discovery based on bipartite graph matching. J. Chin. J. Comput. **31**(8), 1364–1375 (2008). https://doi.org/10.3724/sp.j.1016.2008.01364
14. Shi, M., Liu, J., Zhou, D., Tang, M., Cao, B.: WE-LDA: a word embeddings augmented LDA model for web services clustering. In: 2017 IEEE International Conference on Web Services (ICWS). IEEE Computer Society (2017). https://doi.org/10.1109/icws.2017.9
15. Zhong, Y., Fan, Y., Huang, K., Tan, W., Zhang, J.: Time-aware service recommendation for mashup creation in an evolving service ecosystem. In: 2014 IEEE International Conference on Web Services (ICWS). IEEE Computer Society (2014). https://doi.org/10.1109/icws.2014.17
16. Meng, Y., Rumshisky, A., Romanov, A.: Temporal information extraction for question answering using syntactic dependencies in an LSTM-based architecture (2017)
17. Li, Z., Wang, J., Zhang, N., He, C., He, K.: A topic-oriented clustering approach for domain services. J. Comput. Res. Dev. **51**(2), 408–419 (2014). https://doi.org/10.7544/issn1000-1239.2014.20120776

18. Cai, M., Zhang, W.Y., Zhang, K.: Manuhub: a semantic web system for ontology-based service management in distributed manufacturing environments. J. IEEE Trans. Syst. Man Cybern. Part A-Syst. Hum. **41**(3), 574–582 (2011). https://doi.org/10.1109/tsmca.2010.2076395

19. Lin, J., Yu, H., Shen, Z., Miao, C.: Using goal net to model user stories in agile software development. In: 2014 15th IEEE/ACIS International Conference on Software Engineering, Artificial Intelligence, Networking and Parallel/Distributed Computing (SNPD). IEEE (2014). https://doi.org/10.1109/snpd.2014.6888731

20. Platzer, C., Dustdar, S.: A vector space search engine for web services. In: European Conference on Web Services (2005). https://doi.org/10.1109/ecows.2005.5

A Fast Public Key Searchable Encryption Scheme Against Inside Keyword Attacks

Can Liu[1], Ningjiang Chen[1,2], Ruwei Huang[1,2(✉)],
and Yongsheng Xie[1]

[1] School of Computer and Electronic Information,
Guangxi University, Nanning 530004, China
liucango@163.com, ruweih@126.com
[2] Guangxi Key Laboratory of Multimedia Communications
and Network Technology, Nanning 530004, China

Abstract. With the advent of the era of cloud computing technology, the security and search efficiency of ciphertext retrieval have has become the focus of research. However, in the traditional encryption schemes, most of them only solve the problem of defending against external keyword guessing attacks. We tend to ignore malicious third party cloud server provider, which sometimes tries to guess the user's ciphertext information through the trapdoor or keywords. For the sake of improving the security of ciphertext, an inside keyword attack scheme on the basis of inverted index is proposed. Firstly, When building inverted indexes, the private key of the data owner is added to protect against the attack of malicious server. Secondly, an efficient public key ciphertext search scheme of parallel encryption index structure is introduced to realize the parallel search task of keywords. Compared with the traditional public-key searchable encryption, our scheme not only improves the search efficiency but also improves the security of the system.

Keywords: Inside keyword attack · Malicious server · Parallel search · Ciphertext search

1 Introduction

Driven by the application of Internet technology, enterprises and individuals prefer to store large amounts of data in the cloud to relieve the data maintenance pressure of data owners. In practice, cloud servers do not provide a completely trusted storage space. Therefore, in order to protect the data security of the data owner, the data must be encrypted before being uploaded to the cloud server, which makes the use of data more difficult than traditional storage (saving data without encryption).

Searchable encryption [1] is one of the important solutions to ciphertext retrieval. It allows users to retrieve encrypted documents containing user specified keyword. When the data user uploads the data trapdoor, the cloud server provider can search the data that the user wants to find according to the ciphertext keyword feature. Encryption can be implemented in symmetric searchable encryption (SSE) [13] and asymmetric searchable encryption (ASE) [6]. Although SSE is efficient, the distribution of secret

© Springer Nature Singapore Pte Ltd. 2019
Y. Sun et al. (Eds.): ChineseCSCW 2019, CCIS 1042, pp. 50–64, 2019.
https://doi.org/10.1007/978-981-15-1377-0_4

keys is complex and suitable for one-to-one scenarios. In today's big data sharing, data owners are more eager to outsource the sharing of encrypted data. For the sake of achieving data sharing, Boneh et al. [2] first proposed Public-Key encryption with keyword search (PEKS), enabling users to search for encrypted data in asymmetric encryption. The PEKS system consists of the data owner (DO), the data user (DU) and the cloud service provider (CSP). The mail sender extracts the keyword from the plaintext, generates the encrypted index and ciphertext, and uploads it to the CSP. According to the requirements, the recipient uploads the keyword trapdoor to the cloud server provider. The cloud server provider performs a match according to the data user's ciphertext and keywords trapdoor and returns the result to the recipient. BDOP-PEKS [2] is semantically secure in resisting keyword guess attacks (SS-KGA). This would imply that if malicious cloud server cannot obtain keyword trapdoor, it cannot get any of keywords contained in the meet of SS-KGA secure PEKS ciphertext. This ensures keyword privacy for the data owner. However, CSP is an honest and curious third party. After having the keyword trapdoor, it is inevitable that CSP will obtain the keyword information of ciphertext through keyword guessing attack [9, 11]. The reason for keyword guessing attacks is that keyword tend to be selected from a small space and low entropy. Adversary can be individually for any possible keyword matching, respectively with the key trapdoor to carry on the pair, therefore, the attacker knows which keyword the user is looking for. Otherwise, the attacker continues to traverse the keyword domain then encrypt it, and test it in a loop. If a trapdoor for a keyword satisfies an equality relationship, the attacker will guess the keyword. Therefore, under KGA, even if the keyword space is small, the traditional PEKS model needs to be extended to achieve keyword privacy protection.

The rest of the paper is organized as follows. Section 2 introduces the related works of the research content. In Sect. 3, We analyze the PEKS model which can resist external keyword attacks. In Sect. 4, we propose our PSEFKS model. The construction and verification of the subsequent scheme are described in Sects. 5 and 6. Finally, the experimental results are given, and the conclusions are drawn.

2 Related Works

Boneh et al. [2] is the person who proposed PEKS, which is the first scheme based on bilinear pairings and forward index construction. The research results show that the scheme not only has high computational cost in encryption and retrieval, but also has many security problems. Byun et al. [3] indicate that the keyword space is far smaller than the secret key space, proposed the problem of offline keyword attack, and succeeded in breaking Boneh's plan. For the sake of resisting the offline keyword guessing attack, reference [5] and reference [6] proposed the scheme based on inverted index and certificateless authentication respectively to resist keyword guessing attack. However, there are still opportunities for successful keyword guessing attacks against adversary within the scheme. Reference [20] proposed an encryption scheme based on fuzzy keyword search, which can resist external keyword guessing attacks. Rhee et al. [4] come up with the security concept of "trapdoor indistinguishable", proving that trapdoor indistinguishable is a sufficient condition to prevent keyword guessing attacks.

However, neither of their work has solved the problem of keyword guessing attack (KGA) [1, 10, 21] being a server.

Shao et al. [22] is the first to raise an honest and curious server problem. In order to resist the problem of server keyword attack, reference [22] adopts the signature of information technology and the authentication of authoritative institutions. In the test algorithm, it is not allowed to test the illegally produced encrypted index, so as to resist the inside keyword guessing attack (IKGA). Reference [8, 17] proposes a scheme that requires double servers, but requires a large system overhead. In addition, some works, such as reference [18], dedicate to achieve the SPKE scheme resisting keyword guessing attack. Whereas, based on the forward index, the search efficiency is lower. In order to improve efficiency, the concept of no authentication and double servers was proposed in [7, 21]. The private key of the data owner is added in the index building process. Anyone without the private key of the data owner cannot generate a legal index, and the CSP cannot carry out a normal attack. However, considering the inside keyword attack, the above literatures are all based on the positive index method of "plaintext-keywords", and the retrieval time complexity of ciphertext information is proportional to the total number of ciphertext. For encrypted files, when the ciphertext contains a large number of keywords, this linear linked list index structure will greatly reduce the search efficiency. With the development of big data era, in the face of more and more data volume in cloud servers, how to improve the efficiency of data search is extremely important. Therefore, in order to better protect the privacy of data to improve search efficiency, this paper designed the parallel inverted index based on public key encryption scheme can search quickly and securely using safe and efficient inverted index [15] to realize linear search, the search efficiency is only associated with containing the query keyword, by adding the user private key technologies to resist key attack.

Given the above description of the problem, the paper's contributions are as follows:

- This paper proposes a scheme to resist inside keyword attack. According to the public key algorithm of bilinear pairings, a searchable encryption in a public key environment that can resist inside keyword attacks in a complete public key environment is constructed.
- This paper provides an efficient keyword search scheme. For the sake of improving the security as well as search efficiency of the system, a fast parallel inverted index is adopted and the private key of the data owner is introduced when the index is encrypted.

3   Preliminaries

3.1   Bilinear Pairing

G is multiplicative group order and G_T is multiplicative group order for Q cycle. The size of G, G_T is determined by the security parameter. Then a bilinear mapping $\hat{e} = G \times G \to G_T$, and satisfy the following properties:

(1) Computable: There are valid algorithms to compute $\hat{e}(g, h)$;
(2) Bilinear: Exiting $x, y \in Z$, make $\hat{e}(g^x, h^y) = \hat{e}(g, h)^{xy}$ was established;
(3) Non-degenerate: $\hat{e}(g, g) \neq 1$.

3.2   Public-Key Encryption with Keyword Search

Traditional public key searchable encryption consists of the following components [2]:

1. **KeyGen**(Param): Input common parameters Param, and then generate data user and data owner common public/private key pair (PK, SK).
2. **PEKS** (w, PK): Input the keyword of w and the public key PK of the data user, and output the encrypted keyword and the ciphertex C_w.
3. **Trapdoor** (w', SK): enter a keyword w' and private key of SK, and output a corresponding trapdoor $T_{w'}$.
4. **Test** $(C_W, T_{w'}, PK)$: Get data user's public key PK, the searchable ciphertext of $C_W = \text{PEKS}(w, PK)$ and a trapdoor of $T_w = \text{Trapdoor}(w', SK)$. If $w = w'$, output "1", otherwise output "0".

The keyword guessing attack problems of PEKS is the cloud server provider match between keywords with trapdoors by violence way. if $|w| \leq ploy(k)$, The CSP can effectively test all keywords of ciphertext. After the inside attacker gets the trapdoor of the keywords, perform the following actions:

1. Set $i = 1$;
2. Select from the keyword space w_i, and then run $\text{PEKS}(w_i, PK)$;
3. If the output is "1" then $w_i = w'$, the attacker knew the trapdoor contains;
4. If the output is "0", set $i = i + 1$, and continue with step 2. Otherwise output "$\perp$".

Therefore, based on the analysis of PEKS in literature [4], it can be seen that although the BDOP-PEKS scheme can achieve multi-user sharing, in practical application scenarios, any attacker can obtain the keywords searched by users by means of infinite cyclic measurement and matching after obtaining the keyword trapdoor. And the search encryption scheme is based on a forward index. In practice, where have too many ciphertext keywords, which seriously affects the efficiency of the scheme. Therefore, in this paper, from the aspect of resisting internal keyword attack and improving search efficiency, the private key of DO referenced in the encryption algorithm limits the ability of entities other than users to conduct keyword encryption, so as to realize the problem of resisting internal keyword attack.

4   PSEFKS Scheme

In this section, we describe the process of building the system prototype and index, and formalize the Public-key Security Encryption with Fast Keyword Search (PSEFKS) model.

4.1   System Prototype

In the application scenario of system prototype, it mainly includes uploading data, searching data and matching data, and it is shown in Fig. 1.

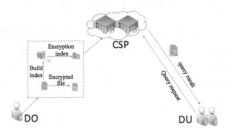

Fig. 1. System prototype

- **Data Owner (DO):** Extract document keywords, encrypt the document, establish inverted index, and upload the ciphertext and encrypted index of the document to the cloud server provider.
- **Data User (DU):** Generate keyword trapdoor and send it to the cloud server. The cloud server matches the keyword trapdoor with the stored ciphertext index and returns the search results to the user.
- **Cloud Server Provider (CSP):** Cloud servers are used to store encrypted data and indexes uploaded by owners of the data. According to the text algorithm, the inverted index is used for calculation, and the ciphertext of the matched document is returned to the data owner.

4.2   Efficient Inverted Index Construction

In inverted index, document and search efficiency are sublinear [5, 12, 16, 17]. In order to achieve rapid retrieval of massive data, this paper adopts efficient inverted index architecture to search ciphertext. Parallel encrypted index structure [15] based on inverted index, each keyword has a calculator to record the amount of ciphertext generated. When $k_w = 1$, this is the first generation of cipher w, w and k_w as input, generating ciphertext C_{w,k_w}. Then C_{w,k_w} and ciphertext document collection uploaded to the server. When the server receives the keyword trapdoor request from the user, it finds the corresponding logical address in the inverted index and gets the encrypted document list. Finally, the user's public key is used to decrypt the encrypted document list one by one to obtain the document ID set. Because this work can be done in parallel, the communication overhead can be reduced based on keyword matching when ciphertext matching. As shown in the Fig. 2:

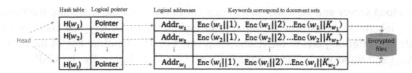

Fig. 2. Efficient inverted index

Suppose ciphertext C_w extract keywords set for $C_w = \{w_1, w_2, \ldots, w_i\}$, H is a pseudo-random function and inverted index is the ciphertext $C_1, C_2, \ldots, C_w$ establish "keywords—document" index. Each keyword w_i have corresponding keywords set of documents for $(w_i, value)$ index set, including *value* corresponding document collection by keywords. The K_{w_i} of $Enc(w_i \| K_{w_i})$ denote the current generated counter namely document keywords w_i number value.

4.3 Definitions of PSEFKS

The scheme of PSEFKS is mainly composed of two hash functions, three objects and seven probabilistic polynomial time algorithms: Setup, $KeyGen_{DO}$, $KeyGen_{DU}$, StrucInit , StrucEnc, Trapdoor, Test. The specific implementation form of each algorithm is as follows:

- **Setup**(λ): System initialization algorithm. The algorithm is input with security parameter λ, the output system common parameter param.
- **KeyGen$_{DO}$**(Param): The data owner generates the secret key algorithm. The algorithm takes the parameter Param as input, and produce the data owner′s public and private key (PK_{DO}, SK_{DO}).
- **KeyGen$_{DU}$**(Param): The data user generated secret key algorithm. The algorithm takes the parameter Param as input, and produce the data user′s public key and private key (PK_{DU}, SK_{DU}).
- **StrucInit**(Param): Hide the relationship structure initialization algorithm. The algorithm is to initialize a hidden relationship structure used to encrypt index generation. The public system parameters Param as input and output hidden relationship structure $HS = (PRI, PUB)$.
- **StrucEnc**(w, PRI, PK_{DU}, SK_{DO}): The hidden relational structure encryption which is called the encryption index generation algorithm. Algorithm with keyword w, hidden relationship structure PRI, data owners of private key SK_{DO} and data users public key PK_{DU} as input, generate a searchable index of encryption CI_w and update PRI, performed by the data owner.
- **Trapdoor**(w', PK_{DO}, SK_{DU}): Keywords trapdoor generation algorithm. Algorithm to generate the trapdoor of keywords w', and then uploaded to the server to search phase matching work. Public key algorithm input data owner PK_{DO}, data user private key SK_{DU}, and output the trapdoor information containing the specified keyword, which is run by data user.
- **Test**($C_w, CI_w, PUB, PK_{DU}, PK_{DO}, T_{w'}$): The test algorithm. When the cloud server receives the trapdoor $T_{w'}$ of data users, it matches the keyword index CI_w stored on

the cloud server and runs the matching algorithm to find the matching the trapdoor $T_{w'}$ ciphertext. Algorithm are input with ciphertext C_w, encrypted index set CI_w, hide the common parts of the relationship structure PUB, public key of the data user PK_{DU} and trapdoor $T_{w'}$ and output "0" or "1", which is run by the CSP.

5   PSEFKS Structure

5.1   Constructing Our Instance

The specific steps of this algorithm are as follows:

- **Setup**(λ): System initialization algorithm. Algorithm output system public system param $= \{p, g, \hat{e}, G, G_T, H_1, H_2\}$. p is a largest prime numbers associated with security parameters. G, G_T are order for p circulation order. $\hat{e} = G \times G \to G_T$ is a bilinear pairing with high efficiency and nondegeneracy. H_1, H_2 are two hash functions which are satisfied with the relationship of $H_1 : \{0,1\}^* \to G$, $H_2 : G_T \to \{0,1\}^{\log^p}$.
- **KeyGen$_{DO}$**(param): The data owner generates the secret key algorithm. The algorithm outputs the public/private key pair of the data owner $(PK_{DO}, SK_{DO}) = (g^x, x)$, among them $x \xleftarrow{R} Z_p^*$.
- **KeyGen$_{DU}$**(param): The data user generates the secret key algorithm. The algorithm outputs the public/private key pair of the data user $(PK_{DU}, SK_{DU}) = (g^y, y)$, among them $y \xleftarrow{R} Z_p^*$.
- **StrucInit**(param): Hide the relationship structure initialization algorithm. The algorithm input for the system of public parameters param and output a hidden relationship structure HS $= (PRI, PUB) = (\alpha, g^\alpha)$. $\alpha \xleftarrow{R} Z_p^*$ and PRI are dynamic list seem as $(\alpha, \{(w, k_w)|w \in W, k_w \in N\})$, $k_w \in N$, initialized (α).
- **KwEnc**($w_w, SK_{DO}, PK_{DU}, PRI$): Encryption index generation algorithm. Algorithm to generate document w keywords $w_i = (w_1, w_2, \ldots, w_t)$, data owners of private key SK_{DO} and data users public key $PK_{DU} = g^y$ as input, generating searchable ciphertext C_{w_i} by the following ways:

1. Judge record (w, k_w) of keyword w whether in PRI;
2. If the record does not exist, set $k_w = 1$ and add(w, k_w) to PRI; otherwise, set $k_w = k_w + 1$ and update PRI;
3. Output searchable ciphertext $C_{w_i} = (C_1, C_2)$, one of them

$$C_1 = H_1(w, PK_{DU})^{k_w \cdot SK_{DO}} \cdot g^r \tag{1}$$

$$C_2 = PK_{DU}^r \tag{2}$$

4. When all keywords calculation is finished, record the index set $CI_w = (CI_{w_1}, CI_{w_2}, \ldots, CI_{w_t})$, and send the ciphertext containing the keywords to the cloud server provider.

- **Trapdoor**(w', PK_{DO}, SK_{DU}): Keywords trapdoor generation algorithm. Algorithm to generate the keyword trapdoor w', and then uploaded to the cloud server provider to search phase matching work. Algorithm take the data owner PK_{DO}, data user private key SK_{DU} as input, calculate the trapdoor of keywords $T_{w'} = \hat{e}(H_1(w')^{SK_{DU}}, PK_{DO})$ and sent $T_{w'}$ to the cloud server provider.
- **Test**$(CI_w, C_w, PUB, PK_{DU}, PK_{DO}, T_{w'})$: Cloud server in the received data users to send the trapdoor $T_{w'}$, according to keywords index CI_w, run the algorithm find matching ciphertext of the trapdoor $T_{w'}$. Algorithm in a ciphertext C_w, encryption index set CI_w, the public ip structure $PUB = g^{\alpha}$, data users' public key $PK_{DU} = g^y$ and trapdoor $T_{w'}$ as input, through the following steps to search matching set of ciphertext:

1. Set $M = j + N * t$, Correctness Certificationand calculate the trapdoor $T_{w',M} = T_{w'}^M$;
2. Calculate $C' = \hat{e}(PUB, PK_{DU}, T_{w'}^M)$;
3. Find the encryption index set $CI_w = (C_1, C_2)$ which meet the needs of $C' = C[i]$. If found, the index of the corresponding ciphertext index set C_w ciphertext $C[i]$ added to C', set $t = t + 1$, and then continue to step 1;
4. If traversal encryption index set CI_w could not find a matching index, the output is C'.

5.2 Correctness Certification

In which the public and private key pairs of data owner $(PK_{DO}, SK_{DO}) = (g^x, x)$, the public and private key of user data $(PK_{DU}, SK_{DU}) = (g^y, y)$. Keywords w with keywords contained in ciphertext C_w, keywords w' are contained in the trapdoor $T_{w'}$. The specific proofs are as follows:

According to **KwEnc:**

$$C_1 = H_1(w_i)^{k_w \cdot SK_{DO}} \cdot g^r = H_1(w_i)^x \cdot g^r \tag{3}$$

$$C_2 = (g^y)^r \tag{4}$$

According to **Trapdoor:**

$$T_{w'} = \hat{e}(H_1(w')^{SK_{DU}}, PK_{DO}) = \hat{e}(H_1(w')^y, g^x) \tag{5}$$

According to **Test:**

$$C' = \hat{e}(g^{\alpha}, H_1(w)^x) = \hat{e}(H_1(w), PK_{DU}^r) = C[1] \tag{6}$$

Therefore, when the first ciphertext containing the keyword w is successfully matched, all ciphertext containing the keyword w can be successfully matched by bilinear pairings.

Therefore, there are two scenarios:

1. When the keyword w is equal to the keyword of trapdoor w':

$$H(w)^M = H(w')^{k'_w} \tag{7}$$

$$C' = \hat{e}\left(g^\alpha, H_1(w)^{M*x}\right) = \hat{e}\left(H_1(w), PK_{DU}^r\right) = \text{C[j]} \tag{8}$$

Therefore, the result of Test algorithm is "1".

2. When the keyword w is not equal to the keyword of trapdoor w':

$$H(w)^M \neq H(w')^{k'_w} \tag{9}$$

By the collision function of hash function

$$C' = \hat{e}\left(g^\alpha, H_1(w)^{M*x}\right) \neq \hat{e}\left(H_1(w), PK_{DU}^r\right) = \text{C[j]} \tag{10}$$

Therefore, the test algorithm result is "0".

From the above proof, we can draw the conclusion that the scheme of this paper is established.

5.3 Safety Certification

The safety certification of PSEFKS index is according to the construction of CBDH problem, that is, when the CBDH problem is difficult to solve, the index encryption part is indistinguishability under chosen keywords attack (IND-CKA).

- **Initialization phase:** A given keyword w and parameters $(p, g, g^a, g^b, g^c, G, G_T, \hat{e}, t)$, the algorithm by simulating the following steps to complete interactions with the enemy A problem:

  1. Initializes an empty list HL $\in w \times G_1 \times Z_p^* \times \{0, 1\}$, Plist $\in Z_p^* \times G_1$;
  2. Establishes $PK = (p = g^a, g, G, G_T, \hat{e})$;
  3. Initializes the hidden relationship N and the implementation is as follows:

  (1) Random selection $u_i \leftarrow Z_p^*$ and $C_i \leftarrow \{0, 1\}$;
  (2) If $C_i = 1$, and then calculate $PUB_i = g^{b*u_i}$;
  (3) Otherwise, calculate $PUB_i = g^{u_i}$;
  (4) The other N hidden structure combining HSET $= (g^{b*u_1} \ldots g^{b*u_N})$, and add multi component system $<w, PUB_i, u_i, C_i>$ to table HL;
  (5) Send (PK, PUB_i) to adversary.

- **Inquiry phase 1:**

Random machine query $\mathcal{O}_{H_1}(w)$**:** When adversary receiving the keyword w, perform random machine query to obtain hash value $H(w)$ of w. The operation as follows:

1. Select $s \xleftarrow{R} Z_p^*$ and $C_i \xleftarrow{\sigma} \{1, 0\}$ and $\xleftarrow{\sigma}$ denotes $pr[C_i = 0] = \sigma$;
2. If Coin $= 1$, add $<w, PUB_i, g^{b*u_1}, u_i, C_i>$ to the list HL and output g^{b*u_1};
3. Otherewise, add $<w, PUB_i, g^{u_1}, u_i, C_i>$ to the list HL and output g^{u_1}.

The trapdoor query $\mathcal{O}_{Trapdoor}(w)$**:** When the choice of the adaptive random Or acle $\mathcal{O}_{Trapdoor}$, and obtain the trapdoor of keyword $w \in \{0, 1\}^*$,$\mathcal{B}$ algorithm is as follows:

1. Judge whether $<w, PUB_i, u_i, C_i>$ is in the table of HL, if not, then the request hash query $\mathcal{O}_{H_1}(w)$;
2. Take out tuple $<w, PUB_i, u_i, C_i>$ of keywords from the list of HL. if $C_i = 0$, output g^{b*u_1}. Otherwise, output $\perp$;
3. Finally, return the trapdoor of the keyword to the attacker.

- **Challenge stage:** Attacker randomly Choose two keywords w_0^*, $w_1^* \leftarrow \{0, 1\}^l$ and Hidden vector PUB_0^*, PUB_1^* send to challenger, and then the challenger does the following to generate the encrypted index:

  1. Respectively determine $<w_0^*, PUB_0^*, u_0^*, C_0^*>$ and $<w_1^*, PUB_1^*, u_1^*, C_1^*>$ whether in the list of the HS. If the keyword w is not in HS, query request hash query $\mathcal{O}_{H_1}(w)$;
  2. If the keywords w_0^* and w_1^* corresponding tuples $C_0^* = C_1^* = 1$, then the algorithm $\mathcal{B}$ returns the challenge failure and output $\perp$;
  3. At least, there is a "0" between C_0^* and C_1^*, randomly selected d $\in \{1, 0\}$ to make $C_d^* = 0$;
  4. Send the ciphertext to attacker.

- **Inquiry phase 2:** As same as query phase 1 but at this stage, not allowed to ask keywords w_0^*, w_1^* and hidden vector PUB_0^*, PUB_1^*.
- **Guessing stage:** The adversary A output a bit b'. If $b' = b$, algorithm $\mathcal{B}$ output "1", otherwise "0".

6 Analysis and Experiments

6.1 Theoretical Analysis

Firstly, through a large number of literature reading and analysis, this paper selects a relatively representative scheme in public key searchable encryption to conduct theoretical comparison with the PSEFKS scheme proposed in this paper. As shown in Table 1, the security situation of each scheme is summarized. Compared with the following schemes, PSEFKS scheme can not only realize the inseparability of ciphertext and trapdoor, but also realize the defense against external keyword attack, and more effectively resist internal keyword attack with higher security.

Table 1. Comparisons of security

	Boneh [1]	Rhee [4]	Shao [22]	PSEFKS
Ciphertext indistinguishable	YES	YES	YES	YES
Trapdoors indistinguishable	NO	YES	YES	YES
Resistance to external attack	NO	YES	YES	YES
Resist internal attack	NO	NO	YES	YES

The comparison of calculation costs of each scheme in different stages is obtained. Table 2 summarizes the time complexity of each scheme in four major stages. Where, E is the time used to run a modular exponential operation, H is the time used to run a hash transformation operation of string and group elements, and P is the time used to calculate a bilinear mapping.

Table 2. Performance analysis

	Boneh	Rhee	Shao	PSEFKS
Setup	2E	3E	(n + 3)E	2E
Index	2E + 2H + P	2E + 2H + P	9E + 3H + 3P	3E + H+2P
Trapdoor	E + H	3E + 2H	2E	3E + H+P
Test	H + P	E + H+P	5E + H+4P	2P

As can be seen from Table 2, from BDOP-PEKS [1] scheme is put forward, for the first time in 2015 Shao [22] first consider the internal key attack, mentioned in their scheme, due to the particularity of forward index, we through the trapdoor, and encryption in matching algorithm to find contain keyword w ciphertext document, the Test algorithm to virtually all of its encrypted index matching again, so the search phase of time complexity is o(n). Therefore, the above comparison scheme only applies to the searchable encryption system with fewer documents, and for the system with more documents, the efficiency will be seriously reduced. However, in the PSEFKS scheme, due to the use of the encryption index structure of parallel inverted index, the cloud server allows parallel keyword search matching tasks to be performed when it receives the keyword search trapdoor, and then finds all matching papers to improve the retrieval efficiency. Therefore, in the cloud storage of big data storage today, this scheme can realize the function of resisting internal keyword attack under effective inverted index, making the scheme more secure and efficient.

6.2 Retrieval Efficiency

This experimental prototype system is developed and tested based on Ubuntu18.04 (64 bit). The experiment used C language, the compiler was CodeLite software library was pbc-0.5.14 [14], and the elliptic curve was selected as type-A. For the sake of evaluating the efficiency and safety of the experiment, this experiment was verified by multiple comparisons, and the average operation time was obtained and the performance of SPCHS [19] scheme was compared.

Because the scheme of SPCHS [19] have representativeness in the problem of implement fast inverted index aspect, therefore, the query efficiency is compared with its experimental results. The comparison is made in three main aspects: data owner generates searchable ciphertext time, cloud server performs keyword search matching time and space occupied by comparison. A total of 100,000 documents were selected in the experiment, and 5 common keywords were extracted from each document for analysis. The extracted data are shown in Table 3.

Table 3. Document keyword analysis

Keywords	Documents contains keywords
Data	7320
Document	2003
Then	8511
Of	9013
Article	4123

(1) **Comparing Search Performance:** In SPCHS scheme, the chain index relation is used to find the next ciphertext based on the previous ciphertext, which is relatively inefficient. But in the scheme of PSEFKS, the new relationship allows a keyword search task to be performed in pallel. Therefore, We can get user's trapdoor in the cloud server and use the test algorithm for comparison and traversal. A pairing operation is required when the plaintext to which the index points contains the keyword w. Figure 3 shows the results of comparing experimental data. There can be found that when the time approach to 8 s, PSEFKS scheme can find approximately16500 ciphertexts, and the comparative scheme of SPCHS can only find approximately 4500 ciphertexts. Therefore, the scheme of PSEFKS have a great enhance in search performance.

(2) **Comparing Encryption Performance:** We upload the generated searchable encryption ciphertext to the cloud server through PSEFKS scheme and SPCHS scheme respectively. Figure 3 shows our results. The SPCHS solution takes about 90 s to generate 10000 searchable ciphertext, while we need just over 60 s. Therefore, to generate keyword-searchable ciphertexts, our new scheme of PSEFKS has a big improvement (Fig. 4).

(3) **Comparing Communication Cost:** This experiment compares the communication cost of our scheme of PSEFKS and the comparative scheme of SPCHS by statistically evaluating the byte size of the generated keyword-searchable ciphertexts. As shown in Fig. 5, The experimental comparison shows that 10000 keyword-searchable ciphertexts, our new scheme of PSEFKS takes approximately 350 KB and the comparative scheme of SPCHS takes approximately 3600 KB. Therefore, compared with the scheme of SPCHS, the communication cost of our new scheme is greatly improved.

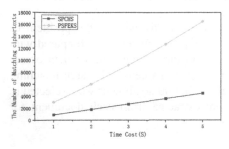

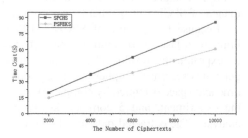

Fig. 3. Comparison of search efficiency

Fig. 4. Index generation time

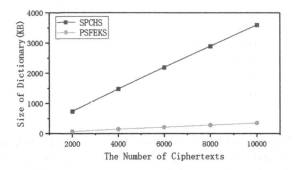

Fig. 5. Comparison of Communication Cost

Above all, comparison of the above three main aspects with the scheme of SPCHS, We can conclude that our scheme can not only resist inside keywords attacks, but also has a great improvement in performance.

7 Conclusions

This scheme of PSEFKS based on parallel inversion index to resist internal keyword attack. In addition, the correctness and security of PSEFKS are verified by analysis and on the premise of obtaining semantic security, it has higher practicability. However, when using inverted index in this paper, the scheme cannot guarantee the dynamic update of encrypted data. In the future, the inverted index structure can be improved to ensure the search efficiency and security of users, and the index structure can be dynamically updated.

Acknowledgment. This work is supported by the National Natural Science Foundation of China (61762008) and the Natural Science Foundation Project of Guangxi (2016GXNSFAA 380115).

References

1. Song, D.X., Wagner, D., Perrig, A.: Practical techniques for searches on encrypted data. In: Proceeding 2000 IEEE Symposium on Security and Privacy, pp. 44–55. IEEE (2000)
2. Boneh, D., Di Crescenzo, G., Ostrovsky, R., Persiano, G.: Public key encryption with keyword search. In: Cachin, C., Camenisch, J.L. (eds.) EUROCRYPT 2004. LNCS, vol. 3027, pp. 506–522. Springer, Heidelberg (2004). https://doi.org/10.1007/978-3-540-24676-3_30
3. Byun, J.W., Rhee, H.S., Park, H.A., Lee, D.H.: Off-line keyword guessing attacks on recent keyword search schemes over encrypted data. In: Jonker, W., Petković, M. (eds.) SDM 2006. LNCS, vol. 4165, pp. 75–83. Springer, Heidelberg (2006). https://doi.org/10.1007/11844662_6
4. Rhee, H.S., Park, J.H., Susilo, W.: Trapdoor security in a searchable public-key encryption scheme with a designated tester. J. Syst. Softw. **83**(5), 763–771 (2010)
5. Wang, B., Song, W., Lou, W., Hou, Y.T.: Inverted index based multi-keyword public-key searchable encryption with strong privacy guarantee. In: 2015 IEEE Conference on Computer Communications (INFOCOM), pp. 2092–2100. IEEE (2015)
6. Yanguo, P., Jiangtao, C., Changgen, P., Zuobin, Y.: Certificateless public key encryption with keyword search. China Commun. **11**(11), 100–113 (2015)
7. Huang, Q., Li, H.: An efficient public-key searchable encryption scheme secure against inside keyword guessing attacks. Inf. Sci. **403**, 1–14 (2017)
8. Wang, C.H., Tu, T.Y.: Keyword search encryption scheme resistant against keyword-guessing attack by the untrusted server. J. Shanghai Jiaotong Univ. (Sci.) **19**(4), 440–442 (2014)
9. Fang, L., Susilo, W., Ge, C., Wang, J.: Public key encryption with keyword search secure against keyword guessing attacks without random oracle. Inf. Sci. **238**, 221–241 (2013)
10. Yau, W.C., Heng, S.H., Goi, B.M.: Off-line keyword guessing attacks on recent public key encryption with keyword search schemes. In: Rong, C., Jaatun, M.G., Sandnes, F.E., Yang, L.T., Ma, J. (eds.) ATC 2008. LNCS, vol. 5060, pp. 100–105. Springer, Heidelberg (2008). https://doi.org/10.1007/978-3-540-69295-9_10
11. Sun, L., Xu, C., Zhang, M., Chen, K., Li, H.: Secure searchable public key encryption against insider keyword guessing attacks from indistinguishability obfuscation. Sci. China Inf. Sci. **61**(3), 038106:1–038106:3 (2018)
12. Zhang, R., Xue, R., Yu, T., Liu, L.: Dynamic and efficient private keyword search over inverted index–based encrypted data. ACM Trans. Internet Technol. **16**(3), 21 (2016)
13. Song, D.X., Wagner, D., Perrig, A.: Practical techniques for searches on encrypted data. In: Proceeding 2000 IEEE Symposium on Security and Privacy, pp. 44–55. IEEE (2000)
14. Lynn, B.: PBC Library. https://crypto.stanford.edu/pbc/
15. Xu, P., Tang, X., Wang, W., Jin, H., Yang, L.T.: Fast and parallel keyword search over public-key ciphertexts for cloud-assisted IoT. IEEE Access **5**, 24775–24784 (2017)
16. Wang, B., Song, W., Lou, W., Hou, Y.T.: Inverted index based multi-keyword public-key searchable encryption with strong privacy guarantee. In: 2015 IEEE Conference on Computer Communications (INFOCOM), pp. 2092–2100. IEEE (2015)
17. Chen, R., Mu, Y., Yang, G., Guo, F., Wang, X.: Dual-server public-key encryption with keyword search for secure cloud storage. IEEE Trans. Inf. Forensics Secur. **11**(4), 789–798 (2015)
18. Saito, T., Nakanishi, T.: Designated-senders public-key searchable encryption secure against keyword guessing attacks. In: 2017 Fifth International Symposium on Computing and Networking (CANDAR), pp. 496–502. IEEE (2017)

19. Xu, P., Wu, Q., Wang, W., Susilo, W., Domingo-Ferrer, J., Jin, H.: Generating searchable public-key ciphertexts with hidden structures for fast keyword search. IEEE Trans. Inf. Forensics Secur. **10**(9), 1993–2006 (2015)
20. Ding, S., Li, Y., Zhang, J., Chen, L., Wang, Z., Xu, Q.: An efficient and privacy-preserving ranked fuzzy keywords search over encrypted cloud data. In: 2016 International Conference on Behavioral, Economic and Socio-cultural Computing (BESC), pp. 1–6. IEEE (2016)
21. Du, M., Wang, Q., He, M., Weng, J.: Privacy-preserving indexing and query processing for secure dynamic cloud storage. IEEE Trans. Inf. Forensics Secur. **13**(9), 2320–2332 (2018)
22. Shao, Z.Y., Yang, B.: On security against the server in designated tester public key encryption with keyword search. Formation Process. Lett. **115**(12), 957–961 (2015)

Grey Fault Detection Method Based on Context Knowledge Graph in Container Cloud Storage

Birui Liang[1], Ningjiang Chen[1,2(✉)], Yongsheng Xie[1], Ruifeng Wang[1], and Yuhua Chen[1]

[1] School of Computer and Electronic Information, Guangxi University, Nanning 530004, China
liangbirui@foxmail.com, chnj@gxu.edu.cn
[2] Guangxi Key Laboratory of Multimedia Communications and Network Technology, Nanning 530004, China

Abstract. In the field of container cloud storage cluster resource scheduling, the activities, such as how to schedule resources according to load changes, and migrate according to resource conditions, are mainly considered. These activities bring about frequent changes in the context and also changes in the application's operating environment. They pose great difficulties in locating fault, especially the location of grey faults, which affect the operation of the application in the containers. Therefore, in order to ensure the normal operation of the application, grey fault detection method is proposed, which establishes a relationship knowledge graph for the relationship between the context change and the grey fault by studying the change of the application attention feature, which are brought by the context change. The method introduces temporal and spatial snapshot group architecture to solve a large number of situational temporal queries caused by too large structure of knowledge graph. The method is validated in the container cluster project and the Google open source dataset, which can effectively detect grey fault scenarios and the accuracy rate has been improved by more than 90%.

Keywords: Fault detection · Context · Grey failure · Cloud storage · Knowledge graph

1 Introduction

With the rise of microservice architecture, container technology such as docker has had a profound impact on the development of cloud computing. In a cloud cluster consisting of containers, the core of cluster attention is how to give applications a stable operating environment. In a cluster, some frequent changes in the

Supported by the Natural Science Foundation of China (No. 61762008), and the Guangxi Natural Science Foundation Project (No. 2017GXNSFAA198141), and Key R&D project of Guangxi (No. GuiKE AB17195014).

© Springer Nature Singapore Pte Ltd. 2019
Y. Sun et al. (Eds.): ChineseCSCW 2019, CCIS 1042, pp. 65–80, 2019.
https://doi.org/10.1007/978-981-15-1377-0_5

context in which the application runs, such as resource scheduling and container migration, will result in changes to the application's operating environment, which makes it difficult to locate grey faults.

Grey failure [1] is the faults that the system can easily ignore under the influence of a large number of fault-tolerant mechanisms of the storage system, while the application can detect it. As shown in Fig. 1. The data server 1 cannot return data to the data manager normally because of severe capacity limitations. But the system fault detector thinks this is not a malfunction.

The application monitor of the application detects this failure because the operation of the application is hindered. The difference between this type of system fault detection and the application's own fault detection mechanism is the source of grey faults. There is also a fault tolerance mechanism such as the clos network [2], which is to tolerate random and silent packet drop problems by using high redundancy in the network. But for applications, this random and silent grouping problem has a serious impact on its normal operation. The fault detection methods in existing container cloud systems are easily unable to detect these faults due to these fault tolerance mechanisms.

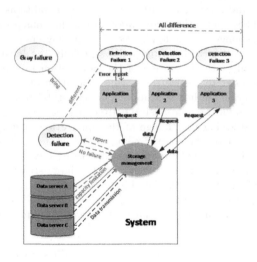

Fig. 1. Grey fault case model.

So far, there is less research on grey fault detection of the container cloud storage system. Kubernetes [3] and Docker Swarm [6], which are the mainstream container cluster framework, do not provide fault detection. And the most of the container cloud storage service systems based on them only provide reactive fault alarms.

The existing detection methods for grey faults are mostly based on the original fault detection method in the cloud storage environment. They have poor observability for differences between systems and applications caused by grey faults [4]. Panorama [3], a framework tool that detects system component failures through API injection, is representative. However, this tool has some incompatibility in the container cloud storage, because the instructions of the container monitoring information depend on the characteristics of the Linux system instructions. And the containers are isolated from each other. It cannot clearly identify a specific container (or the name of the container that contains it) by name.

Therefore, it is necessary to analyze grey faults by studying the changes of the context in the container cloud and the application running, and to help the fault detection and system maintenance of the container cloud storage environment, which is application-centric.

Question 1: How to perceive and analyze context to find out its relationship to grey faults?

Question 2: How to quickly and accurately detect grey faults in a container cluster system based on context changes?

Solution 1: We finds the relationship between context and grey fault by sampling analysis of specific events of context and classifying the situation of grey faults. Specific events are primarily related to system performance changes and application operational failure events.

Solution 2: Through sampling analysis, the knowledge graph of the relationship between context change and grey fault is established to analyze the fine-grained and comprehensive features of the grey fault scene generated by the container context change. For the changes in the context, we take the temporal and spatial snapshot of the scenario based on the current changes. The graph uses the breadth-first search algorithm to quickly compare snapshots with scene snapshots in the knowledge graph.

The given experiments results show that the method effectively discovers the grey fault in the process of container cluster context change.

The rest of this paper is organized as follows. Section 2 introduces the preparation for the establishment of knowledge graph. Section 3 presents the framework results of the knowledge graph. Section 4 conducts an experimental verification and discusses the experimental results. Section 5 introduced the related work of fault detection in container-based cloud storage. Section 6 concludes the paper.

2   Preparation for the Establishment of Knowledge Graph

2.1   Analysis of Context

In order to analysis of the context in the container cloud and the application running, this paper design a plugin in order to collect various of information of performance statistics during the container cloud operation, The plugin could help identify performance changes such as performance bottlenecks and effectively support for memory, CPU scalability analysis. The plugin can also monitor events in the container. It periodically monitors when a particular event that exceeds a predefined threshold occurs. The specificity of these specific events is that:

(1) The event of container access memorizer failing AF;
(2) The event of reporting memorizer measurement MR;
(3) The event that the system fails to recognize the precise instruction pointer during application operation PI;
(4) The event of failed to access valid data address accurately VD.

The cost of sampling memory access events is evaluated based on data access latency in the CPU, which is calculated as follows:

$$MCost = \sum_{i=1}^{n}(CH_i + DT_i) \tag{1}$$

where n is the number of sampling memory access events, CH is the cache hit buffer period of the CPU, and DT is the clock period of the access data DRAM.

The latency information can be used to identify and quantify performance issues such as scaling bottlenecks in the storage load of the container cloud. Given the storage load is L_i $1 \leq i \leq N$ N is the number of containers in the cluster. Only computational overhead and communication overhead are considered here. So L_i is calculated as Fomular (2), where W_i is the calculation cost, C_i is the communication overhead.

$$L_i = W_i + C_i \tag{2}$$

The total load is ML, then the storage load scaling bottleneck $f(n)$ is

$$f(n) = \begin{cases} -1, & \sum_{i=1}^{N} L_i \geq ML \\ 1, & \sum_{i=1}^{N} L_i < ML \end{cases} \tag{3}$$

the symbols are given in Table 1.

Table 1. Symbol description.

Symbol	Description
MemoryDelay	Memory delay
$PrivateM_i$	The private memory of application i
$PrivateB_i$	The bandwidth of application i
ShareM	Shared memory of the cluster system
ShareB	Shared bandwidth of the cluster system

The sampling process algorithm is as follows:

Algorithm:	SamplingContext
Input:	Application set A={ $a_1,a_2,\ldots,a_m$ }, $1 \leq$ m.
Output:	Context sampling set C={AF,MR,PI,VD}
1	Function samplingContext()
2	While A $\neq$ NULL
3	if a_i need more L_i
4	if MemoryDelay ++
5	$AF \leftarrow af_i$
6	$MR \leftarrow mr_i$
7	end if
8	end if
9	if $PrivateM_i + +\&\&PrivateB_i + +$
10	if $ShareM + +\&\&ShareB + +$
11	$MR \leftarrow mr_i$
12	end if
13	end if
14	if a_i recognition pointer failed
15	$PI \leftarrow pi_i$
16	end if
17	if a_i access address failed
18	$VD \leftarrow vd_i$
19	end if
20	end while

2.2 Classification According to the Relationship Between Context and Grey Fault

Based on the detected faulty context, the relationship between the environment and the fault are classified. The relationship between context and grey fault is related to the causality between fault classes, as shown in Fig. 2.

In this paper, the grey faults caused by the context change are mainly divided into behavioral failure, design failure, and content failure.

(1) Behavioral failure are mainly related to application behavior, including transactions in applications, dynamic content and execution events. Assume that application i has a behavioral failure.

$$BF = \{bf_i \mid 1 \leq i \leq m\} \tag{4}$$

(2) Design failure mainly refer to the failure of the rules and transactions in the design of the container cloud system, including load balancing strategy, container migration strategy, resource scheduling algorithm and so on. Assume that there is a design failure at time t,

$$DF_{new} \leftarrow DF_{previous} \cup \{df_t\} \tag{5}$$

(3) Content failure refer to the failure of the process when the static data in container cloud system make interacting with the system, including memory bottleneck, CPU bottleneck and so on. Suppose there is a content failure at time t,

$$CF_{new} \leftarrow CF_{previous} \cup \{Cf_t\} \tag{6}$$

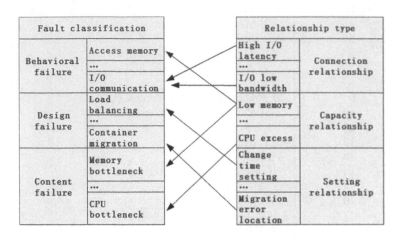

Fig. 2. Classification structure of fault relational patterns.

There are many classifications of associations, such as connection relationships, capacity relationships, setting relationships, terminal relationships, and data usage fault relationships. References from fault classes to association classes determine the relationship between them. In this paper, the reference is implemented by defining identifiers for fault classes. Therefore, this paper obtains a pattern consisting of fault classes, association relational classes, and references from associated relational classes to fault classes. The examples is the following:

(1) The lack of part about the code in the application may lead to the incorrect operation of the application transaction failure;
(2) Inconsistent time between containers may lead to incorrect container migration of container cloud systems;
(3) Memory bottleneck failure may be caused by the increase of a large number of applications.

The vertex set V of the knowledge graph established in this paper is mainly composed of the context in which these faults occur.

$$\{BF, DF, CF\} \in V \tag{7}$$

And the associations form the set E of the edges of the knowledge graph,

$$\{ConR, CapR, SetR, TerR, DuR, \ldots\} \in E \tag{8}$$

In order to accurately describe the relationship between grey fault and context scene transformation, this section conducts targeted sampling analysis on the context, and obtains the main types of correlation between context and grey fault, providing basic conditions for the establishment of knowledge graph and the search of grey fault scene.

3   Establish the Knowledge Graph

3.1   Overview of the Build Process

After successfully sampling the context and classifying the grey fault according to the characteristics of the context, this section start by constructing a knowledge graph that reflects the relationship between the context and the grey fault.

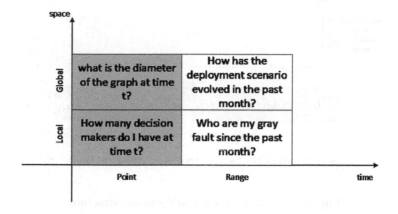

Fig. 3. Classification of the graph work.

The knowledge graph tracks all information related to the evolution of this context and grey faults, including the addition of vertices representing the context or grey fault occurrence scenarios. It also includes additions and deletions of edges (representing associations), editing activities of the entire graph, and its timestamps. Such a graph can store associations between context scenarios that can analyze grey faults in the past. It builds detailed metrics of graphs at specific instances to detect future failure trends.

The classification of the main work on the graph is shown in Fig. 3. It allows to query for points in time, access subsets, or entire graphs. For example, at time t, it can find the shortest path by calculating the shape diameter to construct and traverse a snapshot of the graph at t. The graphical query in this article can span a range of time.

After the knowledge graph is successfully built, the size of the data structure given for the graph is limited. When the knowledge graph saves more and more data/information/knowledge from Sect. 2, the graph will inevitably become

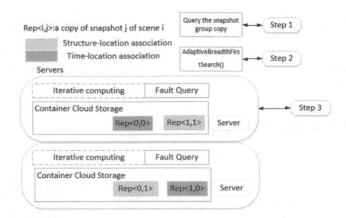

Fig. 4. Overview of knowledge graph architecture.

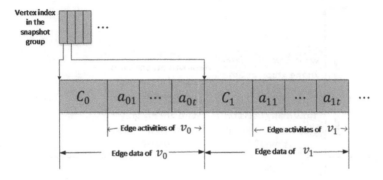

Fig. 5. Time-location relationship layout structure.

larger, and the final structure is too large, which will result in an inefficient iterative calculation. Therefore, in order to improve the performance of iterative calculations, this article focuses on improving memory access and minimizing concurrency control overhead. Therefore, the model will use snapshot groups to simplify the query process.

3.2 Knowledge Graph Structure

Figure 4 shows the structure of the knowledge graph. The rounded rectangle in the figure represents a server that stores a snapshot group copy using a structure-location or a time-location association. The knowledge graph continuously stores multiple snapshot groups in a container cloud persistent store across multiple servers. First, the first step in the knowledge graph is to query a copy of the snapshot group in the time range (step 1). The second step is to perform a fault assessment query, using an adaptive breadth-first search algorithm to select the scene copy with the greatest similarity (step 2). The final step is to detect the result (step 3).

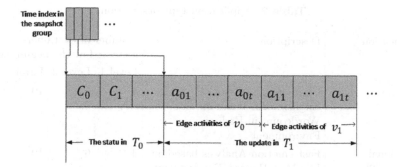

Fig. 6. Structure-location relationship layout structure.

The time-position correlation structure of the edge of the knowledge graph is shown in Fig. 5. Figure 5 begins with each vertex in the snapshot group at the edge of the graph, followed by a series of segments corresponding to the vertices. Here, the starting point of the segment corresponding to the vertex of the particular grey fault scene can be located without sequential scanning. The segment of vertex v_0 consists of checkpoint sector C_0. This sector includes the edge associated with v_0 and its attributes at the beginning of the snapshot group. Following it is the side activity associated with v_0. For example, C_0 includes the $(v_0, v_1, w_1)(v_0, v_5, w_2) \ldots (v_0, v_n, w_m).(v_0, v_1, w_1)$ indicates that the graphical snapshot at the beginning of the snapshot group contains edges (v_0, v_1) with weights of $w_1.(v_0, v_5, w_2)$ indicates that the graphical snapshot at the beginning of the snapshot group contains edges (v_0, v_5) with weights of $w_2.(v_0, v_n, w_m)$ indicates that the graphical snapshot at the beginning of the snapshot group contains edges (v_0, v_n) with weights of w_m. And so on. List $a_{01}, a_{02}, \ldots, a_{0t}$ is the edge activity associated with v_0, sorted by timestamp.

$$a_{01} = \langle addE, (v_0, v_6, w), t_1 \rangle$$
$$a_{02} = \langle addE, (v_0, v_1, w'), t_2 \rangle$$
$$\ldots$$
$$a_{0t} = \langle delE, (v_0, v_5), t \rangle$$

(9)

a_{01} is an activity of adding an edge (v_0, v_6) with a weight of w at time t_1. a_{02} is an activity of adding an edge (v_0, v_1) with a weight of w' at time t_2. a_{0t} is the activity of deleting the edges (v_0, v_5) at time t.

The structure of the structure-location relationship is shown in Fig. 6. Structure-location associations are different from time-location associations. It processes the snapshot group based on the time interval $T_i, 1 \leq i \leq n$. The format of each inspection scenario C_i is the same as the time-location association. The vertex activity within each T_i is continuously stored in timestamp order.

After the knowledge graph is created, the breadth-first algorithm is used to match the current scene with the grey fault scene in the graph. The scenario of successful matching is the detection result of this algorithm.

Table 2. Application type description.

Application	Description	Number of containers allocated at the beginning		
		Level 1	Level 2	Level 3
Counter	Character Statistics of Englishwords based on Wordcount in Hadoop Environment	4	6	10
Emotional snalysis [9]	Fast emotion Analysis based on LSTM in Hadoop Environment	4	6	10
Stock detection [10]	Stock forecasting based on SVM in Hadoop Environment	4	6	10
Scenic spot recommendation [19]	Scenic spot recommendation on CF algorithm and user behavior data in Zookeeper Environment	8	14	18
Keyword extraction [20]	Application of keyword extraction on TF algorithm in Zookeeper Environment	8	14	18
Frequent itemsets calculation	Frequent itemsets calculation on FP-tree in the ZooKeeper environment for the behavior data of tourist attractions purchased by users	8	14	18

Every update activities of graph is stored in the log, which is the change of the context environment and the occurrence of grey faults. The snapshot group $G_{[t_1,t_2]}$ consists of the graphical state of $[t_1, t_2]$ in the time range. It contains the context environment change point and grey fault occurrence point for the entire graph at start time t_1, and all graphics updates up to t_2. Therefore, the time graph consists of a series of snapshot groups of consecutive time ranges. The snapshot group $G_{[t_1,t_2]}$ contains enough information to access the graph snapshot at any point in the time range. So we can look at changes in context across two dimensions of time and space.

4   Experiment

This section will analyze the performance of relationship knowledge graph about context change and grey fault.

4.1   Experimental Design and Configuration

The experiment was carried out in a container cluster composed of 181 container nodes. For consistency, all docker containers use Ubuntu14.04 mirrors. We use

Table 3. Example of the injected grey failure.

ID	Cluster environment	Grey fault	Type of failure for context environment change
1	HDFS	Blockpool could not be initialized but still continues to initialize	Content fault
2	HDFS	Too many dynamic partitions cause the master node to be abnormal	Design fault
3	HDFS	HDFS block lost threshold is too high	Content fault
4	HDFS	The HDFS service is abnormal, and the HDFS health check timeout is 1 h	Design fault
5	Hbase	The client process actively calls multiple User # login methods to log in	Behavior fault
6	Zookeeper	Failed disk lock in leader causes cluster	Design fault
7	Zookeeper	Transient network partitioning causes long failures in service requests	Content fault
8	Zookeeper	Destroyed packets in deserialization	Behavior fault
9	Zookeeper	Transaction thread exception	Design fault

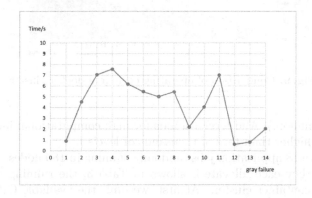

Fig. 7. Grey fault identification.

Kubernetes as container cluster management, and use its own load balancing strategy and resource scheduling strategy. This 181 containers are installed in the same hadoop framework and Zookeeper system for the application to run. One of the containers acts as the primary container and it is responsible for being the primary node for Kubernetes, Hadoop, and Zookeeper. There uses six types of application jobs, as shown in Table 2.

The same type of application separately prepares 1, 2, and 3 levels of the same type of application work of different strengths: level 1 represents low computational strength (The data count is lower than a million levels.). And level 2 represents computational intensity (The data count is higher than a million,

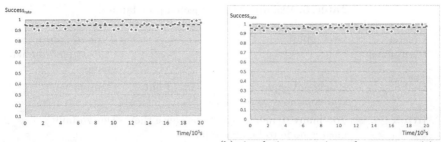

(a) Google cluster recognition success rate. (b) simulation container cluster recognition success rate.

Fig. 8. Identification success rate.

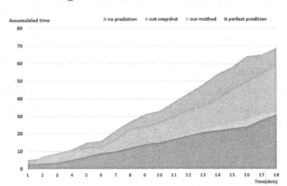

Fig. 9. Comparison of operating costs in a docker cluster.

lower than a million.). And level 3 stands for high computational intensity (The data count is higher than tens of thousands of levels.).

So there are 18 application jobs randomly running in 180 nodes. The number of containers they each allocate is shown in Table 3, the running environment of simulating container clusters. At first, we limit the available CPU resources for each container to 1. And all other configurations remain the default. We started HDFS, HBASE and Zookeeper, and calculated the running time of the application. Then, the available CPU and memory resources are increased one by one to observe the changes in the running time of the application. In this paper, fault injection is used to simulate the experiment. An example of the injected grey failure is shown in Table 3.

Through a 60-day fault injection simulation experiment on the container cluster, a knowledge graph of about 400,000 vertices and 1.2 million edges is stablished. For the grey fault simulated in this section, the identification time of the graph is shown in Fig. 7.

Table 4. The result of fault detection in Google.

	The method of this paper	Isolated forest	LSTM	BLSTM
Precision	94.31%	81.91%	68.88%	78.86%
Recall	73.59%	56.33%	50.33%	55.72%
F1	82.67%	66.75%	58.16%	65.30%

Table 5. The validity of the algorithm in docker cluster.

	The method of this paper	Isolated forest	LSTM	BLSTM
Precision	95.60%	83.48%	75.91%	80.06%
Recall	75.43%	51.23%	44.12%	53.07%
F1	84.33%	63.49%	55.81%	63.83%

4.2   The Result of Fault Detection

In order to prove the universality of the method, Google's public data set [21] is used for experiments. The main record in the Google Cluster public dataset has more than 12,000 nodes clustered in 29 days, 670,000 application jobs and 26 million application tasks.

The operational information contains more than 10 million situational changes with faults and more than 800,000 grey faults. The knowledge graphs created using this dataset have more than 11 million vertices and 42 million edges.

For the previously mentioned injected faults and faults in the Google cluster data, an example of the recognition success rate of each 10^5S of the method is shown in Fig. 8. It can be seen that the recognition success rate is above 90%.

The Fig. 9 illustrates the running cost of this method, with 9 failures and 18 time slots in the Table 3 as examples. As a comparison, three curves are add: "perfect detection", the detection of the method before optimization, and "no detection". "Perfect detection" means that all grey failures can be detected without any wrong detections. "No detection" refers to a system that does not have a detector or any reflection of the scheme. It is defined that in the event of a failure, the cost of execution will increase, which is the overhead of false negation using the SEP [11] method. From the Fig. 9, because of the snapshot structure, the performance curve of the proposed method is close to the complete detection curve, even completely coincident. And it can detect most faults, thereby reducing the cost of fault processing.

In order to evaluate the performance of the knowledge graph, the isolated forest [11], LSTM [12] and BLSTM [13] are used for the comparative experiment. The experiment was compared based on three indicators: Precision, Recall, and F-measure. The Precision is the percentage of correct results in the detection results of the model. The Recall is the percentage of the results of the detected grey fault in the model detection results among all true grey fault results. F-measure is the harmonic mean of Precision and Recall. Because Precision and Recall are two different values, F-measure is a comprehensive indicator of the values of these two indicators. Whether in Google Cluster or in the cluster scenario simulated in this paper, the method achieves high F-measure values and precision, as shown in Tables 4 and 5.

The experimental results show that the knowledge mapping method has better detection accuracy than the comparison method. However, the method of this paper has some limitations. In the temporal local format, all temporary data for the vertices are put together. Therefore, a block of data may contain too much time information beyond a given point in time. The closer the time point to the start time, the more likely the data block will contain useless time data in the time-location layout. This will mislead the search comparison of later scenes to detect the result of grey fault. This is also the optimization work to be carried out later in this paper.

5   Related Work

As Docker is supported by numerous public cloud platforms, it becomes a core cloud business in addition to virtual machines, and container cloud clusters are beginning to emerge. Virtual public cloud vendors such as AWS, Google, Azure, and Docker official cloud services basically support both virtual machine services and container services. They even launched a container cloud business.

The existing container cloud cluster storage service mainly uses the mainstream orchestration tools in containerized cloud storage such as Docker Swarm and Kubernetes, as shown in Table 6. And mainstream orchestration tools in containerized cloud storage like Docker Swarm and Kubernetes do not provide fault detection techniques. They only provide fault tolerance when encountering a fault [15]. Reactive anomaly management like Docker Swarm and Kubernetes is not enough to meet these requirements.

Table 6. Comparison of some containerized cloud storage services.

System	Orchestration tools	Measures for failures
ECS	Choreography component	Distributed key-value database
Tencent Cloud [16]	kubernetes	Multidimensional chart alarm monitoring
Aliyun [17]	Kubernetes Docker Swarm	Multidimensional chart alarm monitoring
BlueDockIBM [14]	Kubernetes	Kubelet-based monitoring components

In the past, research work on failures caused by context changes focused on workload changes, and studied the impact of resource scheduling on application operations. Fadishei et al. [18] analyzed workload tracking from the GridWorkloadArchive project. They found a certain correlation between job failures and performance metrics such as memory usage, CPU utilization, queue utilization, exit time, and job migration.

There are not many studies on grey fault detection. Reference [4] is mentioned, but no specific detection method is given. The literature [5] reproduces the grey faults they collected from ZooKeeper, HDFS, HBase and Cassandra. Each of them causes a serious service interruption.

Grey faults are closely related to the application scenario. Changes in the context, such as container migration, can cause changes in application health. Studying the relationship between grey faults and context scenarios will help with fault detection and system maintenance for application-centric containerized cloud storage environments. Therefore, it could solves the grey fault problem caused by the configuration of the container shielding application context that establish a knowledge graph for the context scenario change and the grey fault association relationship. It helps to support the intelligent operation and maintenance of the system, with application and research significance.

6 Conclusion

In order to carry out fine-grained analysis and comprehensive feature analysis on the grey fault scenarios generated by the container context change, a knowledge graph of the relationship between environmental background changes and grey faults is established. In order to establish and search the scene quickly in the graph, under the premise of accurately classifying the grey fault scene, the graph introduces the snapshot group structure of continuous time and space range to improve the accuracy of the query. Through the monitoring of the resource performance bottleneck inside the container cloud system, the knowledge graph can track the change characteristics of the fault area brought by the resource bottleneck. This will help the operation and maintenance personnel to find a recovery mode to curb the failure propagation in time. The method was test in the container cluster environment. And the experimental results show that the proposed method is fast and effective in grey fault detection.

References

1. Huang, P., et al.: Gray failure: the Achilles' heel of cloud-scale systems. In: Proceedings of the 16th Workshop on Hot Topics in Operating Systems, pp. 150–155. ACM (2017)
2. Miao, Y., et al.: ImmortalGraph: a system for storage and analysis of temporal graphs. ACM Trans. Storage (TOS) 11(3), 14 (2015)
3. Docker: docker (2014). https://docs.docker.com/swarm/
4. Bernstein, D.: Containers and cloud: from LXC to docker to kubernetes. IEEE Cloud Comput. 1(3), 81–84 (2014)
5. Huang, P., Guo, C., Lorch, J.R., Zhou, L., Dang, Y.: Capturing and enhancing in situ system observability for failure detection. In: 13th USENIX Symposium on Operating Systems Design and Implementation (OSDI 2018), pp. 1–16 (2018)
6. Kubernetes: kubernetes (2014). https://www.kubernetes.org.cn/
7. Islam, T., Manivannan, D.: Predicting application failure in cloud: a machine learning approach. In: 2017 IEEE International Conference on Cognitive Computing (ICCC), pp. 24–31. IEEE (2017)

8. Alquraan, A., Takruri, H., Alfatafta, M., Al-Kiswany, S.: An analysis of network-partitioning failures in cloud systems. In: 13th USENIX Symposium on Operating Systems Design and Implementation (OSDI 2018), pp. 51–68 (2018)
9. duoergun0729: nlp. https://github.com/duoergun0729/nlp/blob/master
10. jerry81333: StockPrediction. https://github.com/jerry81333/StockPrediction/
11. Hariri, S., Kind, M.C.: Batch and online anomaly detection for scientific applications in a Kubernetes environment. In: Proceedings of the 9th Workshop on Scientific Cloud Computing, p. 3. ACM (2018)
12. Song, B., Yu, Y., Zhou, Y., Wang, Z., Du, S.: Host load prediction with long short-term memory in cloud computing. J. Supercomput. **74**(12), 6554–6568 (2018)
13. Gupta, S., Dinesh, D.A.: Resource usage prediction of cloud workloads using deep bidirectional long short term memory networks. In: 2017 IEEE International Conference on Advanced Networks and Telecommunications Systems (ANTS), pp. 1–6. IEEE (2017)
14. IBM: IBM cloud private technical community. https://www.ibm.com/developerworks/community/wikis/home?lang=zh#!/wiki/W1559b1be149d_43b0_881e_9783f38faaff
15. Gupta, S., Muthiyan, N., Kumar, S., Nigam, A., Dinesh, D.A.: A supervised deep learning framework for proactive anomaly detection in cloud workloads. In: 2017 14th IEEE India Council International Conference (INDICON), pp. 1–6. IEEE (2017)
16. Tencent: Tencent cloud. https://cloud.tencent.com/document/product/457/9112
17. jianshu: Aliyun cloud. https://www.jianshu.com/p/b7a402c2cf2a
18. Chen, X., Lu, C.D., Pattabiraman, K.: Failure analysis of jobs in compute clouds: a Google cluster case study. In: 2014 IEEE 25th International Symposium on Software Reliability Engineering, pp. 167–177. IEEE (2014)
19. Hwang, S.Y., Yang, W.S.: On-tour attraction recommendation in a mobile environment. In: 2012 IEEE International Conference on Pervasive Computing and Communications Workshops, pp. 661–666. IEEE (2012)
20. Cao, L., Luo, J., Gallagher, A., Jin, X., Han, J., Huang, T.S.: A worldwide tourism recommendation system based on geotagged web photos. In: 2010 IEEE International Conference on Acoustics, Speech and Signal Processing, pp. 2274–2277. IEEE (2010)

Performance Evaluation of Auto Parts Suppliers for Collaborative Optimization of Multi-value Chains

Liangyan Li[1,2], Zhanting Wen[3], Dazhi Wang[4],
and Changyou Zhang[1(✉)]

[1] Laboratory of Parallel Software and Computational Science,
Institute of Software, Chinese Academy of Sciences, Beijing,
People's Republic of China
changyou@iscas.ac.cn
[2] School of Information Science and Technology,
Shijiazhuang Tiedao University, Shijiazhuang, Hebei,
People's Republic of China
[3] China Electronic Science and Technology Network Information
Security Co., Ltd., Chengdu, Sichuan, People's Republic of China
[4] Chengdu Guolong Information Engineering Co., Ltd., Chengdu, Sichuan,
People's Republic of China

Abstract. The performance of auto parts suppliers is becoming an important factor in multi-value chain collaboration. In order to improve the productivity of all links in the auto parts value chain and the competitiveness of the whole value chain, this paper proposes a performance evaluation method for parts suppliers and for the multi-value chain coordination of automobiles. Firstly, from the supplier business data in the auto parts value chain collaboration platform, the relevant description attributes are extracted, and the initial index system of supplier performance evaluation is established. Then, based on the grey system theory and the neighborhood rough set theory, a screening method for the importance of the performance evaluation indexes of auto parts suppliers is designed. Then, the index weights are calculated by the orness measure. Finally, according to the MEOWA idea, the integrated grayscale attribute values. Corresponding weights are used to calculate the comprehensive performance and guide the performance-based accessory supplier optimization. Data from the experimental results on the actual business shows that the supplier evaluation method can correctly reflect the performance of the parts suppliers and provide a quantitative reference for the business synergy of the parts value chain.

Keywords: Multi-value chain · Parts supplier · Performance evaluation · Grey theory · Neighborhood rough set

© Springer Nature Singapore Pte Ltd. 2019
Y. Sun et al. (Eds.): ChineseCSCW 2019, CCIS 1042, pp. 81–91, 2019.
https://doi.org/10.1007/978-981-15-1377-0_6

1 Introduction

China has become the largest producer and owner of automobiles in the world. Competition, among automobile related corporation, has evolved to competition among the value chain of auto-factories. Parts suppliers have a unique position for keeping long-term competitive advantage in the value chain of the automotive industry.

The supplier performance evaluation is the base of supplier selection. It can pre-judge whether the candidate suppliers are qualified to achieve the basic requirements of strategic cooperation or they achieve the established objectives of transaction [1–3]. It can also help suppliers to adapt to the changing market and improve their market competitiveness by analyzing the factors which is affecting the operation of supplier enterprises.

In the aspect of supplier performance evaluation, foreign researchers began research from different perspectives, such as index and weight setting, evaluation methods, etc. The evaluation index is a clear measurement standard. In the automobile industry value chain condition, on the one hand, enterprises can use quantitative methods to evaluate the collaboration effect with upstream partners, on the other hand, they can help enterprises compare performance before and find the right way to compete with other competitors. In the current research literature, the index system of supplier performance evaluation mainly evaluates the performance of suppliers from four aspects: service level, financial situation, internal business and learning ability [4, 5].

According to different condition, various researchers offer diverse evaluation indexes, and some indexes affected by subjectivity, which are ambiguous and difficult to quantify. In the evaluation of parts suppliers, these factors need to be analyzed. Rough set theory and grey system theory are two effective methods to deal with these uncertainty problems.

Rough set theory was proposed by Pawlak who is a Polish mathematician [6]. The theory is based on classification mechanism, which divides data into sets. It aims at describing and analyzing incomplete and uncertain information, and discovering hidden information, so as to make the results objective. Pawlak's original rough set theory is only applicable to dealing with nominal variables. In order to deal with numerical variables, Lin proposed a neighborhood model [7]. The idea of this model is to use the neighborhood of spatial points as basic information particles to granulate the universe space.

Yao [8] and Wu [9] respectively studied the properties of 1-step neighborhood and k-step neighborhood information systems. In 2008, Hu Tsinghua studied a numerical attribute reduction model based on neighborhood granulation and rough approximation, and proposed a neighborhood information system and a neighborhood decision table model. The algorithm is described as follows [10, 11]:

Algorithm
Input: NDT=\<U,A,D\>;
Output: reductive red.
1. $\forall a \in A$: calculate neighborhood relations Na;
2. $\emptyset \rightarrow$red;
3. for any $a_i \in A$-red, calculate SIG(a_i,red,D)=$\gamma_{red \cup a}$(D)- γ_{red}(D)   // defined $\gamma_{\emptyset}$(D)=0;
4. choose a_k, which satisfy: SIG(a_k,red,D)=max(SIG(a_i,red,D));
5. If SIG(a_k,red,D)>0, red$\cup a_k \rightarrow$red, go to Step 3, else returned, end.

From the perspective of probability and statistics, grey system theory provides an effective method to deal with limited and incomplete information by using less data or more variable factors, which has been widely used [12]. The essential goal of the GM (1, 1) model is to process discrete data continuously. Moreover, it can keep its stochastic characteristics when the original time series is weakened, the original data series will be replaced by the generated number series. Grey prediction is based on the grey model GM(1, 1) to carry out quantitative analysis and bring about the prediction of unknown quantities.

Supplier performance evaluation is a comprehensive measure of the overall performance of suppliers. This paper intends to use neighborhood rough set algorithm to bring about supplier comprehensive performance evaluation method.

2 Performance Evaluation Model of Parts Suppliers

2.1 Performance Evaluation Process of Parts Suppliers

Index Selection Process Based on Neighborhood Rough Set. The key point of neighborhood rough set is to select the important index by using variable precision threshold and attributive distance. Then updating the kernel set by iteration. Based on the grey theory and neighborhood rough set [13], as mentioned in the previous section, this section designs a performance evaluation model for automobile parts suppliers. The algorithm steps are shown in Fig. 1.

In Fig. 1, firstly, selecting the index system and constructing the neighborhood decision system. Then, standardizing the neighborhood decision system, finding the neighborhood relation matrix, determining the measurement threshold accuracy and identifying the lower approximation based on each performance index. Furthermore, the attributive dependency is calculated to determine the importance of performance evaluation indexes. Finally, the kernel set and reduction set of performance evaluation are selected and updated.

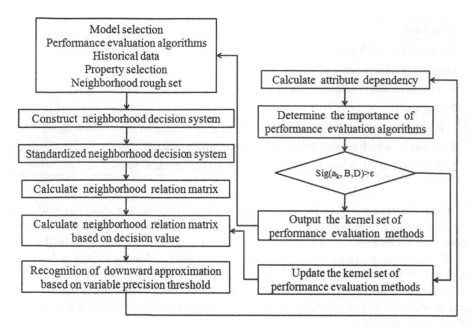

Fig. 1. Index selection process based on neighborhood rough set

Maximum Entropy Evaluation Method Based on OWA Operator. Influenced by MEOWA algorithm with ordered weighted average operator (OWA) weight [14], the maximum entropy evaluation method of OWA operator is integrated to sort suppliers. The algorithm steps are shown in Fig. 2.

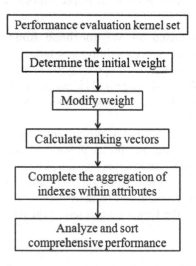

Fig. 2. Maximum entropy evaluation based on OWA operator

In Fig. 2, firstly, according to the neighborhood rough set, the performance evaluation kernel set and reduction set are obtained. Then, the initial weight and the modified weight are determined, and the orness of the weight vector is calculated. The OWA operator collects the information of each index in the attribute and obtains the ranking vector of the comprehensive attribute value. Finally, analyzing the comprehensive performance and according to result put suppliers in numerical order.

2.2 Calculation Model for Performance Evaluation of Parts Suppliers

Construct Neighborhood Decision System. Firstly, select three first-level procurement performance evaluation indexes. Then select two types of business performance evaluation indexes from each category. The results of performance evaluation, as the decision attribute D of suppliers, are expressed by discrete values to distinguish different categories. When the attribute value is 1, the intentional supplier is indicated, and when the attribute value is 2, the qualified supplier is indicated. There are many types of attribute values. To make calculation easier, these values are uniformly expressed by grey numbers through the standardization process.

For the process of converting fixed values into grey numbers, as shown in formula 1:

$$\oplus v_{ij} = \{[\underline{v_{ij}}, \overline{v_{ij}}] | v_{ij} = \underline{v_{ij}} = \overline{v_{ij}}\} \quad (1)$$

The standard grey numbers corresponding to the range of values [VH, H, M, L, VL] can be shown in Table 1.

Table 1. Attribute level grey scale

Level	V_{ij}
VH	[0.8, 1]
H	[0.6, 0.8]
M	[0.4, 0.6]
L	[0.2, 0.4]
VL	[0, 0.2]

The standardization process of grey number is shown in formula 2.

$$\underline{\tilde{v}_{ij}} = \frac{\left|v_{ij} - v_j^{min}\right|}{\left|v_j^{max} - v_j^{min}\right|}, \overline{\tilde{v}_{ij}} = \frac{\left|\overline{v_{ij}} - v_j^{min}\right|}{\left|v_j^{max} - v_j^{min}\right|} \quad (2)$$

$\oplus V_{ij}$ represents the attribute a_j measurement value of performance indexes for supplier i. v_j^{max} and v_j^{min} represent the historical maximum and minimum values of attribute a_j, respectively.

Calculate Neighborhood Relation Matrix. Neighborhood relationship refers to the similarity between parts suppliers expressed by neighborhood distance. Firstly, calculate the relationship matrix. At this time, there is only a single attribute in the performance evaluation index of parts suppliers (only consider the distance between two suppliers with a single attribute).

If the number of attribute cores is more than 1, formula 3 is used to calculate, where $p = \infty$:

$$
\Delta_p(x_i, x_j) = \max \left| \left[\left(\underline{v_{ik}} - \underline{v_{jk}} \right)^2 + \left(\overline{v_{ik}} - \overline{v_{jk}} \right)^2 \right]^{\frac{1}{2}} \right| \tag{3}
$$

Based on the distance relationship matrix calculated from the above formula, the neighborhood relationship among different parts suppliers is obtained. The neighborhood relationship matrix is established according to formula 4.

$$
M_B(N) = (r_{ij})_{n \times n}, \quad \text{where } r_{ij} = \begin{cases} 1, & \Delta(x_i, x_j) \le \delta \\ 0, & \text{otherwise} \end{cases} \tag{4}
$$

Among them, $\delta = 0.4$, because $\Delta(x_i, x_i) = 0 \le \delta$, so $\Delta(x_i, x_i) = 1$, that is, the diagonal element is 1. The assessment of Initial Performance Indexes Atr is an empty set, and attribute set B is a combination of separate performance evaluation indexes.

Calculate the Downward Approximation of Rough Sets. Determining the accuracy of measurement threshold, recognizing the downward approximation of each evaluation index and get the downward approximation of Rough Sets.

Using formula 5 to calculate neighborhood information particles $\delta_B(x_i)$.

$$
\delta_B(x_i) = \{x_j | x_j \in U, \Delta_B(x_i, x_j) \le \delta\} \tag{5}
$$

Establish the neighborhood supply relation matrix $M_B^D(N)$ based on the same decision attribute by formula 6.

$$
M_B^D(N) = (r_{ij})_{n \times n}, \quad \text{where } r_{ij} = \begin{cases} 1, & (\Delta(x_i, x_j) \le \delta, \ D_i = D_j) \\ 0, & \text{otherwise} \end{cases} \tag{6}
$$

Calculate the downward approximation of neighborhood rough sets based on formula 7.

$$
X = \{x_i | I(\delta(x_i), X) \ge k, x_i \in U\}; \ \overline{N^k}X = \{x_i | I(\delta(x_i), X) \ge 1 - k, x_i \in U\} \tag{7}
$$

Kernel Set and Reduction Set of Performance Evaluation. Firstly, calculate the data dependency. For any $a_i \in C - Atr$, formula 8 is used to calculate the dependence of performance result D on performance result Atr and the dependence of D on adding another performance index a_i to Atr. These two dependencies are used to judge the importance of subsequent adding attributes.

$$\gamma_B(D) = \frac{|POS_B(D)|}{|U|} \tag{8}$$

Then calculate the importance of performance evaluation indexes. Definition the dependence of D on Atr is 0. According to formula 9, the importance of evaluation index a_i is calculated based on attribute set B and result D.

$$Sig_1(a_i, B, D) = \gamma_B(D) - \gamma_{B-a_i}(D) \tag{9}$$

Select and update the performance evaluation kernel set and reduction set. Choose an attribute index AI to satisfy formula 10:

$$Sig(a_k, B, D) = \max_i(Sig(a_i, B, D)) \tag{10}$$

Based on the formula mentioned above, get the most important index and update Atr. Always make $Sig_1(a_k, B, D) > \varepsilon(\varepsilon = 0.001)$. The steps are iterated until the condition of this inequality is no longer satisfied. Finally obtain the final reduction set.

Calculate Weight and Orness Measure. Weight calculation and orness measure calculation are shown in formulas 11, 12 and 13.

$$w_i = Q\left(\frac{i}{n}\right) - Q\left(\frac{i-1}{N}\right) \tag{11}$$

$$Q(r) = \begin{cases} 0 & r < a \\ \frac{r-a}{b-a} & a \leq r \leq b \\ 1 & r > b \end{cases} \tag{12}$$

$$orness(w) = \frac{1}{n-1}\sum_{i=1}^{n}(n-i)w_i \tag{13}$$

Q is a Fuzzy Linguistic Quantifier (FLQ).

Supplier Comprehensive Evaluate. Before collecting the information of OWA operator, we should sort the reduction set which is finally selected above. Then the comprehensive attribute value is obtained by multiplying the revised weight. Formula 14 is used to establish the possibility degree matrix of pairwise comparison and formula 15 is used to obtain the ranking vector.

$$p(\tilde{v}_{ij} \geq \tilde{v}_{ik}) = \frac{\min\{l_{\tilde{v}_{ij}} + l_{\tilde{v}_{ik}}, \max(\overline{\tilde{v}}_{ij} - \underline{\tilde{v}}_{ik}, 0)\}}{l_{\tilde{v}_{ij}} + l_{\tilde{v}_{ik}}} \tag{14}$$

$$v_i = \frac{\sum_{j=1}^{n} p_{ij} + \frac{n}{2} - 1}{n(n-1)} \tag{15}$$

Calculate the total score of each supplier and select the best supplier.

3 Evaluation and Verification for the Automotive Industry Collaboration Platform

The parts suppliers use the automobile industry chain collaborative platform to select the evaluation indexes. We choose three primary purchasing performance evaluation indexes. Then we select two business performance evaluation indexes from each category. The "initial" evaluation value of the automobile parts supplier is D. The neighborhood decision system of this paper selected 9 decision attributes of 30 suppliers, and some data are listed in Table 2.

Table 2. Decision makers' assessment of the performance of automobile parts suppliers

Index	Cos1	Cos2	Time1	Time2	Qual1	Qual2	FB1	FB2	D
Su1	10	[1, 5]	[11, 14]	[4, 8]	[0.02, 0.09]	VH	VH	176	2
Su2	10	[2, 8]	[42, 49]	[23, 24]	[0.22, 0.73]	H	VH	198	2
Su3	10	[11, 5]	[13, 16]	[20, 23]	[0.5, 1.21]	VH	VH	186	2
...	...	...	...	...	...	...	...	...	...
Su30	1	[42, 47]	[42, 45]	[2, 5]	[1.5, 1.9]	VL	VH	162	1

We consider the variety of values of the above indexes. We standardize them into gray numbers for ease of calculation. The results of the standardized supplier performance evaluation are shown in Table 3.

Table 3. Standardized automobile parts supplier performance evaluation results

Index	Cos1	Cos2	Time1	Time2	Qual1	Qual2	FB1	FB2	D
Su1	[1, 1]	[0.02, 1]	[0.183, 0.233]	[0.167, 0.333]	[0.002, 0.009]	[0.8, 1]	[0.8, 1]	[0.88, 0.88]	2
Su2	[1, 1]	[0.04, 0.16]	[0.7, 0.817]	[0.958, 1]	[0.022, 0.073]	[0.6, 0.8]	[0.8, 1]	[0.99, 0.99]	2
Su3	[1, 1]	[0.22, 0.3]	[0.217, 0.267]	[0.8333, 0.958]	[0.05, 0.121]	[0.8, 1]	[0.8, 1]	[0.93, 0.93]	2
...	...	...	...	...	...	...	...	...	...
Su30	[0.1, 0.1]	[0.84, 0.94]	[0.7, 0.75]	[0.083, 0.203]	[0.15, 0.19]	[0, 0.2]	[0.8, 1]	[0.81, 0.81]	1

Then execute the algorithm process of Fig. 1, get the performance evaluation kernel set and reduction set.

Orness measure is calculated according to formulas 11, 12 and 13. Based on the orness value, the weights are modified by maximizing the entropy, and the index information is extracted on the premise that the revised weights remain unchanged. According to the simplified formula 16, 17, the revised value is obtained.

$$\sum\nolimits_{i=1}^{n} \left(\frac{n-i}{n-1} - \text{orness(w)}\right)h^{n-i} = 0 \tag{16}$$

$$w_i^* = \frac{h^{n-i}}{\sum_{i=1}^{n} h^{n-i}} \qquad (17)$$

Obtain the results of h and revised weights.

Take supplier Su1 as an example, sort the index set. According to the performance weight W, OWA operator is used to aggregate the information of performance attribute D. Then we get the comprehensive performance Dn.

Select Max (Dn), obtain a result that the highest comprehensive evaluation score of supplier is Su4. It can be selected as the best supplier. The results of selection are consistent with the actual performance of suppliers running on the platform.

4   A Case Study of Multi-value Chain Collaboration

According to the evaluation method of automobile industry collaboration platform proposed above, multi-value chain collaboration is designed in the platform. Service providers select the best parts supplier through the performance evaluation of parts suppliers, and the best parts supplier can directly support the service providers. Those parts which are not from best supplier will transfer to warehouse to examine, Parts which reaching the standard can be supplied to the service provider. As shown in Fig. 3, it achieves multi-value chain business collaboration.

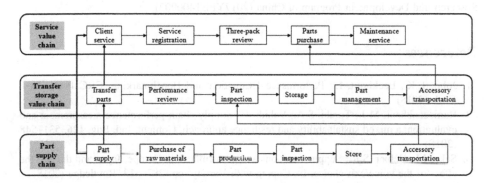

Fig. 3. Multi-value chain collaboration

We evaluate multi-value chain business collaboration through performance evaluation. First, in the service value chain, the service provider directly selects high-quality parts based on the performance evaluation of the part suppliers. This will reduce the swelling inventories and operational risk, and it will increase revenue. Secondly, the transfer library has increased the inspection of part goods. This has played a controlling role in the quality of the parts. It increases the value of the value chain while also reducing costs.

Finally, due to performance evaluation, some parts suppliers have some extra costs in the detection of parts, but it indeed improves the quality of the supply parts and the supplier's benign competitiveness and optimizes its value chain. Performance evaluation Multi-value chain business collaboration can correctly reflect the performance of part suppliers. It provides a quantitative reference for multi-value chain collaboration. It does multi-value chain collaboration. And it increases the productivity of all segments of the value chain and the competitiveness of the full value chain.

5   Conclusions and Prospects

The paper initially analyzed the performance evaluation indexes of parts suppliers. We combined current evaluation algorithms with existing indexes for interpretation and analysis and gave an evaluation index system for part suppliers for multi-value-chain collaboration. Then we studied several major evaluation algorithms. We selected the neighborhood rough set theory algorithm for the characteristics of the part suppliers in the trading system. Finally, after comprehensive consideration, the evaluation index system is given. We designed a performance evaluation algorithm for the part suppliers that incorporate MEOWA's improved neighborhood rough set for the actual situation. It has certain reference value for collaborative optimization of multiple value chains.

Acknowledgment. The author wishes to thank the editor and anonymous referees for their helpful comments and suggested improvements. This paper is supported by The National Key Research and Development Program of China (2017YFB1400902).

References

1. John, L.K., Eeckhout, L.: Performance Evaluation and Benchmarking. CRC Press, New York (2018)
2. Ramezankhani, M.J., Torabi, S.A., Vahidi, F.: Supply chain performance measurement and evaluation: a mixed sustainability and resilience approach. Comput. Ind. Eng. **126**, 531–548 (2018)
3. Sako, M., Helper, S.R.: Supplier relations and performance in Europe, Japan and the US: the effect of the voice/exit choice. In: Coping with Variety, pp. 287–313. Routledge (2018)
4. Ding, R., Ren, P.: The logistics performance evaluation index system in the transportation industry based on big data. In: 2018 IEEE 3rd International Conference on Big Data Analysis (ICBDA). IEEE (2018)
5. Sinha, A.K., Anand, A.: Development of sustainable supplier selection index for new product development using multi criteria decision making. J. Clean. Prod. **197**, 1587–1596 (2018)
6. Pawlak, Z., et al.: Rough sets. Commun. ACM **38**(11), 88–95 (1995)
7. Lin, T.Y.: Granular computing on binary relations I: data mining and neighborhood systems. Rough Sets Knowl. Discov. **1**, 107–121 (1998)
8. Yao, Y.Y.: Relational interpretations of neighborhood operators and rough set approximation operators. Inf. Sci. **111**(1-4), 239–259 (1998)
9. Wu, W.Z., Zhang, W.X.: Neighborhood operator systems and approximations. Inf. Sci. **144** (1-4), 201–217 (2002)

10. Ma, Y., et al.: Selection of rich model steganalysis features based on decision rough set α-positive region reduction. IEEE Trans. Circuits Syst. Video Technol. **29**, 336–350 (2018)
11. Hu, Q.H., Yu, D.R., Xie, Z.X.: Numerical attribute reduction based on neighborhood granulation and rough approximation. J. Softw. **19**(3), 640–649 (2008)
12. Wang, C.N., et al.: Performance evaluation of major asian airline companies using DEA window model and grey theory. Sustainability **11**(9), 2701 (2019)
13. Yang, X., et al.: Pseudo-label neighborhood rough set: measures and attribute reductions. Int. J. Approx. Reason. **105**, 112–129 (2019)
14. Kang, B., et al.: Generating Z-number based on OWA weights using maximum entropy. Int. J. Intell. Syst. **33**(8), 1745–1755 (2018)

Research and Implementation of Flow Table Optimization Strategy for SDN Switches Based on the Idea of "Main Road"

Zhaohui Ma[1,2,3], Zenghui Yang[4], and Gansen Zhao[3(✉)]

[1] Collaborative Innovation Center for 21st-Century Maritime Silk Road Studies, Guangdong University of Foreign Studies, Guangzhou 510006, China
26593978@qq.com
[2] School of Information Science and Technology, Guangdong University of Foreign Studies, Guangzhou 510006, China
[3] School of Computer Science, South China Normal University, Guangzhou 510631, China
gzhao@m.scnu.edu.cn
[4] Crime and Professional Information Battalion of Crime and Case Investigation Department, Nansha District Bureau of Guangzhou Public Security Burea, Guangzhou, China
3291235182@qq.com

Abstract. The seperation of control layer from data layer through SDN (software defined network) enables network administrators to plan the network programmatically without changing network devices, realizing flexible config-uration of network devices and fast forwarding of data flows. The controller sends the flow table down to the switch, and the data flow is forwarded through matching flow table items. However, the current flow table resources of the SDN switch are very limited. Therefore, this paper studies the technology of the latest SDN Flow table optimization at home and abroad, proposes an efficient opti-mization scheme of Flow table item on the main road through the directional flood algorithm, and realizes related applications by setting up experimental topology. Experiments show that this scheme can greatly reduce the number of flow table items of switches, especially the more hosts there are in the topology, the more obvious the experimental effect is. This method can solve the problem of insufficient resources of Flow table items of Open Flow switch, and the experiment proves that the optimization success rate is over 90%.

Keywords: Software defined networking · Main road · Directional flooding · Flow table optimization

1 Introduction

With the rapid development of cloud computing, virtualization, blockchain technology and other network services, great challenges have been posed to traditional network architecture and network equipment. Due to the increasingly complex network service, the data flow is also growing rapidly. Such shortcomings as low efficiency, the

© Springer Nature Singapore Pte Ltd. 2019
Y. Sun et al. (Eds.): ChineseCSCW 2019, CCIS 1042, pp. 92–103, 2019.
https://doi.org/10.1007/978-981-15-1377-0_7

variability and complexity in agreement and poor flexibility existed in the traditional network make it more and more difficult to meet the present network demand. In order to meet the demand of the new services, a special network framework is urgently needed. The network framework consists of a manageable, simple general hardware devices which can program freely. Under such circumstances, SDN began to receive extensive attention from the society and academia [1], and received strong support from operators and related network service enterprises. It was once believed that SDN would be a new technology to subvert the traditional network.

OpenFlow is the first standard southward interface designed for SDN, which proposes the concept of flow. Packets in the network are processed according to the pre-defined matching domain in the flow table and the controller writes all the fields that needed to be matched and the corresponding actions into the flow table by installing the flow table to the switch, which break the concept of two layers and three layers in the traditional network. For OpenFlow switches, the SDN network provides fine-grained flow matching. The frequent interactive access between different hosts and different servers in the network requires the controller to issue a large number of flow tables to ensure the smooth operation of the network. When the switch fails to match many streams, the switch will frequently send packet_in package to the controller, which may lead to overloading of the controller, thus extending the down-time of the flow table and resulting in high network delay. Since the current switch flow table can only load about 1 k–2 k flow table items and the flow table resources are very limited [2, 3], one of the problems that must be solved for the wide application of SDN network is the performance optimization of switch flow table.

At present, some scholars have put forward and studied the scheme of switch flow table optimization [4, 5], including optimization based on the flow table reuse, optimization based on multi-level flow table technology, optimization based on the stagnation timeout time of flow table items and optimization based on caching mechanism.

Optimization scheme based on flow chart resources reuse [6] puts forward flow table storage resources reuse through the analysis of the existing flow table data structure so as to achieve the purpose of matching the data flow as much as possible. The flow table items in the same matching domain will be compressed and stored in a same flow chart space through this method, which greatly improves the efficiency of flow table resources use. But it reduces the fine-grained processing power of flow table items.

In literature [7], a new Open Flow multi-level Flow table structure was proposed. Based on this structure, a mapping algorithm was proposed for multi-level Flow table lookup. In this paper, a single flow table is mapped to a multi-level flow table by mapping algorithm for efficient storage and search. However, it fails to take into account the balance between the series of flow tables and the search time, and its search algorithm has the problem of repeated search for many times.

Literature [8, 9] obtains real-time information such as forwarding rules, actions, stagnant timeout [10, 11] of flow and hard timeout of flow by programming the controller and dynamically adjusts the timeout of switch flow table items. Thus, the switch flow table resources are more effectively utilized. However, there are still some deficiencies in parameter setting of real-time dynamic adjustment.

The design of cache mechanism of Cache Flow is inspired by the author Katta N. In his view, the flow table items which can match most of flow are deployed in the switch (including software switches and hardware switches). In the data layer the remaining flow table items are cached with a data structure called Cache Master so as to increase the matching efficiency of the Flow [12, 13]. However, the cache space of the switch is limited, so it is impossible to pre-deploy all the rules to each switch, and the statically deployed rules cannot adapt to the dynamic changing network state.

In view of the deficiencies in the above optimization of Flow table resources, this paper proposes an efficient Open Flow table item optimization scheme based on the main road through the directional flood algorithm. This method can solve the problem of insufficient flow table item resources of the Open Flow switch well. Experiments show that the optimization success rate is over 90%.

The following arrangements are as follows: Sect. 2 briefly introduces the related SDN technologies, including OpenFlow, Mininet and RYU controllers; Sect. 3 describes the idea and implementation process of "main road"; Sect. 4 demonstrates the realization process of flow table optimization based on the main road; Sect. 5 carries on the simulation experiment and the data comparison analysis; finally, Sect. 6 summarizes the full text and evaluates its prospects.

2 Overview of the SDN Related Technologies

2.1 OpenFlow

OpenFlow is the most widely used protocol in the SDN southward interface (control layer docking data forwarding layer) at present. OpenFlow has developed from the original 1.0 version to the current 1.5 version, of which the 1.0 version and 1.3 version are the most widely used and best-compatible versions. Version 1.3 has more new structures than version 1.0. First of all, the header domain of version 1.0 can only match 12 items of content, while version 1.3 has increased its matching domain to 39 items, including such matching items as source and destination MAC address and source and destination IPV6 address, etc., which is used to continuously improve the compatibility of the matching domain [14].

Then the instruction of version 1.3 is superior to the action of version 1.0. The instruction introduces the concept of action set on the basis of the action, which facilitates the flow to perform a series of actions. The action here is divided into mandatory actions and optional actions, including all actions in the traditional network, such as forwarding, discarding, flood and so on. In addition, some new attributes have been added in version 1.3: table miss flow table items, timeout, and additional information, etc.

2.2 Mininet

Mininet is a real-time lightweight virtual network. It uses virtualization technology to provide virtual simulation platform for operations simulating in real environment. It supports various protocols such as OpenFlow and OpenvSwith. Mininet also provides a

Python API for others to collaborate on. The SDN experiment can be carried out conveniently through the switch, host and controller of Mininet. In addition, Mininet supports multiple controllers. Most importantly, experiments conducted on Mininet can be seamlessly migrated to the real environment, with good hardware portability and high scalability [15].

2.3   Ryu Controller

Ryu is an open source SDN controller, which literally means "Flow" in Japanese. Fully programmed in python language, it is in exact compliant with the Apache open source license and supports all versions of the OpenFlow protocol currently. The goal is to provide a SDN operating system with logically centralized control. Through RYU's extensive API interface, SDN programmers can easily create new management and control applications.

3   Brief Description and Implementation Process of "Main Road" Idea

3.1   Brief Description of the Thought of "Main Road"

This paper proposes a new flow chart storage optimization strategy based on "main road". Aiming at solving the problems of limited flow table resources in the internet, this strategy builds a "main road" in the center area which has the frequent circulation of the network data flow to reduce the total number of network flow table items. The main road is essentially composed of a set of OpenFlow switches. It stipulates that the data flow must go through the main road before arriving at the destination, which makes part of the network flow centralize on the main road. After that the flow table items of the main road will be optimized by using the corresponding flow table algorithm to reduce the number of flow table items on the main road. At the same time, the number of flow table items in other areas of the network is also reduced, so as to optimize the flow table resources. Specific validation scenarios are described in part 3 in detail.

3.2   The Concrete Process of Realizing the Idea of "Main Road"

The flow table optimization design based on the idea of "main road" can be divided into four stages. The first stage is topological initialization. At this stage, the location of the specific switches and hosts in the topology can be obtained. In the second stage, the data entered the main road. The most important thing in this stage is to determine which node on the main road will be selected to enter for data. The selected node is described as the primary node. The third stage is to determine the data transmission path on the main road. In the fourth stage, the data leaves the main road and reaches the terminal.

4 Implementation of the Optimization Design of Flow Table Based on Idea of "Main Road"

The detailed implementation process of flow table optimization based on the idea of "main road" is mainly carried out through the four stages mentioned in the second section, and the operation results are observed, so as to verify the validity of the idea of "main road".

4.1 Topology Discovery

When such events as switch connection/disconnect, add/delete port, port ON/OFF, delete/add link occur in the network topology, a topology is reacquired to accommodate the dynamic changing process of the topology. First of all, the list of switches and links in the topology is obtained through the RYU controller. The switch list creates port mappings between switches and switches, switches and hosts, resulting in three different port collections: i.e., the collection of all ports of the switch; the collection of ports of switch connected to a switch; the collection of ports of switches connected to a host. The source and destination ports of the link are obtained in the link list, forming a mapping, that is, the mapping between the source port of the source switch and the destination port of the destination switch. The shortest path algorithm is then used to calculate the shortest path between all switches in the whole topology and store it for later use.

4.2 The Selection of Main Nodes on the "Main Road"

In Fig. 1 topology, s5-s6-s9-s10 is the main road. Next, the main node from the host to the main road will be determined. The primary node means the nearest node from the host to the main road. First, the shortest path from each host to each node in the main road is calculated, and the shortest path node is selected as the candidate primary node. When there is more than one candidate node, a single primary node needs to be selected from multiple nodes. The selection principle is to choose nodes as close as possible to both ends of the main road. Because the switch that each host accesses is unique, the calculation of each segment host is converted into the calculation of access switch. For example, if the candidate node of S2 switch to the main road is [S6, S9, S10], S10 is determined to be the final primary node by S2 according to the principle of closest endpoints, which means that the final path of S2 switch for data to enter the main road is through S10.

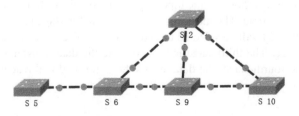

Fig. 1. Node switches on the main road

4.3   Main Road" Data Transmission – "Directional Flood" Algorithm

When the data packet enters into the main road through the main node, it is forwarded based on directional flood algorithm. The description of directional flood algorithm is as follows:

First, during the initialization process, we set S5, S6, S9 and S10 switches as the main node of the main road.

Then wildcard flow table with lower priority will be distributed to four nodes for the transmission of packets on the main road. The flow tables possess a high optimization effect and they just match the ingress port. For example, the data packet from the left port of S6 switch will only be transmitted to its right port. Data packet from all ports except the right port of S5 switch will be sent to S6, which not only saves a lot of flow table resources, but also avoids the occurrance of the loop.

Finally, if the data flow cannot match the exit after entering the main road, it will conduct directional flood at both ends to ensure data transmission in the main road. That is, S5 – S6 – S9 – S10 and S10 – S9 – S6 – S5. When the flow enters the main road from the left main node, the process is shown in Fig. 2 (similar to that of the right end). and when the flow enters the main road from the middle main node, the process is shown in Fig. 3.

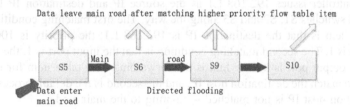

Fig. 2. Flow chart of "directional flood" 1

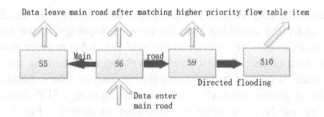

Fig. 3. Flow chart of "directional flood" 2

4.4   Data Leaves the "Main Road"

When the specific conditions within the time limit adjustment mechanism are reached, it can be seen from part C that the data transmission in the main road is by default matched with the directional flood table items at both ends with lower priority. Therefore, After the data is transferred to a node of the main road, it needs to take a higher priority flow table item to leave the main road from this node. Flow table items with different priority issued by the switch can effectively ensure the correct transmission of data. How does the switch install flow table items? As shown in Fig. 4, host h1 enters the main road through the main node S4. When the data of h1 enters the main node S4, the flow table will be sent to the switch along the passing path. The installation of flow table is shown in Fig. 4:

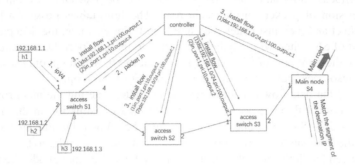

Fig. 4. Installation of flow table

The controller issues 192.168.1.1 as the source IP and destination IP flow table items to all switches S1, S2 and S3 along the way. The first matching condition for the flow table item is that the destination IP is 192.168.1.1, the priority is 100 and the output port is 1. The second matching condition is that the input port is 1, the priority is 10 and the output port is 4. The first is the forwarding flow table item for the access switch s1 to match the destination host IP, and the second is the default processing after the destination host IP is not matched – pointing to the main node.

5 Comparison of Simulation Experiment and Data Analysis

In order to verify the effectiveness of the optimization strategy for the flow table of "main road", under the condition that the network topology is the same as other experimental environments, in contrast to the total flow table number generated in the SDN network based on the "shortest path" forwarding algorithm the feasibility of the idea of "main road" is verified through comparative analysis. The experimental scheme is as follows. In the mininet network simulation environment, a SDN network topology under a single remote Ryu controller is constructed, as shown in Fig. 5.

The application based on the "shortest path" and the application based on the "main road" are respectively operated on the controller. Then ping operation is conducted on all the hosts in the network topology. If the entire network can be connected to each other through ping, it shows that the corresponding flow table item has been issued to all OpenFlow switch. After that check and count the number of the flow table in all switches. At last, do a comparison of the results. In addition, in order to make the experimental results more persuasive, besides horizontal comparison, vertical comparison is also carried out, that is, the experiment is carried out in the network environment of 16 hosts and 40 hosts respectively, to reach the final conclusion.

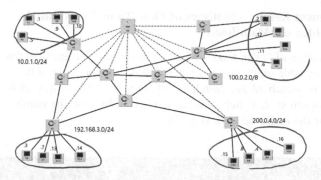

Fig. 5. Experimental topology

5.1 Experiment Results – Statistics of Forwarding Flow Table Items Based on SPF (Shortest Path First) Algorithm

In the environment of Fig. 5, the shortest path application is run on the controller. The number of flow table items in all switches will be checked and counted when all hosts can be connected through ping. Figure 6 shows the partial flow table items generated by switch s2.

```
mininet> dpctl dump-flows
*** s2 ----------------------------------------------------------------
NXST_FLOW reply (xid=0x4):
 cookie=0x0, duration=27.363s, table=0, n_packets=729, n_bytes=43740, idle_age=0, priority=0
 cookie=0x0, duration=410.320s, table=0, n_packets=76, n_bytes=7448, idle_age=9, priority=10
 cookie=0x0, duration=391.376s, table=0, n_packets=76, n_bytes=7448, idle_age=9, priority=10
 cookie=0x0, duration=388.339s, table=0, n_packets=76, n_bytes=7448, idle_age=9, priority=10
 cookie=0x0, duration=334.954s, table=0, n_packets=78, n_bytes=7644, idle_age=9, priority=10
 cookie=0x0, duration=410.317s, table=0, n_packets=57, n_bytes=5586, idle_age=9, priority=10
 cookie=0x0, duration=391.374s, table=0, n_packets=56, n_bytes=5488, idle_age=9, priority=10
 cookie=0x0, duration=388.336s, table=0, n_packets=57, n_bytes=5586, idle_age=9, priority=10
 cookie=0x0, duration=334.953s, table=0, n_packets=57, n_bytes=5586, idle_age=9, priority=10
 cookie=0x0, duration=12.090s, table=0, n_packets=0, n_bytes=0, idle_timeout=15, hard_timeou
 cookie=0x0, duration=12.089s, table=0, n_packets=0, n_bytes=0, idle_timeout=15, hard_timeou
 cookie=0x0, duration=12.089s, table=0, n_packets=0, n_bytes=0, idle_timeout=15, hard_timeou
 cookie=0x0, duration=11.994s, table=0, n_packets=0, n_bytes=0, idle_timeout=15, hard_timeou
 cookie=0x0, duration=11.993s, table=0, n_packets=0, n_bytes=0, idle_timeout=15, hard_timeou
 cookie=0x0, duration=11.993s, table=0, n_packets=0, n_bytes=0, idle_timeout=15, hard_timeou
 cookie=0x0, duration=11.968s, table=0, n_packets=0, n_bytes=0, idle_timeout=15, hard_timeou
 cookie=0x0, duration=11.964s, table=0, n_packets=0, n_bytes=0, idle_timeout=15, hard_timeou
 cookie=0x0, duration=11.964s, table=0, n_packets=0, n_bytes=0, idle_timeout=15, hard_timeou
 cookie=0x0, duration=11.922s, table=0, n_packets=0, n_bytes=0, idle_timeout=15, hard_timeou
 cookie=0x0, duration=11.922s, table=0, n_packets=0, n_bytes=0, idle_timeout=15, hard_timeou
```

Fig. 6. Partial flow table items generated based on SPF

By statistics, the number of flow table items of switch s1 is 110, switch s2 110, switch s3 110, switch s4 110, switch s5 66, switch s6 98, switch s7 34, switch s8 2, switch s9 66, switches s10 2. In total, the number of flow table items of the entire network is 708.

5.2   Experiment Results – Statistics of Flow Table Items Based on the Idea of "Main Road"

Run the application of "main road". When all hosts can be connected through ping the statistics of flow table of all switches are as follows: the number of flow table items of switch s1 is 10, switch s2 10, switch s3 10, switch s4 10, switch s5 6, switch s6 4, switch s7 2, switch s8 2, switch s9 6, switch s10 4. In total, the number of flow table items is 64 for the entire network (Fig. 7).

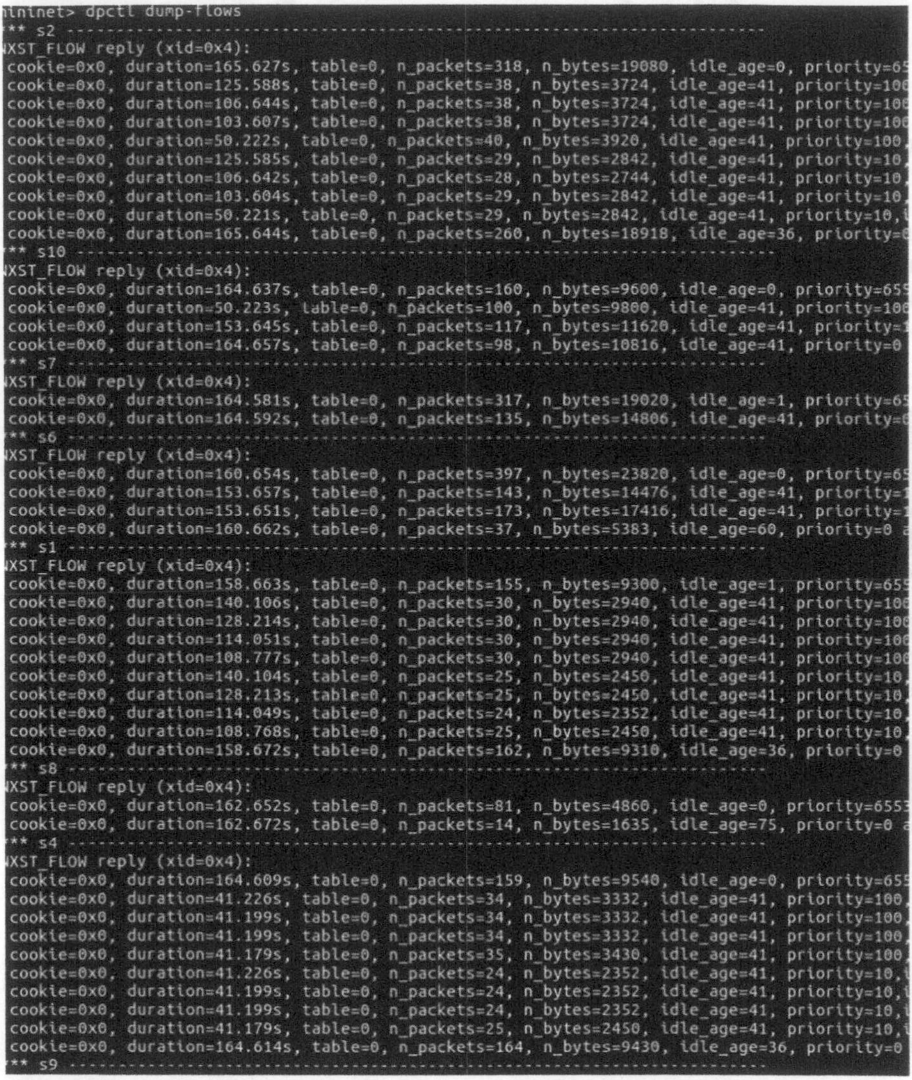

Fig. 7. Partial flow table items based on the "main road"

5.3   Data Analysis and Comparison

Under the condition that 16 hosts can be connected in entire network, the total number of flow table items generated by the whole network is 64 when using the flow table optimization algorithm based on the idea of "main road" while the total number of flow table items generated by the whole network is 708 when using the shortest path forwarding algorithm. The number of the flow table items of the entire network resulted from the application of main road is less than one tenth of the number of flow table items by running on the shortest path forwarding application, which greatly optimizes the performance of the flow table of switch. In another experiment, the entire network involves 40 hosts. The total number of flow table items by running the shortest path forwarding application reached 4,511 while the total number of flow table items by running the application based on the idea of "main road" is only 122.

By comparing the number of flow table items by running two different applications in two topologies, it can be concluded that the more hosts there are in the topology, the more obvious the optimized effect is when using the idea of "main road". The comparative analysis is shown in Fig. 8.

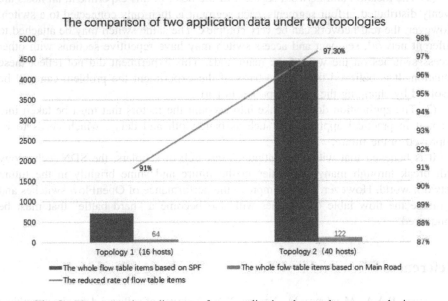

Fig. 8. The comparison diagram of two application data under two topologies

6   Conclusions

SDN is a new kind of network architecture. It realizes the dynamic control of network flow through the separation of traditional network control layer and data forwarding layer. It can carry on the real-time control to the network behavior. It is academia and industry research hot at present and it is valued by both network operators and enterprise. However, before the practical and large-scale application, there are still some key technical problems to be solved, one of which is the optimization of SDN performance. This paper studies the latest SDN flow table optimized technology at home and abroad, analyzes and puts forward the new idea of "main road". Through constructing the experimental topological environment and the realization of related applications by using RYU controller programming, the experiment proves the conclusion that the number of flow table items can be greatly reduced. In particular, the more the number of the host in the topology, the more obvious the optimized effect is, the more the number of flow table items reduces.

Although the experiment got the expected effect, there are still some deficiencies in the following aspects. The paper needs to be further improved in the following aspects on the basis of the existing research:

(1) Selection of main road. In this experiment, the selection of the main road is fulfilled by artificial work. The main road can not be changed according to the real changing network topology, so at present the main road is only applicable to the experiment topology. The improvements in this aspect needs to select representative node for different areas, and connect each representative node as the main road in the network.

(2) The location and network segment of the host. In this experiment, all hosts are evenly distributed in four segments, each segment is then only connected to a switch. However, the real network can be very complex. The same switch may be attached to different network segment and access switch may have repetitive sections with other access switches on the way to the main node. This experiment did not reflect these situation. It is believed that on the basis of the experiment the problem can still be resolved by changing the code in packet in parts.

(3) The application does not take into account the factors that must be taken into account in practical applications, such as bandwidth and delay, which needs to be improved in the future.

It is believed that with the continuous research of scholars, the SDN technology will break through many difficulties in the future and shine brightly in the future network world. However, how to improve the performance of OpenFlow switches and optimize the flow table of switches will also become a "hard battle" that must be conquered.

References

1. Nianes, B.A.A., Mendonca, M., Nguyen, X.N.: A survey of software-defined net working: past, present, and future of programmable networks. IEEE Commun. Surv. Tutor. **16**(3), 1617–1634 (2014)
2. Foysal, M.A., Anam, M.Z., Islam, M.S.: Performance analysis of ternary content addressable memory (TCAM). In: International Conference on Advances in Electrical Engineering (2016)
3. Leng B., Huang L., Wang X.: A mechanism for reducing flow tables in software defined network. In: Proceedings of IEEE ICC (2015)
4. Feghali, A., Kilany, R.: SDN security problems and solutions analysis. In: International Conference on Protocol Engineering (2015)

5. Nadeau, T.D., Gray, K.: Software Definition Network and Openflow Analysis, p. P7. People's Postal and Telecommunications Press, Beijing (2014)
6. Li, X., Ji, M., Cao, M.: OpenFlow flowchart storage optimization scheme based on resource reuse. Opt. Commun. Res. (2), 8–11 (2014)
7. Liu, Z., Li, Y., Su, L.: TCAM-efficient flow table mapping scheme for open flow multiple-table pipelines. J. Tsinghua Univ. (Sci. Technol.) **54**(4), 437–442 (2015)
8. Zhu, H., Fan, H., Luo, X.: Intelligent timeout master: dynamic timeout for SDN-based data centers. In: 2015 IFIP/IEEE International Symposium on Integrated Network Management (IM), pp. 734–737. IEEE (2015)
9. Xie, L., Zhao, Z., Zhou, Y.: An adaptive scheme for data forwarding in software defined network (2014)
10. Kim, T., Lee, K., Lee, J.: A dynamic timeout control algorithm in software defined networks. **3**(5) (2014)
11. He, C.H., Chang, B.Y., Chakraborty, S., Chen, C., Wang, L.C.: a zero flow entry expiration timeout P4 switch. OL (2018)
12. Katta, N., Alipourfard, O., Rexford, J.: CacheFlow: dependency-aware rule-caching for software-defined networks. In: Proceedings of ACM Symposium on SDN Research (SOSR), pp. 1–12 (2016)
13. Curtis, A.R., Mogul, J.C., Tourrilhes, J.: DevoFlow: scaling flow management for high-performance networks. ACM SIGCOMM Comput. Commun. Rev. **41**(4), 254–265 (2011)
14. OpenFlow v1.3 Messages and Structures. http://ryu.readthedocs.io/en/latest/ofproto_v1_3_ref.html
15. Tang, Y., Zhang, Y., Zijian Y., Zhu, G.: Flow table optimization for software-defined networks. J. Xian Jiaotong University (2017)

Two-Layer Intrusion Detection Model Based on Ensemble Classifier

Limin Lu, Shaohua Teng[✉], Wei Zhang, Zhenhua Zhang, Lunke Fei,
and Xiaozhao Fang

School of Computer Science and Technology, Guangdong University of Technology,
Guangzhou, China
shteng@gdut.edu.cn

Abstract. Ensemble classifier can not only improve the accuracy of learning system but also significantly improve its generalization ability by utilizing different deviations of each classifier. Although different classifier ensemble methods are proposed in intrusion field, they are more or less defective and still need further improvement. Aiming at realizing a strong generalization intrusion detection model with high detection rate (DR) and low false positive rate (FPR), a two-layer intrusion detection model based on ensemble classifier (TLMCE) is proposed in this paper. R2L and U2R are classified using JRip classifier in the first layer, and the ensemble classifier is used to classify Normal, DoS, and Probe in the second layer. The stacking optimization strategy is applied to the ensemble classifier using J48, JRip, RandomForest (RF), BayesNet, and SimpleCart as the base classifier. In addition, a modified sequential forward selection method is proposed to select appropriate feature subsets for TLMCE. The experimental results on the NSL-KDD dataset demonstrate that the TLMCE has better performance than some existing ensemble models. It achieved an overall accuracy rate of 89.1% and a FPR of 3.1%.

Keywords: Intrusion detection · Two-layer · Ensemble classifier · Stacking · Feature selection

1 Introduction

In recent years, since new forms of network intrusion has increased rapidly and the attacks have become more and more complicated, the need to defend against various intrusions and protect user systems has increased dramatically [1]. In network security technology, intrusion detection system (IDS) compensates for the shortage of traditional network protection methods such as firewalls, anti-virus. Therefore, IDS can protect the network security effectively. In recent researches, many IDSs based on machine learning or statistical learning techniques(e.g. decision tree(DT) [2], Bayesian [3], support vector machine (SVM) [4], neural network [5], etc) have been developed. Many IDSs have focused on how to achieve

© Springer Nature Singapore Pte Ltd. 2019
Y. Sun et al. (Eds.): ChineseCSCW 2019, CCIS 1042, pp. 104–115, 2019.
https://doi.org/10.1007/978-981-15-1377-0_8

improved performance for single classifier, but it is difficult to achieve high generalization ability and low FPR when applying a single classifier to intrusion detection. Aiming at overcoming the shortcomings of single classifier, ensemble methods using a multi-classifier combination has applied in IDS.

Dietterich [6] once pointed out that ensemble learning has a very important research significance, and it will become one of the future development trends of machine learning. Ensemble learning, also known as the multi-classifier system, first generates multiple learners through certain rules, and then combines them with some integration strategy. It can achieve better classification performance than any single classifier through merging multiple classifiers with certain differences [7]. It can realize a better balance between detection abilities and false alarm [8–10].

In the ensemble approach, the performance of the ensemble learning depends on the base classifier and the combination strategy. For intrusion detection, although many ensemble methods have been proposed, how to select the base classifier and how to combine the base classifiers are key issues and difficult problems. A two-layer intrusion detection model based on ensemble classifier is proposed in this paper, which aims to further improve the DR of intrusion and reduce the FPR. JRip is used to classify R2L and U2R in first layer, and the ensemble classifier is adopted to classify Normal, Dos, and Probe in second layer. The optimized Stacking combination strategy is applied in ensemble classifier. J48, JRip, RandomForest, BayesNet, and SimpleCart are selected as a base classifier in Stacking 0-layer, and BayesNet is used in a stacking 1-layer. Different evaluations and comparative experiments were performed to validate the effectiveness of the TLMCE. The experimental results on the NSL-KDD dataset demonstrate that the TLMCE achieves a high accuracy rate, especially for the DR of both probe and DoS attacks with comparing with the other implemented ensemble model.

Intrusion detection data contains redundant data and irrelevant features, which will lead to the issue of excessive training and predicting time. Hence the data needs to be pre-processed. Feature selection techniques plays an important role in data preprocessing, which is used to reduce the dimension of data [11]. To eliminate redundant and further enhance the overall performance of the TLMCE, a modified sequential forward selection method is proposed.

In this paper, our contributions can be summarized as follows:

1. A modified sequential forward selection method is proposed to eliminate redundant.
2. A two-layer classification structure is proposed. First, the R2L and U2R are classified, and then the rest of classes are classified. On the first layer, JRip almost filtered all other data to the next layer. This structure improves the accuracy of R2L without affecting the further classification of other classes, and realizes the information fusion of different hierarchical classification decisions.
3. In the second layer of the TLMCE model, the ensemble classifier using the Stacking optimization strategy achieves better classification results for the other classes (i.e., Normal, Dos, and Probe).

2 Related Work

Ensemble learning produces a strong learner by combining the results of multiple (homogeneous or heterogeneous) base learners. Therefore, they are also called multi-classifier systems [12,13]. These systems overcome the limitations of traditional classification methods based on a single classifier, and utilize the different deviations of each classifier model to fuse their decision results. Various ensemble methods have been proposed for classification techniques, and developed algorithms to achieve strong generalization capabilities. Ensemble learning algorithms have been widely applied in the fields of image, biology, medicine and finance and so on [14–17]. In recent years, ensemble technology has been also applied more and more widely in intrusion detection [9].

Peddabachigari et al. [18] proposed two hybrid IDS modeling methods. Combining DT and SVM to form hierarchical hybrid intelligent system model (DT-SVM) and classifier-based integration method. Borji [19] proposed a combinatorial classification method for intrusion detection. Three combination strategies are used to fuse the output of four basic classifiers ANN, SVM, KNN and DT: majority vote, bayesian average and belief measure. The results show that this method is superior to single classifier. Hu and Maybank [20] presented an based AdaBoost intrusion detection algorithm, which used decision tree stump as base classifier. The algorithm achieves low complexity and low false alarm rate. Panda and Patra [21] proposed an intrusion detection model based on combination classifier, and REP tree was used for base classifier, which increased the DR of the model and reduced its false alarm rate. Hu et al. [22] proposed two online intrusion detection algorithms based on Adaboost, which use decision stumps and GMMs as base classifiers respectively. Experimental results demonstrate that both algorithms are superior to existing intrusion detection algorithms. Amini et al. [23] propose a new combinatorial classifier, which USES the hybrid combinatorial method to aggregate the prediction results of the weak classifiers (RBF neural network and fuzzy clustering). The results reflect that the ensemble methods is more effective than other methods. Mehdi et al. [24] present an intrusion detection approach based on a combination of GA (Genetic algorithm) and Bagging ensemble MLP (Multilayer Perception) Neural Network. The results reflect that the combined method exceeds other approaches. Teng et al. [25,26], proposed a novel adaptive intrusion detection model based on decision tree, which uses SVM at each layer of the decision tree.

In addition, feature selection technology simplifies the classification process and makes classifier decisions more accurate by eliminating redundant and irrelevant features. In the ensemble method, especially the heterogeneous method, different classifiers have different requirements for features. Fadi et al. [27] combine information gain (IG) and PCA with ensemble classifiers based on SVM, IBK, and MLP. A new intrusion detection hybrid dimension reduction technique is proposed, and the method and related work are compared and analyzed. It is found that presented integration method is more advanced than most existing ones in classification accuracy, DR, and FPR. A good feature selection method can speed up the ensemble process and improve classification accuracy.

Although there are many ensemble technologies, how to choose the optimal system design for a specific application or different data sets needs to be further studied in the absence of unified guiding principles. A two-layer intrusion detection model based on ensemble classifier (TLMCE) is presented in this paper. At the first layer, R2L and U2R are classified using JRip classifier, and at the second layer, other classes (Normal, Dos, Probe) are classified using Stacking method. And a modified sequential forward selection method is employed to select appropriate feature subset for different classifiers. The TLMCE model realizes the information fusion of different layers classification decision.

3 Proposed TLMCE Model

Details of TLMCE model are described in this section. The original data is feature-selected by the modified sequential forward selection method to get a training set and a test set. In view of the imbalance of intrusion detection sample data, Smote oversampling technique is used to increase samples of minority classes. Then, a new training set is generated by the Smote technique. Finally, The TLMCE detector is constructed by the training set and the test set is used to evaluate its performance. The construction process of the TLMCE model is shown in Fig. 1.

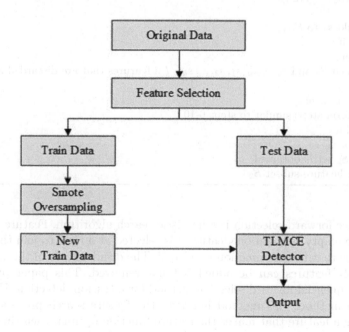

Fig. 1. The construction process of the TLMCE model

3.1   A Modified Sequential Forward Selection Method

Feature selection is a significant means to enhance the performance of learning algorithm and also a crucial step of data preprocessing. Feature selection methods can be broadly divided into two categories: filter method and wrapper method [28]. Filter method relies on training data features to select the independent features of classification model, while wrapper method takes classifier optimization as part of feature subset evaluation and selection process.

Algorithm 1. A modified sequential forward selection method

Input: Training dataset X, Feature set(S).
Procedure:
1:Evaluate each feature in S set using OneRAttributeEval through training set X
2:Sort feature in descending order to obtain ranked $S_1(s_1, s_2, ..., s_m)$
3:Select the first feature from S_1 to S_2
4:Initialize $S_3 = S_2$, $S_{remain} = null$
5:for j=2, , m:
6:    $S_3 = S_2$;
7:    add s_j of S_1 to S_3
8:    if $(f(S_3) > f(S_2))$
9:        $S_2 = S_3$;
10:       $f(S_2) = f(S_3)$;
11:   else
12:       add s_j to S_{remain}
13:   end if
14:End for
15:Obtained S_2 and $S_{remain}(s_1, s_2, ..., s_n)$// n features that are discarded in the last iteration
16:for k =1,..., n:
17:   Perform steps similar to steps 6-10
18:   End if
19:End for
20:Then S_2 set is obtained
Output: Feature subset S_2

Sequence forward selection is a heuristic search algorithm. Feature subset X starts from empty set, and one feature x is selected at a time to join the feature subset X, so that feature function is optimal. The disadvantage of this algorithm is that only features can be added but not removed. This paper presents a modified sequential forward selection method for intrusion detection. Unlike the traditional method, this method improves the feature search process. Instead of selecting a feature that makes the feature function optimal each time, it uses the feature evaluation method to evaluate the contribution of individual features and sort it in descending order. Based on this sequence, features are added to the feature subset one by one. If the added feature causes the value of the feature function to be increased, it is retained, otherwise it is discarded to the remaining set. The specific process of the method is as follows Firstly, OneRAttributeEval is employed to evaluate the contribution of individual features in feature set S, and descending order is carried out to get feature set S_1. Then add one feature at a time from S_1 to S_2 in order and use the classifier of model to evaluate performance of subset S_2. The performance evaluation function is as follows:

$$f(S_i) = \frac{Acc(S_i) + 1}{TTC(S_i)}.$$ (1)

$Acc(S_i)$ and $TTC(S_i)$ represent accuracy and training time of classifier when using S_i feature subset. The larger $f(S_i)$ function value, the better performance of the feature subset. When adding a feature from S_1 to S_2, if the function value f is larger than the value without adding this feature, the feature will be retained in S_2, otherwise it will not be retained, but added to S_{remain}. Repeat this operation until the features in S_1 are iterated over. It is also considered that the discarded features may be related to some features added later in S_2, thereby improving the classification performance. Therefore, after traversing S_1, the features in S_{remain} are re-added to S_2 one by one, if f value becomes larger, this feature is retained in S_2. The algorithm for feature selection is shown in Algorithm 1.

3.2   A Two-Layer Intrusion Detection Model

The design of multi-level structure for IDS has been proposed by some researchers [29–32]. These multi-level models generally classify the R2L and U2R classes in the last layer or later layer, which will make many R2L or U2R be misclassified into other classes in the previous layers and unable to flow to the last layer, resulting in low accuracy of R2L and U2R. In this paper, a two-layer intrusion detection model based on ensemble classifier (TLMCE) is proposed. In the first layer, the minority classes R2L and U2R were first classified, which was helpful to improve their classifier accuracy, and almost all other classes (Normal, DoS, Probe) were filtered to the next layer. At the second level, the other classes (Normal, Dos, Probe) were classified using the stacking strategy. This paper studies and analyzes some literatures [34–36], which compare and analyze the performance of different classifiers used in intrusion detection. In addition, the actual experiment of the influence of each single classifier on the model is also performed. Based on the experimental results and previous experience, the training time, accuracy and differences between them are considered. JRip is used for the first layer classification because it can better detect R2L and U2R categories compared with other classifiers. And the J48, JRip, RandomForest, BayesNet and SimpleCar are used in the second for ensemble. The structure of the TLMCE is shown in Fig. 2.

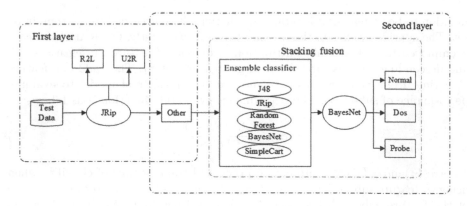

Fig. 2. The structure of the TLMCE model

At second layer, using Stacking combination strategy to integrate base classifiers (J48, JRip, RandomForest, BayesNet, SimpleCart). Unlike traditional Stacking algorithm, It not only takes the prediction category of the sample as the input to the next layer, but also considers the maximum probability that the sample belongs to the category. In other words, the prediction category and the maximum probability of each base classifier in the 0-layer are taken as the input attribute of the 1-layer. Taking into account the probability that the prediction category corresponds to the maximum is equivalent to adding a weight value to the prediction result. More information has been added to facilitate more accurate classification. Thereby, the information fusion of different layers of classification decision is realized.

4   Experiments

Empirical experiments for TLMCE are run in an Intels Core(TM) i5 CPU @1.80GHz computer with 6.00 GB RAM running Windows 10. All experiments are implemented using the Java language and WEKA API was called.

4.1   Dataset

Experiments on NSL-KDD dataset [33], which is obtained after KDD99 has been modified, was conducted to test the TLMCE performance. As the KDD99, the NSL-KDD Dataset provides 41 features with 3 symbolic features and 38 numeric features. Each data instance is either of a normal type or of a concrete attack type (DoS, Probe, U2R, and R2L).

Before building the model, data pre-processing that may determine the classification performance is necessary. As most classifiers only accept numeric values, this paper uses numerical data types to replace symbols according to the appropriate rules. For data imbalance problems, Smote oversampling technology is used to add samples of a few classes. Then use the modified sequential forward selection method to choose optimal feature subset for each classifier.

TLMCE model was trained using 20% of NSL-KDD training set ($Train\_20\%$) and then tested through the given test set (Test+). The distribution of NSL-KDD the dataset is shown in Table 1.

Table 1. The distribution of the dataset on NSL-KDD dataset

Dataset	Total	Normal	DoS	Probe	R2L	U2R
$Train\_20\%$	25242	13499	9234	2289	209	11
$Test+$	22544	9711	7458	2421	2754	200

4.2 Assessment Metrics

The following measurements are adopted to test the TLMCE performance: DR, FPR, OCA, and TT. On the basis of the confusion matrix presented in Table 2, the definitions of DR, FPR are as follows:

Table 2. Confusion matrix

	Test result positive	Test result negative
Actual positive class	True positive(TP)	False negative(FN)
Actual negative class	False positive(FP)	True negative(TN)

(1) $DR(detection\ rate) = TP/(TP + FN)$
(2) $FPR(false\ positive\ rate) = FP/(FP + TN)$
   In addition, the definition of OCA, T-time are as follow:
(3) OCA(Overall classification accuracy) = (samples of test dataset correctly classified)/(samples of test dataset)
(4) TT means the time of training model.

4.3 Experimental Results

The TLMCE places particular emphasis on the selection of base classifiers. Many literatures [34–36] were collected and analyzed, which made detailed comparative analysis on the performance of different classifiers. In the meantime, some actual experiments of the influence of different single classifiers on the model was carried out. Thence, for the consideration of their training time, accuracy and the difference between them, JRip classifier is selected in the first layer and J48, JRip, RF, BayesNet, and SimpleCart are selected for ensemble in the second layer.

The first experiment evaluated the performance of a modified sequential forward selection method. The goal of the feature selection method is to improve the accuracy and minimize the TT of classifiers, thus enhance the performance of the whole model. Table 3 shows that the ACC of the classifiers is improved and TT is shorten as a modified sequential forward selection method is used.

Table 3. The result of experiment for feature selection

Classifier	JRip		Ensemble classifier	
Measurement	ACC	TT(s)	ACC	TT(s)
with feature selection	93.01	4.11	95.47	14.49
without feature selection	87.61	6.58	87.75	16.53

The second experiment aims to evaluate effectiveness of the presented ensemble strategy. The obtained result of TLMCE is compared with other existing technologies. The comparisons of the DR of each classes and FPR between the methods of [23, 24, 37, 38] and the TLMCE model are reported in Table 4. It is obvious that the proposed model provided better performance. The detection rates of Normal, DoS, Probe, R2L, and U2R reached 96.79%, 93.44%, 96.45%, 49.16%, and 15.0% respectively. The TLCEM ensemble model outperforms all other method in two classes DoS, and Probe and maintains a very high DR for the Normal classes. However, The result is slightly disappointing in terms of U2R and R2L as the training samples of R2L (209 samples) and U2R (11 samples) are rare and small in the $Train\_20\%$. Of note, The TLCEM offers an enhancement by improving about 14.3% of R2L compared to the work in [37] that also used $Train\_20\%$. More importantly, it achieves very low FPR (obtained 3.1%).

Table 4. The comparisons of the DR and FPR between other existing methods and the TLMCE model on NSL-KDD

Methods	Dataset	Normal	DoS	Probe	R2L	U2R	FPR
Two-tier model [37]	$Train\_20\%$	94.56	84.68	79.76	34.81	67.16	4.83
GA-Bagging-MLP [24]	Train+	97.90	92.23	76.91	67.14	56.72	2.10
Two-level Ensemble [38]	$Train\_20\%$	N/A	N/A	N/A	N/A	N/A	12.60
RBF-ensemble[23]	Train+	95.70	83.50	86.80	82.10	87.50	N/A
TLMCE	$Train\_20\%$	96.79	93.44	96.54	49.16	15.0	3.1

Aiming to better interpret the result obtained from proposed TLCEM model, the comparison of overall classification accuracy (OCA) between the model with other existing technologies [23, 24, 37, 38] is presented in Fig. 3. As seen from Fig. 3, the OCA of the TLMCE reaches 89.10%. Its performance is outperform other existing methods except RBF-ensemble as it used all train data (Train+) that larger than ours.

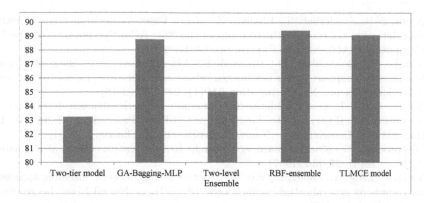

Fig. 3. Comparison of OCA between TLCEM model and other existing methods

5   Conclusion

This paper presents a two-layer intrusion detection model based on ensemble classifier (TLMCE). The experimental results on the NSL-KDD dataset confirm that the TLMCE is superior to some other existing ensemble models (overall accuracy rate obtained 89.1%). In the first layer of the model, JRip is used to classify R2L and U2R. In the second layer, the ensemble classifier with stacking strategy is used to classify Normal, DoS, and Probe. A modified sequential forward selection method is presented to choose the appropriate feature subset for different classifiers, which further improves the performance of the model. The two-layer classifier performs collaborative detection to achieve information fusion of different hierarchical classification decisions.

Considering that different datasets have different requirements for the base classifier of Stacking combining strategy. Our further work is to optimize the stacking strategy according to different data subsets. Data subsets are divided by network protocols (TCP, UDP, and ICMP).

Acknowledgement. This research is supported in part by the National Natural Science Foundation of China (Grant No. 61772141, 61702110, 61603100), Guangdong Provincial Science & Technology Project (Grant No. 2016B010108007), Guangdong Education Department Project (Grant No. [2018] 179, [2018] 1), and Guangzhou City Science & Technology Project (Grant No. 201604046017, 201604020145, 201802030011, 201802010042, 201802010026, 201903010107).

References

1. Liao, H.J., Lin, C.H.R., Lin, Y.C., Tung, K.Y.: Intrusion detection system: a comprehensive review. J. Netw. Comput. Appl. **36**(1), 16–24 (2013)
2. Lee, J.H., Lee, J.H., Sohn, S.G., Ryu, J.H.: Effective value of decision tree with KDD 99 intrusion detection datasets for intrusion detection system. In: International Conference on Advanced Communication Technology. IEEE (2008)

3. Amor, N.B., Benferhat, S., Elouedi, Z.: Naive Bayes vs decision trees in intrusion detection systems. In: Proceedings of the 2004 ACM Symposium on Applied Computing (SAC), Nicosia, Cyprus, pp. 14–17 (2004)
4. Yingjie, T., Mahboubeh, M., Hosseini, B.S.M., Huadong, W., Qiang, Q.: Ramp loss one-class support vector machine; a robust and effective approach to anomaly detection problems. Neurocomputing **310**(1), 223–235 (2018)
5. Zhang, Z., Li, J., et al.: A hierarchical anomaly network intrusion detection system using neural network classification. In: CD-ROM Proceedings of 2001 WSES International Conference on: Neural Networks and Applications (2001)
6. Dietterich, T.G.: Machine learning research: four current directions ai magazine. Ai Mag. **18**(4), 97–136 (1997)
7. Yang, J., Zeng, X., Zhong, S., Wu, S.: Effective neural network ensemble approach for improving generalization performance. IEEE Trans. Neural Netw. Learn. Syst. **24**(6), 878–887 (2013)
8. Aburomman, A.A., Reaz, M.B.I.: A novel SVM-KNN-PSO ensemble method for intrusion detection system. Appl. Soft Comput. **38**(C), 360–372 (2016)
9. Folino, G., Sabatino, P.: Ensemble based collaborative and distributed intrusion detection systems: a survey. J. Netw. Comput. Appl. **66**, 1–16 (2016)
10. De Jongh, A.: Neural network ensembles. IEEE Trans. Pattern Anal. Mach. Intell. **12**(10), 993–1001 (2004)
11. Zhao, Z., Morstatter, F., Sharma, S., Alelyani, S., Anand, A., Liu, H.: Advancing feature selection research. In: ASU Feature Selection Repository, pp. 1–28 (2010)
12. Schapire, R.E.: The strength of weak learnability. Mach. Learn. **5**(2), 197–227 (1990)
13. Woźniak, M., Graña, M., Corchado, E.: A survey of multiple classifier systems as hybrid systems. Elsevier Science Publishers B. V (2014)
14. Takemura, A., Shimizu, A., Hamamoto, K.: Discrimination of breast tumors in ultrasonic images using an ensemble classifier based on the adaboost algorithm with feature selection. IEEE Trans. Med. Imaging **29**(3), 598–609 (2010)
15. Partalas, I., Tsoumakas, G., Hatzikos, E.V., Vlahavas, I.: Greedy regression ensemble selection: theory and an application to water quality prediction. Inf. Sci. **178**(20), 3867–3879 (2008)
16. Korfiatis, V.C., Tassani, S., Matsopoulos, G.K.: A new ensemble classification system for fracture zone prediction using imbalanced micro-CT bone morphometrical data. IEEE J. Biomed. Health Inform. **22**(4), 1189–1196 (2017)
17. Bin, W., Lin, L., Xing, W., Megahed, F.M., Waldyn, M.: Predicting short-term stock prices using ensemble methods and online data sources. Expert Syst. Appl. **112**(2), 258–273 (2018)
18. Peddabachigari, S., Abraham, A., Grosan, C., Thomas, J.: Modeling intrusion detection system using hybrid intelligent systems. J. Netw. Comput. Appl. **30**(1), 114–132 (2007)
19. Borji, A.: Combining heterogeneous classifiers for network intrusion detection. In: Cervesato, I. (ed.) ASIAN 2007. LNCS, vol. 4846, pp. 254–260. Springer, Heidelberg (2007). https://doi.org/10.1007/978-3-540-76929-3_24
20. Hu, W., Hu, W., Maybank, S.: Adaboost-based algorithm for network intrusion detection. IEEE Trans. Syst. Man Cybern. Part B Cybern. **38**(2), 577–583 (2008)
21. Panda, M., Patra, M.R.: Ensemble of classifiers for detecting network intrusion. In: International Conference on Advances in Computing. ACM (2009)
22. Hu, W., Gao, J., Wang, Y., Wu, O., Maybank, S.: Online adaboost-based parameterized methods for dynamic distributed network intrusion detection. IEEE Trans. Cybern. **44**(1), 66–82 (2014)

23. Amini, M., Rezaeenour, J., Hadavandi, H.: A neural network ensemble classifier for effective intrusion detection using fuzzy clustering and radial basis function networks. Int. J. Artif. Intell. Tools **25**(2), 1550033 (2016)
24. Mehdi, M., Khalid, E.Y., Seddik, B.: Mining network traffics for intrusion detection based on bagging ensemble multilayer perceptron with genetic algorithm optimization. Int. J. Comput. Sci. Netw. Secur. **18**(5), 59–66 (2018)
25. Teng, S., Du, H., Wu, N., Zhang, W., Su, J.: A cooperative network intrusion detection based on fuzzy SVMs. J. Netw. **5**(4), 475–483 (2010)
26. Teng, S., Wu, N., Zhu, H., et al.: SVM-DT-based adaptive and collaborative intrusion detection. IEEE/CAA J. Autom. Sinica **5**(1), 108–118 (2018)
27. Fadi, S., Ali, B.N., Aleksander, E.: Dimensionality reduction with IG-PCA and ensemble classifier for network intrusion detection. Comput. Netw. **148**, 164–175 (2019)
28. Li, Y., Xia, J., Zhang, S., Yan, J., Ai, X., Dai, K.: An efficient intrusion detection system based on support vector machines and gradually feature removal method. Expert Syst. Appl. **39**(1), 424–430 (2012)
29. Gogoi, P., Bhattacharyya, D.K., Borah, B., Kalita, J.K.: MLH-IDS: a multi-level hybrid intrusion detection method. Comput. J. **57**(4), 602–623 (2014)
30. Xiang, C., Yong, P.C., Meng, L.S.: Design of multiple-level hybrid classifier for intrusion detection system using Bayesian clustering and decision trees. Pattern Recogn. Lett. **29**(7), 918–924 (2008)
31. Xiang, C., Chong, M.Y., Zhu, H.L.: Design of mnitiple-level tree classifiers for intrusion detection system. In: IEEE Conference on Cybernetics & Intelligent Systems (2004)
32. Lu, H., Xu, J.: Three-level hybrid intrusion detection system. In: International Conference on Information Engineering & Computer Science (2009)
33. Tavallaee, M., Bagheri, E., Lu, W., Ghorbani, A.A.: A detailed analysis of the KDD CUP 99 data set. In: IEEE International Conference on Computational Intelligence for Security & Defense Applications (2009)
34. Chauhan, H., Kumar, V., Pundir, S., Pilli, E.S.: A comparative study of classification techniques for intrusion detection. In: International Symposium on Computational & Business Intelligence. IEEE Computer Society (2013)
35. Aziz, A.A.S., Hanafi, E.O., Hassanien, A.E.: Comparison of classification techniques applied for network intrusion detection and classification. J. Appl. Log. **24**(A), 109–118 (2016)
36. Ahmad, L., Basheri, M.J., Raheem, A.: Performance comparison of support vector machine, random forest, and extreme learning machine for intrusion detection. IEEE Access **6**, 33789–33795 (2018)
37. Hamed, H.P., GholamHossein, D., Sattar, H.: Two-tier network anomaly detection model: a machine learning approach. J. Intell. Inf. Syst. **48**(1), 61–74 (2017)
38. Tama, B.A., Patil, A.S., Rhee, K.H.: An improved model of anomaly detection using two-level classifier ensemble. In: Asia Joint Conference on Information Security. IEEE Computer Society (2017)

A Minimum Rare-Itemset-Based Anomaly Detection Method and Its Application on Sensor Data Stream

Saihua Cai[1], Ruizhi Sun[1,2(✉)], Huiyu Mu[1], Xiaochen Shi[1], and Gang Yuan[1]

[1] College of Information and Electrical Engineering, China Agricultural University, Beijing 100083, China
{caisaih, sunruizhi}@cau.edu.cn

[2] Scientific Research Base for Integrated Technologies of Precision Agriculture (Animal Husbandry), The Ministry of Agriculture, Beijing 100083, China

Abstract. In recent years, the scale of data stream is becoming much larger in real life. However, the anomaly data often exists in the collected data stream, while the existence of anomaly is a main reason for the decrease of the accuracy of data-based operations. The anomaly data have two main characteristics, that is, appear rarely and deviate much from most data elements, thus, the anomaly detection methods should accurately detect the anomaly data by considering these two attributes. Because the data stream is continuously generated and constantly flowing, thus, the previous static anomaly detection methods are not suitable for processing data streams. In addition, the large amount of data stream makes the time consumption and memory occupation of rare itemset mining phase very high. To effectively solve these problems, this paper first proposes an efficient MRI-Mine method for mining minimum rare itemsets, and then proposes an accurately anomaly detection method called MRI-AD based on anomaly index to identify the implicit anomaly data. The experiments indicate the proposed MRI-Mine method can mine the minimum rare itemsets in less time consumption and memory occupation, and the detection accuracy of MRI-AD method is also competitive.

Keywords: Anomaly detection · Minimum rare itemsets · Anomaly index · Sensor data stream · Data mining

1 Introduction

With the sustainable development of sensor technology, more and more sensors are used in daily production and life, such as: agricultural monitoring [1], water quality monitoring [2], etc., the use of sensors can provide data-based scientific decision-making for production and life. However, the security problem in the sensor network is particularly prominent, and the abnormal data is one of the main aspects of the sensor network security problem. Therefore, in order to reduce the problems caused by abnormal data, it is necessary to accurately detect these abnormal data. The data generated by the sensor usually exists in the form of a stream [3], it is composed of a

© Springer Nature Singapore Pte Ltd. 2019
Y. Sun et al. (Eds.): ChineseCSCW 2019, CCIS 1042, pp. 116–130, 2019.
https://doi.org/10.1007/978-981-15-1377-0_9

large number of rapidly generated and continuous data elements. Compared with traditional anomaly detection operation on static datasets, the anomaly detection operation on sensor data stream needs to take the requirement of speed into consideration, it is owing to that once some new sensor data transfer to the computer terminal, the old information will be overwritten immediately.

The appearing frequency of the data element and the anomaly degree of the data element from other data [4] are the key to judging whether there are some anomalies existing in sensor data streams, because anomaly is often rarely occurring and different to most normal data, where only one of the attributes is satisfied is usually due to the existence of concept drift, thus, anomaly detection on sensor data stream requires that two properties of abnormal data are fully considered. For the common used anomaly detection methods, such as clustering-based approaches [5], distance-based approaches [6] and density-based approaches [7], only the different degree between each data and other data were used as the criterion for anomaly detection, but the appearing frequency is not considered for abnormal judgment.

Compared with the three common anomaly detection methods mentioned above, two characteristics of abnormal data are considered in itemset-based anomaly detection methods [8–14], thus, the use of itemset-based anomaly detection method can improve the detection accuracy in maximal degree. For itemset-based anomaly detection methods, the Find FPOF method [8] was the first algorithm that proposed to discover the anomaly data from large-scale datasets. Although FindFPOF method could identify the anomaly data, but the simple anomaly judging index made the detection accuracy is not very high. Then, the MFPM-AD method [9], OODFP method [10], FCI-Outlier method [11] and LFP method [14] were proposed to improve the detection accuracy, where the MFPM-AD method and OODFP method were realized based on maximal frequent itemsets, and the FCI-Outlier method was realized based on frequent closed itemsets. Against the uncertain data stream, an efficient anomaly detection approach, namely FIM-UDSOD [12], was designed for accurately finding the existing anomalies, where the length of each transaction was considered in the anomaly detection process. Based on two anomaly indices, the MWIFIM-OD-UDS approach [13] was used for seeking for the anomalies from weighted data streams, where the *weighted support* value and the length of every minimum weighted infrequent itemset, and the length of each transaction were considered in the anomaly detection process, but it is not suitable for processing precise datasets.

For the itemset-based anomaly detection methods, they are mainly composed of two stages. (1) In the mining stage, the frequent itemsets or rare itemsets are mined from the data sets to reflect the appearing frequency of the data. (2) In anomaly detection stage, the designed anomaly index is used to calculate the anomaly degree of each transaction in the datasets, which is used to reflect the anomaly degree of the data. However, the existing itemset-based anomaly detection approaches still exist next several problems: (1) The scale of sensor data stream is very large, which results the time cost of anomaly detection process is long, thereby difficult adapting to the update speed of data stream; (2) The simple anomaly index (such as: FindFPOF [8], OODFP [10]) makes the accuracy of anomaly detection relatively low. Compared to the frequent-itemset-based anomaly detection methods, because the rare itemsets refer to itemsets that occur less frequently, thus, it is more efficient to use rare itemsets during detection process rather than to use frequent itemsets. However, the huge scale of mined rare itemsets will affect the speed of anomaly detection process. Because the minimum rare itemsets are the generators of rare itemsets (that is, the rare itemsets can be gained by extending the minimum rare itemsets) and its number is much smaller than that of rare itemsets, thus, the use of minimum rare itemsets can reduce the time cost during the anomaly detection process.

Based on the above ideas and sliding window technology [15], we propose a minimum rare-itemset-based anomaly detection method, namely MRI-AD, for accurately finding the anomalies from the collected data stream, and then applied the proposed MRI-AD method to actual sensor data streams. The main contributions can be summarized as follows:

(1) We propose an efficient matrix structure to store the specific information of each data element, which supports mining the itemsets through only one scan of the data streams. Based on the matrix structure, we design a minimum rare itemset mining method, namely MRI-Mine, for efficiently mining the minimum rare itemsets from data streams with the use of "seek minimum" operation and "itemset connection" operation, thereby reducing the time cost on itemset mining process.

(2) We design an efficient anomaly index, namely transaction anomaly index (*TAI*), for evaluating the anomaly degree of each transaction.

(3) Based on the defined anomaly index and the mined minimum rare itemsets, we describe a minimum rare-itemset-based anomaly detection approach, namely MRI-AD, for effectively seeking for the anomalies in the incoming data streams.

The remainder of this article is organized as follows. Section 2 introduces some preliminaries related to this paper. Section 3 presents the minimum rare itemset mining and anomaly detection methods. Section 4 states the empirical studies and experimental analysis. Section 5 concludes this paper and prospects for the future.

2 Preliminaries

Assume that itemset $I = \{i_1, i_2, ..., i_n\}$ is a set of items. Each transaction T in the sliding window is identified by an *ID*, namely *TID*. Let $I_s = \{i_i, i_{i+1}, ..., i_j\} \subseteq I$ and $i, j \in [1,n]$, then, I_s is the sub-set of I and I is the super-set of I_s. If the length of I_s ($|I_s|$) is k, I_s is called a k-itemset. Data stream $DS = [T_1, T_2, ..., T_m)$ ($m \rightarrow \infty$) is composed of an infinite sequence of transactions T_i. For the sliding window (*SW*) model, its size is denoted as $|SW|$, and it only processes the recently transactions. In addition, the minimum *support* threshold (*min_sup*) is used to measure the itemsets belong to rare itemsets or frequent itemsets. To better explain the relevant definitions, we describe them using an example that is shown in Table 1, where the letters in the table can be regarded as the goods in the shopping basket.

Definition 1. *count*: The number of itemset $\{I_i\}$ included in *DS* is denoted as *count*.

For itemset $\{a\}$, it is appearing in T_1, T_3, T_5 and T_6, thus, *count*(a) = 4; For itemset $\{ab\}$, it is appearing in T_1, T_3 and T_5, thus, *count*(ab) = 3.

Definition 2. *support* (*sup*): The appearing frequency of itemset $\{I_i\}$ in *DS* is defined as *support*, i.e. $sup(I_i) = count(I_i, DS)/|SW|$.

Table 1. Transaction dataset

TID	Transaction	TID	Transaction	TID	Transaction
T_1	$\{a, b, c, d\}$	T_2	$\{b, d\}$	T_3	$\{a, b, c, e\}$
T_4	$\{c, d, e\}$	T_5	$\{a, b, c\}$	T_6	$\{a, c, f\}$
...	......	...	......	...	......

For itemset $\{a\}$, its *count* value is 4, thus, $sup(a) = 4/6$; For itemset $\{ab\}$, its *count* value is 3, thus, $sup(ab) = 0.5$.

Given two items $\{X\}$ and $\{Y\}$ $(X \subset Y)$, if item $\{Y\}$ is contained in a transaction, then item $\{X\}$ is must contained in this transaction, thus, $sup(X) \geq sup(Y)$.

Definition 3. *rare itemset (RI)*: The itemset $\{I_i\}$ is an *RI* if its *support* value is small than *min_sup*, that is, $sup(I_i) < min\_sup$.

Definition 4. *frequent itemset (FI)*: The itemset $\{I_i\}$ is a *FI* if its *support* value is large or equal to *min_sup*, that is, $sup(I_i) \geq min\_sup$.

Definition 5. *minimum rare itemset (MRI)*: The itemset $\{I_i\}$ is a *MRI* if it is an *RI* and every subset of $\{I_i\}$ is a *FI*.

For an itemset $I_a = \{i_1, i_2, ..., i_m\}$ that composed by m items, the number of potential extensible itemsets is 2^m [16], thus, the itemset mining operation is a time consuming process. To reduce the time consumption during the *MRI* mining phase, the primary goal is to reduce the scale of extensible itemsets. In addition, although the itemset-based anomaly detection method takes the appearing frequency of each itemset as a criterion for anomaly detection, but the anomaly index in the existing itemset-based anomaly detection method is relatively simple, that is, the detection process does not consider more factors that may affect the anomaly detection results, which also leads to the accuracy of anomaly detection is not competitive. Thus, it is necessary to design an anomaly index for considering more factors that lead to data anomalies, thereby making the detection accuracy much higher.

3   Minimum Rare-Itemset-Based Anomaly Detection

In this section, an efficient *MRI*-based anomaly detection method is proposed for seeking for the anomalies from data stream. Specifically, the first subsection presents the *MRI* mining method; the second section designs the anomaly index and puts forward the outlier detection method.

3.1   Minimum Rare Itemset Mining Method

Based on a two-dimensional matrix structure, we propose a *MRI* mining method, namely MRI-Mine, for mining the *MRIs* from data stream, where the mining operation is conducted by "itemset connection" operation and "seeking minimum" operation.

For the *MRI* mining, the most important problem needs to be solved is the time consumption on mining process. Thus, controlling the scale of extensible itemsets and reducing the times of data stream scanning is the key to solve this problem. In the itemset mining process, the long itemsets are mined by "itemset connection" operation of the short itemsets, thus, the mining efficiency will be improved if the number of short itemsets can be discarded and not participate to the next extending operations.

Theorem 1. *Anti-monotone property*: The supersets of *RIs* are also *RIs*.

Proof. Assume that $\{I_a\}$ is a *RI*, $\{I_b\}$ is a superset of $\{I_a\}$ and $\{I_b\}$ is a *FI*. Because $\{I_a\}$ is the subset of $\{I_b\}$, thus, the transactions that contains $\{I_b\}$ must contain $\{I_a\}$, thus, $sup(I_a) \geq sup(I_b)$. Because itemset $\{I_b\}$ is a *FI*, thus, itemset $\{I_a\}$ is also a *FI*, it is contradicted with the assumption. Thus, the supersets of *RIs* are also *RIs*.

It can be seen from Theorem 1 that if itemset $\{I_k\}$ is a *RI*, then, the extending of $\{I_k\}$ is meaningless because any superset of $\{I_k\}$ is not *MRI* and any superset of $\{I_k\}$ is not *FI*, that is, $\{I_k\}$ can be discarded directly, thereby reducing the meaningless time consumption.

For Apriori-like methods, the itemset mining process needs to scan the entire datasets for several times, thus, the mining efficiency is not competitive. To improve the mining efficiency, the FP-Growth-like methods are proposed by scanning the entire datasets for only two times, while the mining process needs to generate many conditional trees, which will consume large memory usage. Thus, if the entire datasets only need to be scanned once to perform the mining operation, the mining efficiency can be further improved. Based on this idea, the two-dimensional matrix structure [13] is designed for storing the specific information of the items in the transactions, which supports scanning the entire datasets for only once to store the data information, and then the itemset mining is conducted with the use of column vectors from the constructed two-dimensional matrix. The specific construction process of the two-dimensional matrix structure (denoted as matrix A) is like the MWIFIM-UDS method [13], but the last line of the matrix is used to store the *count* value of each item. A_{dk} is recorded as 1 if item i_k is appearing in T_d, otherwise, A_{dk} is recorded as 0.

Definition 6. *itemset connect*: For two frequent $(n + 1)$-itemsets with same prefix, $I_a = \{i_1, i_2, \ldots, i_n, i_{n+1}\}$ and $I_b = \{i_1, i_2, \ldots, i_n, i_{n+2}\}$, "itemset connect" is to connect these two $(n + 1)$-itemsets to $(n + 2)$-itemset, $I_c = \{i_1, i_2, \ldots, i_n, i_{n+1}, i_{n+2}\}$. In particular, for the frequent 1-itemsets, they can be connected to 2-itemset directly without determining whether the prefix is the same.

Definition 7. *seeking minimum* (Λ): For two frequent n-itemsets I_a and I_b, and the extended $(n + 1)$-itemset I_c that extended by I_a and I_b, "seeking minimum" is to find the minimum probability between I_a and I_b in each transaction to form the probability for I_c.

Definition 8. *minimum checking*: For a rare k-itemset ($k > 2$), "minimum checking" is to check if any subset $(k-1)$-itemset of k-itemset is existing in MRIL.

Then, we describe the implementation of the *MRI* mining method (called MRI-Mine) in detail. When the specific information of the items in the transactions in current sliding window is saved in matrix, the *count* value of each item is calculated to seek for the frequent 1-itemsets that can be further extended. For these frequent 1-itemsets, they are saved to the frequent itemset library (FIL), while these rare 1-itemsets are saved to the minimum rare itemset library (MRIL) and they are no longer participating "itemset connect" operation. For the frequent 1-itemsets in FIL, they are connected with each other by "itemset connect" operations, and the *count* value of the extended 2-itemsets is calculated by "seeking minimum" operation. For the extended 2-itemsets, they are saved to the MRIL directly if the *count* value is less than *min_sup*, it is owing to that the subsets of these 2-itemsets are frequent, thus, no subset of them is existing in the MRIL. Then, the frequent $(k-1)$-itemsets that with the same prefix are selected to extend to k-itemsets through "itemset connect" operation, $(k > 2)$, and the *count* value of the extended k-itemsets is also calculated through "seeking minimum" operation. Different to 2-itemsets, the rare k-itemsets $(k > 2)$ cannot saved into MRIL directly because some subsets $((k-1)$-itemset) of the k-itemset may exist in MRIL. For example, $\{a,b,c\}$ can be extended by frequent 2-itemsets $\{a,b\}$ and $\{a,c\}$, but the subset $\{b,c\}$ of $\{a,b,c\}$ is a rare

---

Algorithm 1: MRI-Mine

Input: Data stream, $|SW|$, *min_sup*

Output: *MRIs*

01. **for** $i \leq |SW|$ **do**
02.   construct matrix A
03. **end for**
04. **foreach** item $\{i_n\}$ **do**
05.   **if** $sup(i_n) < min\_sup$ **then**
06.   MRIL$\leftarrow \{i_n\}$ $//\{i_n\}$ is saved to MRIL
07.   **else**
08.   FIL$\leftarrow \{i_n\}$ $//\{i_n\}$ is saved to FIL
09.   **end if**
10. **end for**
11. **foreach** frequent 1-itemsets $\{i_n\}$ and $\{i_p\}$ **do**
12.   connect $\{i_n\}$ and $\{i_p\}$ to $\{i_n,i_p\}$
13.   **if** $sup(i_n,i_p) < min\_sup$ **then**
14.   MRIL$\leftarrow \{i_n,i_p\}$
15.   **else**
16.   FIL$\leftarrow \{i_n,i_p\}$
17.   **end if**
18. **end for**
19. $n=2$
20. **foreach** frequent $\{i_1,...,i_{(n-1)},i_n\}$ and $\{i_1,...,i_{(n-1)},i_{(n+1)}\}$ with same $(n-1)$ prefix **do**
21.   connect $\{i_1,...,i_{(n-1)},i_n\}$ and $\{i_1,...,i_{(n-1)},i_{(n+1)}\}$ to $\{i_1,...,i_{(n-1)},i_n,i_{(n+1)}\}$
22.   **if** $sup(i_1,...,i_{(n-1)},i_n,i_{(n+1)}) < min\_sup$ **then**
23.   minimum checking
24.   **if** any subset of $\{i_1,...,i_{(n-1)},i_n,i_{(n+1)}\}$ in MRIL **then**
25.   discard $\{i_1,...,i_{(n-1)},i_n,i_{(n+1)}\}$
26.   **else**
27.   MRIL$\leftarrow \{i_1,...,i_{(n-1)},i_n,i_{(n+1)}\}$
28.   **end if**
29.   **end if**
30. **end for**
31. n++
32. return *MRIs* in MRIL

itemset, thus, {*a*,*b*,*c*} cannot be saved into MRIL. For this reason, the "minimum checking" operation needs to be conducted before saving the *k*-itemsets to MRIL, thereby guaranteeing the rare itemsets saved in the MRIL are *MRIs*. The specific process of *MRI* mining (called MRI-Mine) is shown in Algorithm 1. Note that, the final mined *MRIs* are {*e*}, {*f*}, {*ad*}, {*bd*} and {*cd*} if *min_sup* is set to 0.4 and |*SW*| is set to 6.

When all *MRIs* in current *SW* are mined and the implicit anomalies are detected, the oldest transactions are covered by the new transactions directly, and then the mining process is conducted repeatedly to effectively mine the *MRIs*.

3.2 Anomaly Detection Method

After the mining process, the anomaly detection operation is conducted to accurately identify the implicit anomaly data from data stream, where the anomaly index is defined to evaluate the anomaly degree of every transaction in *SW*. Thus, in detection process, the quality of anomaly index directly determines the accuracy of anomaly detection. For the FindFPOF method [8], the anomaly index is only the ratio of the number of contained *FIs* to the number of entire *FIs*, while the single anomaly index makes the detection accuracy is not very high. In addition, the time consumption of FindFPOF is relatively high because the *FIs* are used in the anomaly detection process, while the scale of *FIs* is very huge. Then, the maximal-*FI*-based anomaly detection methods, namely OODFP [10] and MFPM-AD [9], and the frequent-closed-itemset-based method, namely FCI-Outlier [11], are proposed for reducing the time consumption in anomaly detection stage and improving the detection accuracy. However, these methods use the *FIs* in the anomaly detection process, while the *RIs* can better represent the anomaly data, thus, the detection accuracy will be improved in theory if the used itemsets can be changed from *FIs* to *RIs*.

In addition, the following factors need to be considered in anomaly detection process. Firstly, similar to FindFPOF method, the number of contained *MRIs* is a core factor to influence the accuracy of anomaly detection, because a transaction is more abnormal if more *MRIs* are contained in this transaction. Secondly, the *support* value of the contained *MRIs* is an important factor to influence the accuracy of anomaly detection, because a transaction is more abnormal if the *support* value of the contained *MRIs* is much small. Thirdly, the length of a transaction will also influence the accuracy of anomaly detection, for two transactions with different length but contain same *MRIs*, the short transaction is more abnormal because the *MRIs* are appearing more frequently.

Based on the above factors that may cause the anomalies, we design an anomaly index for evaluating the anomaly degree of every transaction in *SW*, where the transactions with high anomaly degree are judged as the true anomaly transactions.

Definition 9. *Transaction Anomaly Index* (*TAI*): For each transaction T_i, its length is *len*(T_i). The *support* value of contained *MRI* {*X*} is *sup*(*X*) and the number of contained *MRIs* is *num*(*X*). Then, *TAI*(T_i) value is defined as

$$TAI(T_i) = 1 - \frac{\sum\limits_{X \subseteq T_i, X \subseteq MRIL} num(X)}{|MRIL| * len(T_i)} * \left(\sum_{X \subseteq T_i, X \subseteq MRIL} min\_sup - sup(X) \right) \quad (1)$$

$TAI(T_i)$ is an anomaly index of transaction T_i, and the larger $TAI(T_i)$ value means that T_i is more likely an anomaly transaction. Note that, if no MRI is contained in transaction T_i, its $TAI(T_i)$ value is set to 0 directly because all itemsets are appearing frequently and the contained items are not different from other items.

Then, the anomaly detection process is composed of next three steps. (1) Determine the contained $MRIs$ in the transaction. (2) Calculate the $TAI(T_i)$ value for each transaction T_i. (3) Sort the transactions with decreasing $TAI(T_i)$ values order. Based on the above three steps, the top k transactions with lowest $TAI(T_i)$ value are more likely to be the abnormal transactions because their anomaly degree is much larger than other transactions, where k is freely specified by the users. The detailed process of the MRI-based anomaly detection method, namely MRI-AD, is shown in Algorithm 2.

Algorithm 2: MRI-AD

Input: Data stream, min_sup, $|SW|$, k

Output: Anomaly transactions

01.call Algorithm 1 // mine $MRIs$

02.$num(X)$=0, $TAI(T_i)$=0, $S(X)$=0

03.**foreach** $\{X\} \subseteq T_i$, $\{X\} \in MRIL$ **do**

04. $num(X)$ +=1

05. $S(X)$=min_sup-$sup(X)$

06.**end for**

07. $TAI(T_i) = 1 - \dfrac{num(X)}{| MRIL | * len(T_i)} * S(X)$

08.arrange each T_i by increasing $TAI(T_i)$ value

09.Anomaly transactions ←top k T_i

10.returen anomaly transactions

11.slide the window

12.go to 01

---

4   Experimental Results

For evaluating the efficiency of MRI-AD method and MRI-Mine method, we conduct massive experiments in this section, where the experiments on MRI-AD method is to test the detection accuracy and the experiments on MRI-Mine method is to test the time consumption on MRI mining process. In the experiment on testing the detection accuracy of the MRI-AD method, a public dataset $WBCD$ and an actual sensor data stream that transmitted from agricultural sensors are used as the datasets, while the FindFPOF [8], OODFP [10], FCI-Outlier [11], LFP [14] and KNN [6] are used as the

compared methods. For testing the mining efficiency of the MRI-Mine method, two
public datasets *T10I4D100K* and *pumsb** are used as the datasets, while the Apriori-
Rare [17], MRG-Exp [17] and IFP-min [18] are used as compared methods. Note that
some features of the used datasets are shown in Table 2. For the entire experiments,
they are implemented on a machine with an Intel dual core i3-2020 3.30 GHz processor
and 8 GB RAM, and all the compared methods are realized by python 3.6.

Table 2. Characteristic of the datasets

Datasets	Num. of trans	Avg. trans. size	Download address
WBCD	699	9	UCI Machine Learning Repository[a]
T10I4D100K	100000	10	Frequent Itemset Mining Implementations Repository[b]
pumsb*	49046	74	

[a]https://archive.ics.uci.edu/ml/datasets/Breast+Cancer+Wisconsin+(Diagnostic)
[b]http://fimi.cs.helsinki.fi/data/

4.1 Detection Accuracy of MRI-AD on *WBCD*

This experiment is conducted for evaluating the detection accuracy of MRI-AD method
on dataset *WBCD*, that is, for counting the exact number of selected transactions when
all malignant are detected. On dataset *WBCD*, 458 instances are benign (denoted as
normal data) and 241 instances are malignant (denoted as anomaly data). The exper-
iment is conducted in different *min_sup* values and different |*SW*|, where |*SW*| is chosen
as 20 and 30, while the ratio of *min_sup* to the |*SW*| is chosen as 30% and 40%. The
specific results are shown in Figs. 1 and 2, where the detection accuracy (in y-axis)
means the ratio of the number of malignant in current sliding window to the number of
selected transactions when all malignant are detected.

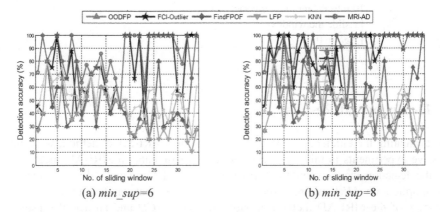

(a) *min_sup*=6   (b) *min_sup*=8

Fig. 1. Detection accuracy on dataset *WBCD* when |*SW*| is 20

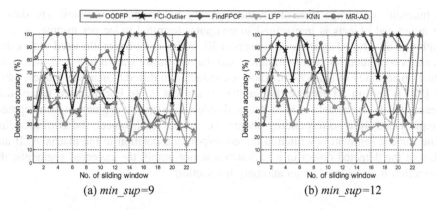

Fig. 2. Detection accuracy on dataset *WBCD* when |*SW*| is 30

It can be seen from Fig. 1(a) that when *min_sup* is assigned to 6, MRI-AD approach is much accurate than other five approaches except in one sliding window (slightly lower than the FCI-Outlier approach), while the detection accuracy of LFP approach, FindFPOF approach and OODFP approach is lower than that of FCI-Outlier approach and MRI-AD approach. As shown in Fig. 1(b) that when *min_sup* is assigned to 8, MRI-AD approach is slightly accurate than that when *min_sup* is assigned to 6, and the appearing times that the detection accuracy of MRI-AD approach is lower than that of FCI-Outlier approach are in three sliding windows, while the detection accuracy of other four approaches is also relatively low. The reason for the detection accuracy of MRI-AD approach is much higher in large *min_sup* values is that scale of *MRIs* is much larger in large *min_sup* values, thus, more *MRIs* are contained in the transactions.

When the |*SW*| is assigned to 30, the detection accuracy of six approaches is shown in Fig. 2. When the *min_sup* is assigned to 9, the detection accuracy of MRI-AD approach can reach to 100% in eleven sliding windows, and the detection accuracy of MRI-AD approach is not less than that of other five compared approaches in all sliding windows. When the *min_sup* is assigned to 12, the 100% detection accuracy of MRI-AD approach is appearing in eleven sliding windows of the 23 sliding windows, while the detection accuracy of MRI-AD approach is slightly lower than that of FCI-Outlier approach in two windows.

In general, the proposed MIFP-Outlier method is more effective in anomaly detection than the FindFPOF approach, OODFP approach, KNN approach, LFP approach and FCI-Outlier approach, and the MRI-AD approach can play a greater advantage in large *min_sup* values.

4.2   Detection Accuracy of MRI-AD on Sensor Data Stream

It can be known from Subsect. 4.1 that the proposed MRI-AD method can effectively find the anomalies. In this subsection, we test the detection efficiency of MRI-AD method on the sensor data stream that collected from agricultural sensor network. The sensor data stream is composed of five attributes, including the concentration of CO_2, the humidity and temperature of air, the humidity and temperature of soil. The data of the monitored objects is transmitted to computer terminal twice per minute, and the number of distributed agricultural sensors is 50. In this experiment, the used sensor data stream is continued for twenty minutes, and the results are shown in Fig. 3, where the detection accuracy of MRI-AD method is conducted in different *min_sup* values, while the |*SW*| is kept in 50 because the number of sensors is 50. For the used sensor data stream, we divide the data of each detected object into several segments (each segment is represented by a corresponding character), and for each object, the corresponding value for that segment is written at the corresponding position in the matrix A. If any value of the detected object in a transaction is not within the expected segments, the transaction is considered as an anomaly transaction.

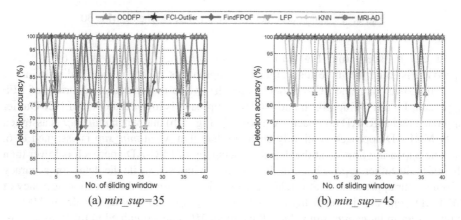

(a) *min_sup*=35   (b) *min_sup*=45

Fig. 3. Detection accuracy of MRI-AD on sensor data stream

It can be seen from Fig. 3(a) that when the *min_sup* is assigned to 35, MRI-AD is the most accurate approach in the six approaches and the FindFPOF approach is most inaccurate. In addition, the detection accuracy of MRI-AD approach that cannot reach at 100% is only in four sliding windows (*SWs*), while the detection accuracy of other five approaches that cannot reach at 100% is in more *SWs*. We can know from Fig. 3(b) that when the *min_sup* is assigned to 45, five compared approaches (except for KNN) is slightly accurate than that when the *min_sup* is assigned to 35, and detection accuracy of MRI-AD approach can reach to 100% in all forty *SWs*. The results illustrate that the detection accuracy of MRI-AD on sensor data stream is very high in large *min_sup* values.

4.3   Time Consumption and Memory Occupation of MRI-Mine

This subsection is to evaluate the mining efficiency of the MRI-Mine method, where the time consumption and memory occupation are tested in the experiments.

Time Consumption of MRI-Mine. For evaluating the time consumption of the MRI-Mine approach, the experiments are conducted on different $|SW|$ and different *min_sup*, and the results are listed in Figs. 4 and 5.

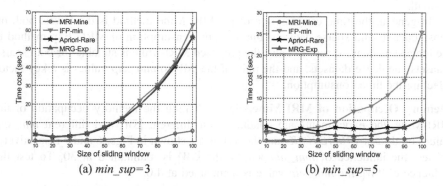

(a) *min_sup*=3   (b) *min_sup*=5

Fig. 4. Time consumption of MRI-Mine on dataset *T10I4D100K*

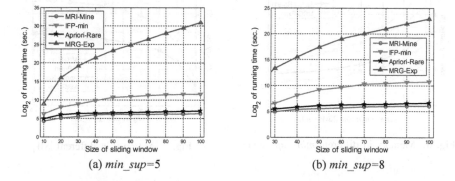

(a) *min_sup*=5   (b) *min_sup*=8

Fig. 5. Time consumption of MRI-Mine on dataset *pumsb**

As shown in Fig. 4(a), when *min_sup* is assigned to 3, the time consumption of the MRI-Mine method on sparse dataset *T10I4D100K* is the lowest, and the time consumption of the compared three methods show a similar trend. When *min_sup* is assigned to 5, the time consumption of MRI-Mine is also the lowest, while the MRG-Exp method is the second lowest and the IFP-min method is the highest. When the $|SW|$ is constant, the time consumption of the compared four methods shows a decrease trend with the increase of *min_sup* values, the reason is that the scale of *FIs* is much small in large *min_sup* values, which results the time consumption in "itemset connect" operation is much less. Compared with the IFP-min method and Apriori-Rare method, the proposed MRI-Mine method is very stable, which indicates the stability of the MRI-Mine method is very good.

On dense dataset *pumsb**, under fixed *min_sup* values, the time consumption of the compared four methods shows an increase trend with the increase of $|SW|$, and the increase trend is becoming smaller while the time consumption of MRG-Exp approach is much larger than that of other three methods. Under fixed $|SW|$, the time consumption of the compared four methods shows a decease trend with the increase of *min_sup* values, the reason is that the scale of potential *FIs* is very large when the *min_sup* value is small.

In general, the time consumption of the MRI-Mine method is the lowest both on dense datasets and sparse datasets, but the time consumption of MRG-Exp method is the longest on dense datasets. Thus, the proposed MRI-Mine is suitable for both sparse datasets and dense datasets, and the ratio of the $|SW|$ to *min_sup* values is a key factor affecting the time consumption.

Memory Occupation of MRI-Mine. For evaluating the memory occupation of the MRI-Mine, the experiments are conducted on different $|SW|$ and different *min_sup* values on dataset *T10I4D100K*, and the results are shown in Figs. 6 and 7 respectively. To test the influence of *min_sup* values, the $|SW|$ is maintained at 30. To test the influence of $|SW|$, the *min_sup* value is maintained at 4.0.

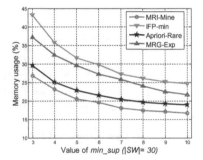 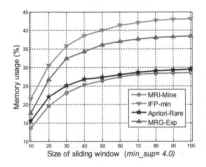

Fig. 6. Memory in different *min_sup* values   **Fig. 7** Memory in different $|SW|$

We can know from Fig. 6 that on dataset *T10I4D100K*, under the fixed $|SW|$, the memory occupation of MRI-Mine, IFP-min, Apriori-Rare and MRG-Exp methods shows a decrease trend with the increase of *min_sup* values, and the decline tendency is gradually slowing down. For these methods, the memory occupation of MRI-Mine is the lowest, while the memory occupation of IFP-min is the highest. The reason for this situation is that IFP-min method constructs many conditional trees during the itemset mining process, while the construction of conditional trees consumes a lot of memory. But for the MRI-Mine method, the rare itemsets are deleted directly before "itemset connect" process, so these rare itemsets are not consuming any memory at all, it also results in a large amount of meaningless "itemset connect" operation not occupy the memory usage.

It can be known from Fig. 7 that under fixed *min_sup* values, the memory occupation of the four methods shows an increase trend with the increase of $|SW|$, the reason is that the scale of *FIs* is very huge in big *SWs*, thus, much memory will be used in "itemset connect" operation. In addition, the increase trend of the compared four methods is gradually slowing down, it is owing to that when the *SW* changes from small to large, the number of *FIs* in the window increases dramatically, but when the window size becomes larger again, the scale of *FIs* in the window increases not much, so only a little memory is consumed for additional "itemset connect" operations. In the compared four methods, the memory occupation of our proposed MRI-Mine method is the lowest and the memory occupation of IFP-min method is the highest.

5 Conclusions

In this paper, we first introduce an efficient minimum rare-itemset-based anomaly detection method, namely MRI-AD, for detecting the anomalies from data stream, and then apply the MRI-AD method to sensor data stream, thereby solving the problem in practice. The proposed MRI-AD method is divided into two phases, that is, *MRI* mining phase and anomaly detection phase. In *MRI* mining phase, we design a matrix structure to store the specific information of items in the transactions, thereby supporting the entire mining operation by scanning the data stream for only once. And then the anti-monotone property, "itemset connect" operation, "seeking minimum" operation and "minimum checking" operation are used to propose the MRI-Mine method, thereby speeding up the mining operation. In anomaly detection phase, an efficient anomaly index is proposed to accurately measure the anomaly degree of all incoming transactions in the sliding window, where the *support* value and length of each contained *MRI* and the length of each transaction are considered in the anomaly index. And then, the transactions having large anomaly degree are judged as anomaly transactions.

The experimental results show that MRI-AD method is more accurate both on public datasets and sensor data stream under different |*SW*| and different *min_sup* values, while the detection accuracy on sensor data stream is not less than 75% and the detection accuracy can reach at 100% in most sliding windows. The experimental results also show that the time consumption and memory occupation of MRI-Mine method is less than that of the compared IFP-min method, Apriori-Rare method and MRG-Exp method both on sparse datasets and dense datasets, it indicates that the proposed MRI-AD method can handle the anomaly detection operation more quickly.

In recent years, the uncertain data stream and weighted data stream are more common in real-life, so the main work in the next stage is to effectively detect the anomaly data from uncertain data stream and weighted data stream.

References

1. Zhang, Z., Wu, P., Han, W., Yu, W.: Remote monitoring system for agricultural information based on wireless sensor network. J. Chin. Inst. Eng. **40**(1), 75–81 (2017)
2. Okazaki, T., Orii, T., Ueda, A., Kuramitz, H.: A reusable fiber optic sensor for the real-time sensing of $CaCO_3$ scale formation in geothermal water. IEEE Sens. J. **17**(5), 1207–1208 (2017)

3. Yuan, J., Wang, Z., Sun, Y., Zhang, W., Jiang, J.: An effective pattern-based Bayesian classifier for evolving data stream. Neurocomputing **295**, 17–28 (2018)
4. Hawkins, D.M.: Identification of Outliers, vol. 11. Chapman and Hall, London (1980)
5. Huang, J., Zhu, Q., Yang, L., Cheng, D., Wu, Q.: A novel outlier cluster detection algorithm without top-n parameter. Knowl.-Based Syst. **121**, 32–40 (2017)
6. Ramaswamy, S., Rastogi, R., Shim, K.: Efficient algorithms for mining outliers from large data sets. In: ACM SIGMOD Record, Dallas, USA, vol. 29, no. 2, pp. 427–438 (2000)
7. Zhang, L., Lin, J., Karim, R.: Adaptive kernel density-based anomaly detection for nonlinear systems. Knowl.-Based Syst. **139**, 50–63 (2018)
8. He, Z., Xu, X., Huang, Z., Deng, S.: FP-outlier: frequent pattern based outlier detection. Comput. Sci. Inf. Syst. **2**(1), 103–118 (2005)
9. Cai, S., Sun, R., Li, J., Deng, C., Li, S.: Abnormal detecting over data stream based on maximal pattern mining technology. In: Sun, Y., Lu, T., Xie, X., Gao, L., Fan, H. (eds.) ChineseCSCW 2018. CCIS, vol. 917, pp. 371–385. Springer, Singapore (2019). https://doi.org/10.1007/978-981-13-3044-5_27
10. Feng, L., Wang, L., Jin, B.: Research on maximal frequent pattern outlier factor for online high dimensional time-series outlier detection. J. Converg. Inf. Technol. **5**(10), 66–71 (2010)
11. Hao, S., Cai, S., Sun, R., Li, S.: An efficient frequent closed itemset-based outlier detecting approach on data stream. In: CCF Conference on Computer Supported Cooperative Work and Social Computing, Guilin, China, pp. 371–385 (2018)
12. Hao, S., Cai, S., Sun, R., Li, S.: An efficient outlier detection approach over uncertain data stream based on frequent itemset mining. J. Inf. Technol. Control **48**(1), 34–46 (2019)
13. Cai, S., Sun, R., Hao, S., Li, S., Yuan, G.: Minimal weighted infrequent itemset mining-based outlier detection approach on uncertain data stream. Neural Comput. Appl. **9**, 1–21 (2018)
14. Zhang, W., Wu, J., Yu, J.: An improved method of outlier detection based on frequent pattern. In: WASE International Conference on Information Engineering (ICIE), Washington, USA, pp. 3–6 (2010)
15. Dallachiesa, M., Jacques-Silva, G., Gedik, B., Wu, K., Palpanas, T.: Sliding windows over uncertain data streams. Knowl. Inf. Syst. **45**(1), 159–190 (2015)
16. Yang, G.: The complexity of mining maximal frequent itemsets and maximal frequent patterns. In: ACM SIGKDD International Conference on Knowledge Discovery and Data Mining, Seattle, WA, pp. 344–353 (2004)
17. Szathmary, L., Napoli, A., Valtchev, P.: Towards rare itemset mining. In: International Conference on Tools with Artificial Intelligence (ICTAI), Patras, Greece, pp. 305–312 (2007)
18. Gupta, A., Mittal, A., Bhattacharya, A.: Minimally infrequent itemset mining using pattern-growth paradigm and residual trees. In: International Conference on Management of Data, Bangalore, India, pp. 1–14 (2011)

A Dynamic Evolutionary Scheduling Algorithm for Cloud Tasks with High Efficiency and High QoS Satisfactions

Xiaoyong Guo and Jiantao Zhou[✉]

Inner Mongolia Engineering Lab of Cloud Computing and Service Software,
College of Computer Science, Inner Mongolia University, Hohhot, China
IM_GuoXY@163.com, cszhoujiantao@qq.com

Abstract. Efficient task scheduling is one of the main ways to increase cloud computing's throughput. In cloud computing, many tasks need to be scheduled on different virtual machines to augment system throughput while satisfying QoS conditions. Task scheduling is an NP-Complete problem, especially dynamic task scheduling in heterogenous cloud environments. This paper presents a dynamic task scheduling algorithm based on heterogeneous cloud environment. The proposed algorithm uses the preponderance of Topological sort, Genetic Algorithm (GA) and NSGA-II. The experimental results show that the scheduling efficiency and QoS satisfaction of the algorithm are significantly better than GA, and the latter is one of the most commonly used heuristic optimization techniques in task scheduling problems.

Keywords: Dynamic task scheduling · Evolution algorithm · Task dependencies · Global task priority · Heterogenous cloud environment

1 Introduction

Cloud computing is one of the hotspots of commercial and scientific research institutions [1]. Through virtualization technology, cloud computing organizes heterogeneous Information Technology (IT) to provides resource-extensible on-demand services [2]. The goal pursued by cloud computing service providers is to maximize benefits while ensuring service quality [3]. Therefore, how to realize efficient scheduling of massive tasks under the premise of satisfying user Quality of Service (QoS) is a research hotspot of cloud computing [4]. However, task scheduling problem is NP-complete, hence seek out an exact solution is a very tricky problem particularly for huge task sizes [5].

Task scheduling strategy in cloud environment focuses on how to establish mapping relationship between resources and tasks [6]. Task scheduling algorithms are divided into static task scheduling algorithms and dynamic task scheduling algorithms. Static task scheduling algorithms are to schedule only existing tasks in the system. Before the static task schedule, user program is first partitioned into interdependent subtasks [7], Then, some static task scheduling

© Springer Nature Singapore Pte Ltd. 2019
Y. Sun et al. (Eds.): ChineseCSCW 2019, CCIS 1042, pp. 131–142, 2019.
https://doi.org/10.1007/978-981-15-1377-0_10

algorithms are proposed in [8–13], all subtasks are displayed as a DAG, where each node represents a subtask, and the communication relationship between two subtasks is exhibited by each edge. Finally,the scheduler assigns all subtasks to the proper resources then determine the order of execution. In dynamic task scheduling algorithms, users can submit new tasks to scheduler at any time. Some algorithms that can be used of dynamic task scheduling are proposed in [14,15], in which the tasks are independent of each other, and each task can have one or more priorities, such as work mode priority, deadline priority and so all. [5,16–20] makes different optimizations for task scheduling in other aspects. According to our research, there is no dynamic task scheduling algorithm considering both global priority of tasks and the dependencies of tasks in cloud environment. However, in cloud environment, user programs arrive at any time and are partitioned into interdependent subtasks. Neither static task scheduling algorithm considering task dependencies nor dynamic task scheduling algorithm considering task priority and not considering task dependencies can satisfy cloud task scheduling. Therefore, it is necessary to design a dynamic task scheduling algorithm that considers both global task priorities and task dependencies.

In order to design a dynamic task scheduling algorithm considering both task dependencies and global task priorities, firstly, we need to design a cloud task scheduling model satisfying the conditions. Secondly, use the appropriate algorithm to solve the problem according to the cloud task scheduling model.

The rest of the paper is organized as follows. The related work is discussed briefly in Sect. 2. Section 3 describes system model and the proposed scheduling algorithm. In Sect. 4, results are discussed. Section 5 is the concluded of this paper.

2   Related Work

Two important goals of cloud task scheduling are high efficiency and high QoS [4]. Hence, the research intends to find algorithms that should satisfy high efficiency or high QoS in the scheduling of tasks. Some static task scheduling algorithms were proposed in [8,10–13], scientists consider the correlation between tasks and static scheduling of dependent parallel tasks. Some dynamic task scheduling algorithms were proposed in [14,15], scientists consider tasks as independent tasks and give different priority for each task to get better scheduling effects. In addition, topological ordering was discussed in [21], NSGA-II was proposed in [22] and optimized in [23].

In terms of static task scheduling algorithms, a cloud workflow tasks scheduling algorithm based on GA was presented in [8], the up-down leveling method was used to assign priority to each task. [10] proposed and proved some standard task graphs, and a novel task scheduler was designed by using Greedy approach. In [11], memetic local optimization based on local pre-evaluation was used to reduce the overall evaluation of individuals and improve the algorithm's search efficiency. At the same time, the cross-mutation operation based on DAG task graph hierarch was used to avoid illegal solutions, and finally reduce the computational cost of the algorithm. Quantum Genetic Algorithm with Rotation Angle Refinement (QGARAR) was highlighted in [12]. N-GA was proposed in [13]. N-GA improves the genetic algorithm which using the advantages of heuristic approach. In order to analyze the correctness of N-GA, it proposed a behavioral modeling approach then maked check by NuSMV and PAT model checkers based on model-checking techniques.

In dynamic task scheduling algorithms, [14] proposed an improve GA, all tasks are regarded as independent tasks, and the priority of each task is divided into two parts: task mode priority and task deadline priority. In addition, improve GA also uses time utilization rate as one of the evaluation indicators of scheduling results. The Artificial Bee Colony (ABC) algorithm was used in [15] to solve the load balanced of preemptive independent tasks on virtual machines and considers the expected remaining completion time and priority of the tasks with an aim to minimize latency and maximize throughput.

There are still some problems in existing methods. Although the static task scheduling algorithm based on task dependencies has a good performance in static scheduling, it cannot meet the requirements of dynamic cloud task scheduling. The dynamic task scheduling algorithm based on independent task priority has good results in terms of QoS and performance, but the scheduling algorithm can not meet the cloud task scheduling requirements with complex dependencies. This paper will try to address the issue of dynamic task scheduling that considering both global task priority and task dependencies in Heterogeneous cloud environment.

3   Cloud Model and Proposed Algorithm

3.1   Cloud Model

The cloud application model's inovation is based on application model coming from [13]. The model in [13] is a simplified model of the cloud environment, and give the formal verification of the algorithm based on this simplified model. In proposed algorithm, cloud system has a set of m Virtual Machines, to make task scheduling uncomplicated, each Virtual machine completely interconnected with a high-speed network, it is hypothesis that the inter virtual machine communications run at the equivalent speed on all links. Each task has randomly number of dependent subtasks and illustrated as follow:

$$G = (V, E) \tag{1}$$

where V nodes indicate subtasks, dependencies and communication cost between the subtasks are represent by E edges. A simple DAG t_1 is shown in Fig. 1.

Before the scheduling, a small quantity of tasks are prepared randomly, and added randomly generated tasks to the scheduling queue during the scheduling until the number of subtask meets the test requirements. Each subtask must be executed on only one virtual machine, and all subtasks should be scheduled. If two dependent subtasks are not allotted to the same virtual machine, the communication cost between them must be considered.

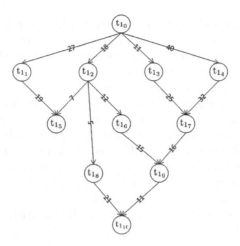

Fig. 1. A simple DAG with 11 nodes and 14 edges

To make a simple example, there are three virtual machines in the system. The computation costs for each subtask in Fig. 1 to execute in each virtual machines is shown in Table 1, where the first column illustrates the application's subtasks from Fig. 1, each subtasks' computation cost is presented by columns 2 up to 4 on vm_1 to vm_3, each subtasks' average computation cost is shown by the last column on all vms. As shown in Table 1, virtual machines may be faster for some subtasks and slower for some other subtasks. In addition, subtasks can be initiated after all of their predecessors have been executed. For example, subtask t_{1_5} cannot start until t_{1_1} and t_{1_2} complete their execution.

Table 1. Computation cost matrix of DAG in Fig. 1

Subtask	vm_1	vm_2	vm_3	$\overline{w}$
t_{1_0}	10	11	12	11
t_{1_1}	11	12	13	15
t_{1_2}	12	8	13	11
t_{1_3}	14	10	18	14
t_{1_4}	27	20	19	22
t_{1_5}	15	12	18	15
t_{1_6}	9	14	19	14
t_{1_7}	19	12	14	15
t_{1_8}	14	10	15	13
t_{1_9}	15	12	15	14
$t_{1_{10}}$	18	10	17	15

3.2 Dynamic Task Scheduling

This paper proposes a dynamic task scheduling algorithm based on GA and NSGA-II for heterogeneous cloud environments that combines task dependencies and global task priority which named TN-GA. TN-GA is divided into three steps. Firstly, topological sorting for subtasks which included in each task. The second step use GA combined with task global priority and task topology ordering for task scheduling. Finally, use the NSGA-II to perform task scheduling in the idle gap generated by the second step task scheduling.

Topologic Sort for NS-GA. In this paper, we use topological sorting to transform the DAG task graph into a sequential execution queue that satisfies subtasks dependencies. For a task in this paper, the steps for topological sort are as follows:

- Get the in-degree and out-degree of all subtasks
- Adding subtask with zero in-degree and max out-degree to topological rank.
- For each subtask traversed, update the in-degree of its successors subtask: reduce the in-degree of the successors subtask by 1.
- Repeat step 2 until all subtasks are traversed.

Task Priority for TN-GA. Before each task scheduling, the scheduler needs to determine the subtasks to be scheduled. In order to achieve higher QoS satisfactions, it is necessary to determine the global priority for each task, and priority scheduling subtasks in higher global priority tasks. In this paper, global task priority is determined by task execution status and deadline. The global task priority is obtained using Eq. (2).

$$priority_i = (N_i - finish\_subtask_i.length) * \frac{deadline_i - time}{deadline_i - insert\_time_i} \quad (2)$$

In Eq. (2), $priority_i$ denotes the global priority of task t_i, N_i denotes the number of subtasks in task t_i, $finish\_subtask_i$ denotes the set of subtasks completed in task t_i, $finish\_subtask_i.length$ denotes the number of subtasks completed in task t_i, $deadline_i$ denotes the deadline of task t_i in system, $insert\_time_i$ denotes the time of task t_i join the system, the $time$ denotes the current time of the system. The low priority value of task represents that is has a higher priority. The scheduler selects from high to low according to priority when deciding to schedule subtasks, until all runnable subtasks are selected or selected to enough subtasks.

Genetic Algorithm for TN-GA. GA is an effective optimization algorithm to choose a better solution in the massive solution space. The purpose of GA is to perform selected subtasks with as little time as possible. GA algorithm comprises three base operations. Parents are selected from population by selection operator. Offspring are produced by the selected parents by crossover operator. The mutation operator changes offspring according to mutation rules.

Based on the cloud model, the GA was described by the following will in detail, which comprises coding, initializing population, designing the fitness function, selection, crossover and mutation.

For the coding and population initialization in this paper, real coding strategy is used by us here, where an individual's each gene is represented by subtask id. The chromosome's length depends on the total number of idle virtual machines, and the chromosome's each gene represents the task assignment to the proper virtual machine. The gene is -1 means that no subtask assignment to the proper virtual machine.

Which chromosomes can survive to engender population's next generation is determined according to the fitness value. The fitness value in the proposed algorithm is the sum of the times when all subtasks are completed during the current scheduling.

Equations (3) and (4) obtain the subtask t_{i_j}'s earliest execution start time on virtual machine vm_k which is symbolized as $EST(t_{i_j}, vm_k)$.

$$EST(t_{entry}, vm_k) = 0 \tag{3}$$

$$EST(t_{i_j}, vm_k) = max_{t_{i_s} \in pred(t_{i_j})}(AFT(t_{i_s}) + c(t_{i_s}, t_{i_j})) \tag{4}$$

where t_{entry} denotes subtask t_{i_j} is the entry of entire task t_i, $t_{i_s} \in pred(t_{i_j})$ denotes subtask t_{i_j} depends on subtask t_{i_s}, $AFT(t_{i_s})$ denotes the actual execution finish time of subtask t_{i_s}.

Equation (5) obtain the subtask t_{i_j}'s actual execution start time on virtual machine vm_k which is symbolized as $AST(t_{i_j}, vm_k)$.

$$AST(t_{i_j}, vm_k) = max(EST(t_{i_j}, vm_k), Avail(vm_k)) \tag{5}$$

Where $Avail(vm_k)$ is the time when the virtual machine vm_k is idle and ready for subtask execution.

Equation (6) obtain the subtask t_{i_j}'s earliest execution finish time on virtual machine vm_k which is symbolized as $EFT(t_{i_j}, vm_k)$.

$$EFT(t_{i_j}, vm_k) = w(t_{i_j}, vm_k) + AST(t_{i_j}, vm_k) \tag{6}$$

Where $w(t_{i_j}, vm_k)$ is the computation cost of the subtask t_{i_j} on virtual machine vm_k.

Equation (7) obtain the subtask t_{i_j}'s actual execution finish time on virtual machine vm_k which is symbolized as $AFT(t_{i_j}, vm_k)$, where vm_k is the appropriate virtual machine for the subtask t_{i_j}.

$$AFT(t_{i_j}, vm_k) = min_{1 \leq l \leq m} EFT(t_{i_j}, vm_k) \tag{7}$$

The runtime of the subtask t_{i_j} on virtual machine vm_k is symbolized as $runtime(t_{i_j}, VM_k)$ which is calculation by Eq. (8):

$$runtime(t_{i_j}, vm_k) = AFT(t_{i_j}, vm_k) - time \tag{8}$$

where *time* is the system's current time.

In this paper, we use tournament selection as selection strategy. Two queues of length $PopSize/10$ ($PopSize$ denote the size of population) and no repeat individuals were randomly selected in the population, and two individuals with the best fitness in the two queues were selected as parents to generate the next generation of individuals.

For crossover, first we randomly generate a crossover point. Second, the gene before the crossover point of a parent chromosome is taken into the offspring. Finally the genes that are not in offspring in another parent chromosome are added to offspring in turn, and two offspring are generated, as shown in Fig. 2.

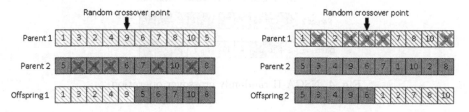

Fig. 2. Single-point crossover operation

In population, each chromosome has 20% probability of mutation. As shown in Fig. 3, during the mutation, a mutation point is randomly generated, and the mutation point is exchanged with a point after the mutation point.

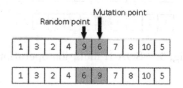

Fig. 3. Mutation operation

NSGA-II for TN-GA. Genetic algorithm may generate idle gap of virtual machine after each scheduling. The purpose of NSGA-II is to use these idle gaps to schedule more subtasks to enhance the throughput and resource utilization of the cloud environment. In this paper, there are two fitness functions as the target parameters of NSGA-II, the tournament selection method is used as the selection function, and the crossover and mutation functions are redesigned.

There are two fitness functions in this paper, one is the sums of idle gaps for scheduled subtasks, and the other is the total time for the subtasks to run.

The running time of subtask t_{ij} scheduled in idle gaps is equal to the computational cost of the subtask t_{ij} on virtual machine VM_k: $C(t_{ij}, VM_k)$.

In this paper, we proposed a new randomly crossover operator to increase the population's diversity. Each chromosome in the offspring is randomly selected form the parents at corresponding gene. As shown in Fig. 4, the first gene of offspring is random from parent1, second gene of offspring is random from parent2. The sixth gene of offspring is random from parent1, but this gene already exists in offspring, so, we give the gene in parent2 to offspring. The ninth gene of offspring is random from parent2, but this gene already exists in offspring and the ninth gene of parent1 also, so, we give the gen -1 to offspring which means that no subtasks are assigned to the corresponding virtual machine idle gap.

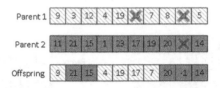

Fig. 4. NSGA-II randomly crossover operation

In population, each chromosome has 20% probability of mutation. As shown in Fig. 5, during the mutation, a mutation point is randomly generated, random selection other subtask substitution which can be completed in this idle gap.

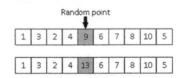

Fig. 5. NSGA-II mutation operation

4   Results and Discussion

To assess the TN-GA approach, we compared it with traditional GA and GA combined with global task priority. We use TGFF [24] to generate a large number of tasks randomly, each of which contains many dependent subtasks. TN-GA is applied in java and execution on desktop with i5, 3.30 GHz CPU and 6 GB RAM. The values of the parameters used in the proposed algorithm are represented by Table 2. Table 3 signifying the randomly generated graphs' values, where the first column signifying the minimum figure of subtasks per test, the second column shows the figure of subtasks on each DAG, third column shows the figure of virtual machine per test, the fourth column shows each task's insert time per test, the fifth column is each task's deadline, the sixth column is the computation cost randomly selecting from 1 to 50, the seventh column signifying the edge's communication cost between two subtasks selecting from 1 to 40 randomly.

Table 2. Parameters' values

Parameters	Description
The population size of GA	Four times the number of idle virtual machine
The mutation probability of GA	0.2
The termination condition of GA	Up to 100 generations or 5 consecutive generations with the same minimum running time
The population size of NSGA-II	Idle virtual machine number*executable subtasks number (up to 200)
The mutation probability of NSGA-II	0.2
The termination condition of NSGA-II	Up to 20 generations

Table 3. Parameters of the random task graphs

Minimum number of subtasks per test	The number of subtasks on each DAG	The number of virtual machine	insert_time	Deadline	Computation cost range	Communication cost range
10000	randomly from (10–1000)	50	randomly from (0–1000)	insert_time $+N_i * 5$	randomly from (1–50)	randomly from (1–40)
20000	randomly from (10–1000)	50	randomly from (0–1000)	insert_time $+N_i * 10$	randomly from (1–50)	randomly from (1–40)
50000	randomly from (10–1000)	100	randomly from (0–1000)	insert_time $+N_i * 15$	randomly from (1–50)	randomly from (1–40)
100000	randomly from (10–1000)	100	randomly from (0–1000)	insert_time $+N_i * 25$	randomly from (1–50)	randomly from (1–40)

Figure 6 shows the turnaround time of traditional GA, GA combined with task global priority and TN-GA on each test. As shown in Fig. 6, in the experiments of different subtasks from 10000 to 100000, the GA combined with task global priority is slightly better than the traditional GA in time efficiency, and TN-GA is obviously better than GA combined with task global priority and traditional GA.

Figure 7 shows the number of task meeting deadline. In Fig. 7, 10000, 20000, 50000 and 100000 subtasks are consist of 26, 39, 94 and 195 random DAG tasks, respectively. In the experiments result, TN-GA and GA combined with task global priority have the same number of tasks meeting the deadline, and they are obviously more than traditional GA.

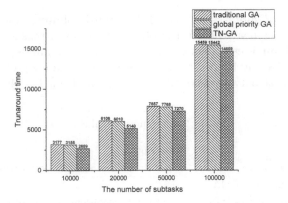

Fig. 6. Turnaround time of three algorithms on each test

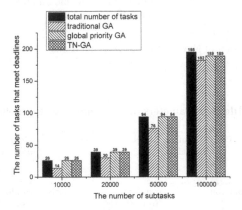

Fig. 7. The number of tasks that meet deadlines of three algorithms on each test

5   Conclusion

In this paper, we presented a dynamic task scheduling algorithm with task dependencies and tasks global priority based on GA and NSGA-II which named TN-GA. Performance of the proposed scheduling algorithm with respect to turnaround time and QoS satisfaction is analyzed empirically by randomly generate DAG tasks and randomly join the scheduler. Further, comparison the scheduling results of TN-GA, GA with combined with task global priority and traditional GA. The performance and QoS satisfaction of our proposed scheduling schemes is better than traditional GA and GA combined with task global priority.

Acknowledgement. The research is supported by Natural Science Foundation of China under Grant No. 61662054, 61262082, Inner Mongolia Science and Technology Innovation Team of Cloud Computing and Software Engineering and Inner Mongolia Application Technology Research and Development Funding Project "Mutual Creation

Service Platform Research and Development Based on Service Optimizing and Operation Integrating" under Grant 201702168, Inner Mongolia Engineering Lab of Cloud Computing and Service Software and Inner Mongolia Engineering Lab of Big Data Analysis Technology.

References

1. Zhao, C.-Y.: Research and Implementation of Job Scheduling Algorithm in Cloud Computing. Beijing Jiaotong University, Beijing (2009)
2. Chen, K., Zheng, W.-M.: Cloud computing: system instances and current research. J. Softw. **20**, 1337–1348 (2010). https://doi.org/10.3724/SP.J.1001.2009.03493
3. Liu, X.-Q.: Research on Data Center Structure and Scheduling Mechanism in Cloud Computing. University of Science and Technology of China (2011)
4. Shen, Q., Xu, M.-Y., Chun-Mao, J.: Review of task scheduling research in cloud computing. Intell. Comput. Appl. **4**, 75–77 (2014)
5. Abdullahi, M., Ngadi, M.A., Abdulhamid, S.M.: Symbiotic Organism Search optimization based task scheduling in cloud computing environment. Future Gen. Comput. Syst. **56**, 640–650 (2016). https://doi.org/10.1016/j.future.2015.08.006
6. Wu, H.: Research of Task Scheduling Algorithm in the Cloud Environment. Nanjing University of Posts and Telecommunications, Nanjing (2013)
7. Dean, J., Ghemawat, S.: MapReduce: simplified data processing on large clusters. Commun. ACM **51**, 107–113 (2008). https://doi.org/10.1145/1327452.1327492
8. Cui, Y., Xiaoqing, Z.: Workflow tasks scheduling optimization based on genetic algorithm in clouds. In: 2018 IEEE 3rd International Conference on Cloud Computing and Big Data Analysis (ICCCBDA), pp. 6–10. IEEE, Chengdu (2018). https://doi.org/10.1109/ICCCBDA.2018.8386458
9. Akbari, M., Rashidi, H., Alizadeh, S.H.: An enhanced genetic algorithm with new operators for task scheduling in heterogeneous computing systems. Eng. Appl. Artif. Intell. **61**, 35–46 (2017). https://doi.org/10.1016/j.engappai.2017.02.013
10. Byrappa, S.D., Hegde, S.N., Rajan, M.A., Krishnappa, H.K.: A novel task scheduling scheme for computational grids - greedy approach. In: 2018 IEEE 32nd International Conference on Advanced Information Networking and Applications (AINA), pp. 1026–1033. IEEE, Krakow (2018). https://doi.org/10.1109/AINA.2018.00149
11. Li, Z.-Y., Chen, S.-M., Yang, B., Li, R.-F.: Multi-objective memetic algorithm for task scheduling on heterogeneous cloud. Jisuanji Xuebao/Chinese J. Comput. **39**(2), 377–390 (2016). https://doi.org/10.11897/SP.J.1016.2016.00377
12. Gandhi, T., Alam, T.: Quantum genetic algorithm with rotation angle refinement for dependent task scheduling on distributed systems. In: 2017 Tenth International Conference on Contemporary Computing (IC3), pp. 1–5. IEEE, Noida (2017). https://doi.org/10.1109/IC3.2017.8284295
13. Keshanchi, B., Souri, A., Navimipour, N.J.: An improved genetic algorithm for task scheduling in the cloud environments using the priority queues: formal verification, simulation, and statistical testing. J. Syst. Softw. **124**, 1–21 (2017). https://doi.org/10.1016/j.jss.2016.07.006
14. Hao, L., Yang, X., Hu, S.: Task scheduling of improved time shifting based on genetic algorithm for phased array radar. In: 2016 IEEE 13th International Conference on Signal Processing (ICSP), pp. 1655–1660. IEEE, Chengdu (2016). https://doi.org/10.1109/ICSP.2016.7878109

15. Shobana, G., Geetha, M., Suganthe, R.C.: Nature inspired preemptive task scheduling for load balancing in cloud datacenter. In: International Conference on Information Communication and Embedded Systems (ICICES2014), pp. 1–6. IEEE, Chennai (2014). https://doi.org/10.1109/ICICES.2014.7033816

16. Zhou, J., Dong, S.-B., Tang, D.-Y.: Task scheduling algorithm in cloud computing based on invasive tumor growth optimization. Jisuanji Xuebao/Chinese J. Comput. **41**(6), 1360–1375 (2018). https://doi.org/10.11897/SP.J.1016.2018.01360

17. Li, J.-F., Peng, J.: Task scheduling algorithm based on improved genetic algorithm in cloud computing environment. J. Comput. Appl. **31**(01), 184–186 (2011)

18. Hamad, S.A., Omara, F.A.: Genetic-based task scheduling algorithm in cloud computing environment. Int. J. Adv. Comput. Sci. Appl. **7** (2016). https://doi.org/10.14569/IJACSA.2016.070471

19. Li, K., Xu, G., Zhao, G., Dong, Y., Wang, D.: Cloud task scheduling based on load balancing ant colony optimization. In: 2011 Sixth Annual Chinagrid Conference, pp. 3–9. IEEE, Liaoning (2011). https://doi.org/10.1109/ChinaGrid.2011.17

20. Zhang, F., Cao, J., Li, K., Khan, S.U., Hwang, K.: Multi-objective scheduling of many tasks in cloud platforms. Future Gen. Comput. Syst. **37**, 309–320 (2014). https://doi.org/10.1016/j.future.2013.09.006

21. Wang, X.Y., Wei, Z.J.: Discussion on the algorithm in topological collating. J. Northwest Univ. (Nat. Sci. Edn.) (2002). https://doi.org/10.16152/j.cnki.xdxbzr.2002.04.007

22. Deb, K., Pratap, A., Agarwal, S., Meyarivan, T.: A fast and elitist multiobjective genetic algorithm: NSGA-II. IEEE Trans. Evol. Comput. **6**(2), 182–197 (2002). https://doi.org/10.1109/4235.996017

23. Chen, J., Xiong, S., Lin, W.: Improved strategies and researches of NSGA-II algorithm. Comput. Eng. Appl. **47**(19), 42–45 (2011)

24. Dick, R.P., Rhodes, D.L., Wolf, W.: TGFF: task graphs for free. In: Proceedings of the Sixth International Workshop on Hardware/Software Codesign, pp. 97–101. IEEE, Seattle (1998). https://doi.org/10.1109/HSC.1998.666245

Improved Bat Algorithm for Multiple Knapsack Problems

Sicong Li[1], Saihua Cai[1], Ruizhi Sun[1,2(✉)], Gang Yuan[1], Zeqiu Chen[1], and Xiaochen Shi[1]

[1] College of Information and Electrical Engineering,
China Agricultural University, Beijing 100083, China
{lsc, caisaih, sunruizhi, chenzq}@cau.edu.cn
[2] Scientific Research Base for Integrated Technologies of Precision Agriculture
(Animal Husbandry), The Ministry of Agriculture, Beijing 100083, China

Abstract. Bat algorithm has been paid more attention because of its excellent conversion ability between global search to local search and its high robustness. To solve the problem of 0–1 single knapsack problem, scholars introduced binary encoding on the basis of bat algorithm and put forward the binary bat algorithm. However, when solving the multiple knapsack problem (MKP), the binary encoding will lead to the emergence of illegal solutions, so it is necessary to use the multi-value encoding to re-model the MKP, thereby applying the bat algorithm to MKP. To improve the entire search ability of the algorithm, we optimized the effective solution in the algorithm using the greedy algorithm, and then proposed a greedy algorithm-based bat algorithm, namely MKBA-GA, for solving the MKP. To further improve the solution ability of the MKBA-GA algorithm, we used Single Running Technique (SRT) to optimize the effective solution, and then proposed an efficient SRT-based bat algorithm called MKBA-SRT. In order to verify the performance of the proposed MKBA-GA algorithm and MKBA-SRT algorithm, we compare them with BBA, IRT and SRT algorithms on twelve datasets, and the experimental results show that the solution ability of MKBA-GA algorithm is stronger than that of BBA algorithm, and the ability of MKBA-SRT algorithm is superior to that of other four compared algorithms on eleven datasets.

Keywords: Bat algorithm · Multiple knapsack problem · Multi-value encoding · Greedy algorithm · Single running technique

1 Introduction

Knapsack problem (KP) [1] is a combinatorial optimization problem, it can be described as: for some items in a group with value and weight, select some items into a knapsack to make the total value of the items loaded into the knapsack is the largest while the weight is not exceeding the weight limit of the knapsack. In real life, many problems can be transformed into KP and its sub-problems [1], such as: resource allocation, investment decision, loading problem, etc. Multiple knapsack problem (MKP) [1] is a less complex and less constrained KP sub-problem, it can be described

© Springer Nature Singapore Pte Ltd. 2019
Y. Sun et al. (Eds.): ChineseCSCW 2019, CCIS 1042, pp. 143–157, 2019.
https://doi.org/10.1007/978-981-15-1377-0_11

as: for n items with their own value and weight, select some items into m knapsacks to make the total value of the items loaded into all knapsacks is the largest while the weight is not exceeding the weight limit of each knapsack.

At present, the algorithms used to solve MKP is divided into two categories: (1) precise algorithm, including exhaustive method, dynamic programming method [2], branching delimitation method [3], etc., (2) imprecise algorithm, including ant colony optimization algorithm [4], artificial fish swarm algorithm [5], genetic algorithm [6], etc. For the precise algorithm, the quality of the solution is much higher, but its high time complexity will lead to the time cost is also very long, so it is not suitable for large datasets. Compared with the precise algorithm, the time complexity of imprecise algorithm is lower and the global optimal solution or its approximate solution can be obtained in a short time, so it is more suitable for solving large datasets.

In this big data era, the size of the datasets to be solved becomes very large, so the imprecise algorithms are more attuned to the current era than the precise algorithms. In addition to the early proposed simulated annealing algorithm (including: particle swarm algorithm, ant colony optimization algorithm and artificial fish swarm algorithm), more efficient imprecise algorithms, such as bat algorithm, fruit fly optimization algorithm, were proposed in recent years. Among them, bat algorithm is a kind of bionic algorithm, it mimics the ability of bats in the biological world to locate and search prey by echo location. Compared with other imprecise algorithms, bat algorithm has the following advantages: (1) bat algorithm can automatically adjust the individual's pulse emission frequency and loudness to transform between global search and local search, so the algorithm has excellent conversion ability between global search and local search; (2) fewer parameters need to be adjusted in the bat algorithm; (3) bat algorithm has internal parallelism and high robustness. Because of these advantages of bat algorithm, it has received wide attention in recent years. In addition, many scholars have further optimized the bat algorithm and applied them to the solution of practical problems, such as 0–1 knapsack problem [7], classification problem [8], three-dimensional positioning [9], workshop job scheduling [10], etc.

The traditional bat algorithm is mainly aimed at solving the continuous problem. In recent years, scholars have improved the bat algorithm into binary bat algorithm by using 0–1 encoding technology, and applied it to the solution of 0–1 KP. Compared with the 0–1 KP, the MKP is more constrained, so using 0–1 encoding to solve the MKP will cause some illegal solutions that produced in the process of solving solution. In addition, the quality of the effective solution obtained after processing the invalid solution in the algorithm is also relatively low, it will reduce the overall search efficiency of the bat algorithm. As far as we know, no bat algorithm is proposed until now for solving the MKP. Based on the above reasons, this paper uses the multi-value encoding to re-model the MKP first, in addition, in the basic idea of bat algorithm, this paper adds two strategies: (1) greedy algorithm and (2) single running technique (SRT) [3], for further improving the quality of effective solutions. The main contributions of this paper can be listed as follows:

(1) We use the multi-value encoding to re-model for the MKP and bat algorithm, thereby making the bat algorithm is suitable for solving the MKP.

(2) We use the greedy algorithm to optimize the effective solution for improving the quality of effective solutions, and then propose a greedy algorithm-based bat algorithm, namely MKBA-GA, for solving the MKP.

(3) We combine the basic bat algorithm and SRT method to further improve the quality of effective solutions, and then propose a SRT-based bat algorithm, namely MKBA-SRT, for solving the MKP.

The remaining parts can be organized as follows. Section 2 reviews the improvements of bat algorithm. Section 3 discusses two algorithms of MKBA-GA and MKBA-SRT. The experimental analysis is stated in Sect. 4. Section 5 concludes the full paper and looks forward to the future work.

2 Improvements of Bat Algorithm

Bat algorithm is a relatively new imprecise algorithm, it has received extensive attention because of it has few parameters need to be adjusted and has excellent conversion ability between global search and local search. In recent years, many scholars improved the bat algorithm and applied it to solve the practical problems.

Gan et al. [11] first proposed a new local search method, that is, the local search is re-done by disrupting the current local optimal method, so as to improve the ability of the algorithm to jump out of the local optimal solution. In addition, the weight factors were added to the speed update formula to improve the diversity and flexibility of bats. Chakri et al. [12] set two sets of pulse emission rules for bats in different directions to enhance the exploration capability and development capability of Bat algorithms, and then they used the position of the best bat to guide the next search direction of the rest bats, which was allowed the bat to move to the solution space near the current optimal solution with a greater probability, thus obtaining a better solution.

The above bat algorithm could only be used for continuous problems, to make the bat algorithm can be used for the solution of discrete problems, many scholars have improved the bat algorithm. Mirjalili et al. [13] discretized the original bat algorithm by setting a new conversion function of speed and position, and then proposed the binary bat algorithm BBA. On the basis of the BBA algorithm, Zhou et al. [14] plural the speed and position of the bat to increase the space search range of the bat. Compared with BBA algorithm, IBBA algorithm [15] first added the dynamic velocity weight factor to the velocity formula, and then added the influence factor of the position of the local optimal bat on the flight direction of the bat to the formula, so as to improve the ability of the algorithm to jump out of the local optimal solution.

3 MKBA-GA and MKBA-SRT Approaches

In this section, we will detailed introduce the proposed two improved bat algorithms, namely MKBA-GA algorithm and MKBA-SRT algorithm. The two algorithms are based on IBBA algorithm [15], which optimizes the effective low quality solution in the algorithm, thereby achieving a better solution effect. Among them, the MKBA-GA algorithm adopts an effective greedy idea-based solution optimization method called

GAVO, which will be described in Subsect. 3.3, while the MKBA-SRT algorithm optimizes the effective solution based on SRT method. The main idea of the two algorithms is consistent, that is, they first generate the initial population by stochastic method, and then seek for the optimal solution by multiple iterations. In addition, in each iteration process, the GAVO or SRT method is used to optimize the effective low quality solution.

Here, we first explain the following three nouns, namely: illegal solution, invalid solution and effective solution.

Definition 1. *Illegal solution***:** There are existing more than one of 1 in a column of binary solution x, that is, an item is placed in multiple knapsacks at the same time. It violates the constraint that each item can only be added into one knapsack.

Definition 2. *Invalid solution***:** The solution is not an illegal solution, but the total weight of the items at least in one knapsack is exceeding its limit weight.

Definition 3. *Effective solution***:** The solution satisfies all the constraints in the problem model.

3.1   Model Reconstruction of MKP

In the original model of the MKP mentioned in the first section, the solution x uses the 0–1 encoding. In this representation of 0–1 encoding, the number of solution spaces of MKP is 2^{m*n}, that is, x is a m*n matrix. The 0–1 encoding [16] form causes the following two problems: (1) the number of solution space shows an exponential increase trend with the increase of m and n, and (2) there are a large number of illegal solutions in this 2^{m*n} solution spaces.

To solve the above two problems, we re-model the MKP and use the multi-value encoding [5] to represent the solution x, the mathematical model is as follows.

$$max f(Y) = \sum_{i=1}^{n} exis(x_i) * p_i \tag{1}$$

$$exis(x_i) = \begin{cases} 0, x_i = 0 \\ 1, x_i \neq 0 \end{cases} \tag{2}$$

$$x_i \in \{0, 1, 2, \ldots, m\}, 1 \leq i \leq n \tag{3}$$

where x_i represents the number of knapsacks that placed by the first item. If x_i is 0, it means that the first item has not been placed in any knapsack, and $exis(x_i)$ is set to 0; otherwise, $exis(x_i)$ is set to 1.

In the case representing the solution x through multi-value encoding, the number of solution spaces changes from 2^{m*n} to $(m + 1)*n$, and the existence of illegal solution is avoided. In the proposed MKBA-GA algorithm and MKBA-SRT algorithm, the new model in used to solve the MKP.

3.2   Location Update Based on Multi-value Encoding

As mentioned in Subsect. 3.1, in order to solve the problems that the much number of solution spaces and the easy appearance of illegal solutions, we abandon the original 0–1 encoding and adopt the multi-value encoding. Thus, the original position formula of the IBBA is no longer applicable to the MKP model because of the use multi-value encoding. Based on the above factors, we give a new position formula for solving the MKP model under multi-value encoding. The formula is as follows:

$$x_i^{t+1} = \begin{cases} \text{rand}(0, m), & \text{if } \text{rand}(0,1) < V(v_i^{t+1}) \\ x_i^t, & \text{else} \end{cases} \tag{4}$$

where m is the number of knapsacks, rand $(0,m)$ means an integer is randomly selected between 0 and m (containing 0 and m), it indicates re-randomly assigning the items to the knapsack.

3.3   Greedy Algorithm-Based Valid Solution Optimization Method

In the process of solving MKP, the algorithm can find some invalid solutions, and an usual way is to delete some items from the invalid solution so that the invalid solution becomes an effective solution, but the quality of such an effective solution is very low (for example, the utilization space of knapsack is low). Thus, we propose an effective greedy algorithm-based valid solution optimization method, namely GAVO, for improving the space utilization rate of the knapsack and improving the quality of effective solution.

The main idea of GAVO method is shown as follows: (1) sorting the remaining items according to the quality value ratio from high to low; (2) putting the items to the knapsacks that still having the remaining space according to the sorting order until all knapsacks can no longer add any items. That is, adding items to the knapsacks to reduce the free space of the knapsacks while ensuring that the load limit of knapsacks is not exceeded.

Algorithm 1: GAVO

Input: an valid solution x

Output: the new solution x

01.calculate the η of all items $//\eta$ is the ratio of value and weight
02.rank the items by the decreasing η values
03.**for** i=1 to m **do**
04. **for** j=1 to r **do**
05.   **if** x_j=0 **then**
06.   **if** sum(weight of items in M_i)<m_i **then**
07. $i{\rightarrow}x_j$
08.   **end if**
09.   **end if**
10. **end for**
11.**end for**
12.output new solution x

The pseudo-code of the GAVO method is shown in Algorithm 1. Where m is the number of the knapsacks, n is the total number of items, $sum()$ is a summation function, M_i represents the collection of items in the ith knapsack, m_i represents the load-bearing limit for the ith knapsack, and w_j represents the weight of the jth item.

Then, an example is given to explain the GAVO method in detail. Assuming that there are two knapsacks with a weight limit of $\{5, 8\}$ (that is $m_1 = 5$, $m_2 = 8$), seven items $\{n_1, n_2, n_3, n_4, n_5, n_6, n_7\}$ with a value of $\{p_1, p_2, p_3, p_4, p_5, p_6, p_7\}$:$\{10, 5, 2, 6, 4, 5, 4\}$, the weight of these seven items $\{w_1, w_2, w_3, w_4, w_5, w_6, w_7\}$ is $\{5, 1, 3, 4, 2, 1, 3\}$. For the first knapsack, the placed item is n_1, that is, the set of items is $M_1 = \{n_1\}$. For the second knapsack, the placed items are n_2 and n_6, that is, the set of items is $M_2 = \{n_2, n_6\}$. The remaining items that have not been placed in any knapsack is $Rem = \{n_3, n_4, n_5, n_7\}$, and solution $x = [1, 2, 0, 0, 0, 2, 0]$.

First, the ratio of value and weight of the items is calculated and sorted in order from high to low, and the result $Rem = \{n_5, n_4, n_7, n_3\}$. In current, there is no remaining space in the first knapsack, and the remaining space in the second knapsack is 6. According to the sorting results, we first try to put item n_5 into the second knapsack, because the weight of n_5 ($w_5 = 2$) is less than the remaining space 6, so it can be put into the second knapsack. At this point, the remaining space changes to 4. Then, item n_4 is attempted to be put into the second knapsack, because the weight of n_4 ($w_4 = 4$) is equal to the remaining space, so it also can be placed into the knapsack. At this point, there is no space left in the second knapsack, so no items can be further placed. The final result becomes $M_1 = \{n_1\}$, $M_2 = \{n_2, n_6, n_5, n_4\}$, $Rem = \{n_7, n_3\}$, and the solution $x = [1, 2, 0, 2, 2, 2, 0]$.

3.4 MKBA-GA Algorithm and MKBA-SRT Algorithm

To improve the quality of MKP, we adjust the location update mode of the bat based on the IBBA algorithm, and then the effective solution of the algorithm is optimized combined with the GAVO method, thereby proposing a greedy algorithm-based bat algorithm called MKBA-GA for solving the MKP.

The main idea of MKBA-GA algorithm is shown as follows:

Step 1: Initialize the parameters of bat population, including: the total number of bats, the maximum value and minimum value of the loudness, the maximum value and minimum value of the pulse frequency, the maximum number of iterations of the algorithm, etc.

Step 2: Initializes the speed, position, pulse emission frequency, pulse frequency, and loudness of each bat, as well as other parameters. When initializing the position of each bat, a random allocation method is used. In this step, the probability that each item is divided into each knapsack is the same as the probability that it is not be placed in the knapsack.

Step 3: Perform an iteration operation, where a random number between $(0, 1)$ is obtained first. If the random number is greater than its own pulse frequency, the bat performs a local search; otherwise, the bat performs a global search.

Step 4: Determine whether the newly obtained solution is an effective solution. If the new solution is an effective solution, jump to the next step directly; otherwise, the effective solution transformation of the invalid solution is carried out first (the items with the lowest current value and quality are removed from the knapsack until they are transformed into effective solutions), and then the newly obtained effective solution is optimized by using the GAVO method.

Step 5: Determine whether the quality of the new solution is higher than the quality of the current global optimal solution. If it is higher than the global optimal solution, the new solution is retained and its loudness and pulse emission frequency are updated; otherwise, go to the next step directly.

Step 6: Update the global optimal solution and determine whether the maximum number of iterations is reached. If the maximum number of iterations is reached, perform the next step directly; otherwise, go to step 3 to perform the next iteration.

Step 7: Output global optimal solution.

Algorithm 2: MKBA-GA

Input: the max capacities of m knapsacks, the value and weight of n items

Output: optimal solution ($f(G_{best})$)

01.initialize parameters //such as bat_num, $iter\_max$, f_{max}, f_{min}, A, $Rmax$
02.**for** i=1 to bat_num **do**
03.   set the position of bat i, $x_i \in [0,m]$
04.   set the f_i of bat i
05. $A \to A_i$, $Rmax \to r_i$
06.   **if** $x_i \in$ invalid solution **then**
07.   optimize the solution x_i by GAVO // or by SRT
08.   **end if**
09.**end for**
10.**if** $iter<iter\_max$ **then**
11.   **for** i=1 to bat_num **do**
12.   adjust the f_i of bat i
13.   get x_{new} according to formula (8)
14.   update the v_i of bat i
15.   **if** rand(0,1)>r_i **then**
16.   change some dimensions of x_{new} according to position of G_{best}
17.   **end if**
18.   **if** $x_{new} \in$ invalid solution **then**
19.   optimize the solution x_{new} by GAVO // or by SRT
20.   **end if**
21.   **if** rand(0,1)<A_i and $f(x_{new})$>$f(G_{best})$ **then**
22. $x_{new} \to x_i$
23.   update A_i and r_i
24.   **end if**
25.   **end for**
26.   update G_{best}
27.   ($iter$+1)$\to iter$
28.**end if**
29.go to 10
30.output fitness(G_{best})

The pseudo-code of the MKBA-GA algorithm is shown in Algorithm 2, where $f(x)$ is an adaptive function of the knapsack problem, it indicates the total value of all items that placed in the knapsack currently; G_{best} represents the global optimal solution;

bat_num represents the number of bats in the bat population; *iter* represents the current number of iterations; *iter_max* represents the maximum number of iterations of the algorithm; f_{max} represents the upper limit of the pulse frequency; f_{min} represents the lower limit of the pulse frequency; A represents the initial loudness value, A_i represents the loudness value of ith bat; *Rmax* represents the initial pulse emission frequency; x_i represents the position of ith bat; f_i represents the pulse frequency of ith bat; r_i represents the pulse emission frequency of ith bat; v_i represents the speed of ith bat.

For optimizing the effective solution, the main problem that is solved by GAVO is to improve the space utilization rate of the knapsack, so its optimization effect is still very limited. To further improve the quality of the effective solution, the solution can be optimized by the SRT method [3]. In other words, by combining the basic bat algorithm with the SRT method, we propose a SRT-based bat algorithm, namely MKBA-SRT, for finding an approximate solution that is closer to the global better solution. The framework of MKBA-SRT algorithm is basically consistent with MKBA-GA algorithm. Note that, the difference of MKBA-SRT algorithm and MKBA-GA algorithm is only that the optimization method of the effective solution. That is, in the MKBA-SRT algorithm, the SRT method is used to find an approximate solution instead of GAVO method. Compared with MKBA-GA algorithm, the differences of the pseudo-code of MKBA-SRT algorithm are only in lines 7 and 19.

4 Experimental Results

To prove the performance of the proposed MKBA-GA algorithm and MKBA-SRT algorithm, we compared it with three classical ant colony optimization algorithms, including BBA [13], IRT [3], SRT [3]. In addition, 12 public datasets[1] are used in the experiment, and according to the correlation between the value and the weight, the 12 datasets [5] can be divided into three categories: strong correlation, weak correlation and no correlation, where the characteristics of the datasets can be found in literature 5. In this experiment, the number of each category of the datasets we selected is four.

All algorithms compared in the experiment are implemented using the C++ programming language in Microsoft Visual C++ 6.0, and they are running on a machine of Intel i7-6700 K 4.0 GHZ, 16 GB memory. In the experiments, we run 30 independent experiments for each algorithm to reduce the influence caused by the randomness of the algorithms. The parameters of each algorithm are set as follows:

(1) BBA algorithm: *bat_num* = 30, $a = 0.25$, *Rmax* = 0.5, f_{max}= 2, f_{min}= 0, *iter_max* = 200;
(2) MKBA-GA and MKBA-SRT algorithms: *bat_num* = 30, $a = 0.25$, *Rmax* = 0.5, f_{max}= 2, f_{min}= 0, δ_{init} = 0.6, w_{max}= 0.9, q_{max}= 50, $M = 50$, $N = 2$, *iter_max* = 200.

---

[1] https://github.com/DBEngine/MKP.

4.1   Experimental Results on No Correlated Datasets

In this subsection, we use four no correlated datasets to test the efficiency of the proposed MKBA-GA algorithm and MKBA-SRT algorithm, where the number of knapsacks is selected from 5 or 50, the number of items is selected from 200 or 500, and the value of R (R is a generation range of the value and the weight) is set to 100. The experimental result is shown in Fig. 1(a) to (d), where the x-axis represents the number of the experiments, and the y-axis represents the total values of the items that putted in all knapsacks.

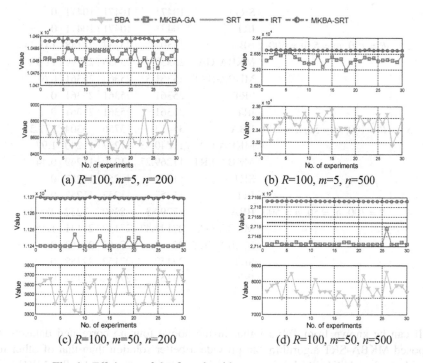

(a) R=100, m=5, n=200

(b) R=100, m=5, n=500

(c) R=100, m=50, n=200

(d) R=100, m=50, n=500

Fig. 1. Efficiency of the five algorithms on no correlated datasets

It can be seen from Fig. 1(a) to (d) that the efficiency of the proposed MKBA-GA algorithm performs better than that of BBA algorithm, but compared with IRT algorithm and SRT algorithm, the efficiency of the MKBA-GA algorithm only performs better in the first dataset. In addition, the efficiency of the proposed MKBA-SRT algorithm is much better than that of other four compared algorithms. That is, in all four no correlation datasets, the solutions found by MKBA-SRT algorithm is always more near to the global optimal solution.

To explain the effect of the MKBA-GA and MKBA-SRT algorithms more clearly, the experimental results of the five compared algorithms after running for 30 times are organized, and then the average and standard difference that can represent the efficiency stability of the algorithm are calculated, the results are shown in Table 1. In Table 1,

$S_{average}$ represents the average value of the 30 experiments, S_{max} represents the optimal solution of the 30 experiments, S_{min} represents the worst solution of the 30 experiments, $S_{variance}$ represents the standard deviation of the 30 experiments.

Table 1. Comparisons of the five algorithms on no correlated datasets

No.	Parameters	Algorithms	$S_{average}$	S_{max}	S_{min}	$S_{variance}$
1	$R = 100, m = 5, n = 200$	BBA	8592.03	8928	8428	109.51
		MKBA-GA	10481.73	10485	10477	2.21
		MKBA-SRT	10488.47	10489	10488	0.5
		SRT	10471	10471	10471	0
		IRT	10471	10471	10471	0
2	$R = 100, m = 5, n = 500$	BBA	23507.7	23737	23148	143.24
		MKBA-GA	25332.4	25355	25299	11.99
		MKBA-SRT	25361.33	25363	25361	0.6
		SRT	25361	25361	25361	0
		IRT	25361	25361	25361	0
3	$R = 100, m = 50, n = 200$	BBA	3541.93	3756	3304	137.2
		MKBA-GA	11240.83	11247	11240	1.97
		MKBA-SRT	11269.27	11270	11269	0.44
		SRT	11257	11257	11257	0
		IRT	11257	11257	11257	0
4	$R = 100, m = 50, n = 500$	BBA	7729.83	8300	7232	256.28
		MKBA-GA	27141.47	27149	27141	1.45
		MKBA-SRT	27163	27163	27163	0
		SRT	27152	27152	27152	0
		IRT	27152	27152	27152	0

It can be known from Table 1 that in the above four no correlated datasets, the proposed MKBA-SRT algorithm can provide a better solution than that of other four algorithms. In addition to the first dataset, the solution provided by the MKBA-GA algorithm is worse than that of IRT and SRT algorithms. But in the above 4 datasets, the solution stability of the proposed MKBA-GA algorithm is better than that of BBA algorithm.

4.2 Experimental Results on Weak Correlated Datasets

In this subsection, we use four weak correlated datasets to test the efficiency of the proposed MKBA-GA algorithm and MKBA-SRT algorithm, and the experimental result is shown in Fig. 2(a) to (d).

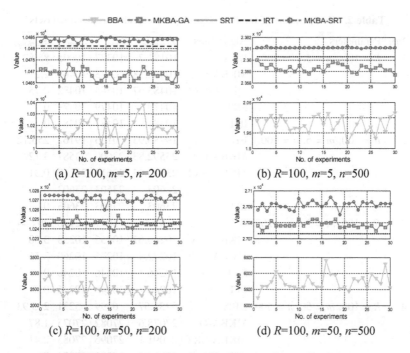

Fig. 2. Efficiency of the five algorithms on weak correlated datasets

It can be seen from Fig. 2(a) to (d) that the ability of the proposed MKBA-SRT algorithm for seeking for the solutions outstands than that of the compared MKBA-GA, BBA, IRT and SRT algorithms. On the third dataset, the performance of the proposed MKBA-GA algorithm is almost consistent with that of the IRT and SRT algorithms, and the performance of the MKBA-GA algorithm is better than that of the IRT and SRT algorithms on the fourth dataset. Compared with BBA algorithm, the performance of MKBA-GA algorithm is much better on all four datasets.

Then, the experimental results of the five compared algorithms after running for 30 times are organized, and the average and standard difference that can represent the efficiency stability of the algorithm are calculated and shown in Table 2.

It can be seen from Table 2 that on the four weak correlated datasets, the solutions solved by the proposed MKBA-SRT algorithm is always the best, and the results are also stable (only on the fourth dataset, the standard deviation of MKBA-SRT is slightly higher than that of MKBA-GA). The solutions solved by MKBA-GA algorithm and its stability are worse than that of MKBA-SRT algorithm, but they are better than that of BBA algorithm. In addition, the solutions solved by MKBA-GA algorithm are worse than that of IRT and SRT algorithms. Specifically, from the view of mean value, maximum value and minimum value, MKBA-GA performs better than IRT and SRT only on the fourth dataset, and the standard deviation of the solutions solved by MKBA-GA algorithm is very small on all four datasets, and on the fourth dataset, the standard deviation of the MKBA-GA algorithm is lower than that of MKBA-SRT algorithm.

Table 2. Comparisons of the five algorithms on weak correlated datasets

No.	Parameters	Algorithms	$S_{average}$	S_{max}	S_{min}	$S_{variance}$
1	$R = 100, m = 5, n = 200$	BBA	10181.7	10377	10005	85.87
		MKBA-GA	10468.53	10473	10465	2.2
		MKBA-SRT	10483.67	10485	10482	0.75
		SRT	10479	10479	10479	0
		IRT	10481	10481	10481	0
2	$R = 100, m = 5, n = 500$	BBA	19783.93	20156	19162	255.72
		MKBA-GA	23592.83	23600	23587	3.12
		MKBA-SRT	23611.03	23612	23610	0.31
		SRT	23603	23603	23603	0
		IRT	23603	23603	23603	0
3	$R = 100, m = 50, n = 200$	BBA	2502.4	3041	2176	202.44
		MKBA-GA	10245.67	10256	10238	3.88
		MKBA-SRT	10272.4	10275	10260	3.56
		SRT	10246	10246	10246	0
		IRT	10246	10246	10246	0
4	$R = 100, m = 50, n = 500$	BBA	5731.57	6399	5231	245.93
		MKBA-GA	27078.6	27082	27075	1.87
		MKBA-SRT	27091.1	27096	27085	2.41
		SRT	27073	27073	27073	0
		IRT	27073	27073	27073	0

4.3   Experimental Results on Strong Correlated Datasets

In this subsection, we use four strong correlated datasets to test the efficiency of the MKBA-GA algorithm and MKBA-SRT algorithm, and the experimental result is shown in Fig. 3(a) to (d).

It can be seen from Fig. 3(a) to (d) that on the four strong correlated datasets, the solutions solved by MKBA-SRT algorithm are better than that of BBA algorithm, while on the three datasets, the solutions solved by MKBA-SRT algorithm are better than that of MKBA-GA, SRT and IRT algorithms. Specifically, on the second dataset, the solutions solved by MKBA-SRT are worse than that of IRT, SRT and MKBA-GA algorithms. Compared with BBA, the solutions solved by MKBA-GA are more advantaged. Compared with IRT, the solutions solved by MKBA-GA are basically the same only on the first dataset, while worse on other three datasets. Compared with SRT, the solutions solved by MKBA-GA algorithm are much better on the first dataset, while on the second and third datasets, the effect of SRT and MKBA-GA algorithms is essentially the same, but on the fourth dataset, the solutions solved by MKBA-GA are worse than that of SRT.

Then, on the four strong correlated datasets, the experimental results of the five compared algorithms after running for 30 times are organized, and the average and standard difference that can represent the efficiency stability of the algorithm are calculated and shown in Table 3.

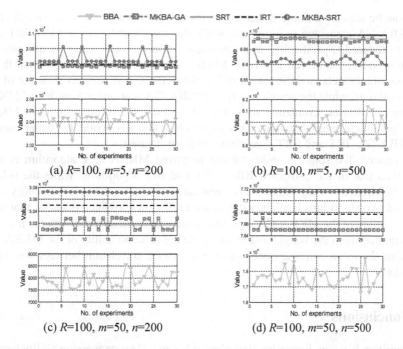

Fig. 3. Efficiency of the five algorithms on strong correlated datasets

Table 3. Comparisons of the five algorithms on strong correlated datasets

No.	Parameters	Algorithms	$S_{average}$	S_{max}	S_{min}	$S_{variance}$
1	$R = 100, m = 5, n = 200$	BBA	20415.67	20544	20238	72.27
		MKBA-GA	20788.47	20806	20775	8.17
		MKBA-SRT	20833.23	20917	20817	36.31
		SRT	20719	20719	20719	0
		IRT	20793	20793	20793	0
2	$R = 100, m = 5, n = 500$	BBA	59451.8	61406	58348	777.14
		MKBA-GA	66796.87	66871	66741	49.56
		MKBA-SRT	66134.2	66491	66088	120.83
		SRT	66858	66858	66858	0
		IRT	66978	66978	66978	0
3	$R = 100, m = 50, n = 200$	BBA	7921	8557	7404	321.43
		MKBA-GA	30189.07	30292	30090	94.27
		MKBA-SRT	30719.67	30730	30711	5.14
		SRT	30176	30176	30176	0
		IRT	30499	30499	30499	0
4	$R = 100, m = 50, n = 500$	BBA	17592.7	18845	16638	543.89
		MKBA-GA	76507.63	76700	76501	35.72
		MKBA-SRT	77156.83	77158	77152	1.49
		SRT	76644	76644	76644	0
		IRT	76770	76770	76770	0

It can be seen from Table 3 that on the strong correlated datasets, the search ability of MKBA-GA algorithm is improved, while the search ability of MKBA-SRT algorithm is reduced. Specifically, from the view of mean value, maximum value and minimum value, the search ability of MKBA-GA algorithm is stronger than that of BBA on four datasets, and the search ability of MKBA-GA is stronger than that of SRT on two datasets, while the search ability of MKBA-GA is stronger than that of MKBA-SRT on one dataset. Compared with other algorithms, the solution effect of MKBA-SRT algorithm is better than that of BBA on all four datasets, while it is better than that of MKBA-GA, IRT and SRT algorithms on three datasets.

In general, the solution ability of our proposed MKBA-SRT algorithm is much better than that of the compared BBA, SRT and IRT algorithms, while the solution ability of MKBA-GA algorithm is very general. In addition, the solution ability of the proposed two algorithms is slightly different on different correlated datasets, that is, the solution ability of the MKBA-GA algorithm shows an increase trend with the increased correlation between the weight and value, while the solution ability of the MKBA-SRT algorithm shows an increase trend with the decreased correlation between the weight and value.

5   Conclusions

Bat algorithm is a new imprecise algorithm, it has excellent conversion ability between global search to local search and its high robustness. In this paper, we propose two improved bat algorithms, namely MKBA-GA and MKBA-SRT, for solving the solutions of MKP. To solve the problems of excessive number of solution spaces in the problem model and the easily occurrence of illegal solutions, we discard the original 0–1 encoding mode and use the multi-value encoding to re-model for the MKP. To solve the problem that the decreased search ability of the algorithm caused by the low quality of effective solution, we propose a GAVO method to optimize the effective solution, so as to improve the quality of the solution. Then, we propose the MKBA-GA algorithm by combining the GAVO method and the improved bat algorithm. To further improve the search efficiency of the algorithm, we adopt another efficient solution optimization method SRT to optimize the low quality effective solution in the algorithm, and then put forward the MKBA-SRT algorithm. And then, we use twelve datasets to verify the validity of the proposed two algorithms, where three algorithms (including BBA, IRT, SRT) are compared in the experiments. The experimental results show that the solution ability and stability of MKBA-SRT algorithm are the strongest, while the effect of MKBA-GA algorithm is general.

At present, the initial solution construction method of MKBA-GA algorithm and MKBA-SRT algorithm adopts stochastic method, so the quality of initial solution cannot be guaranteed. In addition, the optimization of the low quality effective solution has a great influence on the overall search performance of the algorithm. In view of the above problems, in the future, we will continue to explore the new initial solution construction method, as well as invalid solution and effective solution optimization method, for further improving the search efficiency of the algorithm.

References

1. Kellerer, H., Pferschy, U., Pisinger, D.: Knapsack Problems. Springer, Berlin (2004). https://doi.org/10.1007/978-3-540-24777-7
2. Sitarz, S.: Multiple criteria dynamic programming and multiple knapsack problem. Appl. Math. Comput. **228**, 598–605 (2014)
3. Laalaoui, Y.: Improved swap heuristic for the multiple knapsack problem. In: Rojas, I., Joya, G., Gabestany, J. (eds.) IWANN 2013. LNCS, vol. 7902, pp. 547–555. Springer, Heidelberg (2013). https://doi.org/10.1007/978-3-642-38679-4_55
4. Xiong, W., Wei, P., Jiang, B.: Binary ant colony algorithm with congestion control strategy for the 0/1 multiple knapsack problems. In: 8th World Congress on Intelligent Control and Automation, pp. 3296–3301. IEEE, Jinan (2010)
5. Liu, Q., Odaka, T., Kuroiwa, J., Shirai, H., Ogura, H.: A new artificial fish swarm algorithm for the multiple knapsack problem. IEICE Trans. Inf. Syst. D **3**, 455–468 (2014)
6. Fukunaga, A., Tazoe, S.: Combining multiple representations in a genetic algorithm for the multiple knapsack problem. In: 2009 IEEE Congress on Evolutionary Computation, pp. 2423+. IEEE, Trondheim (2009)
7. Rizk-Allah, R., Hassanien, A.: New binary bat algorithm for solving 0–1 knapsack problem. Complex Intell. Syst. **4**(1), 31–53 (2017)
8. Bangyal, W., Ahmad, J., Rauf, H.: Optimization of neural network using improved bat algorithm for data classification. J. Med. Imaging Health Inform. **9**(4), 670–681 (2019)
9. Huang, L., Wang, P., Liu, A., Nan, X., Jiao, L., Guo, L.: Indoor three-dimensional high-precision positioning system with bat algorithm based on visible light communication. Appl. Opt. **58**(9), 2226–2234 (2019)
10. Fusai, D.: The application research of improved bat algorithm based on chaos for job shop scheduling. In: 3rd Workshop on Advanced Research and Technology in Industry Applications, Guilin, China, pp. 353–356 (2017)
11. Gan, C., Cao, W., Wu, M., Chen, X.: A new bat algorithm based on iterative local search and stochastic inertia weight. Expert Syst. Appl. **104**, 202–212 (2018)
12. Chakri, A., Khelif, R., Benouaret, M., Yang, X.: New directional bat algorithm for continuous optimization problems. Expert Syst. Appl. **69**, 159–175 (2017)
13. Mirjalili, S., Mirjalili, S.M., Yang, X.: Binary bat algorithm. Neural Comput. Appl. **25**(3–4), 663–681 (2014)
14. Zhou, Y., Bao, Z., Luo, Q., Zhang, S.: A complex-valued encoding wind driven optimization for the 0–1 knapsack problem. Appl. Intell. **46**(3), 684–702 (2017)
15. Huang, X., Zeng, X., Han, R.: Dynamic inertia weight binary bat algorithm with neighborhood search. Comput. Intell. Neurosci. **8**, 1–15 (2017)
16. Pisinger, D.: An exact algorithm for large multiple knapsack problems. Eur. J. Oper. Res. **114**(3), 528–541 (1999)

Customer Value Analysis Method Based on Automotive Multi-value-Chain Collaboration

Wen Bo[1,2], Yufang Sun[3], Xiaobo Hu[4], Qian Xu[5],
and Changyou Zhang[1(✉)]

[1] Laboratory of Parallel Software and Computational Science,
Institute of Software, Chinese Academy of Sciences, Beijing,
People's Republic of China
changyou@iscas.ac.cn

[2] School of Computer and Information Technology, Shanxi University, Taiyuan,
Shanxi, People's Republic of China

[3] School of Computer Engineering, Chengdu Technological University,
Chengdu, Sichuan, People's Republic of China

[4] Chengdu Guolong Information Engineering Co., Ltd., Chengdu, Sichuan,
People's Republic of China

[5] School of Information Science and Technology,
Southwest Jiaotong University, Chengdu, Sichuan, People's Republic of China

Abstract. From the view point of the automotive multi-value-chain in collaborative businesses, different customers will produce unequal values in the future services. To strive for the continuous contribution of high-value customers, and optimize the collaborative business efficiency in parts supplier value chain and service multi-value-chain, in this paper, we will mention a customer analysis method for service multi-value-chain in automotive industry. Firstly, according to the related data in our automotive service collaborative business platform, we selected some evaluation indicators to describe the customer's current value and potential value. Then, a customer value analysis model was offered based on a multi-classification SVM, which would give a classification recommendation according to the potential value of customs. Finally, we randomly selected 40 customers' 36 months related information from our platform, to verify our method by some experiments. The experimental result shows that the customer classification was close to the actual situation. Among the 40 testing samples, the accuracy rate of customer value classification reaches to 87.5%. The results of customer value forecasts can be used to guide parts supplier and service providers to make a more effective policy, such as giving specific offers for outstanding customers. It will expand the synergy of multi-value-chain outside the Three-guarantee period of serving period of automobile service, and realize the strategic management of service life cycle centered on customer value.

Keywords: Automotive service · Multi-value-chain · Customer analysis · Support vector machine · Collaboration

© Springer Nature Singapore Pte Ltd. 2019
Y. Sun et al. (Eds.): ChineseCSCW 2019, CCIS 1042, pp. 158–169, 2019.
https://doi.org/10.1007/978-981-15-1377-0_12

1 Introduction

In recent years, Chinese auto industries have a rapid development. The study found that loyal customers increase every 5% then the profit will increase about 25% to 85%. And about the cost of winning, a new customer is five times of an existing customer. In an increasingly competitive environment, customers have become the most important strategic resource for the company to enhance its competitiveness. Frederick, a representative of customer value research, proposed and improved the net present value evaluation system for customer value [1, 2]. Conway and other researchers believe that customer value reflects the customer's current contribution to corporate value, directly equating customer value with customer profit, and has not considered customer loyalty [3, 4]. At the same time, Aljawarneh et al. proposed to use a voting algorithm with information gain to filter the data to select important features of the model [5].

Domestic scholars have also implemented a series of studies around customer value. Liu et al. mentioned a customer value analysis model based on dual attributes from the perspective of enterprise attributes and customer attributes [6–8]. Xia analyzed customer value from the three dimensions of customer's current value, customer potential value and customer loyalty, and proposed customer value analysis system and customer retention plan with enterprise [9].

In summary, customer value is an important part of the strategic management of automotive enterprises. The basis of strategic management is the business collaboration around customer value. This paper proposes customer value as the core and differentiates customer value. From the perspective of marketing value chain and service value chain, this paper proposes customer value model. We use SVM to classify customers, predict customer value. The model promotes marketing value chain and optimizes the service multi-value-chain of the automotive industry.

2 Related Works

In order to help enterprises rationally allocate service resources and enhance competitiveness, this paper studies the customer value classification method of enterprise groups. It is used for the multi-value-chain. The customer value classification can be classified as the data mining and it already had accumulated a lot of research results [10].

2.1 Construction of Indicator System

We combined the actual situation of the customer value theory and the multi-value-chain. We studied from the service multi-value-chain, the existing data in the system database and established detailed indicators of customer value analysis are established. Customer current value includes: Vehicle purchase quantity, Total number of services, Total service cost, Purchase type, Average service period, Purchase average cycle, Total number of replacement parts, Education, Gender. And Customer potential value includes: Number of suggestions, Repeat purchases, Service satisfaction, Major failures, Age.

2.2   Data Preprocessing

To First step of the data preprocessing is data cleansing. Eliminating invalid data: The service data is derived from the sales system and the after-sales service system. In the actual business collaboration, there are customers who buy vehicles in the enterprise but don't service in the enterprise. Then processing the missing data: In the action of sales and after-sales service, some original business documents are not verified. During the business process, the operators may don't fill in the actual value, that will result in the lack of data. Finally, correcting the error data: The service personnel inevitably enter the error information during the service process. We can be modified according to the average value.

2.3   Classification Algorithm for Customer Value

The traditional classification algorithms mainly include: Support vector machine, Bayes Rule, Decision tree, neural networks and so on. The recently emerging classification algorithms include deep learning and Graph neural network methods. In general, the traditional algorithm requires relatively small amount of training samples, and the computational power requirements of the underlying hardware are relatively low. For systems with limited computational conditions, solving the problem of suitable scale has unique advantages.

SVM is a learning method based on statistical learning theory and structural risk minimization criterion [11, 12]. The disadvantages are: sensitive to missing data, no general solution to nonlinear problems.

Bayes Rule disadvantage: classification efficiency is poor when the number of attributes is large and the correlation is complex. The decision tree algorithm disadvantages are: there are over-fitting problems, difficulties in dealing with missing data. The neural network disadvantage is that it requires a large number of parameters, a large amount of training data and a long learning time.

The deep learning is based on deep neural networks. It combines low-level features to form more abstract high-level features to discover distributed feature representations of data [13]. For classification problems, deep learning acquires features using unsupervised or semi-supervised feature learning and hierarchical feature extraction algorithms. Recently, more research on graph neural networks (GNN) is a kind of method based on deep learning to process graph information. For graph-based systems such as social networks and chemical molecular structures, they exhibit powerful description capabilities.

2.4   Selection of Customer Value Classification Algorithms

As we mentioned before, SVM is suitable for small sample problems and has advantages in solving nonlinear and high dimensional problems [14, 15].

The number of experimental samples obtained in this paper is small, and the value chain collaborative platform works on information management. It doesn't have enough computing power of large data. Therefore, this paper chooses SVM as the classification algorithm.

3   Customer Potential Value Analysis Models and Methods

3.1   Customer Value Classification

Combine the indicator system construction and the actual situation of the multi-value-chain in the previous chapter. First, we will describe two aspects of the customer value: the current value of the customer and the potential value of the customer.

According to the customer value norms and the existing research findings in the previous part, we construct a customer value rating matrix.

(1) High-value customer. They have large number of vehicles and contribute over half of benefits to those corporations who service the customer. Therefore, it is defined as the high-value customer.

(2) Risk customer. The ratio of profit contribution to business is not high at present. However, there are more opportunities in future to sale. This type of customer has considerable room for profit.

(3) Marginal customers. These customers have high profit contribution but have low earnings in future. This type of customer is currently of considerable value and needs to maintain the existing service level, but does not invest too many resources.

(4) Low-value customer. This type of customer has less profit contribution and business volume. There is a lack of value to be tapped in the future. This type of customer has a low loyalty and the serious churn rate.

3.2   Customer Value Classification

Customer value classification is the process of defining customer value. This paper divides the customer value into 4 levels: high-value, risk, margin and low-value. They correspond to the first, second, third and fourth customers respectively.

3.3   Customer Value Analysis Model Based on Multi-classified SVM

In response to the customer value theory proposed above. This section will use a multi-classified SVM to customer value. The customer value classification is completed by three steps: data discretization, kernel function selection and design based on multi-classified SVM.

Data Preparation. First we discretize the data, and the processing of discretized data is not detailed here. The results of the discretized customer value analysis indicators are shown in Table 1. Repeat purchase, total service cost, total number of services and vehicle purchase quantity are actual value.

Table 1. Discretization results of customer value analysis indicators

Index name	Selection value
Purchase type	1: [0–1] 2: [2, 3] 3: [4, 5] 4: [6–∞]
Average service period	1: Within 1 month: Within the first quarter 3: Within half a year 4: Within one year 5: More than one year
Purchase average cycle	1: Within one year 2: Within two years 3: Within three years 4: More than three years
Total number of replacement parts	1: [0–2] 2: [3–5] 3: [6–8] 4: [9–∞]
Education	1: Elementary school and below 2: Middle school 3: High school 4: College and above
Gender	1: male 2: female
Number of suggestions	1: [0–2] 2: [3–5] 3: [6–8] 4: [9–∞]
Service satisfaction	1: Very satisfied 2: Satisfied 3: General 4: Displeasure 5: Very displeasure
Major failures	1: [0–2] 2: [3–5] 3: [6–8] 4: [9–∞]
Age	1: [18–25] 2: [26–35] 3: [36–50] 4: [50–65]

SVM Framework Design. Let the training set be denoted as $\{(x_1, y_1), (x_2, y_2), \ldots, (x_n, y_n)\}$, where $x_i = (x_{i1}, x_{i2}, \ldots, x_{ir})$ is an r-dimensional input vector, y_i denotes its corresponding category label, and $y_i \in \{1, -1\}$ denotes a positive sample and a negative sample. The classification plane in the two-dimensional space can be expressed as shown in Eq. 1.

$$w \cdot x + b = 0 \tag{1}$$

To find the optimal classification hyperplane, the classification interval needs to be maximized, and we need to define the function: $f(x) = w \cdot x + b$. The distance from the sample point to the hyperplane is as shown in Eq. 2.

$$S = \frac{1}{||w||} |f(x_i)| \tag{2}$$

$||w||$ is the second-order norm of w. In order to maximize the distance S. it can be equivalent to finding the minimum value of $||w||$, and is equivalent to finding the minimum value of $||w||^2/2$. That is, to make the classification interval maximum, it needs to be solved as Eq. 3.

$$\begin{cases} \min \frac{1}{2} ||w||^2 \\ \text{s.t. } y_i(x_i \cdot w + b) \geq 1, i = 1, 2, \ldots, N \end{cases} \tag{3}$$

That is, to make the classification interval maximum. By introducing the Lagrange multiplier a_i, the minimum value of the Lagrange function corresponds to the minimum value of w and b, they are considered 0, which is shown in Eq. 4.

$$\begin{cases} \frac{\partial L}{\partial w} = w - \sum_{i=1}^{N} a_i y_i x_i = 0 \\ \frac{\partial L}{\partial b} = -\sum_{i=1}^{N} a_i y_i = 0 \end{cases} \tag{4}$$

The Lagrange optimization method is used to convert the problem into its dual problem, and the quadratic programming problem in the form of Eq. 5 is obtained.

$$\begin{cases} \max \sum_{i=1}^{N} a_i - \frac{1}{2} \sum_{i,j=1}^{N} a_i a_j y_i y_j (x_i \cdot x_j) \\ \text{s.t.} \sum_{i=1}^{N} a_i y_i = 0, \ a_i \geq 0, i = 1, 2, \ldots, N \end{cases} \tag{5}$$

Equations 6 can be obtained by Eq. 5.

$$\begin{cases} w^* = \sum_{i=1}^{N} a_i y_i x_i \\ b^* = -\frac{\max_{y_i=-1}(w^* \cdot x_i) + \min_{y_i=1}(w^* \cdot x_i)}{2} \end{cases} \tag{6}$$

The resulting optimal classification function is shown in Eq. 7.

$$f(x) = \text{sign}\{(w^* \cdot x) + b\} = \text{sign}\{\sum_{i=1}^{N} a_i^* y_i(x_i \cdot x) + b^*\} \tag{7}$$

Select Kernel Function. The choice of kernel function is critical to the performance of the model, especially for linearly inseparable data. Commonly used non-custom kernel functions are: linear, polynomial, RBF and sigmoid.

The reason is that the number of hyperparameters affects the complexity of the model selection, while the polynomial kernel function has more hyperparameters than the RBF kernel function.

In addition, according to Vapnik's research, the sigmoid kernel is invalid under some parameters (it's not the inner product of two vectors). This paper intends to use the RBF kernel function, as shown in Eq. 8.

$$K(x_i \cdot x_j) = \exp\left(-\gamma \|x_i - x_j\|^2\right), \gamma > 0 \tag{8}$$

Optimal Parameters of the Kernel Function. According to the above analysis, the RBF function is determined as the kernel function of the SVM model, and it needs to determine two very important parameters C and gamma. Where C is the penalty factor, which is the tolerance for the error. The higher the C, the more the over-fitting is easy.

Gamma is a parameter of the RBF kernel function. The larger the gamma is, the smaller the support vector is. The number of support vectors affects the speed of training and prediction. This article uses "grid search" to find the optimal C and g. The so-called grid search is to try out the various possible (C, g) pairs of values, and then cross-validate to find the (C, g) pairs that the most accurate.

SVM Multi-classified Design. The original intention of SVM is to solve the dichotomy problem. It is a typical two-class classifier. In this paper, the customer value levels are divided into 4 categories, so when dealing with the multi-classified problems, we need to construct a suitable multi-class classifier. This paper uses one-to-one method for multi-classification, which is to design an SVM between any two types of samples. So, k categories of samples need to design k (k − 1)/2 classifiers. In the form of voting, the voting process is shown in Fig. 1. If it is classified as Class A, then Class A plus 1, belonging to Class B, then Class B voting plus 1, and so on.

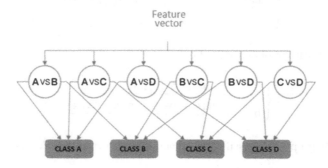

Fig. 1. One-to-one multi-classified voting

Accumulate the scores of each category, select the category with the highest score as the test data category, and finally get a set of results, such as Eq. 9.

$$y = \text{sgn}\left(W_{ij}^{T}\emptyset(x) + b_{ij}\right) \tag{9}$$

4   Verification of the Multi-value-Chain Collaboration

4.1   Customer Value Experiment

In the test verification phase, this paper randomly extracted the basic information service records and consumption records of 400 customers from the data of the automobile industry chain collaborative platform. For these customers, this paper refers to the opinion of service auditor and classifies the customer's value of the sample data. This article lists the 40 data of the test sample. See Table 2 for details.

Table 2. Customer value analysis test sample

BB	SN	SA	CB	AS	AB	AC	E	S	CT	RB	TM	FT	A	Label
1	11	1111.52	1	3	4	1	1	1	1	0	2	1	3	4
1	5	496.82	1	3	4	2	2	1	1	0	2	1	3	4
1	16	1605.60	1	2	4	2	2	1	1	0	2	1	3	4
2	21	12457.12	1	3	4	2	1	1	2	1	3	2	3	3

After testing, the paper uses the model for actual customer value classification prediction. It was finally determined that the RBF kernel function parameter G = 0.89, and the penalty parameter C = 108.4, the accuracy of 40 test samples reached a maximum of 87.5%.

4.2 Analysis of Results

The classification of 40 test samples is shown in Table 3. The experimental results show that the number of primary and secondary customers is small, and the number of low-value level customer is large. According to the current situation of most automobile companies in China, the experimental results are in line.

Table 3. Sample data analysis

Project	High-value	Risk	Margin	Low-value
Actual sample	5	3	11	21
Correct sample	4	3	7	21

5 The Multi-value-Chain Collaboration

In the process of multi-value-chain collaboration, different customer groups are distinguished, so that high-value customers feel the existence of differentiated services, thus creating more benefits for the enterprise.

Taking the standard ¥200,000-¥300,000 cars as an example, the service item pricing, cost price, and income are shown in Table 4.

Table 4. Service standards

Service/once	Price	Cost	Times/year
Whole insurance	8890	4000	1
Whole maintenance	1500	500	2
Oil	450	120	2
Lacquer	800	350	Uncertain
Wash	50	5	Uncertain

5.1   Multi-value-Chain Collaboration

For the collaboration between the parts supply chain and the service multi-value-chain designing differentiated services will optimizes multiple value chains. Therefore, according to the information in the platform, the differentiated services for the automotive enterprise group are designed, as shown in Fig. 2.

In the collaboration of the parts supply chain and the service value chain A, customers can choose the parts that vehicles are outside three-guarantee period and the service provider submits the demand and the order to the parts supplier. For example, most of the high-level customers and risk customers choose the original factory parts are free when cars maintenance. The synergy between the two chains increases the benefits of component suppliers and service providers. Optimized the parts supply chain and extended the service value chain A.

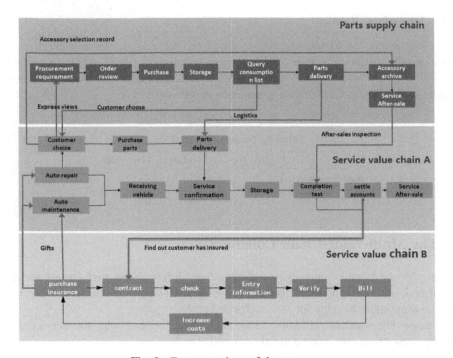

Fig. 2. Future earnings of the company

In the collaboration between the service value chain A and the insurance value chain B, different value customers purchase insurance will be gifted different maintenance products. At the same time, service provider A customizes different services for different customers. They can conduct business monitoring or return visit tracking. And they combine customer value classification information to know whether high-value customers have targeted high-quality services in the service business collaboration. It increases Insurer B's revenue. Service provider A also gets more quality customers at the front end of the life cycle, extending the service value chain. The difference services provided are shown in Table 5.

Table 5. Customer value service customization

Service	High-value	Risk	Marginal	Low-value
Maintenance gift	1 oil/2 times	1 oil/4 times	Non	Non
Wash service	1/1 month	1/2 months	1/3 months	Non
Purchase insurance	2 lacquers	2 lacquers	1 oil	1 oil

5.2 Increased Revenue from Services

About 80% of high-level customers, 60% of risk customers, 20% of marginal customers, and 5% of low-level customers are willing to purchase insurance for vehicles at the factory and maintain them 4th a year. The revenues are:

$$(4890 + 1000 * 4 - 350 * 2 - 120 * 0.5 - 5 * 12) * 80\% = 6456,$$

$$(4890 + 1000 * 4 - 350 * 2 - 120 * 0.25 - 5 * 6) * 60\% = 4878,$$

$$(4890 + 1000 * 2 - 120 - 5 * 3) * 20\% = 1351,$$

$$(4890 + 1000 * 2 - 120) * 5\% = 338.5.$$

If enterprises do not provide differentiated services, and they provide services according to the standards of high-value customers, then, human resources will be wasted and increase corporate expenses. If they are based on the criteria of low-value customers, it is optimistic to estimate that the high-level and risk customers are reduced to 50% and 30% respectively, and the revenues are:

The service provider will lose ¥2078.5 per year for each high-value customer and lose ¥2251.5 per year for each risk customer. At the same time, service providers lost the income from high level customers purchase the vehicles again.

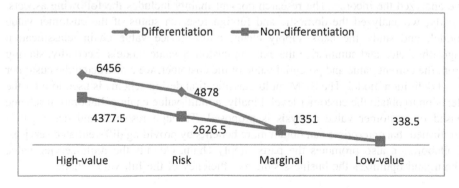

Fig. 3. Future earnings

We analyze the proportion of customers in the company's revenue. As shown in Fig. 3, we can see that the source of revenue mainly depends on two customers with high value.

5.3 Analysis

We analyze the proportion of customers in the company's revenue. As shown in Fig. 4 we can see that the source of revenue mainly depends on two customers with high value.

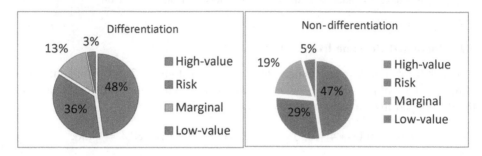

Fig. 4. Proportion of income

The customer value analysis method based on automotive multi-value-chain collaboration, increases the visible benefits for multiple service providers and component suppliers. High-value customer and risky customer create more than three-quarters of the company's revenue. Thus, the customer value analysis model is very important. Multi-value chain collaboration can help the business community create more revenue.

6 Conclusions and Prospects

This paper introduces the customer value analysis method base on automotive multi-value-chain collaboration, and proposes customer value classification model, customer value definition scheme, and multi-value-chain collaborative service process. Finally, we analyzed the income. The research content mainly includes the following aspects: Firstly, we analyzed the domestic and foreign research status of the customer value model, and study the parts supply chain, multi-service value chain, classification algorithm, etc., and summarize the existing customer value models. Secondly, starting from the current value and potential value of the customer, we constructed the customer level definition model. The SVM multi-classification design scheme is used to infer the decision to obtain the customer level. Finally, a multi-value chain collaboration scheme based on customer value analysis is proposed. Through research and design, it is confirmed that enterprises can obtain more benefits by providing differentiated services. Collaboration also promotes the parts supply chain, extends the multi-service value chain, and optimizes the business synergy efficiency of the full value chain.

Acknowledgment. The authors would like to thank the anonymous referees for their valuable comments and helpful suggestions. This paper is supported by The National Key Research and Development Program of China (2017YFB1400902).

References

1. Bachetti, A., Sacanni, N.: Spare parts classification and demand forecasting for stock control: investigating the gap between research and practice. Omega **40**, 722–737 (2011)
2. Frederick, F.R.: The Loyalty Effect: The Hidden Force Behind Growth, Profits, and Lasting Value, pp. 5–18. Harvard Business School Press, Boston (2002)
3. Baines, T.S., Lightfoot, H.W., Benedettini, O., Kay, J.M.: The servitization of manufacturing: a review of literature and reflection on future challenges. J. Manuf. Technol. Manag. **20**, 547–567 (2009)
4. Walter, A., Ritter, T., Gemunden, H.G.: Value creation in buyer-seller relationships. Ind. Market. Manag. **30**, 365–377 (2001)
5. Aljawarneh, S., Aldwairi, M., Yassein, M.B.: Anomaly-based intrusion detection system through feature selection analysis and building hybrid efficient model. J. Comput. Sci. **25**, 152–160 (2018)
6. Bao, Z., Zhao, Y., Zhao, Y., Hu, X., Gao, F.: Segmentation of baidu takeaway customer based on RFA model and cluster analysis. J. Computer Science **45**(S2), 436–438 (2018)
7. Wu, Y.L., Li, E.Y.: Marketing mix, customer value, and customer loyalty in social commerce: a stimulus-organism-response perspective. J. Internet Res. **28**(1), 74–104 (2018)
8. Mahmoud, M.A., Hinson, R.E., Anim, P.A.: Service innovation and customer satisfaction: the role of customer value creation. J. Eur. Innov. Manag. **21**(3), 402–422 (2018)
9. Xia, W., Wang, Q.: Customer segmentation and retention strategy based on customer value. J. Manag. Sci. China **19**(4), 35–38 (2006)
10. Xu, Q.: The Research on Business Collaborative Technology Supporting Service Lifecycle Strategy Management. Southwest Jiaotong University (2017)
11. Zhou, H., Li, S., Sun, J.: Unit model of binary SVM with DS output and its application in multi-class SVM. In: International Symposium on Computational Intelligence & Design. IEEE (2011)
12. Yang, F., Yang, L., Wang, D., Qi, P., Wang, H.: Method of modulation recognition based on combination algorithm of k-means clustering and grading training SVM. J. China Commun. **15**(12), 55–63 (2018)
13. Huang, L., Jiang, B., Lv, S., Liu, Y., Li, D.: The review of research on recommendation systems based on deep learning. J. Chin. Comput. **41**(07), 1619–1647 (2018)
14. Goodfellow, I., Bengio, Y., Courville, A.: Deep Learning. MIT Press, Cambridge (2016)
15. LeCun, Y., Bengio, Y., Hinton, G.: Deep learning. J. Nat. **521**(7553), 436 (2015)
16. Alam, S.: Performance of Alzheimer disease classification based on PCA, linear SVM, and multi-kernel SVM. In: The Eighth International Conference on Ubiquitous and Future Networks. IEEE (2016)
17. Yuanhang, D., Lei, C., Weiling, Z., et al.: Multi-support vector machine power system transient stability assessment based on relief algorithm. In: 2015 IEEE PES Asia-Pacific Power and Energy Engineering Conference (APPEEC), pp. 1–5. IEEE (2015)

Research on Power Distribution Strategy for Bi-directional Energy Cooperation Diamond Channel with Energy Harvesting Nodes

Taoshen Li[1,2], Peipei Chen[1(✉)], Li Sun[1], Zhe Wang[1], and Mingyu Lu[1]

[1] School of Computer and Electronic Information, Guangxi University, Nanning 530004, China
19920091@gxu.edu.cn, 984689507@qq.com, 1160742044@qq.com, lumingyu_26080468@qq.com, designbyyili@163.com
[2] Nanning University, Nanning 530200, China

Abstract. To solve the end-to-end throughput maximization problem of dual-relay channels based on harvested energy and bi-directional energy cooperation, a power distribution strategy for bi-directional energy cooperation diamond channel with nodes of harvested energy is proposed. The strategy extends Gaussian diamond channel model of energy harvesting to the diamond communication network model of bi-directional energy cooperation, and applies the delay policies to decompose the problem into the energy distribution problem and energy transmission problem of each time slot. The two-way water injection algorithm to solve the practical energy consumption distribution, and then the optimization schem of the original problem is obtained by solving solutions of the two separated problems. Simulation results proved that proposed power distribution strategy has obviously improved the system throughput with the power split strategy based on uni-directional energy collaboration and bi-directional energy collaboration when the energy collection of source and relay nodes are very different.

Keywords: Energy harvesting · Energy cooperation · Optimal power distribution · Energy transfer strategy · Diamond channels

1 Introduction

The battery life of devices affects the performance of wireless communication networks. The most of battery-powered devices have a short life span and tend to stop working when the battery runs out, which is expensive even if replacing the battery becomes a possibility [1]. By energy harvesting (EH) technique, wireless networks may continuously gather energy from the environment, which greatly prolongs the life of wireless devices, reduces maintenance costs, and improves the performance of wireless network systems [2]. Due to the great difference between energy harvesting communications and

© Springer Nature Singapore Pte Ltd. 2019
Y. Sun et al. (Eds.): ChineseCSCW 2019, CCIS 1042, pp. 170–182, 2019.
https://doi.org/10.1007/978-981-15-1377-0_13

traditional battery-driven communication, we need to re-establish the model and the optimal transmission strategy, so as to maximize the throughput [1, 3].

Due to environmental protection and energy conservation had been widely concerned by the society, the power split in energy harvesting wireless networks (EH-WN) had become a research craze for scholars. In [4], an energy harvesting network node transmission time minimization scheme equipped with the unlimited capacity battery was proposed, which the optimal off-line scheduling strategy was obtained by optimizing the problem of the packet scheduling in the energy harvesting wireless systems for single-user. [5] utilized energy harvesting transmission nodes with limited battery capacity to communicate in the wireless channel to optimize point-to-point data transmission, by controlling the transmission power time sequence which subjects to energy storage capacity and causality constraints, to optimize the throughput maximization before the deadline, and the optimal off-line and online strategy was presented. In [6], an energy quantum channel allocation algorithm was proposed, which was based on considering several transfer links in the frequency-selective fading channel. [7] presented a point-to-point wireless network of transmitter battery with limited capacity and fading channel to achieve the purpose of harvesting energy storage and data transmission, and proposed a technology to resolve non-convex optimization problems. The research on throughput maximization and its solutions had been extended to multi-terminal models. The authors of [8] considered a cooperative Gaussian multiple-access channel (MAC) for two users, proposed a successive convex approximation method on this basis, which was effective in every step of calculation, and proved that this method converges to the optimal solution. [9] investigated the two-hop network for energy acquisition, which had decoding and forwarding half-duplex relay, and explored the throughput maximization problem. For maximizing the throughput of peer-to-peer system, [10] proposed a transmission strategy suitable for joint scheduling and power distribution in double- hop relay communication system with limited battery capacity. In [11], the diamond channel with the cooperative multi-access capacity area was decomposed into internal and external maximization problem, so as to obtain the optimal rate and power.

Energy harvesting network is a network with intermittent availability, and the energy cooperation between network nodes can improve the performance of the network, which has attracted people's attention and research. Based on the research of energy transfer of energy harvesting wireless nodes, a corresponding management strategy of the multi-users network structures was formulated in [12]. To maximize peer-to-peer throughput, [13, 14] proposed optimal joint off-line energy management strategy for bi-directional, two-hop, and multiple access situations of uni-directional energy collaboration respectively. Under the constraint of throughput maximization, a delay strategy for multi-terminal networks with energy harvesting was proposed in [15], energy was transferred only when it was used immediately, and the transferred energy must be fully utilized before ending the current time slot. The purpose of the delay strategy is to divide the joint optimization problem into the optimal energy transmission problem and the power distribution strategy problem. [16] proved that in the optimal strategy, nodes cannot receive and transfer energy at once. [17] studied the energy allocation problem under the constraint of limited battery energy for multi-users. The original assumption of infinite battery capacity is accurate to the optimal battery capacity and corresponding strategies are formulated. [18] presented a variety of optimizing power distribution and relay selection methods for energy acquisition collaborative wireless network. In [19], the throughput fairness of WPCN was improved, which was solved by multi-user collaboration. Due to the influence of geographical location or time, the energy collection by energy harvesting nodes in the wireless network varies greatly. Although uni-directional energy transfer improves the system end-to-end throughput to some extent, there are still some problems. For example, for nodes with less energy, the energy harvester may be different at different times. For this condition, it is necessary to realize the mutual transfer and supply of energy between nodes through the way of bi-directional energy cooperation. Power allocation, relay selection and energy cooperation strategies

In this paper, the Gaussian diamond channel model of energy harvesting proposed in [11] is extended to the diamond communication network model of bi-directional energy cooperation. For greatly improving the peer-to-peer throughput, we proposes the optimal power distribution and energy transmission methods for off-line transmission. Considering the causality constraints of information and energy between transmission nodes, a system throughput optimization model is constructed. By decomposing the original problem, the optimization solution can be obtained simultaneously. Finally, the effectiveness of the proposed strategy is demonstrated by experimental simulation.

2 System Model

We study a diamond channel based on harvested energy and bi-directional energy cooperation. In Fig. 1, S node represents the source node in the channel, R1 and R2 are two relay nodes, and D node is the destination node. The energy harvested from the environment by source and relay nodes is stored in the corresponding batteries. Supposing that the source and relay nodes are equipped with infinite battery capacity, and the destination node is powered by fixed power. The physical layer of the diamond channel is composed of the broadcast channel and multi-address channel. By constructing the Wireless Power Communication Network (WPCN), it can supply power to multiple communication devices with different physical conditions and service requirements.

The diamond channel proposed in this paper is the Gaussian diamond channel [20]. Specifically, the broadcast channel in the former part and the multiple access channel in the latter part are both the Gaussian channel, and the noise of the channel is additive Gaussian white noise. System takes time slot as the minimum transmission time unit, and a transmission period contains L equal time slots. The strategy described in this paper can be extended to any time slot length.

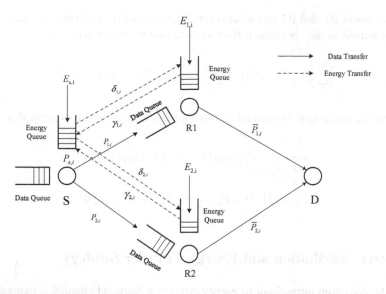

Fig. 1. Diamond channel with energy collection and two-way energy cooperation

Supposing the energy harvested by the nodes in each time slot can be predicted in advance, and the energy transfer efficiency is equal. Let $E_{s,i}$ denotes the energy harvested by the source node in ith time slot, $E_{1,i}$ and $E_{2,i}$ denote the energy harvested by the first and second relay nodes respectively. $\delta_{k,i}$ is the energy transferred through the source to the relay, and $\gamma_{k,i}$ is energy transferred through the relay to the source, where $k = 1$ and 2 represent the first and second relay nodes respectively; $\alpha(0 \leqslant \alpha \leqslant 1)$ is the efficiency coefficient of energy transfer. $P_{s,i}$ is the efficiency of the total transmission at the source, $P_{k,i}$ is the transmission power from S node to $R_k(k = 1, 2)$ node and $\bar{P}_{k,i}$ is the transmission power from $R_k(k = 1, 2)$ node to D node. Furthermore, P_s, P_k, $\bar{P}_k, \delta_k, \gamma_k$ represent the vector of power and energy transfer respectively. The conclusion of this paper also applies to the case of $k = 1$ or $k = 1, 2, ..., j(j > 2)$.

In this paper, the transmission nodes S, R1 and R2 must satisfy the causality constraints of energy, means that, energy not reached within the specified n time slots cannot be used. The causal constraints are as follows:

$$\sum_{i=1}^{n} \bar{P}_{k,i} \leq \sum_{i=1}^{n} \left(E_{k,i} + \alpha\delta_{k,i} - \gamma_{k,i} \right), \forall n, k = 1, 2 \tag{1}$$

$$\sum_{i=1}^{n} \bar{P}_{1,i} + \bar{P}_{2,i} \leq \sum_{i=1}^{n} \left(\begin{array}{c} E_{s,i} + \alpha\gamma_{1,i} + \alpha\lambda_{2,i} \\ -\delta_{1,i} - \delta_{2,i} \end{array} \right), \forall n \tag{2}$$

Relay nodes R1 and R2 should also satisfy the causality constraints of data, that is, data not arrived in the specified n time slots cannot be forwarded, i.e.,

$$\sum_{i=1}^{n} \frac{1}{2} \log \left(1 + \bar{P}_{k,i}\right) \leq \sum_{i=1}^{n} \frac{1}{2} \log \left(1 + P_{k,i}\right), \forall n, k = 1, 2 \tag{3}$$

Then, the maximum end-to-end throughput in systems can be expressed as

$$\max_{\substack{\bar{P}_{1,i}, \bar{P}_{2,i}, P_{1,i}, P_{2,i}, \\ \delta_{1,i}, \delta_{2,i}, \gamma_{1,i}, \gamma_{2,i}}} \sum_{i=1}^{L} \left(\tfrac{1}{2} \log \left(1 + \bar{P}_{1,i}\right) + \tfrac{1}{2} \log \left(1 + \bar{P}_{2,i}\right)\right)$$

$$\text{s.t. (1)-(3)} \quad (P_1, P_2, \bar{P}_1, \bar{P}_2, \delta_1, \delta_2, \gamma_1, \gamma_2) \geq 0 \tag{4}$$

3   Power Distribution and Energy Transfer Strategy

When the maximum throughput of energy collection diamond channel is expressed by the rate, the problem in Eq. (4) is convex optimization [10]. To simplify the problem, we apply strict delay constraints on relays. We delete the data buffer of the relay node, force the relay node to decode the received data and forward the data to destination node immediately. Consequently, the problem is expressed as

$$\sum_{i=1}^{n} \left(\frac{1}{2} \log \left(1 + P_{k,i}\right) - \frac{1}{2} \log \left(1 + \bar{P}_{k,i}\right)\right) = 0, \forall n, k = 1, 2 \tag{5}$$

According to (5), we get $P_{k,i} = \bar{P}_{k,i}, \forall i, k = 1, 2$, then (4) can be converted into

$$\max_{\substack{P_{1,i}, P_{2,i}, \delta_{1,i}, \\ \delta_{2,i}, \gamma_{1,i}, \gamma_{2,i}}} \sum_{i=1}^{L} \left(\frac{1}{2} \log \left(1 + P_{1,i}\right) + \frac{1}{2} \log \left(1 + P_{2,i}\right)\right)$$

$$\text{s.t.} \quad \sum_{i=1}^{n} P_{k,i} \leq \sum_{i=1}^{n} \left(E_{k,i} + \alpha \delta_{k,i} - \gamma_{k,i}\right), \forall n, \ k = 1, 2 \tag{6}$$

$$\sum_{i=1}^{n} P_{1,i} + P_{2,i} \leq \sum_{i=1}^{n} \begin{pmatrix} E_{s,i} + \alpha \gamma_{1,i} + \alpha \gamma_{2,i} \\ -\delta_{1,i} - \delta_{2,i} \end{pmatrix}, \forall n$$

$$(P_1, P_2, \delta_1, \delta_2, \gamma_1, \gamma_2) \geq 0$$

Theorem 1. Define a delay policy that satisfies $\delta_{k,i} \gamma_{k,i} = 0$, that is, two-way energy transfer between two nodes cannot occur in the same time slot. Simultaneously, this delay policy satisfies $\bar{P}_{k,i} \geq \alpha \delta_{k,i} - \gamma_{k,i}, P_{s,i} \geq \alpha \gamma_{1,i} + \alpha \gamma_{2,i} - \delta_{1,i} - \delta_{2,i} (i = 1, \ldots, L;$

$k = 1, 2$). That is to say, energy can only be transferred when it is used immediately, and the transferred energy must be used up in the current slot. Therefore, there at least is a delay strategy that can be solved (6).

Proof: Firstly, it is proved that there is no two-way energy transfer between two nodes in same time slot, i.e., $\delta_{k,i}\, \gamma_{k,i} = 0$. Assuming $\delta_{k,i}\, \gamma_{k,i} \neq 0$, if $\delta_{k,i} \geq \gamma_{k,i} > 0$, then we can replace $\delta_{k,i}$ and $\gamma_{k,i}$ with $\delta_{k,i} - \gamma_{k,i}$ and 0. If $\gamma_{k,i} > \delta_{k,i} > 0$, then we can replace $\gamma_{k,i}$ and $\delta_{k,i}$ with $\gamma_{k,i} - \delta_{k,i}$ and 0. Because $(\gamma_{k,i} - \delta_{k,i})0 = 0$, after transformation, it doesn't decrease the total maximum throughput. This proves that $\delta_{k,i}\, \gamma_{k,i} = 0$ is true. That is to say, no bidirectional transfer between two nodes can be established in the same time slot.

Next, it is proved as follows: when the first condition of the delay strategy is satisfied but the second condition is not satisfied, at least have one k and l, satisfy $P_{k,l}^* < \alpha \delta_{k,l}^*$ or $P_{1,l}^* + P_{2,l}^* < \alpha \gamma_{1,l}^* + \alpha \gamma_{2,l}^*$.

We assume exist the optimal solution $\{P_{1,i}^*, P_{2,i}^*, \delta_{k,i}^*, \gamma_{k,i}^*\}$ that satisfies the first condition of the delay strategy $\delta_{k,i}^*\, \gamma_{k,i}^* = 0$, but not satisfies the second condition of the delay strategy. So there is at least one k and l, make $P_{k,l}^* < \alpha \delta_{k,l}^*$ or $P_{1,l}^* + P_{2,l}^* < \alpha \gamma_{1,l}^* + \alpha \gamma_{2,l}^*$. Assuming that the $P_{k,l}^* < \alpha \delta_{k,l}^*$, when $l < L$, let $\delta_{k,l}^* = P_{k,l}^*/\alpha$, $\delta_{k,l+1}^* = \delta_{k,l+1}^* + \delta_{k,l}^* - P_{k,l}^*/\alpha$; when $l = L, \delta_{k,i}^* = P_{k,l}^*/\alpha$. This adjustment has not changed $\{P_{1,i}^*, P_{2,i}^*\}$, the transfer of remaining energy is delayed to next time slot. Because other variables do not change, the optimal solution remains unchanged. Thus it can be proved that energy can only be transferred when it is used immediately, and the transferred energy must be used up in the current time slot, cannot be left to the next time slot, i.e., $\bar{P}_{k,i} \geq \alpha \delta_{k,i} - \gamma_{k,i}$.

Similarly: $P_{1,l}^* + P_{2,l}^* < \alpha \gamma_{1,l}^* + \alpha \gamma_{2,l}^*$.

Theorem 1 shows that the optimal solution can be found in the delay strategy. According to the delay strategy, the transferred energy must be exhausted in the current slot and cannot be used to the next slot, so has $\bar{P}_{k,i} \geq \alpha \delta_{k,i}$. Using $P_{s,i}'$ and $P_{k,i}'$ to represent actual power consumption of nodes, we get $P_{k,i}' = \bar{P}_{k,i} + \gamma_{k,i} - \alpha \delta_{k,i}$, so has $P_{k,i}' \geq \gamma_{k,i}$. Formula (6) can be converted to

$$\max_{\substack{P_{1,i}', P_{2,i}', \delta_{1,i}, \\ \delta_{2,i}, \gamma_{1,i}, \gamma_{2,i}}} \sum_{i=1}^{L} \left(\tfrac{1}{2}\log\left(1 + P_{1,i}' - \gamma_{1,i} + \alpha \delta_{1,i}\right) + \tfrac{1}{2}\log\left(\tfrac{1 + P_{2,i}' - \gamma_{2,i}}{+\, \alpha \delta_{2,i}} \right) \right) \tag{7a}$$

$$\text{s.t.} \quad \sum_{i=1}^{n} P_{1,i}' \leq \sum_{i=1}^{n} E_{1,i}, \forall n$$

$$\sum_{i=1}^{n} P_{2,i}' \leq \sum_{i=1}^{n} E_{2,i}, \forall n \tag{7b}$$

$$\sum_{i=1}^{n} P_{1,i}' + P_{2,i}' \leq \sum_{i=1}^{n} E_{s,i} + (\alpha + 1)\left(\begin{matrix} \gamma_{1,i} + \gamma_{2,i} \\ -\delta_{1,i} - \delta_{2,i} \end{matrix} \right), \forall n \tag{7c}$$

$$P'_{1,i} \geq \gamma_{1,i}, P'_{2,i} \geq \gamma_{2,i}, \forall n \tag{7d}$$

$$(\delta_1, \delta_2, \gamma_1, \gamma_2) \geq 0 \tag{7e}$$

Where, the energy constraint of the $\{\delta_{1,i}, \delta_{2,i}, \gamma_{1,i}, \gamma_{2,i}\}$ is the (7c), (7d) and (7e), thus we can define

$$g\left(P'_{1,i}, P'_{2,i}\right) = \max_{\delta_{1,i}, \delta_{2,i}, \gamma_{1,i}, \gamma_{2,i}} \sum_{i=1}^{L} \begin{pmatrix} \frac{1}{2}\log\left(1 + P'_{1,i} - \gamma_{1,i} + \alpha\delta_{1,i}\right) + \\ \frac{1}{2}\log\left(1 + P'_{2,i} - \gamma_{2,i} + \alpha\delta_{2,i}\right) \end{pmatrix}$$

$$\text{s.t.} \quad \sum_{i=1}^{n} P'_{1,i} + P'_{2,i} \leq \sum_{i=1}^{n} E_{s,i} + (\alpha+1)\begin{pmatrix} \gamma_{1,i} + \gamma_{2,i} \\ -\delta_{1,i} - \delta_{2,i} \end{pmatrix}, \forall n \tag{8}$$

$$P'_{2,i} \geq \gamma_{2,i}, \ P'_{2,i} \geq \gamma_{2,i}, \quad \forall n$$

$$(\delta_1, \delta_2, \gamma_1, \gamma_2) \geq 0, \forall n$$

The original problem is equivalent to

$$\max_{P'_{1,i}, P'_{2,i}} g\left(P'_{1,i}, P'_{2,i}\right)$$

$$\text{s.t.} \quad \sum_{i=1}^{n} P'_{1,i} \leq \sum_{i=1}^{n} E_{1,i}, \quad \forall n$$

$$\sum_{i=1}^{n} P'_{2,i} \leq \sum_{i=1}^{n} E_{2,i}, \quad \forall n \tag{9}$$

$$\sum_{i=1}^{n} P'_{s,i} \leq \sum_{i=1}^{n} E_{s,i}, \quad \forall n$$

$$P'_{1,i} \geq 0, P'_{2,i} \geq 0, \quad \forall n$$

We can seen from the problem (9), the tarbet formula is the joint concave function, and the constraint conditions are all linear conditions. (9) is known as a convex optimization problem and its solution only depends on $P'_{1,i}$ and $P'_{2,i}$, so it is equivalent to initializing the transferred energy to 0. Using the directional water injection algorithm proposed in [5], we obtain the optimal power distribution scheme without energy transfer. Next, the optimal power collaboration of equivalent energy collaboration is briefly explained.

Assume that L_1, L_2 represent two time periods in Fig. 2. We observe that if the water (energy) in units E is injected into a rectangle with a bottom width of L, the injection level is E/L. E_{max} denotes maximum battery power, E_i represents the harvested energy in the ith. If $E_0/L_1 > E_1/L_2$, in order to make the water injection level of the two intervals equal, let some energy flows from slot 1 to slot 2, as shown in Fig. 2 (a). However, if $E_0/L_1 < E_1/L_2$, there is no energy flowing from right to left, which is because of the causal relationship of energy use. That is, energy cannot be used before it is harvested. As can be seen in Fig. 2(b), the injection level in two intervals is not equal at this time.

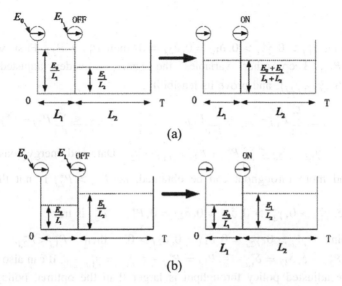

(a)

(b)

Fig. 2. Directional water injection algorithm with right permeable faucet in two time periods

$\{P'_{s,i}, P'_{1,i}, P'_{2,i}\}$ can be obtained by directional water injection algorithm, put $P'_{s,i}, P'_{1,i}, P'_{2,i}$ into (8), accord to the constraint conditions, in order to get the maximum value of the (8), we have $P'_{1,i} + P'_{2,i} \leq P'_{s,i} - (1+\alpha)(\gamma_{1,i} + \gamma_{2,i} - \delta_{1,i} - \delta_{2,i})$. By the Theorem 1, $\delta^*_{k,i}\gamma^*_{k,i} = 0$. In the $\gamma_{1,i}, \gamma_{2,i}, \delta_{1,i}, \delta_{2,I}$ at least two of them are zeros. If there are only two zeros, there are four cases, case 1: $\gamma_{1,i} = 0, \gamma_{2,i} = 0, \delta_{1,i} > 0, \delta_{2,i} > 0$; case 2: $\gamma_{1,i} > 0, \gamma_{2,i} > 0, \delta_{1,i} = 0, \delta_{2,i} = 0$; case 3: $\gamma_{1,i} = 0, \gamma_{2,i} > 0, \delta_{1,i} > 0$, $\delta_{2,i} = 0$; case 4: $\gamma_{1,i} > 0, \gamma_{2,i} = 0, \delta_{1,i} = 0, \delta_{2,i} > 0$. If there are three zeros, the transfer amount of non-zero is less than or equal to $\left| P'_{s,i} - P'_{1,i} - P'_{2,i} \right| / (1+\alpha)$.

As shown in reference [11], for the optimal solution $P^*_{1,i}, P^*_{2,i}, \delta^*_{1,i}, \delta^*_{2,i}, \gamma^*_{1,i}, \gamma^*_{2,i}$, at a certain time, if $\gamma^*_{1,i} = 0, \gamma^*_{2,i} = 0, \delta_{1,i} > 0, \delta_{2,i} > 0$, then $P^*_{1,i} = P^*_{2,i}$. That is to say, when $\gamma^*_{1,i} = 0, \gamma^*_{2,i} = 0, \delta_{1,i} > 0, \delta_{2,i} > 0$, the energy transferred through source node to relay nodes must be equal, i.e., $\delta_{1,i} = \delta_{2,i} \leq \left| P'_{s,i} - P'_{1,i} - P'_{2,i} \right| / 2(1+\alpha)$.

Theorem 2. For the optimal solution $P_{1,i}^*, P_{2,i}^*, \delta_{1,i}^*, \delta_{2,i}^*, \gamma_{1,i}^*, \gamma_{2,i}^*$, at a certain time, if $\gamma_{1,i}^* > 0, \gamma_{2,i}^* > 0, \delta_{1,i} = 0$, and $\delta_{2,i} = 0$, then $P_{1,i}^* = P_{2,i}^*$; if $\gamma_{1,i}^* > 0, \gamma_{2,i}^* = 0, \delta_{1,i} = 0$, and $\delta_{2,i} > 0$, then $P_{1,i}^* \geq P_{2,i}^*$; end if $\gamma_{1,i}^* = 0, \gamma_{2,i}^* > 0, \delta_{1,i} > 0$, and $\delta_{2,i} = 0$, then $P_{1,i}^* \leq P_{2,i}^*$.

Proof:

(1) Assuming $\gamma_{1,i}^* > 0, \gamma_{2,i}^* > 0, \delta_{1,i} = 0, \delta_{2,i} = 0$, then $P_{1,i}^* \neq P_{2,i}^*$. First, we assume $P_{1,i}^* > P_{2,i}^*$, keep other variables the same, consider adjusted strategy $(\bar{P}_{1,i}, \bar{P}_{2,i}, \bar{\gamma}_{1,i}, \bar{\gamma}_{2,i})$, and prove its feasibility.

$$\sum_{i=1}^{n} \bar{P}_{1,i} - \bar{\gamma}_{1,i} \leq \sum_{i=1}^{n} P_{1,i}^* - \gamma_{1,i}^* \leq \sum_{i=1}^{n} E_{1,i}, \qquad \sum_{i=1}^{n} \bar{P}_{2,i} - \bar{\gamma}_{2,i} \leq \sum_{i=1}^{n} P_{2,i}^* - \gamma_{2,i}^* \leq \sum_{i=1}^{n} E_{2,i},$$

$\sum_{i=1}^{n} \bar{P}_{1,i} + \bar{P}_{2,i} - \bar{\gamma}_{1,i} - \bar{\gamma}_{2,i} \leq \sum_{i=1}^{n} P_{1,i}^* + P_{2,i}^* - \gamma_{1,i}^* - \gamma_{2,i}^*$. Data and energy constraints are satisfied, and more throughput can be obtained, so $P_{1,i}^* > P_{2,i}^*$ is not the optimal solution.

Therefore, $\gamma_{1,i}^* > 0, \gamma_{2,i}^* > 0, \delta_{1,i} = 0, \delta_{2,i} = 0, P_{1,i}^* = P_{2,i}^*$ is true.

(2) Assuming $\gamma_{1,i}^* > 0, \gamma_{2,i}^* = 0, \delta_{1,i} = 0, \delta_{2,i} > 0$, then $P_{1,i}^* < P_{2,i}^*$. Adjusting $\bar{P}_{2,i} = P_{2,i}^* - \varepsilon, \bar{\delta}_{2,i} = \delta_{1,i}^* - \frac{\varepsilon}{\alpha}, \bar{P}_{1,i} = P_{1,i}^* - \varepsilon, \bar{\gamma}_{1,i} = \gamma_{1,i}^* - \varepsilon$. It can also be proved that the adjusted policy throughput is larger than the optimal policy, so it is inconsistent with the hypothesis. Therefore, if $\gamma_{1,i}^* > 0, \gamma_{2,i}^* = 0, \delta_{1,i} = 0$, and $\delta_{2,i} > 0$, then $P_{1,i}^* \geq P_{2,i}^*$

By symmetry, we can prove $\gamma_{1,i}^* = 0, \gamma_{2,i}^* > 0, \delta_{1,i} > 0$, and $\delta_{2,i} = 0$, then $P_{1,i}^* \leq P_{2,i}^*$.

By Theorem 2 $\gamma_{1,i}^* > 0, \gamma_{2,i}^* > 0, \delta_{1,i} = 0, \delta_{2,i} = 0$. The energy transferred from the relay node to the source node must be equal, i.e., $\gamma_{1,i} = \gamma_{2,i} \leq \left| P_{s,i}' - P_{1,i}' - P_{2,i}' \right| / 2(1+\alpha)$. If $\gamma_{1,i}^* > 0, \gamma_{2,.}^* = 0, \delta_{1,i} = 0, \delta_{2,i} > 0, \delta_{2,i} \leq \left(\alpha \left| P_{s,i}' - P_{1,i}' - P_{2,i}' \right| \right) / (1+\alpha)^2$, then $\gamma_{1,i} = \left(\left| P_{s,i}' - P_{1,i}' - P_{2,i}' \right| (2\alpha+1) \right) / (1+\alpha)^2$; if $\gamma_{1,i}^* = 0, \gamma_{2,.}^* > 0, \delta_{1,i} > 0, \delta_{2,i} = 0, \delta_{1,i} \leq \left(\alpha \left| P_{s,i}' - P_{1,i}' - P_{2,i}' \right| \right) / (1+\alpha)^2$, then $\gamma_{2,i} = \left(\alpha \left| P_{s,i}' - P_{1,i}' - P_{2,i}' \right| (2\alpha+1) \right) / (1+\alpha)^2$.

Combined with the above theoretical analysis, the implementation process of the proposed power distribution algorithm based on energy collection and two-way energy cooperation diamond channel is described as follows:

Algorithm 1. Power distribution algorithm based on energy collection and two-way energy cooperation diamond channel

Input: energy $E_{s,i}$ and $E_{k,i}$ harvested at each time slot;

Output: Maximum and throughput are calculated by proposed optimal power split and energy transfer strategies.

1. Initialize $\delta_{k,i}=0$, $\gamma_{k,i}=0$, $E_{s,i}^0 = E_{s,i}$, $E_{k,i}^0 = E_{k,i}$ ($1 \leq i \leq L$, $n=0$, $sum=0$; $k=1,2$);

2. $\{P_{s,i}^{'0}, P_{1,i}^{'0}, P_{2,i}^{'0}\}$ is obtained by using directional waterflooding algorithm;

3. calculate: $C = \sum_{i=1}^{L} \left(\frac{1}{2}\log\left(1+p_{1,i}^{'0}\right) + \frac{1}{2}\log\left(1+p_{2,i}^{'0}\right) \right)$

4. while $C > sum$ do

   $n++$, $sum = C$;

   use theorem in [10] and theorem 2 to calculate $\{\delta_{1,i}^n, \delta_{2,i}^n, \gamma_{1,i}^n, \gamma_{2,i}^n\}$

   update the following values:

   $$E_{s,i}^n = E_{s,i}^{n-1} + \alpha\gamma_{1,i}^n + \alpha\lambda_{2,i}^n - \delta_{1,i}^n - \delta_{2,i}^n,$$

   $$E_{k,i}^n = E_{k,i}^{n-1} + \alpha\delta_{k,i}^n - \gamma_{k,i}^n$$

   calculate $C = \sum_{i=1}^{L} (\frac{1}{2}\log(1+P_{1,i}^n) + (\frac{1}{2}\log(1+P_{2,i}^n))$

   End while

5. Output the maximum total throughput.

The (9) is a problem of convex optimization, which has an optimal solution. In Algorithm 1, the results obtained by each energy transfer and directional water injection algorithm will gradually increase until they converge to the optimal value.

4 Simulation Experiment and Performance Analysis

The experiment was done in Matlab simulation environment. The running environment is VSCode, Matlab version R2017a_win64. The simulation data were selected from the experimental data of [14–16].

In the experiment, we assume that simulated time slot L = 10 s, duration of each time slot is 1 s, bandwidth Bw = 1 MHz, and the spectral density of noise power in each node is $N_0 = 10^{-19}$ W/Hz [14–16]. The energy harvested by source and relay nodes is evenly distributed between $[0, E_s^{\max}]$ and $[0, E_k^{\max}]$. Suppose that the maximum energy collected by source and relay nodes in each slot is no more than 10 mJ. The purpose of the simulation experiment in this paper is to compare and analyze the optimal power distribution and energy transfer strategies of one-way energy transfer, no-energy transfer diamond channel, the end-to-end total throughput of the power distribution, energy transfer strategies of diamond channel based on energy collection and two-way energy cooperation proposed in this paper under different variables.

Figure 3 demonstrates that end-to-end throughput curves of the three strategies as the energy transfer efficiency between nodes changes. We can see from the Fig. 3, in given power split and the strategy of energy transfer, the end-to-end throughput of our strategy is better than that the strategy of the power distribution based on unidirectional energy collaboration and energy transfer strategy of the diamond channel without energy cooperation. As the augment of energy transfer efficiency, end-to-end throughput of power distribution and energy transfer strategies based on non-energy collaboration remained unchanged, while the end-to-end throughput of power distribution and energy transfer strategies based on one-way and two-way energy collaboration increased significantly.

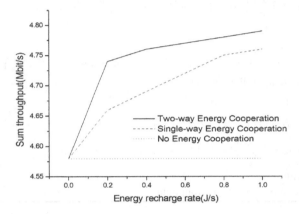

Fig. 3. System throughput comparison of the three strategies under different energy transfer efficiency

Figure 4 demonstrates the curve of end-to-end throughput of three strategies changing with the maximum collected energy of the source node E_s. Let $\alpha = 0.5$.

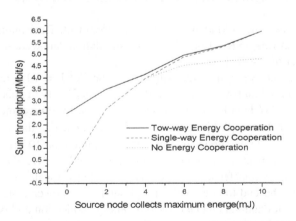

Fig. 4. System throughput comparison of three strategies under the maximum collection energy of different source nodes

From Fig. 4, we can see when the source node harvests less energy, the throughput performance of the power distribution and energy transfer strategies based on one-way energy cooperation and non-energy cooperation channels are similar. The throughput of our strategy is higher than that the power distribution and energy transfer strategies based on one-way energy cooperation and non-energy cooperation. With the increase of the maximum harvested energy of source nodes, the throughput performance gap of power distribution and energy transfer strategies based on one-way energy collaboration and two-way energy collaboration gradually decreases. However, the throughput performance of our strategy is still better than that of power distribution and energy transfer strategies based on one-way energy collaboration and non-energy collaboration.

According to the experimental results, bidirectional energy cooperation is more suitable for the situation where there is a large gap between the energy collection ability of source and relay nodes. If energy collected by source is much larger than that collected by the relay node, the unidirectional energy cooperation performance of source to relay nodes is better. If the relay node collects much more energy than the source node, the relay node has excellent one-way energy cooperation performance to the source node. Therefore, there is a big gap between source and relay nodes in collecting energy, the bi-directional cooperative energy distribution and transmission strategies are proposed for improving the performance.

5 Conclusion

We proposes a power splits and transfer strategy by maximizing end-to-end throughput in this paper. The system throughput optimization model is constructed according to the energy and information causality constraints, and the optimal solution is obtained under these constraints. The problem is decomposed into optimal power split and slot-by-slot energy transfer problems by using delay strategy and solved by directional water injection algorithm and inequality analysis respectively. Simulation results show that comparing with the power allocation strategies on the basis of energy-free collaboration and uni-directional energy collaboration, the proposed power distribution strategy significantly improves the system throughput when the energy collection from source and relay nodes is quite different.

Acknowledgments. These works are supported by the NNSF of China (No. 61762010).

References

1. He, Y., Cheng, X., Peng, W., et al.: A survey of energy harvesting communications: models and offline optimal policies. IEEE Commun. Mag. **53**(6), 79–85 (2015)
2. Atallah, R., Khabbaz, M., Assi, C.: Energy harvesting in vehicular networks: a contemporary survey. IEEE Wirel. Commun. **23**(2), 70–77 (2016)
3. Ulukus, S., Yener, A., Erkip, E., et al.: Energy harvesting wireless communications: a review of recent advances. IEEE J. Sel. Areas Commun. **33**(3), 360–381 (2015)

4. Jing, Y., Ulukus, S.: Optimal packet scheduling in an energy harvesting communication system. IEEE Trans. Commun. **60**(1), 220–230 (2012)
5. Ozel, O., Tutuncuoglu, K., Yang, J., et al.: Transmission with energy harvesting nodes in fading wireless channels: optimal policies. IEEE J. Sel. Areas Commun. **29**(8), 1732–1743 (2011)
6. Wang, Z., Wang, X.D., Aggarwal, V.: Transmission with energy harvesting nodes in frequency-selective fading channels. IEEE Trans. Wirel. Commun. **15**(3), 1642–1656 (2016)
7. Shafieirad, H., Adve, R.S., Shahbazpanahi, S.: Throughput maximization with an energy outage constraint for energy harvesting links. In: Wireless Communications & Networking Conference Workshops, pp. 1–6. IEEE Press, San Francisco (2017)
8. Gurakan, B., Kaya, O., Ulukus, S.: Energy harvesting cooperative multiple access channels with data arrivals. In: 2016 IEEE International Conference on Communications, pp. 1–6. IEEE Press, Kuala Lumpur (2016)
9. Orhan, O., Erkip, E.: Energy harvesting two-hop communication networks. IEEE J. Sel. Areas Commun. **33**(12), 2658–2670 (2015)
10. Li, P.Q., Senior, M.: Optimal transmission policies for relay communication networks with ambient energy harvesting relays. IEEE J. Sel. Areas Commun. **34**(12), 3754–3768 (2016)
11. Gurakan, B., Ulukus, S.: Cooperative diamond channel with energy harvesting nodes. IEEE J. Sel. Areas Commun. **34**(5), 1604–1617 (2016)
12. Gurakan, B., Ozel, O., Yang, J., et al.: Energy cooperation in energy harvesting communications. IEEE Trans. Commun. **61**(12), 4884–4898 (2013)
13. Gurakan, B., Ozel, O., Yang, J., et al.: Energy cooperation in energy harvesting two-way communications. In: 2013 IEEE International Conference on Communications, pp. 3126–3130. IEEE Press, Budapest (2013)
14. Gurakan, B., Ulukus, S.: Energy harvesting diamond channel with energy cooperation. In: 2014 IEEE International Symposium on Information Theory, pp. 986–990. IEEE Press, Honolulu (2014)
15. Tutuncuoglu, K., Yener, A.: Energy harvesting networks with energy cooperation: procrastinating policies. IEEE Trans. Commun. **63**(11), 4525–4538 (2015)
16. Tutuncuoglu, K., Yener, A.: Multiple access and two-way channels with energy harvesting and bi-directional energy cooperation. In: 2013 Information Theory and Applications Workshop, pp. 1–8. IEEE Press, San Diego (2013)
17. Tutuncuoglu, K., Yener, A.: The energy harvesting and energy cooperating two-way channel with finite-sized batteries. In: 2014 IEEE Global Communications Conference, pp. 1424–1429. IEEE Press, Austin (2014)
18. Baidas, M.W., Alsusa, E.A.: Power allocation, relay selection and energy cooperation strategies in energy harvesting cooperative wireless networks. Wirel. Commun. Mob. Comput. **16**(14), 2065–2082 (2016)
19. Lei, M., Zhang, X., Yu, B.: Max-min fairness scheme in wireless powered communication networks with multi-user cooperation. In: Chellappan, S., Cheng, W., Li, W. (eds.) WASA 2018. LNCS, vol. 10874, pp. 211–222. Springer, Cham (2018). https://doi.org/10.1007/978-3-319-94268-1_18
20. Bertsekas, D.: 6.253 convex analysis and optimization, spring 2004. Athena Sci. **129**(2), 420–432 (2004)

Research on Cloudlet Placement in Wireless Metropolitan Area Network

Zhanghui Liu[1,2], Yongjie Zheng[1,2], Bing Lin[3(✉)], Xing Chen[1,2],
Kun Guo[1], and Yuchang Mo[4]

[1] College of Mathematics and Computer Science, Fuzhou University,
Fuzhou 350116, China
[2] Fujian Key Laboratory of Network Computing and Intelligent Information
Processing, Fuzhou 350116, China
[3] College of Physics and Energy, Fujian Normal University, Fujian, China
WheelLX@163.com
[4] Fujian Province University Key Laboratory of Computational Science,
Huaqiao University, Quanzhou, China

Abstract. With the development of 5G technology, the Internet of Everything is becoming possible, and the communication traffic of mobile terminals will explode. Meanwhile, the requirements of mobile terminals in terms of fluency and the computing power of applications are also becoming more stringent. However, the computing power of mobile devices is always limited due to their portability. Edge computing can offer a timely manner by offloading tasks of the mobile device to nearby cloudlet. Therefore, the computing tasks can be processed quickly nearby network edge, which can effectively reduce the system delay. Although there are many researches on cloudlet placement technology, how to optimize the cloudlet placement in a given network to improve the performance of mobile applications is still an open issue. This paper mainly proposes a particle swarm optimization algorithm based on genetic algorithm (PSO-GA) to optimize the cloudlet placement in a wireless metropolitan area network, aiming at reducing the average response time for users to process tasks. The simulation results show that the PSO-GA approach performs better in user service quality and reduces system average response time compared with other cloudlet placement schemes.

Keywords: Edge computing · Wireless metropolitan area network · Cloudlet placement · System average response time

1 Introduction

With the gradual advancement of the Internet of Everything, the network connection objects are expanding from people to things. Mobile phones, pad and other devices are very important in social life, learning, social contact, work and other fields. Meanwhile, smart devices, such as Sensors, smart meters and smart cameras, are widely used in industry, agriculture, medical care, education, transportation, smart home,

© Springer Nature Singapore Pte Ltd. 2019
Y. Sun et al. (Eds.): ChineseCSCW 2019, CCIS 1042, pp. 183–196, 2019.
https://doi.org/10.1007/978-981-15-1377-0_14

environmental protection and other industries [1]. According to IDC's statistics, China's internet data traffic will reach 8806EB, with a compound annual growth rate of 49% in 2020 [2].

The communication traffic of mobile terminals shows explosive growth. At the same time, users and enterprises are increasingly demanding the fluency of the terminal and the user experience. Mobile applications are becoming more computationally intensive, requiring more computing power. However, the computing power, storage capacity and battery life of mobile devices are limited by their size [3].

Edge Computing came into being. OEC defines edge computing: edge computing provides small data centers, namely edge nodes. Cloudlet is a cluster of trusted computers that can connect to high-speed Internet and serve mobile devices. The use of cloudlet technology can alleviate the battery pressure of mobile devices, improve the storage and processing capabilities of data, and strive to achieve ultra-low delay transmission of information [4]. Compared with mobile edge computing (MEC) and fog computing, Cloudlet is primarily used for mobile enhancements and provides rich computing resources. Especially, video analysis application focusing on edges can extract labels and metadata of edge data and transfer them to the cloud to achieve efficient global search [5].

By offloading tasks computationally to the nearest cloudlet, users can significantly reduce delay and costs. User access cloudlets have the advantages of the short delay, single hop, high bandwidth and low cost, and can get a real-time response. It is one of the effective ways to reduce the system response time. However, compared with cloud computing centers, cloudlets have the following disadvantages: cloudlet can only access a small range of Wi-Fi APs, and computing resources are significantly insufficient compared to cloud data centers. Therefore, in the wireless metropolitan area network (WMAN), it is important to study how to place several cloudlets and make full use of the resources of these cloudlets.

However, there is little focus on cloudlet placement. How to optimize cloudlet placement to improve mobile application performance remains to be further studied. This paper focuses on the research of cloudlet placement schemes for complex network environments in WMAN. We focus on solving the following optimization problem: select K $\geq$ 1 APs in the WMAN to place the cloudlet. Users can offload tasks to the cloudlet with the smallest transmission and waiting time so that the average waiting time for completing the user task processing is minimum under the placement scheme of the K cloudlets. In this paper, we mainly propose a particle swarm optimization algorithm based on a genetic algorithm (PSO-GA). Simulation experiments show that the proposed algorithm can effectively optimize the cloudlet placement.

2   Related Work

First, the resource-Rich cloud center is a good choice to improve the performance of mobile applications. Cloud computing has enormous computing resources and processing abilities. However, it is far away from mobile devices, and the data delay may be lengthy. As more and more devices access to the network, the remote cloud is easy to cause congestion, resulting in high access delay, which seriously affects the network performance. Therefore, someone proposes the cloudlet model where a smart device can offload computing tasks.

The cloudlet is composed of computers and usually placed on APs. It has more computing power and processing power than mobile devices [6]. Mobile users often offload the tasks to a virtual machine (VM) and then loads it into the cloudlet for execution. Therefore, users can offload more tasks to the VM in the cloudlet, and once the cloudlet executes the tasks on the VM, will return the result to users [7].

In the paper [8, 9], the existing frameworks and algorithms for task offloading are mainly involved in the cloud, but cloudlet quickly replaces the cloud as the offloading destination [10, 11]. However, most of the research focused on how to allocate resources for the received requests, pay little attention to cloudlet placement [12].

Paper [13] effectively reduce data transmission time by combining edge computing and cloud computing. Moreover, paper [14] offers an adaptive offloading framework for android applications in mobile edge computing. Tan [15] considered the cloud with unlimited capacity and used a perspective model to provide a progressively optimal cloud choice for each request.

Cloudlet has limited resources and coverage. Capacity and user movement limit the stability of connections. In [4], in the case of access to multiple cloudlets, the author considers human mobility behaviour and cloudlet resource usage to optimize distributed computing offload, reduce terminal energy consumption, and efficiently solve computational offload problems.

Besides, some researchers have observed that cloudlet can communicate with another one. Jia [16] propose an optimal algorithm to offload tasks from overloaded cloudlets to other under-loaded cloudlets instead of the cloud.

In the paper, we studied the cloudlet placement, which is similar to facility location issues, but they are essentially different. We define cloudlet is a cluster of computers coexisting with AP. Besides, users within the coverage of the AP are free to choose the appropriate AP, and users can be self-sufficient without a cloudlet.

3   Models and Problem Formulation

3.1   WMAN System Model

WMAN system consists of several APs that can connect to each through the internet and sets of users can access the network through AP. We define graph G = (V, E) represent the relationship between the users and APs. In addition, V = AP ∪ User. E contains two types of edges: One is the direct connection between two APs with a low-latency high-speed transmission delay, any AP contained in G can visit another AP through the high-speed internet; another is a wireless connection between the user and AP, through which the user can access the network.

Figure 1 is a simple diagram of the WMAN system. All APs communicate with each other through a wired high-speed network, and AP can directly connect to all users within its coverage [17]. Therefore, we define that users can connect to multiple APs. As shown in Fig. 1, users can select either AP 4 or AP 6. In a WMAN system, the amount of tasks generated by mobile users is floating and cannot be accurately predicted, especially when multiple applications are running at the same time. We define the average task amount of user in a certain period as the number of tasks of the user at this moment. We assume that each user can produce the task and entry the network with arrival rate λi.

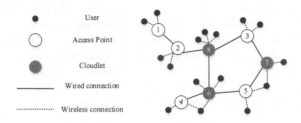

Fig. 1. A simple WMAN

Let $W = \{\omega_{ij}|0 \leq i < n, 0 \leq j < m\}$ be the delays between the user and his several wirelessly connected APs. Let $D = \{D_{ij}|0 \leq i, j < m\}$ be the transmission delays between APs.

In the network, the user passes the task requirements and offloads his task to the cloudlet for execution through the directly connected AP. The transmission is the first delay. We assume that the delay of tasks transmitted between the same APs through the network is equal. We define the matrix is $D \in R^{m*m}$, where D_{jk} represents the transmission delay generated by the task between the AP_j and AP_k. First, $user_i$ transmits the task to the AP_j, which requires the wireless delay ω_{ij}. At this point, if the task needs to be scheduled to the cloudlet on the AP_k, then the task needs D_{jk} delay to transfer from AP_j to AP_k.

3.2 Offloading System Model

We also define the tasks into the cloudlet as a queuing network. Then tasks can be executed on one of the K cloudlets. We assume that in the same WMAN system, each cloudlet can offer c servers with service rate μ. When the user needs to offload the task to the cloudlet, the waiting time is composed of the transmission and queue time.

In this part, we introduce the *FuncQ*, which can return the average queue time according to its service and arrival rate λ

$$Func_Q(\lambda) = C(c, \lambda/\mu)/(c\mu - \lambda) \tag{1}$$

Where

$$C(c, \rho) = \frac{\left(\frac{(c\rho)^c}{c!}\right)\left(\frac{1}{1-\rho}\right)}{\sum_{k=0}^{c-1}\frac{(c\rho)^k}{k!} + \left(\frac{(c\rho)^c}{c!}\right)\left(\frac{1}{1-\rho}\right)} \tag{2}$$

Formula (2) is Erlang's formula [18].

According to formula (1), the higher arrival rate to the cloudlet, the longer queuing time it is. If the arrival rate is too high, the queue time will be very long, which will seriously increase the average response time of the entire WMAN system, resulting in bad user experience.

$$\Lambda(k) = \sum\nolimits_{u_i \in User_j} \lambda_i \tag{3}$$

Where user$_j$ represents the set of users that are scheduled to the cloudlet$_k$.

The average queue waiting time on the cloudlet placed at AP$_k$ is:

$$t_{cloudlet}(k) = Func_Q(\Lambda(k)) + 1/\mu \tag{4}$$

The average waiting time of user$_i$ in WMAN is the sum of transmission delay and queue waiting delay.

$$t_i = \omega_{ij} + D_{jk} + t_{cloudlet}(k) \tag{5}$$

Therefore, the average response time of the system is:

$$\text{ave\_t} = \frac{1}{n} \sum\nolimits_{i=1}^{n} t_i \tag{6}$$

3.3   Cloudlet Placement

We define three sets of $Cloudlet = \{cloudlet_i | 0 \leq cloudlet_i < n\}$, $ConnectAp = \{connectAp_i | 0 \leq connectAp_i < n\}$ and $RunAp = \{runAp_i | 0 \leq runAp_i < n\}$, which represent the placement location of the cloudlet, the AP selected while user task scheduling, and the cloudlet location where the task runs.

We define the cloudlet placement in a network as follows: Given an integer K and system model parameters $(G, \Lambda, W, D, \mu, c)$, we need to find the placement of cloudlets among the APs such that the system response time in Eq. (6) is minimized.

$$\mathop{min}\limits_{cloudlet}\text{ave\_t} \tag{7}$$

4   Cloudlet Placement Scheme Based on PSO-GA

In this part, we first discuss two heuristic algorithms for dealing with cloudlet placement and then propose a PSO-GA algorithm to solve the problem. Related researchers have proposed two heuristic algorithms, Heaviest-AP First (HAF) algorithm and Density-Based Clustering (DBC) algorithm [19]. Compared with our algorithm in simulation experiments, it shows that using our PSO-GA algorithm to place the cloudlet can effectively reduce the average response time.

4.1   Heuristic Algorithm

The Heaviest-AP First (HAF) algorithm places the cloudlet directly to the AP where the user's arrival rate is the heaviest. The algorithm first counts the arrival rate of the AP by calculating the task arrival rate of all users directly connected to the AP. Then take the top K APs for the cloudlet placement. However, the HAF algorithm has a major disadvantage. The AP with the largest arrival rate may not be necessarily the AP closest to the user it serves. On the other hand, although some APs only connect some users wirelessly, the AP is not far from most users. The AP should be a better cloudlet placement point.

Density-Based Clustering (DBC) algorithm places APs with a relatively large working density as cloudlet. When the arrival rate is too high, the cloudlet will overload resulting in a long time queuing time.

4.2   PSO-GA Algorithm

We proposed an algorithm combining the PSO algorithm and the GA operator (PSO-GA). By improving the PSO algorithm and adding the genetic operator rules, the cloudlet placement was optimized.

GA. Genetic algorithm (GA) [20] is a probabilistic search algorithm derived from the evolution of biological evolution. The basic idea comes from the evolution process of the population. Each evolution selects the adaptable individual to generate new individuals, and the new population replaces the old population. It is not a simple random optimization. According to the principle of survival of the fittest, the algorithm can generate better and better offspring by mutating and crossing the previous generation of outstanding individuals and gradually find better offspring.

PSO. Particle Swarm Optimization (PSO) algorithm [21] is a random search algorithm based on group collaboration developed by simulating bird foraging behaviour. In the PSO algorithm, each solution is a "bird", recorded as a particle. All particles evaluate their fitness by function and use velocity to guide particle flight. Particles search the answer space by following the current optimal particle. The features are simple and easy to implement, there are not many parameters to adjust, and there is a wide range of applications.

PSO-GA. We propose a PSO-GA algorithm based on the PSO algorithm and adding the rules of genetic algorithm operation. The PSO-GA algorithm combines the advantages of the PSO and the GA algorithm. The algorithm uses the crossover operator, and the mutation operator can avoid the premature convergence of the PSO algorithm, which enhanced the diversity of population evolution and effectively reduced the system average response time.

Particle Coding Strategy. The coding method will affect the search efficiency and performance, so we need a good coding strategy to solve the cloudlet placement. Rules: In WMAN, if you need to place K cloudlets, the encoding length of the particle is K. If the AP is selected, the AP number is recorded in the particle. In particle, the value is incremented, and the value does not exceed the maximum value of the AP sequence number.

If K = 5 cloudlets need to place, and APs with numbers 2, 3, 5, 8, and 10 are selected, the particle code is shown in Fig. 2.

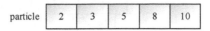

| particle | 2 | 3 | 5 | 8 | 10 |

Fig. 2. An example of particle encoding

Fitness Function Strategy. The fitness of particles is the main indicator for evaluating the quality of particles. Generally, particles with smaller fitness correspond to better solutions. We use the formula (7) to calculate the average response time as the fitness of the particle. If the time of the particle is shorter, the particle is better.

Particle Iteration Strategy. Firstly, we randomly initialize several particles and calculate the fitness of each particle by the function, and record it as the optimal value of the particle itself. Then record the individual with the best fitness. In each iteration, the particles are obtained through the mutation operator and the intersection of the excellent individuals to obtain new particles, and the optimal value of each particle is selected, and the optimal value of the population particles is entered into the next generation.

Traditional PSO particle update strategy:

$$V_i^{t+1} = w * V_i^t + c_1 r_1 \left(p\_best_i^t - X_i^t \right) + c_2 r_2 \left(g\_best^t - X_i^t \right) \tag{8}$$

$$X_i^{t+1} = X_i^t + V_i^{t+1} \tag{9}$$

Where t represents the current number of iterations. V_i^t and X_i^t respectively represent the velocity and position of the i-th particle at the t-th iteration. w is the inertia weight, indicating that the particle can maintain the current speed. $p\_best_i^t$ and g_best^t respectively represent the optimal historical value of the particle itself and the historical value of the population after t iterations. r_1 and r_2 are two random factors. c_1 and c_2 are learning factors, which can control the ability of the particle to learn its optimal historical value and the historical value of the population. However, the traditional PSO particle update strategy is suitable for solving continuous problems. It is easy to converge prematurely in the iterative process and fall into local optimum so that a more satisfactory result cannot be obtained. The improved PSO-GA algorithm particle introduces the crossover operator of the genetic operator, and the update strategy formula is:

$$X_i^{t+1} = c_2 \oplus C_g \left(c_1 \oplus C_p \left(w \oplus Mu(X_i^t), P_{best_i}^t \right), g\_best^t \right) \tag{10}$$

During the process of updating, the particles undergo the three processes of individual mutation, individual crossover and social population crossover, so that the particles search toward the optimal solution.

As shown in Fig. 3, a particle has a w-probability individual mutation in the iterative process, and w is the inertia weight. As the iteration proceeds, the particle's mutation probability becomes lower and lower.

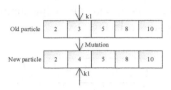

Fig. 3. An example of particle mutation

The mutation of the particle individual itself:

$$A_i^{t+1} = w \oplus Mu(X_i^t) \tag{11}$$

For example, Fig. 4 is the crossover of the individual (social) cognitive part. K_1 and K_2 respectively represent the two positions of the randomly selected coded particles in the crossover operator and replace the code between the two positions with the same position in p_best and g_best.

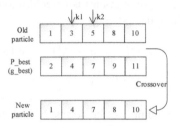

Fig. 4. An example of particle crossover

Individual cognitive crossover formula:

$$B_i^{t+1} = c_1 \oplus C_p(A_i^{t+1}, p\_best_i^t) \tag{12}$$

Social population cognitive crossover formula:

$$X_i^{t+1} = c_2 \oplus C_g(B_i^{t+1}, g\_best^t) \tag{13}$$

Parameter Settings. The learning factor $c_1(c_2)$ of the individual (the population), and the original inertia weight w: gradually decrease with the iteration, which means that the individual tends to maintain his original particle, and the optimal particle learning ability for individual history and population is reduced:

$$c = (c_{start} - c_{end}) * i/\text{iteration} \tag{14}$$

$$w = w_{max} - (w_{max} - w_{min}) * i/\text{iteration} \tag{15}$$

However, the original inertia weight w is suitable for linear problem-solving. In this paper, cloudlet placement is a nonlinear problem, so the new inertia weight formula is:

$$w = w_{max} - (w_{max} - w_{min}) * exp^{\frac{d(X_i^{t-1})}{d(X_i^{t-1})-1.01}} \qquad (16)$$

The meaning of formula (17) is the rate of the difference between the current particle and the best placement.

$$d(X_i^{t-1}) = \frac{diff(X^{t-1}, g\_best^{t-1})}{K} \qquad (17)$$

Algorithm Flow.

Algorithm: PSO-GA
Input: G, Λ, W, D, μ, c, K
Output: Cloudlet
1:        Initialize particles :
2:        **for** i ← 1 to iterations **do:**
3:            generate a new particle Randomly   // Each particle represents a cloudlet placement
4:            fitness ← ave_t   //use Eq. (6) evaluate the fitness of particle
5:            **if** fitness < g_best.fitness **then:**// the particle is the g_best
6:                g_best ← particle   g_best.fitness ← fitness
7:            **end if**
8:        **end for**
9:        Update iterative particles with genetic operators:
10:        **for** j ← 1 to iterations **do:**
11:            Update particle[j]   // Update particles based on genetic operator rules
12:            **if** particle[j].fitness < particle[j].p_best.fitness **then:**

13:                particle[j].p_best.fitness ← particle[j].fitness
                particle[j].p_best ← particle[j]
            // the particle[j] is the p_best

14:                **if** particle[j].fitness < g_best.fitness **then:**

15:                    g_best.fitness ← particle[j].fitness
                    g_best ← particle[j]
                // The particle is the g_best

16:                **end if**
17:            **end if**
18:            **Output** g_best and g_best.fitness(cloudlet and ave_t)

5   Simulation

In this section, we define some parameters related to the system (Table 1).

Table 1. System parameters

Symbol	Definition	Default value
n	Users number	100–150
m	APs number	50
D_{ij}	Communication delay of the link AP_i, AP_j	/
K	Cloudlets number	/
λ_i	Task arrival rate of $user_i$	$0 \le \lambda_i \le 2.99$
ω_{ij}	Wireless data-rate between $user_i$ and AP_j	$0.1 \le \omega_{ij} \le 0.4$
μ	Servers service rate	10
c	Cloudlet servers numbers	8

We simulate the WMAN system environment through simulation experiments. First, there are m APs, and the APs form an undirected connection diagram. We calculate the shortest path between two APs through the dijkstra algorithm and record it into the matrix D. We use the average user location and arrival rate replaces the moment. Then we use the placement algorithm to get the best cloudlet placement in this case.

We also adopt the setting of [19]. In each experiment, the arrival rate λ_i of the task is suitable for a normal distribution with an average of 2.0 and a variance of 0.5. λ_i is not more than 2.99. ω_{ij} is the wireless delay that the user connects to the AP through Wi-Fi, and delay of the user to the AP by simulating the real physical distance from the user to the AP. And satisfy $0.1 \le \omega_{ij} \le 0.4$. Besides, the server service rate in the cloudlet is 10, and the cloudlet consists of 8 servers.

5.1   Benchmark Algorithm

To evaluate our proposed PSO-GA algorithm, we introduced three benchmark algorithms. The first benchmark algorithm is Random algorithm. It works by randomly selecting K APs to place the cloudlet. The results of the Random experiment are obtained from an average of 100 Random algorithm. The other two-benchmark algorithms are the two heuristic algorithms introduced in this paper HAF algorithm and DBC algorithm.

5.2   Algorithm Performance Comparison

First, in the WMAN system, we calculate the average response time generated by the different cloudlet placement scheme and then draw a line graph to compare.

By looking up the data and the parameter settings in reference [22], the relevant parameters of our PSO-GA algorithm are set to: Maximum number of iterations 1000, initial population size 100, the max of w is 0.9, the min of w is 0.4, $c1_{start} = 0.9$, $c1_{end} = 0.3$, $c2_{start} = 0.2$, $c2_{end} = 0.9$. The inertia factor w, the self-cognitive factor c1 and the population cognitive factor c2 are set using Eqs. (14) and (16).

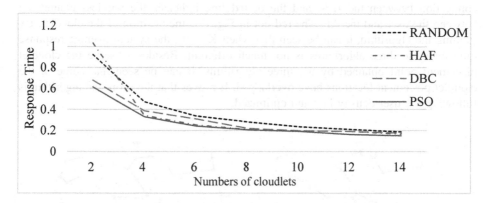

Fig. 5. System average response times by different numbers of cloudlets

In Fig. 5, we show several experiments in the case of different cloudlet placements in the WMAN with APs m = 50 and users n = 150 and as the number of cloudlet placement changes, the system average response time changes. We can see that under all algorithms, the system time always decreases with the increase of the cloudlets. However, it turns out that as the cloudlets increases, the decline rate is slow down. We can say that we should consider the scale of WMAN, put a suitable number of cloudlets, do not waste cloudlet resources, and use effective algorithms to place cloudlets properly to achieve maximum cost performance.

In Fig. 5, the proposed algorithm is always better than other algorithms, and when K > 8, there are enough cloudlets in the network, so that all user tasks can be offloaded to the nearby cloudlet. Therefore, each user can save transmission and waiting time, and the system response time will tend to be stable, no longer significantly reduced.

Since our algorithm is an intelligent search algorithm, under a suitable parameter setting, we can always find a better solution that approximates the optimal solution under a global range. The essential meaning of the random algorithm is the response time on average. The DBC algorithm performs close to the PSO-GA algorithm, but sometimes it takes into account that the user task is too dense and some cloudlets are overloaded, which affects the system response time. The performance of the HAF algorithm is worse than DBC algorithm. The result of HAF algorithm at K = 2 is longer than the other three placement methods. The cloudlet placed on two APs with the largest arrival rate, but the APs may not be the closest to the other users, the average response time is higher than other algorithms.

Meanwhile, some APs are only connecting to some users, but most users in the network are not far from the AP. Therefore, the AP should be a better choice to place cloudlet. Then, when K = 4, with the increase of the cloudlet numbers, the HAF algorithm approaches the DBC algorithm, and the system average response time is greatly reduced.

5.3   Different Algorithms for Cloudlet Placement

For example, Fig. 6 shows the cloudlet placement algorithms used in WMAN. Figure 6 is a WMAN instance of 50 APs and 150 users. The solid line indicates the wired connection between the APs, and the dotted line indicates the wireless connection between the user and the AP. The red dot in Fig. 6 is the position of the cloudlet. By running the algorithm, it can be seen that when K = 10, the system average response time by these three algorithms is not much different. Besides, observe the cloudlet placement point obtained by the three algorithms. It can be seen that some of the cloudlet placement locations have overlapped. It proves that the proposed algorithm is following the user density law and optimized.

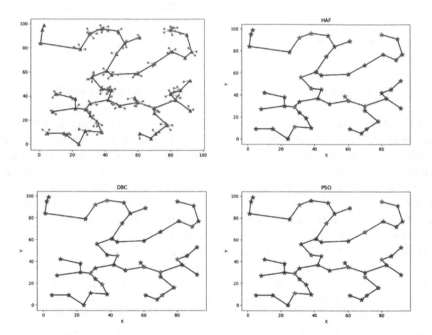

Fig. 6. Cloudlet placement of three algorithms when K = 10 (Color figure online)

6   Conclusion

Mobile devices with limited computing power can reduce the system response time through cloudlet technology. Although there have been many researches on cloudlet offloading technology, optimizing the cloudlet placement in a given network is usually ignored. In this study, a particle swarm optimization algorithm based on a genetic algorithm (PSO-GA) was proposed to optimize the cloudlet placement in WMAN, aiming at reducing the average response time for users to process tasks. We found that a reasonable and appropriate cloudlet placement solution could reduce the system response time with better cost performance. Compared with the HAF and DBC algorithms, our PSO-GA algorithm can optimize user service quality efficiently.

Since we focus on the cloudlet placement, the impact of task scheduling is ignored in our model. Meanwhile, we do not consider the tasks redirecting to nearby idle cloudlets, resulting in a large increase in system response time with the overloaded cloudlets. In the future, we will consider the impact of task scheduling and the tasks redirecting in a new model.

References

1. Liu, X., Yang, Z., Hu, Z., et al.: Seamless service handoff based on delaunay triangulation for mobile cloud computing. In: International Symposium on Wireless Personal Multimedia Communications. IEEE (2013)
2. Wang, F.-S., Wang, G.-C., Peng, Y.: Study of energy consumption minimization data transmission strategy in mobile cloud computing. J. Chin. Comput. Syst. **40**(3), 560–566 (2019)
3. Shi, Y., Xu, X., Lu, C., Chen, S.: Distributed and weighted clustering based on d-hop dominating set for vehicular networks. KSII Trans. Internet Inf. Syst. **10**(4), 1661–1678 (2016). https://doi.org/10.3837/tiis.2016.04.011
4. Xu, X.: Research on computation offloading strategy based on mobility behavior analysis in distributed mobile cloud computing. Beijing University of Posts and Telecommunications (2017)
5. Li, Z., Xie, R., Sun, L., Huang, T.: A survey of mobile edge computing. Telecommun. Sci. **34**(1), 87–101 (2018)
6. Clinch, S., Harkes, J., Friday, A., Davies, N., Satyanarayanan, M.: How close is close enough? Understanding the role of cloudlets in supporting display appropriation by mobile users. In: 2012 IEEE International Conference on Pervasive Computing and Communications (PerCom), pp. 122–127. IEEE (2012)
7. Ha, K., Pillai, P., Richter, W., Abe, Y., Satyanarayanan, M.: Justin-time provisioning for cyber foraging. In: Proceeding of the 11th Annual International Conference on Mobile Systems, Applications, and Services, pp. 153–166. ACM (2013)
8. Kemp, R., Palmer, N., Kielmann, T., Bal, H.: Cuckoo: a computation offloading framework for smartphones. In: Gris, M., Yang, G. (eds.) MobiCASE 2010. LNICST, vol. 76, pp. 59–79. Springer, Heidelberg (2012). https://doi.org/10.1007/978-3-642-29336-8_4
9. Zhang, Y., Liu, H., Jiao, L., Fu, X.: To offload or not to offload: an efficient code partition algorithm for mobile cloud computing. In: 2012 IEEE 1st International Conference on Cloud Networking (CLOUDNET), pp. 80–86. IEEE (2012)
10. Kosta, S., Aucinas, A., Hui, P., Mortier, R., Zhang, X.: ThinkAir: Dynamic resource allocation and parallel execution in the cloud for mobile code offloading. In: 2012 Proceedings IEEE INFOCOM, pp. 945–953. IEEE (2012)
11. Shiraz, M., Abolfazli, S., Sanaei, Z., Gani, A.: A study on virtual machine deployment for application outsourcing in mobile cloud computing. J. Supercomput. **63**(3), 946–964 (2013)
12. Tong, L., Li, Y., Gao, W.: A hierarchical edge cloud architecture for mobile computing. In: IEEE INFOCOM 2016 (2016)
13. Lin, B., et al.: A time-driven data placement strategy for a scientific workflow combining edge computing and cloud computing. IEEE Trans. Ind. Inform. https://doi.org/10.1109/tii.2019.2905659
14. Chen, X., Chen, S., Ma, Y., Liu, B., Zhang, Y., Huang, G.: An adaptive offloading framework for android applications in mobile edge computing. Sci. China Inf. Sci. https://doi.org/10.1007/s11432-018-9749-8

15. Tan, H., Han, Z., Li, X.-Y., Lau, F.C.: Online job dispatching and scheduling in edge-clouds. In: IEEE INFOCOM 2017 (2017)
16. Jia, M., Liang, W., Xu, Z., Huang, M.: Cloudlet load balancing in wireless metropolitan area networks. In: IEEE INFOCOM 2016 (2016)
17. Cai, X., Kuang, H., Hu, H., Song, W., Lü, J.: Response time aware operator placement for complex event processing in edge computing. In: Pahl, C., Vukovic, M., Yin, J., Yu, Q. (eds.) ICSOC 2018. LNCS, vol. 11236, pp. 264–278. Springer, Cham (2018). https://doi.org/10.1007/978-3-030-03596-9_18
18. Kleinrock, L.: Queueing systems, volume i: theory, pp. 101–103 (1975)
19. Jia, M., Cao, J., Liang, W.: Optimal cloudlet placement and user to cloudlet allocation in wireless metropolitan area networks. IEEE Trans. Cloud Comput. **PP**(99), 1 (2015)
20. Holland, J.H.: Adaptation in Natural and Artificial Systems. University of Michigan Press, Ann Arbor (1975)
21. Kennedy, J., Eberhart, R.: Particle swarm optimization. In: IEEE International Conference on Neural Networks, pp. 1942–1948. IEEE (2002)
22. Shi, Y., Eberhart, R.: A modified particle swarm optimizer. In: IEEE International Conference on Evolutionary Computation Proceedings. IEEE World Congress on Computational Intelligence, pp. 69–73 (1998)

Research on Customer Segmentation Method for Multi-value-Chain Collaboration

Lei Duan[1,2], Wen Bo[1,3], Qing Wen[4], Shan Ren[5],
and Changyou Zhang[1(✉)]

[1] Laboratory of Parallel Software and Computational Science,
Institute of Software, Chinese Academy of Sciences, Beijing,
People's Republic of China
changyou@iscas.ac.cn
[2] School of Computer and Communication Engineering,
University of Science and Technology Beijing, Beijing,
People's Republic of China
[3] School of Computer and Information Technology, Shanxi University,
Taiyuan, Shanxi, People's Republic of China
[4] Chengdu Guolong Information Engineering Co., Ltd., Chengdu,
Sichuan, People's Republic of China
[5] School of Information Science and Technology,
Southwest Jiaotong University, Chengdu, Sichuan,
People's Republic of China

Abstract. For multi-value-chain collaborative business in automotive industry, the value-based customers segmenting has become an important method to improve the synergy efficiency of automobiles. In order to accurately discover the value of potential customers, this paper proposes a customer segmentation method for multi-value-chain collaboration. Firstly, we screened evaluation index with high degree of customer value relevance, and establish a value-based customer data representation model according to customer's information we collected on the collaborative marketing platform of the automobile marketing value chain; Then, according to the distribution characteristics of the customer information, we used improved initial centroid selection method for k-means algorithm to establish customer segmentation method. Finally, based on the customer data accumulated on the car collaborative business platform, design an experiment to verify the accuracy of customer segmentation. The result of experiment shows that the customer segmentation method effectively reduces the computational complexity. This method can guide the designing of multi-value chain coordination mechanism for customer segmentation and create more value of both automobile production value chain and sales value chain.

Keywords: Multi-value-chain · Customer segmentation · Customer representation model · Cluster analysis · K-means algorithm

© Springer Nature Singapore Pte Ltd. 2019
Y. Sun et al. (Eds.): ChineseCSCW 2019, CCIS 1042, pp. 197–211, 2019.
https://doi.org/10.1007/978-981-15-1377-0_15

1   Introduction

The automobile services include three stages: pre-sales service, in-sales service and after-sales service. Pre-sales service as the link between marketing and sales is very important for the entire automobile service chain. Automobile pre-sales service is a series works that stimulate customer's purchase desire before they contact the product. The main purpose of pre-sales is to accurately grasp and satisfy customer's actual needs. The international competition in automobile consumer market makes the traditional pre-sales service method hard to acquire the subtle needs of users. It needs to be improved by modern information technologies. The competition between automobile companies has gradually become the competition between industry chains. Pre-sales service is at the forefront of the entire sales value chain, and is also the window between automotive products and customer service brands. It is the closing point of the automotive product life cycle. Therefore, automotive pre-sales services are critical to understand user's ability and build a user-centric product service strategy. Moving the focus to customer service is helpful to keep existing customers and discover potential customers. Provide different customers with personalized services can create value for customers and maximize enterprises' profits. It becomes an effective method to enhance the competitiveness of enterprises, and enhance the value creation ability of the automobile sales value chain. In 1950s, American scholar Wendell Smith proposed the theory of customer segmentation. Customer segmentation refers to the company formulating precise strategic strategies based on its own business operation model, classifying customers according to various factors such as customers' natural characteristics, value, purchasing behavior, needs and preferences in certain marketing, and providing targeted products, services and marketing models to customers within the limited resources [1–3]. Customer segmentation in automobile pre-sales services can help companies develop precise strategic strategies based on their business models. On the one hand, customers are classified according to their natural characteristics, value, purchasing behavior, needs and preferences to provide targeted products, services and marketing models. On the other hand, it needs to identify potential customers' acquirements and provide them with accurate and satisfactory services.

2   Related Works

2.1   Related Research on Customer Segmentation

In the research of customer segmentation, scholars from various countries have proposed a variety of methods [4]. Commonly, the enterprise selects the necessary customer segmentation variables and indicator systems from many segmentation variables according to the background of the enterprise customer base. The normal customer segmentation methods include: simple statistics-based customer segmentation, behavior-based customer segmentation, value-based customer segmentation, data mining-based customer segmentation, etc. [5].

The simple statistics-based customer segmentation method is based on demographic characteristics such as age, gender, and region, or based on social characteristics such as occupation, education, marital status, and occupation. Behavior-based customer segmentation methods classified customers primarily based on customer's behavior characteristics. Value-based customer segmentation is the way that companies use for themselves, and in most cases, it used with behavior-based customer segmentation. The data mining-based customer segmentation extracts the segmentation variables to describe customers from their basic attributes and purchase behavior. Using the relevant data mining technology to establish the customer segmentation model [6], which divides customers into different categories [7].

2.2   Cluster Analysis

The traditional customer classification methods either divide customers into high, medium and low categories according to the purchase price, or divide into new customers and old customers based on purchase date. Both two methods are rough and subjective, hard to accurately predict consumption behavior of customer bases. Clustering is unsupervised learning, no label in advance, compute cluster based on the similarity between data. The objects in same class have a relatively high similarity, while the objects in different classes have lower similarity [8]. The cluster analysis can be divided into many kinds [9, 10].

Partitioning based-clustering. For a given data set and the value of categories k, the data set can be divided into k different categories according to a similarity measure, ensuring that each category has one data item at least and each data item can only belong to one category [11]. Hierarchical-based clustering. This is also a common method in clustering, which is to hierarchically decompose a given data set and form each data object into a clustering tree [12]. Density-based clustering. The main idea of this method is to divide data objects into similar categories according to the density of the data set. If the density in the "neighborhood" exceeds the threshold, the merge will continue [13]. Model-based clustering. This method can be used to locate each category using the density function of the data space, or can determine the number of clusters from standard statistics [14]. Network-based clustering. This type of method is applied to any attribute data, and the method is fast in processing, can identify categories of any shape, and input fewer parameters, less affected by isolated points.

According to the data and practicality provided by the enterprise, this paper chooses the k-means algorithm based on partition [15], which is a relatively classic unsupervised clustering algorithm, also called distance-based clustering algorithm. The method is easy to implement, and the algorithm has good efficiency and scalability on large data sets. Therefore, it has been widely used in customer segmentation [16–18].

2.3   The Main Idea of K-means

The k-means algorithm uses Euclidean distance [19] to measure the similarity, the closer the two objects are, the higher the similarity. The basic idea is to first determine the number k of clusters and select k initial cluster centers from the data sets. Then calculate the distance from the dataset to the k initial cluster centers, and merge the dataset to the cluster where is closer. Re-adjust the cluster center of each cluster until the cluster centers are not changing, indicating that the clustering is ended and get k classes [20].

This algorithm works best if the objects of the dataset is dense and the difference between the classes is significantly larger. Even facing the relatively large dataset, the algorithm is relatively scalable and efficient, and the complexity of the algorithm is O (nKI). Calculating the new cluster center is the same time complexity as calculate the clustering criterion function value, and the time complexity required is O(nd).

3 The Customer Segmentation Model

3.1 Construction of Customer Segmentation Model

In this paper, the customer value segmentation model is established for automobile sales depend on the improved k-means algorithm. The constructed indicators of this model are basic attributes and consumption behavior of the customer. The basic attributes include customers' gender, age, annual income, education level, etc. Consumer behavior is the automobile model they purchased and the spent. Taking each customer's basic attributes and consumption behavior attributes as a dimension, the basic data about the customer becomes a multi-dimensional space, and then using the clustering algorithm to build the customer segmentation model.

Represent customer information as a customer space $\{C1, C2, \ldots, Cn\}$, where each C represents a customer's attribute dimension. A concept class $\{cl_{s1}, cl_{s2}, \ldots, cl_{si}\}$ can be determined according to the customer space, where $cl_{si}(1 \ll i < k)$ represents a concept cluster, a group of customers [22], such as "High-income people". If the customer belongs to a concept class, then all attribute dimensions in the customer space can determine which cluster it belongs to in the concept class. This process completes the customer segmentation and finally performs a functional analysis of the results of the classification.

According to the model established above, the customer segmentation mainly uses the clustering algorithm to analyze the historical customer data of the enterprise, and constructs a model that can predict the future category of the customer and applies it to the enterprise marketing. The managers of the enterprise analyze the results of the subdivision to formulate targeted marketing strategies, provide suggestions for hierarchical services for different segments of the customer segmentation, and dig out the maximum value of customers.

3.2 The Improvement of Customer Segmentation Algorithm

The improvement of this paper is based on the theory that the farthest point is the least likely to be assigned to the same cluster [23]. This method can effectively prevent the objective function from falling into local optimum, also avoids that the selected points are too close, causing multiple selected cluster centers in the same class. The process of selecting the initial clustering center for the k-means algorithm by this improved method is as follows: first select the two points p and q which are the farthest distance as the first two initial cluster centers, and record them as $V_1 = o_p$, $V_2 = o_q$, $d_1 = d_{pq}$.

Then, all the remaining points are classified into V1 and V2 as the center [24], and two classes are recorded as d_{21} and d_{22}. For any point i $\in$ {1, 2, ..., n} (excluding p, q), if $|o_i - V_1| < |o_i - V_2|$, then o_i is attributed to the V_1 class, otherwise it belongs to V_2. Then calculate the distance from the data in d_{21} to V_1 and the distance from the data in d_{22} to V_2:

$$\begin{cases} d_{21} = max\{|o_i - V_1|, \, o_i \in D_{21}\} \\ d_{22} = max\{|o_i - V_2|, \, o_i \in D_{22}\} \end{cases} \tag{1}$$

$d_2 = max\{d_{21}, d_{22}\}$ then the corresponding data is recorded as V_3 as the third initial cluster center. And so on, it is executed repeatedly until k initial cluster center points are selected.

3.3   The Effect of Improved Clustering Algorithm

To verify the feasibility and effectiveness of the initial clustering center selection, in this paper we selected three sample sets which are Balance scale, Iris and Wine datasets from UCI [25] for comparative analysis. Table 1 indicates the size, category numbers, and dimensions of the three sample sets, while Table 2 lists each cluster's data number of three sample sets.

Table 1. The basic attribute of sample set.

Sample set name	Sample size	Number of clusters	Dimension
Balance-scale	625	3	4
Iris	150	3	4
Wine	178	3	13

Table 2. The cluster number of sample set.

Clustering category	Balance-scale	Iris	Wine
First cluster	49	50	59
Second cluster	288	50	71
Third cluster	288	50	48
Total	625	150	178

Using three sample sets to finish experiments, the comparison is mainly based on the stability and accuracy of the clustering results. The accuracy of the clustering result is calculated as follows:

$$P = \frac{n}{N} * 100\% \tag{2}$$

P is accuracy, n is the correct number of classification, and N is the total number of samples. The clustering results of two methods are shown in Table 3.

Table 3. Experimental simulation results.

Number of clusters	Experimental data					
	Balance-scale data		Iris data		Wine data	
	Initial center	Accuracy	Initial center	Accuracy	Initial center	Accuracy
1	2585864	0.4276	1915119	0.5612	5016174	0.5213
2	230416549	0.6140	14514142	0.6183	8950173	0.5725
3	10630509	0.6178	13982139	0.7010	15017857	0.6500
4	375138437	0.6193	1410963	0.5633	991453	0.6123
5	91389179	0.4789	10587145	0.7988	16715170	0.5217
6	1651889	0.5988	11940139	0.6944	104140154	0.6210
7	509145387	0.5990	1723147	0.5074	167106140	0.7011
8	1642530	0.4856	1409645	0.5744	1096184	0.7011
9	220517439	0.4956	7913250	0.6700	40121125	0.5997
10	360116509	0.4891	1146769	0.6597	9079127	0.6590
Average accuracy		0.5358		0.6359		0.6160
New method accuracy		0.6048		0.8900		0.6730

As we can see from the above table, the method of randomly selecting the initial cluster center makes the clustering result unstable and a relatively low accuracy. The method we adopted to select the initial clustering center in this paper is better than the original method and is relatively stable. It can be applied to handle a large amount of data generated in enterprise production, and can produce relatively good customer segmentation results in the automotive industry.

4 Verification of Automotive Industry Collaboration Platform

4.1 The Choice of Customer Segmentation Variables

For different industries, there are many attributes can describe the characteristics and differences of customers. So it results different customer segmentation variables and numbers. According to the data of automotive industry and internal business of the enterprise, this paper considered various factors of customers. Based on customer's basic data provided by the enterprise and the data quantifiable principle, selected the segmentation variables from two aspects: the customer's basic attributes and the car purchase behavior. Customer's consumption amount and purchase frequency can be used to measure the current value to the enterprise, and basic attributes such as gender, income, age and education can be used to measure its future value to the enterprise. In customer's profile, the information includes the vehicle purchase records and service personnel information. Before clustering, we need to find some basic attributes of customers according to the kind of business and the quality of historical data. The basic attributes include name, age, region, gender, income, car brand, car model, color, purchase amount, vehicle use, etc. In the first step, we need to extract customer-related data and integrated into a customer information table as the data source for establishing customer segments.

For raw data, first filter out useless attributes such as name, customer number, and so on. Obtain a data source table that can be analyzed, and select the following attributes as the cluster sample attributes as shown in Table 4.

Table 4. Cluster sample attributes.

Attribute name	Attribute value range	Type of data
Gender	{Male, Female}	Class attribute
Age	{0–100}	Numerical type
Income (ten thousand yuan)	{≥0}	Numerical type
Address	{Provinces, Municipalities, Autonomous regions}	Class attribute
Transaction amount (ten thousand yuan)	{Integer greater than 0}	Numerical type
Number of transactions	{Integer greater than 0}	Numerical type
Brand	{A B C D E}	Class attribute
Color	{Company's existing model color}	Class attribute
Model	{The models produced by the company}	Class attribute
Education	{Primary school, Junior high school, High school, Bachelor, Master's degree, Doctor}	Class attribute
Whether to buy again	{0, 1}	Numerical type

4.2 The Discretization and Normalization of Data

According to the information in the enterprise's existing database, we extracted 11 variables from the database. These data not only have numeric attributes but also other types of attributes. However, the k-means algorithm is a distance-based clustering algorithm and cannot handle non-numeric data. Therefore, in practical applications, the customer data needs to be extracted, cleaned, converted, and also needs to be discretized [26], which means some non-numeric-type attributes are converted into integers by using some data encoding method. Some of the attributes are handled as follows in Table 5.

Data discretization does not consider the importance and relevance of the relevant attributes, so the discretized data needs to be normalized. At the same time, it can reduce the number of iterations and improve the convergence speed. The normalization is to map the data value of a certain attribute to a specific range, and eliminate the deviation of the clustering result due to the difference size of numerical value. This method is mainly used in neural networks and distance-based classification and cluster mining. K-means algorithm we used is a distance-based algorithm. Therefore, the normalization process can eliminate the unfair clustering effect caused by different value ranges of each attribute.

Table 5. Cluster sample attributes.

Indicator name	Indicator value
Gender	1: Male 2: Female
Age	1: [0–30] 2: [30–40] 3: [40–50] 4: [50–65]
Education	1: Primary school and below 2: Junior high school 3: High school 4: Associate and above
Model	1: A 2: B 3: C 4: D 5: D
Address	0001: Chengdu 0101: Mianyang 0102: Deyang
……	……

We used a common method for data normalization which is Z-score normalization [27]. The method is to normalize the data set's mean to 0 and the variance to 1. The processed data conforms to the standard normal distribution. The normalization method is shown in following equation:

$$x' = \frac{x - \mu}{\sigma} \tag{3}$$

The x' represents the value of x after normalization, μ represents the mean, and σ represents the standard deviation.

4.3   The Result of Customer Segmentation

According to the past business experience of AA companies on the automobile industry chain collaboration platform, the company's customers are divided into 4 categories, and the value of K equals 4. We summarized the results for each cluster, as shown in Table 6. The area counted in this table have relatively good sales. The top three digits of the car model number in the enterprise are used to classify the product categories as A, B, C, D, and E, and the color of the model is relatively fixed.

Table 6. Cluster sample attributes.

Cluster number	1	2	3	4
Number of clients	7894	15675	10396	5851
Client ratio	19.83%	39.39%	26.11%	14.70%

4.4   Results Analysis

According to the different subdivision results, the categories are divided as Table 7:

Table 7. Analysis of clustering results.

Cluster number	1	2	3	4
Number of clients	7894	15675	10396	5851
Client ratio	19.83%	39.39%	26.11%	14.70%
Average consumption	4.6w	8.8w	6.8w	11.6w
Average age	37	42	50	40
Male	86.86%	87.02%	86.28%	82.53%
Female	13.14%	12.98%	13.72%	17.47%
Main area	Sichuan 14.4% Shandong 7.9% Henan 7.5%	Sichuan 19.5% Chongqing 9.6% Guizhou 8.9%	Sichuan 15.3% Hebei 6.8% Henan 9%	Sichuan 17.8% Chongqing 13% Henan 6.4%
Main product category	B 25.67% C 20.98%	B 35.78% D 12.45%	A 35.78% B 19.24%	C 29.47% E 16.29%
Color	Plain white 32.0%	Pearl White 40.3%	Pearl White 26.1%	Pearl White 40.1%
Best seller	SQJ6451B	SQJ6460C	F16-T01	B60X-T02

Evaluate the different characteristics of each category, we can realize that the second and third types of customers accounted for 65.5% of the total customers, and the transaction amount accounted for 78.69% of the total transaction, which indicates that these two types are the core customers of the enterprise. From the average age of customers, most of them from 37 to 50 years old, and male customers' accounts for a large proportion. The color of the automobile is mainly white. From the aspect of customer's area distribution, customers are concentrated in Sichuan, Chongqing, Henan, Shandong and other regions. From the product categories that customers mainly purchase, A, B, and C are the main products of the company, while B is the best-selling product. It can be seen from the Table 7 that the customer group of the company is monotonous, and the scope of the covered consumer groups is very narrow. Most customers are men, lack of female customers, and the colors are mainly dominated by white. Based on the above statistics, each type of customer has the characteristics of its consumption behavior. Different consumption behaviors can be viewed through the customer information attribute in the clustering result. Relevant personnel in the enterprise can formulate marketing policies according to the attributes of customers, and maximize the value created by the customers. In addition, the manufacturer can analyze customer's purchase characteristics, understand their target customer's needs, and seize the market to provide relevant service.

5 Supporting Service Quality Based on Customer Segmentation

After completing the segmentation of customers, the statistical analysis method is used to analyze the data between different customer groups. Enterprises need to use limited resources, making corresponding marketing strategies and implementation, to achieve the actual value of customer segmentation, improve the service quality of enterprises. According to customer's commonality and characteristics, sales personnel made corresponding marketing plans for different customer groups to achieve differentiated marketing services. Combined the potential customer management module to predict the purchase will of potential customers, optimize the original sales business process and create more value.

5.1   Differentiate Marketing Strategies

Customers can be divided into four types through customer segmentation: core customers, high-end customers, normal customers, and potential customers.

The core customers made greatest contribution to the company and is the main source of enterprise's profits. In normal times, it is necessary to improve their service quality and let them experience a sense of superiority. The measures include set up a VIP area and point strategy, or give a certain repair discount, etc. To ensure that core customers can be retained for a long term, and enhanced customer loyalty. The high-end customers have relatively high purchasing power and most of them are male. According to that, the enterprise can provide them certain products or maintenance services to keep them and create greater value for the company. The normal customer's transaction amount is low, but their amount is large. They have the characteristics of dissociation, always make comparison of corresponding products from multiple enterprises, and pay attention to the sales activities. Therefore, enterprise can develop some promotion strategies to increase the consumption propensity of normal customers. The potential customers are also a huge market. According to the car model recommended by the system, set a tracking plan to provide them relatively correct requirements in an accurate and timely manner, making them become the real customer of enterprise.

5.2   Evaluation and Analysis of Service Quality

The customer segmentation can improve service quality, differentiated services can be provided for different types of customers to improve their satisfaction. Develop more potential customers while retaining existing customers, thereby maximizing the profits and enhancing corporate competitiveness. To evaluate the effect of customer segmentation in pre-sales service, this paper proposes a pre-sales service evaluation system based on the enterprise service capability model. The service evaluation system includes service operation and service performance, as shown in Fig. 1.

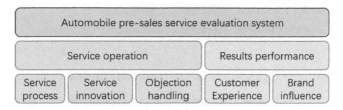

Fig. 1. Automobile pre-sales service evaluation system.

The service operation factors are used to evaluate the specific service operation of the enterprise, including customer evaluation of service processes, service innovations, and the disputes in services. Service performance factors are the qualitative or quantitative evaluation used to evaluate service outcomes, including customer experience and brand impact. The system default weight of each evaluation factor item given by the enterprise in Table 8 (the weight of each evaluation element can be adjusted according to the needs of different enterprises):

Table 8. Service quality evaluation element weight.

Service process	Service innovation	Objection handling	Customer experience	Brand influence
2	2	1	4	1

Based on the automobile pre-sales service evaluation system, we randomly selected 2,000 customers from AA companies to finish service quality assessment studies. After segmenting 2000 customers, the result shows that core customers accounted for about 43%, high-end customers accounted for about 27%, normal customers accounted for about 18%, and potential customers accounted for about 12%. Dividing 2,000 customers into two parts (maintaining the same customer's proportion of four types), 1000 of them provide unified pre-sales services, and another 1000 customers provide differentiated services according to their type. In the group which provide unified pre-sales service, we collect and summarize points of the service quality scored by customers, and shown in Fig. 2.

Fig. 2. Traditional service quality score table.

The result is, 235 customers have purchased the car of this company, and effective customers account for 23.5%. In the customer group with differentiated marketing strategy, provide targeted customer service for different customer types, and summarize points of the service quality scored by customers, and shown in Fig. 3.

Fig. 3. Differentiated service quality score table.

The result is, 371 customers have purchased the car of this company, and effective customers account for 37.1%.

5.3   Collaboration of Multiple Value Chain

Customers evaluate the service quality of multiple sale providers, and multiple enterprises achieve the collaboration of marketing value chain through the evaluation system. From the above part, we can know that the service quality score table is based on the service quality assessment and the conversion rate of effective customer. After customer segmentation, using differentiated marketing strategies, the service satisfaction of potential customers and general customers has been slightly improved, but high-end customers and core customers have been greatly improved. It effectively aroused more purchase desires of high-end and core customers, which stimulated healthy competition among enterprises and prompted the quality of service. At the same time, it has effectively improved the satisfaction of customers, converted more potential customers into effective customers, and brought more customer resources to entire enterprises group. The customer segmentation reduced the waste of human resources, improves the efficiency of enterprises, and promoted the collaboration efficiency of the automobile multi-marketing value chain.

There are two main tasks of the manufacturing factory, one is the development of new cars, and the other is automobiles production. The development of new cars must first go through market research, figure the model and market goals, and then start the designing. Finally, the quality inspection begins after a series of adjustments, and after the quality inspection is passed, new car's data is entered into the database and arranged by the company's business department for production. The business department issues the production task to the production workshop, and the purchasing department submits the applications to purchase the materials and parts. After the production workshop produces a car, it has to pass the quality inspection, and then register and put into storage. When receiving the shipping order, the logistics department is responsible for transporting the car out of the warehouse. The above process constitutes the production value chain of the manufacturing factory. The sales department provides a series of services such as registration information, product introduction, test drive, and handles the delivery procedures after customer placed the order.

To improve the operation process of internal value chain, the company strengthens the links between various parts and establishes the industrial collaboration value chain as we shown in Fig. 4.

Based on large amount of customer data, dealers finished customer segmentation to integrate customer's feedback and market information. Sharing customer's feedback data to the new vehicle develop department of the manufacturer, it will provide a reference for the future development, which will help the manufacturer to design a car that better meets the market demand and bring a greater benefit to the manufacturers and dealers. At the same time, the dealers sharing the data with business management department of the manufacturer. Through the research on customer's feedback and market requirement, the business department adjusts the promotion and inventory of each type of car in the next quarter. Reduce the possibility of inventory backlog of vehicles while ensuring sufficient sales volume. The business department will share the main production type of next quarter and part of publicly available sales plan to the dealer, and guarantee to provide sufficient cars to dealers. When introducing the product to customers, the dealers will focus on the main production type. This measure will reduce the decrease of customer satisfaction caused by the lack of stock and the effect of brand reputation. Through the interaction process, the collaborative relationship between the manufacturing value chain and the marketing value chain is established to achieve a win-win goal.

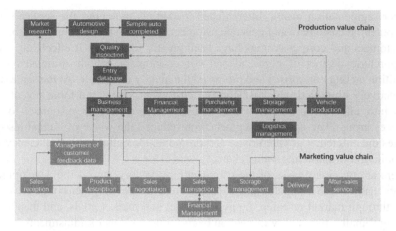

Fig. 4. Value chain collaboration.

6 Conclusion and Prospect

Verification experiment shows that cluster analysis using K-means algorithm can achieve a well effect of customer segmentation in automotive multiple value chain collaborative services. The results of customer segmentation and the consumption behavioral characteristics of each type of customer will effectively improve the collaborative efficiency between the marketing value chain and the manufacturing value chain, provide individualized and accurate services for customers. This method has a strong practical guiding significance. The successful application of cluster analysis will promote the development of customer relationship management in automotive service industry and optimize the collaboration between multiple value chains.

Acknowledgment. The authors would like to thank the anonymous referees for their valuable comments and helpful suggestions. This paper is supported by The National Key Research and Development Program of China (2017YFB1400902).

References

1. Parsell, R.D., Wang, J., Kapoor, C.: Customer Segmentation, WO/2014/099928 (2014)
2. Wang, J., Yin, L., Xu, J.: Research and Application of LNG Customer Segmentation Based on Cluster Analysis. Zhonghai Petroleum Gas and Electricity Group Co., Ltd. (2019)
3. Chen, Z.-Y., Fan, Z.-P., Sun, M.: Multi-kernel support tensor machine for classification with multitype multiway data and an application to cross-selling recommendations. Eur. J. Oper. Res. (11), 34–37 (2016)
4. Ou, J., Cao, X., Zhang, J., Ding, C.: Research on power customer segmentation based on hybrid neural network. Comput. Digit. Eng. (2019)
5. Zhang, H.: Research and Application of Clustering ensemble Algorithm in Customer Segmentation. Anhui University (2016)

6. Zhang, L., Ma, Y.: Research on aviation customer churn and segmentation based on data mining technology and implementation of R language program. Math. Pract. Theory (2019)
7. Fan, X.: Research on Customer Segmentation and Marketing Strategy of H Bank Based on Cluster Analysis. East China University of Science and Technology (2016)
8. Zhang, Y., Zhou, Y.: A review of cluster algorithms. J. Comput. Appl. (2019)
9. Li, Z., Zhang, Y.: Analysis and evaluation of clustering analysis algorithm. Electron. Technol. Softw. Eng. (2019)
10. Wang, X., Wang, H., Wang, J.: Comparison of clustering methods in data mining. Comput. Technol. Dev. **16**(10), 20–22 (2006)
11. Lai, X., Gong, X., Han, L.: Genetic algorithm based K-medoids clustering within MapReduce framework. Comput. Sci. **44**(3), 23–26 (2017)
12. Madan, S., Dana, K.J.: Modified balanced iterative reducing and clustering using hierarchies (m-BIRCH) for visual clustering. Formal Pattern Anal. Appl. 1–18 (2015)
13. Li, W.: The Study of Clustering Algorithm based on Density. Hunan University (2010)
14. Sharma, V.P., Sharma, S, Kumar, S., et al.: Cob web and dry bubble diseases in Lentinula edodes cultivation-a new report. International Society for Mushroom Science (2016)
15. Guo, D.: Research on Algorithms of Recommendation in Personalized Recommendation Systems. Beijing Jiaotong University (2017)
16. Wang, J.: The Application Research of Cluster Analysis in Caifupai Customer Segmentation. University of Electronic Science and Technology of China (2016)
17. He, L.: Customer segmentation of China telecom. Southwest Jiaotong University (2016)
18. Huang, Y.: Analysis of airline customer value based on k-means clustering method. Commun. World, 303–305 (2018)
19. Ma, C., Wu, T., Duan, M.: Clustering algorithm based on membership degree of K-nearest neighbor. Comput. Eng. Appl. **52**(10), 55–58 (2016)
20. Mokdad, F., Haddad, B.: Improved infrared precipitation estimation approaches based on k-means clustering: application to North Algeria using MSG-SEVIRI satellite data. Adv. Space Res. (2017)
21. Viegas, J.L., Vieira, S.M., Melício, R., Mendes, V.M.F., Sousa, J.M.C.: Classification of new electricity customers based on surveys and smart metering data. Energy (09), 123–125 (2016)
22. Lu, H.: Study on value-added service for power customer segmentation. Guangdong University of Technology (2016)
23. Huang, M., Lin, H.: A new method of k-means clustering algorithm with events based on variable time granularity. In: Web in formation Systems and Applications Conference. IEEE (2017)
24. Fan, Z., Sun, Y.: Clustering of college students based on improved K-means algorithm. In: Computer Symposium. IEEE (2017)
25. UCI Data Set. [EB/OL]. http://archive.ics.uci.edu/ml/
26. Wu, C., Dong, S., Li, C.: Research on the discretization algorithm of association classification based on Gaussian mixture model. J. Chin. Comput. Syst. (2018)
27. Liu, J., Zhang, K., Wang, G.: Comparative study on data standardization methods in comprehensive evaluation. Digit. Technol. Appl. (2018)

A Data-Driven Agent-Based Simulator for Air Ticket Sales

Yao Wu[1], Jian Cao[1($\boxtimes$)], Yudong Tan[2], and Quanwu Xiao[2]

[1] Department of Computor Science and Engineering, Shanghai Jiao Tong University,
Shanghai, China
{wuyaoericyy,cao-jian}@sjtu.edu.cn
[2] Department of Air Ticket Business Ctrip.com International Ltd Shanghai,
Shanghai, China
{ydtan,qwxiao}@ctrip.com

Abstract. In order to better design sales strategy, air companies or travel agents would predict sales of air tickets. In this paper, we propose an agent-based ticket sales simulator based on data analysis. The features of air tickets and passengers are extracted by analyzing real data. A Long Short-Term Memory recurrent neural network is used to forecast the daily customer search volume. Then a purchase decision tree is designed and embedded into the customer agent to simulate the decision process when a customer tries to find and buy an air ticket. Experimental results show that our prediction model achieves better prediction accuracy than three compared approaches. Moreover, through the simulation experiment on the historical real data, we obtain good simulation results, and verify the validity and practicability of our ticket sales simulator.

Keywords: Simulator · Agent-based · Big data · Deep learning · Air ticket sales

1 Introduction

Currently, more and more people choose to travel by plane, and customers inquire and buy tickets through different platforms. At the same time, air companies and travel agents adjust their sales strategies dynamically in order to promote their sails since the competitions in the airline market are becoming fierce.

However, to predict the effect of sales strategies is difficult. It is very complex if not possible to find appropriate mathematical models to reflect the relationship between market effect and sails strategy since there are so many factors can affect the sales. It is also not possible to try one strategy and change it to another when the performance is not satisfying. Therefore, how to accurately simulate the sales environment of air tickets and the behavior of customers is very important.

The simulation of ticket sales also faces many challenges. First, the air ticket prices are affected by various factors, such as the number of days in advance to buy the ticket, the remaining seats, fuel prices and so on. These make the price

© Springer Nature Singapore Pte Ltd. 2019
Y. Sun et al. (Eds.): ChineseCSCW 2019, CCIS 1042, pp. 212–226, 2019.
https://doi.org/10.1007/978-981-15-1377-0_16

of air tickets show a certain periodicity and randomness. In addition, there are many types of customers taking different strategies for purchasing air tickets, so it is difficult to simulate the purchasing behaviors of customers. Besides, air companies or travel agents don't only care the profits but also customer stickiness and so on. At last, due to the large number of air companies and travel agents today, customers' purchasing behaviors are also influenced by competitive ticket providers, but the data obtained from other companies are also very limited.

If the above problems can be effectively solved, the companies can execute a variety of strategies on the simulated market environment, and select the optimal solution by comparing the results. In addition, they can make forward-looking plans and make some corresponding adjustment according to the change of virtual market. It can not only bring more substantial profits to the companies, reduce their risks, but also can provide customers with more affordable services.

In this paper, we design and implement an agent-based flight sales simulator driven by big data. Firstly, we analyze the real data to extract the features of air tickets and users. Then, by counting the distribution of data, we model the different features to characterize customer-agents and air ticket sales environment. Under demand of future environment simulation, we predict the search volume, the input of the simulator, and the prediction is based on the Long Short-Term Memory (LSTM) neural network model. On the basis of feature extraction and search volume prediction, a customer purchase decision model is designed to simulate the interaction between the customer and the purchase environment.

Experimental results show that our prediction model achieves better prediction accuracy than compared approaches. Through experiment on tickets sales simulation, we obtain good simulation results, which verifies the validity and practicability of our ticket sales simulator.

The paper is organized as follows. Section 2 discusses related work, highlighting the difference between our proposed approach and the existing ones. Section 3 introduces the structure of the simulator and the implementation principles and details of different modules. Section 4 shows the process of experiments and the analysis of the experimental results. We conclude the paper in Sect. 5.

2 Related Work

In the research field of simulators, the earliest researches are to simulate the problem from a macro point of view. Regression models are often used in academic literature to analyze the impact of certain factors on the results in the simulation process (Kim et al. 2008). In addition, with the development and maturity of neural networks, some literatures adopted neural networks and deep learning methods to study the influence of various factors on simulation results (Choua et al. 2010).

Recent studies have increasingly implemented agent-based models from an individual perspective in order to improve existing solutions and explore new ways to support more functionality. Different from the model from a macro perspective, the agent-based simulator simulates the problem by judging the behavior of each individual agent at the micro level. Janssen and Jager, who first proposed this model in the literature, used agent-based models to simulate the process leading to "deadlock" in the consumer market (Janssen and Jager 1999). One of the most successful models used in practice is the agent-based consumer goods market simulator developed by North and Macal for P&G (North et al. 2010). Okada and Yamamoto used agent-based simulation model to study the influence of online communication on the buying habits of B2C website customers (Okada and Yamamoto 2009). Most of the above literatures are based on the study of individual phenomena in a field through simulators, which take fewer factors into consideration and the data set is not large enough.

Some scholars proposed simulators for more macro problems in a more complex environment and with bigger data sets. By modeling e-commerce environment and users, Sava and Aleksander provide a set of simulation framework for customer performance of B2C websites (Sava and Aleksander 2017). By introducing the knowledge of data mining and combining the agent-based model, Duarte and Hugo established the simulator of user performance in the e-commerce environment (Duarte et al. 2018). Through the analysis of big data and combining with the classification and clustering technology of data mining, David and Chidozie built the agent-based mobile user performance model to study the problem of user loss (David and Chidozie). Although the above literatures studied the simulators under the general environment, most of them only provided a framework and direction for implementation, with little introduction to the specific implementation.

There is little research on simulators in the field of airline ticket sales. The PODS model, developed by Boeing and the Massachusetts institute of technology, is a prime example. Based on the demand forecast and inventory control model established by Boeing company, this model introduces passenger behavior, ticket product and other information, and simulates the market performance under the general environment of ticket sales by modeling the demand, passengers and purchase decisions. However, as this simulator was born in the 1990s, and the Internet technique is not developed enough, most of the analysis and decision-making are based on the questionnaire survey of aviation experts. Due to trade secrets and other factors, no specific implementation of each module is provided, which is not suitable for simulation in the internet era.

From what has been discussed above, the simulator research on the air ticket sales market based on the big data environment is a research topic to be explored.

3   Agent-Based Ticket Sales Simulator

The simulator implemented in this paper is agent-based simulator. It simulates the market by judging the behavior of each individual customer agent at the micro level. Through data distribution, we generate individual customer agent data that conform to the distribution characteristics, make purchase judgment on each customer agent individually, and finally integrate individual results to macro results, and then evaluate the simulator at the macro level. The whole simulator is divided into four modules: data analysis module, customer agent generation module, customer agent decision module and evaluation module. The structure diagram is shown in Fig. 1.

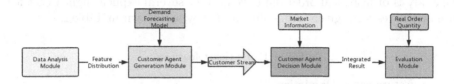

Fig. 1. Simulator structure

In the data analysis module, the first module, we analyze a large amount of data extracted from the big data platform, extract the features of customer agents and air tickets, and analyze the distribution of features. The second module, customer agent generation module, is responsible for generating the daily customer agent stream according to the demand model and customer agent feature distribution from the first module. Each customer agent in the stream is generated separately, and the values of features are randomly assigned according to the probability distribution. In the third module, the user decision module, we build a comprehensive purchase decision tree according to the actual process of the purchase of air tickets by the user, combining the characteristic values of each customer agent and the air ticket information obtained when the user requests. Through a series of complex decision-making process, we finally obtain the probability of the customer agent purchasing the air ticket. Because the purchase strategies and goals of different categories of customer agents are not the same, each type of customer agents is modeled separately. In the fourth module, evaluation module, we randomly get a certain result for each customer agent purchasing air tickets according to the probability obtained from the second module. With the simulation results, the total number of customer agents who purchased air tickets in the user stream are counted as the final simulated order quantity every day, and compared with the actual value to evaluate the performance of the simulator through a series of evaluation indicators.

3.1 Data Analysis Module

In this module, we extract the features of air tickets and customer agents, as well as the distribution of these features, based on the prior knowledges in the field of air tickets and analysis of the data from the big data platform of a large online travel network in China.

Features. Unlike books, scenic spot tickets and some other static commodities with relatively fixed properties, air ticket prices are unstable, and with time sensitivity. So, it is difficult to treat airfare of a certain flight as a fixed commodity. At the same time, the ticket also has some other explicit characteristic attributes, each of which may affect the purchase decision of users.

Based on the information from the flight ticket search result pages and with the analysis of historical order data, we extract several explicit flight ticket feature attributes for a given airline line, and they are shown in Table 1.

Table 1. Air ticket features

Feature name	Discription
dcity	The departure city of the flight
acity	The arrival city of the flight
price	Airfare
take_off_time	Departure time of the flight
order_date	Date of purchase
airline	The airline of the flight
class	The class of the ticket
subclass	The subclass of the ticket
dport	Departure airport (some cities have multiple airports)
aport	Arrival airport

The travel purpose of passengers on different air routes has bigger differences, for example, passengers on the Shanghai-Beijing route are mostly business travelers, while those on the Shanghai-Hainan route are mostly tourists, which results in different characteristics and purchase strategies of users on different routes. In order to simplify the problem, we use the data of a single airline, Shanghai-Los Angeles airline, for research in this paper. The Shanghai-Los Angeles airline is an international airline with a very large passenger flow, and the travel purposes of passengers are more abundant than those of domestic airlines, which meets our demand of simulation for a variety of passengers.

As for customer agent features, different customers have different purchase strategies in the air ticket purchase. For example, students with limited income may pay more attention to the price of tickets, while business people with work plans may pay more attention to the departure time and delay rate of flights, and those with higher income may pay more attention to the service and comfort level of airlines. Therefore, we need to establish different customer agent selection models to simulate the purchase behaviors of different categories of users. Similarly, feature extraction also needs to be done for different categories of customer agents.

Customer classification is also a complex problem, and there are many clustering algorithms for implementation. However, what we adopt is the existing customer classification within the company's big data platform. The classification is based on the company's long-term business experience in the field of ticket sales and is updated regularly. The classification process relies heavily on the user's previous purchases. The categories are individual user, business user, small group, lower ticket agent, and distributor.

After finishing the classification of customer agents, the features related to air ticket purchase are extracted from the user portrait provided by the company's big data platform. According to the prior knowledge in the aviation field, four customer agent features are extracted, which are enough to show the differences between different types of customer. The features are shown in Table 2.

Table 2. Customer agent features

Feature name	Discription
Time preference	The departure time of the flight preferred by the customer
Price preference	The customer's preferred airfare
Airline preference	The preferred airline
Class preference	The preferred class of airplane seats

Although there are corresponding values of these four features for each customer in the user portrait, we can't match each simulated customer agent to a specific user in the simulation process, and can only generate the values for each customer agent randomly through historical information. Therefore, we further extract the corresponding feature distribution of different categories of customers from the past order information, and analyzed and modeled each feature.

Distribution of Features. In this problem, we can only get the distribution of features through the analysis of the user's past purchase behavior. Therefore, we extract a large amount of historical order information from the company's big data platform, and try to analyze orders from different dimensions to obtain distribution characteristics of different features.

For the time preference, the time preference of different types of customer agents is converted into the probability of users appearing in different time periods, and it is measured through the proportion of orders of different categories of customer agents.

For price preference, it is basically the summary of the price of air tickets purchased by users in the past. But for a simulated customer agent, we cannot simulate the agent's past purchase record. As mentioned above, different categories of customer agents have different purchasing habits, so the price preference of individual customer agents can be determined according to the agent's category. Therefore, the evaluation of individual customer agents' price preference is transformed into the evaluation of different types of customers' price preference.

For airline preference and class preference, we also adopt the idea of time preference and convert the preference into the proportion of orders of various customers. In the analysis of the change of order quantity over time before, it has been found that the order quantity changes regularly with the change of weekday and takes one week as the cycle. Therefore, we mainly analyze how these preferences change over different weekdays in this part.

3.2 Customer Agent Generation Module

In the part, we get the size of the customer agent stream on the simulated day according to the demand forecasting model. Here, we take the users who have searched the air tickets of that day as the hidden air ticket purchasing customer agents. Feature values are randomly assigned to each individual customer agent according to the feature distributions from data analysis module. Then the generated customer agents will be sent to the decision model for decision simulation. The flow of the customer agent generation module is shown in Fig. 2.

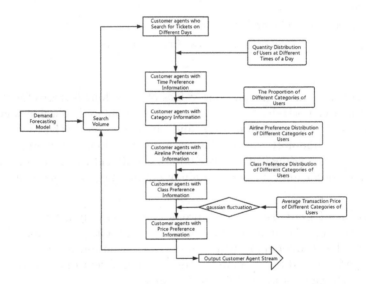

Fig. 2. The flow of the customer agent generation module

The demand forecasting model in this part is to predict the input of the simulator at a certain time in the future, that is, the prediction of the number of customer agents who may buy air tickets. In this question, we convert the demand into the number of users' searches.

Each search volume data is classified by two variables of time, search date and departure date. However, we need to reduce the time dimension as time series prediction can only have one axis on time. In air ticket sales, we usually consider the search data within 180 days before departure as reasonable. Therefore, for each search date, only the number of searches in the 180 days following the search date are counted. Then, we take the search date as the timeline of the prediction problem, and take the search quantity of tickets with different departure dates on each day as the features of each time node, which solves the problem of time dimension reduction.

In this study, Long Short-Term Memory (LSTM) deep learning framework is adopted to make predictions. It was proposed in 1997 and has been widely used to solve the problem of long-term dependence. This algorithm is very suitable to our problem for the number of searches per day being related to the number of searches in a short period and also related to a long time, such as the number of searches on the same date last month or the number of searches on the same date last year. Using LSTM algorithm can help us learn the potential correlation between the past data and the present data to make the prediction better.

3.3   Customer Agent Decision Module

After obtaining the customer agent stream from the customer agent generation module, we simulate the customer agent's purchase behavior by combining the extracted air ticket information. The start probability of each agent to buy an air ticket is assumed to be 100%, and the purchase probability of different air tickets is simulated according to the decision tree in Fig. 3. We choose the ticket with highest purchase probability as the target ticket, generate the purchase results of agents randomly according to the simulated probability. The decision process takes all the user features into account and combines some of the typical purchase behaviors of a customer agent purchasing an airline ticket. The interaction process between a customer agent and the environment is shown in Fig. 4.

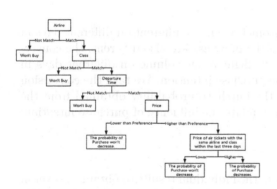
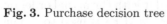

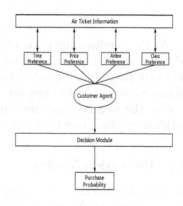

Fig. 3. Purchase decision tree

Fig. 4. Interaction between a customer agent and the environment

Table 3 shows the evaluation criteria of different indicators in the decision tree. In terms of price preference, we not only compare the air ticket price with the customer agent's price preference, but also compare it with the air ticket price of the same class and airline in recent three days to simulate the decision-making psychology of users who wait and see the ticket regularly.

Table 3. Decision indicators

Decision indicators	Decision criteria
Airline	If the airline of the ticket does not match the airline preference of the user, the user will not buy the ticket
Class	If the class of the ticket does not match the class preference of the user, the user will not buy the ticket
Departure time	If the departure time of the ticket is not within the time preference period of the user, the user will not buy the ticket
Price	Select the lowest price ticket of the same airline and class ticket for judgment; If the price of the ticket is lower than the price preference of the user, the probability of the user buying the ticket will not decrease; If the price of the air ticket is the lowest price of the same class air ticket of the same airline in the last three days, the probability of the user buying the air ticket will not decrease; If the price of the air ticket is higher than the price preference of the user, the probability of the user buying the air ticket will be reduced according to the price difference, which is realized by the Sigmoid function

As the purchase intention of customer agents is different on different days in advance, and the criteria for judging the purchase are also different, the conversion rate from flight search volume to flight order volume on different days in advance is calculated to measure the purchase intention. We take the conversion rate as the standard and multiply the purchase probability obtained from the decision tree by a new coefficient to simulate the influence of purchase intention.

3.4 Evaluation Module

In the evaluation module, we integrate the individual results to obtain the overall order quantity on the macro level and compared it with the actual order quantity on that day. The mean absolute percent error (MAPE) is used to evaluate the accuracy of the simulation results.

4 Experiment

Experiments are carried out separately according to the modules of the simulator. First, we extract the distribution of user characteristics by analyzing the data within the big data platform. Then, the demand of the simulator is forecasted based on the historical data of flight search volume. Finally, according to the experimental results of the first two parts, we generate the output data of the user generation module, and judge the user stream according to the user decision module, and finally evaluate the simulation results.

4.1 User Feature Distribution Extraction Based on Big Data Platform

In this experiment, according to the ideas in the previous chapter, we extract the historical order data from the company's big data platform and count the orders in different dimensions.

Time Preference. We calculate the proportion of orders in different time periods for different types of user from 2017 to 2018. The statistics are divided into two dimensions, namely, different weekdays and different time periods in each day, to analyze the changes in the order proportion of different types of users over time. Taking 2017 data as an example, Fig. 5 shows the proportion of various types of users in the passenger stream in different time periods within a day. Figure 6 shows the proportion of different users on different weekdays.

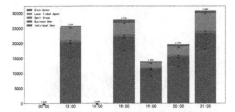

Fig. 5. Order proportion at different time periods of a day

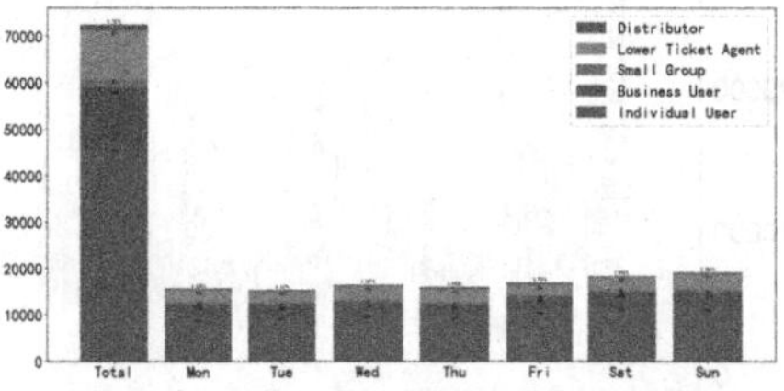

Fig. 6. Order proportion on different weekdays

As shown in Fig. 5, this airline has passengers in five time periods, and the proportion of different types of passengers in different time periods is very stable. As shown in Fig. 6, the total number of orders on this route is almost the same on every weekday, and the proportion of different categories of users on each weekday is also very stable. We also conduct cross statistics of the two dimensions and find that the proportion of all kinds of passengers is very stable no matter in the hour dimension or weekday dimension.

Therefore, here comes to the following conclusion: in different time periods, the proportion of different types of user orders remains basically unchanged. Based on the above conclusion, the proportion characteristics of the order quantity of different categories of users over time can be converted into a static proportion distribution. Then, based on the total number of users in different time periods of each day and according to the proportion of different categories of users, we assign users a category one by one according to the probability, and the number of times allocated is the total number of users. Since it is generated at the individual level, and the number of users per day is not too large, the total number of users from different category generated each time will not be the same, so this method can also simulate the data fluctuation over time.

Price Preference. We calculate the air ticket price distribution of different types of users from 2017 to 2018, and hope to find the characteristics of the data distribution. From the analysis of the price data, it shows that the price distribution is chaotic and there is no obvious rule, but it can be noted that the purchase price of the same type of users is relatively concentrated. Therefore, in order to reflect the differences in price preferences of various users, we take the average ticket price of different types of users as the measurement of price preferences. Taking the data of 2017 as an example, the average transaction price of all kinds of users is shown in Fig. 7.

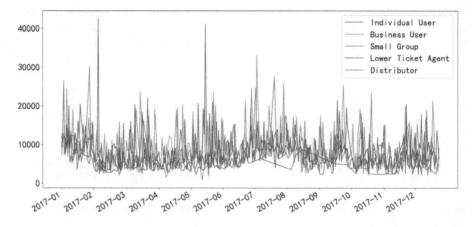

Fig. 7. Average daily ticket price distribution for all types of users in 2017

As can be seen from Fig. 7, individual users' ticket prices fluctuate less over time and are in a lower position among the five categories of users, while the ticket prices of other users also vary greatly and have their own characteristics. Therefore, these performances verify that we can effectively simulate the difference of price preference characteristics between different categories of users by taking average transaction price as the benchmark of price preference. For the generation of each specific user, a gaussian fluctuation is added to the average airfare to simulate the differences of different individuals.

Airline Preference and Class Preference. We adopt the same idea of time preference, and the result is also satisfactory. All kinds of users' choice of class and airline in different time is also stable. So airline preferences and class preferences are modeled in the same way that we model time preferences.

4.2   Demand Forecasting

In this experiment, we extract the search volume data from the big data platform from 2017 to 2018 for the experiment. These data are divided into training set and test set to evaluate the accuracy of demand prediction on the real data.

As mentioned earlier, we are considering a search with a maximum lead time of 180 days. So based on the previous design, for each day, we already have 181 features which are the total numbers of searches within 180 days before takeoff. In addition, some new features are added to the training sets. In the data analysis of the change of order quantity over time before, we knew that the order quantity had a certain periodicity every week, every month and every year. Therefore, we also add four characteristics: the year, month, date and weekday of the search date, trying to obtain more information according to the periodicity of the order quantity to make the predicted result more accurate. Before using the LSTM algorithm to train the data, the data is further preprocessed to improve the effect of the model.

In order to make the prediction more accurate, we use the method of single-step rolling prediction, that is, after each step prediction, the real result of the new day is added as a new training sample to the training set, so that the model can learn the information of the new day in the next round of learning, which makes the model obtain more information.

The effects of the model are evaluated through comparing with two classical time series prediction algorithms: Double Exponential Smoothing (Holt-Winters), Autoregressive Integrated Moving Average (ARIMA) and a classical machine learning algorithm k-nearest Neighbors (KNN). We adopt the commonly used valuation indexes for regression problems: Mean Absolute Error (MAE), Mean Squared Error (MSE) and Root Mean Squared Error (RMSE) to evaluate the predicted results. The experimental results are shown in Table 4.

Table 4. Prediction experiment results

Algorithm	Evaluation index		
	RMSE	MES	MAE
LSTM	17.94	322.14	7.21
ARIMA	71.24	5075.74	34.02
Holt-Winters	61.33	3760.92	30.11
RNN	75.96	5769.36	37.51

According to the comprehensive comparison of the three evaluation indexes, the performance of the LSTM method in the test set is far better than that of the other three methods, and each evaluation index is far less than that of other methods. The performance of the holt-winters method is slightly better than that of the ARIMA method and KNN method, while the performance of KNN is the worst. Thus, we verify the accuracy of our prediction model and verify that it can generate user search volume prediction data close to the real value for our simulator.

4.3 Purchase Simulation

According to the ideas in the previous sections, the simulator is implemented using Python programming language, and the 2017–2018 data on the Shanghai-Los Angeles airline have been used as the source data

We combine the predicted search volume, the characteristic distribution obtained by analysis the historical data, and the real flight ticket data as the input of the simulator. The output daily simulated order quantity is used to verify the simulation effect on the air ticket sales process of our simulator. Daily simulated order quantity and real order quantity are shown in Fig. 8.

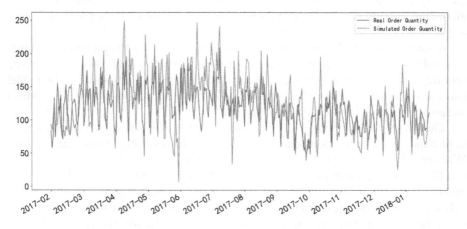

Fig. 8. Daily simulated order quantity and real order quantity

It can be seen from the image that the variation trend of broken line of the simulated order quantity is basically the same as that of the real order quantity, and the cycle fluctuation mentioned above is well simulated. The cycle of fluctuation is basically consistent with that of the real order quantity. But the fluctuation range of the simulator is obviously larger than the real data, which may be related to the lack of flight information.

We also compare the numerical values of simulated orders and real orders, and use mean absolute percent error (MAPE) between them on the entire data set as the measurement indicator. The experimental results are that the overall deviation between the result of the simulator and the real data is 27.0654%, which means 72.9346% accuracy. The high accuracy can verify the high reliability of our simulator.

5   Conclusion

In this paper, we design and implement a big data-driven agent-based flight sales simulator. We extract the features of users and airline tickets, and analyze the data distribution of features in historical data to model the customer agent features. LSTM deep learning network is used to realize the accurate prediction of ticket demand. Those above provide reliable input data for the simulator. A relatively complete decision process for purchasing air tickets is also designed to simulate the interaction between customer agents and the environment. The experimental results on real data verify the authenticity and reliability of our simulator, which makes our simulator play a guiding role in the market regulation for air ticket agents and air companies.

References

Greasley, A., Owen, C.: Modelling people's behaviour using discrete-event simulation: a review. Int. J. Oper. Prod. Manag. **38**(5), 1228–1244 (2018)

Box, G.E., Jenkins, G.M., Rrinsel, G.C.: Time Series Analysis: Forecasting and Control, vol. 734. Wiley, Hoboken (2011)

Brockwell, P.J., Davis, R.A.: Introduction to Time Series and Forecasting. Springer, Heidelberg (2006)

Kim, B.S., Kang, B.G., Choi, S.H., Kim, T.G.: Data modeling versus simulation modeling in the big data era: case study of a greenhouse control system. Simul. Trans. Soc. Model. Simul. Int. **93**(7), 580–594 (2017)

Changa, M., Cheungb, W., Laib, V.: Literature derived reference models for the adoption of online shopping. Inf. Manag. **42**, 543–559 (2004)

Choua, P.H., Lib, P.H., Chenc, K.K., Wua, M.J.: Integrating web mining and neural network for personalized e-commerce automatic service. Expert Syst. Appl. **37**(4), 2898–2910 (2010)

Ctrip Flight. http://flights.ctrip.com/. Accessed 10 May 2018

Bell, D., Mgbemena, C.: Data-driven agent-based exploration of customer behavior. Simul. Trans. Soc. Model. Simul. Int. **94**(3), 196–212 (2017)

Yang, F., Cao, J., Milosevic, D.: An evolutionary algorithm for column family schema optimization in HBase. In: IEEE International Conference on Big Data Computing Service and Applications (Big Data Service), pp. 439–445 (2015)

Forsythe, S., Liu, C., Shannon, D., Gardner, L.: Development of a scale to measure the perceived benefits and risks of online shopping. J. Interact. Mark. **20**, 55–75 (2006)

Gilbert, N., Troitzsch, G.: Simulation for the Social Scientist. Open University Press McGraw-Hill Education, London (2005)

Godes, D., Mayzlin, D.: Using online conversations to study word of mouth communication. J. Mark. Sci. Arch. **23**, 545–560 (2004)

Ferreira, H.S., Azevedo, J.: Framework for multi-agent simulation of user behaviour in E-commerce sites. Faculdade de Engenharia da Universidade do Porto (2016)

Duarte, D., Ferreira, H.S., Dias, J.P., Kokkinogenis, Z.: Towards a framework for agent-based simulation of user behaviour in E-commerce context. In: De la Prieta, F., et al. (eds.) PAAMS 2017. AISC, vol. 619, pp. 30–38. Springer, Cham (2018). https://doi.org/10.1007/978-3-319-61578-3_3

Hummel, A., Kern, H., Kuhne, S., Dohler, A.: An agent-based simulation of viral marketing effects in social networks. In: European Simulation and Modelling Conference (2012)

Janssen, M., Jager, W.: An integrated approach to simulating behavioural processes: a case study of the lock-in of consumption patterns. J. Artif. Soc. Soc. Simul. **2**, 2 (1999)

Wilson, J.L.: The Value of Revenue Management Innovation in a Competitive Airline Industry. Cornell University, New York (1993)

Kalekar, P.S.: Time series forecasting using holt-winters exponential smoothing, pp. 1–13. Kanwal Rekhi School of Information Technology (2004)

Kim, D., Ferrin, D., Raghav Rao, H.: A trust-based consumer decision-making model in electronic commerce: the role of trust, perceived risk, and their antecedents. Decis. Support Syst. **44**(04), 544–564 (2007)

Liu, X., Tang, Z., Yu, J., Lu, N.: An agent based model for simulation of price war in B2C online retailers. Adv. Inf. Sci. Serv. Sci. **5**, 1193–1202 (2013)

Moe, W.: Buying: differentiating between online shoppers using in-store navigational clickstream. J. Consum. Psychol. **13**, 29–39 (2003)

Alotaibi, M.B.: Adaptable and adaptive E-commerce interfaces: an empirical investigation of user acceptance. J. Comput. **8**(8), 1923–1933 (2013)

North, M., et al.: Multiscale agent-based consumer market modelling. Complexity **15**(5), 37–47 (2010)

Okada, I., Yamamoto, H.: Effect of online word-of-mouth communication on buying behavior in agent-based simulation. In: 6th Conference of the European Social Simulation Association (2009)

Said, B., Drogoul, A.: Multi-agent based simulation of consumer behavior: towards a new marketing approach. In: International Congress on Modelling and Simulation, MODSIM (2001)

Sava, C., Aleksandar, M.: Agent-based modelling and simulation in the analysis of customer behaviour on B2C ecommerce sites. J. Simul. **11**(04), 335–345 (2017)

Lee, S.: Seoul National University, Seoul, South Kerean (2000)

Understanding LSTM Networks. http://colah.github.io/posts/2015-08-Understanding -LSTMS/. Accessed 24 May 2017

Chen, Y., Cao, J., Feng, S., Tan, Y.: An ensemble learning based approach for building airfare forecast service. In: International Conference on Big Data. IEEE (2015)

Zhang, T., Zhang, D.: Agent-based simulation of consumer purchase decision-making and the decoy effect. J. Bus. Res. **60**, 912–922 (2007)

RETRACTED CHAPTER:
A New Information Exposure Situation Awareness Model Based on Cubic Exponential Smoothing and Its Prediction Method

Weijin Jiang[1,2], Yirong Jiang[3(✉)], Jiahui Chen[1], Yang Wang[1],
and Yuhui Xu[1]

[1] Institute of Big Data and Internet Innovation, Mobile E-Business Collaborative
Innovation Center of Hunan Province, Hunan University of Technology
and Business, Changsha 410205, China
jlwxjh@163.com, 18508488203@163.com,
810663304@qq.com, 363168449@qq.com
[2] School of Computer Science and Technology,
Wuhan University of Technology, Wuhan 430073, China
[3] Tonghua Normal University, Tonghua 134002, China
307553803@qq.com

Abstract. A lot of information in the social network is accompanied by the continuous transmission of users, and there are many forms of propagation, fermentation, evolution, emergence and outbreaks, which make it difficult for analysts to predict the information dissemination situation at the next moment. However, if the information dissemination can be effectively predicted and perceived, it plays a very important role in hot event discovery, personalized information recommendation, bad information early warning and so on. Therefore, the study of this problem is of great practical value. This paper first study of situational awareness information transmission method, including the definition of information dissemination situational awareness problem and expounds the basic thought, and analyzes the information dissemination situation and level of the modularity, the relationship between the three exponential smoothing is used for information dissemination model for situational awareness, and to evaluate the application effect of the model has carried on the detailed; In addition, this chapter also studies the prediction method of information spread outburst, including the definition of information explosion, the analysis of related factors that affect the prediction of information dissemination, and the modeling and evaluation of the information outburst prediction model. In addition, some issues related to which features are more sensitive to information explosion prediction are also studied.

Keywords: Social network information communication · Situational awareness · Propagation forecast

The original version of this chapter was retracted: The retraction note to this chapter is available at
https://doi.org/10.1007/978-981-15-1377-0_61

© Springer Nature Singapore Pte Ltd. 2019, corrected publication 2020
Y. Sun et al. (Eds.): ChineseCSCW 2019, CCIS 1042, pp. 227–242, 2019.
https://doi.org/10.1007/978-981-15-1377-0_17

1   Information Dissemination Situational Awareness Method

The basic idea of information dissemination situation awareness method is described, and the relationship between information dissemination situation and Modularity will be analyzed. At the same time, the three exponential smoothing method is used to model the information spread situational awareness model, and the model is carefully evaluated.

1.1   Basic Idea of Information Communication Situation Perception

The information released by celebrities or socialites can soon cause social hot spot, such as the 2014 Oscar Lee DeGeneres, a self photographed push, published on Twitte: with many film and television stars, and the 35 and 2 h after the release of 810 thousand and 1 million 670 thousand times. However, most grassroots users' information is rarely forwarded [1–3]. In addition, even if the two information is the same, its macro transmission situation may be significantly different, for example, the forwarding amount of the ripple and fireworks information propagation model mentioned in the previous chapter is likely to be at an order of magnitude [4–6].

In addition, one or more outbreaks may be formed during the propagation of information in a social network, which can cause two or more large-scale forwarding of information, so these two or multiple large-scale forwarding users can also be called an outbreak group. If t_1 has formed a certain scale of outbreaks at some time in the process of communication, then at some point in the future, it is likely that t_2 will form more large-scale outbreaks and lead to the outbreak of information, of which $(t_1 < t_2)$. Therefore, the basic idea of information dissemination situational awareness can be shown in Fig. 1 [7].

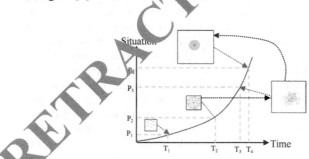

Fig. 1. Basic idea for perceiving information propagation situation

As shown in Fig. 1, as time goes on, the trend of information dissemination is also changing. At the time of T_1, the message has only 6 forwarding nodes; however, with the continuous attention of the social network users, the information has formed a certain scale of forwarding nodes at the time of T_2, but the overall situation of the information dissemination is still in ripple and does not have some characteristics of the information explosion; however, the information has been caused at the time of T_3. Therefore, the information dissemination situation awareness method based on time series can be designed based on the schematic diagram of Fig. 1, and the historical data in the network can be used to predict the future propagation situation [8, 9].

1.2 The Relationship Between Information Dissemination Situation and Modularity

Real Data Observation. Modularity is originally proposed by NEWMAN, which is used to evaluate the accuracy of community detection algorithm in social network [10–12].

Since the information transmission situation is dynamic with the change of time, this paper is based on the time sequence of information forwarding to carry on the average slice of the information transmission process. The time value calculation formula of the I part time slice is:

$$t(i) = t(start) + \frac{(t(end) - t(start))i}{k} \quad (1)$$

In which the $t(start)$ represents the initial release time of the information, and $t(end)$ represents the end of the information transmission. The k representative divides the entire information propagation tree into k portions. In this paper, the case 1 (3443749588269364) and case 2 (3436314995649160) are divided into 20 copies of the life cycle of the information propagation process [13], and the Modularity values of each information propagating subtree are compared, and the comparison results are shown in Fig. 2.

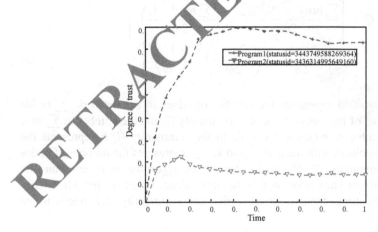

Fig. 2. Comparison with modularity of Case1 and Case2

From Fig. 2, we can see that the difference between the Modularity value of case 1 and case 2 is very obvious as time goes on. At first time, there is only one node (the publication node of original information), so both of the Modularity are 0; between second and sixth, the Modularity value of case 1 and case 2 increases gradually, and the

Modularity value of case 1 increases from 0.321 to 0.745. The Modularity value of case 2 increased from 0.153 to 0.202 and fell to 0.153, and then from seventh to twentieth, the Modularity values were also stable, around 0.727 and about 0.121, respectively. It can be seen that the information dissemination situation and the Modularity value have a very obvious mapping relationship [14–16].

Theoretical Verification. The physical meaning of Modularity is to calculate the difference between the proportion of the inner side of the same community in the social network and the expected value of the proportion of the internal network of the reference network under the same degree distribution [17–19]. Therefore, the higher the Modularity value, the better the result of community division. The specific calculation method of Modularity is:

$$Q = \frac{1}{2m} \sum_{vw} [A_{vw} - \frac{k_v k_w}{2m}] \delta(c_v, c_w) \tag{2}$$

In which, m represents the number of edges in the network, A represents the adjacency matrix of the social network, and k_v represents the degree value of the node v, so $k_v = \sum_w A_{vw}$; m represents the number of edges in the network; $\frac{k_v k_w}{2m}$ represents the probability of appearing in the base network. In addition, c_v represents the community of the partitioned node v, if v = w, that is, node v and node w belong to the same community, then $\delta(c_v, c_w) = 1$, otherwise $\delta(c_v, c_w) = 0$.

Because computational Modularity belongs to NP-complete problem, so a heuristic algorithm is used to calculate Modularity in this paper. The specific formula is:

$$Q = \left[\frac{\sum inside + 2k_{i.inside}}{2m} - \left(\frac{\sum total + k_i}{2m} \right)^2 \right]$$
$$- \left[\frac{\sum inside}{2m} - \left(\frac{\sum total}{2m} \right)^2 - \left(\frac{k_i}{2m} \right)^2 \right] \tag{3}$$

In the form (3), m will represents the number of edges in the network, $\sum inside$ represents the number of internal edges in the community C, and the number of $\sum total$ representing the number of edges to the node in the community C, k_i represents the number of edges associated with the node i, and $k_{i.inside}$ represents the number of nodes from node to community C. The main idea of the algorithm is to combine the community first and see each node as a community alone, based on the Modularity incremental maximization standard to determine the community that needs to be merged [20].

1.3   Information Dissemination Situational Awareness Model

The three exponential smoothing method is one of the nonlinear prediction methods, which can predict the trend of numerical change based on time series better [22]. Although the three exponential smoothing prediction method is more complex than the one and two exponential smoothing forecasting methods, they have the same objective of correcting the predicted values, and can better track the nonlinear variation trend of the time series [23]. In the process of information dissemination, we suppose $X = \{X_1, X_{12}, \ldots, X_i\}$ represents a time series vector, and S represents an exponential smoothing value, so $S_t^{(1)}$ represents an exponential smooth value of the t time and $S_t^{(2)}$ and $S_t^{(3)}$ represent the two and three exponential smooth values of the t moment, respectively. The method of calculating the first, two and three exponential smoothing values is:

$$
\begin{cases}
S_t^{(1)} = \alpha X_i + (1 - \alpha)S_{t-1}^{(1)} \\
S_t^{(2)} = \alpha S_t^{(1)} + (1 - \alpha)S_{t-1}^{(2)} \\
S_t^{(1)} = \alpha S_t^{(2)} + (1 - \alpha)S_{t-1}^{(3)}
\end{cases}
\tag{4}
$$

In which, X_i represents the actual value of the moment, the α represents the exponential smoothing factor, and the α satisfies $(0 < \alpha < 1)$. In addition, the three exponential smoothing prediction model can be expressed as:

$$
\hat{X}_{t \mid T} + \lambda \cdot \rho = T + \delta T^2
\tag{5}
$$

In which, $\hat{X}_{t+T}$ represents the current t time and the prediction period is T time prediction value. λ, ρ and δ represent the smoothing coefficients of exponential smoothing. The specific calculation method is:

$$
\begin{cases}
\lambda = 3S_t^{(1)} - 3S_t^{(2)} + S_t^{(3)} \\
\rho = \frac{\alpha}{(1-\alpha)^2}\left[(6 - 5\alpha)S_t^{(1)} - 2(5 - 4\alpha)S_t^{(2)} + (4 - 3\alpha)S_t^{(3)}\right] \\
\delta = \frac{\alpha}{2(1-\alpha)^2}\left(S_t^{(1)} - 2S_t^{(2)} + S_t^{(3)}\right)
\end{cases}
\tag{6}
$$

Based on the above analysis, this paper proposes an information spread situational awareness algorithm, which combines the three exponential smoothing methods with the computational Modularity method. Information propagation situational awareness algorithm such as Algorithm 1.

Algorithm 1 Information Communication Situation Awareness Model Simulation
1: Obtain information propagation nodes vector $N(t_1)$ and edges vector $E(t_1)$ by *sid* at time
$t_1 \rightarrow N(t_1) = \{ t_1, t_2,..., t_i \}$, $E(t_1) = \{ e_1, e_2,..., e_j \}$
2: Initialize modularity vector $M(t_k) = null$
3: Initialize the target vector for perceiving information propagation model $v = null$
4: $t_k \leftarrow t_1$
5: while $t_k < t_{final}$ do
6: Generate information propagation tree by both $N(t_k)$ and $E(t_k) \rightarrow Tr(t_k)$
7: Compute the value of modularity $\omega(t_k)$ by $Tr(t_k)$
8: if $\omega(t_k) != null$ && $\omega(t_k) != 0$ then
9:      $\Delta M(t) \leftarrow \omega(t_k)$
10:    if $k=1$ then
11:        Generate Modularity vector $M(t_k) = \Delta M(t)$
12:    else
13:        Generate Modularity vector $M(t_k) = M(t_{k-1}) + \Delta M(t)$
14:    endif
15: endif
16: Compute the target vector for triple exponential smoothing model vector by $M(t_k) \rightarrow v$
17:    if k=final then
18:        return v
19:    endif
20:    k +=1
21: endwhial
22: return v
End

1.4 Verification of Information Communication Situational Awareness Model

Verifying Accuracy of Information Communication Situational Swareness Model Fig. 3, the blue curve ("+" symbol curve) represents the actual value of the information propagation situation. The purple red curve ("O" sign curve) represents the predicted value calculated using the information spread situation awareness model. It can be seen that the two curves are roughly fitted to the state, and with the gradual increase of the exponential smoothing factor alpha. Large, subsequent time slice prediction results will be more accurate, but the initial time segments will also fluctuate. Therefore, when the exponential smoothing factor alpha satisfies $0.3 < \alpha < 0.5$, the prediction accuracy of the information spread situational awareness model is higher when the exponential smoothing factor alpha satisfies the index alpha.

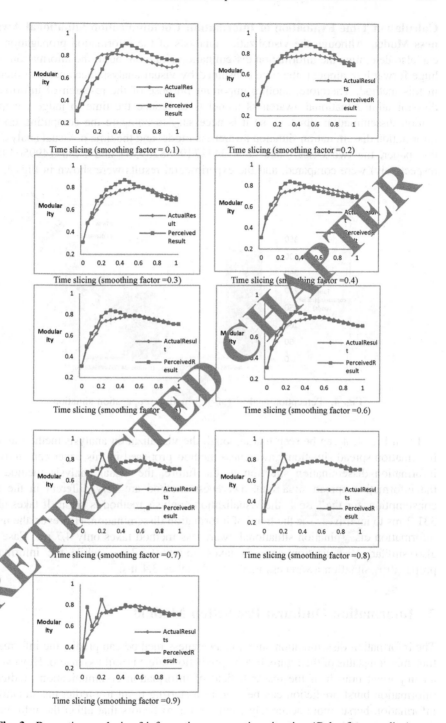

Fig. 3. Perception analysis of information propagation situation (Color figure online)

Calculation Time Evaluation of Information Communication Situational Aware-ness Model. Although the visualization analysis of the information propagation tree can also determine the information dissemination situation, but if the information has a huge forwarding amount, the time consumed by visual analysis becomes a bottleneck in this method. Therefore, another important purpose of the research of information dissemination situational awareness model is to reduce the time to judge the infor-mation dissemination situation, so it is necessary to compare the computing time of information dissemination situation awareness model and the time of visual analysis. In this paper, the two cases (statusid is 3443749588269364 and 3436314995649160, respectively) were compared, and the experimental results were shown in Fig. 4.

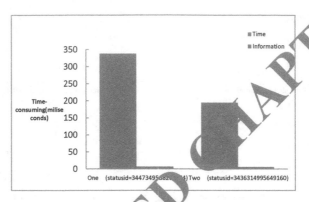

Fig. 4. Perception analysis of information propagation situation

From Fig. 4, it can be seen that although the visualization analysis method and the information spread situational awareness method proposed in this paper can study the information dissemination situation in the future, the visual analysis method and the information spread situational awareness method are very different in the time consumption, for the case 1, the visualization analysis method is spent. It takes about 331.2 ms to intuitively see the trend of information dissemination. However, the use of information dissemination situational awareness method takes only 4.7 ms. Case 2 is also similar. Visualization analysis takes about 196.7 ms, while the information propagation situation awareness method only takes 3.4 ms.

2   Information Outburst Prediction Method

The information dissemination situation awareness method can predict the information transmission status of the future, but the prediction effect is still too macro. However, as an important branch of the research field of information communication prediction, information burst prediction can be used as a starting point to predict the situation of information bursts more accurately from the micro factors that affect the information dissemination [24–26].

2.1   The Definition of Information Outburst Prediction

In epidemiology, outbreaks are usually defined as sudden increases in the number of people infected by diseases at certain time and place. However, in the study of social networks, there is no clear and recognized definition of the outbreak, and some researchers have also tried to define the outbreaks of information. Some scholars have used information forwarding times to define the definition of information explosion in social networks [27].

In order to explore the rationality of the threshold value, the correlation between Modularity calculation results and the degree of information explosion is compared. Firstly, three representative and highly distinguished information propagation trees (the statusid of the three is 3801498036877201, 38014850899489 5 and 3801511685531606) are visualized, and the Modularity values of each information propagation tree are calculated at steady state, respectively (Figs. 5, 6 and 7).

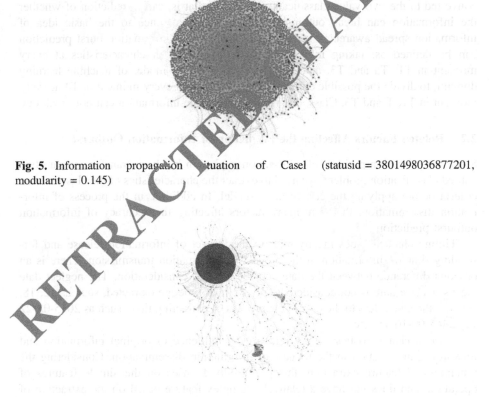

Fig. 5. Information propagation situation of Case1 (statusid = 3801498036877201, modularity = 0.145)

Fig. 6. Information propagation situation of Case2 (statusid = 3801485089948925, modularity = 0.263)

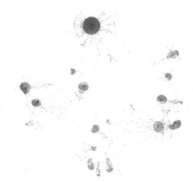

Fig. 7. Information propagation situation of Case2 (statusid = 380151_85531606, modularity = 0.856)

Based on this definition, the information burst prediction problem can be further converted to the two value classification problem, that is, early prediction of whether the information can break out in the future. With reference to the basic idea of information spread awareness in Fig. 1, the problem of information burst prediction can be defined as: taking information outburst prediction characteristics at every moment in T1, Ta and T3, and using the classification model of machine learning domain, to divide the possible outbreak of information at every moment of T4 at every moment in T1, T and T3. Class, that is, at T4 time, the information can not break out.

2.2 Related Factors Affecting the Prediction of Information Outburst

Since the prediction problem of information explosion has been transformed into two valued classification problem, we need to extract the characteristics of the classification criteria before applying the classification model. In addition, in the process of information dissemination, there are many factors affecting the accuracy of information outburst prediction.

Timing characteristics mainly refer to the impact of information release and forwarding time of information in the process of information transmission. There is an obvious difference between the date and the date of consideration. The neglect date means that the date is not considered, only 24 h of time are extracted, such as 12, 18, etc., and the date refers to the date in the process of extracting time, such as 2015-08-22 12, 2015-09-16 18, etc.

Content characteristics mainly refer to the influence of original information and forwarding information in the process of information dissemination. Considering the timeliness of feature extraction, this paper only focuses on the simple features of operation, and does not have a relatively complex feature based on the extraction of Natural Language Processing technology. It mainly includes whether the original information contains URL, whether the original information contains question marks (considering the difference between punctuation marks in Chinese and English, here we consider "?" Or "?"), and whether the original information contains exclamation marks ("!"). The number of "@" symbols in the original information, the length of original information (according to bytes), the entropy of all information length distribution in the information propagation network (according to bytes), the standard deviation of all information length distribution in the information propagation network (by bytes), and the only forwarding of all the users of the information propagation network. The proportion of content is not released, and the proportion of users who transmit information and publish content is distributed.

2.3   Evaluation of Information Outburst Prediction Model

From Fig. 8, we can see that the least forward amount in 340 information propagation trees is 61 times, the maximum forwarding amount is 146838 times, and the forwarding amount is random distribution, which again confirms the non subjectivity in the process of data selection. In addition, we set the threshold of information explosion $\delta = 0.2$, and calculate the Modularity value in ϕ_{20} when information propagating steady state. We use the 44 features to predict the information explosion under different classifiers, as shown in Fig. 9.

Fig. 8. Distribution of the number of forwarding nodes

From Fig. 9, we use the 44 feature information burst prediction with high accuracy. Under different classifiers, the accuracy rate is maintained at 79% to 97%, while the recall rate stays between 74% and 97%. In addition, as time goes on, all indicators (Precision, Recall, F1-Measure and AUC) have a significant upward trend. In the first time slice ϕ_1, the F values under different classifiers are kept in the range of 74% to 83%, while the F values under the different classifiers are kept in the range of 86% to 97% in the last time slice ϕ_{20}. It is not difficult to find that Ada Boost and Random Forest classifier are most suitable for information burst prediction from different curves. In addition, the prediction accuracy of the above algorithms can be sorted from high to low to Ada Boost $\geq$ Random Forest > C4.5 > Bayes Net > Naive Bayes.

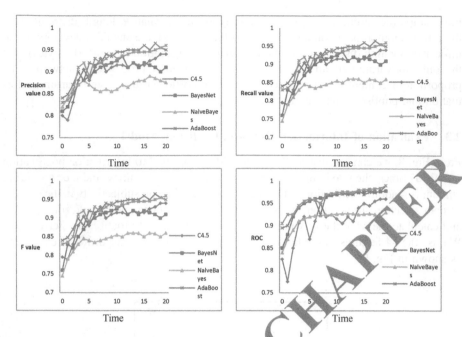

Fig. 9. Results from predicting outbreaks using different classifiers

2.4   Feature Selection and Evaluation of Information Outburst Prediction

From the previous section, we can learn that the above 44 features can accurately predict the information burst. Therefore, in this paper, feature selection algorithm (Cfs Subset combined with Best First's feature selection scheme) is applied to each time slice data ($\Phi1$, $\Phi2$, ..., $\Phi20$).

From Fig. 10, the feature Entropy Eigenvector Centrality and Entropy Page Rank all appear in the 20 time slice, that is, the feature selection algorithm is selected in all the process of information propagation, which also indicates that the probability of the two features to be selected by the feature selection under the 20 time slice data set is 100%. In the same way, the number of three features of Standard Deviation Betweenness, No Content Proportion, and Average Path Length appeared in 20 time slices were 14, 13 and 13 respectively, which also indicated that the probability of the three features to be selected by the feature selection under the 20 time slice data set was greater than 50%.

From Fig. 11, we can see the comparison of information burst prediction using all characteristics, five characteristics and two features, among which, under the C4.5, Ada Boost and Random Forest classifiers, the F value curves of the three are basically stable and similar; Under the Bayes Net and Naive Bayes classifier, the F value predicted by information explosion using fewer features is significantly higher than that of F values after using all features.

Therefore, this paper finds that the five features of Entropy Eigenvector Centrality (the entropy of the eigenvector distribution of the information propagation network),

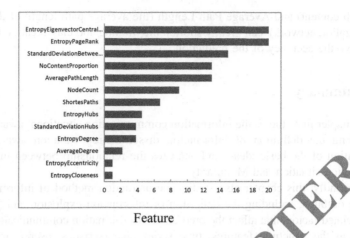

Fig. 10. Results from combining various selected features

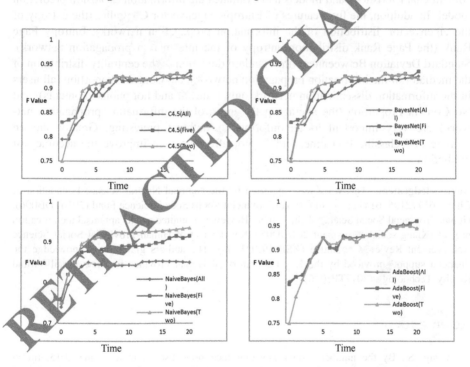

Fig. 11. Comparative results from using all 44 features, 5 features, and 2 features

Entropy Page Rank (the Page Rank distribution entropy of the information propagation network), Standard Deviation Betweenness (the standard deviation of the centrality distribution of the media in the information propagation network), No Content Proportion (all users in the information dissemination network only transmit and not publish content) and Average Path Length (the average path length of the information propagation network), can greatly shorten the training and classification time, and even improve the accuracy of the prediction.

3   Summary

This chapter first studies the information communication situational awareness method, including the definition of information dissemination situation awareness and the exposition of the basic ideas, and analyzes the relationship between information dissemination situation and Modularity.

Secondly, this chapter also studies the prediction method of information dissemination outburst, including the definition of information explosion and the analysis of four related factors that affect the prediction of information communication (44 factors, covering the structural features, time series characteristics, user characteristics and content characteristics), and models and evaluates the information explosion prediction model. In addition, the five features of Entropy Eigenvector Centrality (the entropy of the eigenvector distribution of the information propagation network), Entropy Page Rank (the Page Rank distribution entropy of the information propagation network), Standard Deviation Betweenness (the standard deviation of the centrality distribution of the mediator of the information propagation network), No Content Proportion (all users in the information dissemination network only transmit and not publish contents), and No Content Proportion (the average path length of the information propagation network) are very important to the information outburst pretesting. Greatly shorten training and classification time, and do not affect or even improve the accuracy of prediction.

Acknowledgement. This work was supported by the National Natural Science Foundation of China (61772196; 61472136), the Hunan Provincial Focus Social Science Fund (2016ZDB006), Hunan Provincial Social Science Achievement Review Committee results appraisal identification project (Xiang social assessment 2016JD05), Key Project of Hunan Provincial Social Science Achievement Review Committee (XSP 19ZD1005). The authors gratefully acknowledge the financial support provided by the Key Laboratory of Hunan Province for New Retail Virtual Reality Technology (2017TP1026).

References

1. Craig, S.: By the numbers: 200+ amazing face book user statistics, June 2015. http://expandedramblings.com/index.php/by-the-numbers-17-amazing-facebook-stats/. Accessed 29 June 2015, 19 July 2015
2. Statista: Leading social networks worldwide as of March 2015, ranked by number of active users (in millions) [OL], 15 April 2015. http://www.statista.com/statistics/272014/global-social-networks-ranked-by-number-of-users/. Accessed 19 July 2015
3. Statista: Number of monthly active WeChat users from 2nd quarter 2010 to 1st quarter 2015 (in millions) [0L], 30 April 2015. http://www.statista.com/statistics/255778/number-of-active-wechat-messenger-accounts/. Accessed 19 July 2015

4. Craig, S.: By the Numbers: 40 Amazing Weibo Statistics [OL], 18 April 2015. http://expandedramblings.com/index.php/weibo-user-statistics/. Accessed 19 July 2015
5. Luchina, F.: Ellen's Oscar Selfie Most Retweeted Tweet Ever [OL], 02 March 2014. http://abcnews.go.com/blogs/entertainment/2014/03/ellens-Oscar-selfie-most-retweeted-tweet-ever/. Accessed 19 July 2015
6. Wikipedia: Socialnetwork [OL], 29 July 2015. http://en.wikipedia.org/wild/Social. Accessed 29 July 2015
7. Lumbreras, A., Gavalda, R.: Applying trust metrics based on user interactions to recommendation in social networks. In: Proceedings of the 2012 International Conference on Advances in Social Networks Analysis and Mining, ASONAM 2012, pp. 1159–1164 (2012)
8. Marin, A., Wellman, B.: Social network analysis: an introduction. In: The SAGE Handbook of Social Network Analysis, pp. 11–25 (2017)
9. Kwak, H., Lee, C., Park, H., et al.: What is Twitter, a social network or a news media? In: Proceedings of the 19th International Conference on World Wide Web, pp. 591–600. ACM (2010)
10. Ren, X., Lu, L.: Overview of sorting methods for important nodes in networks. Chin. Sci. Bull. **13**, 004 (2014)
11. Abdullah, S., Wu, X.: An epidemic model for news spreading on Twitter. In: 2011 23rd IEEE International Conference on Tools with Artificial Intelligence (ICTAI), pp. 163–169 (2017)
12. Xiong, F., Liu, Y., Zhang, Z., et al.: An information diffusion model based on retweeting mechanism for online social media. Phys. Lett. A **376**(30), 2103–2108 (2016)
13. Lilt, D., Chen, X.: Rumor propagation in online social networks like Twitter—A simulation study. In: 2011 Third International Conference on Multimedia Information Networking and Security (MINES), pp. 278–282 (2016)
14. Zaman, T., Fox, E.B., Bradlow, E.: A Bayesian approach for predicting the popularity of tweets. Ann. Appl. Stat. **8**(3), 583–1611 (2014)
15. Suh, B., Hong, L., Pirolli, P., et al.: Want to be retweeted? Large scale analytics on factors impacting retweet in Twitter network. In: 2016 IEEE Second International Conference on Social Computing (SocialCom), pp. 177–184 (2016)
16. Peng, H.K., Zhu, J., Piao, D., et al.: Retweet modeling using conditional random fields. In: 2015 IEEE 11th International Conference on Data Mining Workshops (ICDMW), pp. 336–343 (2015)
17. Hong, L., Dan, O., Damson, B.D.: Predicting popular messages in Twitter. In: Proceedings of the 20th International Conference Companion on World Wide Web, pp. 57–58. ACM (2016)
18. Bao, Y., Yi, C., Xue, Y., Dong, Y.: Precise modeling rumor propagation and control strategy on social networks. In: Kazienko, P., Chawla, N. (eds.) Applications of Social Media and Social Network Analysis. LNSN, pp. 77–102. Springer, Cham (2015). https://doi.org/10.1007/978-3-319-19003-7_5
19. Wang, J., Jiang, C., Qian, J.: Robustness of interdependent networks with different link patterns against cascading failures. Phys. A: Stat. Mech. Appl. **393**, 535–541 (2014)
20. Wang, J.: Mitigation strategies on scale-free networks against cascading failures. Phys. A **392**(9), 2257–2264 (2013)
21. Dou, B.L., Wang, X.G., Zhang, S.Y.: Robustness of networks against cascading failures. Phys. A: Stat. Mech. Appl. **389**(11), 2310–2317 (2016)
22. Qiao, X., Yang, C., Li, X., et al.: A user context based trust degree calculation method in social network services. J. Comput. Sci. **34**(12), 2403–2413 (2011)

23. Jiang, J., Yi, C., Bao, Y., et al.: Online community perceiving method on social network. In: 1st International Workshop on Cloud Computing and Information Security. Atlantis Press (2013)
24. Beatty, P., Reap, I., Dick, S., et al.: Consumer trust in e-commerce web sites: a meta-study. ACM Comput. Surv. (CSUR) **43**(3), 14 (2011)
25. Sherchan, W., Nepal, S., Paris, C.: A survey of trust in social networks. ACM Comput. Surv. (CSUR) **45**(4), 47 (2013)
26. Jure, L.: Stanford Large Network Dataset Collection [0L]. http://snap.Stanford.edu/data/. Accessed 26 August 2015
27. Jiang, W., Zhong, L., Zhang, L., Shi, D.: Multi-agent dynamic collaboration model based on active logical sequence of complex systems. Chin. J. Comput. **36**(5), 1115–1124 (201.
28. Jiang, W., Xu, Y., Gu, H., Zhang, L.: Dynamic trust calculation model and management mechanism of online trading. Sci. Sin. Inform. **44**(9), 1084–1101 (2014)

Budget Constraint Bag-of-Task Based Workflow Scheduling in Public Clouds

Pengfei Sun, Zhicheng Cai$^{(\boxtimes)}$, and Duan Liu

School of Computer Science and Engineering,
Nanjing University of Science and Technology, Nanjing 210094, China
caizhicheng@njust.edu.cn

Abstract. Bag-of-Tasks (BoT) workflows have appeared in distributed computing platforms such as Spark, MapReduce, and Pegasus. Budget constraints usually exist for these applications. It is crucial to design scheduling algorithms to minimize makespans under budget constraints for BoT workflows. However, most existing workflow algorithms are tailored for general workflows without considering batch structures of Bot-workflows. The main challenge for scheduling BoT workflows is to distribute the budget to different BoTs appropriately considering BoT structures. In this paper, a configuration-and-serialization iterative adjusting based heuristic algorithm (CSIA) is proposed to minimize the makespans under budget constraints. CSIA allocates VM configurations and serial degrees to different BoTs appropriately to decrease the makespan. Experimental results illustrate that the proposal gets shorter makespans on several types of workflow instances than existing algorithms under budget constraints.

Keywords: Cloud computing · Bag of tasks · Workflow scheduling · Budget constraint · Makespan

1 Introduction

Bag-of-Tasks(BoT) workflows are widespread in Cloud computing [6]. Different from general workflows which only consist of single task, a BoT workflow is composed of multiple bag-of-tasks. Most of applications on the widely used distributed computing platform Spark [1, 19] can be modeled as BoT workflows. For example, Fig. 1 is an BoT workflow which finds functional dependencies from distributed data sets on the Spark platform. This BoT workflow consists of 7 BoTs and 14 tasks. In order to accelerate the execution, there is a trend to deploy BoT workflows to virtual machines (VM) rented from public clouds elastically. Various types of virtual machines are provided by public Clouds with different charging modes such as On-demand (fixed prices) and Spot (dynamic prices) VMs. There are usually budget constraints for Cloud users to rent resources [10]. Cloud users are confronted with the problem of allocating appropriate numbers, types of virtual machines at proper time [20]. Therefore, it is crucial to design BoT-workflow scheduling algorithms to minimize makespans under budget constraints.

© Springer Nature Singapore Pte Ltd. 2019
Y. Sun et al. (Eds.): ChineseCSCW 2019, CCIS 1042, pp. 243–260, 2019.
https://doi.org/10.1007/978-981-15-1377-0_18

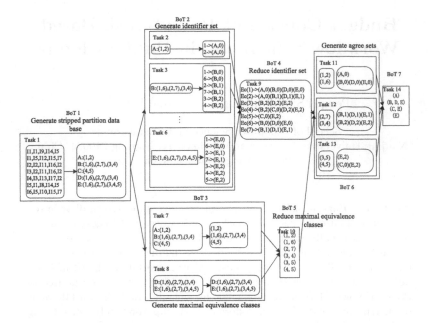

Fig. 1. A BoT workflow which finds functional dependencies from distributed data sets

The optimal workflow scheduling is usually NP-hard [11]. Most existing algorithms are developed for general workflows without considering BoT structures. Only a few algorithms consider the BoT-workflow scheduling [6,11]. However, they are tailored for rental cost minimization under deadline constraints rather than the makespan minimization under budget constraints considered in this paper. The main challenge of makespan minimization under budget constraints is how to allocate the limited budget to competitive BoTs. The Serialization Degrees (SD) is the number of tasks of the BoT allowed to be executed on each VM. Given different VM configurations and Serialization Degrees (SD) to a BoT, there are different execution times and rental costs. Better configurations usually mean shorter makespans and higher costs while larger SDs usually means longer makespans and lower costs. Therefore, allocating budgets appropriately means selecting proper configurations and SDs for each BoT. This problem can be modeled as the traditional NP-hard Discreet Time-Cost Tradeoff Resource Problem (DTCTP) [15] with additional resource sharing.

In this paper, a configuration-and-serialization iterative adjusting based heuristic algorithm (CSIA) is proposed. CSIA first allocates appropriate configurations and SDs to BoTs by performing SD-increasing and configuration-updating operations iteratively. Then, tasks are executed according to the final configuration and SD allocation plan. Meanwhile, it is important to evaluate the rental cost and makespan of VM configuration and SD allocation plans efficiently during the planning stage. In the literature, the rental cost and makespan of allocation plans are usually evaluated by considering BoTs separately without considering the sharing of idle slots on rented VMs leading to inaccurate estimation. In order to estimate the rental cost and makespan of each SD and configuration allocation plan more precisely, an earliest start time based BoT-schedule generating method is developed first by considering each BoT as a super task with multiple modes which considers the sharing of rented intervals among different BoTs. The main contributions of this paper are as follows:

(1) Earliest start time based BoT-schedule generating method for estimating the makespan and rental cost of each configuration and SD allocation plan.
(2) A configuration-and-serialization iterative adjusting based heuristic algorithm for minimizing the makespan under the budget constraint.

The rest of the paper is organized as follows. Sections 2 and 3 are the related work and problem description. Preliminaries and proposed CSIA are given in Sects. 4 and 5 respectively. Section 6 compares the proposal with an existing algorithm and Sect. 7 concludes the paper.

2   Related Work

Objectives of cloud workflow scheduling mainly consist of makespan and rental cost. For traditional workflows, scheduling algorithms have been designed to minimize the makespan under the budget constraint [13, 14, 20], minimize the rental cost under the deadline constraint [3, 8–10, 12, 16, 21] or find the Pareto front of the makespan and the rental cost with/without deadline and budget constraints [17, 18]. For example, Constrained Budget-Decreased Time (CB-DT) was developed by Ghafouri et al. [10] to minimize the makespan of traditional workflow under the budget constraint. CB-DT decreases the makespan as much as possible by allocating faster VMs to tasks on critical paths and slower VMs to non-critical tasks.

Existing scheduling algorithms for traditional workflows are not suitable for BoT workflows. When algorithms designed for traditional workflows are applied to BoT-workflow directly, it is likely to spend too much cost on long critical tasks without considering the number of tasks of each BoT. Moreover, frequent changes of virtual machines for the same BoT's tasks can incur additional cost when compared to the scheduling algorithms for BoT workflows. Fot BoT workflows, both the length and SD of each BoT have an impact on the final makespan. For example, BoTs with larger SDs usually consume more cost to update VM configurations in order to decrease the makespan.

No algorithms have been designed for minimizing the makespan of BoT workflow under budget constraints. Several algorithms have been developed for scheduling bag-of-task workflows [6, 11]. However, they are tailored for minimizing the rental cost of cloud resources under deadline constraints. Therefore, in this paper, CSIA is proposed to allocate the budget to BoTs appropriately based on the proposed definitions of VM-configuration-updating effectiveness and SD-increasing effectiveness.

3   Problem Description

Public cloud providers provision different types of virtual machines with different configurations. For each configuration, there are two types of virtual machines, On-demand VM and Spot VM respectively. The price of On-demand VM instance is constant, while Spot VM instances of different zones have dynamic prices changing over time [6]. In addition, the price of the On-demand VM instances is usually higher than that of Spot VM instances. Therefore, only Spot instances are used in this paper. The rental cost C_m of a Spot VM instance m of type δ is calculated as follows:

$$C_m(T_m^b, T_m^f) = \sum_{k=0,1,\ldots,\lfloor \frac{T_m^f - T_m^b}{L} \rfloor} P_{\delta, T_m^b + \mathbf{L} \times k} \tag{1}$$

where T_m^b is the start renting time, T_m^f is the release time, $\mathbf{L}$ is the length of pricing interval and $P_{\delta, T_m^b + \mathbf{L} \times k}$ is the price of Spot type δ per interval at time $T_m^b + \mathbf{L} \times k$. A BoT workflow is modeled by a directed acyclic graph $G(U, T, E)$. $u_i = \{t_k | k = n_1, n_2, \ldots, n_{u_i}\} \in U$ denotes the ith BoT of the BoT workflow in which t_k is a task and U is the set of all BoTs. Tasks in the same BoT have the same depth from the source node and the same type of functions. $T = \{t_0, t_1, \ldots, t_N\}$ is the set of all tasks in the workflow. $E = \{(t_i, t_j) | i < j\}$ represents precedence constraints that task t_j must execute after task t_i.

When a task is scheduled to execute on a virtual machine, the task need to receive the transfer data from its parents tasks. The $Tran_{t_i}$ represents the maximum time the task t_i takes to receive transfer datas from its parent tasks. The transfer time is calculated through the Eq. 2.

$$Tran_{t_i} = (\sum_{I_{i,k} \in I_i} (1 - x_{i,k,m,S_{i,m}}) \times Z_{i,k})/W \tag{2}$$

in which $I_{i,k}$ is the k-th input data file of t_i and $Z_{i,k}$ is the size of $I_{i,k}$. W is the bandwidth of the cloud system. If the task needs the output file of the other tasks which have been executed on the same virtual machine, file transfer time will be zero for those files. Hence, $x_{i,k,m,S_{i,m}}$ is 1 if $I_{i,k}$ is already in the local storage of VM m at time t, otherwise $x_{i,k,m,S_{i,m}}$ is 0.

ET_{t_i} represents the execution time of task t_i. The estimated task processing time of t_i on VM m including software and data preparation time is

$$ET_{t_{i,m}} = E[t_{i,m}] + \sqrt{V(t_{i,m})} + (1 - y_{i,m,S_{i,m}}) \times T_u^s + Tran_{t_i} \tag{3}$$

in which $t_{i,m}$ is a stochastic variable which represents the pure task execution time of t_i on VM m excluding the data and software preparation time, $E[t_i, m]$ and $V(t_i, m)$ are the expectation and the standard deviation of $t_{i,m}$ respectively. $S_{i,m}$ is the start time of t_i on VM m. T_u^s is the software setup time of BoT u ($t_i \in u$ and all tasks of the same BoT needs the same type of software). $y_{i,m,S_{i,m}}$ is 1 if software has been installed on VM m at time t_i, otherwise $y_{i,m,S_{i,m}}$ is 0.

Let Bt be the budget of the BoT workflow and M be the set of rented VMs. The objective of this paper is to minimize the makespan of the BoT workflow under the budget constraint.

$$\min S_{t_{end}} + ET_{t_{end}} \tag{4}$$

subject to,

$$\sum_{m \in M} C_m(T_m^b, T_m^f) <= Bt$$

$$S_i + ET_{t_i} \leq S_j, \forall(i, j) \in E$$

where t_{end} is the end task of the BoT workflow and S_i is the start time of task t_i. Important notations of this paper are listed in Table 1.

4 Preliminaries

Serialization Degree (SD) of a BoT. It refers to the maximum number of tasks of the BoT that can be performed on a single virtual machine. For example in Fig. 2, when a BoT contains five tasks and the serialization degree is two, three VMs are needed to execute the BoT as shown in Fig. 2(b). Figure 2(c) shows that two VMs are consumed when SD is three.

Remaining Cost (RC) is the left rental cost during the schedule process which is equal to the budget Bt minus the cost already been consumed.

Estimated Wasted Cost (EWC) has been proposed in [6] which is defined to measure the waste of cost assigning different sub-deadlines to a BoT. For a BoT u, given VM type δ and a sub-deadline σ (called mode $M_{\delta,\sigma}$), the EWC of u is the total wasted rental cost of unused fractions on the rented intervals and the software setup cost, which is computed as follows.

$$W_{u,\delta,\sigma} = \frac{\{g_\sigma[N_{u,\delta,\sigma}^m(I_{u,\delta,\sigma}^m L - T_u^s) - T_{u,\delta}^e)] + T_u^s N_{u,\delta,\sigma}^m\}P_\delta}{L} \tag{5}$$

where $T_{u,\delta}^e$ is the total processing time of BoT u, T_u^s is the software setup time needed by BoT u, $N_{u,\delta,\sigma}^m$ and $I_{u,\delta,\sigma}^m$ are the minimum number of VM instances and rented intervals for $M_{\delta,\sigma}$ respectively. g_σ is a binary variable. The L represents the rent interval. If $\sigma \leq L$, $g_\sigma = 1$. Otherwise, $g_\sigma = 0$. Because most tasks last less than one time interval.

5 Proposed Algorithm

Configuration-updating of a BoT usually means decreasing of the makespan and increasing of cost while the serialization degree increasing (SD-increasing) of a BoT means increasing of the makespan and decreasing of the cost. Figure 3(a) shows a workflow and Fig. 3(b) is an initial schedule. When VM configurations of BoT 2, 3, 4 and 5 are upgraded to faster VMs, the makespan is reduced and the rental cost grows as shown in Fig. 3(c). On the contrary, increasing the SD

Table 1. Notations

Notations	Descriptions
δ	A type of virtual machine configuration
u	A bag-of-task in a workflow
σ_u	The deadline of a BoT u
δ_u	The Vm type δ of BoT u
$C_m(T_m^b, T_m^f)$	The rental cost of spot VM m during the duration from T_m^b to T_M^f
T_m^b	The start renting time of spot VM m
T_m^f	The release time of spot VM m
$P_{\delta,t}$	The price of spot type δ per interval at time t
P_δ	The average price of spot type δ per interval
$Z_{i,k}$	The size of $I_{i,k}$
t_i	The i-th task of the workflow
$t_{i,m}$	The task t_i which executes on Vm m
$I_{i,k}$	The k-th input date file of t_i
$x_{i,k,m,t}$	1, if $I_{i,k}$ is already in the local storage of VM m at time t, otherwise 0
W	The bandwidth of the cloud system
T_u^s	The setup time of the software needed by BoT u
$y_{i,m,t}$	1, if software has been installed on VM m at time t, otherwise 0
S_i	The start time of task t_i
$S_{i,m}$	The start time of task t_i on Vm m
t_{end}	The end task of a BoT workflow
ET_{t_i}	The execution time of task t_i
$N_{u,\delta,\sigma}$	Minimum number of VM instances of $M_{\delta,\sigma}$ in BoT u
$M_{\delta,\sigma}^m$	The mode with given VM type δ and a sub-deadline σ
$I_{u,\delta,\sigma}^m$	The rented intervals for $M_{\delta,\sigma}$ in BoT u
$T_{u,\delta}^e$	The total processing time of unit u with given VM type δ
L	The rental interval of virtual machine
g_σ	1, if $\sigma \leq L$, otherwise 0
$T_{u,\delta}^m$	The processing time of the longest task of u on VM type δ
$|u|$	The task number of of BoT u
$C(p)$	The cost of schedule plan p
$M(p)$	The makespan of schedule plan p
U_{cp}	The set of BoTs on critical path
RC	The left available cost during the schedule
Bt	The user-given budget during the workflow schedule
ξ_u	The set of Vms to which BoT u is assigned in the BoT schedule

of a task-unit will increase the makespan and have a probability to reduce the rental cost. For example, when the serialization degrees of BoT 2, 3, 4 and 5 are increased, the rental cost decreases and the makespan increases as shown in Fig. 3(d). It is critical to select appropriate configurations and serialization degrees (SD) for different BoTs considering the tradeoff of makespan and cost. If VMs are not shared among different BoTs, this problem can be modeled as a

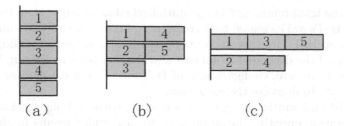

Fig. 2. An example of SD adjusting of a BoT

discrete time-cost tradeoff problem (DTCTP) which is NP-hard [15]. However, VM sharing makes the BoT workflow scheduling more complex.

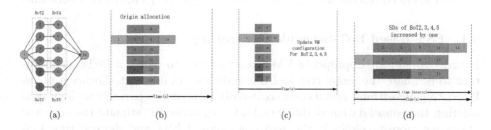

Fig. 3. A example of adjusting VM-configuration and serial-degree allocation plans

In this paper, a configuration-and-serialization iterative adjusting based heuristic algorithm (CSIA) is proposed which mainly consists of three steps as shown in Algorithm 1. In the first two steps, each BoT is considered as a super task with multiple alternative execution modes (different VM configurations and SDs) and appropriate modes are allocated to BoTs to minimize the makespan through updating VM configurations and increasing SDs of BoTs iteratively. In the last step, tasks of BoTs are scheduled according to the BoT-schedule generated based on selected VM configurations and SDs for BoTs.

Algorithm 1. Configuration-and-Serialization Iterative Adjusting Based Heuristic Algorithm (CSIA)

---

1. Call SDEBS to generate an initial BoT schedule;
2. Call CSDCA to by performing configuration updating and SD adjusting iteratively;
3. Call BSTE to execute tasks according to the final BoT schedule;

---

(1) Initial BoT-Schedule generation. The cheapest-slowest VMs are assigned to BoTs to generate an initial schedule by setting the SDs of BoTs to be one.

When the total rental cost of the initial schedule is larger than the budget, SDs of BoTs are increased to decrease rental costs. In order to minimize the makespan, BoTs are selected to increase the SDs in the descending order of the ratio of the decreased rental cost and the increased makespan. When RC is larger than zero, configurations of BoTs in the critical path are updated one by one to decrease the makespan.

(2) SD-and-Configuration cooperative adjusting without budget constraints. A configuration-updating operation is performed which results in $RC < 0$. In order to fulfilling the budget constraint, multiple SD-increasing operations are done to save rental cost. If there is no enough cost can be saved fulfilling $RC \geq 0$, the previous configuration-updating and SD-increasing operations are undo. Otherwise a new BoT is selected to update the configuration.

(3) Generate a BoT-Schedule according to configurations and SDs of BoTs in which BoTs are assigned to VMs based on ESTs. Then, tasks are scheduled to VMs in consistent with positions and orders of BoTs in the BoT-schedule.

5.1 EST Based BoT-Schedule Generating

In order to allocate appropriate VM types and SDs to different BoTs, it important to estimate the rental cost and the makespan of different allocation plans. In existing algorithms, rental costs are usually estimated separately without considering the consolidation of different BoTs. In order to estimate the cost and makespan more precisely, in this paper, a smallest EST and shortest time first BoT schedule method (SESTF) is proposed.

Different plans have different SDs and VM configurations for different BoTs. Scheduling each task of BoTs separately to estimate the rental cost and makespan is time consuming. Therefore, in this paper, a BoT schedule is generated by considering each BoT as a super-task to estimate the rental cost and makespan efficiently for each configuration and SD allocation plan. For a BoT u, the length σ_u of the super-task is:

$$\sigma_{u,\delta,SD} = T_u^s + T_{u,\delta}^m \times SD \tag{6}$$

where T_u^s is the setup time of software needed by BoT u, $T_{u,\delta}^m$ is the processing time of the longest task of u on VM type δ and SD is the SD of u. Meanwhile, the super task consumes $\left\lceil \frac{|u|}{SD} \right\rceil$ VMs. In other words, VM configurations and SDs determine execution times and consumed VM counts of BoTs. Then, super tasks are scheduled in the ascending order of the earliest start time (EST) and length. Dependency constraints are kept and only super tasks with the same VM configuration can be consolidated to share rented VM intervals. BoTs are scheduled one by one in the ascending order of ESTs. If multiple BoTs have the same EST, the earliest finished BoT (which has the shortest execution time) is scheduled first. Idle time slots on previous VMs of the same configuration can be shared by later BoTs (super tasks). After all BoTs are scheduled, a BoT schedule is generated for the current VM configuration and SD allocation plan. Assuming p is a SDs and configurations allocation plan, the rental cost and makespan of

p estimated by SESTF is $C(p)$ and $M(p)$ respectively. Formal description of SESTF is shown in Algorithm 2.

Algorithm 2. Smallest EST and Shortest Time First BoT Schedule Method(SESTF)

Input: The set U of BoTs in the workflow, VM configuration and SD allocation plan p;

Output: Estimated rental cost, Makespan;

1. Update ready super-task set $\mathcal{R}$ by adding super tasks (BoTs) whose predecessors have been scheduled;
2. **while** $\mathcal{R} \neq \emptyset$ **do**
3.   Get a ready super-task u with the earliest EST and finish time from $\mathcal{R}$;
4.   **if** $u \neq null$ **then**
5.   Calculate the requied VM number $N_u \leftarrow \left\lceil \frac{|u|}{SD_u} \right\rceil$ of u according to p, $|u|$ is the number of tasks in u;
6.   Try to find N_u number of existing VMs of the configuration δ_u defined in p.
7.   **if** N_u^e number of VMs are found and $N_u^e < N_u$ **then**
8.   Allocate $N_u - N_u^e$ number of new VMs of type δ_u;
9.   **end if**
10.   Schedule the super task u to the N_u number of VMs;
11.   Update ready task set $\mathcal{R}$;
12.   **end if**
13. **end while**
14. Calculate $C(p) \leftarrow$ the total cost of allocated VMs and $M(p) \leftarrow$ the finish time of the last super task;
15. Return $C(p)$ and $M(p)$;

5.2 SD-Increasing and Configuration-Updating Effectiveness Based Initial BoT Schedule Generating

A SD-increasing effectiveness based method (SDEBS) is proposed to generate an initial schedule. Firstly, all BoTs are assigned the cheapest-slowest VMs and SD is initialized to be 1. Then, SDs of BoTs are increased to decrease cost when $RC < 0$. SD-increasing for different BoTs usually leads to different saved costs and increased makespans. In order to decrease the cost as much as possible and increase the makespan as short as possible, BoTs are ordered and selected based on the ratio of estimated saved cost and the increased makespan which is defined as the SD-increasing effectiveness (SDE). When the SD of a BoT increases from SD to SD' base on a configuration and SD allocation plan p, the estimated execution costs change too. The **Estimated Saved Cost** (ESC) is

$$ESC_{u,\delta}(SD, SD') = C(p) - C(p + \{SD'\}) \tag{7}$$

Where $C(p + \{SD'\})$ and $C(p)$ are the estimated execution cost of the whole workflow after and before SD-increasing respectively. ESC represents saved cost by increasing the SD. Then, **SD-increasing effectiveness** (SDE) is

$$SDE_u = \{(\mathcal{W}_{u,\delta,\sigma} - \mathcal{W}_{u,\delta,\sigma'}) * \gamma + ESC_{u,\delta}(SD, SD') * \eta\}/T_{u,\delta}^m \qquad where \ \gamma + \eta = 1 \tag{8}$$

in which ESC are cooperated with EWC to improve utilization and to decrease rental cost. Then, BoTs are ordered in the descending order of SDE and SD-increasing operations are executed to increase SD by one each time.

Updating configurations of BoTs are beneficial to decrease makespan. If there is still available cost for updating configurations of some BoT after SD-increasing operations, configuration-updating operations are needed to decrease the makespan further. In order to decrease the makespan as much as possible under limited budget, a **Configuration-Update Effectiveness** (CUE) is defined as follows. For each BoT u, when the VM type is changed from δ to δ', CUE is

$$CUE_u = \frac{(\sigma_{u,\delta,SD} - \sigma_{u,\delta',SD})}{\{(\mathcal{W}_{u,\delta',\sigma_{u,\delta'},SD} - \mathcal{W}_{u,\delta,\sigma_{u,\delta,SD}}) * \alpha + (C(p + \{\delta'\}) - C(p)) * \beta\}} \tag{9}$$

where $\alpha + \beta = 1$ and $C(p + \{\delta'\})$ is the rental cost after configuration updating based on p. Higher CUE means longer makespan are decreased and lower cost are used. The current critical path is generated based on the current SDs and configurations. CUEs of BoTs on the critical path is calculated based on Eq. (9). Then, the configuration of the BoT with the maximum CUE is updated iteratively until $RC < 0$ (no available budget to update). The last configuration-updating is undone to ensure $RC > 0$.

5.3   Configuration and SD Cooperative Adjusting

In this step, Vm configurations of BoTs on critical paths are updated to decrease the makespan as much as possible. Because the budget is nearly used up in the initial step, configuration updating of critical BoTs usually leads to budget violation $RC < 0$. In order to make up for the cost consumed by configuration updating, SDs of some BoTs are adjusted to save cost for the previous VM configuration updating operation. Different SDs of different BoTs leads to different costs and makespans of final schedule. In order to save cost and avoid makespan increasing as much as possible of SD adjusting operations, **SD-Adjusting Effectiveness** (SDAE) is defined as follows. When SD_u is updated to SD'_u based on the current VM configuration and SD allocation plan p, $C(p)$ and $M(p)$ are the rental cost and makespan got by SESTF before SD adjusting. $C(p + \{SD'_u\})$ and $M(p + \{SD'_u\})$ are the rental cost and makespan got by SESTF after SD adjusting.

$$SDAE_u(p + SD'_u) = \frac{C(p) - C(p + \{SD'_u\})}{M(p + \{SD'_u\}) - M(p)} \tag{10}$$

where $p + \{SD_u\}$ means SD and VM configuration plan after adjusting SD of u to be SD_u based on p.

Algorithm 3. SDE Based Initial BoT Schedule Generating (SDEBS)

Input: Budget Bt

1. Allocate the cheapest-slowest VM to each BoT
2. Initialize SD of each BoT to be one;
3. Initialize U to be the set of BoTs;
4. Calculate SDE for all BoTs;
5. $u' \leftarrow \text{argmax}_{u \in U}\{SDE_u\}$;
6. **while** $u' \neq null$ **do**
7.   **if** $RC > 0$ **then**
8.   Break;
9.   **end if**
10.   Calculate $ESC_{u',\delta}(SD_{u'}, SD_{u'} + 1)$ according to Equation(7);
11.   **if** $ESC_{u',\delta}(SD_{u'}, SD_{u'} + 1) < 0$ **then**
12.   Update $U \leftarrow U - \{u'\}$;
13.   **else**
14.   Increasing SD of u' by one;
15.   **end if**
16.   Update SDE for all BoTs;
17. $u' \leftarrow \text{argmax}_{u \in U}\{SDE_u\}$;
18. **end while**
19. **if** $RC < 0$ **then**
20.   No solution can be generated;
21. **end if**
22. $U_{CP} \leftarrow$ the current critical path based on BoTs;
23. **for** each $u \in U_{CP}$ **do**
24.   Calculate CUE_u with Equation (9);
25. **end for**
26. **while** $RC > 0$ **do**
27. $u' \leftarrow \text{arg max}_{u \in U_{CP}}\{CUE_u\}$;
28.   **if** $u' \neq null$ **then**
29.   Find the next cheapest-slowest VM type δ'_u for the selected u';
30.   Update the configuration of u' from δ_u to δ'_u;
31.   RC$\leftarrow Bt - C(p + \{\delta'_u\})$;
32.   **end if**
33.   Update U_{CP} and CUE_u of BoTs;
34. **end while**
35. Undo the latest updation of configuration;

A configuration and SD cooperative adjusting method (CSDCA) is proposed to decrease makespan further as shown in Algorithm 3. Initially, critical paths are calculated based on execution times of BoTs estimated based on the current SDs and VM configurations, and U_{CP} is the set of Bots on the critical path. Then, CUE of each BoT in critical path is calculated based on Eq. (9) and the BoT u' with the maximum CUE is selected of which the VM configuration will be updated. If the configuration of the one BoT can not be updated any more, the CUE is set to be $-\infty$. M is the current makespan before configuration updating and a check point is set. Next, the configuration of the selected u' is updated without considering the budget constraint and multiple SD adjusting operations will be carried out to make up for the budget violating. Before selecting a BoT to adjust the SD, the rental cost $C(p)$ and the makespan $M(p)$ of current plan p is estimated by SESTF. $SDAE$s of different SDs of each BoT are calculated to select the most effective combination of BoT and SD. If the selected SD adjust plan can decrease the rental cost while guaranteeing that the new makespan is smaller than M, the SD adjust is carried out. The SD adjusting is iterated to save more costs until $RC > 0$ which means enough rental cost is saved to make up for the budget violating produced by the previous configuration updating operation. Then, the critical path and CUE of Bots on the critical path are updated. If no SD adjusting can be performed and $RC > 0$, the configuration updating and SD adjusting operations after the latest check point are undone. u' is removed from the critical path set U_{CP}, when the cooperative adjusting operation fails. Finally, a new BoT on critical path is chosen to update configuration as mentioned above until no BoT can be adjust in terms of VM configuration to decrease the makespan.

5.4 Schedule Tasks According to the Final BoT Schedule

After all BoTs are assigned appropriate VM configurations and SDs, a final allocation plan is obtained based on which a final BoT schedule is generated by SESTF. The final BoT schedule determines to which VMs each BoT is assigned and the order of BoTs on each VM. Because each BoT may consume multiple VMs, each VM has a list of BoTs with different starting times and different VMs may contain the same BoT with the same starting time. Then tasks are scheduled according to the order of BoTs and VM allocations in the BoT schedule as follows. Ready task set is first updated by adding tasks whose predecessors have been scheduled. Let ξ_u is the set of VMs to which BoT u is assigned in the BoT schedule. A task $t_k \in u$ from the ready task set can be scheduled only when BoTs prior to u on each VM of ξ_u have been completed. If some VMs in ξ_u do not exists, they are rented from public Cloud elastically. t_k is scheduled to the VM of ξ_u with the earliest finish time to minimize the makespan as much as possible. Then, another task from the ready task is tested as mentioned above until no task can be scheduled. Next, the ready task set is updated and traversed until there is no task to be scheduled. The details of BoT-Schedule based task execution (BSTE) is shown in Algorithm 5.

6 Performance Evaluation

In this section, the proposed CSIA is compared with the Constrained Budget-Decreased Time (CB-DT) developed by Ghafouri et al. [10]. In the literature, there is no BoT workflow scheduling algorithm on Spot VM instances minimizing the makespan under budget constraints. CB-DT was proposed to schedule general workflows on On-demand VMs which gets the best results. In this paper,

---

Algorithm 4. Configuration and SD Cooperative Adjusting (CSDCA)

Input: Budget Bt, U to be the set of BoTs;
1. $U_{CP} \leftarrow$ calculate critical path based on current CDs and configurations;
2. **for** each $u \in U_{CP}$ **do**
3.   Calculate CUE_u with Equation(9);
4. **end for**
5. $u' \leftarrow \arg max_{u \in U_{CP}}\{CUE_u\}$
6. **while** $u' \neq null$ **do**
7.   Update $M \leftarrow M(p)$ and set CheckPoint;
8.   Find the next cheapest-slowest VM type δ'_u for the selected u';
9. $p \leftarrow$ update the configuration of u' from δ_u to δ'_u;
10. $RC \leftarrow Bt - C(p + \{\delta_{u'}\})$;
11. $State_{sd} \leftarrow true$
12.   **while** $RC < 0$ **do**
13.   Estimate the current makespan $M_c \leftarrow M(p)$ and the current cost $C_c \leftarrow C(p)$
    by SESTF;
14.   **for** each $u \in U$ **do**
15.   **for** SD'_u in $[1, |u|]$ **do**
16.   Call SESTF to estimate the makespan $M(p + \{SD'_u\})$ and the cost
$C(p + \{SD'_u\})$;
17.   **end for**
18.   **end for**
19. $(u, SD'_u) \leftarrow \arg max_{\{u \in U, SD'_u \in [1, |u|]\}}\{SDAE_u(p + SD'_u)|M(p + \{SD'_u\}) \leq M, C(p + \{SD'_u\}) \leq C_c\}$;
20.   **if** $(u, SD_u) \neq null$ **then**
21. $p \leftarrow$ adjust SD of u to be SD_u
22.   Update RC;
23.   **else**
24. $p \leftarrow$ undo the last configuration-updating and SD-changeing operations
    after CheckPoint;
25. $State_{sd} \leftarrow false$
26. $U_{CP} = U_{CP} - \{u'\}$
27.   Break;
28.   **end if**
29.   **end while**
30.   **if** $State_{sd} = true$ **then**
31.   Update critical path, U_{CP} and CUE_u;
32.   **end if**
33. $u' \leftarrow \arg max_{u \in U_{CP}}\{CUE_u\}$
34. **end while**
35. **while** true **do**
36.   **if** Existe BoT u in critical path which can decrease SD with RC>0 **then**
37.   Do the SD-decreasing operation on u
38.   Update the critical path;
39.   **else**
40.   Exit the algorithm;
41.   **end if**
42. **end while**

Algorithm 5. BoT-Schedule Based Task Execution (BSTE)

Input: A given simulation scheme;
1. Update the ready task set Γ;
2. **while** $\Gamma \neq \emptyset$ **do**
3.   **for** each $t_k \in \Gamma$ **do**
4.   **if** all tasks of the BoT u are in Γ, $t_k \in u$ **then**
5.   Update ξ_u to be the set of VMs to which BoT u $(t_k \in u)$ is assigned in the BoT schedule;
6.   Ensure that VMs in ξ_u exist by renting new VMs from public Cloud;
7.   **if** BoTs prior to u on each VM of ξ_u have been completed **then**
8.   Schedule t_k to the VMs of ξ_u with the earliest finish time;
9.   **end if**
10.   **end if**
11.   **end for**
12.   Wait for the finish of a task;
13.   Update the ready task set Γ;
14. **end while**

CB-DT has been modified to be suitable to Spot VMs by adding the support to rent the cheapest VMs every time. Realistic workflows [4] such as Montage, CyberShake, Genomic, LIGO and Sipht which are produced by the Workflow Generator [2] randomly. These workflows have BoT structures in nature and workflows with small (400), middle (600) and large (800) number of tasks are selected. Task execution times have different probability distribution types and variances which are generated by a stochastic method proposed by Cai et al. [7]. The budget of a BoT workflow is set to be $\lambda \times C_{cheapest}$ where λ is the *budget factor* taking values from $\{1, 1.1, 1.3, 1.5, 2, 6\}$ and $C_{cheapest}$ is the renal cost on the cheapest and most effective VMs.

6.1   Experimental Results

The proposal is evaluated on ElasticSim [5]. For different combinations task number, budget factor and workflow type, five different types of workflow instances have been executed. Figure 4 shows the mean plots of makespans (in seconds) for five types of Bot workflows with different budget factors and sizes. In total, the proposed CSIA gets shorter makespans on Montage and CYBERSHAKE workflows and longer makespans on LIGO workflows than the CB-DT. Meanwhile, CSIA has similar results with CB-DT on SIPHT and GENOME workflows.

For MONTAGE workflows, the CB-DT can not generate schedule plan when the budget is middle and tight. On the contrary the proposed CSIA can get schedule successfully for middle and tight budget constraints. Meanwhile, CSIA gets shorter makespans than CB-DT when the budget becomes loose. When the budget is large enough, the makespan of CSIA converges to a value and the fluctuations are small. The makespan of CB-DT also decreases as the budget increases. However, the makespan of CB-DT is always larger than that of CSIA. The reason is that the MONTAGE workflow consists of several BoTs with huge

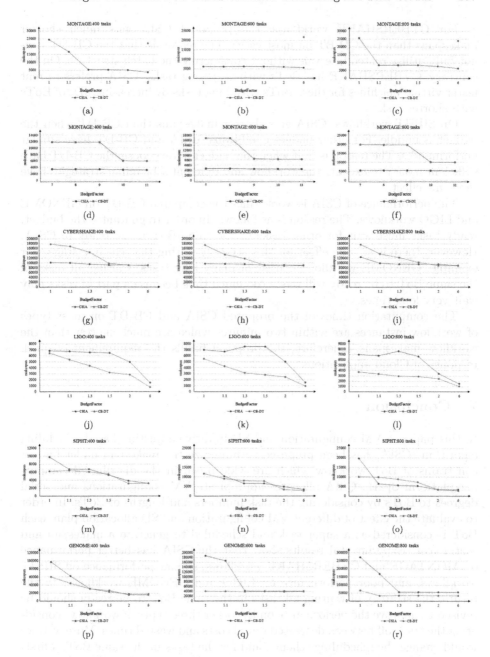

Fig. 4. Makespans of CSIA and CB-DT on different types of workflows

numbers of short tasks. The CSIA prefers to assign large SDs to these BoTs to save cost which can be used to decrease the execution times of longer tasks by updating VM configurations.

For CYBERSHAKE workflows, the proposed CSIA gets much shorter makespans than the CD-BT on most cases. The reason is that the CB-DT cannot fully utilize rented intervals without considering the batch structure. On the contrary, CYBERSHAKE has BoTs with very long tasks, CSIA prefers to rent faster virtual machines for these BoTs with long tasks by increasing SDs of BoTs with shorter tasks.

On SIPHT workflows, CSIA gets shorter makespans than CB-DT when the budget is middle. For other budget conditions, CSIA and CB-DT have similar performances. The reason is that when the budget is very loose, the CB-DT have sufficient money to execute the workflow and different scheduling strategies have the same effect.

The performance of CSIA is worse than the compared CB-DT on GENOME and LIGO workflows. The reason is as follows. In order to guarantee the budget, ready tasks must wait for other tasks of the same BoT to be ready in CSIA. However, there is an side effect for this strategy during the schedule stage. For example, GENOME workflows consist of many long tasks, waiting all tasks of the same BoT to be ready will waste too much time because a part of tasks may wait very long times.

The computation times of the proposed CSIA and CB-DT on most types of workflow instances are within two minutes which are much shorter than the workflow makespans. Therefore, the proposal fulfills the requirement of quick response of Cloud applications.

7   Conclusion

In this paper, a VM configuration and SD iterative adjusting based scheduling algorithm (CSIA) has been proposed to minimize the makespan under budget constrains of BoT workflow which are widespread in the distributed computing platform Spark. CSIA allocates appropriate VM configurations and serial degrees to BoTs by considering the task numbers and lengths of BoTs. In order to evaluate the effect of different VM configuration and SD allocation plan, each BoT is considered as a super task and scheduled to generate a makespan and rental cost. Experimental results show that the CSIA has better performance on MONTAGE and CYBERSHAKE workflows, similar performance on SIPHT workflows, and poor performance on LIGO and GENOME workflows. In the future, the strategy of waiting all tasks in the same BoT to be ready should be revised to improve the performance on the other three types of workflows considering the trade off between decreased rental costs and wasted times. Moreover, we would change the scheduling scheme, and for the tasks in the same BoT, a first-come-first-execution strategy can be applied to further shorten the makespan. Finally, when the budget cost is high, considering shortening SDs of BoTs on critical paths is also a promising future work.

Acknowledgements. Zhicheng Cai is supported by the National Natural Science Foundation of China (Grant No. 61602243), the Natural Science Foundation of Jiangsu

Province (Grant No. BK20160846), the Fundamental Research Funds for the Central Universities (No. 30919011235) and the Fundamental Research Funds for the Central Universities (No. 30920120180101). Duan Liu is supported by the Postgraduate Research and Practice Innovation Program of Jiangsu Province (No. KYCX18_0434).

References

1. Spark lightning-fast unified analytics engine. http://spark.apache.org. Accessed 14 May 2019
2. A workflow generator. https://confluence.pegasus.isi.edu/display/pegasus/ WorkflowGenerator. Accessed 30 June 2016
3. Abrishami, S., Naghibzadeh, M., Epema, D.H.J.: Deadline-constrained workflow scheduling algorithms for infrastructure; as a service clouds. Future Gener. Comput. Syst. **29**(1), 158–169 (2013)
4. Bharathi, S., Chervenak, A., Deelman, E., Mehta, G., Su, M.H., Vahi, K.: Characterization of scientific workflows. In: Third Workshop on Workflows in Support of Large-Scale Science, pp. 1–10. IEEE, Austin (2008)
5. Cai, Z., Li, Q., Li, X.: ElasticSim: a toolkit for simulating workflows with cloud resource runtime auto-scaling and stochastic task execution times. J. Grid Comput. **15**, 1–16 (2016)
6. Cai, Z., Li, X., Ruiz, R.: Resource provisioning for task-batch based workflows with deadlines in public clouds. IEEE Trans. Cloud Comput. **7**(3), 814–826 (2019)
7. Cai, Z., Li, X., Ruiz, R., Li, Q.: A delay-based dynamic scheduling algorithm for bag-of-task workflows with stochastic task execution times in clouds. Future Gener. Comput. Syst. **71**(C), 57–72 (2017)
8. Calheiros, R.N., Buyya, R.: Meeting deadlines of scientific workflows in public clouds with tasks replication. IEEE Trans. Parallel Distrib. Syst. **25**(7), 1787–1796 (2014)
9. Chopra, N., Singh, S.: Heft based workflow scheduling algorithm for cost optimization within deadline in hybrid clouds. In: Fourth International Conference on Computing, pp. 1–6 (2014)
10. Ghafouri, R., Movaghar, A., Mohsenzadeh, M.: A budget constrained scheduling algorithm for executing workflow application in infrastructure as a service clouds. Peer-to-Peer Netw. Appl. **3**, 1–28 (2018)
11. Li, X., Cai, Z.: Elastic resource provisioning for cloud workflow applications. IEEE Trans. Autom. Sci. Eng. **14**(2), 1195–1210 (2017)
12. Li, Z., Ge, J., Hu, H., Wei, S., Luo, B.: Cost and energy aware scheduling algorithm for scientific workflows with deadline constraint in clouds. IEEE Trans. Serv. Comput. **11**(1), 713–726 (2018)
13. Lin, X., Wu, C.Q.: On scientific workflow scheduling in clouds under budget constraint. In: IEEE International Conference on Parallel Processing, vol. 46, no. 1, pp. 90–99 (2013)
14. Mao, M., Humphrey, M.: Scaling and scheduling to maximize application performance within budget constraints in cloud workflows. In: IEEE International Symposium on Parallel and Distributed Processing, pp. 67–78 (2013)
15. Nudtasomboon, N., Randhawa, S.U.: Resource-constrained project scheduling with renewable and non-renewable resources and time-resource tradeoffs. Comput. Ind. Eng. **32**(1), 227–242 (1997)

16. Sahni, J., Vidyarthi, D.: A cost-effective deadline-constrained dynamic scheduling algorithm for scientific workflows in a cloud environment. IEEE Trans. Cloud Comput. **6**(1), 2–18 (2018)
17. Shi, J., Luo, J., Fang, D., Zhang, J.: A budget and deadline aware scientific workflow resource provisioning and scheduling mechanism for cloud. In: IEEE International Conference on Computer Supported Cooperative Work in Design, pp. 672–677 (2014)
18. Shi, J., Luo, J., Fang, D., Zhang, J., Zhang, J.: Elastic resource provisioning for scientific workflow scheduling in cloud under budget and deadline constraints. Cluster Comput. **19**(1), 167–182 (2016)
19. Zaharia, M., Chowdhury, M., Franklin, M.J., Shenker, S., Stoica, I.: Spark: cluster computing with working sets. In: USENIX Conference on Hot Topics in Cloud Computing, pp. 1–10 (2010)
20. Zeng, L., Veeravalli, B., Li, X.: ScaleStar: budget conscious scheduling precedence-constrained many-task workflow applications in cloud. In: IEEE International Conference on Advanced Information Networking and Applications, pp. 534–541 (2012)
21. Zhao, H., Sakellariou, R.: Scheduling multiple DAGs onto heterogeneous systems. In: International Parallel and Distributed Processing Symposium, pp. 1–14 (2006)

Function-Structure Collaborative Mapping Induced by Universal Triple I Systems

Yiming Tang$^{(\boxtimes)}$, Jingjing Chen, Fuji Ren, Xi Wu, and Guangqing Bao

Anhui Province Key Laboratory of Affective Computing and Advanced Intelligent Machine, School of Computer and Information, Hefei University of Technology, Hefei 230601, China
tym608@163.com

Abstract. The and/or/not function tree is one of the principal functional models of product concept design. How to model and solve the and/or/not function tree is a key issue in this field. In the past, it was mainly to establish a function tree by artificial means, and it was hard to get the desired effect. Aiming at this problem, a novel function tree modeling method is proposed based on the idea of fuzzy reasoning. Firstly, the function tree is extended to the scope of the fuzzy function tree. Secondly, starting from the universal triple I algorithm of fuzzy reasoning, the fuzzy system for fuzzy function tree modeling is constructed, which is infiltrated into the function-structure mapping of function tree modeling. Therefore, the function-structure collaborative mapping method based on the triple I system is proposed. Through the application examples of the magnetic levitation train, the effectiveness of the proposed method is confirmed, and the problem of function tree modeling is effectively solved, which promotes the development of product concept design.

Keywords: Product conceptual design · Collaboration · Fuzzy reasoning · Function tree

1 Introduction

At present, the market competition in the manufacturing industry is becoming more and more fierce, and the competitiveness of products is largely due to its innovation. Moreover, innovation is mainly reflected in the product concept design stage [1]. At the same time, the research results show that the product concept design determines 75% of the total cost of product design and production [2].

The core of conceptual design is the establishment of functional models and functional solutions [3]. Currently, in the field of product conceptual design, there are a large number of functional models. The function tree containing and/or/not (referred to as the and/or/not function tree, or directly called a function tree) is one of the most common functional models [4]. Therefore and/or/not function tree is studied here.

A number of experts around the and/or/not or function tree have conducted some research. On the basis of axiomatic design theory, Zhu et al. [5] represented the tortuosity mapping of functional domains to domains as product structure trees. The

© Springer Nature Singapore Pte Ltd. 2019
Y. Sun et al. (Eds.): ChineseCSCW 2019, CCIS 1042, pp. 261–274, 2019.
https://doi.org/10.1007/978-981-15-1377-0_19

tree was only decomposed by "and". Zhang et al. [6] constructed a conceptual design cyclic mapping tree model including required domain, functional domain, principle solution domain. Based on the similarity theory and the extenics theory in [7], a similar expansion study was conducted for the and/or/not function tree. In [8], for the and/or/not function tree, Tang et al. investigated the function solving algorithm of the and/or/not function tree by solving the function set family [9]. In [10], a lossless simplification strategy was proposed by reducing the redundancy by logic simplification and facing the and/or/not function tree. Hua et al. [11] established the FFM function solving algorithm by alternately expanding and optimizing. Gan [12] proposed an extension model for innovative design of mechanical products for human-machine collaboration under the environment of human-machine collaboration. Duan et al. [13] proposed a module partitioning method combining functional decomposition and structural clustering for software radio.

The previous function tree was more manually created by domain experts, which was time-consuming and laborious, and the effect was not good. In this regard, we are trying to propose a new function tree modeling method.

In order to describe the function tree more delicately, we extend from the non-zero to 1 hard function tree to the fuzzy function tree, that is, the value of each node of the tree is within the interval [0, 1].

The key to the establishment of fuzzy function trees is functional structure mapping. This is essentially a category from function to structural reasoning. From this we think of introducing fuzzy reasoning methods for research.

For fuzzy reasoning, the core problem is the FMP (fuzzy modus ponens):

$$\text{FMP: For rule } A \to B \text{ and input } A^*, \text{ find output } B^*.$$

where A, A^* are the fuzzy subsets on the input domain X and B, B^* are the fuzzy subset Y on the output domain. There are many methods of fuzzy reasoning, such as the CRI method proposed by Zadeh [14], the triple I algorithm proposed by Wang [15], the universal triple I method proposed by Tang and Liu [16]. Among them, the universal triple I method is a well-recognized method in the field [17, 18].

To this end, we will use the universal triple I algorithm as the kernel and extend it to the fuzzy system, that is, the universal triple I system for functional structure mapping. Based on this point, a functional structure collaborative mapping strategy based on the universal triple I system is proposed.

2   Fuzzification of the Function Tree

First, we introduce the and/or/not function tree. The modeling strategy is as follows. We start from the top down with the total function (derived from the design requirements) as the starting point. The child node of each node is an implementation of its parent node or a further detailed description. Moreover, the type of the node may be a function, or may be a behavior, a structure, or the like. It can be seen that the function tree is actually established by means of deductive methods.

From the function tree modeling strategy, the function tree essentially reflects the idea of classical propositional logic. Specifically, "and", "or", and "not" respectively take "$\wedge$", and extract "$\vee$" and negate "$'$". The meanings see [8]. Let "$\sum$" represent the compound operation of extracting "$\vee$" multiple times, and let "$\prod$" represent the compound operation of "$\wedge$" multiple times.

Definition 1. The atomic proposition x_i is used to represent the leaf nodes in the function tree. Here $x_i = 1$ indicates that the node is established (where the demand of the node is satisfied), and $x_i = 0$ indicates that the node is not established. The proposition G_i used to indicate the gate node (where the meaning of the assignment is similar to the leaf node).

The atomic propositions and propositions are collectively referred to as tree propositions below. Let $H(G_i)$ denote the function tree whose top node is G_i, and let $X(G_i)$ (or X) be the set of all tree propositions.

Definition 2. For $H(G_i)$, let $\phi_{G_1}(X) = \phi(x_1, \cdots, x_n, G_1, \cdots, G_p)$ be its logical function (where n, p represent the atomic proposition and the number of propositions, respectively). $\phi_{G_1} = 1$ indicates that the top node of the function tree is established (that is, the design requirement is fully realized). $\phi_{G_1}(X) = 0$ indicates that the top node is not established. If no proposition exists in the logic function, then it is called complete logic function.

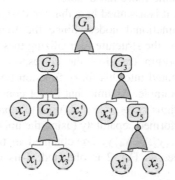

Fig. 1. An example of a function tree

With reference to Definition 2, the logic function of the function tree of Fig. 1 can be obtained, as shown in the following equation:

$$\phi_{G_1}(X) = G_1 = G_2 \vee G_3 = (x_1 \wedge G_4 \wedge x_2') \vee (x_4' \vee G_5')$$
$$= (x_1 \wedge x_2') \vee (x_1 \wedge x_2' \wedge x_3') \vee (x_4 \wedge x_4') \vee (x_4 \wedge x_5)$$

For the function tree, it was previously portrayed in a regular language. But in fact, this often makes the analysis of the problem not detailed enough. For example, in the conceptual design of the teacup, the use of "insulation" to describe the functional requirements is very general. Specifically, the boiling water is poured into the cup, and it cools in 20 min or cools in 60 min, and the degree is different. And this degree of difference leads to different implementations. Therefore, a more detailed analysis of the concept of "insulation" is needed. Then, using fuzzified metrics, this problem can often be better handled.

To this end, the nodes in the function tree are added with specific degree values, and the function tree is called a fuzzy function tree (a fuzzy function tree also referred to as a function tree when not obfuscated).

3 Function-Structure Mapping Based on the Triple I Algorithm

3.1 Macroscopic Process

One of the core steps in functional solution is the mapping of functions to structures (function-structure mapping). In the past, the study of function trees often assumed that the designer had completed a function tree containing information. But in fact, the decomposition of functions and the decomposition of structures are relatively natural, and they often have no a fundamental impact on the design results. The impact of function-structure mapping on the design is direct and essential, which is extremely important. Therefore, the function-structure mapping is separated here. And the function-structure mapping is processed by the universal triple I algorithm to make the selection of the design scheme more reasonable.

For functional solution, it is assumed here that the designer has established a fuzzy function tree with only functional nodes. Since the functional requirements have quantized values at this time, the structure for realizing these functions can be obtained by universal triple I algorithm (requires the corresponding fuzzy system). These structures are then incorporated into this fuzzy function tree.

We take Fig. 1 as an example. Assume that Fig. 1 contains only functional nodes. For several atomic propositions x_1, x_2, x_3, x_4, x_5 (used to represent leaf nodes) contained in it, fuzzy reasoning is performed separately (using the universal triple I algorithm and its fuzzy system). Then y_1, y_2, y_3, y_4, y_5 are obtained in turn, and are added in the original fuzzy function tree. So Fig. 2 is obtained. This is the process of function-structure mapping.

It should be pointed out that for the sake of convenience, functional nodes are generally decomposed into leaf nodes (often inseparable basic functions) before conversion. Then multiple structural nodes may be obtained. The obtained structural nodes may need to be enriched. This will lead to a complete fuzzy function tree.

3.2   The Universal Triple I System for Function Tree Modeling

Here, the universal triple I algorithm is used as the kernel, and the universal triple I system for function tree modeling is proposed. Specifically, it is presented from the framework of the fuzzy system. According to the structure of knowledge base, it includes fuzzer, fuzzy reasoning and defuzzifier.

The function-structure mapping rule is stored in the knowledge base. For example, for the functional requirements of "insulation". In the sub-knowledge database, various

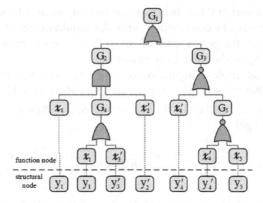

Fig. 2. Consider the function-structure mapping fuzzy function tree

mapping methods of which "insulation" implies "insulation realization structure" have been saved in advance. When the calculation is needed, the corresponding sub-knowledge base is first located, and then the later processing is performed.

Furthermore, due to the using of fuzzy inference methods. Therefore, the input and output of the rules in the knowledge base are fuzzy sets. Considering that the triangular fuzzy set is an extremely common and simple fuzzy set, it is also suitable for the field of innovative conceptual design. Therefore, a triangular fuzzy set is used here as a tool for fuzzy characterization. That is, the input and output of the rules in the knowledge base adopt a triangular fuzzy set.

Figure 3 is a diagram of a common configuration of multiple triangular fuzzy sets. A set with such a structure (consisting of fuzzy sets) is called a fuzzy set family consisting of a triangular fuzzy set, which is simply referred to as a fuzzy set family. And it is assumed that the peaks of the fuzzy set in the fuzzy set are equidistant.

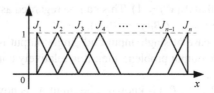

Fig. 3. Fuzzy set family

For input x^*, we use a single point fuzzy set $A_{x^*}^* = A^*(x) = \begin{cases} 1, & x = x^* \\ 0, & x \neq x^* \end{cases}$.

Here we utilize fuzzy reasoning based on the universal triple I algorithm. Suppose there are n given rules $A_i \rightarrow B_i$, and now the input A^* (previously converted x^* to A^* by a single point fuzzier) is used to calculate the output B^*. When $\rightarrow_2 = I_L$, $\rightarrow_1 = I_{La}$, the universal triple I algorithm has better properties in terms of both reductive and responsive performance. In combination with the requirements of the field of innovative concept design, the general idea of using the universal triple I algorithm when $\rightarrow_2 = I_L$ and $\rightarrow_1 = I_{La}$ is used for fuzzy reasoning.

For the universal triple I algorithm of $\rightarrow_2 = I_L, \rightarrow_1 = I_{La}$, because the MinP-universal triple I solution is $B^*(y) = \sup_{x \in E_y} \{A^*(x) + R_1(x, y) - 1\}$,

where $R_1(x, y) = \vee_{i=1}^n (A_i(x) \rightarrow_1 B_i(y)) = \vee_{i=1}^n (A_i(x) \times B_i(y))$.

For input x^*, we get

$$A_{x^*}^* = A^*(x) = \begin{cases} 1, & x = x^* \\ 0, & x \neq x^* \end{cases}, \quad B^*(y) = \begin{cases} R_1(x^*, y), & x^* \in E_y. \\ 0, & x^* \notin E_y \end{cases}$$

Then, for any fixed x^*, suppose $x^* \in E_y$. Two fuzzy sets A_j and A_{j+1} satisfying $A_j(x^*), A_{j+1}(x^*) > 0$ can be found. Then we get:

$$\begin{aligned} B^*(y) &= R_1(x^*, y) = (A_j(x^*) \rightarrow_1 B_j(y)) \vee (A_{j+1}(x^*) \rightarrow_1 B_{j+1}(y)) \\ &= (A_j(x^*) \times B_j(y)) \vee (A_{j+1}(x^*) \times B_{j+1}(y)) \\ &= (a_j \times B_j(y)) \vee (a_{j+1} \times B_{j+1}(y)) \\ &= \begin{cases} a_j \times B_j(y), & a_j \times B_j(y) \geq a_{j+1} \times B_{j+1}(y) \\ a_{j+1} \times B_{j+1}(y), & else \end{cases} \end{aligned} \quad (1)$$

Here $j, j+1 \in \{1, 2, \cdots, n\}$, $A_j(x^*), A_{j+1}(x^*) > 0$, $A_1(x^*) = \cdots A_{j-1}(x^*) = A_{j+2}(x^*) = \cdots = A_n(x^*) = 0$, and let $a_j \cong A_j(x^*)$, $a_{j+1} \cong A_{j+1}(x^*)$.

Here A_j, A_{j+1} are generally two connected fuzzy sets. For example $A_j = J_2, A_{j+1} = J_3$ in Fig. 3. However, B_j, B_{j+1} are obtained by the two rules $A_j \rightarrow B_j, A_{j+1} \rightarrow B_{j+1}$ in the knowledge base. So B_j, B_{j+1} are not necessarily the two fuzzy sets that are connected (such as the case where $B_j = J_3, B_{j+1} = J_1$ may occur). The following are all similar cases.

There is another possibility here. There is only a fuzzy set A_j such that $A_j(x^*) > 0$ (where it is easy to find that $A_j(x^*) = 1$). This can be regarded as a special case of (1). It is similar to obtain $B^*(y) = a_j \times B_j(y) = B_j(y)$.

The fuzzy inference case of single input and single output is given above. If it is a single input and multiple output problem, then it is relatively complicated. The essence of the idea is as follows.

The rule $A \rightarrow (B_1, B_2, \cdots, B_p)$ is known. The input A^* is now given, and the output $(B_1^*, B_2^*, \cdots, B_p^*)$ is sought.

The computing process is as follows. First of all, B_1^* is obtained by A^* and $A \rightarrow B_1$. Then B_2^* is computed by $A \rightarrow B_2$ and A^*. Following that, $(B_1^*, B_2^*, \cdots, B_p^*)$ can be similarly achieved. Finally, the integrated solution $(B_1^*, B_2^*, \cdots, B_p^*)$ is obtained.

This is discussed for the three types of defuzzifiers.

(i) If the center of gravity method is used to solve the ambiguity, we get

$$y^* = \frac{\int_Y yB^*(y)dy}{\int_Y B^*(y)dy} = \frac{\int_{Y_1} yB^*(y)dy + \int_{Y_2} yB^*(y)dy}{\int_{Y_1} B^*(y)dy + \int_{Y_2} B^*(y)dy}$$
$$= \frac{\int_{Y_1} y \times a_j \times B_j(y)dy + \int_{Y_2} y \times a_{j+1} \times B_{j+1}(y)dy}{\int_{Y_1} a_j \times B_j(y)dy + \int_{Y_2} a_{j+1} \times B_{j+1}(y)dy}$$

and $Y_1 = \{y \in Y \,|\, a_j \times B_j(y) \geq a_{j+1} \times B_{j+1}(y)\}$, $Y_2 = Y - Y_1$.

(ii) Note that when $B^*(y)$ is the sum or intersection of M fuzzy sets, then the central average defuzzifier can be used (an approximation of the ambiguity of the center of gravity method, see [19]). At this time $y^* = \frac{\sum_{l=1}^M \bar{y}^l \omega_l}{\sum_{l=1}^M \omega_l}$. Here $\bar{y}^l$ is the center of the l-th fuzzy set, and the weight ω_l is the height ($l = 1, \cdots, M$) of the l-th fuzzy set.

Furthermore, from (1), it is obtained:

$$y^* = \frac{\bar{y}^j \omega_j + \bar{y}^{(j+1)} \omega_{j+1}}{\omega_j + \omega_{j+1}} = \frac{\bar{y}^j a_j + \bar{y}^{(j+1)} a_{j+1}}{a_j + a_{j+1}}.$$

(iii) According to the previous center of gravity method, we get

$$y^* = \frac{\int_Y yB^*(y)dy}{\int_Y B^*(y)dy} \approx \frac{\sum_{i=1}^n y_i B^*(y_i)h_i}{\sum_{i=1}^n B^*(y_i)h_i}.$$

Thus, a defuzzifier can also be obtained, and the result is:

$$y^* = \frac{\sum_{i=1}^n y_i B^*(y_i)h_i}{\sum_{i=1}^n B^*(y_i)h_i}.$$

Here, since the triangle fuzzy set is used for processing, and the peak points are equidistant between them, then:

$$y^* = \frac{\sum_{i=1}^n y_i B^*(y_i)h_i}{\sum_{i=1}^n B^*(y_i)h_i} = \frac{\sum_{i=1}^n y_i B^*(y_i)}{\sum_{i=1}^n B^*(y_i)}.$$

From (1), we know:

$$B^*(y) = (a_j \times B_j(y)) \vee (a_{j+1} \times B_{j+1}(y)),$$

$$y^* = \frac{\sum_{i=1}^{n} y_i B^*(y_i)}{\sum_{i=1}^{n} B^*(y_i)} = \frac{\bar{y}^j B^*(\bar{y}^j) + \bar{y}^{j+1} B^*(\bar{y}^{j+1})}{B^*(\bar{y}^j) + B^*(\bar{y}^{j+1})} = \frac{\bar{y}^j a_j + \bar{y}^{j+1} a_{j+1}}{a_j + a_{j+1}}.$$

Here $\bar{y}^j, \bar{y}^{j+1}$ are the centers of the fuzzy sets B_j, B_{j+1} respectively. It can be seen that this defuzzifier is in fact equivalent to the central average defuzzifier.

Because the actual problems are often complicated, the amount of calculation is huge. The central average defuzzifier is essentially an approximation of the center of gravity defuzzifier. Therefore, in order to facilitate the calculation, the center average defuzzifier is used instead of the center of gravity method for processing.

After that, only the general function and structure are needed here, so after getting y^*, it can be considered to be attributed to a certain class. We calculate the distance $(i = 1, 2, \cdots, n)$ between y^* and the center y_i of each $B_i(y)$. We take the closest one, that is $d = \min\{|y_i - y^*|, \quad i = 1, \cdots, n\}$. Let

$$y^{**} = \max\{y_i \,|\, d = |y_i - y^*|, \quad i = 1, \cdots, n\}. \tag{2}$$

If it is the case of multi-output $y_1^*, \ldots, y_r^*$, then let $y^* = y_1^*, \ldots, y_r^*$. We process $(j = 1, \ldots, r)$ for each y_j^* according to (2), and get the corresponding y_j^{**}, and finally obtain $y^{**} = (y_1^{**}, \ldots, y_r^{**})$.

3.3 Functional Structure Cooperative Mapping Algorithm

Because the key to innovative conceptual design is the collaborative mapping from functional requirements to structural information. Therefore, it is first necessary to establish a knowledge base of functional information and structural information. Furthermore, considering that the step mapping is implemented by fuzzy inference in the later stage, it is necessary to blur the function and structure. Specifically, a standard fuzzy set (such as a 7-fuzzy set family) is selected for each function (or structure). Then, based on the domain knowledge, the corresponding knowledge base of fuzzy inference rules is established.

The function of this module is as follows. The designer enters an initial fuzzy function tree (mainly only with function nodes). Then, the function-structure mapping is performed by fuzzy reasoning (using the universal triple I algorithm and its fuzzy system), so that the structural nodes corresponding to the respective function nodes are obtained. Based on this, a complete fuzzy function tree is formed. Thus, the functional structure collaborative mapping algorithm driven by the universal triple I system is obtained. The main process of its overall processing is Algorithm 1.

Algorithm 1. Functional structure collaborative mapping algorithm.

Input: Initial fuzzy function tree.

Output: Extended fuzzy function tree.

Step 1: Functional structure collaborative mapping algorithm driven by universal triple I system.

Step 2: If the current leaf is a function node to be processed, then continue; otherwise, go to Step 7.

Step 3: We convert the current leaf node into a fuzzy set $A^*(x)$.

Step 4: We locate the corresponding sub-knowledge database in the fuzzy inference rule base, and extract the relevant fuzzy rules in the sub-knowledge base to perform fuzzy reasoning (i.e., mapping $A^*(x)$ to $B^*(y)$).

Step 5: We convert $B^*(y)$ to output y^* by the center average defuzzifier.

Step 6: We perform the classification of the output result and save it.

Step 7: If the current node has the next leaf node in the initial tree, then we let the current node take the next leaf node and go to Step 2; otherwise, continue.

Step 8: Add the obtained structure nodes to the fuzzy function tree.

End procedure

4 Applications

Nowadays in the field of magnetic levitation trains, "floating" and "driving" are two core functions, which are mainly done on the "sensing interface". For example, Japanese large-gap "repulsive" maglev trains, German "small gaps", and long-stack "suction-type" maglev trains use this approach. However, these require high-precision, high-tech control methods, which results in high cost of floating.

This section describes the process of functional solution by taking the innovative concept design of the maglev train as an example (in order to obtain a low-cost design). The core function of the maglev train has two parts. One is the magnetic levitation function and the other is the driving function. In terms of magnetic levitation, it mainly includes two modes of "magnetic repulsion" and "magnetic absorbing". Here we only show the implementation of the "magnetic repulsion". At the same time, given the integrity, the implementation strategy of the driver is also given. Furthermore, there are mainly two types of implementations of the magnetic repulsion function. One is magnetic repulsion based on magnetic-magnetic method, and the other is magnetic repulsion based on magnetic-inductive method.

Around these functions, it is assumed that the initial fuzzy function tree as shown in Fig. 4 is obtained.

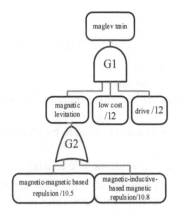

Fig. 4. Initial fuzzy function tree of maglev train

4.1   Functional Requirements Analysis and Fuzzy Inference Rule Base

To begin with, we show magnetic repulsion function based on magnetic-magnetic method. In order to realize the magnetic-magnetic magnetic repulsion function, the bottom of the train and the rail can be made magnetic. Moreover, the magnetic poles on the opposite side of the train and the rail should be the same (mutually exclusive). At the same time, the cost is guaranteed to be low (taking the 7-fuzzy set, see Fig. 3, which shows low cost). The input to this function uses a 7-fuzzy set family whose output is $B_1^*, B_2^*, B_3^*, B_4^*$. Where $B_1^*, B_2^*, B_3^*, B_4^*$ respectively denotes magnetism at the bottom of the magnet-magnetic train (taking 7-fuzzy set family), rail magnetism (taking 5-fuzzy set family), mutual exclusion (taking 2-fuzzy set family), low cost (taking 3 - Fuzzy set family). Table 1 is the fuzzy inference rule base of the relevant part.

Table 1. The rule base for magnetic repulsion function based on magnetic-magnetic case

	Magnetism for the bottom with magnetic-magnetic case	Rail magnetic	Mutually exclusive	Low cost
Magnetic-magnetic based repulsion F_2	F_1	E_1	C_1	D_1
Magnetic-magnetic based repulsion F_2	F_1	E_2	C_1	D_1
Magnetic-magnetic based repulsion F_2	F_1	E_0	C_1	D_1
Magnetic-magnetic based repulsion F_2	F_1	E_2	C_1	D_0
Magnetic-magnetic based repulsion F_2	F_1	E_1	C_1	D_{-1}
Magnetic-magnetic based repulsion F_2	F_1	E_2	C_1	D_{-1}
...	...	...	...	...

Furthermore, we provide magnetic repulsion function based on magnetic-inductive method. In order to realize the magnetic repulsion function based on the magnetic-inductive method, the bottom of the train can be made magnetic. At the same time, the rails are sensed. Similarly, the magnetic properties of the opposite faces should be the same, and the cost needs to be low. The input to this function uses a 5-fuzzy set family whose output is $B_1^*, B_2^*, B_3^*, B_4^*$. Where $B_1^*, B_2^*, B_3^*, B_4^*$ denotes the magneticity of the bottom of the train for magnetic-sensing (taking the 7-fuzzy set family), the inductivity of the rail (employing the 5-fuzzy set family), the mutual exclusion (taking the 2-fuzzy set family), and the cost is low (taking the 3-fuzzy set family). Table 2 is the fuzzy inference rule base of the relevant part.

Table 2. The rule base for magnetic repulsion function based on magnetic-inductive case

	Magnetism for the bottom with magnetic-induction case	Rail inductivity	Mutually exclusive	Low cost
Magnet-sensitive magnetic repulsion E_1	F_1	E_0	C_1	D_1
Magnet-sensitive magnetic repulsion E_1	F_2	E_1	C_1	D_0
Magnet-sensitive magnetic repulsion E_2	F_3	E_1	C_1	D_{-1}
Magnet-sensitive magnetic repulsion E_2	F_3	E_2	C_1	D_{-1}
...	...	...	...	...

4.2 Function-Structure Mapping Process Based on Fuzzy Reasoning

On the one hand, we analyze magnetic repulsion function based on magnetic-magnetic method. For the input "magnetic-magnetic based magnetic repulsion/10.5", i.e., $x^* = 10.5$. The fuzzy reasoning process is shown below.

For x^*, we use $A_{x^*}^* = A^*(x) = \begin{cases} 1, & x = 10.5 \\ 0, & x \neq 10.5 \end{cases}$.

It is easy to know that j (satisfying $A_j(x^*), A_{j+1}(x^*) > 0$) is equal to 2. We get $a_j = A_j(x^*) = F_2(x^*) = 0.75$, $a_{j+1} = A_{j+1}(x^*) = F_3(x^*) = 0.25$. Referring to Table 1, it can be seen that $A_j = F_2$, $A_{j+1} = F_3$ has three output schemes, which are respectively denoted as $B_1^{(1)}, B_1^{(2)}, B_1^{(3)}$ and $B_2^{(1)}, B_2^{(2)}, B_2^{(3)}$. Therefore, in general, $3 \times 3 = 9$ schemes are available.

(i) First we take $A_j \to B_1^{(1)}$, $A_{j+1} \to B_2^{(1)}$ as an example. Comparing to Table 1, we have $B_1^{(1)} = (B_{11}, B_{12}, B_{13}, B_{14}) = (F_1, E_1, C_1, D_1)$, $B_1^{(1)} = (B_{21}, B_{22}, B_{23}, B_{24}) = (F_2, E_2, C_1, D_0)$

Here we need to calculate separately to get $(B_1^*, B_2^*, B_3^*, B_4^*)$. That is, B_1^* is calculated by $A_j \to B_{11}$, $A_{j+1} \to B_{21}$. B_2^* is calculated by $A_j \to B_{12}$, $A_{j+1} \to B_{22}$. From (1), we get:

$$B_1^*(y) = (a_j \times B_{11}(y)) \vee (a_{j+1} \times B_{21}(y)) = (a_j \times F_1(y)) \vee (a_{j+1} \times F_2(y)).$$

$$B_2^*(y) = (a_j \times B_{12}(y)) \vee (a_{j+1} \times B_{22}(y)) = (a_j \times E_1(y)) \vee (a_{j+1} \times E_2(y)).$$

$$B_3^*(y) = (a_j \times B_{13}(y)) \vee (a_{j+1} \times B_{23}(y)) = (a_j \times C_1(y)) \vee (a_{j+1} \times C_1(y)).$$

$$B_4^*(y) = (a_j \times B_{14}(y)) \vee (a_{j+1} \times B_{24}(y)) = (a_j \times D_1(y)) \vee (a_{j+1} \times D_0(y)).$$

Then by the central average defuzzifier, we get the following results. If B_1^*, then $y_1^* = \frac{\bar{y}^j a_j + \bar{y}^{(j+1)} a_{j+1}}{a_j + a_{j+1}} = \frac{8 \times 0.75 + 10 \times 0.25}{0.75 + 0.25} = 8.5$. If B_2^*, then $y_2^* = 9.75$. If B_3^*, then $y_3^* = 12$. If B_4^*, then $y_4^* = 10.5$. So get the output vector $y^* = (y_1^*, y_2^*, y_3^*, y_4^*) = (8.5, 9.75, 12, 10.5)$.

Finally, considering the classification of y^*, we need to convert y^* to $y^{**} = (y_1^{**}, y_2^{**}, y_3^{**}, y_4^{**})$. According to (2), we can get $y_1^{**} = \max_{i \in \{1, \cdots, n\}} \{y_i \mid \min_{j \in \{1, \cdots, n\}} \{|y_j - y_1^*|\}\} = |y_i - y_1^*|\} = 8$, $y_2^{**} = 9$, $y_3^{**} = 12$, $y_4^{**} = 12$. Then one has $y^{**} = (8, 9, 12, 12)$.

(ii)   For $A_j \to B_1^{(1)}$, $A_{j+1} \to B_2^{(2)}$, where $B_1^{(1)} = (F_1, E_1, C_1, D_1)$, $B_2^{(2)} = (F_3, E_1, C_1, D_{-1})$. After calculation, similarly we get $y^{**} = (10, 9, 12, 12)$.

(iii) For $A_j \to B_1^{(1)}$, $A_{j+1} \to B_2^{(3)}$, we have $y^{**} = (10, 9, 12, 12)$.

(iv)   For $A_j \to B_1^{(2)}$, $A_{j+1} \to B_2^{(1)}$, we get $y^{**} = (8, 12, 12, 12)$.

(v)   For $A_j \to B_1^{(2)}$, $A_{j+1} \to B_2^{(2)}$, we have $y^{**} = (10, 12, 12, 12)$.

(vi)   For $A_j \to B_1^{(2)}$, $A_{j+1} \to B_2^{(3)}$, we get $y^{**} = (10, 12, 12, 12)$.

(vii) For $A_j \to B_1^{(3)}$, $A_{j+1} \to B_2^{(1)}$, we have $y^{**} = (10, 9, 12, 12)$.

(viii) For $A_j \to B_1^{(3)}$, $A_{j+1} \to B_2^{(2)}$, we get $y^{**} = (10, 6, 12, 12)$.

(ix)   For $A_j \to B_1^{(3)}$, $A_{j+1} \to B_2^{(3)}$, we have $y^{**} = (10, 9, 12, 12)$.

Here, schemes (ii), (iii), (vii), and (ix) are the same, and schemes (v) and (vi) are the same, so that 5 schemes are actually obtained.

On the other hand, we analyze magnetic repulsion function-structure mapping based upon magnetic-magnetic method. For the input "magnetic-inductive-based magnetic repulsion/10.8", i.e., $x^* = 10.8$, a single point fuzzy set is used. Similar to the above, we can get the corresponding conclusion. Finally, four options are available. They are $(10, 9, 12, 6)$, $(10, 9, 12, 6)$, $(12, 9, 12, 0)$, $(12, 12, 12, 0)$.

Then we obtain a complete fuzzy function tree. Considering the integrity of the design, the implementation strategy of the driver is given directly here. That is, the drive can be completed in several ways, such as a linear motor, a DC motor armature, and a drive plate of the primary winding. At this point, the structural nodes obtained by the function-structure mapping are added to the fuzzy function tree to obtain a complete fuzzy function tree, as shown in Fig. 5. Referring to Definition 2, the logic function of the fuzzy function tree shown in Fig. 5 is given below:

$$G_1 = G_2 \wedge x_1 \wedge G_3 = (G_4 \vee G_5) \wedge x_1 \wedge (x_2 \vee x_3 \vee x_4 \vee G_6)$$
$$= \{[(x_6 \wedge x_7 \wedge x_8 \wedge x_1) \vee (x_9 \wedge x_7 \wedge x_8 \wedge x_1) \vee (x_6 \wedge x_{10} \wedge x_8 \wedge x_1) \vee (x_9 \wedge x_{10} \wedge x_8 \wedge x_1)$$
$$\vee (x_9 \wedge x_{11} \wedge x_8 \wedge x_1)] \vee [(x_{12} \wedge x_{13} \wedge x_8 \wedge x_{14}) \vee (x_{15} \wedge x_{13} \wedge x_8 \wedge x_1')$$
$$\vee (x_{15} \wedge x_{16} \wedge x_8 \wedge x_1')]\} \wedge x_1 \wedge [x_2 \vee x_3 \vee x_4 \vee ((x_{17} \vee x_{18}) \wedge x_5)].$$

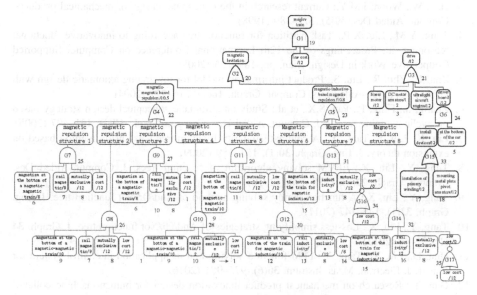

Fig. 5. Complete fuzzy function tree of train

5 Summary and Outlook

This paper introduces the universal triple I algorithm for the problem of function tree modeling. The functional structure collaborative mapping driven by the universal triple I system is proposed. First, we extend the hard function tree to the scope of the fuzzy function tree. That is, the value from non-zero or 1 is extended to the entire [0, 1] interval. Second, based on the universal triple I algorithm of fuzzy reasoning, a fuzzy system for modeling fuzzy function trees (i.e., the universal triple system) is constructed. It is infiltrated into the functional structure mapping of the function tree modeling. Then, the functional structure collaborative mapping method driven by the universal triple I system is proposed. Finally, the effectiveness of the proposed method is verified by an application example of a maglev train.

In the future research, other methods of fuzzy reasoning will be further tried. For example, the symmetric implication algorithm [20] can be used as a new kernel for function tree modeling. And, we will try to infiltrate fuzzy reasoning into the whole process of conceptual design.

Acknowledgement. This work was supported by the National Natural Science Foundation of China (Nos. 61673156, 61672202, 61432004, U1613217, 61877016).

References

1. Shai, O., Reich, Y., Rubin, D.: Creative conceptual design: extending the scope by infused design. Comput. Aided Des. **41**(3), 117–135 (2009)
2. Chakrabarti, A., Bligh, T.P.: A scheme for functional reasoning in conceptual design. Des. Stud. **22**(6), 493–517 (2001)
3. Hsu, W., Woon, I.M.Y.: Current research in the conceptual design of mechanical products. Comput. Aided Des. **30**(5), 377–389 (1998)
4. Tang, Y.M., Liu, X.P.: Task partition for function tree according to innovative functional reasoning. In: Proceedings of the 12th International Conference on Computer Supported Cooperative Work in Design, Xian, pp. 189–195 (2008)
5. Zhu, L., Zhu, R., Liu, S.: Product information model for integrating axiomatic design with DFA. J. Comput.-Aided Des. Comput. Graph. **16**(2), 216–221 (2004)
6. Zhang, S., Feng, P., Pan, S.X., et al.: Study on automatic conceptual design strategy based on cyclic mapping model. J. Comput.-Aided Des. Comput. Graph. **17**(3), 491–497 (2005)
7. Liu, X., Liu, J., Tang, Y.: A contrast similarity function-tree's extension method based on extension theory. J. Eng. Graph. **30**(1), 153–159 (2009)
8. Tang, Y., Liu, X.: EFVM solving algorithm of function trees. J. Comput.-Aided Des. Comput. Graph. **22**(9), 1578–1586 (2010)
9. Tang, Y., Liu, X.: A function family solving method of And/Or/Not function tree. J. Eng. Graph. **32**(1), 143–147 (2011)
10. Tang, Y., Liu, X.: Lossless simplifying strategies of And/Or/Not function tree. J. Graph. **34**(1), 31–40 (2013)
11. Hua, D., Liu, X., Tang, Y.: FFM function-solving and its application in control system for sensor. J. Electron. Meas. Instrum. **30**(6), 975–981 (2016)
12. Gan, Y.: Research on mechanical product innovation design for human-machine collaboration. China Plant Eng. **8**, 107–108 (2019)
13. Duan, T., Huang, Y., Wang, Y.: Research on module partition for reconfiguration in software radio. Electron. Des. Eng. **27**(9), 92–99 (2019)
14. Zadeh, L.A.: Outline of a new approach to the analysis of complex systems and decision processes. IEEE Trans. Syst. Man Cybern. **3**(1), 28–44 (1973)
15. Wang, G.: On the logic foundation of fuzzy reasoning. Inf. Sci. **117**(1), 47–88 (1999)
16. Tang, Y., Liu, X.: Differently implicational universal triple I method of (1, 2, 2) type. Comput. Math. Appl. **59**(6), 1965–1984 (2010)
17. Tang, Y., Ren, F.: Fuzzy systems based on universal triple I method and their response functions. Int. J. Inf. Technol. Decis. Making **16**(2), 443–471 (2017)
18. Tang, Y., Pedrycz, W.: On continuity of the entropy-based differently implicational algorithm. Kybernetika **55**(2), 307–336 (2019)
19. Wang, L.X.: A Course in Fuzzy Systems and Control. Prentice-Hall, Englewood Cliffs (1997)
20. Tang, Y., Pedrycz, W.: On the $\alpha(u, v)$-symmetric implicational method for R- and (S, N)-implications. Int. J. Approx. Reason. **92**, 212–231 (2018)

A Feature Selection Based on Network Structure for Credit Card Default Prediction

Yanmei Hu[✉], Yuchun Ren, and Qiucheng Wang

Chengdu University of Technology, Chengdu, China
huyanmei@cdut.edu.cn,
renyuchunCS@126.com,
wqc230@126.com

Abstract. The problem of credit card default prediction is important in finance and electronic commerce, thus it has been attracting more and more attention. Generally, the existing research work on credit card default prediction directly applies a classification model to the historical data and train a predictor, but rarely deeply explores the data. In this paper, we research the problem of credit card default prediction in an unconventional way. First, we study the records of consumption by credit card from the perspective of network to uncover the relationships between features and the ones between features and label. Second, based on the network structure we propose a new feature selection algorithm named as NSFSA. Finally, we apply the NSFSA to five machine learning models to train predictors over the real dataset of consumption records by credit card, and also compare with four existing feature selection algorithms. Experimental results show that the proposed NSFSA performs excellently, which demonstrates the potentials of our way to research the credit card default problem.

Keywords: Network structure · Credit card default prediction · Feature selection · Community structure

1 Introduction

Currently, credit card is widely used in people's daily consumption, but on the other hand the credit card default is obviously increasing along with the popularity of credit card. This brings about big challenges to the financial institutions that issue credit cards. Thus an increasing number of researchers from different fields, including finance, economics and computer science, have been paying attention to the prediction of credit card default.

Generally, the existing research work uses machine learning models to solve the problem of credit card default. For example, to estimate credit risk parameters Sun and Jin present two ensemble learning methods by random forest and stochastic gradient boosting (Sun and Jin 2016); Ajay et al. applied several machine learning models such as random forest, naïve bayesian classifier and support vector machine to train predictors of credit card default (Ajay et al. 2016). In (Leow and Crook 2016), credit card exposure at default is estimated by a new mixture model constructed from panel models

© Springer Nature Singapore Pte Ltd. 2019
Y. Sun et al. (Eds.): ChineseCSCW 2019, CCIS 1042, pp. 275–286, 2019.
https://doi.org/10.1007/978-981-15-1377-0_20

and survival models. It is surprising and interesting that almost all the work directly applying the machine learning models is considered as practical on the problems related to credit card.

However, the work above applies a somewhat black-box approach, i.e., it directly takes the data of consumption records by credit card to machine learning models as inputs, and cannot produce more except the prediction result. We believe that more can be done if more information could be studied from the consumption behaviors, e.g., the patterns hidden in the consumption behaviors and the effect of the patterns on the prediction, and the most and least informational features to the prediction. For instance, when the critical features are identified, more attention can be paid on their collection to reduce mistakes; exploring hidden patterns can make us deeply insight the consumption behaviors by credit card; moreover, exploring the effects of those patterns on prediction can produce further potentially valuable information. On the other hand, when the size of instances is small but the dimensionality of each instance is high, the problem of over-fitting is appearing (Bermingham et al. 2015). In this case, we can pick the features based on the knowledge mined from the data of consumption records, and take them to learning models as input to overcome the problem of over-fitting. Even in the case that the feature dimensionality is not that high, only using the important features to learn models is still desirable, e.g., ignoring the data noise and reducing computational cost. Therefore, we continue our previous work (Wang et al. 2017, 2018) and develop the research in this paper. Consequently, we proposed an improved algorithm of feature selection for the prediction of credit card default.

A lot of work about feature selection (Li et al. 2016; Alelyani et al. 2016) and credit card default prediction (Leow and Crook 2016; Abdou et al. 2016) has been witnessed. We continue this research line in a way different from our previous work. First, we observe the data of consumption records by credit card from the perspective of network, and find that the feature network presents an obvious community structure, which means that the features are clustered into different groups. Further, in order to consider both of the relationships among features and the ones between features and the label, we weight each feature based on the network structure and its relevance to label. Then, based on the observation and the feature weight, we proposed a feature selection algorithm and apply it to learning of prediction models. Finally, we experiment the effect of the proposed feature selection on the prediction of credit card default. In this paper, our contributions are as follows:

(1) We weight each feature based on the structure of the feature network and its relevance to label, in order to consider both of the relationships among features and the ones between features and the label.
(2) We proposed a new algorithm of feature selection by integrating the observed feature pattern and the feature weight.
(3) We use the proposed algorithm to select important and informational features to the prediction and then take those features to several widely used machine learning models as input to obtain several prediction models.
(4) We test the prediction models on real dataset of consumption records by credit card. To further test the effectiveness of the proposed feature selection, we also compare it with other feature selection algorithms which are widely used in the

literature. Experiments show that the proposed algorithm yields excellent performance and has potential value to the problem of credit card default prediction.

The following is the organization of the rest of this paper. Related work is reviewed in Sect. 2. The dataset used and the problem statement are described in Sect. 3. The observation over the data of consumption records is presented in Sect. 4, followed by the proposed network structure based feature selection algorithm. The machine learning models which are used to train predictors are described in Sect. 5. Experimental results are presented in Sect. 6, and the conclusions are presented in Sect. 7.

2 Related Work

Presently, the prediction of credit card default is a critical issue in finance and has witnessed a lot of research work from both academia and industry. Mainly, the machine learning models (particularly, the classification models) are used to train predictors based on historical data, see (Wang et al. 2015; Hon and Bellotti 2016; Abdou et al. 2016; Evangelista and Artes 2016) for example. In this line of work, a classification model directly takes almost all the features of instances as input to train the predictor.

The above work has been demonstrated to be useful to credit card default prediction, but directly and simply use the machine learning models to train predictors can only produce the prediction result and may causes other problems such as lack of interpretability. There are probably valuable patterns hidden in data. It would be meaningful if we can explore them and further use them to improve the solution of a problem. For instance, in the work of (Yang and Leskovec 2014; Hu et al. 2016; Hu and Yang 2017) novel patterns related to community structure are observed and further applied to improve the solution of community detection; in (Nie et al. 2009; Zhao et al. 2015), patterns hidden in the consumption data of credit card are also mined.

Our work here is closely related to the work on feature selection (Li et al. 2016; Alelyani et al. 2016; Zhang et al. 2016). Feature selection is a necessity when the feature dimensionality is high and the sample size is small (Kung and Mark 2009; Boln-Canedo et al. 2015; Asir Antony Gnana Singh et al. 2016). It is also very useful even the feature dimensionality is not that high, such as eliminating noise and reducing computational cost. There is a huge literature of feature selection and it is impossible to list all here, but the exiting algorithms of feature selection can be traditionally categorized into the following three classes: filter method, wrapper method and embedded method. Both of wrapper and embedded methods are dependent on the predictor, and filter method directly evaluates features from data itself. Thus, filter method costs much less computational time than wrapper and embedded methods. See (Tallón-Ballesteros et al. 2016; Peng et al. 2015), (Ruiz et al. 2006; Ma et al. 2017) and (Mejía-Lavalle et al. 2006; Lin et al. 2012; Fu et al. 2009) for examples of filter, wrapper and embedded methods, respectively.

The work in (Butterworth et al. 2005; Zhou et al. 2014; Han and Kim 2015) is most closely related to ours. They select features by incorporating the technique of clustering, but in our work we select features according to the feature pattern explored from consumption behaviors.

3 Preliminaries

In this work, we focus on mining and analyzing a set of credit consumption data from a bank in China, and further aim to find a solution to the prediction of credit card default. Thus, we first briefly describe the data and then discuss the problem of credit card default following in this section.

3.1 Data

The dataset is consisted of 9393 instances with each instance corresponding to one credit card and containing 87 features. The features are extracted from the records of consumption behaviors such as the number of transactions, the maximum consumption amount, and the number and the amount of cash withdrawals by credit card. Moreover, each credit card corresponds to one instance which is explicitly marked a label representing whether it has the records of default or not. Thus, the set of instances are classified into two groups in accordance with the label, i.e., default or not default.

3.2 Problem Statement and Notations

Based on the historical consumption records of a set of credit cards, the problem of credit card default prediction is to predict whether a credit card holder will be default in the near future. Formally, assuming that the dataset of n instances $X = \{x_1, x_2, ..., x_n\}$, where $x_i = \{f_i, y_i\}$ is the ith instance which has a d-dimensional feature vector $f_i = \{f^1, f^2, ..., f^d\}$ with a label $y_i = 1$ or $y_i = 0$ (the value 1 corresponds to default and 0 corresponds to non-default). Then the task is to forecast the label y_j of a new instance x_j, on the basis of the dataset X.

There are many classification models such as Logistical regression, SVM and Decision Tree which have mature theories behind to support their practical applications and are widely used in many fields. A straight way for the credit card default prediction is to apply these models to train predictors on X. However, there are usually useless features to the prediction, and some features may be redundant. If those features are taken as input to the predictor, not only the accuracy of prediction will be decreased, but also the time cost by training predictor will be increased. This problem will be critical when the feature dimensionality is high, which causes the "curse of dimensionality". Thus, proceeding feature selection before applying these models to train predictors can often improve the prediction. Generally, the goal of feature selection is to select a subset of features F' from the original set of features F which is useful and non-redundant to the prediction of credit card default.

On the other hand, along with the dramatically increase of data's scale and complexity, the relationships between features and the ones between features and the label become complex. The relationships between features are probably not linear and the one between features and the label are not straightly useful or useless. We believe that mining the complex relationships and capturing the patterns hidden in data can give us a new insight to feature selection. Thus, there are two main goals in this paper: (1) exploring the pattern to represent the complex relationships between features and the ones between features and the label; (2) proposing an algorithm of feature selection on the basis of the pattern, aiming at exploring the effect of the pattern on the prediction.

4 The Network Structure Based Feature Selection

In this section, we first describe the observation by exploring the pattern hidden in data, and then present the proposed feature selection algorithm which is on the basis of the observation.

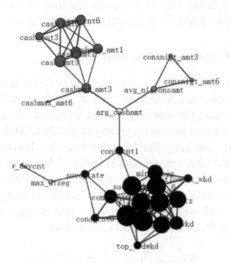

Fig. 1. A part of feature network. The edges whose weight is lower than 0.6 are ignored. The feature network presents obvious community structure (nodes marked as the same color are in one community, the nodes not in any community are marked as white color, and the size of each feature node is proportional to its weighted degree) (Wang et al. 2017, 2018). (Color figure online)

4.1 Observation

For the sake of observation, we represent the features and the relationships between them as network since network is a simple and natural representation for complex systems. Particularly, we consider each feature as a node and the relationship between each pair of features as an edge between the corresponding nodes, and then for each pair of features we evaluate their relationship by the correlation coefficient and assign it as the weight of the corresponding edge. For the correlation coefficient, here we use the Pearson product-moment correlation coefficient, which is widely known as Person's r (Pearson 1895). It is noted that in the literature several metrics such as kendall's tau (Lapata 2016), Person's r (Prion and Haerling 2014) and the Rank correlation (Sedgwick 2014) can be used to measure the correlation and dependence between variables. In those metrics, Person's r is the most widely used one, thus we apply it in this paper. To avoid noise caused by weak relevance, we ignore the edges whose weight is lower than a given threshold. After that we obtain a weighted network named as feature network.

By observing the feature network, we find that it present obvious community structure, i.e., feature nodes clustered into different communities. For illustration, Fig. 1 shows a part of the feature network built from the data. It can be seen that feature nodes cluster together to form several communities, which implies that the corresponding features are clustered into different groups. Features in one group can be treated as strongly correlated and features in different groups can be treated as weakly correlated. It can also be seen that there are feature nodes that do not belong to any community (see the white feature nodes in Fig. 1). For these features, we can easily infer that there are no strong correlations between them and other features.

4.2   The Proposed Feature Selection Algorithm

The observation above introduces us a new direction to select features to fulfill the task of credit card default prediction. Simply, we can choose from each community the most representative features (i.e., the one with largest weighted degree) to obtain the final feature set as done in our previous work (Wang et al. 2017, 2018). In this way, we can avoid redundancy and also keep the most information. However, the relationships between features and the label are not considered, which cannot exclude the irrelevant features to the prediction task and even may exclude some relevant features. To overcome this issue, we assign a weight to each feature node by considering both the correlation between it and other features and the one between it and the label. Particularly, the weight for feature node v_i is calculated as

$$\mathrm{W}(v_i) = \sum\nolimits_{v_j \in N(v_i)} W(e_{ij}) \cdot C_i, \tag{1}$$

where $N(v_i)$ is the neighbor set of v_i and $W(e_{ij})$ is the edge weight between v_i and v_j, and C_i is the correlation coefficient between feature i and the label.

By weighting each feature node, the feature network not only represents the relationships between features, but also considers the relationship between each feature and the label. Then, we choose features based on the community structure and the node weight, and form our feature selection algorithm which is named as Network Structure based Feature Selection Algorithm (NSFSA). Table 1 shows the main steps of the proposed algorithm.

There are four main steps in the proposed network structure based feature selection algorithm. In Step 1, the relationship between each pair of features and the one between each feature and the label is evaluated by Person's r. In Step 2, the feature network is constructed by taking each feature as a feature node and the Person's r value between each pair of features as the edge with weight between the corresponding feature nodes. To get rid of the noise introduced by weakly related features, only the edges with Person's r value higher than σ are kept. Further, for each feature node we calculate its node weight by Eq. (1). In Step 3, community detection is performed to discover communities in the feature network. Particularly, we apply the weighted LPA (Hu and

Yang 2017) as the community detection algorithm here. In Step 4, we choose from each community the feature corresponding to the highest node weight into the final feature set.

Table 1. The network structure based feature selection algorithm

---

Input: S_o: the original set of features; σ: the threshold for correlation coefficient.
Output: S_f: the final set of features

---

1. Let $S_f = \emptyset$, then evaluate the relationship $r(a_i, a_j)$ between each pair of features a_i and a_j in S_o and the one $r(a_i, label)$ between each feature a_i and the label by the Person's r;
2. Construct the feature network **G** as follows:
   a. Construct a feature node n_i for each feature $a_i \in S_o$;
   b. For each pair of features a_i and a_j in S_o:
      if $r(a_i, a_j) \geq \sigma$,
      then link the corresponding feature nodes n_i and n_j with weight $r(a_i, a_j)$;
   c. Calculate the node weight by Eq. (1) for each feature node n_i;
3. Discover communities **C** in the feature network **G**;
4. For each community $c_i \in$ **C**:
   a. Choose the feature a_i corresponding to the node with the highest node weight from c_i;
   b. $S_f = S_f \cup a_i$.

---

5 Models

To explore the effectiveness of the proposed NSFSA on the prediction of credit card default, we use the NSFSA to choose features from the original feature set and then take these features as input to five popular models including Naïve Bayesian, Logistic Regression, Support Vector Machine, K Nearest Neighbor and Random forest to train predictors. The following is the description of the models.

Naïve Bayesian (NB): NB is a classifier on the basis of the Bayes' theorem and the assumption of conditional independence (Friedman et al. 1997; Langley et al. 1992). Given a training set, NB first learns the joint probability distribution of the data, presumed that all the features are independent conditioned by the category; then for a new instance, the learned model computes the posterior probability of each class, and the new instance is categorized to the class with the largest posterior probability. In our case, there are two classes representing the default and not default respectively.

Logistic Regression (LR): LR assumes that the log odd of one event occurring is the linear combination of feature values, then the probability of one class (e.g., c_k) given a instance as a condition (e.g., $\mathbf{x}$), $p(c_k|\mathbf{x})$, is inferred to be an exponent function of the linear combination of feature values (Kim and Wright 2016). The parameters of the feature values are optimized by maximizing the loss function which is the likelihood of the observed data. Given a new instance, LR computes the conditioned probability

$p(c_k|\mathbf{x})$ for each class and puts the instance into the class corresponding to the largest conditioned probability.

Support Vector Machine (SVM): SVM classifies the data by searching for a plane that separates the data at the maximum extent, that is the plane has the largest distance to the nearest data points from each class. For separating the data better or dealing with data that is not linearly separable, SVM usually projects the data into higher dimensional space and then searches the hyperplane (Byun and Lee 2002). The parameters are obtained by maximizing the distance of the nearest data points to the plane. For a new instance $\mathbf{x}$, SVM classify it by checking its position to the hyperplane in the higher dimensional space.

K Nearest Neighbor (KNN): KNN is a classifier which puts a new instance into the class that most of its k nearest neighbors belong to. Usually, the default value of k is set to be 3.

Random Forest (RF): RF is a collection of basic classifiers which are decision trees. Each decision tree is constructed by maximally fitting the observed data. For a new instance, each decision tree produce its prediction result and the instance is classified into the class which most of the decision trees predict (Ajay et al. 2016).

6   Preliminary Experiments

In this section, we test the effectiveness of NSFSA by experiments on real dataset. We also compare our algorithm with four widely used feature selection algorithms from the field of filter method, considering that NSFSA follows the frame of filter method which are also more efficient. Next, we first describe the compared algorithms and the evaluation metrics, and then present the experimental results and analysis.

Community Based (CB). It is also based on the observation of community structure over the feature network, but it doesn't consider the correlation of each feature to the label (Wang et al. 2017, 2018).

Mutual Information (MI). Basically, MI evaluates the mutual information of each feature with other features and the label, and then iteratively picks the features that has the minimal mutual information with other features and the maximal mutual information with the label to form the subset of feature (Deng and Runger 2015).

Chi-square Test (CT). For each feature, CT measures its relevance to the label using chi-squared. Then, the feature selection is fulfilled by picking the features that are highly relevant to the label (Sharma et al. 2015).

Correlation Coefficient (CC). CC aims for obtaining a subset of features that are uncorrelated with each other but highly correlated with the label. The correlation is particularly measured by Pearson's r and the final subset of features is obtained by heuristic search (Peng et al. 2015).

Here we apply the metrics of accuracy, precision, recall and F1-score to evaluate the experimental results. The accuracy is the proportion of the correct prediction. The precision is the proportion of actual defaults among the ones that are predicted as default, while the recall is the proportion of defaults that are predicted correctly among the actual defaults. The F1-score comprehensively considers both precision and recall, and is the harmonic mean of them. It is noted that higher value means better prediction for all of the four metrics.

All the experiments are conducted by independently running 10 trials and in each trial 75% data is randomly selected to form train set and the remained 25% data forms the test set. Then the metrics are calculated on the results of each trial and the average values of the metrics are shown in Figs. 2 and 3 (in this case the parameter σ is empirically set to be 0.85). From Fig. 2 it can be seen that the NSFSA leads to the best accuracy on NB, LR, KNN and RF followed by CB, but is a little worse than CT on SVM. In term of precision, the NSFSA produces the best result on all of the models, and CB is the second best feature selection.

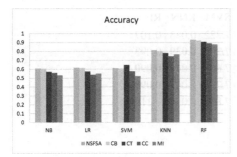 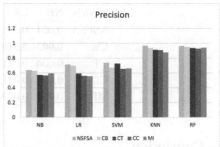

Fig. 2. The accuracy and precision caused by different algorithms of feature selection on different models.

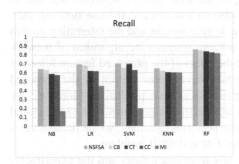 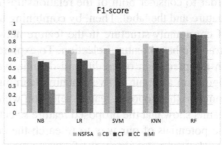

Fig. 3. The recall and F1-score caused by different algorithms of feature selection on different models.

On the metric of recall, see Fig. 3, NSFSA performs best on all of the models, but the superiority to CT on the model of SVM is marginal. Comprehensively considering precision and recall, we apply the metric of F1-score. It can be seen that NSFSA performs best on all of the models in comparison with other algorithms of feature selection, and the following one is CB. Overall, NSFSA works best, which is followed by CB and CT, and MI works worst.

Running times are also compared among different algorithms of feature selection on different models, see Table 2. CB costs the least running time on the models of NB, LR, SVM and KNN, followed by NSFSA. On the model of NB, CT and CC cost running time equal to NSFSA, which is a little more than CB. On all of the models, NSFSA costs more running time than CB, but the difference is very marginal. On all of the models, MI costs the most running time. Overall, NSFSA is less efficient than CB, but is more efficient than other feature selection algorithms. This is reasonable since that compared with CB, NSFSA adds the step of weighting nodes according to the correlation of features to the label.

Table 2. The running time caused by different algorithms of feature selection on different models.

Models	NB	LR	SVM	KNN	RF
NSFSA	0.004	0.021	7.222	0.043	0.107
CB	**0.003**	**0.018**	**7.205**	**0.039**	0.093
CT	0.004	0.021	7.661	0.044	**0.090**
CC	0.004	0.023	7.796	0.048	**0.090**
MI	0.004	0.023	7.838	0.054	0.110

7 Conclusion

In this paper, we transform the consumption records of many credit cards to feature network and then study the dataset from the perspective of network. Moreover, we weight each feature based on its relationship to the label and the network structure, in order to consider both of the relationships between features and the one between each feature and the label. Then, by combining the feature weight and the interesting pattern of community structure in the feature network, we propose the algorithm of network structure based feature selection. To explore the effect of the proposed feature selection on the prediction of credit card default, we take the features chosen by the proposed algorithm as input to five machine learning models to train predictors. Moreover, we compare the proposed feature selection with four existing algorithms. The experimental results show that the proposed feature selection performs excellent, which demonstrates the potentials of our way to research the credit card default problem.

However, we obtain the feature pattern only from one dataset and the experiments are preliminary. In the future work, we will collect more datasets and study the generality of the observed pattern, and improve the code and complete the experiments.

Acknowledgements. This work is supported by Natural Science Foundation of China under Grant No. 61802034.

References

Ajay, A., Venkatesh, A., Gracia, S., et al.: Prediction of credit-card defaulters: a comparative study on performance of classifiers. Int. J. Comput. Appl. **145**(7), 36–41 (2016)

Leow, M., Crook, J.: A new Mixture model for the estimation of credit card Exposure at Default. Eur. J. Oper. Res. **249**(2), 487–497 (2016)

Sun, S.H., Jin, Z.: Estimating credit risk parameters using ensemble learning methods: an empirical study on loss given default. J. Credit Risk (2016, Forthcoming)

Bermingham, M.L., Pongwong, R., Spiliopoulou, A., et al.: Application of high-dimensional feature selection: evaluation for genomic prediction in man. Sci. Rep. **5**, 10312 (2015)

Wang, Q., Hu, Y., Li, J.: Community-based feature selection for credit card default prediction. In: Cherifi, C., Cherifi, H., Karsai, M., Musolesi, M. (eds.) Complex Networks & Their Applications VI. SCI, vol. 689, pp. 153–165. Springer, Cham (2018). https://doi.org/10.1007/978-3-319-72150-7_13

Li, J., Cheng, K., Wang, S., et al.: Feature selection: a data perspective. arXiv:1601.07996 (2016)

Alelyani, S., Tang, J., Liu, H.: Feature selection for clustering: a review. Encycl. Database Syst. **21**(3), 110–121 (2016)

Abdou, H.A., Tsafack, M., Ntim, C.G., et al.: Predicting creditworthiness in retail banking with limited scoring data. Knowl. Based Syst. **103**(1), 89–103 (2016)

Wang, H., Xu, Q., Zhou, L., et al.: Large unbalanced credit scoring using Lasso-logistic regression ensemble. PLoS ONE **10**(2), e0117844 (2015)

Hon, P.S., Bellotti, T.: Models and forecasts of credit card balance. Eur. J. Oper. Res. **249**(2), 498–505 (2016)

Evangelista, R.D., Artes, R.: Using multi-state markov models to identify credit card risk. Production **26**(2), 330–344 (2016). The Scientific Electronic Library Online

Yang, J., Leskovec, J.: Structure and overlaps of ground-truth communities in networks. ACM Trans. Intell. Syst. Technol. **5**(2), 26 (2014)

Hu, Y., Yang, B., Wong, H.: A weighted local view method based on observation over ground truth for community detection. Inf. Sci. **355–356**, 37–57 (2016)

Hu, Y., Yang, B.: Characterizing the structure of large real networks to improve community detection. Neural Comput. Appl. **28**(8), 2363 (2017)

Nie, G., Wang, G., Zhang, P., Tian, Y., Shi, Y.: Finding the hidden pattern of credit card holder's churn: a case of China. In: Allen, G., Nabrzyski, J., Seidel, E., van Albada, G.D., Dongarra, J., Sloot, P.M.A. (eds.) ICCS 2009. LNCS, vol. 5545, pp. 561–569. Springer, Heidelberg (2009). https://doi.org/10.1007/978-3-642-01973-9_63

Zhao, B., Wang, W., Xue, G., Yuan, N., Tian, Q.: An empirical analysis on temporal pattern of credit card trade. In: Tan, Y., Shi, Y., Buarque, F., Gelbukh, A., Das, S., Engelbrecht, A. (eds.) ICSI 2015. LNCS, vol. 9141, pp. 63–70. Springer, Cham (2015). https://doi.org/10.1007/978-3-319-20472-7_7

Zhang, C., Kumar, A., Ré, C.: Materialization optimizations for feature selection workloads. ACM Trans. Database Syst. **41**(1), 2 (2016)

Kung, S.Y., Mak, M.W.: Feature selection for genomic and proteomic data mining, Chap. 1. In: Machine Learning in Bioinformatics. Wiley, Hoboken (2009)

Boln-Canedo, V., Snchez-Maroo, N., Alonso-Betanzos, A.: Feature Selection for High-Dimensional Data. Springer, Heidelberg (2015). https://doi.org/10.1007/978-3-319-21858-8

Asir Antony Gnana Singh, D., Appavu alias Balamurugan, S., Jebamalar Leavline, E.: Literature review on feature selection methods for high-dimensional data. Methods **136**(1) (2016)

Tallón-Ballesteros, A.J., Riquelme, J.C., Ruiz, R.: Merging subsets of attributes to improve a hybrid consistency-based filter: a case of study in product unit neural networks. Connection Sci. **28**(3), 242–257 (2016)

Peng, H., Ding, C., Long, F.: Minimum redundancy-maximum relevance feature selection and its applications. Feature Selection (2015)

Ruiz, R., Riquelme, J.C., Aguilar-Ruiz, J.S.: Incremental wrapper-based gene selection from microarray data for cancer classification. Pattern Recogn. **39**(12), 2383–2392 (2006)

Ma, L., Li, M., Gao, Y., et al.: A novel wrapper approach for feature selection in object-based image classification using polygon-based cross-validation. IEEE Geosci. Remote Sens. Lett. **99**, 1–5 (2017)

Mejía-Lavalle, M., Sucar, E., Arroyo, G.: Feature selection with a perceptron neural networks. In: International Workshop on Feature Selection for Data Mining, pp. 131–135 (2006)

Lin, X., Yang, F., Zhou, L., et al.: A support vector machine-recursive feature elimination feature selection method based on artificial contrast variables and mutual information. J. Chromatogr. B Anal. Technol. Biomed. Life Sci. **910**(23), 149–155 (2012)

Fu, H., Xiao, Z., Dellandréa, E., Dou, W., Chen, L.: Image categorization using ESFS: a new embedded feature selection method based on SFS. In: Blanc-Talon, J., Philips, W., Popescu, D., Scheunders, P. (eds.) ACIVS 2009. LNCS, vol. 5807, pp. 288–299. Springer, Heidelberg (2009). https://doi.org/10.1007/978-3-642-04697-1_27

Butterworth, R., Piatetskyshapiro, G., Simovici, D.A., et al.: On feature selection through clustering. In: 5th International Conference on Data Mining, pp. 581–584 (2005)

Zhou, X., Hu, Y., Guo, L., et al.: Text categorization based on clustering feature selection. Proc. Comput. Sci. **31**, 398–405 (2014)

Han, D., Kim, J.: Unsupervised simultaneous orthogonal basis clustering feature selection. In: IEEE Conference on Computer Vision and Pattern Recognition, pp. 5016–5023 (2015)

Pearson, K.: Note on regression and inheritance in the case of two parents. Proc. Roy. Soc. Lond. **58**, 240–242 (1895)

Lapata, M.: Automatic evaluation of information ordering: Kendall's tau. Comput. Linguist. **32**(4), 471–484 (2016)

Prion, S., Haerling, K.A.: Making sense of methods and measurement: Pearson product-moment correlation coefficient. Clin. Simul. Nurs. **10**(11), 587–588 (2014)

Sedgwick, P.: Spearman's rank correlation coefficient. BMJ (2014)

Friedman, N., Geiger, D., Goldszmidt, M., et al.: Bayesian network classifiers. Mach. Learn. **29**, 131–163 (1997)

Langley, P., Iba, A.W., Thompson, K., et al.: An analysis of Bayesian classifiers. In: International Conference on Artificial Intelligence, pp. 223–228 (1992)

Kim, T., Wright, S.: PMU placement for line outage identification via multinomial logistic regression. IEEE Trans. Smart Grid **9**, 122–131 (2016)

Wang, W., Lin, W., Zhang, R., et al.: Research on human face location based on Adaboost and convolutional neural network. In: IEEE 2nd International Conference on Cloud Computing and Big Data Analysis (2017)

Byun, H., Lee, S.-W.: Applications of support vector machines for pattern recognition: a survey. In: Lee, S.-W., Verri, A. (eds.) SVM 2002. LNCS, vol. 2388, pp. 213–236. Springer, Heidelberg (2002). https://doi.org/10.1007/3-540-45665-1_17

Deng, H., Runger, G.: Feature selection via regularized trees. In: International Joint Conference on Neural Networks (2015)

Sharma, A., Imoto, S., Miyano, S.: A top-r feature selection algorithm for microarray gene expression data. IEEE/ACM Trans. Comput. Biol. Bioinf. **9**(3), 754–764 (2015)

Trust-Based Agent Evaluation in Collaborative Systems

Yin Sheng[1], Wenting Hu[2(✉)], and Xianjun Zhu[3,4]

[1] State Key Laboratory of Air Traffic Management System
and Technology, Nanjing 210000, Jiangsu, China
[2] Jiangsu Open University, Nanjing 210000, Jiangsu, China
xiaohu_nanjing@163.com
[3] Jingling Institute of Technology, Nanjing 210000, Jiangsu, China
[4] Software Testing Engineering Laboratory of Jiangsu Province,
Nanjing 211169, China

Abstract. Agent evaluation is an important topic in collaborative systems. Agent evaluation is to accurately assess the ability of the agent to complete a task. The ability of the agent is demonstrated by the result of the task, and the ability changes over time. Thus, this paper studies the evaluation of agent in collaborative systems. Considering the independent and interaction abilities of an agent for a role, as well as the historical and current evaluation value, we propose a trust-based evaluation approach that promotes the effectiveness of evaluation. The experimental results show that this approach can obtain more accurate results when the agent capability changes fast and there are only a few known evaluation values.

Keywords: Collaborative system · Agent evaluation · Multi-agent system

1 Introduction

Agent evaluation is a very important task in collaborative systems. Agent evaluation is to measure the value, ability, importance and other indicators of each agent based on some criteria [1]. In this paper, an agent is defined as an entity (including person, robot [2], program, etc.) that can complete a task. Before tasks are assigned, agent evaluation can understand the intrinsic characteristics of the capabilities and reliability of each agent, and provide administrators with the basis of agent assignment. After the task is executed for a period of time, the agent evaluation can get the performance of each agent in time, so that the agents with good performance can continue to work, while the agents with poor performance should be replaced. For example, a company can evaluate newly recruited employees based on the resume they receive to determine what position they should be assigned to. After one period of work, the performance of the employees can be reassessed. The underperforming employee may be considering changing his position because he does not fulfill his potential.

Agent evaluation is a complex task. This paper assumes that the ability of the agent is manifested by the effect of completing the task. The effect obeys the normal distribution with the agent ability as the mean. There are many similar examples in reality.

© Springer Nature Singapore Pte Ltd. 2019
Y. Sun et al. (Eds.): ChineseCSCW 2019, CCIS 1042, pp. 287–296, 2019.
https://doi.org/10.1007/978-981-15-1377-0_21

A basketball player scores differently in several games, but all scores fluctuate around his ability. A student has different grades in each subject, but each grade will fluctuate near the ability of his subject. In addition, the capabilities of one agent change over time. Players may make progress, scores of many games fluctuates around a higher level; it may also regress. The same is true for students. Therefore, this paper studies the method of evaluating agent's ability in the case where the effect of the agent's completion of the task fluctuates around the ability and the ability changes over time.

In recent years, Gershoff et al. [3] discusses the topic of agent evaluation. Although the agent is different from the definition adopted in this paper, the method proposed still effective. The opinions of the customers in the article become an important indicator for evaluating the agent. Jensen [4] presents the difficulties and challenges of assessing the overall performance of an employee. The author believes that evaluation is the most difficult and time consuming task in management. Some assessment techniques, such as papers, interviews, behavior lists, etc., are also discussed. These methods are both qualitative and quantitative and have been widely used in management practices. Moore et al. [5] proposes a series of questions about how to select agents for specific roles, as well as preliminary ideas for considering role requirements and Agent capabilities, but they do not fully address the agent evaluation problem. Neely et al. [6] designs a framework for measuring organizational performance and point out some criteria for performance evaluation methods, such as evaluation methods should be easy to understand, concise, objective, clear definition. These standards provide a reference for the scientific nature of the proposed method. Zhu et al. [7] analyzes the abilities of the agents and the requirements of a role, and determined the evaluation value of the agent by the multi-attribute decision-making method.

Other literature explores agent evaluation from the perspective of the trust model. Osman [8] describes on how to choose the appropriate agent to participate in the collaboration problem, proposes a trust model based on historical information, and proposed an algorithm to evaluate the trust of the agent and select the appropriate one. Wang et al. [9] proposes a trust model based on evidence, and used statistical laws to determine whether the agent is credible and can handle situations where the evidence conflicts. Literature [10–12] evaluated the agent using cloud theory, network analysis and grey theory.

The basic assumption of the above literature is that the ability of the agent does not change over time, which does not reflect the actual situation. These papers only consider one aspect of the agent's own attributes and interactive capabilities, and do not consider both. In addition, there is no integration of historical evaluation and current evaluation values.

This paper believes that we cannot simply say whether an agent is good or not, but whether this agent is suitable for a certain role [13]. A role is a collection of rights and obligations. In this paper, a role refers to a position, and teachers, principals, etc. For example, it is not appropriate to let employees who are good at the market write programs and conclude that this employee's ability is not strong. Therefore, this paper proposes a trust-based agent evaluation method for roles, taking into account the historical evaluation value and the current evaluation value. When the agents' abilities change, the proposed approach will get more accurate result than those considering only the current evaluation values.

2 Trust-Based Agent Evaluation Approach

2.1 Introduction to Trust-Based Agent Evaluation Approach

Trust has many meanings [9]. This paper focuses on the objective characteristics of trust, that is, through the behavior to judge whether the agent is reliable, rather than subjective opinions of managers. Intuitively the high trust of an agent means that other agents believe that this agent may perform a task well [14]. When a role requires an agent, the manager selects an agent with a higher degree of trust.

In the traditional P2P (Peer to Peer) multi-agent system, there is only mutual cooperation and communication between agents, and there is no need to play a specific role [9]. In a role-based multi-agent system, the agent not only needs to complete the role requirements, but also interacts with other agents through roles. On the football field, the player who plays the role backward needs to defense and pass the ball to his own team. The ability to defense is an independent ability and has no direct interactive with other players. Passing the ball to the player of his team belongs to the interactive ability. The quality of the pass is evaluated by other players.

The trust of the agent is influenced by the current and historical trust. If someone performs well in the last evaluation period and the current evaluation period does not perform well, it is not easy to say that the agent's ability has decreased.

Therefore, the agent evaluation needs to comprehensively consider the independent ability and interactive ability, and the ability needs to be obtained from the performance of each time period. That is, historical evaluation values need to be fully utilized to make the assessment more accurate.

2.2 Independent Satisfactory of an Agent

The independent satisfactory of an agent for a role refers to the ability level to play a role, and this ability does not be affected by other agents' evaluations. For example, in a university, being a professor requires a variety of abilities such as research, teaching and instructing students. Among them, the scientific research should complete the task independently, and does not need to interact with other agents.

This paper assumes that the time intervals between two assessments are equal, such as a day, a month, or a year. Between the two evaluation points, the agent plays a role and needs to complete multiple tasks. The quality of each task completed has an evaluation value. If the year is used as the evaluation time unit, a professor may publish multiple articles, and each article has a good or bad evaluation result.

Definition 1: During time period $t - 1$ to t, plays, and the satisfaction of agent i for task m of role j at time t is $RpSat^t(a_i, r_j, m) \in [0, 1]$, where 0 means completely dissatisfied and 1 completely satisfied.

There are several ways to evaluate the completion of a single task $RpSat^t(a_i, r_j, m)$. For example, to evaluate the quality of a professor's paper, the impact factor of the journal can be used as the evaluation criteria. If the impact factor is high, $RpSat^t(a_i, r_j, m)$'s value is large, otherwise the value is small. The assessment of a single task is not the focus of this paper. We assume that the $RpSat^t(a_i, r_j, m)$ is known.

Between the two evaluations, the agent will complete multiple tasks, and the completion effect of these tasks will determine the satisfaction of the agent's role in independent ability during this time.

Definition 2: During time period $t - 1$ to t, Agent i plays the role j. The satisfaction of the independent ability at time t is the weighted average of all tasks in this time period:

$$RSat^t(a_i, r_j) = \sum_{k=1}^{m} w_k \times RpSat^t(a_i, r_j, k), \sum_{k=1}^{m} w_k = 1, w_k \in [0, 1] \tag{1}$$

In general, the evaluation interval is not too long, only related to the importance of the task, and not to the order in which the tasks are completed. For example, when evaluating a professor's work for a year, we may adopt a rule that a good paper has a greater weight of the evaluation value, but generally does not allow the order of good papers and general paper publications to influence the final evaluation value.

2.3   Interactive Satisfactory of an Agent

When the agent plays a role, it constantly interacts with other agents through roles. The satisfaction of other agents on the evaluated Agent will affect its final evaluation value. In the example of Sect. 2.2, the ability to teach and the ability to teach students is reflected in the interaction with the student, and is evaluated by the student. Another example, when we shop online, the buying and selling behavior is that the seller constantly interacts with the buyer. After the transaction is over, the buyer will rate the seller. We tend to think that merchants with high scores are more credible, and the definition of satisfaction with interactive capabilities is similar to the traditional P2P multi-agent system.

Definition 3: During time period $t - 1$ to t, Agent k and Agent i perform one interaction, and Agent k's satisfaction with Agent i as role j is $InoSat^t(a_k, a_i, r_j)$ $\in [0, 1]$, where 0 means completely dissatisfied, 1 means completely satisfied.

Definition 4: Agent k's satisfaction with Agent i's role j being interactive at time t is:

$$InSat^t(a_k, a_i, r_j) = \sum_{l=1}^{p} \alpha_l \times InoSat^t(a_k, a_i, r_j) \tag{2}$$

where p is the number of interactions, $\sum_{l=1}^{p} \alpha_l = 1$, $\alpha_l \in [0, 1]$ indicates the weight of the lth interaction.

In the interval between the two evaluations, the Agent may interact multiple times, and the satisfaction of the Agent to another Agent is the weighted average of all interactive satisfaction. In general, the importance of each interaction is the same $\alpha_l = 1/p$. The superscript t indicates that the interaction occurs during the period from $t - 1$ to t.

In the following, $IS(a)$ is used to represent all the set of Agents that interact with Agent a, and the number of elements of the set $IS(a)$.

Definition 5: Agent i plays the role j. The satisfaction of the interactive ability at time t is the average of the satisfaction of Agent to Agent i in all $IS(a)$:

$$ISat^t(a_i, r_j) = \frac{\sum_{a_k \in IS(a_i)} InSat^t(a_k, a_i, r_j)}{|IS(a_i)|} \tag{3}$$

This paper believes that all the agents that interact with Agent i give the same evaluation value. In reality, the importance of all students who evaluate the level of professorship is the same.

2.4   Trend of Role Satisfaction

The ability of the agent to play a role is closely related to the historical assessment. The trend to introduce capacity changes is to integrate historical estimates with current estimates.

Definition 6: If a straight line pair, $RSat^0(a_i, r_j)$, $RSat^1(a_i, r_j)$, ..., $RSat^t(a_i, r_j)$, is fitted by the least squares method, Agent i plays the role of the role j at the time t The change trend of the satisfaction of the independent ability DRt is the slope of this line.

Definition 7: If a straight pair, $ISat^0(a_i, r_j)$, $ISat^1(a_i, r_j)$, ..., $ISat^t(a_i, r_j)$, is used to fit together, the Agent i plays the role of the role j. The trend of the satisfaction of the interactive ability at time t is the slope of this line.

2.5   Agent's Trust of Independent Ability

Definition 8: The trust of the agent's role-independent ability is composed of the satisfaction of the current independent ability and the predicted value derived from the satisfaction of the historical independent ability. The weight of each part is λ and $1 - \lambda$.

$$RTr^t(a_i, r_j) = \lambda \times RSat^t(a_i, r_j) + (1 - \lambda) \times (RSat^{t-1}(a_i, r_j) + DR^{t-1} \times \Delta t) \tag{4}$$

Where Δt is the time interval between the two assessments.

2.6   Agent's Trust of Interactive Ability

Definition 9: The agent's trust of interactive ability is composed of the satisfaction of the current interactive ability and the predicted value derived from the satisfaction of the historical interactive ability. The weight of each part is μ and $1 - \mu$.

$$ITr^t(a_i, r_j) = \mu \times ISat^t(a_i, r_j) + (1 - \mu) \times (ISat^{t-1}(a_i, r_j) + DI^{t-1} \times \Delta t) \tag{5}$$

Where Δt is the time interval between the two assessments.

2.7   Trust of an Agent for a Role

The trust of an agent for a role is derived from the combination of independent and interactive trust.

Definition 10: During the time period $t - 1$ to t, the confidence degree of agent i acting as role j at time t is the weighted average sum of the independent capability and the reliability of the interactive ability:

$$Tr^t(a_i, r_j) = \beta \times RTr^t(a_i, r_j) + (1 - \beta) \times ITr^t(a_i, r_j), \ \beta \in [0, 1], \tag{6}$$

In particular, $\beta = 0$ indicates that the trust of the agent consists only of the interactive capability, and the role does not contain the content that needs to be completed independently by the agent. The trust of the agent is only affected by the evaluation values of other agents. When $\beta = 0$ is indicated that the agent does not interact with other agents, the trust depends only on the tasks that are completed independently.

2.8   Trust-Based Agent Evaluation Steps

The trust-based agent evaluation step is shown in Fig. 1. Firstly, the satisfaction of Agent's role independent ability and interactive ability is calculated. Then, the trend of the two satisfactions at each evaluation time is linearly fitted to the trend. The trust of independent capabilities and interactive capabilities can be calculated from trends. Combining these two trust value can lead to the evaluation of the agent's role.

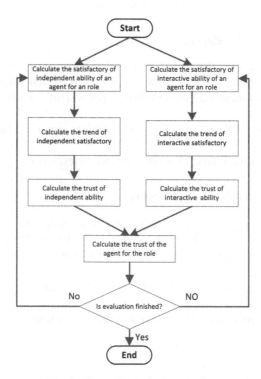

Fig. 1. Process of agent evaluation

3 Experiment

The experimental hardware environment is a laptop computer, the CPU model is Intel Core i5, the clock speed is 2.5 GHz, and the memory is 4 GB. The operating system is Windows 7, and the development environment is Matlab R2011b.

In order to verify the performance of the method of Agent evaluation in the article, the experiment simulates 200 Agents, which are divided into 10 categories according to the initial capacity, 20 classes in each class. The satisfaction of the initial role-playing ability of each type of Agent is 1, 2, ..., 10, the initial satisfaction of role interactive ability is 2, 4, ..., 20. For each type of Agent, there are 10 Agents that gradually increase with time, and the other 10 gradually decline. $w_k = 1/m$, λ and μ are both 1/2, $\beta = 0.6$. The experiment simulates the true value of the agent's ability value at each evaluation time, and generates a normal distribution random number with the true value as the mean value, and simulates the evaluation value with the random number. Then use the method of this paper to calculate the trust, the sum of the deviations of all Agent trust and the real value to measure the performance of the evaluation method:

$$Er = \sum_{i=1}^{200} \left| Tr^t(a_i, r_j) - \overline{Tr^t(a_i, r_j)} \right|, \text{ where } \overline{Tr^t(a_i, r_j)} \text{ is a real value, } Tr^t(a_i, r_j) \text{ is evalu-}$$

ation value. The smaller the deviation between the two values, the better.

In the first experiment, the independent and interactive abilities of the agent change by 3% of the initial value, that is, if the initial value was 5, the ability increased or decreased by 0.15 at each evaluation. The randomly generated evaluation value variance is 1. The number of agents completed between the two evaluations is subject to a Poisson distribution with a mean of 5, and the number of interacting agents and the number of interactions with each agent are subject to a Poisson distribution with a mean of 2 and 3.

The solid line in Fig. 2 is the evaluation error of the approach only considering the current evaluation value, and the dotted line shows the error of the proposed approach. The error of our approach is smaller.

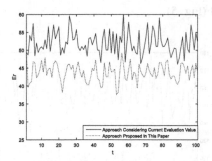

Fig. 2. The evaluation error of these two evaluation approaches in the first experiment

In the second experiment, the independent and interactive abilities change by 5% of the initial value, and other settings are the same as the first experiment. When the abilities change rapidly, our method does not perform well (Fig. 3).

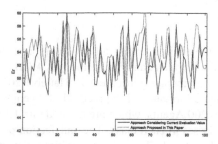

Fig. 3. The evaluation error of these two evaluation approaches in the second experiment

In the third experiment, the number of agents completed between the two evaluations is subject to a Poisson distribution with a mean of 15, and other settings are the same as the first experiment. When there are many evaluations during an interval of time, our method does not perform well, either (Fig. 4).

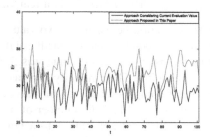

Fig. 4. The evaluation error of these two evaluation approaches in the third experiment

In the third experiment, evaluation value variance is 9, and other settings are the same as the second experiment. When the evaluation fluctuate significantly, our approach performs well (Fig. 5).

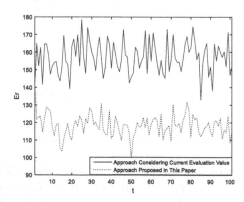

Fig. 5. The evaluation error of these two evaluation approaches in the fourth experiment

4 Conclusion

The approach proposed in this paper performs well when there are a small number of evaluation values between the two evaluations, the agent's abilities change slowly, or the evaluation values fluctuates greatly. For the second and third experiments, the proposed approach is not as good as the one considering only the current evaluation value. In the real world, in many cases, the ability value of the agent between the two evaluations generally does not occur, and the agent's abilities do not change rapidly. Because the time interval of the evaluation can be adjusted, the shorter the evaluation interval, the more timely the ability information of the agent is obtained. Therefore, there will not be too many evaluation values between the two evaluations. For example, in the American professional basketball league, the player is evaluated once a week, and the best player of the week is selected instead of one year. There are usually three games in a week. Thus, in the actual evaluation, if the number of evaluation values between the two evaluations is small and the agent's abilities change rapidly, the approach of this paper can be directly applied. If the interval between the two evaluations is long, the time interval can be shortened and then the proposed approach can be applied. The proposed approach not only ensures the timeliness of the assessment, but also improves the accuracy of the assessment.

Our approach only uses the values of the previous evaluation period, and more historical information can be used in future research. In addition, it is also possible to evaluate the fluctuation of the ability of the agent.

Acknowledgement. The authors wish to thank Natural Science Foundation of the Jiangsu Higher Education Institutions of China under Grant 18KJB520007 and 18KJB520018, the High-Level Talent Foundation of Jinling Institute of Technology under Grant JIT-B-201703, the High-Level Talent Foundation of Jiangsu Open University under Grant 19RC-2.

References

1. Zhu, H.: Maximizing group performance while minimizing budget. IEEE Trans. Syst. Man Cybern. Syst. 1–13 (2017). https://doi.org/10.1109/tsmc.2017.2735300
2. Haiyan, G.U., Chi, X.U.: Cooperative operational technologies for manned/unmanned teaming. Command Inf. Syst. Technol. **8**(6), 37–45 (2017). https://doi.org/10.15908/j.cnki.cist.2017.06.006
3. Gershoff, A.D., Mukherjee, A.: Few ways to love, but many ways to hate: attribute ambiguity and the positivity effect in agent evaluation. J. Consum. Res. **33**(4), 499–505 (2007). https://doi.org/10.1086/510223
4. Jensen, J.: Employee evaluation: it's a dirty job, but somebody's got to do it. Grantsmanship Center News **8**(4), 36 (1980)
5. Moore, J.P., Inder, R., Chung, P.W.H., Macintosh, A., Stader, J.: Who does what? Matching agents to tasks in adaptive workflow. In: International Conference on Enterprise Information Systems, pp. 181–185 (2000)
6. Neely, A., Richards, H., Mills, J., Platts, K., Bourne, M.: Designing performance measures: a structured approach. Int. J. Oper. Prod. Manag. **17**(11), 1131–1152 (1997). https://doi.org/10.1108/01443579710177888

7. Zhu, H., Feng, L., Pickering, R.: Agent evaluation in distributed adaptive systems. In: Proceedings of the 2013 IEEE International Conference on Systems, Man, and Cybernetics, pp. 752–757 (2013). https://doi.org/10.1109/smc.2013.133

8. Osman, N., Sierra, C., Mcneill, F., Pane, J., Debenham, J.: Trust and matching algorithms for selecting suitable agents. ACM Trans. Intell. Syst. Technol. 5(1), 1–39 (2013). https://doi.org/10.1145/2542182.2542198

9. Wang, Y., Singh, M.P.: Evidence-based trust: a mathematical model geared for multiagent systems. ACM Trans. Auton. Adapt. Syst. 5(4), 1–28 (2010). https://doi.org/10.1145/1867713.1867715

10. Shi, F.L., Yang, F., Xu, Y.P.: Power index method for operational capability evaluation of weapon equipment based on ANP and simulation. Syst. Eng.-Theory Pract. 31(6), 1086–1094 (2011). https://doi.org/10.1090/S0002-9939-2011-10775-5

11. Yao, C.M., Wang, Q.Y., Xie, R.S.: Multi-target threat assessment method for tank. Command Inf. Syst. Technol. 9(1), 68–72 (2018). https://doi.org/10.15908/j.cnki.cist.2018.01.012

12. Xu, T., Xie, B., Liu, B., Jin, X.: Effectiveness evaluation for collaborative planning pattern across services and arms. Command Inf. Syst. Technol. 10(2), 29–33 (2019)

13. Zhu, H., Zhou, M.C.: Roles in information systems: a survey. IEEE Trans. Syst. Man Cybern. Part C: Appl. Rev. 38(3), 377–396 (2008). https://doi.org/10.1109/tsmcc.2008.919168

14. Castelfranchi, C., Falcone, R., Marzo, F.: Being trusted in a social network: trust as relational capital. In: Stølen, K., Winsborough, W.H., Martinelli, F., Massacci, F. (eds.) iTrust 2006. LNCS, vol. 3986, pp. 19–32. Springer, Heidelberg (2006). https://doi.org/10.1007/11755593_3

Adaptive Graph Planning Protocol:
An Adaption Approach to Collaboration
in Open Multi-agent Systems

Jingzhi Guo[1,2(✉)], Wei Liu[1,2], Longlong Xu[1,2], and Shengbin Xie[3]

[1] School of Computer Science and Engineering, Wuhan Institute of Technology,
Wuhan, China
{guojingzhi, liuwei}@wit.edu.cn
[2] Hubei Province Key Laboratory of Intelligent Robot, Wuhan, China
[3] Enshi No. 1 Senior Middle School of Hubei, Enshi, China

Abstract. The adaptive system requires each agent to provide effective adaptive scheme in runtime according to dynamic changes in the environment. This paper offers an Adaptive Graph Planning Protocol (AGPP) that uses the Goal-Capability-Commitment (GCC) meta-model to dynamically reconstruct. The method uses the concept of capability to represent the executable capabilities possessed by the Agent, and introduces the concept of context state to represent the dynamic environment in the adaptive system. The adaptive graph planning protocol generation method is optimized by calculating the semantic matching degree of the context state. To evaluate the effectiveness of our approach, we provide an experimental scheme based on intelligent robot parking system (IRPS). This scheme verifies the execution time efficiency of this method and the adaptive efficiency of offline in case of emergency.

Keywords: Open multi-agent system · Graph planning · Adaptive collaboration · Goal-Capability-Commitment model

1 Introduction

Open Multi-Agent System (OMAS) is composed of several autonomous agents, which use the available resources in the system and collaborate with each other to achieve their goals [1]. In OMAS, it is impossible to have a central coordinator controlling all agents, and the agents join and leave systems is frequently and unpredictably [2]. The characteristic to adaptively adjust or collaborate to achieve system goals based on their capabilities in a dynamic environment is adaption [3]. In an open environment, agents can adaptively adjust or collaborate to achieve system goals based on their capabilities. However, in many cases, predefined context state and agent executable operations during system design do not comply with the changing environmental requirements of the system.

One of the solutions to this problem is to transform the predefined information at design time into control adaptation at runtime. In order to adapt to changing user requirements and environments, Floch et al. [4] proposed using an architecture model at runtime to implement an adaptive solution. The scheme uses general middleware

© Springer Nature Singapore Pte Ltd. 2019
Y. Sun et al. (Eds.): ChineseCSCW 2019, CCIS 1042, pp. 297–303, 2019.
https://doi.org/10.1007/978-981-15-1377-0_22

component reasoning and control adaptation. When it detects changes in the system context, it infers the change and makes the decision to perform the adaptation, and finally implements the adaptation choice. Built on the scheme, Blair et al. [5] proposed the software model runtime adaptive mechanism models@run.time to extend the model-driven engineering technology to the runtime environment. However, these runtime models are the most likely to propose solutions on the model architecture and do not propose effective runtime technical solutions. To solve this problem, Morin et al. [6] proposed a technique that combines model-driven and face-oriented aspects to identify and execute a dynamic adaptive software system. This approach not only provides a high degree of automation and verification, but also controls the complexity of the adaptive system due to dynamic changes.

The GCC model [7] models the three core concepts of goals, agent capabilities, and dynamically generated commitments in an open system. Given the agent specification and protocol, we can use semantics to verify whether the protocol supports a specific agent goal, whether the Agent's specification meets a specific commitment, etc. Günay et al. [8] proposed a three-phase framework that generates a commitment protocol at runtime, where the dynamic creation of a shared protocol is achieved by generating, sorting, and negotiating protocols. The previous work has contributed to achieve its goals, assuming that the collaborative participant agent has a commitment or a commitment to creating another collaborative participant. The promise of these methods utilizes only the knowledge and beliefs available locally. However, in the real-world case, each agent needs to understand and communicate with each other. Therefore, we need to propose flexible diagnostic and compensation methods at runtime, not only to determine which agent can achieve the goal, but also to dynamically determine the agent.

This paper makes a contribution and proposes a method based on the GCC model, called Adaptive Graph Planning Protocol (AGPP), which mainly achieves adaptive collaboration at runtime by the capability commitment cooperation of heterogeneous agents under the open multi-agent system.

Our approach is based on two main ideas. The basic idea is to use tools to improve GCC modeling of domain scenarios, match goals and capabilities, and dynamically generate commitments [9]. Collaborate with capabilities and commitments to achieve system goals, generate commitments in the event of a failure, and collaborate in both capabilities and commitments [10]. The second idea of AGPP is to perform semantic similarity calculation on concept vectors [11], and quantify the choice of collaboration by calculating the matching degree between capabilities, goals and commitments. By calculating the matching degree between capabilities and goals, commitments and goals, collaboration with superior matching degree is selected to make the optimal decision.

Built on the collaboration of capability commitments, AGPP quantifies the choice of collaboration by computing the degree of concept matching, presented in the form of graph plans. We use the Intelligent Robot Parking System (IRPS) used by Yee-fung robotics to establish an ontology model [12] for the scenario to implement and verify AGPP.

2 Semantic Matching Degree Calculation Between Concepts

In our approach, Agent collaboration needs to be selected when encountering multiple paths. We perform the optimization based on the capability and goal and the semantic matching result between the commitment and the goal.

Definition 1 (Capability and goal semantic matching degree). Given two concepts of GCC model C_i *{In-constraints, Out-constraints}* and G_j *{Trig-conditions, Final-states}*, the semantic matching degree of a capability and a goal is represented as:

$$SMD(Capability, Goal) = \frac{\sum_{i=1}^{a} sim(cs_i, cs_j) + \sum_{q=1}^{b} sim(cs_p, cs_q)}{a+b} \qquad (1)$$

Where $cs_i \in$ *In-constraints*, $cs_j \in$ *Trig-conditions*, a is equal to the number of context states in *In-constraints*; $cs_p \in$ *Final-states*, $cs_q \in$ Out-constraints, b is the number of context states in *FinalStates*.

Definition 2 (Commitment and goal semantic matching degree). Given two concepts of GCC model Co_i *{Antecedent, Consequent}* and G_j *{Trig-conditions, Final-states}*, the semantic matching degree of a commitment and a goal is represented as:

$$SMD(Commitment, Goal) = \frac{\sum_{i=1}^{m} sim(cs_i, cs_j) + \sum_{q=1}^{n} sim(cs_p, cs_q)}{m+n} \qquad (2)$$

Where $cs_i \in$ *Antecedent*, $cs_j \in$ *Trig-conditions*, m is equal to the number of context states in *Antecedent*; $cs_p \in$ *Final-states*, $cs_q \in$ *Consequent*, n is the number of context states in *FinalStates*.

3 Our Approach

The Agent performs task assignment and collaboration-based on system requirements to achieve specific system social goals. An adaptive graph planning protocol based on capability commitment collaboration is to find an optimal path from the initial state to the goal state. In the process of graph planning, if a goal can be completed by the capability alone, the capability is used for planning; if a goal requires multiple capabilities to collaborate to complete, and the premise of the commitment meets the trigger condition of the goal, the Commitment to replace the capability to collaborate to achieve the goal.

In the adaptive graph planning, the semantic matching degree is utilized to judge whether the capability cooperation is needed. We divide the matching of the capability and the goal into three types according to the calculated matching value range: Complete match, Contained match, and Disjoint match. Similarly, when there is capability collaboration in the graph plan, it is judged whether the commitment can achieve the goal according to the semantic matching degree of the commitment and the goal.

The flow chart of the adaptive graph planning protocol method is illustrated in Fig. 1. The input parameters to the algorithm are: initial state set CS_{Init} and goal state set CS_{Goal}, and the output is the plan graph *GraphPlanning*.

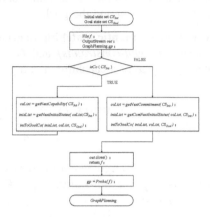

Fig. 1. Flow chart of the adaptive graph planning protocol method

The method is mainly composed of three steps. Step 1: Define the file f, the byte output stream *out*, and the graph planning *gp*. Then, traverses the context state to determine which collaboration to utilize. Step 2: Perform capability collaboration graph planning. If the context state is not in a position to commit to collaboration, use the capability for collaborative graph planning. Otherwise, conduct capability commitment collaboration diagram planning. If there is a commitment to collaboration in this context state, use capability and commitment to collaborating on graph planning. Then, close the byte output stream and return the file f. Step 3: uses the *Proba(f)* method to parse and draw the Dot Diagram Plan statement. Finally, returns the graph planning *gp* drawn using Graphviz.

4   Experiments and Discussion

In order to verify the efficiency of our proposed plan-based collaborative commitment collaborative planning method, we conducted two sets of experiments to verify the efficiency of AGPP, and observed the efficiency through experiments.

Experiment 1 analyzes the performance at the execution time of Capability Graph Planning (CG) and Capability Commitment Graph Planning (CCG). At the same time, the Similarity Degree (SD) calculation method and not using Similarity Degree (NSD) calculation method is compared, and the execution time of the String-Matching-Method (SMM) is used to verify the effectiveness of the proposed algorithm. Experiment 2 simulates the efficiency of system adaptive graph planning in the event of a sudden power outage by comparing the efficiency of the system under online and offline conditions. As showed in Fig. 2, we verify that the selected capabilities and goals can be fully planned. We use capability collaboration (a) and capability commitment collaboration (b) to generate adaptive graph planning.

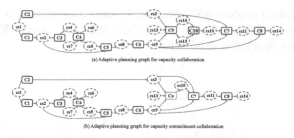

(a) Adaptive planning graph for capacity collaboration

(b) Adaptive planning graph for capacity commitment collaboration

Fig. 2. Capability commitment collaboration generated adaptive graph planning

As showed in Fig. 3, capabilities in the domain scene model are raised to 20, 30, and 40. When the number of capabilities in the scene model is 30, the time consumption of using the CCG method is 105.4 ms less than that of the CG method, and when the number of capabilities is increased to 40, the efficiency of the CG is higher than that of the CCG. It can be observed that the best application scenario of this method is that the capability is about 20–30. Another group, it can be seen that the method of using similarity have significantly lower execution time than the other two methods. It can be seen from the experimental results that adding similarity to the graph planning can dramatically improve the efficiency of graph planning, and efficiency can be improved in a simpler and more complicated scenario.

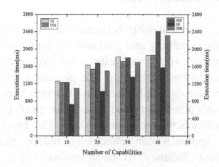

Fig. 3. Comparison of execution time based on the number of capabilities

In the subsequent experiment, we analyzed the matching rate of the AGPP method in both online and offline situations, and observed the change of the matching rate in both online and offline cases by increasing the number of goals. Experiments consider the execution time and matching rate performance of the online generation commitment and the prestored commitment in the local case. As showed in Fig. 4, when the number of goals is 8, the time consumption in the online case is about 1200 ms, and in the case of offline, it is about 500 ms, and the matching rate is 88.90%. It can be seen that in the offline case, the generation time of the online commitment is decreased, and the execution time is significantly reduced. We increase the number of goals exponentially. When the number of goals increases to 32, it can be seen that the execution time in the online case has soared to about 4500 ms, and the time consumption in the offline case is about 1500 ms, and the matching rate in the offline case is only 52.8%, which is significantly lower than the case of online.

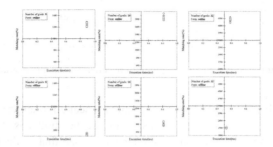

Fig. 4. Comparison of matching rate and execution time based on the number of goals

Through two sets of comparative experiments, we can find that our proposed capability commitment collaborative planning method can effectively reduce the decision time in graph planning and improve the efficiency of graph planning.

5 Conclusion and Future Work

In this paper, we describe the relationship between hidden concepts in the ontology tree structure, and calculate the capability and goal and the degree of commitment and goal matching based on the concept vector similarity calculation, and use the graph planning to optimize the collaboration-based planning. We propose a collaborative approach to collaborative planning based on graph planning. Through experimental analysis, we find that the adaptive graph planning protocol based on graph planning can effectively improve the planning efficiency of execution time. The capability to commit to a collaborative planning approach can effectively reduce planning time, thereby increasing decision making efficiency. At the same time, our approach can be utilized not only in online situations, but also in the offline situation.

Next, we are working on finding a way to introduce the concept of time in context state modeling, so that in a perfect adaptive scenario, the context state changes over time, triggering dynamic programming. The dynamic collaborative planning based on this will be more suitable for the needs of adaptive systems.

Acknowledgment. Project supported by the National Natural Science Foundation of China under Grant (No. 61502355), supported by Scientific Research Project of Education Department of Hubei Province (No. Q20181508), supported by Graduate Innovative Fund of Wuhan Institute of Technology (No. CX2018203).

References

1. Gottifredi, S., Tamargo, L.H., García, A.J., Simari, G.R.: Arguing about informant credibility in open multi-agent systems. Artif. Intell. **259**, 91–109 (2018). https://doi.org/10.1016/j.artint.2018.03.001

2. Wang, D.D., Zhou, Q.H., Zhu, W.: Adaptive event-based consensus of multi-agent systems with general linear dynamics. J. Syst. Sci. Complexity **31**(1), 120–129 (2018). https://doi.org/10.1007/s11424-018-7360-0

3. Albrecht, S.V., Stone, P.: Autonomous agents modelling other agents: a comprehensive survey and open problems. Artif. Intell. **258**, 66–95 (2018). https://doi.org/10.1016/j.artint.2018.01.002

4. Floch, J., Hallsteinsen, S., Stav, E., Eliassen, F., Lund, K., Gjorven, E.: Using architecture models for runtime adaptability. IEEE Softw. **23**(2), 62–70 (2006). https://doi.org/10.1109/MS.2006.61

5. Blair, G., Bencomo, N., France, R.B.: Models@run.time. Computer **42**(10), 22–27 (2009). https://doi.org/10.1109/mc.2009.326

6. Morin, B., Barais, O., Jezequel, J., Fleurey, F., Solberg, A.: Models@run.time to support dynamic adaptation. Computer, **42**(10), 44–51 (2009). https://doi.org/10.1109/mc.2009.327

7. Liu, W., Li, S., Wang, J.: Goal-capability-commitment based context-aware collaborative adaptive diagnosis and compensation. In: Cong Vinh, P., Ha Huy Cuong, N., Vassev, E. (eds.) ICCASA/ICTCC -2017. LNICST, vol. 217, pp. 79–89. Springer, Cham (2018). https://doi.org/10.1007/978-3-319-77818-1_8

8. Günay, A., Winikoff, M., Yolum, P.: Dynamically generated commitment protocols in open systems. Auton. Agent. Multi-Agent Syst. **29**(2), 192–229 (2015). https://doi.org/10.1007/s10458-014-9251-7

9. Krupitzer, C., Roth, F.M., VanSyckel, S.: A survey on engineering approaches for self-adaptive systems. Pervasive Mob. Comput. **17**, 184–206 (2015). https://doi.org/10.1016/j.pmcj.2014.09.009

10. Zhao, T.Q., Zhao, H.Y., Zhang, W., Jin, Z.: Survey of model-based self-adaptation methods. J. Softw. **29**(1), 23–41 (2018)

11. Liu, H.Z., Bao, H., Xu, D.: Concept vector for similarity measurement based on hierarchical domain structure. Comput. Inform. **30**(5), 881–900 (2012)

12. Farias, T.M.D., Roxin, A., Nicolle, C.: SWRL rule-selection methodology for ontology interoperability. Data Knowl. Eng. **105**, 53–72 (2016). https://doi.org/10.1016/j.datak.2015.09.001

Dynamic Adaptive Bit-Rate Selection Algorithm Based on DASH Technology

Taoshen Li[1,2(✉)], Zhihui Ge[2], and Junkai Zeng[2]

[1] Nanning University, Nanning 530200, China
tshli@gxu.edu.cn
[2] School of Computer, Electronics and Information, Guangxi University,
Nanning 530004, China
{tshli,zhihuige}@gxu.edu.cn, zengjunkai@foxmail.com

Abstract. Aiming at the existing problems of the dynamic adaptive bit-rate selection algorithm, an improved dynamic adaptive bit-rate selection algorithm based on DASH technology is proposed. To solve the optimal allocation of resources in the process of streaming media transmission, the algorithm reduces the number of video re-buffering by dynamically adjusting the buffer's key value, and improves the broadcasting quality of video by effectively reducing the startup time of video playback and switching frequency between videos with different quality. Simulation results show that the proposed algorithm can better adjust the playback bit-rate and increase the quality and stability of video playback under various bandwidth conditions. It can optimal configuration of DASH service and provide users with a good video playback experience.

Keywords: Streaming media · Bit-rate selection algorithm · Buffer · Dynamic adaption · Quality of experience (QoE)

1 Introduction

With the increasing of streaming media services, mobile terminals have become the first choice for many users to watch videos. Because the traditional streaming media technology cannot meet the demand of exponential growth of audio data transmission traffic, new streaming media technology is urgently needed to solve the some new problems [1]. Dynamic streaming over HTTP(DASH) is an open source video streaming technology, which has become the international standard of streaming media [2]. It can provide users with better quality of experience (QoE). In DASH, clients can select video clips with appropriate bit-rate for playback according to current or predicted network conditions, so dynamic bit-rate adaptive selection algorithm is the core technology to enables high quality streaming of media content over the Internet delivered from conventional HTTP web servers [3].

At present, most of the existing rate adaptive selection algorithms take the current available channel bandwidth or client buffer state as decision criteria [4–7]. Different adaptive strategies can be used to adapt to different bit-rates. However, most of dynamic rate adaptive selection algorithm only considers the server and network conditions, and do not considers many factors synthetically and user's QoE.

© Springer Nature Singapore Pte Ltd. 2019
Y. Sun et al. (Eds.): ChineseCSCW 2019, CCIS 1042, pp. 304–310, 2019.
https://doi.org/10.1007/978-981-15-1377-0_23

In recent years, the research of QoE model has attracted attention of researchers. Some QoE models for mobile terminals have been proposed [8, 9]. In DASH, [10] proposed a QoE evaluation strategy for video service based on DASH and data mining, and [11] presented a QoE evaluation model considering average video bit- rate, average pause time and times, average switching times and switching amplitude. The authors of [12] made a comprehensive balance between the overall QoE and playback continuity, and proposed a rate adaptation algorithm to select the appropriate request rate. However, the existing DSAH rate adaptive algorithm does not consider the optimizing the algorithm from the aspect of user's QoE. Therefore, how to optimize the adaptive selection algorithm in DASH technology and effectively improve the user's QoE is an urgent problem for streaming media services.

This paper optimizes the allocation of DASH services from technology and user's subjective feelings, and proposes an improved dynamic adaptive rate selection algorithm for DASH technology.

2 Description of the Algorithm

2.1 Design Ideas of Algorithm

There are many factors affecting the QoE of DASH service. According to the quantitative analysis results in [13], the most important factors affecting the QoE of DASH service are the average video quality, the number of video re-buffering, and the switching frequency between video with different qualities. For these three key factors, we have made the following improvements to the algorithm.

(1) When video starts playing, the lowest quality video is sent to the user to reduce the initial delay of viewing video by users. Then, to improve the quality of video quickly, we use a quick start mechanism to change the bit-rate of video slices to the rate suitable for the current network environment.

(2) To decrease the bit-rate switching frequency and buffer time, we use the media presentation description file(MPD) to better predict video download time, and video rate can be dynamically adjusted by combined with bandwidth download rate and buffer state.

(3) In order to reduce the frequency of video re-buffering, the buffer threshold parameters are dynamically adjusted to increase the buffer length in time. This can reduce the rate switching frequency and ensure the stability of video playback.

2.2 Design of Algorithm

Set $Buff_{min}$, $Buff_{low}$, $Buff_{high}$ and $Buff_{max}$ represent the shortest playback time, shorter playback time, longer playback time and maximum playback time of the buffer respectively; $Buff_{curr}$ is the current video buffer grade at client; VR_{curr} is the current downloaded video rate, VR_{next} is the next download video rate, VR_{max} is the maximum downloaded video rate; Δt is time length of slicing. The design of the algorithm is described below.

Buffer Stream Storage Calculation. Usually, in order to maintain the continuous play of video, the client's buffer data can be used to express the situation that the video play speed is faster than the download speed. In DASH, the videos in buffer have different bit-rates, so we use the playable time of the buffer data to express the grade of the buffer data. At time t, the bit stream storage $B(t)$ of the buffer is calculated as follows:

$$B(t) = B(t_{k-1}) \frac{\int_{k-1}^{t} c(t) - p(t)dt}{R(i)} \tag{1}$$

Quick Start Mechanism. When the video starts playing, in order to enable users to watch the video in the shortest time, we first send the lowest quality video to the users to reduce the initial delay to watch the video. Then, according to the buffer's status and the needs of playback, we use a quick start mechanism to change the video to a higher bit-rate video. Based on the current buffer condition and the video slice which grade is higher than the current download slice, this mechanism determines if switch to the next higher bit-rate video. The specific steps are as follows:

Step1: If $Buff_{curr} < Buff_{min}$ and $R_{next} < \alpha_1 \cdot \mu(\Delta t)$, then switch to the next higher rate video, otherwise keep the current video playback rate. Where, α_1 represents the current bandwidth weight, the calculation of $\mu(\Delta t)$ is as follows:

$$\mu(\Delta t) = \frac{\int_{t-\Delta t}^{t} \rho(t)dt}{\Delta t} \tag{2}$$

Where, $\rho(t)$ is the download rate of video at time t.

Step2: If $Buff_{min} < Buff_{curr} < Buff_{low}$ and $R_{next} < \alpha_2 \cdot \mu(\Delta t)$ then switch to the next higher rate video, otherwise keep the current video playback rate. Where, α_2 $(0 < \alpha_2 < 1)$ is a current bandwidth weight factor.

Step3: If $Buff_{curr} > Buff_{low}$ and $R_{next} < \alpha_3 \cdot \mu(\Delta t)$ then switch to the next higher rate video, otherwise keep the current video playback rate. Where, α_3 $(0 < \alpha_3 < 1)$ also is a current bandwidth weight factor.

By fast start mechanism, the algorithm can effectively switch to the rate video suitable for the current network condition in short time.

Adjustment of Video Rate. To ensure the stability of video playback, the algorithm needs to adjust the video rate based on the network state and buffer occupancy. The adjustment is mainly includes the following situations:

(1) If $Buff_{curr} < Buff_{min}$ and the current video download rate is less than the current video download rate, the lowest bit-rate is selected.

(2) If $Buff_{curr} < Buff_{low}$ and the buffer data is decreasing, the bit-rate of the next video slice is lowered by one level; otherwise, the current bit-rate remains unchanged.

(3) When $Buff_{low} < Buff_{curr} < Buff_{high}$, it shows that the current network environment can better support the current rate video playback. If $r_{i-1} < VR_{max}$ && $VR_{next} < \mu(\Delta t)$, the bit-rate of the next video slice is switched to the next higher bit-rate video slice; otherwise, keep the current bit-rate video unchanged.

(4) When $Buff_{high} < Buff_{curr} < Buff_{max}$ and $\mu(\Delta t) \not> r_{i-1}$, the current bit-rate remains unchanged.

Dynamic Adjustment of Buffer Threshold Parameters. After fast starting, if $Buff_{curr} > Buff_{low}$ and $Buff_{curr} = Buff_{max}$, it means that the beginning stage of video playback has passed and users may be satisfied with the quality of video playback, and the algorithm's task is to maintain the stability of video playback. When the current network conditions and buffer occupancy are good, to stabilize video playback and improve the ability to resist network changes, the algorithm will adjust the threshold parameters $Buff_{low}$, $Buff_{high}$ and $Buff_{max}$. By increasing the value of these three parameters, the capacity of the buffer is properly enlarged and the frequency of video re-buffering is reduced. If the video click phenomenon occurs, the three threshold parameters of the buffer are restored to the initial state to restore the number of video segments in the buffer to normal state as quickly as possible.

2.3 Algorithm's Description

The main pseudocode of the improved algorithm is described as follows:

Algorithm 1. The improved dynamic adaptive bit-rate selection algorithm based on DASH technology

---

Input: Number of buffers $Buf(i)(i \in [1,n])$

Output: Rate r_i of the *i-th* downloaded video slice.

if $(VR_{max} > r_{i-1}$ && $Buf(t_1) < Buf(t_2))$ {

  if $((Buff_{low} < Buff_{curr}$ && $VR_{next} < \alpha_3 \cdot \mu(\Delta t))$ || $(Buff_{curr} < Buff_{min}$ && $VR_{next} < \alpha_1 \cdot \mu(\Delta t)))$

    $r_i := r_{i-1}$;

  else if $(Buff_{curr} <= Buff_{low}$ && $Buff_{curr} >= Buff_{min}$ && $VR_{next} < \alpha_2 \cdot \mu(\Delta t))$ $r_i := r_{i-1}$;

}

else {

  if $(Buff_{curr} >= Buff_{high}$ && $r_{i-1} < VR_{max}$ && $r_{i-2} < \mu(\Delta t))$ $r_i := \mu(\Delta t)$;

  else if $(Buff_{curr} >= Buff_{low})$ && $r_{i-1} < VR_{max}$ && $r_{i-2} < \mu(\Delta t))$ $r_i := Buf(i)$;

  else if $(Buff_{curr} >= Buff_{min}$ && $r_{i-1} > \mu(\Delta t))$ $r_i := Buf(i-2)$;

  else if $(Buff_{min} > Buff_{curr})$ $r_i := VR_{min}$;

}

if $((Buff_{low} < B_{curr})$ && $(VR_{curr} == VR_{max}))$ {

    $Buff_{low} = 2*Buff_{low}$; $Buff_{high} = Buff_{high} + Buff_{low}$; $Buff_{max} = Buff_{max} + Buff_{low}$;   }

  else if $(Buff_{curr} == 0)$ {

    $Buff_{low} = Buff_{low}/2$; $Buff_{high} = Buff_{high} - Buff_{low}$; $Buff_{max} = Buff_{max} - Buff_{low}$; }

}

---

3 Performance Analysis of Algorithm

3.1 Experimental Environment

To analyze the algorithm's performance, we build an adaptive video playback proto-type system based on DASH technology on Apache. The system consisted of three parts: DASH server, network simulator and DASH client. The video source used in the experiment was Big Buck Bunny, an open source movie. We used DASH encoder to encode and slice the video source, and generated 20 MPD files with different bit-rates. The total length of the experimental video is about 600 s and is divided into 150 segments. In the experiment, the transmission protocol is TCP, and the existing HTTP Web server is used to transmit video information.

3.2 Contrast Experiments of Algorithm

The comparison algorithm is a rate selection algorithm in Ref. [4]. This algorithm mainly chooses video rate according to network throughput, which is universal. In the experiment, $B_{min} = 2$ s, $B_{low} = 5$ s, $B_{high} = 10$ s, $B_{max} = 12$ s, $\alpha_1 = 0.3$, $\alpha_2 = 0.4$, $\alpha_3 = 0.6$.

Figure 1 shows the comparative experimental results in the case of 4 Mbps network bandwidth. The experimental results show that comparison algorithm is vulnerable to the influence of network bandwidth. When the bandwidth changes greatly, the video with different bit-rates is switched frequently, which makes the video playback unstable and affects the playback quality. In addition to the initial stage, the proposed algorithm can choose a higher bit-rate in other playback stages, which can reduce the switching times of different quality videos, ensure the stability of video playback, and maintain a higher average video quality. The experimental results show that improved algorithm can increase the user's QoE.

Figure 2 is the comparative experimental results in the case of 8 Mbps network bandwidth. From the experimental results, the performance of two algorithms is better in the case of large network bandwidth. But on the whole, the average bit-rate of the improved algorithm is higher than the contrast algorithm.

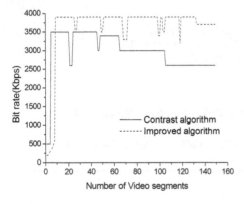

Fig. 1. Comparison results under 4 Mbps network bandwidth

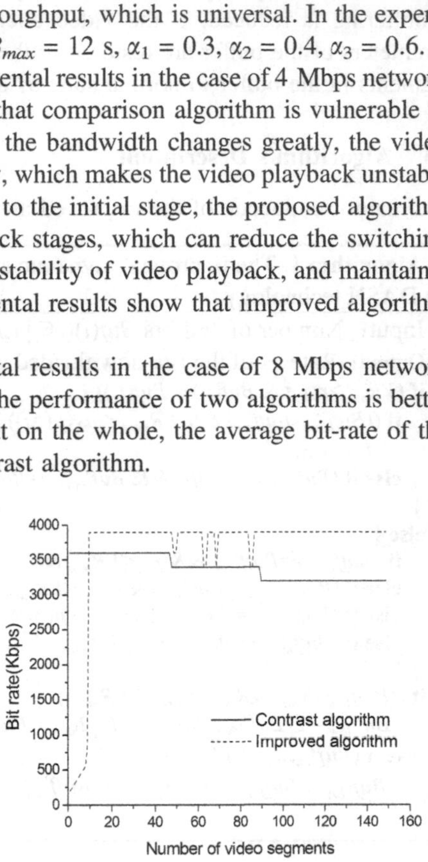

Fig. 2. Comparison results under 8 Mbps network bandwidth

3.3 Experiments of Video Play Quality

This experiment is mainly to verify the applicability of the improved algorithm in the video playback adaptive prototype system based on DASH technology. The experiment is divided into two groups. The first group is a performance comparison experiment of playing video on PC and mobile devices (mobile phones), and the second group is the comparison experiment of different bit-rate levels.

Comparing Experiment Between PC Terminal and Mobile Terminal. In the first group of experiment, the Videos are divided into five levels, that is 50 Kbps, 200 Kbps, 400 Kbps, 700 Kbps, 1500 Kbps. To decrease the impact of uncertain changes in the network environment, many experiments have been done and the average value of the experiment has been taken. Figures 3 and 4 are the rate switching change results of PC terminal and mobile terminal at 5 bit-rates level. The experimental results show that the adaptive performance of video playback on PC is better than mobile terminal.

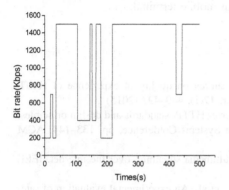
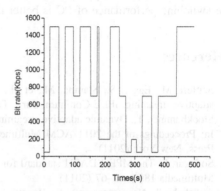

Fig. 3. Changes of bit-rate switching of PC.   **Fig. 4.** Changes of bit-rate switching of mobile terminal.

Comparing Experiment of Bit-Rate Levels. The second group of experiments played two group slices of the same video with different bit rates on PC. The first slice combination is composed of bit-rate slices with 200 Kbps, 600 Kbps, 1500 Kbps respectively. The second combination is composed of bit-rate slices with 200 Kbps, 300 Kbps, 400 Kbps, 500 Kbps, 600 Kbps, 700 Kbps, 800 Kbps, 1000 Kbps, 1200 Kbps, 1500 Kbps respectively. Figures 5 and 6 are the experimental results. Experimental results show that the smaller the bit-rate levels, the more favorable the stable play of video.

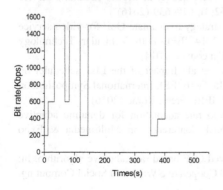
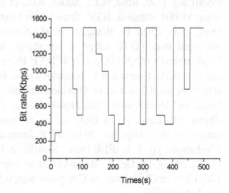

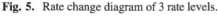

Fig. 5. Rate change diagram of 3 rate levels.   **Fig. 6.** Rate change diagram of 10 rate levels

4   Conclusion

In this paper, an dynamic adaptive bit-rate selection algorithm based on DASH technology is proposed. The algorithm uses a quick startup method to reduce the initial delay of video playback. By dynamically adjusting the frequency of video re-buffering and the average switching times of the bit rate, it can improve the average quality of video playback. The Experimental results show that the improved algorithm can reduce the number of video re-buffers and the switching frequency between different quality videos, and can effectively improve the user's QoE for DASH service. The adaptive rate switching performance of PC is better than mobile terminal.

References

1. Seufert, M., Egger, S., Slanina, M., et al.: A survey on quality of experience of HTTP adaptive streaming. IEEE Commun. Surv. Tutor. **17**(1), 469–492 (2015)
2. Stockhammer, T.: Dynamic adaptive streaming over HTTP- standards and design principles. In: Proceedings of the 2011 ACM Multimedia Systems Conference, pp. 133–144. ACM Press, New York (2011)
3. Sodagar, I.: The MPEG-DASH standard for multimedia streaming over the internet. IEEE Multimedia **18**(4), 62–67 (2011)
4. Akhshabi, S., Narayanaswamy, S., Begen, A.C., et al.: An experimental evaluation of rate-adaptive video players over HTTP. Signal Process.: Image Commun. **27**(4), 271–287 (2012)
5. Egger, S., Reichl, P., HoBfeld, T., et al.: "Time is bandwidth"? Narrowing the gap between subjective time perception and quality of experience. In: 2012 IEEE International Conference on Communications, pp. 1325–1330. IEEE Press, Ottawa (2012)
6. Park, J., Chung, K.: Client-side rate adaptation scheme for HTTP adaptive streaming based on playout buffer model. In: The 30th International Conference on Information Networking, pp. 190–194. IEEE Press, Kota Kinabalu (2016)
7. Juluri, P., Tamarapalli, V., Medhi, D.: SARA: segment aware rate adaptation algorithm for dynamic adaptive streaming over HTTP. In: 2015 IEEE International Conference on Communication Workshop, pp. 1765–1770. IEEE Press, London (2015)
8. Zhou, C., Lin, C.W., Guo, Z.: mDASH: a Markov decision-based rate adaptive approach for dynamic HTTP streaming. IEEE Trans. Multimedia **8**(4), 738–751 (2016)
9. Rodriguez, D.Z., Rosa, R.L., Alfaia, E.C., et al.: Video quality metric for streaming service using DASH standard. IEEE Trans. Broadcast. **62**(3), 628–639 (2016)
10. Deng, X.L., Chen, L., Wang, F., et al.: A novel strategy to evaluate QoE for video service delivered over HTTP adaptive streaming. In: 2014 IEEE 80th Vehicular Technology Conference (VTC 2014), pp. 1–4. IEEE Press, Vancouver (2014)
11. Zahran, A.H., Quinlan, J.J., Ramakrishnan, K.K., et al.: Impact of the LET scheduler on achieving good QoE for DASH video streaming. In: 2016 IEEE International Symposium on Local and Metropolitan Area Networks, pp. 1–7. IEEE Press, Rome (2016)
12. Zhang, H., Jiang, Z.: A QOE-driven approach to rate adaptation for dynamic adaptive streaming over http. In: 2016 IEEE International Conference on Multimedia & Expo Workshops, pp. 1–6. IEEE Press, Seattle (2016)
13. Li, T., Zheng, D., Ge, Z.: Research on an improved QoE-based rate-adaptive algorithm. In: 12th Chinese Conference on Computer Supported Cooperative Work and Social Computing, pp. 1–6 (2017)

KBCBP: A Knowledge-Based Collaborative Business Process Model Supporting Dynamic Procuratorial Activities and Roles

Hanyu Wu[1,2,3], Tun Lu[1,2,3(✉)], Xianpeng Wang[1,2,3], Peng Zhang[1,2,3], Peng Jiang[4], and Chunlin Xu[4]

[1] School of Computer Science, Fudan University, Shanghai, China
whyiot39@163.com, {lutun,18210240196}@fudan.edu.cn,
zhpll@126.com
[2] Shanghai Key Laboratory of Data Science, Fudan University, Shanghai, China
[3] Shanghai Institute of Intelligent Electronics and Systems, Shanghai, China
[4] TongFang SaiWeiXun Information Technology Co., Ltd.,
Chengdu, Sichuan, China
420857890@qq.com, cd_xcl@sina.com

Abstract. In recent years, the focus of business process management has gradually shifted from quantitative assessment to quality assessment. Business processes are no longer limited to the explicit rules, and the creativity and flexibility of the process have become more and more attractive. The data-driven process model drills this creativity and flexibility by collecting the data characteristics of actual instances and has been fully developed over the past decade. The knowledge-intensive process makes use of explicit and implicit knowledge, which is fit for procuratorial scenario. This paper combines the idea of the data-driven process model and knowledge-intensive process to propose a knowledge-based collaborative business process model KBCBP supporting dynamic procuratorial activities and roles. The mechanism of the model is based on the procuratorial background.

Keywords: Business process management · Knowledge-intensive process · Procuratorial scenario · Dynamic process model

1 Introduction

There are two types of flow in business processes models: control flow and data flow. The control flows are generated according to the process logic structure, defining rules or human participation, and are used to control process activities or process tasks. Control flow is an essential component of business processes. The data flow is composed of related data used in process activities or process tasks, and the flow direction is divided into two types: between the process external (such as an external database) and the process node (representing process activities or process tasks) and between the process node and the process node. For process participants, data flows to improve the interpretability of business processes. The combination of control flow and data flow is the foundation of process collaborative management.

© Springer Nature Singapore Pte Ltd. 2019
Y. Sun et al. (Eds.): ChineseCSCW 2019, CCIS 1042, pp. 311–319, 2019.
https://doi.org/10.1007/978-981-15-1377-0_24

There are two kinds of knowledge in business processes: explicit knowledge and implicit knowledge. Explicit knowledge can be recorded or described to others and can be organized, distributed [1]. To understand and use explicit knowledge, people should share the same context [2]. A typical example of explicit knowledge can be found in process models, such as the structure of the business process model, the rules of the execution of the business process model, and so on. Implicit knowledge, on the other hand, is something that people know but usually don't record [1]. Implicit knowledge is more difficult to communicate and understand than explicit knowledge, but it can be externalized to some extent through problem-solving [2]. Typical implicit knowledge such as know-how, understanding of individuals or disciplines. From the perspective of the knowledge dimension, business process collaboration is essentially the collaborative sharing of two types of knowledge. The design and management behavior of the process can also be regarded as the process of knowledge integration and management.

To solve the challenges, a business process should be able to combine control flow and data flow. At the same time, it also needs to support collaborative sharing of explicit knowledge and implicit knowledge, so that explicit knowledge can be sketched. This process skeleton and implicit knowledge flexibly implement specific process tasks. For the problem of complex knowledge understanding, there should be corresponding implicit knowledge externalization component mechanism in the process to meet the persistent storage, reuse and rationality judgment requirements of implicit knowledge.

This article will be divided into the following sections: The second section will introduce some developments in business process management; the third section will briefly introduce the first instance public prosecution case process; the fourth section will introduce a structure and role configurable collaboration model for procuratorial scenario; the fifth section will introduce related experiments; The last section is a conclusion.

2 Related Work

Business Process Management (BPM) is a research area for business process design, operation, analysis, and other related issues. People's needs and expectations for business processes are the starting point for business process management technology research. Typical business processes use activities as atomic elements in the process which are also core elements. Such a process is known as activity-oriented or activity-centric business processes. The main indicators it focuses on are the speed and throughput of process execution. Process-oriented business processes focus on process elements and are oriented toward control flow. The process structure can be represented by a business process model and fully specified at design time [3]. The task sequence is simple and predictable, with clear coordination rules [4] and supports repetitive process operations.

Traditional process-oriented business process management has several problems. First, the activity is an atomic element. Second, focusing on the control flow, moving the process context (such as data related to the entire case instance) into the background can lead to errors and reduce the efficiency of the process. Due to the lack of data flow mechanism, the intelligibility of business processes is reduced. Furthermore, business process routing is used for work distribution and authorization [5], so that work distribution and authorization are coincident, and the organization and coordination of business processes can only be performed on a coarse-grained structure. Finally, business process routing focuses on what should be done rather than what can be done [5]. This push-oriented perspective makes business processes less flexible. In response to these problems, the Case-handling process model [5] was proposed. It provides all the information available to avoid contextual deficiencies; deciding which activities to enable based on available information rather than implemented activities; assignment and authorization are separated; allows staff to view, add, and modify data before and after performing the corresponding activity [5]. Each business process instance has a customer-facing officer responsible for the processing. Case parallel processing is logically independent of each other. Because Case-facing officers are separate, and sometimes knowledge becomes a competitive advantage, they hardly share knowledge.

People are beginning to realize the importance of process quality and flexibility. Flexibility is closely related to implicit knowledge in business processes, so knowledge management concepts are introduced into business process management, resulting in a practice-oriented [1] or knowledge-centric business process, which is also known as knowledge-intensive business processes. It is based on participant collaboration and information exchange. The key elements are human knowledge, experience, and creativity [1]. The predictability of task sequences is weak, with only partially defined coordination rules. The rules of design can only be roughly constrained to not go beyond the organizational boundaries of the process [2], but are not used to specify participant behavior or execution patterns. The process is dynamically adjusted during the specific runtime, and the process tasks appear dynamically at runtime, which is difficult to standardize and lack of the reusability.

How to better regulate the data flow mechanism is also a focus of business process management development. Business Process Modeling and Notation (BPMN) is a widely used business process modeling notation specification. Although BPMN is activity-oriented [6], it can describe the data in the process. Data has two kinds of identification in BPMN, one is data objects and the other is data storage. The data object is used to describe the data (input) used in the activity or the newly generated data (output). The data store is used to describe the database used in the process. However, BPMN cannot describe the constraints of data, the dependencies between data, the specific process of data-driven activities, and so on. Meyer A [7] and others extend the data objects of BPMN: data objects include identifiers, attribute sets, life cycles, and a set of expression fields. The Artifact-centric process model [8] describes the various key elements of the process as Artifacts. Each type of Artifact includes an information model and a life cycle model. The lifecycle model uses a finite state machine or a Guard-Stage-Milestone (GSM) metamodel to describe the life cycle of an Artifact. Wei Xu et al. [9] designed a data-driven process model that can be dynamically extended based on the Artifact-centric process model. The Object-aware process model [10] includes multiple sub-models.

3   Example: First-Instance Public Prosecution Business Process

In this section, we will introduce the business process of the first-instance public prosecution of the provincial institute.

3.1   Overview

As is depicted in Fig. 1, the first-instance public prosecution process consists of many activities. The instance begins with a Start Node. Then it goes to Accept Node, which means accept cases sending from another department. Next node is Distribute, where leaders distribute cases to specific undertakers. The Review node represents undertakers' review activities. They check related files such as case cards, case documents. In this part, they can also delay the deadline, or pause the case. There is a synchronized node named Supplement next to Review, where undertakes can send the case back to the original department to ask for more evidence. After reviewing undertakes have to decide: to get to Deny node, which means the case is not going to the court; or to get to Cancel node, which means there is not criminal and the case should be returned to original department and canceled; or to get to Change node, which means they have not right to review it and send it to upper department; or to get to Prosecute node to make a full prosecution and send it to court. After the prosecution, the case can be returned to the Review node by Recall node activity, or go on to the Attend node. Attend node means the case is sent to the court. If undertakes have found mistakes before the trial is done they can make a correction or supplement (which is not shown in Fig. 1). Then undertakes should review the judgment, if they discover something wrong, they should start the Protest node. The Superior node is a synchronize node to Prosecute node and Review node, which represents the Superior department's supervisions on the corresponding node events. The merge node is to merge two cases that should have the same state. The Split node is to split one case to two or more cases.

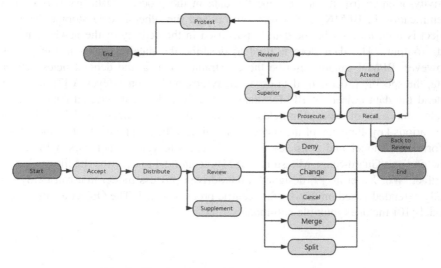

Fig. 1. First-instance public prosecution process

3.2   Problems

Firstly, this model cannot handle dynamic adjustment well. Designing the first-instance public prosecution business process model requires the cooperation of procuratorial business experts and process technology experts. Procuratorial business experts have a large amount of implicit knowledge of the first-instance public prosecution business process, and process technology experts that design the first-instance public prosecution business process needs to transform implicit knowledge into explicit knowledge. However, business experts and technology experts have gaps, so the transmission of information cannot be lossless. The dynamic adjustment is depending on implicit knowledge. It is difficult to cover all dynamic parts into one workflow chart. For one thing, it's too complex and low efficient because some of the dynamic parts are rarely used. For another, it is difficult to record all the special situations since implicit knowledge cannot fully transfer to implicit knowledge.

Secondly, this model cannot deal with the cycle structure. As we can see from Fig. 1, the Recall node will push the case back to Review node, which creates a "cycle" and cannot be depicted by the model.

Thirdly, role authority is rather static. Once the Distribute activity has finished, there is no activity to adjust the role of authority. However, for the prosecution process, prosecutors have their own working groups. Although they have the authorities of all the tasks, they do not need to work on these all by themselves. So they need to assign these authorities to their group mates, which is equivalent to create new roles.

4   KBCBP: Configurable Procuratorial Process Collaborative Model

To solve the problems proposed in the previous section, this section discusses a configurable collaborative process model KBCBP for the procuratorial scenario.

From the perspective of process execution, process configurable refers to the ability of a process instance to dynamically adjust a process model based on the actual scenario. From the process design perspective, process configurable refers to the process model covering the "variants" of all different scenario [11]. We will discuss the dynamic process structure and role configuration. Dynamic means this happens in the execution of the process instance.

4.1   Process Structure Configuration

To configure the process, we should figure out the variants of the procuratorial scenario process. For the procuratorial scenario, there are several variants in the process:

- The case is sent to another department, the current case instance is blocked, waiting for awakening (or ending). For example, in the Review section, the contractor may return the case of unclear facts and insufficient evidence to the public security organ or the self-reporting department of the court for supplementary investigation.

- The case is sent to other departments and is synchronized with the current case instance. For example, the prosecution in court and the review of the results of the referee can be reported to the higher authorities for simultaneous review of the case.
- The case goes back to the previous activity. For example, the Recall activity will push the case back to previous Review activity.
- Consolidation and splitting of cases, i.e. mergers and demolitions, and current case instance will end after the merger or split.

Four types of dynamic modification mechanisms are introduced to meet the above requirements. Dynamic modification refers to modifying the process structure during the execution of a process instance.

Add Mechanism. The add mechanism is to call a new execution event on the current execution event. The addition mechanism is divided into two types: synchronization type and blocking type. Since the process model generates process nodes according to the transfer of processing rights, the addition mechanism for generating new process nodes will involve the transfer of processing rights. The case card model and the case document model need to be processed accordingly. For synchronous addition, the added event has the read operation permission of the case card and the case document of the original process instance. The added event is executed synchronously with the original event. For blocking additions, the added event has the read and write permissions of the case card and the case document of the original process instance. The original event will continue to execute after the added event ends, so we need to save the execution entry information to return before blocking.

Cycle Mechanism. The cycle mechanism is to go back to the previous activity. This operation will refresh the state of the case, which refers to the case documents. They will be cleared and drawn back to the ready state for the previous activity. Case card also will be recovered to previous contents.

Merge Mechanism. The merge mechanism is the process in which the current instance is merged with other instances to generate a new instance. The premise of instance consolidation is that two instances have the same state. The merging of instances includes the merging of the case card and the merging of the case document. For the case card, the merge mechanism will combine the criminal attributes and criminal facts. Data attribute will take the early one. The case document only copies the index of the original document that can be accessed because the status of the pre-merger case does not affect the status of the merged case.

Splitting Mechanism. The splitting mechanism is the process of splitting the current instance into multiple instances and ending. The splitting of the instance includes the splitting of the case card and the splitting of the case document. For the case card, the merge mechanism will split the criminal attributes and criminal facts. The case document will be shared by every instance.

4.2   Process Role Configuration

In addition to the process structure configuration, the role configuration in the process is also an integral part of the process configuration. We propose a simple method to generate new roles which are feasible even when the process instance is running. We mainly focus on the role's authority, so the dynamic configuration is all about authority configuration.

To start with, we will have an original root role that has all the authorities in one activity or the whole process. In the procuratorial process, the prosecutor is the root role. Then we use two mechanisms to generate new type roles.

Inherit Mechanism. The inherit mechanism is the process in which the first role gives one of his authority to the second role (meanwhile the first role can still hold the authority). The first role can decide whether the second role can give this authority to other roles or not.

Hybridization Mechanism, The hybridization mechanism is to combine the authorities of two roles and generate a new role.

Based on the inherit mechanism and hybridization mechanism, the root role can evolve into various new roles. And a prosecutor can design and choose his needed roles for his working group. For every activity, this procedure can be different and is decided by prosecutors, which depends on their implicit knowledge.

5   Experiment

We have tested our process model in a simulated data set because the real procuratorial data cannot be accessed. All these were done by using the Activity workflow framework with BPMN as the graphics.

Figures 2, 3 and 4 are the results of simulated first-instance public prosecution cases. In Fig. 2 we use the Add mechanism to add a parallel Supplement activity when reviewing, which add a parallel gateway that is depicted by diamond. Figure 3 also shows the Add mechanism, the parallel node has changed to Superior. Figure 4 shows that the process using Cycle mechanism to redo the Review activity.

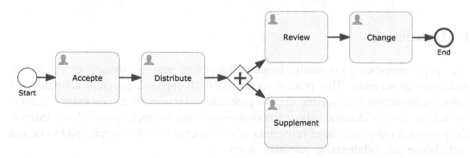

Fig. 2. Result 1 of process modeling

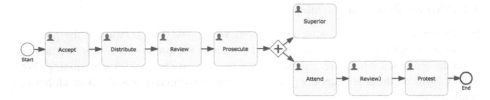

Fig. 3. Result 2 of process modeling

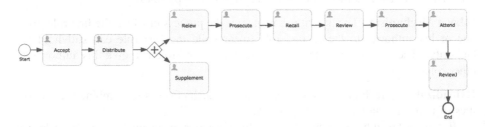

Fig. 4. Result 3 of process modeling

We evaluate the process model through two indicates. One is called process efficiency, which is calculated by the ratio between the numbers of activities in process instances and process model. The other is process complexity. We use the sum of nodes and edges of the process model (Excluding Start and End) to represent process complexity. A process is more efficient when the process efficiency is large, and a process is more complex when process complexity is larger. We choose the model in Fig. 1 as our baseline. The result is shown in Table 1. Our model is superior in both indications.

Table 1. Efficiency and complexity

	Average process efficiency	Average process complexity
Baseline model	0.4286	156
KBCBP	1	57

6   Conclusion

This paper proposes a knowledge-based collaborative business process model for the procuratorial scenario. This process model KBCBP supports dynamic activities and roles configuration. Considering that the procuratorial process is a knowledge-intensive workflow, such mechanisms allow prosecutors to better use their implicit knowledge in the process, to improve the effectiveness of the process model structure, and to design and choose the collaborative activities' roles.

Future work will focus on how to put the model into practice, and validate the model from actual procuratorial scenario and data.

Acknowledgments. This work was supported by the National Key Research and Development Program of China under Grant No. 2018YFC0381402.

References

1. Marjanovic, O., Seethamraju, R.: Understanding knowledge-intensive, practice-oriented business processes. In: Proceedings of the 41st Annual Hawaii International Conference on System Sciences (HICSS 2008), p. 373. IEEE (2008)
2. Marjanovic, O., Skaf-Molli, H., Molli, P., et al.: Collaborative practice-oriented business processes Creating a new case for business process management and CSCW synergy. In: 2007 International Conference on Collaborative Computing: Networking, Applications and Worksharing (CollaborateCom 2007), pp. 448–455. IEEE (2007)
3. Aureli, S., Giampaoli, D., Ciambotti, M., et al.: Key factors that improve knowledge-intensive business processes which lead to competitive advantage. Bus. Process Manag. J. **25**(1), 126–143 (2019)
4. Rychkova, I., Nurcan, S.: Towards adaptability and control for knowledge-intensive business processes: declarative configurable process specifications. In: 44th Hawaii International Conference on System Sciences, pp. 1–10. IEEE (2011)
5. van der Aalst, W.M.P., Weske, M., Grünbauer, D.: Case handling: a new paradigm for business process support. Data Knowl. Eng. **53**(2), 129–162 (2005)
6. Steinau, S., Marrella, A., Andrews, K., et al.: DALEC: a framework for the systematic evaluation of data-centric approaches to process management software. Softw. Syst. Model. **18**, 1–38 (2019)
7. Meyer, A., Pufahl, L., Fahland, D., Weske, M.: Modeling and enacting complex data dependencies in business processes. In: Daniel, F., Wang, J., Weber, B. (eds.) BPM 2013. LNCS, vol. 8094, pp. 171–186. Springer, Heidelberg (2013). https://doi.org/10.1007/978-3-642-40176-3_14
8. Hull, R., et al.: Business artifacts with guard-stage-milestone lifecycles: managing artifact interactions with conditions and events. In: 5th ACM International Conference on Distributed Event-based Systems (DEBS), pp. 51–62. ACM, New York (2011)
9. Xu, W., Su, J., Yan, Z., Yang, J., Zhang, L.: An artifact-centric approach to dynamic modification of workflow execution. In: Meersman, R., et al. (eds.) OTM 2011. LNCS, vol. 7044, pp. 256–273. Springer, Heidelberg (2011). https://doi.org/10.1007/978-3-642-25109-2_17
10. Künzle, V.: Object-aware process management. Ph.D. Thesis, University of Ulm (2013)
11. Kang, G., Yang, L., Xu, W., et al.: Artefact-centric business process configuration. Int. J. High Perform. Comput. Netw. **9**(1–2), 93–103 (2016)

Social Computing (Online Communities, Crowdsourcing, Recommendation, Sentiment Analysis, etc.)

An Analysis Method for Subliminal Affective Priming Effect Based on CEEMDAN and MPE

Min Zhang[1,2], Bin Hu[1,2(✉)], Yuang Zhang[1,2], and Xiangwei Zheng[1,2]

[1] School of Information Science and Engineering, Shandong Normal University,
Ji'nan 250014, China
binhu@sdnu.edu.cn
[2] Shandong Provincial Key Laboratory for Distributed Computer Software
Novel Technology, Ji'nan 250014, China

Abstract. The Subliminal Affective Priming Effect (SAPE) can be accomplished on the arousal level, which manifests that the effect appears when the subjects judge the arousal values of the target stimulus. Based on Electroencephalogram (EEG) signals, this paper analyzes whether the SAPE appears on the arousal level. For effectively solving the problems of the modal aliasing and reconstruction error for EEG study, we introduce the Complete Ensemble Empirical Mode Decomposition with Adaptive Noise (CEEMDAN) algorithm. And Multi-scale Permutation Entropy (MPE) is a common method for the time series. According to the nonlinear characteristics of EEG signals, this paper combines CEEMDAN with MPE to the judgement of SAPE. Firstly, this paper introduces the principle and computational procedure of the analysis method of SAPE. Then, we perform a one-way repeated-measures ANOVAs (ORANO-VAs) for the arousal values. At last, the experimental results are analyzed. The experimental results show that the SAPE occurs in the negative priming group, but SAPE is not observed in the positive priming group, which demonstrates that the analysis method can be used to judge whether the SAPE exists on the arousal level. We also further verify this conclusion by performing the ORANOVAs.

Keywords: Subliminal affective priming effect · CEEMDAN · MPE · Signal reconstruction · Feature extraction

1 Introduction

The subliminal affective priming effect (SAPE) means that this effect can still be observed when the priming stimulus presentation time is decrease to tens of milliseconds, a dozen milliseconds, or even several milliseconds in the affective priming experiment [1]. Since the presentation time of the priming stimulus is very short, the priming stimulus cannot be perceived consciously by the subject, namely, the effect that occurs at this time is called "SAPE" [18].

Emotion plays an indelible role in daily life. In recent years, many researchers have done a lot of study in emotion recognition. The basic emotion theory and emotional dimension theory were two universal standards, and they were proposed by Ekman and Izard respectively [2]. Ekman et al. thought that all our emotions were expanded by the

© Springer Nature Singapore Pte Ltd. 2019
Y. Sun et al. (Eds.): ChineseCSCW 2019, CCIS 1042, pp. 323–334, 2019.
https://doi.org/10.1007/978-981-15-1377-0_25

basic emotion set [3]. Meanwhile, the emotions were divided into several dimensions in the emotional dimension theory, and these dimensions included all human emotions [3]. DEAP, a multimodal data set, is used to analyze human emotional states, and it adopts the dimensional model. And the data set [18] is used in this study which is also based on the dimensional model.

Common emotion recognition methods were mainly divided into two categories [4]. On the one hand, facial expression and language expression are the emotion recognition of non-physiological signals. The advantage is that the operation is simple, and no special equipment is needed. However, the disadvantage is that it cannot guarantee the reliability of the emotion recognition. On the other hand, common physiological signal recognition methods include fMRI and EEG. Due to the fMRI equipment is huge and expensive, so it is not suitable for practical application. However, the technology of EEG has broad application prospects with the merits of simple use and few environmental constraints [19]. Therefore, nowadays, the emotion recognition based on EEG is a more commonly used method.

In recent years, the empirical mode decomposition (EMD) was a common method in the signal processing field [5, 6]. Although EMD had good self-adaptability, it was easy to appear modal aliasing in the process of decomposition, thus which affected the effect of decomposition [7]. The ensemble empirical mode decomposition (EEMD) was an improvement based on EMD, which solved the modal aliasing phenomenon to a certain extent. But its reconstructed component still contained a certain residual noise. Although the reconstruction error could be reduced, which also increased the computational complexity [8]. To completely solve the problems of reconstruction error and modal aliasing, the complete ensemble empirical mode decomposition with adaptive noise (CEEMDAN) was proposed. And the CEEMDAN decomposition process has completeness, and its reconstruction error was almost zero [9].

This paper combines CEEMDAN with multi-scale permutation entropy (MPE) to analysis the SAPE. It is a brand-new field to study the SAPE by using the analysis method of combining CEEMDAN with MPE. Currently, there are few related researches at domestic and abroad. So, the proposed work should be regarded as a kind of application innovation on EEG analysis. Through the analysis of simulation experiments, the method can judge the occurrence of SAPE well.

2   Related Work

The CEEMDAN has been generally applied in the signal processing for many years. Its main application fields include physics, mechanical engineering, biomedicine, etc., and has achieved good results. However, MPE is suitable for the linear and nonlinear signal processing, which has been widely applied in climate, image processing medicine, biology and other fields, and the effects are obvious.

CEEMDAN based physical applications. Han et al. found that CEEMDAN not only achieved good denoising effect in the high SNR Raman spectral signal, but also had obvious advantages in the low SNR Raman spectral signal [10]. For assessing the property of the filter, Zhan et al. found that the filtering method based on CEEMDAN and fuzzy entropy was superior to other filtering methods [11].

The applications of CEEMDAN in mechanical engineering field. Zhu et al. verified that CEEMDAN could eliminate most of the noise and interference signals, and it was successful to apply the diagnosis of vibration screen bearing fault [12]. Li and Zhao come up with a fault diagnosis method of combining CEEMDAN with support vector machine, the experimental results of hydraulic pump fault diagnosis proved the effectiveness and superiority of the method [13].

MPE was first used in medical research on account of its excellent adaptability and it is gradually extended to others. Georgiou et al. extracted the PE of EEG from normal people and patients with epilepsy as the characteristics and found that the correct rate of PE algorithm was the highest [14]. Aziz et al. used the MPE as features to differentiate the patients from healthy people [15].

The applications of MPE in biological field. Zou et al. made use of the PE algorithm to study the data of gene expression and found the permutation entropy algorithm could be used to identify time gene expression profiles of different complexity [16]. Liu et al. calculated MPE for EEG signals and successfully classified different emotions [17].

The above studies show that CEEMDAN can accurately reconstruct the signal and MPE can dispose non-stationary and random time series well. In this paper, the method of combining CEEMDAN with MPE is introduced into the analysis of the SAPE, and it is a kind of application innovation on EEG analysis.

3 An Analysis Method for SAPE Based on CEEMDAN and MPE

This analysis method consists of eight steps, which are briefly introduced in Fig. 1.

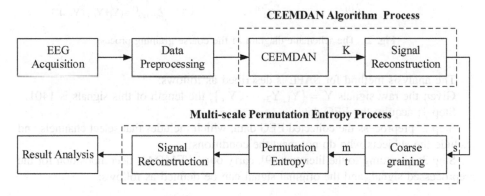

Fig. 1. The overall experimental framework of SAPE analysis method

CEEMDAN is an improvement based on EEMD, and EEMD is an improved EMD. So, CEEMDAN not only solves EMD's modal aliasing problem, but also avoids EEMD's two major problems, namely large reconstruction error and incomplete reconstruction process. CEEMDAN can make the reconstruction error to 0, and the reconstructed signal is almost the same as the original signal. PE is a nonlinear dynamic method to measure the complexity of time, and it can enlarge weak changes in time series and achieve fast calculation, strong noise immunity and high real-time performance. The MPE is a combination of PE and multi-scale, we perform coarse graining processing before calculating the PE, which is that giving a time series y(n), according to a certain multi-scale rule (s is a scale factor), we can get Z(s) and reconstruct it to z (i), the coarse graining process is shown in Fig. 2, then calculate PE of the z(i) of the scale s. Therefore, we propose an analysis method for SAPE based on CEEMDAN and MPE, and this method not only can achieve accurate signal reconstruction, but also can make the calculated PE more accurate, which is more conducive to our analysis of experimental results.

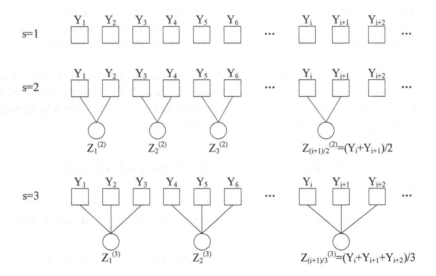

Fig. 2. The schematic diagram of the coarse graining process

The analysis method for SAPE is described as follows:

Given the raw signals $Y = \{Y_1, Y_2, \cdots, Y_N\}$, the length of this signals is 1401.

Step 1: acquire the EEG data;

Step 2: preprocess the collected EEG data, which includes that select channels and exclude the subjects who do not meet the conditions;

Step 3: according to the literature [9], carry out CEEMDAN decomposition for the preprocessed signal, and the original signal can be defined as follows:

$$y(n) = \sum_{k=1}^{K} \widetilde{IMF}_k + R(n) \tag{1}$$

Where $\widetilde{IMF}_k$ is the kth modal component produced by CEEMDAN, and R(n) is the final residual signal;

Step 4: reconstruct the first K IMFs into the high precision signal, and we set $K = 7$, the reconstructed signal is as follows:

$$Y(n) = \sum_{k=1}^{7} \widetilde{IMF}_k \tag{2}$$

Step 5: coarse graining of y(n), the process is as follows:

$$Z_t^{(s)} = \sum_{i=(t-1)s+1}^{ts} Y_i, 1 \leq t \leq \frac{N}{s} \tag{3}$$

Where s is a scale factor. When s = 1, the time series remains unchanged.

Step 6: restructure $Z_t^{(s)}$ to get z(i);

Step 7: according to the literature [14], calculate the PE for z(i) and normalize it, the normalized PE can be expressed as:

$$H_p(m) = -\frac{1}{\ln(m!)} \sum_{j=1}^{J} p_i \ln(p_i) \tag{4}$$

Step 8: the experimental results are analyzed, and the conclusion is drawn.

The analysis method for SAPE is a combination of CEEMDAN and MPE. We first perform CEEMDAN decomposition and reconstruct the first seven IMFs into the new signal, and then calculate the MPE for the reconstructed signal. According to the literature [9], the coefficient ε can be used for selecting the appropriate SNR at each stage of decomposition. When reconstructing the IMFs, we select the first seven IMFs. When calculating the MPE, s and m together determine the size of the MPE, where s represents the scale factor for coarsing graining, and m represents the embedding dimension of computed PE, namely, embedding the signal into the m-dimensional space.

4   Experiments and Discussion

4.1   Data Set

This paper uses a self-collecting dataset and it is derived from the literature [18]. According to the principle of the random sampling, 17 students from two universities in Tianjin were selected as experimental subjects. The subjects were 8 males and 9 females, who were aged between 20 and 26, and each subject completed the experiments of the positive priming group and the negative priming group respectively.

The affective faces pictures were derived from the Chinese Affective Picture System (CAPS). For the high and low priming group of experiments, we selected 10 affective face pictures (half men and half women) as the priming stimulus materials and 80 affective face pictures (half men and half women, do not repeat with the priming stimulus of medium arousal) of medium arousal as the target stimulus material after balancing gender and valance.

The experimental procedure was compiled by E-Prime software. Figure 3 shows the experimental flowchart. In each experiment, the first step was to appear the black cross gaze point (1000 ms) on a white screen. The second step was to display priming stimulus (low or high arousal, 12 ms). The third step was to present shelter stimulus (200 ms). The fourth step was blank screen (300 ms). The fifth step was to present the target stimulus (medium arousal), after the subjects made responding, and then the target stimulus disappeared. Finally, the blank screen of 1200 ms was presented, and the next trail was entered.

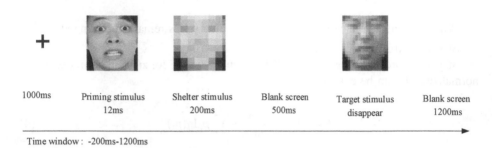

Time window : -200ms-1200ms

Fig. 3. The flowchart of the data generation process

The 64-channel cap of international 10–20 system was used in the experiment, scan4.3 was used to record the EEG signal. The experiment used AC sampling and the sampling rate sets 1000 Hz. In our study, the experimental data is imported into MATLAB for further analysis and processing.

4.2 Data Preprocessing

According to the visibility test results of subjects on the priming stimulus, the subjects who do not meet the conditions are excluded. In the negative priming group, 3 subjects who do not meet the conditions are excluded, and the data of 14 subjects is valid. In the positive priming group, 4 subjects who do not meet the conditions are excluded, and the data of 13 subjects is valid.

In this study, according to the total average graph and relevant literature, we choose the data of channels such as F3, Fz, F4, FC3, FCz, FC4, C3, Cz, C4, CP3, CPZ, CP4, P3, PZ, and P4 when studying the emotional face affective priming effect. The distribution of the channels is as shown in Fig. 4. When selecting the experimental parameters, the data of F3 and C4 channels is selected according to the principle of reducing the computational complexity as much as possible.

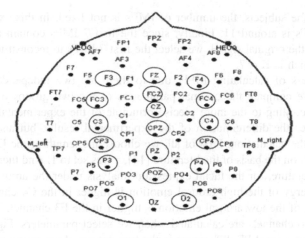

Fig. 4. The experimental channels distribution map (the channels for the experimental analysis in the circle)

4.3 Feature Extraction

In this experiment, each channel contains 1401 data points. First, we take the data of the same emotion pictures on average as the input signal of CEEMDAN. Second, we decompose each row of the input signal into several IMFs and select the first K IMFs to reconstruct the high-precision signal. Finally, we calculate the MPE for the high-precision signal. When calculating the MPE, the value of the MPE depends on the scale factor s, the embedding dimension m, the delay time t, and the data length N. When N and t are fixed, s and m together determine the size of the MPE.

Experimental Parameter Selection. In the process of CEEMDAN decomposition, firstly, the data is decomposed by CEEMDAN, and several IMFs are obtained. The process of decomposition is shown in Fig. 5. Secondly, because of the individual

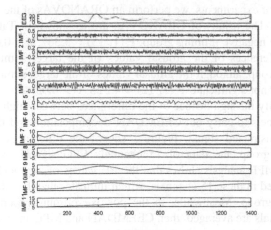

Fig. 5. Example of CEEMDAN decomposition

differences of the subjects, the number of IMFs is not fixed. In this experiment, the number of IMFs is around 11. Finally, since the first 7 IMFs contain almost all the information of the original signal, we select the first 7 IMFs to reconstruct them into a new signal, which is $K = 7$.

In the process of calculating MPE, due to there are two independent groups of experiments, we compare the optimal parameters of the two groups of experiments respectively according to the more precise principle of the experimental results. This not only ensures the differentiation of the experimental results, but also ensures the complexity of the calculation. First of all, we choose the length of the EEG signals is 1401. Secondly, on the basis of the literature [14], τ was set to 1. And then according to the relevant literature, for the facial expression analysis under the arousal dimension, the average energy of the high arousal emotion is higher in the C4 channel, and the average energy of the low arousal emotion is higher in the F3 channel. Therefore, the MPE of the two channels are calculated when we select parameters. Figure 6. shows the high and low arousal PE differences in the two groups of experiments when s or m are different. The study found that in the negative priming group, when the scale factor $s = 2$ and the embedding dimension $m = 6$, the differentiation degree of PE can well represent the change trend. Therefore, for the experiment of the negative priming group, we set $s = 2$ and $m = 6$. In the positive priming group, when the scale factor $s = 2$ and the embedding dimension $m = 5$, the differentiation degree of PE can well represent the change trend. Therefore, for the experiment of the positive priming group, we set $s = 2$ and $m = 5$.

4.4 Analysis of Experimental Results

ANOVAs is the most widely used statistical method in ERP studies. Its main function is to judge whether different experimental treatments have a significant impact on ERP [19]. The ORANOVAs is whether the different levels of a single independent variable have a significant effect on the dependent variable. The purpose of repeated-measures is to reduce or remove the variation caused by individual differences. The ORANOVAs improves the sensitivity of differences in F-test processing.

In two groups of experiments, we perform an ORANOVAs of the values of arousal of target stimulus. The results of ORANOVAs are as shown in Table 1, the results indicate that in the negative priming group, $F (1, 13) = 5.367, P < 0.05$, namely there are significant differences of the arousal values of the target stimulus. And in the positive priming group, although the arousal values of the target stimulus are different, the differences are not significant, $F (1, 12) = 4.328, P > 0.05$.

In order to verify the significance of the differences between high and low arousal, we select 15 channels to carry out CEEMDAN decomposition and then calculate the MPE of the reconstructed signal. Figure 7 shows the differences of MPE. In the negative priming group, the data of 15 channels EEG signals are averaged, then CEEMDAN and MPE analysis shows that the PE values of high and low arousal are clearly distinguished in these 15 channels, which indicates that the SAPE occurs in the negative priming group. Meanwhile, in the positive priming group, the data of 15 channels EEG signals are averaged, then CEEMDAN and MPE analysis shows that the

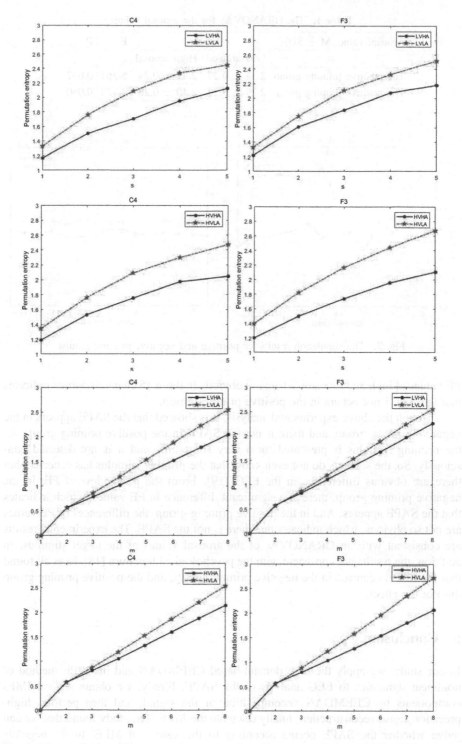

Fig. 6. The results for different parameters s and m (from top to bottom, LV/s, HV/s, LV/m, HV/m)

Table 1. The ORANOVAs for the arousal values

Arousal value (M ± SD)			F	P
	Low arousal	High arousal		
The positive priming group	2.34 ± 0.27	2.41 ± 0.24	5.367	0.037
The positive priming group	2.29 ± 0.26	2.40 ± 0.28	4.328	0.060

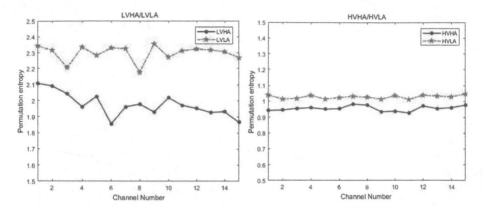

Fig. 7. The simulation results for positive and negative priming groups

PE values of high and low arousal are not obvious in these 15 channels, which indicates that the SAPE not occurs in the positive priming group.

Through the above experimental analysis, it is showed that the SAPE appears in the negative priming group, and there is not the SAPE in the positive priming group. As the priming stimulus is presented for a very short time and it is not detected consciously. So, the subjects do not even know that the priming stimulus has occurred, but there are obvious differences in the EEG [19]. From the point view of PE, in the negative priming group, there is a significant difference in PE values, which indicates that the SAPE appears. And in the positive priming group, the differences of PE values are not so obvious, which indicates that there is not the SAPE. The experimental results are consistent with an ORANOVAs of the arousal values of the target stimulus. In addition, the results are consistent with the psychological literature [18], Li et al. found that the SAPE occurred in the negative priming group, and the positive priming group did not the effect.

5   Conclusion

In our study, we apply the time domain-based CEEMDAN and the MPE method of nonlinear dynamics to EEG analysis of the SAPE. Firstly, we obtain several IMFs components by CEEMDAN decomposition of the signal, and then perform high-precision signal reconstruction, finally calculate the MPE. The study found that we can judge whether the SAPE occurs according to the values of MPE. In the negative priming group, there is the SAPE, and in the positive priming group, there is not the SAPE. This is because the essence of the priming effect is to make people in a state of preparation, which has an evolutionary meaning to the preparation of "danger". From the point of view of biological evolution, negative messages and things are closely related to human subsistence and life, and can play a certain role in prompting. Since positive messages and things are not as strong as negative ones, they will not achieve the same level of priming effect. Through the ORANOVAs, it can be known that in the negative priming group, P is less than 0.05, which indicates that there is the SAPE. Meanwhile, in the positive priming group, P is greater than 0.05, which indicates that the differences between high and low arousal are not obvious, which is that there is not the SAPE. From the point view of the computer, this paper verifies the study of the SAPE under arousal condition, which can provide a new feature extraction method and idea for further research of the SAPE in the future.

Acknowledgements. National Natural Science Foundation of China (61373149) and the Taishan Scholars Program of Shandong Province, China.

References

1. Liu, R.H., Wang, L.: Subliminal affective priming effect. Psychol. Sci. **23**, 97–110 (2000). https://doi.org/10.3969/j.issn.1671-6981.2000.03.024
2. Luo, Y.J., Wu, J.H.: Emotional psychological control and cognitive research strategies. J. Southwest Norm. Univ. **31**, 26–29 (2005). https://doi.org/10.3969/j.issn.1673-9841.2005.02.005
3. Luo, Y.J., Huang, Y.X., Li, X.Y., et al.: Effects of emotion on cognitive processing: Series of event-related potentials study. Adv. Psychol. Sci. **14**, 505–510 (2006). https://doi.org/10.3969/j.issn.1671-3710.2006.04.005
4. Nie, D., Wang, X.W., Duan, R.N., et al.: A survey on EEG based on emotion recognition. Chin. J. Biomed. Eng. **31**, 595–606 (2012). https://doi.org/10.3969/j.issn.0258-8021.2012.04.018
5. Huang, N.E., Shen, Z., Long, S.R., et al.: The empirical mode decomposition and the Hilbert spectrum for nonlinear and non-stationary time series analysis. Proc. A **454**, 903–995 (1998). https://doi.org/10.1098/rspa.1998.0193
6. Bajaj, V., Pachori, R.B.: Classification of seizure and nonseizure EEG signals using empirical mode decomposition. IEEE Trans. Inf Technol. Biomed. **16**, 1135–1142 (2012). https://doi.org/10.1109/titb.2011.2181403
7. Gao, J., Deng, J.W.: Empirical mode decomposition and analysis of its evaluation criteria. Sci. Technol. Rev. **33**, 108–112 (2015). https://doi.org/10.3981/j.issn.1000-7857.2015.02.016
8. Wu, Z.: Ensemble empirical mode decomposition: a noise assisted data analysis method. Adv. Adapt. Data Anal. **01**, 1–41 (2009). https://doi.org/10.1142/S1793536909000047
9. Torres, M.E., Colominas, M.A., Schlotthauer, G., et al.: A complete ensemble empirical mode decomposition with adaptive noise. In: Acoustics: 2011 IEEE International Conference on Acoustics, Speech, and Signal Processing (ICASSP), pp. 4144–4147 (2011). https://doi.org/10.1109/icassp.2011.5947265

10. Han, Q.Y., Sun, Q., Wang, X.D., et al.: Application of CEEMDAN in Raman spectroscopy denoising. Laser Optoelectron. Prog. **52**, 274–280 (2015). https://doi.org/10.3788/lop52. 113003
11. Zhan, L.W., Li, C.W.: A comparative study of empirical mode decomposition-based filtering for impact signal. Entropy **19**, 13–26 (2016). https://doi.org/10.3390/e19010013
12. Zhu, M., Duan, Z.S., Guo, B.L., et al.: Application of CEEMDAN combined with LMS algorithm in signal de-noise of bearings. Noise Vib. Control **38**, 144–149 (2018). https://doi.org/10.3969/j.issn.1006-1355.2018.02.028
13. Li, F., Lin, Y., Zhao, H., et al.: Fault diagnosis of hydraulic pump based on CEEMDAN-SVM. Hydraul. Pneum. **01**, 125–129 (2016). https://doi.org/10.11832/j.issn.1000-4858. 2016.01.026
14. Nicolaou, N., Georgiou, J.: Detection of epileptic electroencephalogram based on permutation entropy and support vector machines. Expert Syst. Appl. **39**, 202–209 (2012). https://doi.org/10.1016/j.eswa.2011.07.008
15. Aziz, W., Arif, M.: Multiscale permutation entropy of physiological time series. In: 2005 Pakistan Section Multitopic Conference 2005, pp. 10–23 (2005). https://doi.org/10.1109/inmic.2005.334494
16. Sun, X., Zou, Y., Nikiforova, V., et al.: The complexity of gene expression dynamics revealed by permutation entropy. BMC Bioinform. **11**, 607 (2010). https://doi.org/10.1186/1471-2105-11-607
17. Liu, X.F., Hu, B., Zheng, X.W.: Facial expression awareness based on multiscale permutation entropy of EEG. Int. J. Data Min. Inform. **21**, 287–300 (2018). https://doi.org/10.1504/ijdmb.2018.098936
18. Li, T.T., Lu, Y.: The subliminal affective priming effects of faces displaying various levels of arousal: an ERP study. Neurosci. Lett. **583**, 148–153 (2014). https://doi.org/10.1016/j.neulet. 2014.09.027
19. Wang, X.W., Shi, L.C., Lu, B.L.: A survey on the technology of dry electrodes for EEG recording. Chin. J. Biomed. Eng. **29**, 777–784 (2010). https://doi.org/10.3969/j.issn.0258-8021.2010.05.022

Cultivating Online: Understanding Expert and Farmer Participation in an Agricultural Q&A Community

Xiaoxue Shen, Adele Lu Jia(✉), and Ruizhi Sun

China Agricultural University, Beijing, China
{xiaoxueshen,sunruizhi}@cau.edu.cn, adele.lu.jia@gmail.com

Abstract. Nowadays, due to the shortage of offline agricultural experts, hundreds of thousands of farmers in China seek online for cultivation advices. A key design issue here is to provide users with timely and high quality answers, which, for general Community-based Question and Answering (CQA) platforms such as Quora, has been extensively studied before. However, answering questions raised in agricultural CQA platforms requires domain knowledge and professional experience, and users are expected to behave differently. In this article, we conduct a case study on a agricultural CQA platform named Farm-Doctor. We obtain the whole knowledge repository of Farm-Doctor, we investigate the behavioral differences between experts and farmers, and we analyze the factors that influence the users' answering behavior. Our results show that there exists obvious behavioral differences between experts and farmers, and differences also exist in the factors that influence user's answering behavior. While *recognition*, i.e., whether answers are well received by the community, has a positive impact on the answering behavior of both experts and farmers, *reciprocation*, i.e., how their own questions are treated, works differently. Experts acknowledge the "effort" of the community, i.e., the number of answers they get for their own questions has a positive effect on the number of answers they provide whereas the quality of the answers does not. Farmers, on the other hand, care both the number and the quality of the answers. Our research provides valuable information for the community to motivate the users to answer questions and therefore helps farmers solving their cultivation problems, in hoping to improve their lives.

Keywords: Community-based Question and Answering · Agriculture CQA · Behavioral analysis · Linear regression

1 Introduction

Community-based Question and Answering (CQA) systems, such as Quora, Yahoo! Answers, and Stack-Overflow are successful platforms where knowledge is shared through question asking and answering. The success of CQA platforms

© Springer Nature Singapore Pte Ltd. 2019
Y. Sun et al. (Eds.): ChineseCSCW 2019, CCIS 1042, pp. 335–350, 2019.
https://doi.org/10.1007/978-981-15-1377-0_26

lies in timely and high quality answers provided by the users and therefore the fundamental research question for CQA platforms is to identify the factors that impact the answering behavior of the users, so that measures could be taken to encourage more users to answer, so as to maintain the community prosperity.

While general CQA platforms have been extensively studied before [3,13–15,26,30], in this article, we focus on a CQA platform that is exclusive for agricultural knowledge. Due to the shortage of offline agricultural experts, a significant amount of farmers in China are now seeking online for cultivation advices. By finding out which factors influence the user's answering behavior in agricultural CQA platforms, we can motivate more users to answer questions and therefore help more farmers solving their problems, which can potentially improve their lives. The motivations behind knowledge sharing have extensively studied before. Qualitative analyses have pointed out that reward mechanisms [5,22], collective reciprocity [24,25,28], altruism [4,19], self-efficacy [18,29], and social attributes [4,11] encourage users to generate positive behavior and continuously contribute to the community. Quantitatively, a number of big data analyses have shown that the overall reputation of the experts and how their answers are received by the community have positive impact on users' willingness to answer questions [7,17,21]. Different from general CQA platforms, proper answers to the agricultural questions require specific domain knowledge and experience. Given the difference in the context, we believe that experts and farmers behave differently in agricultural CQA platforms, and therefore, the same set of factors may have different impacts on their answering behavior. Further, there exists factors that are uniquely provided in agricultural CQA platforms, for example, whether the asker has adopted and carried out the cultivation advices given in the answer. Such factors have not been analyzed before.

In this paper, we conduct a case study on a agricultural CQA platform named *Farm-Doctor*, and obtain the whole knowledge repository of it that consists of over 750 thousand questions and over 3 million answers. Following our motivations as discussed above, we seek to answer the following research questions:

– *What are the behavioral differences between the experts and the farmers?*
– *What are the factors that influence users' answering behavior, and do they have similar effects for the experts and for the farmers?*

We conducted one tailed Kolmogorov-Smirnov test to answer the first problem. To answer the second question, we build linear regression models, separately for experts and for farmers, to identify factors that influence their answering behavior. We consider traditional factors as explored in previous works [7,17,21], as well as factors that have not been explored before. Particularly, we include a set of factors that reflect how the questions are treated by the community. We intend to examine whether reciprocation could be one of the motivations behind. Our analysis makes the following findings:

– *Firstly,* we observe significant behavioral differences between experts and farmers, in terms of the number and the quality of the answers they provide, and the number of questions they ask. We find that, consistent with our intuition, farmers in general ask more questions whereas experts provide a larger number of high quality answers.

- *Secondly,* consistent with previous analyses, we find that the number of answers the users provide is positively correlated with how well their answers are received by the community, reflected by the number of thumbs-up they receive and the number of their answers adopted by the askers.
- *Thirdly,* through analyzing the impact of how the questions are treated by the community, we find that experts are more concerned with the number of answers he received than the quality of the answers, i.e., experts care more about the "effort" the community has made in helping them. On the other hand, farmers concern about both the quality and quantity of answers they receive, suggesting that their priority purpose here is to seek practical help.

The rest of this article is organized as follows. In Sect. 2, we give a brief review of previous related research. In Sect. 3, we develop the hypotheses. In Sect. 4, we give a brief introduction of Farm Doctor and the dataset. In Sect. 5, our evaluation models and analysis are given. Finally, we conclude our work in Sect. 6.

2 Related Work

CQA platforms have been extensively studied before. Comprehensive surveys can be found in [23]. We summarize the related works for the main research topics in CQA as follows.

User Activity Prediction. The most closely related work to our research is user activity prediction. Keeping the user active in CQA platforms is vital to maintaining the community prosperity, and a number of previous works have shed lights on this issue. Dror *et al.* [7] utilized several features including activity rate, personal information, and social interactions with other users to predict whether users will leave the community. The results show that the number of answers and positive responses (such as upvotes) are most relevant to the newcomer user churn. For both experts and new users, Pudipeddi *et al.* [21] utilized different classifiers to identify significant factors that affect users churn. They concluded that the time gap between subsequent posts is the most important indicator of newcomer users and experts churn. More recently, Liu *et al.* [17] proposed an algorithm named PALP that takes into consideration user behavior change over time to predict whether an expert will post in the future.

Most of these studies focus on predicting whether a user will be active in the future, whereas our work takes one step ahead and analyze how active the users will be, in terms of the number of answers they are going to provide. Meanwhile, previous analyses have considered features related to the activity of the experts, how their answers are received by the community, and the overall reputation of the experts. We believe that how questions are treated by the community (e.g., whether they get a sufficient number of answers timely and of good quality) also impact users willingness to answer questions raised by others, and the impacts could vary for different kinds of users, e.g., experts and normal users. This factor, however, has not been analyzed before. In this article, we conduct a detailed analysis on this based on linear regression models tailored for different kinds of users.

Motivations for Knowledge Sharing. Many works have reasoned the motivations for question answering through qualitative analyses. Lou *et al.* [18] proposed the self-determination theory to study the motivational factors that influence the quantity and quality of knowledge contribution in the online CQA platform. They found that rewards, as a manifestation of external rules, is more effective in promoting knowledge contribution quantity than quality. Knowledge self-efficacy, as a manifestation of intrinsic motivation, is more relevant to the quality of knowledge contribution. Choudhury *et al.* [4] utilized Exploratory Factor Analysis to analyze user behaviour and their latent needs. The results shown that users are driven primarily by four needs: social interaction, altruism cognitive need and reputation. Based on social capital theory, social exchange theory and social cognition theory, Jin *et al.* [11] explored why users continuously contribute knowledge to online social Q&A communities. They found a user self-presentation, peer recognition, and social learning have a positive impact on his knowledge-contribution behaviors.

Expert Finding is another line of research on CQA platforms that aims to rank users based on their expertise. Pal *et al.* [20] utilized classification and ranking models to predict whether or not a user will be an expert in the future according to his behavior during the first 2 weeks. For improving authority ranking. Yang *et al.* [15] proposed a probabilistic Topic Expertise Model which used tagging information and voting information to learn topical expertise estimation. They also proposed CQArank model that combines user topical expertise estimation and user authority derived from link analysis to find experts with both similar topical preference and high topical expertise. Zhao *et al.* [30] formed the problem of expert finding as missing value estimation and leveraged graph regularized matrix completion to estimate missing value.

CQA Platform Measurement. In recent years, Yahoo! Answer, Stack Overflow, Quora and other general CQA platforms have received extensive attention from scholars at home and abroad. For Yahoo! Answer, Li *et al.* [13] analyzed 18,895 resolved questions under the Computers & Internet category. [2,6,12,31] have further expanded the datasets, and the number of questions has reached the million level. For Stack Overflow, Yang *et al.* [15] analyzed 8,904 questions and 96,629 answers posted by 663 users. Yang *et al.* [26] further added tag information for analysis. For Quora, Zhao *et al.* [30] analyzed the 444,138 questions posted between September 2012 and August 2013 and 95,915 users who answered these questions. Chen *et al.* [3] studied 50,451 questions, 4,415 users and their following relationships.

The above measurement and public datasets are mainly concentrated in the general CQA platforms. Most of them are limited to a particular type of questions instead of the whole corpus. In contrast, we have obtained and offer public access to the whole knowledge repository of a agricultural CQA platform, i.e., Farm-Doctor.

3   Hypotheses Development

3.1   Behavioral Differences Between Experts and Farmers

In this article, the first research question to be investigated is whether agricultural experts and farmers behave differently in the agricultural CQA platforms. Previously studies on user behavior analysis in CQA platforms mainly focus on general CQA platforms and on highly active users or new users. For instance, Liu et al. [17] made predictions on the activity level for experts, and Dorr et al. [7] studied the problem of churn prediction for new users. Different from questions raised in general CQA platforms such as Quora, proper answers to the agricultural questions require specific domain knowledge and experience, and therefore one would expect that the agricultural experts and farmers behave differently. We make the following hypothesis:

Hypothesis 1. Agricultural experts and farmers behave differently in question answering, in terms of the number and the quality of the answers they provide, and the number of questions they ask.

3.2   Impact of How the Users' Answers Are Received
by the Community

Previous studies have shown that answers being well received by the community has a positive impact on users' willingness to answer more questions [5,7,22]. Being well received can be shown through receiving badges (e.g., Stack Exchange) and reputation points (e.g., Yahoo! Answers). Reputation points encourage users to provide high quality answer for getting "point". The points are cumulative and the users get a "level" according to the total points they received. User's points indicate their social status in the community [5]. Badges are often used in conjunction with reputation scores to reward positive behaviour by relating a users site identity with their perceived expertise and respect in the community [1,9].

Different from the above mentioned general CQA platforms, in agricultural CQA platforms such as Farm-Doctor, experts are selected first through self application and then manually checked by the community administrators. Therefore, badges and reputation points are not needed to improve their social status. Thumbs-up is a common mechanism used both in general CQA platforms and Farm-Doctor, which has been shown to be informative for predicting the activity level of the users [17]. In addition, in Farm-Doctor question askers can claim that they have adopted the solutions given in the answers to solve their planting issues, as a way to show that the answers are valuable and treated seriously by the askers. These considerations lead to the following two hypotheses:

Hypothesis 2. The number of answers a user provides is positively correlated with the number of his answers adopted by the askers.

Hypothesis 3. The number of answers a user provides is positively correlated with the number of thumbs-up the user receives.

3.3 Impact of How the Users' Questions Are Treated by the Community

As mentioned above, most previous works on user behavior analysis have considered features related to the activity of the experts, how their answers are received by the community, and the overall reputation of the experts. We believe that how questions are treated by the community (e.g., whether they get a sufficient number of answers timely and of good quality) also impact users willingness to answer questions raised by others, as users who have received help from the community might be encouraged to contribute as well. We make the following hypothesis:

Hypothesis 4. The number of answers a user provides is positively correlated with the average number of answers received by his own questions.

Hypothesis 5. The number of answers a user provides is negatively correlated with the average delay for the first answer received by his own questions.

Hypothesis 6. The number of answers a user provides is positively correlated with the quality of the answers received by his own questions.

3.4 Impact of Social Relationship

Many CQA platforms have introduced social features, for example, user following, to maintain the community prosperity. The impact of social relationships on the CQA community have been extensively studied before. For instance, [16] showed that number of followers the askers attract have a positive effect on the number of answers received. [10] observed that the answer price has positive and significant correlations with the number of followers. In this study, we analyze how the social relationships of users impact the number of answers they provide. We have the following hypothesis:

Hypothesis 7. The number of answers a user provides is positively correlated with the number of followers the user attracts.

Hypothesis 8. The number of answers a user provides is positively correlated with the number of users he follows (named *followees*).

4   Methodology and the Farm-Doctor Dataset

In this section, we give a brief introduction of Farm Doctor and the dataset used throughout this article.

4.1 An Overview of Farm-Doctor

Farm-Doctor is a CQA platform exclusive for agricultural questions. It is one of the largest CQA platforms in China that provide farmers with advices on cultivation, such as crop disease detection and treatment recommendation. As in general CQA platforms, users in Farm-Doctor can raise and answer questions, vote to the answers, and follow each other. In addition, they can specify the crops that are interested to them and tag their questions with related crops.

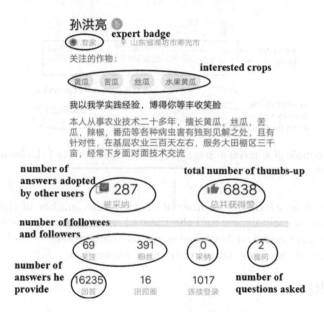

Fig. 1. An example of a user profile page in Farm-Doctor and the basic information we can obtain.

4.2 Dataset

Farm-Doctor identifies each of its user with a unique numerical number in the increasing order. Each identifier corresponds to a webpage with detailed user information that can be obtained with crawlers. We have obtained the whole knowledge base (until April 2019) of Farm-Doctor with detailed information of 344,899 users, 750,831 questions, and 3,282,723 answers.

Figure 1 shows an example of a user profile page in Farm-Doctor and the basic information we can obtain. For each user, we obtain the number of answers adopted by askers, the total number of thumbs-up he received, the number of users that follows he, the number of users that he follows, the number of questions he asked and the number of answers he answered. Figure 2 shows an example of a question page, along with a list of its answers. For each question, we obtain the

Fig. 2. An example of a question page with a list of answers and the basic information we can obtain.

time when it was asked, the number of answers, the number of the thumbs-up for and down for each answer and the time when each answer was left. In total, 78,594 users have raised at least one question (named *askers*) and 45,150 users have answered at least one question (named *answerers*).

5   Results

5.1   Descriptive Analysis

For our analysis, we have only considered users who have answered and have asked at least one question, corresponding to 2,969 experts and 22,412 farmers, respectively. The statistical results of key variables of experts and farmers are described in Table 1. Specifically, variables p2 and p3 reflect how the answers are received by the community, p4, p5, and p6 reflect how the questions are treated by the community, and p7 and p8 reflect the social status of the users.

Comparing the mean values, we find that in general experts provide a larger number of answers than the farmers, and their answers are also of higher quality, in terms of the number of thumbs-up received and the number of answers adopted by the question askers. Experts are also more active in forming social relationships, i.e., compared to the farmers, experts are more active in following others and they have also attracted more followers.

5.2   Results of One Tailed KS Test

To test whether experts and farmers behave differently (Hypothesis 1), we conducted one tailed Kolmogorov-Smirnov (KS) test [8,27] (suitable for both nor-

Table 1. Statistics results of the variables

Variable		Expert	Farmer
The number of answers they provide (p0)	Mean	858	11
	Median	36	2
	Max	164,313	9,638
The number of questions they asked (p1)	Mean	16	20
	Median	6	8
	Max	723	1,227
The number of answers adopted by the askers (p2)	Mean	10	0
	Median	0	0
	Max	1,913	104
The number of thumbs-up they receive (p3)	Mean	487	4
	Median	12	0
	Max	98,785	6,908
The average number of answers received by their own questions (p4)	Mean	4	4
	Median	4	3
	Max	33	47
The delay of the first answer (p5)	Mean	79,614	109,110
	Median	1,185	1,346
	Max	18,205,010	69,214,650
The number of thumbs-up answers received by their own questions (p6)	Mean	89	49
	Median	28	11
	Max	3,953	5,406
The number of followers (p7)	Mean	68	4
	Median	3	2
	Max	5,506	3,702
The number of followees (p8)	Mean	38	4
	Median	4	1
	Max	4,029	4,171

mal and abnormal distribution) on four aspects, namely the number of answers they provide, the number of questions they asked, the number of their answers adopted by askers, and the number of thumbs-up they receive. The results are shown in Table 2.

We observe significant behavioral differences in the above four aspects between experts and farmers. For experts, the number of questions they asked are relatively smaller, while the other features exhibit stronger strength compared with the farmers. Hypotheses 1 is supported. These results indicate that, consistent with our intuition, in general farmers ask more questions whereas experts provide a larger number of high quality answers. We conjecture that, as Farm-Doctor is exclusive for agricultural questions that highly demand domain knowledge and experience, the agricultural experts are more qualified for the questions and hence there exists the significant difference in question answering behaviors between experts and farmers.

5.3   Results of Linear Regression Models

In this section, we propose linear regression models to examine the factors that potentially impact users' willingness to answer questions (Hypotheses 2–8). As discussed above, there exists significant behavioral differences between experts and farmers. Therefore, for our analysis we differentiate experts and farmers, and we build linear regression models for them separately. The results of the regression model for experts and farmers are reported in Tables 3 and 4, respectively. F-statistic indicates the significance of the regression model. The large R^2 indicates that the proposed independent variables explain much of the answer behavior. For experts, the proposed model is statistically significant (F = 2270) and good (adjusted R^2 = 88.5%). And For farmers, the model is significant (F = 3,557) and suitable (adjusted R^2 = 61.4%). According to the results of VIF test, there is no multiple collinearity problem. We have several interesting findings as following.

Table 2. Results of the one tailed KS test: Experts are in the former

Variable	p	Type
The number of answers they provide (p0)	***	Larger
The number of questions they asked (p1)	***	Smaller
The number of their answers adopted by askers (p2)	***	Larger
The number of thumbs-up they receive (p3)	***	Larger

[1]*: $p < 0.05$; **: $p < 0.01$; ***: $p < 0.001$

Impact of How the Answers Are Received by the Community. The results report a positive effect of the number of answers adopted by askers (experts: parameter = 7.9425, $p < 0.001$, farmers: parameters = 24.5022, $p < 0.001$) and the number of thumbs-up received (experts: parameter = 1.3913, $p < 0.001$, farmers: parameters = 0.9211, $p < 0.001$) on the number of answers users provide. This means that the number of answers the users provide is positively correlated with how well the answers are received by the community. Therefore, Hypotheses 2 and 3 are supported.

Table 3. Results of linear regression for experts (R-squared: 0.885, F-statistic: 2270)

Variable	Parameter	Std.error	P value	VIF
p2	7.9425***	0.756	<0.001	2.648
p3	1.3913***	0.017	<0.001	2.813
p4	16.0371*	7.876	<0.05	1.222
p5	8.864e−06	5.35e−05	0.868	1.016
p6	0.0651	0.148	0.439	1.175
p7	0.6398***	0.136	<0.001	2.043
p8	−1.2993***	0.246	<0.001	1.658

[1]*: $p < 0.05$; **: $p < 0.01$; ***: $p < 0.001$

Impact of How the Questions Are Treated by the Community. For experts, the average number of answers received by their own questions has a positive relationship with the number of answers they provide (parameter = 16.0371, $p < 0.05$). But the delay of the first answer (parameter = 0.0000, p = 0.868) and the number of thumbs-up answers received by their own questions (parameter = 0.0651, p = 0.439) show no significant effect on the number of answers they provide. It means that the experts are more concerned with the number of answers he received than the time and quality of the answers, i.e., they focus more on the "effort" of the help from the community, rather than the quality. Therefore, for the experts, hypotheses 4 is supported, and hypotheses 5 and 6 are unsupported. For farmers, on the other hand, the average number of answers received (parameter = 0.4842, $p < 0.001$) and the number of thumbs-up answers received (parameter = 0.0307, $p < 0.001$) both have a positive relationship with the number of answers they provide. However, the delay of the first answer shows no significant effect with the dependent variable (parameter = 0.0000, p = 0.240). This means that, different from the experts, farmers focus on both the quality and quantity of answers they receive. Therefore, for the farmers, hypotheses 4 and 6 are supported, and hypotheses 5 is unsupported.

Table 4. Results of linear regression for farmers (R-squared: 0.614, F-statistic: 3557)

Variable	Parameter	Std.error	P value	VIF
p2	24.5022***	0.297	<0.001	1.210
p3	*0.9211***	0.012	<0.001	1.209
p4	0.4842***	0.066	<0.001	1.205
p5	3.215e−07	2.73e−07	0.240	1.002
p6	0.0307***	0.002	<0.001	1.198
p7	−0.0818***	0.010	<0.001	1.584
p8	0.0486***	0.008	<0.001	1.559

[1]*: $p < 0.05$; **: $p < 0.01$; ***: $p < 0.001$

In addition, in Fig. 3 we show the CDFs of the delay of the first answer for experts and farmers. We observe that the 71% questions asked by farmers and 76% questions asked by experts are answered in 1,000 s (i.e., 17 min). So in Farmer-Doctor, most users can receive timely answers. We conjecture that this is the reason why time has no significant effect on the answering behavior.

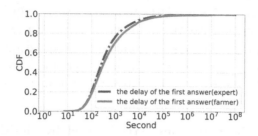

Fig. 3. CDFs of the delay of first answer for experts and farmers

Impact of the Social Relationships. We observe that, for experts, the number of followers has a positive relationship with the dependent variable (parameter = 0.6398, p < 0.001), whereas for farmers the number of followees does so (parameter = 0.0486, p < 0.001). Intuitively, a large number of followers indicates that the experts are well received by the community and therefore encourage them to answer more. On the other hand, a large number of followees shows that the farmers are very active and have followed many other users. These active farmers are also possible to be active in answer questions.

However, we also find that, for experts, the number of followees has a negative effect on the number of answers they provide (parameter = −1.2993, p < 0.001), whereas for farmers, the number of followers shows a negative effect (parameter = −0.0818, p < 0.001). Together with the above results, It is interesting to see that the social features in fact have different impacts on the behavior of the experts and of the users. We conjecture that this is due to the fundamental differences in the roles of the experts and of the farmers in Farm-Doctor. However, reasoning on this argument requires qualitative analyses such as interviews and surveys, which we leave as our future work.

We summarize our hypotheses testing results in Table 5.

Discussion. In this article, we have analyzed the potential factors that affect the number of answers the uses provide. For experts and farmers, we observed that the answer attributes have positive relationships with the answering behavior. This result is in accordance with previous works [5, 22] which states that incentives can encourage users to generate more positive behavior. We have also analyzed the impact of how the users' questions are treated by the community on the answering behavior. We find that the experts are more concerned about the number of answers he received and farmers care both the quality and quantity of answers they received. We speculate that most farmers join the community

Table 5. Hypotheses testing results

Hypotheses	Expert	Farmer
H1	Supported	Supported
H2	Supported	Supported
H3	Supported	Supported
H4	Supported	Supported
H5	Unsupported	Unsupported
H6	Unsupported	Supported
H7	Supported	Unsupported
H8	Unsupported	Supported

to seek help, so they care about both the quality and the quantity of answers, whereas experts have strong expertise, so they do not really care about the quality of the answers they receive, but they care more about the community's effort in helping them, reflected by the number of answers they receive.

6 Conclusion

In this article, we conducted an in-depth analysis on the participation behavior of experts and farmers in a agricultural CQA platforms. Based on the case study on Farm-Doctor, which is one of the major agricultural CQA platforms in China, we revealed the behavioral differences between experts and farmers, in term of the number and the quality of answers that they provide, and the number of questions they ask. We found that farmers in general ask more questions whereas experts provide a larger number of high quality answers. We further analyzed and discussed the factors that influence the users' answering behavior, and observed that the number of answers the users provide is positively correlated with how well their answers are received by the community for both experts and farmers. On the other hand, how the questions are treated by the community has different effects on the experts compared to on the farmers, i.e. experts are more concerned with the number of answers he received than the quality of the answers, whereas farmers concern about both the quality and quantity of answers they receive. Our research can provide a reference for the fundamental research question in the agricultural CQA platforms, i.e., identifying factors that influence users' answering behavior, based on which incentive policies can be tailored to motivate users to answer questions, so as to maintain the community prosperity.

References

1. Anderson, A., Huttenlocher, D., Kleinberg, J., Leskovec, J.: Steering user behavior with badges. In: Proceedings of the 22nd International Conference on World Wide Web, WWW 2013, pp. 95–106. ACM, New York (2013). https://doi.org/10.1145/2488388.2488398

2. Bouguessa, M., Dumoulin, B., Wang, S.: Identifying authoritative actors in question-answering forums: the case of Yahoo! answers. In: Proceedings of the 14th ACM SIGKDD International Conference on Knowledge Discovery and Data Mining, KDD 2008, pp. 866–874. ACM, New York (2008). https://doi.org/10.1145/1401890.1401994

3. Chen, Z., Zhang, C., Zhao, Z., Yao, C., Cai, D.: Question retrieval for community-based question answering via heterogeneous social influential network. Neurocomputing **285**, 117–124 (2018). https://doi.org/10.1016/j.neucom.2018.01.034

4. Choudhury, S., Alani, H.: Exploring user behavior and needs in Q & A communities. In: Rospigliosi, A., Greener, S. (eds.) Proceedings of the European Conference on Social Media: ECSM 2014, pp. 80–89. Academic Conferences and Publishing International Limited, July 2014. http://oro.open.ac.uk/40658/

5. DeVaro, J., Kim, J.H., Wagman, L., Wolff, R.: Motivation and performance of user-contributors: evidence from a CQA forum. Inf. Econ. Policy **42**, 56–65 (2018). https://doi.org/10.1016/j.infoecopol.2017.08.001

6. Dror, G., Koren, Y., Maarek, Y., Szpektor, I.: I want to answer; who has a question?: Yahoo! answers recommender system. In: Proceedings of the 17th ACM SIGKDD International Conference on Knowledge Discovery and Data Mining, KDD 2011, pp. 1109–1117. ACM, New York (2011). https://doi.org/10.1145/2020408.2020582

7. Dror, G., Pelleg, D., Rokhlenko, O., Szpektor, I.: Churn prediction in new users of Yahoo! answers. In: Proceedings of the 21st International Conference on World Wide Web, WWW 2012 Companion, pp. 829–834. ACM, New York (2012). https://doi.org/10.1145/2187980.2188207

8. Fasano, G., Franceschini, A.: A multidimensional version of the Kolmogorov-Smirnov test. Mon. Not. R. Astron. Soc. **225**(1), 155–170 (1987). https://doi.org/10.1093/mnras/225.1.155

9. Grant, S., Betts, B.: Encouraging user behaviour with achievements: an empirical study. In: 2013 10th Working Conference on Mining Software Repositories (MSR), pp. 65–68, May 2013. https://doi.org/10.1109/MSR.2013.6624007

10. Jan, S.T., Wang, C., Zhang, Q., Wang, G.: Pay-per-question: towards targeted Q & A with payments. In: Proceedings of the 2018 ACM Conference on Supporting Groupwork, GROUP 2018, pp. 1–11. ACM, New York (2018). https://doi.org/10.1145/3148330.3148332

11. Jin, J., Li, Y., Zhong, X., Zhai, L.: Why users contribute knowledge to online communities: an empirical study of an online social Q & A community. Inf. Manag. **52**(7), 840–849 (2015). https://doi.org/10.1016/j.im.2015.07.005. Novel applications of social media analytics

12. Jurczyk, P., Agichtein, E.: Discovering authorities in question answer communities by using link analysis. In: Proceedings of the Sixteenth ACM Conference on Conference on Information and Knowledge Management, CIKM 2007, pp. 919–922. ACM, New York (2007). https://doi.org/10.1145/1321440.1321575

13. Li, B., King, I.: Routing questions to appropriate answerers in community question answering services. In: Proceedings of the 19th ACM International Conference on Information and Knowledge Management, CIKM 2010, pp. 1585–1588. ACM, New York (2010). https://doi.org/10.1145/1871437.1871678

14. Li, B., King, I., Lyu, M.R.: Question routing in community question answering: putting category in its place. In: Proceedings of the 20th ACM International Conference on Information and Knowledge Management, CIKM 2011, pp. 2041–2044. ACM, New York (2011). https://doi.org/10.1145/2063576.2063885

15. Yang, L., et al.: CQArank: jointly model topics and expertise in community question answering. In: Proceedings of the 22nd ACM International Conference on Information & Knowledge Management, CIKM 2013, pp. 99–108. ACM, New York (2013). https://doi.org/10.1145/2505515.2505720

16. Liu, Z., Jansen, B.J.: Factors influencing the response rate in social question and answering behavior. In: Proceedings of the 2013 Conference on Computer Supported Cooperative Work, CSCW 2013, pp. 1263–1274. ACM, New York (2013). https://doi.org/10.1145/2441776.2441918

17. Liu, Z., Xia, Y., Liu, Q., He, Q., Zhang, C., Zimmermann, R.: Toward personalized activity level prediction in community question answering websites. ACM Trans. Multimedia Comput. Commun. Appl. **14**(2s), 41:1–41:15 (2018). https://doi.org/10.1145/3187011

18. Lou, J., Fang, Y., Lim, K.H., Peng, J.Z.: Contributing high quantity and quality knowledge to online Q & A communities. J. Am. Soc. Inf. Sci. Technol. **64**(2), 356–371 (2013). https://doi.org/10.1002/asi.22750

19. Nam, K.K., Ackerman, M.S., Adamic, L.A.: Questions in, knowledge in?: a study of Naver's question answering community. In: Proceedings of the SIGCHI Conference on Human Factors in Computing Systems, CHI 2009, pp. 779–788. ACM, New York (2009). https://doi.org/10.1145/1518701.1518821

20. Pal, A., Farzan, R., Konstan, J.A., Kraut, R.E.: Early detection of potential experts in question answering communities. In: Konstan, J.A., Conejo, R., Marzo, J.L., Oliver, N. (eds.) UMAP 2011. LNCS, vol. 6787, pp. 231–242. Springer, Heidelberg (2011). https://doi.org/10.1007/978-3-642-22362-4_20

21. Pudipeddi, J.S., Akoglu, L., Tong, H.: User churn in focused question answering sites: characterizations and prediction. In: Proceedings of the 23rd International Conference on World Wide Web, WWW 2014 Companion, pp. 469–474. ACM, New York (2014). https://doi.org/10.1145/2567948.2576965

22. Resnick, P., Zeckhauser, R., Swanson, J., Lockwood, K.: The value of reputation on eBay: a controlled experiment. Exp. Econ. **9**(2), 79–101 (2006). https://doi.org/10.1007/s10683-006-4309-2

23. Srba, I., Bielikova, M.: A comprehensive survey and classification of approaches for community question answering. ACM Trans. Web **10**(3), 18:1–18:63 (2016). https://doi.org/10.1145/2934687

24. Wasko, M.M., Faraj, S.: Why should I share? Examining social capital and knowledge contribution in electronic networks of practice. MIS Q. **29**(1), 35–57 (2005). http://dl.acm.org/citation.cfm?id=2017245.2017249

25. Wu, P.F., Korfiatis, N.: You scratch someone's back and we'll scratch yours: collective reciprocity in social Q & A communities. J. Assoc. Inf. Sci. Technol. **64**(10), 2069–2077 (2013). https://doi.org/10.1002/asi.22913

26. Yang, B., Manandhar, S.: Tag-based expert recommendation in community question answering. In: 2014 IEEE/ACM International Conference on Advances in Social Networks Analysis and Mining (ASONAM 2014), pp. 960–963, August 2014. https://doi.org/10.1109/ASONAM.2014.6921702

27. Young, I.: Proof without prejudice: use of the kolmogorov-smirnov test for the analysis of histograms from flow systems and other sources. J. Histochem. Cytochem. **25**(7), 935–941 (1977). https://doi.org/10.1177/25.7.894009. Official Journal of the Histochemistry Society

28. Yu, J., Jiang, Z., Chan, H.C.: The influence of sociotechnological mechanisms on individual motivation toward knowledge contribution in problem-solving virtual communities. IEEE Trans. Prof. Commun. **54**(2), 152–167 (2011). https://doi.org/10.1109/TPC.2011.2121830

29. Zhao, L., Detlor, B., Connelly, C.E.: Sharing knowledge in social Q&A sites: the unintended consequences of extrinsic motivation. J. Manag. Inf. Syst. **33**(1), 70–100 (2016). https://doi.org/10.1080/07421222.2016.1172459
30. Zhao, Z., Zhang, L., He, X., Ng, W.: Expert finding for question answering via graph regularized matrix completion. IEEE Trans. Knowl. Data Eng. **27**(4), 993–1004 (2015). https://doi.org/10.1109/TKDE.2014.2356461
31. Zhu, H., Cao, H., Xiong, H., Chen, E., Tian, J.: Towards expert finding by leveraging relevant categories in authority ranking. In: Proceedings of the 20th ACM International Conference on Information and Knowledge Management, CIKM 2011, pp. 2221–2224. ACM, New York (2011). https://doi.org/10.1145/2063576.2063931

Characterizing Urban Youth Based on Express Delivery Data

Dong Zhang and Zhiwen Yu[$\boxtimes$]

Northwestern Polytechnical University, Xi'an 710072, China
zhiwenyu@nwpu.edu.cn

Abstract. The urban youth has emerged as a new concept in recent years, which reflects the degree of rejuvenation of a city. The fine-grained urban youth characterization has the potential value of multi-industry development orientation and business configuration optimization. However, there is no formal definition and structured characterizing system for urban youth between the academic field and the business field. In addition, the express delivery industry has ushered in explosive growth driven by e-commerce. The scale and value of express delivery data increase accordingly. This paper attempts to characterize the urban youth based express delivery data. Along this line, we first propose a concept of Youth Index (YI) to quantify the urban youth. Then, we construct the Youth Index Assessment Model (YIAM) to calculate urban YI, where a Youth Index Dictionary (YID) is constructed based on relevant sociological studies as an auxiliary tool. Furthermore, the YI of urban areas is presented visually combined with a road network-based urban functional area division strategy, which characterizes a fine-grained urban YI. Finally, experiments on Xi'an (a Chinese provincial city) show that urban youth can be characterized excellently according to the comparison with the actual situation of the experimental samples.

Keywords: Urban computing · Urban youth · Urban youth index · Big data · Sociological research

1 Introduction

The Urban Youth has appeared in the public vision as a new concept in recent years. Intuitively, it reflects the degree of the youthfulness of a city. With the development of society, the young degree of the city receives more and more attention. On the one hand, the city's youth reflects its attraction to young people. On the other hand, the urban youth reflects the urban vitality, which is of great reference value to the industrial layout of urban development. However, academic research related to urban youth has not yet been proposed.

In this paper, we have developed a variety of methods to depict urban youth. We propose the concept of Urban Youth Index (UYI) and use relevant data to quantify it. For example, the user information of the social network is used to measure the urban

© Springer Nature Singapore Pte Ltd. 2019
Y. Sun et al. (Eds.): ChineseCSCW 2019, CCIS 1042, pp. 351–362, 2019.
https://doi.org/10.1007/978-981-15-1377-0_27

youth index, the review data of the life consumption platform is mined to analyze the urban youth level, the express delivery information is mined to evaluate the urban youth level, and so on. However, the authenticity of social network user information is low and the reliability is poor. The consumer review data is not well regulated, and the association rule between the data itself and the urban youth is not easy to establish. On the contrary, the express data does not have the problems mentioned above. As a significant global industry, the express delivery industry has ushered in an explosive development period under the driven by e-commerce and keeps being a research hotspot in various countries and regions [1]. The industrial data produced have the characteristics of "large scale, multiplex, wide time span". These characteristics make the data itself have a uniform and dense distribution in space and time. In addition, the express delivery data contains the detailed address information of the recipient and the sender, which enables the research target to be analyzed at the level of urban area youth index (precise to block), so as to achieve fine-grained characterizing results.

The online shopping business accounts for a very high proportion in the express delivery industry. E-consumers buy around 13.5 parcels per person per year for a total of over 400 million parcels. Across France, the road express delivery market is valued at over 43 billion Euros [2]. Therefore, express data is able to reflect the characteristics of online shopping in a specific region. According to existing relevant sociological theories, online shopping behavior can reflect the youngness of consumers. On this basis, this paper proposes a fine-grained method based on express delivery of big data to characterize urban youth.

In summary, our research has made the following contributions:

1. **A new concept of urban youth index is introduced.** The concept of the YI is first proposed, which is supposed to be an indicator to characterize urban youth.
2. **A novel model profiling proposed youth index based on express delivery big data is proposed.** Through putting an insight view of express delivery big data and combining the existing sociological theory, a novel and complete Youth Index Assessment Model (YIAM) is constructed to calculate the proposed YI. This is a brand-new application based on express big data.
3. **Visualization of urban youth based on the proposed youth index is presented.** The results of the hue diagram are used to display the experimental results, and based on this, a typical area is selected for analysis to evaluate the model performance.

2   Related Work

In existing academic research, the study on UYI almost remains a blank. The studies are mostly at the level of the age structure of the national population, rather than studying the degree of youth in cities or even regions. In addition, independent research on urban youth has not yet been done.

As mentioned before, the rapid development of the express industry makes it a research hotspot for multiple-domain research. Li et al. propose a systematic housing demand inference method, named Housing Demand Inference Model (HDIM) to estimate housing demand by exploiting the residential mobility of communities based on express delivery data [3]. Ye et al. implement the Logistics delay prediction model for freight based on Artificial Neural Network [4]. Duan et al. characterize the material flows and environmental implications of post-consumer packaging waste from express delivery in China to provide policy suggestions for the government [5]. Fan et al. estimated an environmental load of express packaging materials consumed in the processes of production and distribution of express delivery based on statistics of the types and quantities of packaging materials consumed by China's express delivery industry [6]. Ding et al. propose an optimized blockchain technology shared express mode based on the technical characteristics of the blockchain to guarantee the efficient operation of the shared express platform [7]. Li et al. focus on privacy issue and proposed improved attribute-based encryption with a hidden access tree to enforce fine-grained access control on the logistic data [8]. Researches mentioned above involve different social fields related to the express industry, but few of them focus on potential user information and application value of express delivery data, let alone dig deep into the data that can characterize the user.

So far, our research has created a mapping model from express delivery data to the YI of urban areas. Unlike existing research on express data, we are committed to discovering the intrinsic link between express data and social hotspots (i.e. YI). And based on the completed work, the direction of program optimization and in-depth research is proposed.

3 Definitions and Sociological Foundations

3.1 Youth Index

The academic community does not currently propose an authoritative definition of the "Youth Index" (YI). The relevant concepts involved in this paper are defined as follows:

Youth Index. Refers to the degree that the age is close to the range of 18 to 35. The value is a fraction distributed over a closed interval of 0 to 1. The closer to 1, the higher the youthfulness. The youth level measured by the YI can be understood as the youthfulness of the mentality, that is, the psychological youth level embodied by specific behaviors (e.g. online shopping);

Item(commodity) Youth Index. Refers to the degree that item whose purchaser's age is close to 18 to 35;

Urban(regional) Youth Index. Mainly refers to the degree of the youngness of the ideology embodied by the groups in the city (region) by analyzing the purchasing behavior of the population in the city.

3.2 Product Category

We classify commodity categories into three levels based on the granularity of division:

1. **Aggregation classification.** A union set of different first-level classification of the same key value. It is combined in the process of constructing the Youth Index Dictionary and is also the category of goods returned by the search operation.
2. **First-level classification.** A collection of classification information directly obtained from the Taobao Express Car Keyword Dictionary, totally 113 species.
3. **Second-level classification.** A collection of all "/" separated subcategories in the first-level classification, which is obtained after string segmentation and data integration of the first-level classification and counts 267 species in total.

3.3 Sociological Foundations

According to the definition of the YI, we conducted a survey on online shopping behaviors of consumers of different ages. The survey found that:

1. Online shopping crowds are mainly concentrated in 18–35 years old
   According to the report [9], among online shopping users in 2016, consumers aged 17–36 accounted for 66.6%. We believe that with the development of mobile Internet and smart devices, people except the young are exposed to online shopping, but neither the freedom, the frequency, nor the reliability of online shopping is less than young people. In view of this, we use the online shopping category preferences in the report as the basis for assessing the ranking of young indices of commodities.
2. Young consumers online shopping preferences
   According to the report [10], between 2013 and 2015, the number of single-use online shopping products continued to expand in the price range, from the external supplies (clothes, shoes, and hats) to the internal supplies (food and health products). In general, the single-user online shopping products category is more and more complete. The report analyzes the purchase behavior of online shopping users. Observing Fig. 1. Comparing the distribution of mobile phone-side online shopping categories with the distribution of overall online shopping categories, it is found that the categories like "food, health products" and "book audio-visual products" are different. We believe that in the online shopping group, users who use mobile phones to shop show greater youthfulness in the consumption behavior pattern, and the purchase preferences of mobile phone users are more valuable for this research. Therefore, we selected the distribution of merchandise categories purchased by mobile online shopping users in 2015 as the basis for ranking the young indices of commodities (categories) in this study.

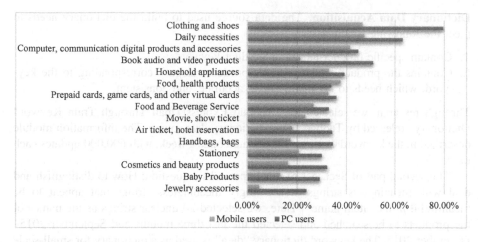

Fig. 1. Distribution of goods purchased by online shoppers in 2015 [9]

4   Youth Index Calculation

4.1   Model Overview

According to the survey results, the relationship between the commodity classification and the YI is established to determine the youth index level of different commodity categories, and the YI of the commodity category can be obtained by processing the grade using a standardized method. Then calculate YI of the commodity by youth index dictionary (YID). After that, a road network-based urban area division method is implemented to map the express data to the corresponding area, finally, analyze the concentration trend of the goods youth index in the area as the YI of the area.

4.2   Youth Index Dictionary

Structural Design. The YID aims to obtain a result set by querying the name of the item and obtain the category of the item and the YI of the item from the result set. Based on this, the dictionary contains three fields, wherein the key is used as a dictionary index, and the other two fields (category and index) are merged into one data tuple, and the corresponding search is performed in the form of a tuple. Structure of YID is shown in Table 1.

Table 1. Structure of youth index dictionary

Field	Description
Key	Key is used to match the input product name string
Category	Category The category to which the item belongs, a string separated by "/"
Index	The youth index of the commodity, which is a fraction of 9 digits after the decimal point in the interval [0, 1]

Dictionary Data Acquisition. The data source used to build the dictionary needs to meet the following conditions:

1. Contain specific product names or product keywords;
2. Contains the product category or the product category corresponding to the keyword, which needs to be authoritative, accurate, and consistent.

Through research, we selected the "Taobao Express/Tmall Through Train Keyword Dictionary" released by Taobao as the dictionary data source. The information module described in the keyword dictionary is updated once a week, with 600,000 updates each time.

The second part of Sect. 5.1 of this paper asks a question: How to distinguish and deal with meaningless strings or seemingly meaningless strings that appear to be meaningful in the "item name"? Here, we selected 20 unclear strings as the names of the products to be searched and selected the last three months and September 2015–December 2015. The keyword dictionary "data" is used as a dictionary for small-scale search experiments. The experimental results are shown in Table 2. We found that dictionary 2 has a better matching performance for ambiguous strings than dictionary 1. The reasons are analyzed as follows:

Table 2. Comparison of small-scale search test results

Dictionary number	A	B
Data time interval	2018.02–2018.05	2015.09–2015.12
Matching degree = 1	6	10
Matching degree ≥ 0.8	12	17

Strings with unknown meanings, if they have practical meaning, mostly refer to the "model" of a certain brand of goods. In this way, we select the keyword dictionary that is close to the time of the express data as the data source of the young index dictionary. The Specific build process is shown in Sect. 5.2.

Determination of the Youth Index of Goods (commodities). Based on Fig. 1, combined with the life experience, the secondary classifications are divided into 15 commodity categories here, and YI ratings were performed on 15 categories. In this paper, we refer to paper [11] to establish a 0–1 transformation method to standardization the data as YI corresponding to the second-level classification.

$$YI^{ac} = \frac{\sum_{i=1}^{n}\left\{YI_i^{sc} * freq(SC_i) | SC_i \in \bigcup_{j=1}^{m}\{split(FC_j)\}\right\}}{\sum_{i=1}^{n}\left\{freq(SC_i) | SC_i \in \bigcup_{j=1}^{m}\{split(FC_j)\}\right\}} \tag{1}$$

Next, the YI of the aggregated classification is calculated based on the secondary classification YI by formula (1). YI^{ac} means YI of aggregate classification. YI_i^{sc} means YI of second-level classification i. SC_i is second-level classification i. $freq(SC_i)$ is the number of repetitions of SC_i when the calculation has a union of the first-level classification and segmentation results. FC_j is first-level classification j.

At this point, the construction of a YID containing three fields has been completed.

4.3 Calculation of Regional Youth Index

The YI for urban areas is determined by all courier data in the region, and its value should have the effect of characterizing the centralized trend of young indices for all courier items in the region. A centralized trend measure is to find a representative or central value of the data level [12]. Three important indicators have the following meanings in this study:

(1) **Mean.** The average of the young indices of all points in the area;
(2) **Mode.** The YI of all points in the region is mapped to the [0, 1] interval with 0.02 as the subinterval of the group distance, and then calculated by Jin's interpolation method using the subinterval with the largest number of data;
(3) **Median.** The median YI of all points in the region.

Since the skewness state of the data distribution in each region is unknown, we use the above three indicators to evaluate the regional YI and compare the results obtained to select the excellent indicators. In the statement of experimental results in Sect. 5.3, in order to avoid tedious narration, we use the symbol YI^r_{mean} to represent regional YI with Mean quantification, YI^r_{mode} to represent regional YI with Mode quantification, and YI^r_{median} to represent regional YI with Median quantification.

5 Experimental Results

5.1 Data Preprocessing

Data Description. The express data is obtained from Xi'an Branch of SF Express Co., Ltd. The data set covers a total of 14,345,165 data from January 2015 to May 2016 with the receiving address or the sending address in the jurisdiction of Xi'an City, Shaanxi Province. Each piece of data contains 30 fields covering order information, recipient and sender information, and item information. In order to protect the privacy of users, the ID number, postal code, delivery time, receipt time, item type and volume of the sender and receiver in the data are replaced by "0".

Data Pre-processing. Literature [13] puts forward ten kinds of data quality problems that need to be solved in the data cleaning process. On this basis, we have developed cleaning rules for this research data as follows:

Field Splitting. Extracts the composition structure from a free-form string, and separates and processes each part.

Verification and Filtering. According to the external lookup table to verify the correctness of the field values, the data of the error or non-compliant data is filtered.

Standardization of Data. Standardization data is mainly to solve the problem of inconsistent type, dimension inconsistency and normalization of indicators, such as telephone number, address information, time and so on.

It should be emphasized that in the process of data pre-processing, the item name is a free string with no structure and no specification, and its processing is the key point and a difficulty. The types of pending fields of this field are shown in Table 3. The method of external dictionary-assisted filtering is used for partially meaningless strings, for "meaningless strings" and "strings that seem meaningless and meaningful". We proposed two approaches to address the problem:

Table 3. Types of items to be processed

Type to be treated	Example
Meaningful strings that do not have (or have, but are not obvious) the young exponential properties defined in this study	Goods, mailing, consignment, returned, file
Meaningless string	FSHKLIVE0394, null
A seemingly meaningless, actually meaningful string	M530AB37.5, 28-TC

1. Expand the size of the YID, rational selection of the dictionary data source to optimize the processing, as much as possible to improve the hit rate of the query.
2. Optimize the data source for building the dictionary by building an auxiliary dictionary. We count the frequency of the item names in the original item list, and select names with higher frequency and no young index attribute (like Table 3) such as the 'attached item', 'item', 'file', 'gift', 'null', 'unrecognized', etc. At the same time, we merge the recipient's mobile phone number and the fixed telephone into the phone number field and format it, and we merge the address info into the recipient address field then convert it to latitude and longitude info using API provided by Baidu map.

5.2 Construction of the Young Index Dictionary

We selected the keyword list that Taobao released on January 28, 2015, May 20, 2015, September 23, 2015, January 20, 2016, and May 18, 2016, as dictionary source data, totally count 1,122,738.

According to Fig. 1, the youth index level of the commodity category is divided as Table 4. 267 second-level classifications are divided into fifteen commodity categories. Then normalized the youth index level as the YI of the second-level class. After that, according to the formula (1), the YI of the aggregated category in each item in the dictionary is calculated, that is, the YI of each group of keywords in YID. At this point, the complete YID has been constructed.

Table 4. The youth index level of commodity category

Commodity category	Level	Commodity category	Level
Clothing and shoes	15	Books and audiovisual products	12
Daily necessities	14	Stationery	6
Computer, communication digital products, and accessories	13	Air ticket, hotel reservation	5
Food, health products	10	Handbags, bags	4
Household appliances	9	Cosmetics and beauty products	3
Food and Beverage Service	8	Baby Products	2
Movie, show ticket	7	Jewelry accessories	1
Prepaid cards, game cards, and other virtual cards	11		

5.3   Regional Youth Visualization

By the urban functional area division based on road network information method proposed by literature [14], we implement the regional division of Xi'an. Since this is not our main contribution, execution steps are not given here. According to the results of regional division, we mapped all express data to each region to observe the overall distribution, as shown in Fig. 2.

After calculating, regional YI is visualized with mean quantification, mode quantification, and median quantification, which are defined as YI^r_{mean}, YI^r_{mean}, and YI^r_{mean} in Sect. 4.3. It was observed that YI^r_{mean} of most areas is close to the 0.5–0.6 from Fig. 3. For better observation and comparison, we adjusted the gradation interval shown as Fig. 4(a) and (b). From these two figures, we selected several universities as observation points where young people congregate in large numbers and found that YI in these regions are significantly higher than the surrounding area.

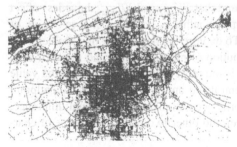 

Fig. 2. Scatter plot of express delivery

Fig. 3. Distribution gradation diagram of YI^r_{mean} in the interval [0, 1]

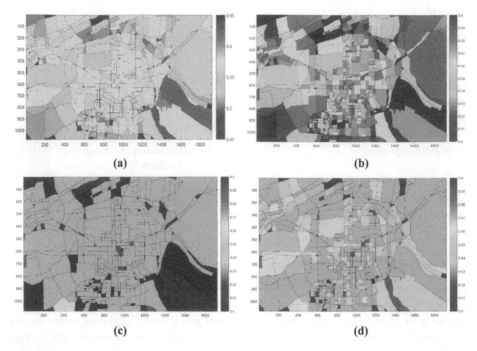

Fig. 4. Gradation diagram of (a) YI^r_{mean} in the interval [0.45, 0.65] (b) YI^r_{mean} in the interval [0.5, 0.6] (c) YI^r_{mode} in the interval [0.5, 0.6] (d) YI^r_{median} in the interval [0.5, 0.6] (Color figure online)

On the other hand, in areas with dense express data, such as the south of Erhuan North Road, Yanta District, and South University City, the results are more in line with the actual situation. Observe the blue part of the map, it's known that residential area (e.g. Qujiang Tiandiyuan residential area), working community, and some urban villages (e.g. Jixiang Village) all show a low YI, which is consistent with life experience.

In addition, for YI^r_{mode} and YI^r_{median}, we analyze the two methods in the same way as YI^r_{mean}, as shown in Fig. 4(c) and (d). It's found that YI^r_{mode} and YI^r_{median} have less effect on the regional YI than YI^r_{mean} in this scheme, whether in accuracy or level. None of them are excellent, and YI^r_{mode} is the worst. To explore the reasons for this problem, we conducted a statistical analysis of the distribution of mean, median, and mode of each region. Seeing Fig. 5, distribution of YI^r_{mean} is the most uniform and the extreme value is small, the distribution of YI^r_{mode} is too centralized and greatly affected by endpoint value, leading to poor characterization performance and insignificant visual difference, and distribution feature of YI^r_{median} is between the other two, thus it has general characterization performance.

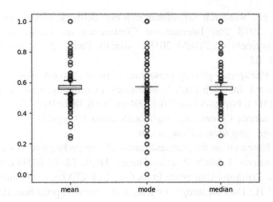

Fig. 5. Distribution box diagram of YI^r_{mean}, YI^r_{mode}, YI^r_{median}

6 Conclusion and Future Work

In this paper, a structured fine-grained assessment model was proposed based on express delivery data to characterize urban youth. A Chinese provincial capital is selected as an example to implement the model, and the optimal evaluation of the regional YI was selected by visual analysis, which shows excellent characterizing performance. Our future work will consider optimizing the model from two directions. One is to conduct more in-depth sociological research and improve the dictionary construction strategy with a larger scale of the data source to improve the performance of the existing model. The other is to adjust the overall framework of the model and apply machine learning algorithm such as the naive Bayesian classification method to do the classification of express items to achieve better performances.

References

1. Economics, Oxford: The impact of the express delivery industry on the global economy. J. September, Oxford (UK) (2009)
2. Export. Gov. https://www.export.gov/article?id=France-Express-Delivery
3. Li, Q., Yu, Z., Guo, B., Lu, X.: Inferring housing demand based on express delivery data. In: 2018 IEEE International Conference on Big Data (Big Data). IEEE (2018). https://doi.org/10.1109/bigdata.2018.8621904
4. Ye, B., Zuo, J., Zhao, X., Luo, L.: Research on the express delivery delay prediction based on neural network in the background of big data. In: 6th International Conference on Electronic, Mechanical, Information and Management Society. Atlantis Press (2016). https://doi.org/10.2991/emim-16.2016.294
5. Duan, H., Song, G., Qu, S., Dong, X., Xu, M.: Post-consumer packaging waste from express delivery in China. Resour. Conserv. Recycl. **144**, 137–143 (2019). https://doi.org/10.1016/j.resconrec.2019.01.037
6. Fan, W., Xu, M., Dong, X., Wei, H.: Considerable environmental impact of the rapid development of China's express delivery industry. Resour. Conserv. Recycl. **126**, 174–176 (2017). https://doi.org/10.1016/j.resconrec.2017.07.041

7. Ding, Y., Xu, H.: Research on shared express delivery mode based on block chain technology. In: 2018 2nd International Conference on Economic Development and Education Management (ICEDEM 2018). Atlantis Press (2018). https://doi.org/10.2991/icedem-18.2018.103

8. Li, T., Rui, Y.: Priexpress: privacy-preserving express delivery with fine-grained attribute-based access control. In: 2016 IEEE Conference on Communications and Network Security (CNS). IEEE (2016). https://doi.org/10.1109/cns.2016.7860501

9. E-Commerce Research Center. http://b2b.toocle.com/zt/16zgxfz/

10. CNNIC homepage. http://www.cnnic.net.cn

11. Li, G., Wu, Q.: Research on the standardization of comprehensive evaluation data based on consistent conclusions. J. Math. Practice Theory **41**(3), 72–77 (2011). (in Chinese)

12. Jia, J.: Statistics. Tsinghua University Press Co., Ltd. (2006). (in Chinese)

13. Rahm, E., Do, H.H.: Data cleaning: problems and current approaches. IEEE Data Eng. Bull. **23**(4), 3–13 (2000)

14. Yuan, N.J., Zheng, Y., Xie, X.: Segmentation of urban areas using road networks. Technical report, MSR-TR-2012–65 (2012)

A Local Dynamic Community Detection Algorithm Based on Node Contribution

Kun Guo[1,2,3], Ling He[1,2,3], Jiangsheng Huang[4],
Yuzhong Chen[1,2,3(✉)], and Bing Lin[5]

[1] College of Mathematics and Computer Sciences, Fuzhou University,
Fuzhou 350116, China
guknl23@163.com, heling_fzu@163.com, yzchen@fzu.edu.cn
[2] Fujian Provincial Key Laboratory of Network Computing and Intelligent
Information Processing, Fuzhou, China
[3] Key Laboratory of Spatial Data Mining and Information Sharing,
Ministry of Education, Fuzhou 350116, China
[4] Power Science and Technology Corporation State Grid Information
and Telecommunication Group, Fuzhou 351008, China
huangjiangsheng@sgitg.sgcc.com.cn
[5] College of Physics and Energy, Fujian Provincial Key Laboratory of Quantum
Manipulation and New Energy Materials, Fujian Normal University,
Fuzhou 350117, China
wheellx@163.com

Abstract. The existence of communities in various complex networks is ubiquitous in all aspects of people's living. Hence, it is crucial to uncover communities accurately, which is one of the hottest research areas in the field of network analysis. Particularly, complex networks are usually in continuous change so that it is more realistic to uncover dynamic communities. In this study, an algorithm based on node contribution for uncovering dynamic communities is proposed. Firstly, the seed nodes are selected via node local fitness in the network, thus guaranteeing that the selected seeds are central nodes of communities. Secondly, a static algorithm is used to obtain communities in initial snapshot of the network. Finally, node contribution is proposed to incrementally uncover communities in non-initial snapshots of the network. The experimental results reveal that our method outperforms all other comparison algorithms in both artificial and real datasets.

Keywords: Complex network · Dynamic community detection · Node local fitness · Node contribution

1 Introduction

In reality, people usually use complex networks to model complex systems. In complex networks, each node represents an individual and interactions between individuals are presented by edges. The existence of community structure is a basic feature of complex networks. Individuals within one community are closely connected, while the connections among communities are relatively sparse [1]. A number of methods have been

© Springer Nature Singapore Pte Ltd. 2019
Y. Sun et al. (Eds.): ChineseCSCW 2019, CCIS 1042, pp. 363–376, 2019.
https://doi.org/10.1007/978-981-15-1377-0_28

proposed for uncovering communities over the past few decades [2], but few of them are appropriate for dynamic networks.

Complex systems in reality are constantly changing, so uncovering communities in dynamic network has attracted increasing attention of researchers. Scholars divided community detection algorithms in dynamic networks into three categories, namely, traditional clustering [3], evolutionary clustering [4] and incremental clustering [5]. In the methods based on traditional clustering, each snapshot of dynamic network applies a static community detection algorithm to uncover communities which cannot take account the structural history. In the methods based on evolutionary clustering, communities at a particular snapshot should be similar to that at the previous snapshot of the dynamic network, which can accurately reflect the changing network structure during that time. In the methods based on incremental clustering, it incrementally updates communities via the structural history and the changing part of the network only, thus greatly reduces the computational cost.

Existing local dynamic community detection algorithms based on incremental strategy may have shortcomings in seed selection or incremental expansion. For this reason, we propose a Dynamic Community Detection algorithm based on Node Contribution (DyCDNC). Our main contributions are given in the following: (1) the seeds are selected in the network via node local fitness, thus guaranteeing that the selected seeds are central nodes of communities; (2) the concept of node contribution is proposed to measure nodes' contribution to the formation of communities. The node contribution and Jaccard coefficient are used to incrementally expand communities, which greatly improved the accuracy of community detection.

Following shows the organization of the rest of this paper: Sect. 2 represents the related work. The description of DyCDNC algorithm is given in detail in Sect. 3. We discuss the experimental results and give the conclusion in Sects. 4 and 5 respectively.

2 Related Work

Methods based on local expansion perform well in discovering communities in complex networks. Compared to methods based on global extension, they do not require the topology of entire network and are easy to be parallelized. In the following, related work about incremental clustering in uncovering dynamic communities is given.

Zakrzewska et al. [6] proposed a seed-expansion-based algorithm in 2015, which incrementally updates communities as the underlying graph changes. This algorithm uses a classic local expansion algorithm to obtain communities in initial snapshot of the dynamic network. The algorithm uses a sequence to store the order in which nodes were added and calculates the fitness of each node. When incrementally processing nodes in non-snapshots, it keeps the sequence sorted via nodes' fitness in non-ascending order and removes the node whose fitness is negative. The algorithm removes nodes behind the changed node from the sequence and uses the static algorithm to recover communities among the removed nodes. The algorithm can obtain high quality communities, but with a low time efficiency. In our method, node contribution is proposed to incrementally precess nodes behind the changed node in the sequence which can improve the algorithm's efficiency.

Hu et al. [7] proposed the LDM-CET algorithm which can not only uncover communities, but also track their evolution in 2016. LDM-CET uses the approximate personalized PageRank method to obtain communities at the initial snapshot. At each non-initial snapshot, LDM-CET only explores local views of the changed nodes by approximate personalized PageRank. Moreover, the evolution of communities is tracked via a partial evolution graph. The algorithm can effectively track communities' evolution when the networks do not change dramatically. However, when communities' boundaries are fuzzy, it is hard for LDM-CET to obtain high quality communities.

In 2017, DiTursi et al. [8] proposed PHASR algorithm which is a filter-and-verify framework used temporal conductance to measure communities of stable memberships. PHASR can be cast as detecting the subgraph and interval of the smallest temporal conductance. The algorithm performs well in dynamic networks in which the nodes interact exclusively.

In 2018, an algorithm using the concept of leadership for uncovering dynamic communities was proposed in [9]. The algorithm considered that defining and selecting the leaders in the network had a significant effect on detecting desired clusters. It is effective in uncovering communities in both real and artificial dynamic network. However, when the size of dynamic networks increases, its time efficiency drops dramatically.

3   The Proposed Algorithm

3.1   Basic Concepts

We use a sequence of snapshots to present dynamic networks, such as $\{G_0, G_1, \ldots, G_T\}$ where $G_t = (V_t, E_t)$ presents the snapshot of the dynamic network at time t. Before showing the details of our method, the definitions and basic concepts are given as follows.

Let m_i presents the i_{th} node added to one community. If m_i is the node v, we say $p(v) = i$. This set of sequence is presented by $\Omega = \{m_i, com_i | 0 \leq i \leq T\}$ shown in Table 1.

Table 1.   The community detection sequence Ω.

Position	0	1	$\cdots$	n
Community nodes	m_0	m_1	$\cdots$	m_n
Node contribution	con_0	con_1	$\cdots$	con_n

Definition 1 (Node Neighbors). The neighbors of node v refer to the nodes those directly connect with it in G. Its specific formula is given as follows:

$$N(v) = \{u | u \in V, (u, v) \in E\} \tag{1}$$

It can be seen from the above definition that $N(v)$ does not include the node v itself. Thus, we define that $\Gamma(v) = N(v) \cup \{v\}$, which is used to represent the neighbors of node v containing itself.

Definition 2 (Community Neighbors). The community neighbors $Ns(c)$ represents the nodes those connect with community c.

$$Ns(c) = \bigcup_{v \in c} N(v) - c \tag{2}$$

Definition 3 (Jaccard Coefficient). The similarity of two is measured by Jaccard coefficient. The larger value of it means that the two nodes are more similar. It is defined as:

$$J(u, v) = \frac{|N(u) \cap N(v)|}{|N(u) \cup N(v)|} \tag{3}$$

Definition 4 (Fitness Function). It is used to measure the tightness of a group nodes. The larger value of it means more closer the connection among the group nodes those are more likely to form a community. Its specific definition is as follows:

$$F_c = \frac{F_{in}^c}{\left(F_{in}^c + F_{out}^c\right)^{\alpha}} \tag{4}$$

Where, F_{in}^c and F_{out}^c represent the internal and external degree of community c, respectively. The scale of communities detected is controlled by parameter α.

Definition 5 (Node Fitness to Community). It is used to decide if one node can be added to the community. It is defined as follows:

$$F_c^v = F_{c \cup \{v\}} - F_c \tag{5}$$

If the value of F_c^v is positive, then the node v can be added to the community c; vice versa.

Definition 6 (Node Local Fitness). The node local fitness is defined as the fitness of its neighbors containing itself minus the fitness of its neighbors. Its specific calculation formula is as follows:

$$F_v = F_{\Gamma(v)} - F_{N(v)} \tag{6}$$

Where, $F_{\Gamma(v)}$ and $F_{N(v)}$ are the fitness of $\Gamma(v)$ and $N(v)$, respectively. The larger value of it means it is more suitable to be the seed node for extending community.

Definition 7 (Node Contribution). It is used to measure the nodes' contribution to the formation of communities. Intuitively, it should be proportional to the node's fitness to the community, but inversely proportional to the distance between the node and the seed node in the community. Its specific definition is as follows:

$$Con_c^v = \frac{F_c^v}{d\left(v, v_c^{seed}\right)} \tag{7}$$

Where, v_c^{sees} represents the seed node in community c and $d\left(v, v_c^{seed}\right)$ represents the shortest path length between node v and the seed node. For the seed node, we consider its local fitness as its contribution to the community.

3.2 Algorithm Details

Static Community Detection. Given a dynamic network, we use a static algorithm to obtain communities at the initial snapshot. In the paper, we use LFM algorithm [10] as the static algorithm but with slight modification. We consider the node with the largest value of local fitness among nodes that are not expanded in the network as the seed node. The procedure of the static algorithm is summarized in Algorithm 1.

Algorithm 1. Static Community Detection

Input: The initial snapshot $G_0(V_0, E_0)$, the parameter α.

Output: the initial community set *communities*, Ω.

1.   Initialize set *communities* and the community sequence Ω ;
2.   **For** each $v \in V_0$ **do**
3.   Calculate F_v;
4.   **End For**
5. $V_{unextended} = V_0$;
6.   Sort $V_{unextended}$ via F_v in non-ascending order;
7.   *communityID* $= 1, i = 0$;   //i represents the position in Ω.
8.   **While** $V_{unextended} = \emptyset$ **do**
9.   Initialize the current community Cs and $Ns(Cs)$;
10. $v_{seed} = v_{max}$;   //v_{max} is the node with the maximum local fitness in $V_{unextended}$.
11. $Cs = Cs \cup \{v_{seed}\}$;
12. $\Omega[communityID][v_{seed}] = F_{v_{seed}}$;
13.   Update Ns;
14.   **While** $Ns \neq \emptyset$ **do**
15.   **For** each v in Ns **do**
16.   Calculate F_{Cs}^v;
17.   **End For**
18.   **If** $f_{v_{max}} > 0$ **then**
19. $Cs = Cs \cup \{v_{max}\}$;
20.   Update Cs;
21. $i \mathrel{+}= 1$;
22.   **else**
23.   break;
24.   **End If**
25.   Update $Ns(Cs)$;
26.   **End While**
27.   *communities* $=$ *communities* $\cup Cs$;
28. $V_{unextended} = V_{unextended} - Cs$;
29.   *communityID* $\mathrel{+}= 1$;
30. **End While**

Dynamic Community Detection. For the non-initial snapshots of the network, $G_t(0 < t \leq T)$ we update communities in an incremental way. There are four events in the dynamic network: (1) edge addition; (2) edge disappearance; (3) node addition; (4) node disappearance. Obviously, addition or disappearance of nodes usually causes the addition or disappearance of edges; otherwise, the changed node is temporarily placed in $V_{unextended}$ or $V_{disappearance}$. We mainly deal with the addition and disappearance of edges in the network.

(1) Edge addition
Obviously, we first determine whether both endpoints of the added edge are newly appearing nodes. If so, we just put the two nodes into $V_{unextended}$. If one endpoint is a newly appearing node, we put the new node into $V_{unextended}$. When the two endpoints of the edge are not newly appearing nodes and are in the same community, the addition of the edge will obviously make the its structure more compact. We do nothing on the two endpoints at this time; otherwise, the two endpoints are in different communities, which will definitely lead to a reduction in the internal compactness of relevant communities as shown in Fig. 1. The processing of endpoints is summarized in Algorithm 2 at the same time.

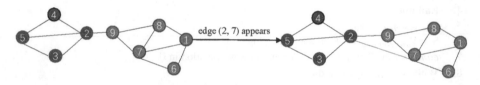

Fig. 1. A new edge appears between two different communities.

Algorithm 2 PrecessingOfEndpoints

Input: node v, communities, Ω, $V_{unextended}$.
Output: updated communities set communities and Ω.
1.   Calculate F_c^v and Δ; // community c is which v belong to.
2.   Update Ω.
3.   **For** $i = p(v) + 1$ to T **do**
4. $con_i = con_i - J(m_i, v) \times \Delta$;
5.   **End For**
6.   **For** $i = p(v)$ to T **do**
7.   **if** $con_i \leq 0$ **then**
8.   Remove node v from community c;
9.   Update communities and Ω.
10. $V_{unextended} = V_{unextended} - \{v\}$;
11.   **else**
12.   Continue;
13.   **End if**
14. **End For**

In algorithm 2, we suppose one of the endpoints is node v which belongs to community c. First, v's contribution to community c is calculated by Eq. 7. Obviously, its value is smaller than that at the previous snapshot. Second, the reduction of node v's contribution is calculated and record as Δ. Next, DyCDNC traverses the subsequent nodes in Ω and subtracts the value of $\Delta * J(v, u)$ (supposing node u added to community c after node v) from the contribution of each node at the previous time step, which means that the node u that is more similar with v, the grater the reduction in its contribution. Finally, for the nodes whose position is behind node v (including itself), if one node whose contribution to community c is negative and then it will be removed from the community c. We add the nodes those no longer belong to any community to $V_{unextended}$. The community will automatically disappear when all nodes are moved. In total, the processing of edge addition is summarized in Algorithm 3.

Algorithm 3 EdgeAddition

Input: $G_t(V_t, E_t), G_{t-1}(V_{t-1}, E_{t-1}), \Omega, communities, V_{unextended}$.
Output: Updated community set $communities$ and Ω.

1.   **For each** $edge \in E_t$ **do**
2.   **if** $edge \notin E_{t-1}$ **then**
3.   **if** $u \notin V_{t-1}$ and $v \notin V_{t-1}$ **then**
4. $V_{unextended} = V_{unextended} \cup \{u, v\}$; // u and v are the edge's endpoints.
5.   **else**
6.   **if** $u \notin V_{t-1}$ and $v \in V_{t-1}$ **then**
7. $V_{unextended} = V_{unextended} \cup \{u\}$;
8.   **else**
9.   **if** $u \in V_{t-1}$ and $v \notin V_{t-1}$ **then**
10. $V_{unextended} = V_{unextended} \cup \{v\}$;
11.   **else**
12.   **if** u and v are in the same community **then**
13. $communities, \Omega = $ ProcessingOfEndpoints($u, communities, \Omega, V_{unextended}$);
14. $communities, \Omega = $ ProcessingOfEndpoints($v, communities, \Omega, V_{unextended}$);
15.   **else**
16.   continue;
17.   **End if**
18.   **End if**
19.   **End if**
20.   **End if**
21.   **End if**
22.   **End For**

---

(2) Edge disappearance.

The way to process the edge disappearance in the network is similar to edge addition. When at least one of the two endpoints of the edge disappear from the network,

DyCDNC uses Algorithm 2 to process both of them. When two endpoints of the edge are both in the current network, if they belong to different communities, no processing for them is performed. While they are in the same community as shown in Fig. 2. They will be processed by Algorithm 2 at this time. The processing of edge disappearance can be seen in Algorithm 4.

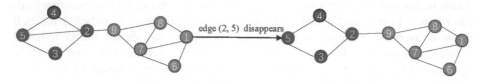

Fig. 2. An edge disappears within the community.

Algorithm 4 EdgeDisappearance
___

Input: $G_t(V_t, E_t), G_{t-1}(V_{t-1}, E_{t-1}), \Omega, communities, V_{unextended}$.
Output: Updated community set *communities* and Ω.
1.   **For each** *edge* ϵ E_{t-1} **do**
2.   **if** *edge* $\notin E_t$ **then**
3.   **if** $u \notin V_t$ and $v \notin V_t$ **then**
4.   *communities*, Ω = ProcessingOfEndpoints(*u, communities, Ω,V$_{unextended}$*);
5.   *communities*, Ω = ProcessingOfEndpoints(*v, communities, Ω,V$_{unextended}$*);
6.   **else**
7.   **if** u and v are in the same community **then**
8.   *communities*, Ω = ProcessingOfEndpoints(*u, communities, Ω,V$_{unextended}$*);
9.   *communities*, Ω = ProcessingOfEndpoints(*v, communities, Ω,V$_{unextended}$*);
10.   **else**
11.   **End if**
12.   **End if**
13.   **End if**
14. **End For**

After we use Algorithm 3 and 4 process the edge added and disappeared, respectively. There may be some nodes those do not belong to any community. For each node in $V_{unextended}$, we calculate the node fitness to every community in *communities* and add the node with the maximum value of node fitness to corresponding community.

In conclusion, the procedures of the DyCDNC algorithm are presented in Algorithm 5.

Algorithm 5 DyCDNC

Input: $G_t(V_t, E_t), G_{t-1}(V_{t-1}, E_{t-1}), \alpha$.
Output: Community set *communities* and Ω.
1. **if** $t == 0$ **do**
2.   Static Community Detection (G_0, α);
3. **else**
4. $V_{unextended} = \emptyset$;
5.   *communities*, Ω = EdgeAddition $(G_t, G_{t-1}, \Omega$ *communities*);
6.   *communities*, Ω = EdgeDisappearance $(G_t, G_{t-1}, \Omega$ *communities*);
7.   **if** $V_{unextended} = \phi$ **then**
8.   continue;
9.   **else**
10.   Static Community Detection $(V_{unextended}, \alpha)$;
11.   Update *communities* and Ω;
12.   **End if**
13. **End if**

3.3 Algorithm Details

For the static algorithm, when communities detected are small, it is almost linear. However, its worst-case complexity is O(n^2).

For DyCDNC algorithm, the time complexity of processing edges addition and disappearance is O($(m_a + m_d) \times k$), where m_a and m_d represent the number of added and disappeared edges, respectively, k presents the average size of communities detected. For processing the nodes unexpanded, its time complexity is O($n_u \times d \times q$), where, n_u represents the number of unexpanded nodes, while d and q present the nodes' average degree and the number of communities detected, respectively. In summary, DyCDNC's time complexity is O($(m_a + m_d) \times k + n_u \times d \times q$) which can be reduced to O(m_c) that represents the number of edges changed.

4 Experimental Results

4.1 The Description of Datasets

(1) Real datasets
Two real datasets are used in the experiments. They are *Enron* dataset [11] and *AS-733* dataset [12].

(2) Artificial datasets
We use the *Dynamic Benchmark Network Generator* [13] to generate dynamic artificial datasets. Four groups of datasets are generated and their parameters are set as follow:

D1: $N = \{1000, 5000, 10000\}$, $muw = 0.2$, $p = 0.2$, $t = 10$;
D2: $N = 1000$, $muw = \{0.1, 0.3, 0.5\}$, $p = 0.2$, $t = 10$;
D3: $N = 1000$, $muw = 0.2$, $p = \{0.1, 0.3, 0.5\}$, $t = 10$;
D4: $N = 1000$, $muw = 0.2$, $p = 0.2$, $t = \{5, 10, 15\}$;

The rest parameters are set by default: $k = 10$, $maxk = 20$, $minc = 20$, $maxc = 100$, $on = 100$, $om = 2$.

4.2   Experimental Schemes and Evaluation Metrics

In the experiments, DyCDNC is compared with three classic approaches for dynamic community detection, namely, AFOCS [14], GreMod [15] and QCA [16]. We mainly compare the accuracy of all algorithms on each dataset.

We chose two widely used evaluation metrics to verify our method: the extension of modularity (EQ) [17] and the Normalized Mutual Information (NMI) [18]. The EQ is defined as follows:

$$EQ = \frac{1}{2m} \sum_i \sum_{v \in c_i, w \in c_i} \frac{1}{O_v O_w} \left(A_{vw} - \frac{k_v k_w}{2m} \right) \delta(c_v, c_w) \qquad (8)$$

Where, A is the adjacency matrix, O_v is the number of communities those node v belongs to and k_v is node v's degree. If node v and node w are connected, then the value of $\delta(v, w)$ is 1; otherwise, it is 0. The larger value of EQ means the better quality of community detection.

NMI uses information entropy to estimate the difference between communities detected and the ground-truth. The larger value of it also means the better quality of community detection. Its specific calculation formula is given as follows:

$$NMI = \frac{-2 \sum_{i=1}^{C_A} \sum_{j=1}^{C_B} C_{ij} log \frac{C_{ij} N}{C_i C_j}}{\sum_{i=1}^{C_A} C_i log \frac{C_i}{N} + \sum_{j=1}^{C_B} C_{ij} log \frac{C_j}{N}} \qquad (9)$$

Where, C_A (C_B) is the number of communities detected and ground-truth, respectively. C_i (C_j) is the number of elements of community C in row i (column j) and N represents the number of nodes in the network.

4.3   Experiments on Real Datasets

Figure 3a and b show the EQ results on Enron and AS-377 dataset, respectively. The experimental results indicate that the value of EQ of DyCDNC algorithm is only second to QCA algorithm. QCA is a module-based optimization algorithm. When incrementally processing nodes in the network, its goal is to maximize the overall modularity. GreMod uses the Louvain algorithm for community detection at the initial time step, so its value of modularity is sightly larger than that of DyCDNC. However, the value of modularity of GreMod and AFOCS will continue to decline over time, while DyCDNC and QCA almost remain unchanged.

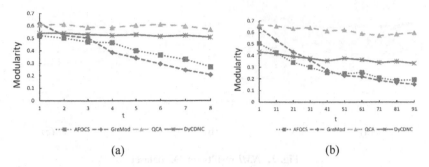

Fig. 3. *EQ* results on real datasets

4.4 Experiments on Artificial Datasets

Figure 4 shows the NMI results on the *D1* artificial dataset. We can find that the number of nodes in the dynamic network can not affect the algorithms' performance. DyCDNC algorithm has the highest accuracy on NMI evaluation metric. Because DyCDNC uses the node local fitness to select high-quality central nodes in the network and expand communities around them. In addition, as the time steps of the dynamic network goes by, the accuracy of AFOCS and GreMod algorithm will decline, while QCA and DyCDNC algorithm can keep it steady.

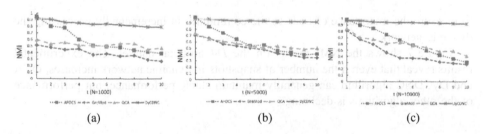

Fig. 4. NMI results on D1 dataset

Figure 5 shows the *NMI* results on *D2* dataset. The experimental results reveal that if the value of *muw* increases, the accuracy of all algorithms decrease. Overall, DyCDNC algorithm performs best. Because DyCDNC algorithm is based on local expansion and suitable for overlapping community detection. However, Fig. 5(c) show that when the value of *muw* reaches 0.5, the performance of DyCDNC becomes to be unstable. At this time, the network structure is more complicated and the mix between communities is higher.

Figure 6 shows the NMI results on the D3 artificial dataset. We can find that when the value of p is less than 0.5, all algorithms' performance will show a slight decline. While the value of p reaches 0.5, all algorithms' performance will drop dramatically. Because there are large changes in the dynamic network at each time step. Overall, DyCDNC algorithm performs best at all values of p, because it uses node contribution

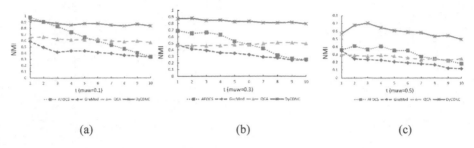

(a)   (b)   (c)

Fig. 5. *NMI* results on D2 dataset

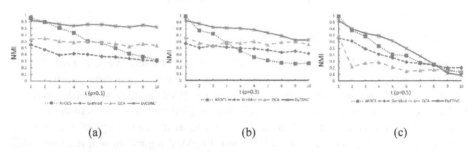

(a)   (b)   (c)

Fig. 6. *NMI* results on D3 dataset

to measure the importance of node to the community in incrementally processing the dynamic network.

Figure 7 shows the NMI results on the *D4* artificial dataset. The experimental results reveal that even if the number of snapshots in dynamic network increases, QCA and DyCDNC algorithm can generally maintain stable performance, but performance of the other algorithms is declining.

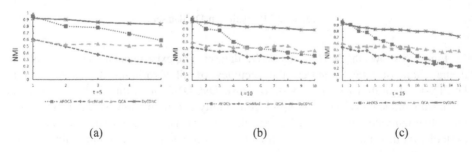

(a)   (b)   (c)

Fig. 7. *NMI* results on D4 dataset

4.5 Running Time Analysis

We can not compare the running time of all algorithms in this paper because the source codes of them were implemented with different languages. Therefore, all algorithms' theoretical time complexity is given in Table 2.

Table 2. All algorithms' theoretical time complexity.

AFOCS	GreMod	QCA	DyCDNC
$O(E)$	$O(E)$	$O(E^2)$	$O(E)$

5 Conclusions

In the paper, we put forward an algorithm for uncovering communities in dynamic networks. On one hand, we select seed nodes via node local fitness, thus guaranteeing that the selected seeds are central nodes of communities. On the other hand, node contribution is proposed to incrementally uncover communities in non-initial snapshots. From the experimental results we can know that our method can accurately uncover communities in both real and artificial networks.

Acknowledgments. This work is partly supported by the National Natural Science Foundation of China under Grant No. 61300104, No. 61300103 and No. 61672159, the Fujian Province High School Science Fund for Distinguished Young Scholars under Grant No. JA12016, the Fujian Natural Science Funds for Distinguished Young Scholar under Grant No. 2015J06014, the Fujian Industry-Academy Cooperation Project under Grant No. 2018H6010 and No. 2017H6008, and Haixi Government Big Data Application Cooperative Innovation Center.

References

1. Newman, M.E.J.: Detecting community structure in networks. Eur. Phys. J. B **38**(2), 321–330 (2004)
2. Fortunato, S., Hric, D.: Community detection in networks: a user guide. Phys. Rep. **659**, 1–44 (2016)
3. Mitra, B., Tabourier, L., Roth, C.: Intrinsically dynamic network communities. Comput. Netw. **56**(3), 1041–1053 (2012)
4. Chakrabarti, D., Kumar, R., Tomkins, A.: Evolutionary clustering. In: Proceedings of the 12th ACM SIGKDD International Conference on Knowledge Discovery and Data Mining, pp. 554–560. ACM, Philadelphia (2006)
5. Li, X., Wu, B., Guo, Q., et al.: Dynamic community detection algorithm based on incremental identification. In: 15th IEEE International Conference on Data Mining Workshop (ICDMW), pp. 900–907. IEEE, Atlantic City (2015)
6. Zakrzewska, A., Bader, D.A.: A dynamic algorithm for local community detection in graphs. In: Proceedings of the 2015 IEEE/ACM International Conference on Advances in Social Networks Analysis and Mining, pp. 559–564. ACM, Paris (2015)
7. Hu, Y., Yang, B., Lv, C.: A local dynamic method for tracking communities and their evolution in dynamic networks. Knowl.-Based Syst. **110**, 176–190 (2016)

8. DiTursi, D.J., Ghosh, G., Bogdanov, P.: Local community detection in dynamic networks. In: 2017 IEEE International Conference on Data Mining (ICDM), pp. 847–852. IEEE, New Orleans (2017)

9. Javadi, S.H.S., Gharani, P., Khadivi, S.: Detecting community structure in dynamic social networks using the concept of leadership. In: Amini, M.H., Boroojeni, K.G., Iyengar, S.S., Pardalos, P.M., Blaabjerg, F., Madni, A.M. (eds.) Sustainable Interdependent Networks. SSDC, vol. 145, pp. 97–118. Springer, Cham (2018). https://doi.org/10.1007/978-3-319-74412-4_7

10. Lancichinetti, A., Fortunato, S., Kertesz, J.: Detecting the overlapping and hierarchical community structure in complex networks. New J. Phys. **11**(3), 033015 (2009)

11. Klimt, B., Yang, Y.: The Enron Corpus: A New Dataset for Email Classification Research. In: Boulicaut, J.-F., Esposito, F., Giannotti, F., Pedreschi, D. (eds.) ECML 2004. LNCS (LNAI), vol. 3201, pp. 217–226. Springer, Heidelberg (2004). https://doi.org/10.1007/978-3-540-30115-8_22

12. Leskovec, J., Kleinberg, J., Faloutsos, C.: Graphs over time: densification laws, shrinking diameters and possible explanations. In: Proceedings of the eleventh ACM SIGKDD International Conference on Knowledge Discovery in Data Mining, pp. 177–187. ACM, Chicago (2005)

13. Greene, D., Doyle, D., Cunningham, P.: Tracking the evolution of communities in dynamic social networks. In: 2010 International Conference on Advances in Social Networks Analysis and Mining, pp. 176–183. IEEE, Odense (2010)

14. Nguyen, N.P., Dinh, T.N., Tokala, S., et al.: Overlapping communities in dynamic networks: their detection and mobile applications. In: Proceedings of the 17th Annual International Conference on Mobile Computing and Networking, pp. 85–96. ACM, Las Vegas (2011)

15. Shang, J., Liu, L., Xie, F., et al.: A real-time detecting algorithm for tracking community structure of dynamic networks. arXiv preprint arXiv:1407.2683 (2014)

16. Nguyen, N.P., Dinh, T.N., Shen, Y., et al.: Dynamic social community detection and its applications. PLoS ONE **9**(4), e91431 (2014)

17. Shen, H., Cheng, X., Cai, K., et al.: Detect overlapping and hierarchical community structure in networks. Phys. A: Stat. Mech. Appl. **388**(8), 1706–1712 (2009)

18. Danon, L., Diaz-Guilera, A., Duch, J., et al.: Comparing community structure identification. J. Stat. Mech: Theory Exp. **2005**(09), P09008 (2005)

A Recommendation Method of Barrier-Free Facilities Construction Based on Geographic Information

Peng Liu[1,2,3], Zhenghao Zhang[1,2,3], Tun Lu[1,2,3(✉)],
Dongsheng Li[1,2,3], and Ning Gu[1,2,3]

[1] School of Computer Science, Fudan University, Shanghai, China
Liudongping0202@163.com,
{18210240021,lutun,dongshengli,ninggu}@fudan.edu.cn
[2] Shanghai Key Laboratory of Data Science, Fudan University, Shanghai, China
[3] Shanghai Institute of Intelligent Electronics and Systems, Shanghai, China

Abstract. Public accessibility facilities play a vital role in the life and travel of the disabled and the elderly. With the increasing population of the disabled and the elderly, the number and types of barrier-free facilities are relatively insufficient and the construction location is not reasonable in most cities of China. This paper collects the geographic location information of public facilities, relevant units of disabled persons, traffic stations and barrier-free facilities in the "National Barrier-Free Construction Demonstration Zone" of Shanghai; divides a single building into several local areas; extracts eigenvalues from environmental information in the local areas and transforms them into eigenvectors to construct accessibility facilities construction model basing on the geographic information. And after experiment validation, the accuracy of the model reaches more than 95%, and part of the error of the number pf accessibility facilities could be controlled within 0.003. The application of this model can provide guidance and suggestions for the government in the construction of public barrier-free facilities, and improve the rationality of the construction of barrier-free facilities.

Keywords: Barrier-free · Accessibility facilities · City planning · Geography information · Recommendation

1 Introduction

Currently, China has begun to step in the aging society, with the increasing number of disabled people, the elderly and other vulnerable groups year by year, more and more attention from all walks of life has been paid to them. Barrier-free facilities in public environment are very significant in the travel of the disabled, the elderly and other vulnerable groups, and helping the disabled groups integrate into society. They are the precondition for them to enjoy social public resources and receive fair treatment. However, there are still some problems in the construction of barrier-free facilities in the public environment. The city barrier-free level is relatively low, which is especially obvious in some cities with relatively backward economy. From the point of view of

© Springer Nature Singapore Pte Ltd. 2019
Y. Sun et al. (Eds.): ChineseCSCW 2019, CCIS 1042, pp. 377–389, 2019.
https://doi.org/10.1007/978-981-15-1377-0_29

vulnerable groups, the more barrier-free facilities in public environment, the better their travel experience. However, urban construction is a gradual process instead of one day project, many old public facilities are lack of supporting barrier-free facilities construction due to historical problems, land-use problems. On the other hand, the vulnerable groups are minorities compared with the general public. The excessive construction of barrier-free facilities will also cause the problem of low utilization rate and high occupation of social resources. Hence, it is difficult to achieve a comprehensive barrier-free city at the current stage. In order to solve the above realistic problems, this paper proposes a recommendation method of barrier-free facilities construction based on geographic location information. Firstly, this method collects the geographic location of public facilities such as stadiums, parks, sunshine centers, transportation stations, wheelchair ramps, barrier-free toilets and other barrier-free facilities in national barrier-free demonstration cities, and divides them into local areas, transform the building information in the local area into eigenvectors. The types of barrier-free facilities need to be constructed are obtained by Bagging, Extra Trees and Random Forest Three Integrated Classification Models based on Decision Tree through voting. The number of barrier-free facilities can be obtained by Extra Trees integrated learning algorithm. The proposed recommendation method of barrier-free facilities construction based on geographic location information can provide planning guidance for the construction of barrier-free facilities in cities.

There are six parts in this paper. The second part mainly introduces the research and problems of barrier-free facilities in public environment. The third part mainly introduces the sources of data acquisition and data preprocessing. The fourth part is mainly about the recommendation method of barrier-free facilities based on geographic location information. The fifth part is experimental evaluation. The sixth part is the summary and discussion of this work.

2   Relevant Research at Home and Abroad

At present, the number of disabled people in China has been increasing year by year. The total of which has exceeded 85 million, and the number of disabled families has exceeded 70 million, accounting for 17.8% of the total number of families in China. But in our daily life, we seldom see the disabled in public. According to statistics, more than 71% of the disabled people in our country have never been to libraries or other cultural venues; more than 68% of the disabled people have never been to sports venues; nearly half of the disabled people have never been to parks [1]. There are two reasons causing this phenomenon: one is the disabled groups themselves, the other is the low level of urban barrier-free facilities of most cities in China, which makes it inconvenient for the disabled groups to travel. The barrier-free construction of foreign cities is also facing the problem of how to achieve full barrier-free construction efficiently with limited funds in barrier-free transformation of old public buildings. [2]. In the field of barrier-free technology for the disabled, the current research focus is on barrier-free aids for the disabled [3] and barrier-free home for the disabled [4]. There are not many studies on barrier-free facilities in public environment, mainly on barrier-free Design of single facilities [5, 6].

Since the concept of barrier-free facilities was put forward in 1959, hundreds of countries and regions have formulated barrier-free design norms and standards, which play an important role in ensuring the successful construction of barrier-free facilities in public [7]. For example, "Codes for Accessibility Design" [8] published by the Ministry of Housing and Urban-Rural Construction of the People's Republic of China, and "National Barrier-free Design Regulations and Standards" formulated by the National Standards Society of the United States [9], etc. The main contents of these standardized standard documents are the basic requirements and minimum standards for the design and construction of barrier-free facilities for each type of barrier-free facilities and different public facilities, without considering the requirements of other factors in the actual public environment for the scope and quantity of barrier-free facilities, meanwhile, within a certain range, the interaction of various types of public buildings will also have an impact on the demand for barrier-free facilities. In view of the above problems and shortcomings, this paper proposes a method of barrier-free facility construction recommendation based on geographic location information. When recommending a new public building for barrier-free facility construction, on the basis of following the "Codes for Accessibility Design" for single buildings, to recommend the types and quantities of barrier-free facilities by taking the location information of public buildings, urban green space, sports venues for the disabled, traffic stations and existing barrier-free facilities around the building as the characteristics of barrier-free facilities. The detailed recommendation method will be introduced in subsequent chapters.

3 Data Acquisition and Preprocessing

As the earliest cities to carry on barrier-free transformation in China, Shanghai promulgated the "Measures for the Construction and Application of Barrier-free Facilities" in 2003, striving to build an all-round barrier-free urban system. In 2015, all 16 districts in the city passed the acceptance of the "barrier-free environment construction" research group of the CPPCC National Committee, and were appraised as the "national barrier-free construction demonstration zone" [10, 11]. Shanghai, as an exemplary role, is leading in the construction of barrier-free public environment in China. Every year, people from all over the country come to visit, inspect and study in Shanghai, such as the Disabled People's Federation and the Housing and Construction Commission, [11]. Therefore, this paper chooses Shanghai's public environment barrier-free construction data as the data set of building barrier-free facilities recommendation model.

According to the "Codes for Accessibility Design" of China, the construction of barrier-free facilities in public environment is related to public buildings such as urban traffic, urban square green space, stadiums and gymnasiums. At the same time, the main service object of the construction of barrier-free facilities is the disabled, hence, the data set of this method also considers some main places related to the disabled groups, such as Sunshine Center, Rehabilitation Guidance Center for the Disabled, Disabled Assistant Instruments Service Society, etc. Based on the above reasons, the data set of this method is composed of 14 kinds of public facilities: vehicle repair shop, auxiliary service club, rehabilitation guidance station, rehabilitation center, park, museum, exhibition hall, community service center, gymnasium, parking lot, sunshine base, sunshine heart garden, "Half-Way House" and public transport stations, which are obtained from the "Shanghai Municipal Government Data Network" [12]. According to the Barrier-Free Design Specification, all recommended types of barrier-free facilities in public environment are classified into 12 categories: barrier-free toilets, wheelchair ramps, handrails, low service desks, barrier-free elevators, barrier-free parking spaces, barrier-free lifts, voice facilities for the blind, wheelchair seats, blind lanes and barrier-free guest rooms, braille Tour Guide Map. The data of barrier-free facilities above are mainly from "Shanghai 2016 Barrier-free Facilities Map" [11]. In order to obtain more comprehensive data of barrier-free facilities, we designed and developed a mobile application of barrier-free facilities. The program uses the design idea of "crowdsourcing" to facilitate users to mark in the online maps through pictures or texts of the barrier-free facilities in the travel process, while the database generates the types of barrier-free facilities and latitude and longitude information. Disabled users can use the system to obtain the specific barrier-free facilities near them, and grade their satisfaction of the barrier-free facilities (see Fig. 1).

Fig. 1. A collaborative accessible facility tagging system

The data of public buildings mentioned above are relatively primitive. Most of the data only contain the name of the building and the type of the building. Therefore, further processing of these data is needed to add the latitude and longitude information of the building to express their geographic location information. In this research, most of the buildings' longitude and latitude information is acquired by application programming interface of GIS which provided by Baidu corporation [13]. The buildings cannot get accurate information through API, it can be acquired by manual labeling through real-time labeling system of accessible facilities.

4 The Recommendation Method of Barrier-Free Facilities Construction Based on Geographic Information

In the previous chapter, we have constructed the data set of public building facilities and accessible facilities, but these data are discrete, and the building has the unique geographical location characteristics. It is impossible to have two different buildings in the same place, and single location information can not express the relationship between public buildings, public buildings and barrier-free facilities. Based on this, the recommended method of accessibility facilities based on geographical location information is mainly composed of three steps: geographical location-based regional division, regional public environment feature extraction, and barrier-free facilities construction recommendation. The content of each part will be described in detail below.

4.1 Regional Division Based on Geographical Location

As can be seen from the above, when building barrier-free facilities, only one public building is considered, thus ignoring the impact of the whole public environment on the demand for barrier-free facilities, which may result in the waste and imbalance of public resources. In order to solve this problem, this paper presents a method to divide public areas according to the distance between buildings. The method begins with the recommended building as the center of a circle O_i, and the fixed distance R as the radius to form a circular area. Then, it calculates whether other public buildings and barrier-free facilities are in the area, and records the number of different types of public buildings and barrier-free facilities in the area as the public environment characteristic data. Through this method, we transform the recommended building into a public area to be recommended, and describe the public environment in the recommended area. The specific algorithm is as follows:

Algorithm1: Division of Area

Input: Public_Facilities set $P = \{P_1, P_2, .., P_n\}$
  Public_Facilities P_i
Output: Public_Areas PA

1: Public_Areas PA
2: int r = 6371, R = 1000
3: for Ti in P.types:
4:   Public_Facilities set pt = P[P.type = Ti]
5:   c=0
6:   for j=1 to n do
7:     float d = haversine(P_i, P_j)
8:     if d<=R
9:       c +=1
10:       PA.add(Ti , c)
11:     end if
12:   end for
13: end for
14: return PA

The haversine function is to calculate the distance between two latitude and longitude coordinate points, R is the average radius of the earth, in kilometers, and R is the radius of the range. The angle of longitude and latitude needs to be converted to radian when calculating, as follows:

$$Ra = degrees * \pi/180 \tag{1}$$

The formula for calculating the distance between two points of the sphere is as follows:

$$d = 2 * r * \sin^{-1}\left(\sqrt{haversin(d/r)}\right) \tag{2}$$

$$haversin\left(\frac{d}{r}\right) = haversin(\varphi_1 - \varphi_2) + \cos(\varphi_1)\cos(\varphi_2)haversin(\Delta\lambda) \tag{3}$$

Where, $\varphi_1\varphi_2$ is the latitude of two points; $\Delta\lambda$ is the difference of longitude between two points.

4.2 Extracting Public Environment Features Within Areas

After converting individual public building data into public area data, we can describe the relationship between public environment and accessibility facilities in a region. However, it is impossible to show the relationship and intensity of public buildings in the region. In a certain range, the intensity of buildings and facilities will also affect the demand for accessible facilities. For example, many bus stations within a small range, comparing with the same number of bus stations within a large range, the needs for wheelchair ramps would inevitably be less. To solve this problem, we calculate the average distance between the buildings in the area based on the 4.1 division of the area to express their density in the area. The concrete formulas are as follows:

$$D_a = \frac{\sum_{i=1}^n d_i}{\sum_{j=1}^n n_j} \tag{4}$$

Among them, di is the calculated spherical distance between the recommending building and building I at the time of area division, nj is the calculated amount of the jth building in the designated area.

As the types of public buildings are defined in accordance with the "Codes for Accessibility Design", there is no relationship between the types of buildings, which is disordered and discrete. In order to express which types of public buildings are included in a public area, we need to digitize this feature. Here we choose to use one-hot encoding, because if we use ordinary numbers from 1 to 14 to represent a building type sequential relationship, which will interfere with the machine learning model, because when the model needs to calculate the average value, the larger the type, the higher average value will be calculated. One-hot coding, also known as one-efficient coding, can represent a classification variable in the form of binary vectors. Specific coding rules are as follows:

$$Building\ Type - > Integer - > Binary\ Vector \tag{5}$$

After applying one-hot coding, different building types are mapped to a point in European space, so that when calculating the distance between features by Euclidean distance, the errors caused by using common numeric representation can be avoided, making the calculation of similarity more reasonable.

Therefore, a set of geographic location features in a recommended area consists of three parts: the types of buildings in the area, the number of public buildings and the average distance between buildings.

$$PA = \{PType\_Code, PType_1\_num, PType_2\_num, \dots, PType_{14}\_num, D_a\} \tag{6}$$

4.3   Barrier-Free Facilities Construction Recommendation

Not only the geographical location of public buildings, but also the people in the area and the internal structure of the building facilities should be considered in the selection of the specific location of the barrier-free facilities construction. Due to the complexities, those factors will not be considered in this paper. Moreover, different from the traditional recommendation system which can predict the content or behavior that the users are interested in the future on the basis of their historical information, the barrier-free facilities construction recommendation cannot predict the barrier-free facilities construction according to the historical situation of a building. Based on the above analysis, the barrier-free facilities construction recommendation in the public environment proposed in this paper mainly refer to the recommended types and the number of barrier-free facilities construction, that is, transforming into the construction of a classification model of barrier-free facilities based on geographic location features within the region and a regression model of barrier-free facilities based on the geographic location features within the region.

Currently the classification models mainly include the linear classification model, the classification model based on the feature distance, and the classification model based on the decision tree. The linear classification model classifies mainly through the linear combination of eigenvalues. It is more efficient in binary classification. The representative algorithm is Logistic Regression [14], etc.; the classification model based on feature distance classifies mainly by calculating the distance between a sample and the other samples in the feature space, but it is not very friendly to the unbalanced data set. The representative algorithm is KNN [15], etc.; the classification model based on the decision tree is to make the data set mapped to a tree structure decision model according to the characteristic attribute and characteristic value, and the decision is thus made. When the target variable is a discrete value, it is a classification tree. When the target variable is a continuous value, it is a regression tree. The representative algorithm mainly includes CART, C4.5, etc. [16].

According to the Codes for Accessibility Design, the barrier-free facilities can be defined as 12 types, so it is multi-class problem to classify the barrier-free facilities based on geographical location features within the region. Furthermore, since the urban public buildings in the city center and urban suburbs are unbalanced in number, the CART decision tree-based integration method is applied [17–21] when we choose to construct a barrier-free facilities classification model and a barrier-free facilities regression model. The integrated prediction is achieved by establishing multiple CART decision trees. The CART decision tree calculates the Gini coefficient of each feature set in the process of constructing the tree nodes and selects the minimum Gini coefficient subset as the child node of the tree. For the training set S with K classifications, the Gini coefficient can be shown as follows:

$$Gini(S) = \sum_K \frac{|C_k|}{|S|} \left(1 - \frac{|C_k|}{|S|}\right) = 1 - \sum_K \left(\frac{|C_k|}{|S|}\right)^2 \tag{7}$$

C_K is the K-class subset and $|C_K|$ represents the value of C_K. The set S can be divided into Subset S1 and Subset S2 by any feature A, and then when the feature A is used as the node splitting condition, the Gini coefficient can be expressed as:

$$Gini(S, A) = \frac{|S_1|}{|S|} Gini(S_1) + \frac{|S_2|}{|S|} Gini(S_2) \tag{8}$$

After the decision tree model is constructed, when a feature set X of a common region is given, the mathematical expression of the model classification process is as follows:

$$G(x) = q_1(x) \cdot g_1(x) + q_2(x) \cdot g_2(x) + \ldots + q_t(x) \cdot g_t(x)$$
$$= \sum_{t=1}^{t} q_t(x) \cdot g_t(x) \tag{9}$$

t is the number of leaf nodes of the tree, namely, the number of paths for classification.

When establishing the regression model, the mean square error is used as the judgement criterion for dividing the data set, and the smallest mean square error value is selected as the child node. The loss function formula is as follows:

$$MSE = \frac{1}{N} \sum_{t=1}^{N} (observed_t - predicted_t)^2 \tag{10}$$

Observed is the estimator, and predicted is the estimated amount. The smaller the MSE is, the better the accuracy of the model is proved.

Based on the three steps, the complete recommendation algorithm for barrier-free facilities construction recommendation based on geographic location information is as follows:

Algorithm 2: Recommendation algorithm

Input: Public_Facilities set $P = \{P_1, P_2, .., P_n\}$,
    Public_Facilities P_i
Output: Accessiblity Facility type set At,
    Accessiblity Facility num set An

1: Public_Areas set PAs
2: for p in P:
3:    Public_Areas PA = Division of Area(p,P)
4:    PA.average_distance = Average_Distance(p,P)
5:    PA.types = One_hotCoder(PA)
6:    PAs.add(PA)
7: end for
8: Public_Area PAi = Division of Area(Pi,P)
9: PAi = Average_Distance(Pi,P)
10: PAi.types = One_hotCoder(PAi)
11: C_model = Classification_train(PAs)
12: R_model = Regression_train(PAs)
13: Accessiblity Facility type set At = C_model(Pi)
14: Accessiblity Facility num set An = R_model(Pi)
15: return At,An

---

5 Experiments

To evaluate the approach for barrier-free facilities construction recommendation based on geographic information proposed in this paper, we collected data of public buildings, transportation sites and barrier-free facilities of Shanghai in 2016. The collection methods have been introduced in the third part. There are 30,662 traffic sites, 1,278 disabled service venues, 6,247 public buildings, and 6,451 barrier-free facilities in the data. These data are preprocessed into feature vectors of regions by the methods mentioned in 4.1 and 4.2. A total of 35562 regions are divided, and the data set is divided into a test set and a training set by 3:7.

The computer used in the experiment has an Intel Core i5-6200U CPU and 8 GB of RAM. First, we select 500 m, 1000 m, and 2000 m as the radius of regions. The CART-based random forest model is used as the classifier to investigate the influence of different radius on the classification accuracy. The results are shown in the Fig. 2.

The results indicate there is no significant correlation between the radius of the region and the classification accuracy. The classifier performs the best when the radius of the region is 1000 m. In the real world, the distance between bus stations is about 1000 m in general, and the public facilities such as stadiums, museums, and cultural centers usually cover areas of about 1 km^2. As a result, when the radius is 500 m, many related public facilities will be missed. And when the radius is 2,000 m, unrelated public facilities may degrade models. The results confirm the idea of this work.

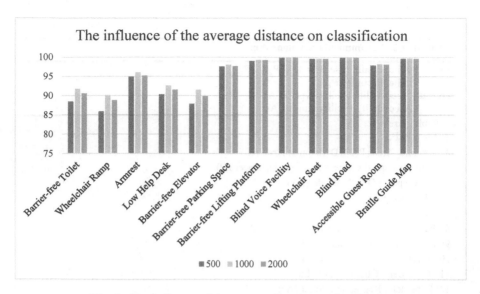

Fig. 2. The influence of the average distance on classification

In order to investigate the influence of the average distance between buildings in an area on the number of barrier-free facilities construction, a data set with no average distance and a data set with average distance is used, and a random forest model is trained to predict the number of barrier-free facilities construction. MSE (mean square error) is used as the loss function. The results are shown in Table 1. It shows that the model corresponding to the data set with the average distance has lower MSE.

Table 1. The influence of the average distance on regression results.

Barrier-free facility	No average distance	With average distance
Barrier-free toilet	0.95575	0.462286
Wheelchair ramp	8.96021	4.14023
Armrest	0.51167	0.278159
Low help desk	1.64514	0.783637
Barrier-free elevator	3.03969	1.6933
Barrier-free parking space	0.03798	0.0304187
Barrier-free lifting platform	0.01718	0.0117006
Blind voice facility	0.00194	0.00135
Wheelchair Seat	0.00713	0.00506
Blind road	0.00134	0.00093
Accessible guest room	0.03648	0.02151
Braille guide map	0.00465	0.00284

Based on the experimental results above, set r = 1000, the classification task and regression task are evaluated using features of regions proposed in this paper. Logistic regression, KNN, decision tree, MLP and CART-based bagging, random forest, Extra Trees, gradient boosting are used in the classification task. The results are shown in Table 2.

Table 2. Results from different classifiers

Classifier	Barrier-free Toilet	Ramp	Wheelchair	Armrest	Low Help Desk	Barrier-free Elevator	Barrier-free Parking Space	Barrier-free Lifting Platform	Blind Voice Facility	Wheelchair Seat	Blind Road	Accessible Guest Room	Braille Guide Map	Mean
Bagging	0.91424	0.89720	0.95499	0.92567	0.90904	0.97946	0.99182		0.99855	0.9954	0.99915	0.98218	0.99727	0.96208
Random-Forest	0.91706	0.90146	0.95900	0.92993	0.91109	0.98056	0.99267		0.99889	0.99531	0.99923	0.98244	0.99702	0.96372
ExtraTrees	0.91765	0.90350	0.95780	0.93411	0.91501	0.98218	0.99361		0.99881	0.99599	0.99932	0.98406	0.99702	0.96492
GradientBoosting	0.81519	0.81570	0.89344	0.82090	0.81545	0.94408	0.98304		0.99685	0.99173	0.99855	0.95891	0.99369	0.91896
LogisticRegression	0.78050	0.79524	0.87461	0.78058	0.78331	0.93146	0.98014		0.99625	0.99011	0.99787	0.94706	0.98909	0.90385
KNeighbors	0.82738	0.81732	0.90103	0.83974	0.82278	0.94843	0.98278		0.99659	0.99062	0.99839	0.9578	0.99148	0.92286
MLPClassifier	0.77922	0.77555	0.88057	0.80104	0.79507	0.93402	0.98142		0.99591	0.99071	0.99812	0.94553	0.98815	0.90544
DecisionTree	0.89864	0.88151	0.94357	0.91075	0.89745	0.97136	0.98977		0.99795	0.99369	0.99864	0.97511	0.99616	0.95455
EnsembleVote	**0.92422**	**0.91032**	**0.96258**	**0.93666**	**0.91919**	**0.98252**	**0.99284**		**0.99889**	**0.99608**	**0.99923**	**0.98500**	**0.99736**	**0.96707**
Stacking	0.91851	0.90581	0.96190	0.93436	0.91706	0.98176	0.99275		0.99889	0.99557	0.99932	0.98508	0.99702	0.96567

The results show that CART-based ensemble models (bagging, random forest and ExtRa Trees) have higher accuracy. Their average accuracies are significantly higher than other models and reach 0.96177, 0.96388 and 0.96551 respectively. The classification accuracy could be improved by ensemble these three models again. We explore two ensemble strategies: (1) voting ensemble: the majority output of the three models are chosen as the final classification output; (2) Logistic Regression ensemble: the output of the three models are passed through a logistic regression model. Train the logistic regression model using the output of three models, and generate the final result by it. The results show that in most tasks, the accuracy of the voting ensemble is slightly higher than that of logistic regression ensemble. This is because the logistic regression ensemble involves a training process, which is prone to over-fitting. The average accuracy rate is 96.71%.

6 Conclusion and Discussion

In view of the shortcomings of barrier-free facilities construction in the existing public environment, this paper proposes a method of barrier-free facilities construction recommendation based on geographic location information. Which not only considers a single public building to be recommended, but also combines the public environmental factors around the public buildings to be recommended, and transform the public buildings to be recommended into the areas to be recommended. The geographic information feature set in the region based on the "Barrier-Free Design Specification" proposed in this paper shall realize the recommendation of the types and quantities of barrier-free facilities in the region. The validity of the feature set based on geographic location information and the feasibility of the recommendation method are proved by experiments. However, there are some shortcomings in this method, such as inaccurate recommendation of the construction location of barrier-free facilities; feature set is only composed of geographic location information, which cannot fully describe other features in the region; at present, it can only realize the recommendation of new barrier-free facilities, but not to propose suggestions for existing barrier-free facilities etc. Based on the above problems and the research in this paper, it is planned to use the barrier-free facilities marking system developed in this paper to grade the existing barrier-free facilities, and propose an improvement method for the design of existing barrier-free facilities according to the evaluation opinions.

Acknowledgment. This work is supported by the Joint Fund of National Natural Science Foundation of China and the China Academy of Engineering Physics (NSAF) under Grant No. U1630115 and the National Natural Science Foundation of China under Grant No. 61332008.

References

1. Zhou, L.: The disabled social security system and public service system construction. Chin. J. Popul. Sci. **02**, 93–101 (2011)
2. Waenlor, W., Wiwanitkit, V., Suwansaksri, J., et al.: Facilities for the disabled in the commercial districts of Bangkok–are they adequate? SE Asian J. Trop. Med. Public Health **33**(3), 164 (2002)
3. Shinohara, K., Tenenberg, J.: A blind person's interactions with technology. Commun. ACM **52**(8), 58–66 (2009)
4. Yatani, K., Banovic, N., Truong, K.: SpaceSense: representing geographical information to visually impaired people using spatial tactile feedback. In: Proceedings of the SIGCHI Conference on Human Factors in Computing Systems, pp. 415–424. ACM (2012)
5. Hirano, K., Kitao, Y.: A study on connectivity and accessibility between tram stops and public facilities. Am. J. Epidemiol. **171**(2), 247–264 (2009)
6. Arbournicitopoulos, K.P., Ginis, K.A.M.: Universal accessibility of "accessible" fitness and recreational facilities for persons with mobility disabilities. Adap. Phys. Activity Q. **28**(1), 1 (2011)
7. Jing, L.: On the development of barrier-free design at home and abroad. Archit. Anhui **1**(1), 26–27 (2002)

8. Ministry of Housing and Urban-Rural Construction of the People's Republic of China: Codes for Accessibility Design. China Architecture& Building Press (2012)
9. Jia, W., Wang, X.: A comparative study on the development of barrier-free design laws and regulations between China, America and Japan. Res. Mod. Cities **2014**(4), 116–120 (2014)
10. XINMIN. CN. http://www.xinmin.cn/. Accessed 14 May 2019
11. Shanghai Disabled Person's Federation. http://www.shdisabled.gov.cn/clwz/clwz/index.html. Accessed 14 May 2019
12. Shanghai Government Data Service. http://data.sh.gov.cn/home!toHomePage.action. Accessed 14 May 2019
13. Baidu Map Platform. http://lbsyun.baidu.com/index.php?title=homepage. Accessed 20 May 2019
14. Hosmer Jr., D.W., Lemeshow, S., Sturdivant, R.X.: Applied Logistic Regression. Wiley, Hoboken (2013)
15. Altman, N.S.: An introduction to kernel and nearest-neighbor nonparametric regression. Am. Stat. **46**(3), 175–185 (1992)
16. Singh, S., Gupta, P.: Comparative study ID3, cart and C4. 5 decision tree algorithm: a survey. Int. J. Adv. Inf. Sci. Technol. (IJAIST) **27**(27), 97–103 (2014)
17. Freund, Y., Schapire, R., Abe, N.: A short introduction to boosting. J. Japan. Soc. Artif. Intell. **14**(771–780), 1612 (1999)
18. Breiman, L.: Bagging predictors. Mach. Learn. **24**(2), 123–140 (1996)
19. Liaw, A., Matthew, W.: Classification and regression by randomForest. news. R News **2**(3), 18–22 (2002)
20. Geurts, P., Damien, E., Louis, W.: Extremely randomized trees. Mach. Learn. **63**(1), 3–42 (2006)
21. Friedman, J.H.: Greedy function approximation: a gradient boosting machine. Ann. Stat. **29**(5), 1189–1232 (2001)

Empirical Research on the Relationship Between Teaching and Scientific Research Based on Educational Big Data in China

Yuyao Li[1,2,3], Hailin Fu[1], Yong Tang[1(✉)], and Chengzhou Fu[1]

[1] School of Computer Science of South China Normal University, Guangzhou 510631, Guangdong, China
everybit@163.com, {190260995,ytang4}@qq.com,
fucz@m.scnu.edu.cn
[2] Collaborative Innovation Center for 21st-Century Maritime Silk Road Studies of Guangdong, University of Foreign Studies, Guangzhou 510420, Guangdong, China
[3] School of Information Science and Technology of Guangdong, University of Foreign Studies, Guangzhou 510006, Guangdong, China

Abstract. The traditional empirical researches on the relationship between teaching and scientific research has been trapped in a dilemma in which lots of conflicting findings coexist because teachers' work performances are used as the main source of research data. Based on the big data, this paper proposes different research hypothesis from the traditional researches, takes the detailed data of teacher's teaching and scientific research behavior recorded in the educational big data system as the research object, adopts the user portrait method to extract the teacher's teaching and scientific research behavior characteristics, and takes association rule mining algorithm to calculate the correlation between teacher's teaching and scientific research behavior. The experimental results effectively support the research hypothesis proposed in this paper, and can effectively present the real relationship between teaching and scientific research.

Keywords: University teacher behavior model · Teaching and scientific research relationship · SCHOLAT · Educational big data · User behavior modeling · Association rule mining · User portrait

1 Introduction

The relationship between teaching and scientific research is a fundamental issue in modern universities. Burton R. Clark even believes that "In modern university education, there is nothing more fundamental than the relationship between teaching and scientific research" [1].

Traditional empirical researches assume that the relationship between teaching and scientific research is reflected in the results of their work, using teachers' working results such as teaching performance evaluation and the number of published papers as research data, trying to find a linear or non-linear quantitative relationship between the

© Springer Nature Singapore Pte Ltd. 2019
Y. Sun et al. (Eds.): ChineseCSCW 2019, CCIS 1042, pp. 390–398, 2019.
https://doi.org/10.1007/978-981-15-1377-0_30

two. Amount of researches based on this traditional method reached its peak in the 1990s and then moved downwards [2].

This paper proposes new research hypothesis from the perspective of big data: teaching and scientific research are two different aspects of the work, and teachers use different methods to coordinate these two tasks in daily work. The association between the two may exist in the working results, or exist in the details of the work process. Therefore, the correlation between the two can be detected under the support of relevant online systems and big data technologies.

Based on this hypothesis, this paper uses the detailed behavior data of 547 teachers distributed in 41 universities in 11 provinces in China to extract the teaching and research behavior characteristics of teachers, and uses association rule mining algorithm to calculate the correlation between teacher's teaching and scientific research behavior. Different from those traditional researches that are trying to make a clear conclusion, the experimental results in this paper demonstrate the complex and diverse relationship between teaching and scientific research.

2 Related Research

2.1 Traditional Empirical Research and Its Inadequacies

Malcolm Tate [2] reviewed the empirical researches on the relationship between teaching and scientific research of many countries in Europe and mainly in America, and discussed the contradictory results in many studies. Genshu et al. [3] in their paper classified traditional research results at home and abroad into 3 conflicting categories, as shown in Table 1:

Table 1. Different correlation models based on traditional researches

Negative correlation model	The scarcity model
	The differential personality model
	The divergent reward system model
Positive correlation model	The conventional wisdom model
	The G model
Zero correlation model	The different enterprises model
	The unrelated personality model
	The bureaucratic funding model

From above table we can draw a conclusion that the research results obtained from these assumptions are often contradictory and even will result in more controversies, thus making traditional empirical research criticized and trapped in dilemma.

2.2   New Research Trends Under Big Data

With the support of big data technology, it has become a reality to discover the people's similarities and differences from the human behavior data of large samples, thus extracting the daily behavior patterns of specific groups, and using quantitative analysis in the mining and analysis of people's characteristics of certain group [4].

With the rapid development of online systems such as academic social network and online education platform, it has become a reality to record and store the detailed data of teachers' online teaching and scientific research process, thus providing rich data details and research possibilities for the research on the relationship between teaching and scientific research in colleges and universities [5, 6]. On the other hand, unfortunately, since most academic social networks and online education websites are independent from each other and have independent user groups, thus data storage standards, logical meanings and functional definitions differ greatly. In this kind of situation, it is difficult to sort and match data, especially to verify the authenticity and validity of user matching in two separate systems, which greatly reduces the credibility of the analysis results.

SCHOLAT (scholat.com) is an online system that integrates the functions of academic social networking and online courses, and provides teaching and scientific research collaboration services for teachers and students in universities and researchers in research institutes [7]. SCHOLAT's users can switch between academic social networking and online teaching functions freely, and its data have unified storage standards, functional definitions and logical meanings. After years of development and accumulation, the system has a large number of teachers and users who have been using the online teaching and academic social networking functions, leaving a lot of operational details. Therefore, objectively, the SCHOLAR provides a unique and semantically consistent data for the research on the relationship between teaching and scientific research.

Based on the user data of the SCHOLAT, this paper extracts the characteristics of teaching and scientific research behaviors from the data of 547 teachers from 41 universities and colleges in 11 provinces in China, and uses association rule mining algorithm to calculate their correlation.

3   Research Design and Data Source

3.1   Research Design

In this paper, the user behavior modeling method is used to establish the behavior model of college teachers, and the association rule mining method is used to verify the model, as shown in Fig. 1:

As shown in Fig. 1, this paper extracts user behavior characteristics based on user portrait method and calculates their correlation through association rule mining algorithm after preliminary processing of original data, such as pre-processing (Processing errors and missing values and conversion, etc.) and normalization processing.

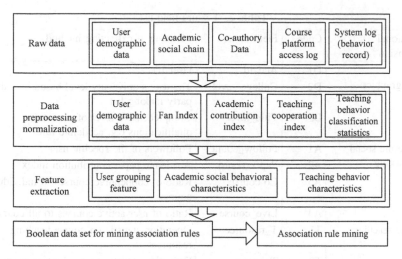

Fig. 1. Research design

3.2 Data Source

The experimental data set of this paper was collected from the user access records of the SCHOLAT from September 15, 2017 to December 31, 2017, including 27,951,571 log records and 138,269 course platform access log records. After pre-processing such as data cleaning, sorting and transformation, detailed academic and teaching behavior data of 547 teachers distributed in 41 universities in 11 provinces in China were obtained, most of which were concentrated in Guangdong Province.

3.3 Feature Extraction

Features are extracted from three dimensions: user demographics, academic social networking, and online teaching. Each dimension has several indicators, as shown in Table 2:

3.4 Sub-features and Boolean Transformation

In order to further process the data, each feature indicator is subdivided into three value ranges: [0–0.29] for low range values, [0.3–0.59] for middle range values, [0.6–1] represents a high range value, and uses "_L", "_M", and "_H" to represent each sub-features of the subdivision. For example, T1_L indicates that a teacher's T1 feature is at a low range value, and the rest are similar.

In addition, Apriori belongs to the Boolean association rule mining algorithm, of which the processing object is a Boolean data set, therefore, it is necessary to map the values of the high, medium and low sub-features in Table 3 to a specific 0–1 relationship. For example, if a teacher's A1 index value is 0.58, then the teacher's A1_H, A1_M, and A1_L are respectively 0, 1 and 0. The rest of the features can be mapped accordingly.

Table 2. Feature field list

Functional dimension	Code	Field name	Meaning or extracting method
User demographics	D1	Acc_id	User ID
	D2	College_level	The user's university level based on a third-party ranking list
	D3	Title_level	Comprehensive level of academic qualifications e.g. academic title
Academic social networking	A1	Follow_count	Followers of the specific user
	A2	Paper_contri	Academic paper contribution index
	A3	Co-course	Ratio of cooperative courses to individual courses
	A4	Live_course	Ratio of user active courses to all courses
Online teaching	T1	T_resource	Operation statistic on course teaching resources
	T2	T_interact	Operation statistic on teaching interaction
	T3	T_assignment	Operation statistic on assignments
	T4	T_attendance	Operation statistic on student attendance
	T5	T_parameter	Operation statistic on setting course parameters

Table 3. Sub-features (SF) and proportion (P)

SF	P	SF	P	SF	P	SF	P	SF	P	SF	P
D2_H	28%	D3_H	14%	A1_H	5%	A2_H	3%	A3_H	5%	A4_H	2%
D2_M	13%	D3_M	8%	A1_M	12%	A2_M	5%	A3_M	77%	A4_M	11%
D2_L	59%	D3_L	78%	A1_L	84%	A2_L	91%	A3_L	18%	A4_L	86%
T1_H	11%	T2_H	0%	T3_H	12%	T4_H	1%	T5_H	3%		
T1_M	18%	T2_M	1%	T3_M	27%	T4_M	2%	T5_M	10%		
T1_L	71%	T2_L	98%	T3_L	60%	T4_L	97%	T5_L	86%		

In the end, the Apriori algorithm is set to have a minimum support threshold of 0.3, a minimum confidence threshold of 0.75, and a minimum lift threshold of 1. For the selection of mining calculation results, the rule with the highest lift degree is generally preferred, followed by the comprehensive consideration of three indicators. In the subsequent experiments in this paper, the rules with the highest lift degree are adopted. Due to the limitation of space, for its algorithm and indicators meaning, please refer to relevant references and here will be no further description.

4 Experiments and Results

4.1 Overall Association Experiment

The current experiment runs the Apriori algorithm directly on the entire data set and gets several very similar rules R101–R104 (support: 0.33, confidence: 1(max), lift: 2.37 (max)), where the sub-features that were set with shading indicate that the rule is different from other rules:

R101: A1_L⇒A2_L⇒A3_M⇒A4_L⇒D2_L⇒D3_L⇒T1_L⇒T4_L⇒T5_L⇒T2_L
R102: A2_L⇒A3_M⇒A4_L⇒D2_L⇒D3_L⇒T1_L⇒T2_L⇒T4_L⇒T5_L⇒A1_L
R103: A1_L⇒A3_M⇒A4_L⇒D2_L⇒D3_L⇒T1_L⇒T2_L⇒T4_L⇒T5_L⇒A2_L
R104: A1_L⇒A2_L⇒A3_M⇒D2_L⇒D3_L⇒T1_L⇒T2_L⇒T4_L⇒T5_L⇒A4_L

In addition, there are rules with a slightly lower confidence R105 ~ R107 (support:0.33, lift:2.37(max)):

R105:A1_L⇒A2_L⇒A3_M⇒A4_L⇒D2_L⇒D3_L⇒T1_L⇒T2_L⇒T5_L⇒T4_L (conf:0.99)
R106:A1_L⇒A2_L⇒A4_L⇒D2_L⇒D3_L⇒T1_L⇒T2_L⇒T4_L⇒T5_L⇒A3_M (conf:0.96)
R107:A1_L⇒A2_L⇒A3_M⇒A4_L⇒D2_L⇒T1_L⇒T2_L⇒T5_L⇒T4_L⇒D3_L (conf:0.91)

The overall association rules R101–R107 are relatively consistent. The following characteristics or conclusions can be found in these rules:

1. Generally speaking, this pattern is basically the set of sub-features with low range values of all features in Table 3 (except A3_M), so the rule R01 conforms to the distribution of data features in Table 3;
2. In rules R101, R105 and R107, A/T features are not directly correlated, but correlated through group features (D-Type features), which means that there is no correlation between teachers' academic social networking and teaching behavior characteristics in the experimental data set;
3. In rules R102, R103, R104 and R106, A-Type features appear after T-type features respectively, which indicates that there is a weak correlation between teachers' teaching behaviors and academic social behaviors.
4. To sum up, from the second the third conclusions above, it's clear that 4 of the 7 rules support the third conclusion, and 3 of the 4 rules with highest lift degree support the third conclusion. Therefore, this paper holds that the overall correlation experiment results based on this data set show that there is a weak correlation between teachers' teaching behaviors and academic social behaviors.

4.2 Group Association Experiment

The current experiment is based on two sub-features of the two sub-group features (D2/D3), which are divided into nine groups for correlation analysis. The results are shown in Table 4:

Table 4. Association rules mining results of each group

Groups	Rules (Support; Confidence; Lift(max))	Code
D2_H-D3_H	A2_L⇒A3_M⇒A4_L⇒T1_L⇒T2_L⇒T5_L⇒T4_L (0.31;1;3.11)	R201
D2_H-D3_M	A1_M⇒T3_M⇒T4_L⇒T5_L (0.31; 1; 2.79)	R202
D2_H-D3_L D2_M-D3_H D2_L-D3_L	A1_L⇒A2_L⇒A3_M⇒A4_L⇒T1_L⇒T4_L⇒T5_L⇒T2_L (0.55;1;1.45) (There are subtle differences in support, confidence, and lift index for each group.)	R203
D2_M-D3_M	A1_M⇒A2_M⇒A3_M⇒A4_M⇒T1_M⇒T2_L⇒T3_L⇒T4_M⇒T5_L (0.5;1;16)	R204
D2_M-D3_L D2_L-D3_H	A1_L⇒A2_L⇒A3_M⇒A4_L⇒T1_L⇒T2_L⇒T3_M⇒T4_L⇒T5_L (0.3;1;1.9) (There are subtle differences in support, confidence, and lift index for each group.)	R205
D2_L-D3_M	A2_L⇒A4_M⇒T5_M⇒T2_L (0.32; 1; 1.54)	R206

In the case of uneven distribution of experimental data, the variation tendency of the relationship between teaching and scientific research reflected by the above experimental results is weak, but it is still helpful to explain the influence of the relationship between teaching and research selected by different types of universities based on practical considerations on specific teacher groups. That is, higher-level universities pay more attention to scientific research, so the group of teachers who work in them has more obvious characteristics of academic social behavior. While lower-level schools pay more attention to teaching, and thus their teachers have more obvious characteristics of teaching behavior. And middle-level schools are "doing two jobs at once and attaching equal importance to each", which shows a balanced relationship between teaching and scientific research and that is exactly what the rule R204 expresses.

5 Discussion

From the verification results based on experimental datasets, there is a relatively weak correlation between teaching and scientific research of teachers in domestic colleges and universities and the relationship between teaching and scientific research has different changes in different groups of teachers. Among them, teachers with middle-level academic titles in middle-level universities are more active in academic social networking and teaching and have a high degree of positive correlation between them. With the above group behavior pattern as the boundary, the academic behavior characteristics of the middle-level teachers in higher-level universities are more obvious, while teachers in the middle and lower level universities have more prominent characteristics of teaching behavior. To some extent, the tendency of these behavior characteristics reflects the current understanding and realistic choices of the relationship between teaching and scientific research in different types of colleges and universities. In general, the experimental results of this paper "reproduce" the complex and diverse characteristics of the relationship between teaching and scientific research in colleges and universities, and effectively verify the research hypothesis proposed in this paper.

This research also has some problems that need to be improved or discussed in the follow-up research. Imbalanced and imperfect datasets is the first problem that needs to be solved in the subsequent study. Since the choices of functions of the SCHOLAT users completely depend on the users' individual intention, from the perspective of research, the collected data sets are imperfect. For example, the proportion of users who take the initiative to label their professional title, education background and other professional qualification information is relatively low. Similarly, in terms of academic paper sharing, the proportion of people who actively and completely share their own data of academic works is not very high. Therefore, more accurate results will be obtained with more perfect datasets. In the follow-up research, similar problems need to be solved in a targeted way, such as obtaining more accurate teacher information through questionnaires or getting relevant academic achievements of teachers through Web Crawling. Secondly, clearer job correlation characteristics should be focused. Experiments discussed in this paper focus on the correlation between behaviors, but the correlation between teaching content and academic content has no relevant characteristics. In the follow-up research, semantic correlation features can be proposed based on semantic similarity calculation and other methods. Thirdly, teachers' working results discussed in traditional researches may as well be included as one of the characteristics. This paper assumes that the relationship between teaching and scientific research exists in the process. If the working results are included as one of the indicators of the model, will it increase the accuracy of the calculation? The above questions and considerations will be further discussed and verified in the follow-up research.

6 Conclusion

Based on the review of traditional empirical researches on the relationship between teaching and scientific research, this paper points out that traditional empirical research takes teachers' working results as the research object and thus ignores the subtle, long-term and non-dominant characteristics existing in the interpenetrative field between teaching and scientific research. Based on the new research hypothesis, this paper takes the educational big data provided by the SCHOLAT which has semantic consistency in teaching and scientific research and records the details of teachers' teaching and scientific research work as the research object. which records the details of the teacher's teaching and scientific research work, adopts the user portrait method to extract the characteristics of teachers' academic social networking and teaching behaviors, and use classical association rules mining algorithm to calculate the teacher's behavioral differences in teaching and scientific research. The experimental results effectively verify the research hypothesis. Finally, the problems that need to be solved and the ideas that will be verified in the following work of this paper are also discussed. This paper introduces the big data processing method into the research on the relationship between teaching and scientific research, and believes that more excellent research results will emerge from the empirical research in this field.

References

1. Clark, B.R.: The modern integration of research activities with teaching and learning. J. High. Educ. **68**(3), 241–255 (1997)
2. Tight, M.: Examining the research/teaching nexus. Eur. J. High. Educ. **6**(4), 293–311 (2016)
3. Genshu, L., Lina, G., Lei, L.: The relationship between research and teaching in chinese higher education institute: empirical analysis. Res. Teach. **28**(4), 286–290 (2005, in Chinese)
4. Duan, H.: Analysis of users' communication behavior based on social network., Ph.D., Huazhong University of Science and Technology, Wuhan (2015, in Chinese)
5. Xianhong, L., Gang, L.: Comparative analysis of recommender systems of research social networking service. Libr. Inf. Serv. (2016, in Chinese)
6. Yang, Z., Luqi, L.: Review and reflection on the status quo of academic social network research at domestic and abroad. Inf. Documentation Serv. **11**, 41–47 (2016, in Chinese)
7. Yuyao, L., Yong, T., Yonghang, H., Rui, D., Chunying, L.: Research on network teaching platform based on scholars' social model. Comput. Educ. **24**, 112–115 (2015, in Chinese)

A New Trust-Based Collaborative Filtering Measure Using Bhattacharyya Coefficient

Xiaofan Qin[1], Wenan Tan[1,2(✉)], and Anqiong Tang[2]

[1] College of Computer Science and Technology,
Nanjing University of Aeronautics and Astronautics, Nanjing 211100, China
qinxfan@foxmail.com, wtan@foxmail.com
[2] School of Computer and Information Engineering,
Shanghai Polytechnic University, Shanghai, China
{watan,aqtang}@sspu.edu.cn

Abstract. With the rapid growth of network data and demands of users, the concept of AI in recommendation system has become a hot academic topic. However, in sparse data, it is difficult for the current user to obtain his efficient neighbors and for some cold-start users, it doesn't do anything. Therefore, we constructed a new measure of trust between users for neighborhood based on Collaborative filtering(CF) which uses a pair of users common ratings and exploits Bhattacharyya similarity to finds relevance of each pair of rated items. We also have measured the validity of the proposed model through accuracy, recall rate and F1 measures. The results show that although some recall rates will be lost, the precision is greatly improved. Overall, it achieved good results.

Keywords: Collaborative filtering · Trust method · Bhattacharyya coefficient · Sparsity problem

1 Introduction

The recommendation system could be used to assist users filtering out the products they are interested in massive amounts of data when users' needs are not clear. We mainly discuss CF. Many experts have studied and improved it due to the data sparsity problem (DS) [1], cold start problem(CS) [2] and other problems that appear in the application process.

For the User-based approach, data sparsity problem occurs owing to the massive increase in products but the user is only interested in a little number of products, which is called hot products, causing the users have very few ratings for items on the whole. CS problem is generally for newly added users, who don't rate any items or have very small number of ratings so we can't effectively recommend them. At the same time, they also affect our recommendation to other users. These problems also exist for the Item-based approach.

© Springer Nature Singapore Pte Ltd. 2019
Y. Sun et al. (Eds.): ChineseCSCW 2019, CCIS 1042, pp. 399–407, 2019.
https://doi.org/10.1007/978-981-15-1377-0_31

We propose a novel measure to estimate the trust between a pair of users based on the [13] in sparse data. This measure is named Bhattacharyya Coefficient in Trust (BCT) owing to using Bhattacharyya coefficient to obtain the similarity between a pair of rated items, which is aimed to improve precision of recommendation in sparse data and make the calculations relatively simple. Our contributions are summarized as follows:

- Compared with the traditional filtering algorithm, it only adds two vital procedures. The one is to calculate the Bhattacharyya coefficient between users. The other is to combine the trust of two any users and the corresponding Bhattacharyya coefficient. In addition, our measure still utilize the traditional cogitation so it is relatively simple to compute and better understood.
- The Bhattacharyya measure is utilized in BCT which plays a major role in our approach. The link of a pair of rated items can be calculated. The BCT combines the trust between users with the link of each rated items, which enhances the trust between users to a certain extent, so even if there is no trust between users, it can be simulated by Bhattacharyya distance of a pair of rated items.

2   Related Work

In order to address problems within traditional CF of the recommendation system such as DS, CS, many experts and scholars have proposed different proposals. Liu et al. [3] leveraged user-owned rating information and behavioral preferences to present a new similarity measure. Najafabadi et al. [4] proposed a mining method based on association rules to make recommendations. Meanwhile, its a hot spot to use the trust between users to solve these problems. For the first time, Jennifer and Hendler [5] integrated a social network based on the semantic web to enhance trust to create predictive movies recommendation. However, the effect is generally not very satisfactory only using the trust between users to recommend. Some scholars have combined it with many other methods. Jamali and Martin [6] proposed a random walk algorithm based on the trust between each of users and they once again combined matrix factorization techniques and trust social network to recommend [7], which performed well on some indicators compared with other existing trust-based methods. Moreover, Jin and Chen [8] proposed a Top-K recommendation based on the social tagging network. Although the recommendation accuracy has improved a lot after the improvement, the overall recommendation efficiency is reduced due to the computational complexity. Jia et al. [9] measured credibility of user ratings from three dimensions. Forsati et al. [10] also did such a similar research. Reshma et al. [11] introduced a semantic-based trust network for recommendation. Patra et al. [12] first applied the Bhattacharya Coefficient to the recommendation solving the user cold start problem. Then they continued to study and proposed a new similarity calculation method for recommendation [13].

3   The Proposed Trust Measures

In the traditional collaborative filtering system, Let $U = \{u_1, u_2, ..., u_m\}$ as the user set that are composed of all users. Let $I = \{i_1, i_2, ..., i_n\}$ as the item set that are composed of all items. Thus, the user-item rating dataset can be denoted by $m \times n$ order matrix R. The rating of user u_k for item i_j can be represented by R_{kj} and if the user u_k has no ratings on item i_j, the $R_{kj} = 0$.

3.1   Bhattacharyya Measure

In our approach, let p_i and q_j indicate the two items i and j obtained from user-item rating dataset. The BC similarity can be indicated the relevance of two items which is expressed as

$$BC\,(i,j) = BC\,(p_i, q_j) = \sum_{h=1}^{m} \sqrt{p_{ih} q_{jh}} \tag{1}$$

where m is the maximum of the current item being rated and $p_{ih} = h/i$, where h is number of users rated the current item i with rating value of h and i is the number of users rated the item i.

3.2   User-Based Implicit Trust Calculation

Trust Derivation. We first obtain the reliability of the given user by appraising the accuracy of the user as the recommender to the active user in the past.

$$P_{aj} = \overline{r_a} + (r_{bj} - \overline{r_b}) \tag{2}$$

Where P_{aj} represents appraised rating of user a on item i_j by the neighbor b and r_{bj} represents the rating of user b on item i_j. The $\overline{r_a}$ and $\overline{r_b}$ denotes the mean rating of a and b respectively.

In the light of prediction errors based on their co-rated items, the normalized Mean Squared Differences (MSD) as same as literature [14] is utilized to measure the credibility of user b relative to user a. The specific calculation formula is as follows:

$$MSD_{ab} = \left(1 - \frac{\sum_{j=1}^{|I_{ab}|} (P_{aj} - r_{aj})}{|I_{ab}|} \right) \tag{3}$$

Where MSD_{ab} indicates that the credibility of the user a relative to the user b and I_{ab} is co-rated item set of the user a and b, r_{aj} represents the actual rating of user a on item i_j.

However, There are still a lot of deficiencies in MSD_{ab}. Because users who rated the very few ratings of items can achieve an extremely high degree of trust with almost all other users in this way. We utilize the user-based Jaccard coefficient to handle this problem, which is calculated as follows:

$$\text{UJaccard}_{ab} = \frac{|I_{ab}|}{|I_a| + |I_b| - |I_{ab}|} \tag{4}$$

Where I_{ab} is the set of co-rated items for users a and b, I_a and I_b are sets of rated items for user a and user b, respectively.

The implicit trust is as follows:

$$DTrust_{ab} = MSD_{ab} \times \text{UJaccard}_{ab} \tag{5}$$

Trust Propagation. As for the users who don't have direct relation in the trust social network, we use the trust propagation to calculate the implicit trust between users. Therefore, we can probably figure out a new relation which can be seen as the implicit trust between users without direct trust. We adopt the Weighted Mean Aggregation Method [14] to calculate the propagation implicit trust. The specific calculation is as follows:

$$PTrust_{ac} = \frac{\sum_{b \in adj(a)} DTrust_{ab} \times (DTrust_{bc} \times \beta_d)}{\sum_{b \in adj(a)} DTrust_{ab}} \tag{6}$$

Where $adj(a)$ represents the set of trusted adjacent neighbors of user a that trust c including the user b.

The MoleTrust is exploited as the weighting scheme in the proposed trust propagation metric, which is shown in Eq. (7).

$$\beta_d = (MPDist - dist + 1) / MPDist \tag{7}$$

where $\beta_d (0, 1]$ represents a weight, the $MPDist$ represents the maximum distance traveled from the source user to where the trust spreads. It is adjustable in different datasets. The parameter $dist$ is the distance of trust propagation from the source user to other recommended users. It can be explained in Fig. 1.

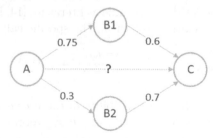

A→ C:(0.75×(0.6×0.5)+0.3×(0.7×0.5))/(0.75+0.3)=0.3

Fig. 1. An example of the trust propagation

3.3   BCT: A New Trust-Based Measure Using Bhattacharyya Coefficient

For improving the accuracy of recommendations, the BCT measure is proposed by utilizing the trust-based collaborative filtering algorithm of Bhattacharyya Coefficient, which calculated as follows:

$$T(a,b) = \sum_{i \in I_a} \sum_{j \in I_b} BC(i,j) \times Trust_{ab} \tag{8}$$

Where the $Trust_{ab}$ is the implicit trust between the user a and b which is calculated by trust derivation or trust propagation.

3.4   Prediction Computation

Generally speaking, the calculation of the final rating is an vital step in the process of recommendation. Thus, we predict a rating P_{ti} of the i-th item using the following Eq. (9).

$$P_{ti} = \overline{r_t} + \frac{\sum_{k \in N^{tru}} T(t,k) \times (r_{ki} - \overline{r_k})}{\sum_{k \in N^{tru}} |T(t,k)|} \tag{9}$$

Where $\overline{r_t}$ is the average of ratings for the target user t, $T(t,k)$ represents the trust between the target user t and his or her k-th neighbor k, r_{ki} represents the ratings of the k-th neighbor on item i and N^{tru} represents the nearest neighbor set of the target user t, including the top k most trusted neighbors.

4   Prediction Computation

In this section, we make comparison with the traditional user-based similarity measure Pearson Correlation (UCFPC) [14] and the similarity measure BCFcor and BCFmed constructed using the Bhattacharyya coefficient proposed in [13].

4.1   Datasets

We adopt the movielens100k dataset (https://movielens.umn.edu/) in our experiment, which includes 10,000 ratings from 1,943 users for 1,682 movies. The ratings scale is from 1 to 5.

4.2   Evaluation Metrics

The precision and recall are better evaluation metrics in Top-N recommendations [4]. Sometimes, the rise of precision causes a decrease in recall. Thus, an evaluation metric called F1_measure is equilibrium for these two metrics [4,13].

Assuming that L_r is the set of the Top-N recommendations for an user and L_{rec} is the set of items that the user's rated item is greater than the relevance threshold which is set as θ. The precision is calculated as follows:

$$\text{Precision} = \frac{|L_r \cap L_{rec}|}{|L_r|} \tag{10}$$

The recall metric is the ability to recommend the system to recommend all relevant items, which computes the ratio of relevant items that are recommended. It is written as follows:

$$\text{Recall} = \frac{|L_r \cap L_{rec}|}{|L_{rec}|} \tag{11}$$

F1_measure combines these two metrics as the best trade-off, which is shown as Eq. (12). Equally to the above two metrics, the larger the F1_measure metric, the better the recommended performance.

$$\text{F1\_measure} = \frac{2 \times \text{Precision} \times \text{Recall}}{\text{Precision} + \text{Recall}} \tag{12}$$

4.3   Experimental Results and Analysis

Our new method called BCT, measuring user trust using Bhattacharyya coefficients, which performs well on precision compared to UCF(PC), BCF_{med} and BCF_{cor} in [13].

We generally prefer to recommend better-rated items to users when recommending to users. Thus, we chose to experiment with θ of 3.5 and 4 respectively. It is specifically shown in Figs. 2 to 3 that even if the value of θ is different, the precision has increased by more than 20 % points, especially when the value of θ is at 4 shown in Fig. 2, it has almost increased by half compared to the other three algorithms. Unfortunately, the substantial increase in the precision of the algorithm we proposed has occurred a certain impact on the recall rate. The recall rate has dropped a lot in Fig. 3. When the value of θ is 3.5, it drops by about 10 % points and when the value of θ is 4, it drops by about 15 % points.

The F1_measure is used to ensure the quality of our proposed algorithm with others shown in Fig. 4. It is obviously that no matter what θ value takes, our algorithm performs well than the others. Although the conversion of the F1_measure metric when θ is taken at 3.5 is not as obvious as the θ is taken at 4. Therefore, the algorithm is more inclined to recommend the better rated item to the user.

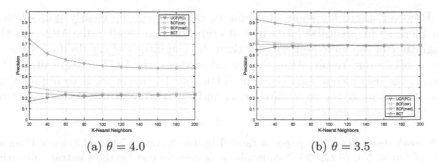

Fig. 2. The Precision of the BCT is compared with other algorithms

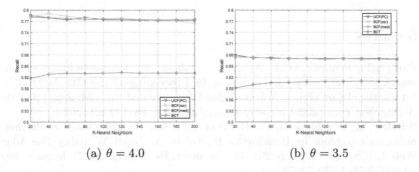

Fig. 3. The Recall of the BCT is compared with other algorithms

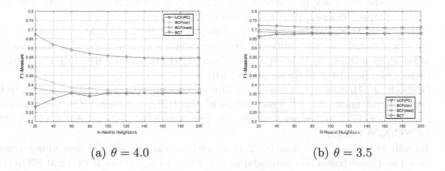

Fig. 4. The F1_Measure of the BCT is compared with other algorithms

5 Conclusion and Further Work

The BCT algorithm mainly aims to solve the data sparseness problem of CF. There are two foremost functions in using Bhattacharyya coefficients. The one is to achieve the purpose of enhancing the trust between users and the other is to simulate a very small number of users who cannot obtain their trust among any other users through trust propagation.

However, there are some problems in BCT. It can be clearly noticed that the increase in precision leads to a decrease in the recall rate. Although the magnitude of the increase is greater than the magnitude of the decline, it is still not a satisfactory result. After theoretical analysis, the BCT is also suitable for solving the user's cold start problem. Thus, our further work is to improve the algorithm to make the recall rate better and explore the utility of this algorithm on the cold start problem.

Acknowledgement. The paper is funded by the National Natural Science Foundation of Grant No. 61272036. Meanwhile, it is also funded by the Central University Fundamental Research Fund and the Key Discipline of Shanghai Second Polytechnic University. The grant numbers are NZ2013306 and XXKZD1604 respectively.

References

1. Park, S.T., Pennock, D.M.: Applying collaborative filtering techniques to movie search for better ranking and browsing. In: Proceedings of the 13th ACM SIGKDD International Conference on Knowledge Discovery and Data Mining, pp. 550–559. ACM, New York (2007) https://doi.org/10.1145/1281192.1281252
2. Schafer, J.B., Frankowski, D., Herlocker, J., Sen, S.: Collaborative filtering recommender systems. In: Brusilovsky, P., Kobsa, A., Nejdl, W. (eds.) The Adaptive Web. LNCS, vol. 4321, pp. 291–324. Springer, Heidelberg (2007). https://doi.org/10.1007/978-3-540-72079-9_9
3. Liu, H., Hu, Z., Mian, A., Tian, H., Zhu, X.: A new user similarity model to improve the accuracy of collaborative filtering. Knowl.-Based Syst. **56**, 156–166 (2014). https://doi.org/10.1016/j.knosys.2013.11.006
4. Najafabadi, M.K., Mahrin, M.N.R., Chuprat, S., Sarkan, H.M.: Improving the accuracy of collaborative filtering recommendations using clustering and association rules mining on implicit data. Comput. Hum. Behav. **67**, 113–128 (2017). https://doi.org/10.1016/j.chb.2016.11.010
5. Golbeck, J., Hendler, J.: FilmTrust: Movie recommendations using trust in web-based social networks. In: Proceedings of the IEEE Consumer Communications and Networking Conference, vol. 96, pp. 282–286(2006). https://doi.org/10.1109/CCNC.2006.1593032
6. Jamali, M., Ester, M.: TrustWalker: a random walk model for combining trust-based and item-based recommendation. In: Proceedings of the 15th ACM SIGKDD International Conference on Knowledge Discovery and Data Mining, pp. 397–406. ACM, New York (2009). https://doi.org/10.1145/1557019.1557067
7. Jamali, M., Ester, M.: A matrix factorization technique with trust propagation for recommendation in social networks. In: Proceedings of the fourth ACM Conference on Recommender Systems, pp. 135–142. ACM, Barcelona (2010). https://doi.org/10.1145/1864708.1864736
8. Jin, J., Chen, Q.: A trust-based Top-K recommender system using social tagging network. In: 2012 9th International Conference on Fuzzy Systems and Knowledge Discovery, pp. 1270–1274. IEEE, Sichuan (2012). https://doi.org/10.1109/FSKD.2012.6234277
9. Jia, D., Zhang, F., Liu, S.: A robust collaborative filtering recommendation algorithm based on multidimensional trust model. JSW **8**(1), 11–18 (2013)

10. Forsati, R., Barjasteh, I., Masrour, F., Esfahanian, A.H., Radha, H.: PushTrust: an efficient recommendation algorithm by leveraging trust and distrust relations. In: Proceedings of the 9th ACM Conference on Recommender Systems, pp. 51–58. ACM, Vienna (2015). https://doi.org/10.1145/2792838.2800198
11. Reshma, M., Pillai, R.R.: Semantic based trust recommendation system for social networks using virtual groups. In: 2016 International Conference on Next Generation Intelligent Systems (ICNGIS), pp. 1–6. IEEE, Kottayam (2016). https://doi.org/10.1109/ICNGIS.2016.7854045
12. Patra, B.K., Launonen, R., Ollikainen, V., Nandi, S.: Exploiting Bhattacharyya similarity measure to diminish user cold-start problem in sparse data. In: Džeroski, S., Panov, P., Kocev, D., Todorovski, L. (eds.) DS 2014. LNCS (LNAI), vol. 8777, pp. 252–263. Springer, Cham (2014). https://doi.org/10.1007/978-3-319-11812-3_22
13. Patra, B.K., Launonen, R., Ollikainen, V., Nandi, S.: A new similarity measure using Bhattacharyya coefficient for collaborative filtering in sparse data. Knowl.-Based Syst. **82**, 163–177 (2015). https://doi.org/10.1016/j.knosys.2015.03.001
14. Shambour, Q., Lu, J.: A trust-semantic fusion-based recommendation approach for e-business applications. Decis. Support Syst. **54**(1), 768–780 (2012). https://doi.org/10.1016/j.dss.2012.09.005

An Online Developer Profiling Tool Based on Analysis of GitLab Repositories

Jing Wang, Xiangxin Meng, Huimin Wang, and Hailong Sun[(⊠)]

School of Computer Science and Engineering,
Beihang University, Beijing, People's Republic of China
sunhl@buaa.edu.cn

Abstract. There are more and more developers as software is playing increasingly important roles in today's economic and social development. As a result, evaluating developers' expertise scientifically has become an urgent need for both Internet companies and developers. However, it seems that there is no satisfactory method to meet this demand currently. In this paper, we propose a solution to profile developers by analyzing their source code. We conduct the analysis of developers in terms of code quantity, code quality, skills, contribution, personalized commit time, and projects they participated in based on the GitLab code repositories. And we comprehensively evaluate developers' expertise from four perspectives, which are code quantity, code quality, contribution and score of projects they participated in. Compared with existing methods, our evaluation indicators are more comprehensive. We design and implement an online tool that can provide developer searching and profiling. Our tool has been used in Neusoft and Wonders Group to characterize the expertise and performance of their software developers.

Keywords: GitLab · Developer expertise · Developer profiling · Program analysis

1 Introduction

The number of developers has increased year by year with the rise of the Internet industry. And the question how to evaluate and rank millions of developers has gotten more and more attention.

Many Internet companies are using GitLab systems to develop and manage projects. From study of Marlow et al. [1], we can know that due to the transparency of data and the difficulty in falsifying activity trajectories, the expertise of job seekers inferred from GitLab data is more reliable than the description on the resume. The activities and contribution history of developers can demonstrate their knowledge of the code [2–4]. Therefore, we can use these data to learn about the developers' expertise in software development. So we believe that we can gain a deep understanding of developers' expertise by analyzing the data on GitLab.

Y. Sun et al. (Eds.): ChineseCSCW 2019, CCIS 1042, pp. 408–417, 2019.
https://doi.org/10.1007/978-981-15-1377-0_32

A comprehensive analysis and a reasonable evaluation of the developers' expertise based on GitLab code repositories are of great significance. In this paper, we analyze GitLab developers from code quantity, code quality, skills, contribution, personalized commit time, and projects they participated in. And we use four indicators, which are code quantity, code quality, contribution and score of projects they participated in, to measure the developers' expertise. Moreover, we design and implement an online tool that can provide developer searching and profiling. Our tool can bring the following benefits: (1) Optimize task assignment. Team leaders can allocate relatively urgent or important projects to employees who can complete tasks with higher quality and efficiency. (2) Improve the employee evaluation mechanism. The management can conduct employee assessments based on the results our tool gives. (3) Enrich the developer selection method. Companies can judge job seekers' expertise according to their code and finally find excellent candidates. Our tool has been used in both our own laboratory and enterprises such as Neusoft and Wonders Group to characterize the software developers' expertise and performance.

The rest of the paper is organized as follows. Section 2 presents related work. In Sect. 3, we provide a detailed introduction of our proposed analysis method and evaluation indicators. Section 4 describes our use cases. We conclude this paper in Sect. 5.

2 Related Work

The research on GitHub is relatively mature. Since GitLab is very similar to GitHub, we learn some methods from GitHub's research and use them to analyze GitLab data. We mainly pay our attention to the following two aspects:

1. The developers' contribution. By analyzing contributing.md researchers found a method to judge how much every developer contributes to the project [5]. Gousios et al. [6] added the number of non-comment lines of code and the contribution factor function to evaluate the developer's contribution.
2. The developers' ability assessment. Li et al. [7] used fuzzy analytic hierarchy process to evaluate developer's capabilities from both social and technical attributes. Ke et al. [8] proposed a fuzzy DEA model, which used the information of code quantity, development time, number of commits, number of bugs, and leader's opinion, to evaluate the efficiency of developers. Constantinou et al. [9] quantified every developer's skills by using his commit activities and ranked the developers by programming languages.

3 Analysis of Developers on GitLab

In this section, we introduce our analysis of GitLab developers from code quantity, code quality, skills, contribution, personalized commit time, and projects they participated in.

In order to make different kinds of data have the same order of magnitude, the following equation is used to normalize data:

$$y' = \frac{y - y_{min}}{y_{max} - y_{min}} \tag{1}$$

where y is an element in the set, y_{min} is the minimum value in the set, and y_{max} is the maximum value in the set.

3.1   Analysis of Code Quantity

The most basic indicator for evaluating a developer's expertise is his code quantity, which visually shows his workload.

We use the number of non-comment lines of code written by the developer to measure his code quantity and define the indicator $dloc$ to evaluate every developer's code quantity as

$$dloc = \frac{ncloc - ncloc_{min}}{ncloc_{max} - ncloc_{min}} \tag{2}$$

where $ncloc$ is the number of the non-comment lines of code developed by the current developer, $ncloc_{min}$ and $ncloc_{max}$ are respectively the minimum and maximum value in the set of developers' number of non-comment code lines.

3.2   Analysis of Code Quality

Code quality is an important manifestation of developer's expertise. For newcomers, whose code quantity is small, the quality of the code can measure their expertise more reasonably. In addition, the quality rather than length of experience is a strong correlate of job performance after the first two years [10].

We use the SonarQube to analyze the code quality. SonarQube is an open source code quality assessment system that supports analysis of more than 20 programming languages. It is widely used in the industry [11,12].

We evaluate the developer's code quality in terms of bugs (reliability), vulnerabilities (security), code smells (maintainability), complexity, and duplicated lines, and propose $dqua$ as an indicator of the developer's code quality. We recount the number of bugs, vulnerabilities, and code smells so that each violation is counted up to once per file. We introduce the following five sub-indicators to measure the quality of every developer's code more scientifically:

$$k\_rate = \frac{16 * blo\_k + 8 * cri\_k + 4 * maj\_k + 2 * min\_k + inf\_k}{ncloc} \tag{3}$$

$$complexity\_rate = \frac{complexity}{ncloc} \tag{4}$$

$$duplicated\_lines\_density = \frac{duplicated\_lines}{lines} \tag{5}$$

Equation (3) calculates sub-indicators for evaluating bugs, vulnerabilities, and code smells, where k can be replaced by *bugs*, *vulnerabilities*, and *code_smells*. The blo_k, cri_k, maj_k, min_k, inf_k respectively represent five levels of violations: blocker, critical, major, minor, and info. And *ncloc* represents the number of non-comment lines of code. Because different levels of violations have different impacts on code, we give different weights to different levels of violations to measure the quality of the code better.

Equation (4) calculates the complexity rate of the code, where *complexity* is the overall complexity of the developer's code, and *ncloc* is the number of non-comment lines of code of the developer.

Equation (5) calculates the density of duplicated lines of the code. In the equation, *duplicated_lines* is the number of duplicated lines of the code developed by the developer, and *lines* is the number of lines developed by the developer.

Finally, we normalize the five sub-indicators with Eq. (1) and evaluate the developer's code quality using the Eq. (6).

$$dqua = \frac{\sum_k k\_rate' + complexity\_rate' + duplicated\_lines\_density'}{5} \quad (6)$$

3.3 Analysis of Skills

We demonstrate every developer's skills from both programming language distribution and keywords of programming fields. After understanding the skills of the developers, the team leaders can assign each member the tasks he is good at, which can help the tasks be completed more efficiently.

We keep a file collection for each developer. The programming language distribution of each developer is obtained according to the file extensions in each collection. And the programming languages that the developer is good at are known.

We treat the content of the files modified by the developer as an article, and use the LDA model [13] to analyze the programming fields that each developer is good at.

We connect a developer's changed lines in all the commits into a string as an article. Then we preprocess the data based on natural language processing, which includes eliminating noise, word segmentation, and deleting the stopwords in the text. We choose the LdaMallet model, which can automatically find the optimal values of α and β. The value of iterations is 700. By comparing some LdaMallet models with different number of topics, we make K be 3.

The trained model's topics and their keywords are shown in Table 1. It is obvious that the topic 1 is a combination of front and back ends, the topic 2 is related to the back end, and the topic 3 is related to the front end.

By entering each developer's changes into the trained model, we can find out the programming fields he does well in.

Table 1. The topics and their keywords of the LDA model

Topic id	Keywords of the topic
1	string, span, class, public, cm, int, import, type, return, java, id, user, org, userid, pagesize, url, div, pageid, github, px
2	file, selection, false, true, line, dir, column, path, option, type, project, info, id, start, caret, state, end, provider, entry, editor
3	div, class, data, var, span, li, col, href, function, id, type, px, text, fa, option, pageid, style, return, content, model

3.4 Analysis of Contribution

We define the indicator *dcon* to evaluate the contribution of each developer to the projects he participated in. It can show the importance of him to the team.

Metrics, such as the commit frequency, lines of changed code, and number of resolved issues, are used by many employers to evaluate their employees' performance [10]. In this paper, the developer's contribution is evaluated in terms of the proportion of added lines, deleted lines, and commits. We calculate three sub-indicators called *add_prop*, *delete_prop*, and *commit_prop*, respectively, indicating the developer's addition, deletion, and commit proportion in a project he participated in. The calculation method is shown in Eq. (7).

$$item\_prop_j = \frac{item_j}{\sum_{i=1}^{m} item_i} \tag{7}$$

where *item* can be replaced by *add*, *delete*, and *commit*, *j* is the index of the current developer, and *m* is the number of developers in this project.

We use *dpcon* to show the contribution of the developer in a project, which can be expressed as the sum of the three sub-indicators.

$$dpcon = add\_prop + delete\_prop + commit\_prop \tag{8}$$

The contribution of the developer, which is named *dcon*, can be expressed as the average of his projects' *dpcon*.

$$dcon = \frac{\sum_{i=1}^{n} dpcon_i}{n} \tag{9}$$

where $dpcon_i$ is the current developer's contribution in the i-th project he participated in, and *n* is the number of projects he participated in.

3.5 Analysis of Personalized Commit Time

A personalized analysis of developers' commit time can help team leaders understand when their team members are more accustomed to developing and committing. We change the date of each commit to the day of the week, then count how many times a developer commits each day from Monday to Sunday and get a personalized analysis of each developer's working time.

3.6   Analysis of Participated Projects

Participated projects can reflect the developer's experience and level of development, which is also a perspective for evaluating developer's expertise. We firstly evaluate the projects in terms of development efficiency, code quality, and collaboration degree, and get the score of each project. Then we use the sum of scores as an indicator measuring the developer's expertise.

Analysis of Development Efficiency. Development efficiency is the simplest evaluation of the development team. After arranging a task, the leader definitely wants the team to complete the project as soon as possible.

According to the number and frequency of commits, we find the last commit time of the project during development, which is noted as $dlcommit\_time$.

The calculation of development efficiency is shown in Eq. (10).

$$peff = \frac{lines}{dlcommit\_time - created\_time} \tag{10}$$

where $lines$ is the number of lines of the project, $dlcommit\_time$ is the date of the last commit during development, and $created\_time$ is the date the project is created.

Analysis of Code Quality. The code quality of the project is measured using calculation method likes Eq. (6), which is shown in Eq. (11). The sub-indicators in the equation are about the values of a project.

$$pqua = \frac{\sum_k k\_rate' + complexity\_rate' + duplicated\_lines\_density'}{5} \tag{11}$$

Analysis of Collaboration. We get the co-authors of each file in the project, count the number of files that every developer has modified together with other developers, and save the result as an adjacency matrix. The row number and column number of the matrix represent the developers' index, and the value of each element is the number of files modified together by these two developers (values on the diagonal are set to 0), as shown in Fig. 1.

$$\begin{bmatrix} 0 & 5 & 3 \\ 5 & 0 & 1 \\ 3 & 1 & 0 \end{bmatrix}$$

Fig. 1. Project collaboration

We introduce collaboration *average* and collaboration *variance* to quantify collaboration degree of a project. The calculation of these two sub-indicators are as follows:

$$average = \frac{2\sum_i c_i}{n(n-1)} \tag{12}$$

where c_i is the i-th element at the top right of the diagonal in the matrix, and n is the number of developers of the project.

$$variance = \frac{2\sum_i (c_i - average)^2}{n(n-1)} \tag{13}$$

where c_i is the i-th element at the top right of the diagonal in the matrix, $average$ is the collaboration average of the project, and n is the number of developers of the project.

The $average$ and $variance$ are normalized using Eq. (1) to obtain $average'$ and $variance'$ respectively. Finally, the collaboration indicator of the project, which is called $pcol$, is introduced.

$$pcol = average' - variance' \tag{14}$$

Project Evaluation. The indicators $peff$, $pqua$ and $pcol$ are normalized to $[0, 1]$ using Eq. (1). Then the normalized indicators are calculated to obtain a comprehensive score of the project. As shown in Eq. (15).

$$pscore = \frac{peff' + 1 - pqua' + pcol'}{3} \tag{15}$$

Finally, considering the number and the quality of the projects together, the evaluation indicator $dsop$ is defined, and the calculation is shown in Eq. (16).

$$dsop = \sum_i pscore_i \tag{16}$$

where $pscore_i$ is the score of the i-th project that the current developer participated in.

3.7   Comprehensive Evaluation of Developers

We select code quantity, code quality, contribution and score of projects they participated in to comprehensively evaluate developers. We normalize the four indicators, which are called $dloc$, $dqua$, $dcon$, $dsop$, to $[0, 1]$ using Eq. (1). Finally, the normalized indicators are weighted and summed to obtain the comprehensive score of each developer.

$$dscore = \alpha * dloc + \beta * (1 - dqua) + \gamma * dcon + (1 - \alpha - \beta - \gamma) * dsop \tag{17}$$

$0 < \alpha, \beta, \gamma, 1 - \alpha - \beta - \gamma < 1$ in the equation are weight parameters, and here we set $\alpha = \beta = \gamma = 0.25$.

4   Use Cases

We collected the data from our laboratory's GitLab system, and got the information of 132 developers, involving in 49 projects. Using the data, we designed and implemented an online tool to show the results of our work.

The developer's page demonstrates the analysis results of each developer in terms of development quantity, contribution, personalized commit time, skills, and code quality, which is shown in Fig. 2.

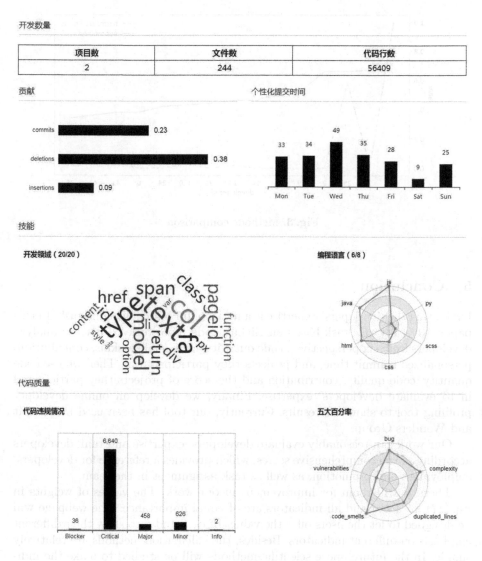

Fig. 2. The developer's page

We compared our method with a method using ncloc as the only metric to evaluate developers. Since ncloc used to be popular. The result is shown in Fig. 3. For developers whose id is 7, 2, 39, 21, 14, their ncloc values are very small, indicating that their code sizes are small. Actually, most of them are new students who mainly changed some code in a few good projects. And their scores are in the top seven using our method. In other words, our method considers four indicators when evaluating developers, which is more reasonable.

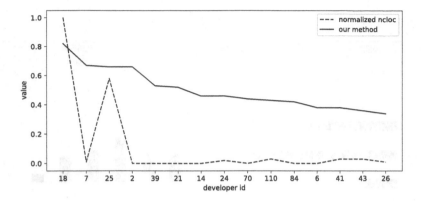

Fig. 3. Methods comparison

5   Conclusion

Evaluating the developers' expertise is a meaningful but challenging task. In this paper, we study this task based on GitLab code repositories. First, we analyze developers from six perspectives: code quantity, code quality, skills, contribution, personalized commit time, and projects they participated in. Then we use code quantity, code quality, contribution and the score of projects they participated in to evaluate developers' expertise. Finally, we develop an online developer profiling tool to show the results. Currently, our tool has been used in Neusoft and Wonders Group.

Our work can reasonably evaluate developers' expertise and rank developers according to the comprehensive scores, which provides a reference for developers' employment and promotion, as well as task assignment in the team.

There is still room for improvement in our work. The values of weights in Eq. (17) are fixed and all indicators are of equal importance. The webpage will be designed to let the users offer the values of α, β and γ to show their different emphasis on different indicators. Besides, the calculation methods are relatively simple. In the future, more scientific methods will be studied to make the indicators more reasonable.

References

1. Marlow, J., Dabbish, L.: Activity traces and signals in software developer recruitment and hiring. In: Proceedings of the 2013 Conference on Computer Supported Cooperative Work, CSCW, pp. 145–156. ACM, New York (2013)

2. Thomas, F., Gail, C.M., Emily, H.: Does a programmer's activity indicate knowledge of code? In: Proceedings of the the the 6th Joint Meeting of the European Software Engineering Conference and the ACM SIGSOFT Symposium on The Foundations of Software Engineering, ESEC-FSE, pp. 341–350. ACM, New York (2007)
3. Constantinou, E., Kapitsaki, G.M.: Developers expertise and roles on software technologies. In: 2016 23rd Asia-Pacific Software Engineering Conference (APSEC), pp. 365–368 (2016)
4. Kagdi, H., Hammad, M., Maletic, J.I.: Who can help me with this source code change? In: 2008 IEEE International Conference on Software Maintenance, pp. 157–166 (2008)
5. Kobayakawa, N., Yoshida, K.: How github contributing.md contributes to contributors. In: 2017 IEEE 41st Annual Computer Software and Applications Conference (COMPSAC), vol. 1, pp. 694–696 (2017)
6. Gousios, G., Kalliamvakou, E., Spinellis, D.: Measuring developer contribution from software repository data. In: Proceedings of the 2008 International Working Conference on Mining Software Repositories, MSR, pp. 129–132. ACM, New York (2008)
7. Li, J., Liu, J., Wu, Z., He, L.: Evaluation method of developers in GitHub based on fuzzy analytic hierarchy process. Appl. Res. Comput. 33(1), 141–146 (2016)
8. Ke, Q., Wu, S.: Evaluation of developer efficiency based on improved DEA model. Comput. Telecommun. 6, 60–62 (2017)
9. Constantinou, E., Kapitsaki, G.M.: Identifying developers' expertise in social coding platforms. In: 2016 42th Euromicro Conference on Software Engineering and Advanced Applications (SEAA), pp. 63–67 (2016)
10. Baltes, S., Diehl, S.: Towards a theory of software development expertise. In: Proceedings of the 2018 26th ACM Joint Meeting on European Software Engineering Conference and Symposium on the Foundations of Software Engineering, ESEC/FSE, pp. 187–200. ACM, New York (2018)
11. Raibulet, C., Fontana, F.A.: Collaborative and teamwork software development in an undergraduate software engineering course. J. Syst. Softw. 144, 409–422 (2018)
12. Kosti, M.V., Ampatzoglou, A., Chatzigeorgiou, A., Pallas, G., Stamelos, I., Angelis, L.: Technical debt principal assessment through structural metrics. In: 2017 43rd Euromicro Conference on Software Engineering and Advanced Applications (SEAA), pp. 329–333 (2017)
13. Blei, D.M., Ng, A.Y., Jordan, M.I.: Latent dirichlet allocation. J. Mach. Learn. Rese. 3(Jan), 993–1022 (2003)

Research on Community Discovery Algorithm Based on Network Structure and Multi-dimensional User Information

Liu Wang, Yi He, Chengjie Mao[✉], Dan Mao, Zuoxi Yang, and Ying Li

South China Normal University, Guangzhou 510631, China
{lwang,yihe}@m.scnu.edu.cn, maochj@qq.com,
1052761446@qq.com, 2345903@qq.com, 964663267@qq.com

Abstract. Recently, most of the community discovery algorithms are based on the structural information of undirected networks, and the social characteristics of users are less considered. Based on the academic social network, we propose a label propagation algorithm that integrates the network structure and multi-dimensional user information (LPA-NU). Through the fusion of multi-dimensional social networks, the algorithm firstly uses the LDA model to mine the similarity of user research directions to derive the hidden social edges between users. Secondly, it constructs a comprehensive directed weighted network, and then classifies the community according to the initial sub-group information. In order to evaluate the quality of community discovery, this paper proposes the definition of overlapping modules of directed networks. We conduct relevant experiments on real social network datasets (SCHOLAT). Experiments show that the LPA-NU algorithm can better divide the structure of the community, and the quality of community division is higher.

Keywords: Academic social networks · Community discovery · LDA · Label propagation · SCHOLAT

1 Introduction

In recent years, with the growth of social big data, the structure of the network composed of nodes and the connections between nodes has become more and more complicated. In order to more accurately mine the structural characteristics of complex networks, reveal the aggregation behavior and evolution law of the network, and predict the relationship between individuals, more and more scholars have researched and improved the algorithms for community discovery [1].

For the research of complex network community discovery algorithms, the traditional algorithms are mainly based on the topology of the network, such as, graph-based partitioning algorithm [2], seed diffusion algorithm [3] and random walk [4], etc. Traditional community discovery algorithms are mostly non-overlapping community discovery algorithms, while in real life, people tend to belong to more than one community. Therefore, the current community discovery research is aimed at overlapping community discovery research, and more and more scholars have found that

© Springer Nature Singapore Pte Ltd. 2019
Y. Sun et al. (Eds.): ChineseCSCW 2019, CCIS 1042, pp. 418–428, 2019.
https://doi.org/10.1007/978-981-15-1377-0_33

the nodes in many complex social networks also have rich semantic information, and the interaction behavior between nodes is also multi-dimensional. Considering the characteristics of a certain aspect alone cannot accurately discover the structure of the community. Therefore, it is necessary to integrate multi-dimensional information structures for overlapping community discovery, which has become a current research trend.

With the rapid development of various online social platforms, there are more and more relevant researches on community discovery of online social networks. Academic social network is a special complex network for scholars, the community division of academic social networks can provide appropriate recommendations for scholars and promote academic exchanges and cooperation among scholars. At present, many relevant researches only consider the connection of scholars in a certain aspect, such as only based on the co-authored information of the scholar's thesis, or the scholar's friend relationship, the user's similarity, etc. However, the social connection between scholars in complex academic social networks involves many aspects, and only by considering multiple social connections can we build a more comprehensive social network structure. Considering that some users who are very similar but lack of interaction are likely to be the same community, by analyzing their similarity, we can derive a hidden social edge to expand the structure of social network among scholars.

Based on the topology of academic social network, combined with user interaction information and semantic content analysis of users, this paper proposes a label propagation algorithm that integrates the network structure and multi-dimensional user information (LPA-NU).

The main contributions of this paper are:

1. By integrating the heterogeneous graphs composed by the user's academic cooperation network, the user's attention network, and the user interaction network, the social connections between users are enriched, the user's directed weighted social network is constructed, so as to solve the sparse problem of the user link matrix.
2. Through the LDA [5] model to calculate the similarity of research directions among users, and the hidden social side of the user is derived to solve the problem that some users in the academic social network are very similar, but the lack of social interaction is likely to belong to the same community.
3. For the complex network with directed weighting, the label propagation algorithm is improved by defining the initial label according to the initial sub-group information and measuring the propagation probability of the label by combining the multi-dimensional trust weights, solving the instability problem of the label propagation algorithm and improving the accuracy rate.

2 Related Work

Due to the simple and efficient characteristics of the label propagation algorithm, many scholars have improved the label propagation algorithm, and these improvements include the improvement of its label initialization, label selection, label propagation and so on. Literature [6] proposed COPRA algorithm based on LPA, which uses membership degree to determine whether a node absorbs the label of its neighboring node when the label is updated, so it can be used to discover overlapping communities. Aiming at the problem that the label propagation algorithm has great randomness in the node update sequence and label propagation process, which leads to unstable partitioning results, the literature [7] proposed a two-stage community discovery algorithm based on label propagation. In literature [8], the structural characteristics of the network are used to predict the edge of the network, and a similarity-based label propagation prediction strategy is proposed. Literature [9] proposed an overlapping community discovery algorithm based on multi-label propagation.

In order to improve the quality of community discovery, many scholars have improved the community discovery algorithm by integrating multi-dimensional features. Literature [10] first calculates the influence of user nodes and the propagation characteristics between networks through LeaderRank, and then calculates the user similarity by using the number of neighbors and the attribute characteristics of user, and combines these and the structure of the network to calculate the probability of label propagation for label propagation. Literature [11] first cluster users according to their interests, so as to obtain the initial interest-based community, and then calculate the correlation between users through the random walk algorithm with restart mechanism, and divide users into the communities with the largest correlation. Literature [12] combined user semantic characteristics with social relationship characteristics, proposed a semantic overlapping community discovery algorithm based on local semantic clustering, and proposed an SQ index which is more suitable for evaluating the quality of semantic community division. Literature [13] extended the definition of modularity to the discovery of overlapping communities with directed networks, extending the applicability of modularity.

The above algorithms do not fully consider the characteristics of the network, mainly for undirected networks or directed unprivileged networks, without considering that with the continuous development of the social platform, the information of the user nodes in the real social network is more and more rich, and the users also have multi-dimensional and directed connections. Considering one aspect of social connection alone cannot accurately show the structure of the real complex network, and the influence among users is not reciprocal. For example, considering the relationship of concern alone, the user follows a professor and an undergraduate user. Based on the identity information, the user accepts the professor's label to a greater extent, and based on other multidimensional features, such as interaction behavior, research direction, the user is more likely to accept the label of the user with frequent interaction and high similarity.

In summary, this paper firstly integrates multi-dimensional connections and user characteristics to build a more realistic directed weighted complex network, and then improves the shortcomings of the label propagation algorithm. Finally, this paper proposes a community discovery algorithm LPA-NU for directed weighted complex networks based on label propagation, which improves the quality of community discovery.

3   Algorithm Design and Implementation

3.1   Building Multidimensional Social Network

Based on the scholar's social network, the more connections between users, the more likely we think they belong to the same community. Considering that the social contact of users involves various aspects, it is impossible to accurately describe the complex social relationship between users by considering only the connection of the topological structure of the social network formed by a certain dimension. For example, if we only consider that A follows B, we can define them as only weak connections [14], and we need to integrate all aspects of the relationship to determine the user's social connection strength. In this paper, only three kinds of social connections are considered: academic cooperation connections, concern connections and some behavioral interaction connections among users. These three connections constitute the user's academic cooperation network, the user's attention network, and the user interaction network. By integrating the heterogeneous graph composed by the three networks, the directed weighted social network of users is constructed and the strength of social trust among users is calculated. Considering that the algorithm proposed in this paper is directed to a directed social network, the mutual trust between users is not the same.

Academic Cooperation Network. The most representative information of scholars is their academic achievements. If there is academic cooperation between two users, we define that there is a two-way edge between the two users. For user j who has worked with user i, the number of cooperation between the two users is $coauthor(i,j)$, and the academic trust of user i to user j is defined as:

$$AT(i,j) = \frac{coauthor(i,j)}{max\{coauthor(i,k)\}} \tag{1}$$

and the academic trust of user j to user i is defined as:

$$AT(j,i) = \frac{coauthor(j,i)}{max\{coauthor(j,k)\}} \tag{2}$$

User Attention Network. User i chooses to follow user j, which indicates that user i is interested in user j and has a certain degree of trust in user j. Therefore, according to the relationship of concern, we define that when user i follows user j, there is an edge of i pointing to j, A_{ij} is the node adjacency matrix, and the weight of the edge is defined as:

$$FT(i,j) = \begin{cases} 0, A_{ij} = 0 \\ 1, A_{ij} = 1 \end{cases} \tag{3}$$

User Interaction Network. The interaction behavior between users can also reflect the intimate trust relationship between users, $comment(i,j)$ is the number of comments by user i on user j dynamics, $transpond(i,j)$ is the number of forwards, and $like(i,j)$ is the number of likes, $sum(i,j)$ is the number of interactions between user i and user k, and k is the user who interacts most with user i. Calculate the interaction trust of user i to user j as:

$$IT(i,j) = \frac{\{comment(i,j) + transpond(i,j) + like(i,j)\}}{max\{sum(i,k)\}} \tag{4}$$

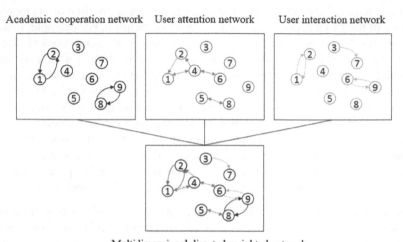

Academic cooperation network   User attention network   User interaction network

Multidimensional directed weighted network

Fig. 1. A comprehensive social network built by converging the heterogeneous graphs

As shown in Fig. 1, a comprehensive directed weighted social network is obtained by fusing the three-layer heterogeneous graph. Whether user i has a social edge to user j should consider the three heterogeneous graphs at the same time. If one is satisfied, there is a pointing edge, and the real social trust of user i to user j is defined as:

$$ST(i,j) = \alpha * AT(i,j) + \beta * FT(i,j) + (1 - \alpha - \beta) * IT(i,j) \tag{5}$$

All the above three kinds of connections belong to the real social connections of users, without considering that although some users lack social connections but have high similarities, there is a strong tendency of interaction between them, and they are likely to belong to the same community.

3.2 Extend the Network by Combining Edges Derived from User Similarity

In order to compensate for the lack of social interaction of some users, we introduce the calculation of similarity to add edges and weights between users with high similarities. At present, the commonly used methods for analyzing the similarity of texts composed of natural language are TF-IDF [15], LDA [5], and word2vec [16]. In this paper, the LDA topic model is used to mine the user's research direction from the information of natural language composed by scholars' papers, projects, etc. Then we calculate the similarity of the user's research direction by cosine similarity [17]. The hidden social edges among similar users are derived from this result, and the value of the similarity is assigned to these edges. The weight of edges is defined as:

$$RT(i,j) = RT(i,j) = cos(D_i, D_j) = \frac{\sum_{k=1}^{m}(W_{ik} \times W_{jk})}{\sqrt{\sum_{k=1}^{m}(W_{ik})^2} \times \sqrt{\sum_{k=1}^{m}(W_{jk})^2}} \quad (6)$$

$D_i = \{w_{i1}, \ldots, w_{ik}, \ldots, w_{im}\}$ is the research interest vector of user i, m is the number of topics, and w_{ik} is the probability that user i is interested in topic k. Set the threshold value as α and only when the user's similarity is higher than the threshold, the hidden social edges among similar users are derived and the directed weighted network is reconstructed. As shown in Fig. 2 below, the weight of the directed weighted social network is redefined as:

$$W(i,j) = \lambda RT(i,j) + (1 - \lambda)ST(i,j) \quad (7)$$

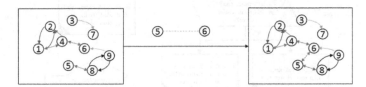

Fig. 2. Extend the network by combining edges derived from user similarity

3.3 Label Propagation Algorithm that Integrates the Network Structure and Multi-dimensional User Information

The traditional label propagation algorithm has strong randomness, the results are unstable, and the accuracy of community partitioning is not high. For the complex academic social networks, this paper proposed a label propagation algorithm LPA-NU that integrates network structure and multi-dimensional user information. This algorithm builds a new type of social network and improves the traditional tag propagation algorithm in the aspects of label initialization, label selection rules and label propagation. In this paper, we first construct a directed weighted social network by combining user semantic information and user interaction behavior, and define the propagation probability of the label according to the trust weight of the social edge. Considering that the user's rich attribute information contains social team information, and because of that the user's contact in the social team is relatively close, the user contact between the groups is sparse, we can define the initial label for users in the team based on the initial subgroup information. In the real life, users tend to accept the labels of users they trust, therefore, in terms of label propagation, we define that the user only accepts the label of the user that he or she points to, that is, only considers the outgoing edge of the user node. And because the user may belong to different communities, in terms of label selection, we build the user's label set based on the propagation probability of the labels.

Each label of a node is represented by a label pair (l,p), l is a specific label, p is the probability that the node has the label, and the probability of the node label in the initial subgroup is defined as 1. According to the structure and edge weight of the weighted social network, the LPA-NU algorithm continuously performs label propagation iteration until the label converges. The node's label set is expressed as:

$$L(i) = L(i) + \sum\nolimits_{j \in S(i)} L(j) * ST(i,j) \qquad (8)$$

Delete the label whose weight is less than the threshold value λ, and the final user label set is obtained. The label set of node a is expressed as $label(i) = \{(l_1,p_1), (l_2,p_2), \ldots, (l_n,p_n)\}$, and the nodes with the same label are divided into the same community, and n is the number of communities to which the node i belongs. The flow chart of the algorithm proposed in this paper is shown in Fig. 3:

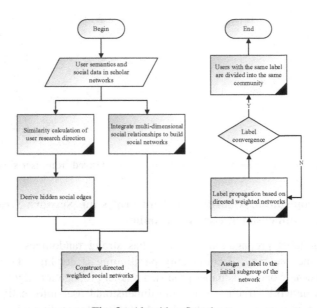

Fig. 3. Algorithm flowchart

4 Experiment

4.1 Data Set

The data set used in this paper is the real data of the academic social networking site – SCHOLAT (www.scholat.com). SCHOLAT is a scientific research collaboration platform for scholars, and provide scholars with a series of services, such as academic achievement management, curriculum and team management, news dynamic management, and user online interaction, etc. Users of the SCHOLAT can establish mutual social contacts, such as mutual attention, social interaction, co-authored papers, etc. At present, the number of users of the scholar network has exceeded one hundred thousand, and the users themselves also have rich semantic information.

This paper selects the cooperation information of users' papers, patents, projects and publications, as well as users' concern information and users' interaction information. These kinds of data are used to construct complex social networks, and then mine users' academic achievement data, calculate the similarity of users' research direction, and derive the hidden edge of similar users. The network information composed of these types of data is shown in Table 1:

Table 1. Network data set

Type	Node	Edge
Academic cooperation network	856	3060
User attention network	29391	96188
User interaction network	1858	19716
A network derived from user similarity	4626	728956
A network that integrates multidimensional social connections	29979	101027
Comprehensive directed weighted social network	30400	820779

4.2 Evaluation Index

At present, a lot of researches use the overlapping modularity EQ [18] to evaluate the quality of overlapping community partitions. According to the above definition of overlapping modularity, this paper proposes the overlapping modularity based on directed networks, which is defined as follows:

$$DEQ = \frac{1}{R} \sum_{c_k \in C} \sum_{i,j \in c_k} \left(A_{ij} - \frac{O_i O_j}{R} \right) \frac{1}{S_i S_j} \tag{9}$$

In this formula, R is the total number of directed edges of the social network, C is the set of communities divided by the algorithm proposed in this paper, K is the total number of communities, A is the adjacency matrix of the nodes, A_{ij} judges whether node i points to node j with connected edge. The value is 1 if it is a connected edge, otherwise the value is 0. O_i is the degree of the node i, and S_i is the number of communities which the nodes i belong. The higher the value of the overlapping module, the better the quality of the community partition.

4.3 Results

In order to verify the effectiveness of this algorithm and explore the influence of the different social network weights and network weights derived from the similarity of user research directions, we compare our proposed LPA-NU algorithm with the following algorithm by using the above DEQ evaluation index.

(1) Label propagation algorithm based on user attention network (LPA-UA): this algorithm just considering the user concerned information in directed unweighted network.

(2) Label propagation algorithm based on multidimensional social network (LPA-MS): This algorithm adds the user multi-dimensional social network information as the weight and perform the label propagation in the directed weighted network.

(3) Label propagation algorithm based on network derived from user similarity (LPA-US): This algorithm adds the user research direction similarity based on the academic achievement information as the weight and perform the Label propagation in the directed weighted network.

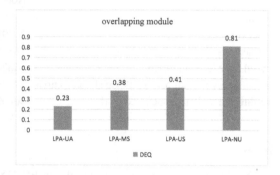

Fig. 4. Overlapping module of various algorithms

Figure 4 respectively shows the results of the different algorithm in DEQ evaluation index. The DEQ by 0.23 and 0.38 in LPA-UA and LPA-MS. That means integrating multi-dimensional connection can get better effect of community division than single social connection. The comprehensive and real interaction information can express more accurate community structure. The LPA-US get the better result than the LPA-MS. We hypothesize the particularity of academic social network is the main reason. For these academic users, the research direction is a more important factor to decide the community of them. The result of the LPA-NU get significantly improvement. Therefore, the synthesis method with multi-dimensional social connections and semantic similarity of users is more suitable for complex academic social networks.

We also discuss the effect of different weights λ on results. in multi-dimensional social connections and semantic similarity of users. Figure 5 shows the DEQ with different weights parameters λ The λ is the weight of semantic similarity of users and the $(1 - \lambda)$ is the weight of multi-dimensional social connections. We can get the best quality of community partitioning when the value of λ is 0.8. The experimental results show that user research direction similarity has greater impact than user interaction behavior in academic social networks. And the users with similar research are more likely in the same community is also accord with the characteristics of academic social network.

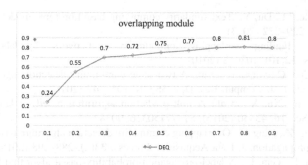

Fig. 5. Influence of λ index on overlapping module degree.

5 Conclusion

We proposed the LPA-NU algorithm, a new label propagation algorithm by combing network structure and multi-dimensional academic information in academic social network. This algorithm builds a multi-dimensional social network to alleviate the problem that traditional algorithm cannot accurately define the strength of social trust. At the same time, for these scholar users with much similarity information and little socialization information, we also used the LDA model to exploit the academic achievement information of them and constructed the hidden social edges by the calculated similarity of the research direction. This algorithm also improves the label propagation partition to solve the instability problem of the algorithm and improves the quality of community partition obviously. For the LPA-NU algorithm, we only consider the initial subgroup in the definition process of the initial label. And the propagation probability of the label only considers the influence between users. In the future, we will try to construct the initial core group by social structure and user attribute information. Measuring the probability of the label propagation and conduct community discovery by user global influence is also one of the major areas of continuing research.

Acknowledgements. Our works were supported by the National Natural Science Foundation of China (No. U1811263, No. 61772211) and Innovation Team in Guangdong Provincial Department of Education (No. 2018-64/8S0177).

References

1. Ying, K., Gu, X., Bo, Y., et al.: A multilevel community detection algorithm for large-scale social information networks. Chin. J. Comput. (1), 169–182 (2016)
2. Li, J., Zhou, Z, R.: Community discovery of P2P resources based on bipartite graph. In: International Conference on Computational Intelligence & Software Engineering (2009)
3. Qi, J., Liang, X., Yi, W.: Overlapping community detection algorithm based on selection of seed nodes. Appl. Res. Comput. (12), 20–23 + 54 (2017)
4. Lai, D., Lu, H., Nardini, C.: Finding communities in directed networks by PageRank random walk induced network embedding. Phys. Stat. Mech. Appl. 389(12), 2443–2454 (2010)

5. Wang, Z., He, M., Du, Y.: Text similarity computing based on topic model LDA. Comput. Sci. **40**(12), 229–232 (2013)
6. Gregory, S.: Finding overlapping communities in networks by label propagation. New J. Phys. **12**(10), 103018 (2010)
7. Zhen, W., Che, C., Qian, Y., et al.: A two-stage community detection algorithm based on label propagation. J. Comput. Res. Dev. **55**(09), 135–147 (2018)
8. Liu, J., Xu, B., Xu, X., et al.: A link prediction algorithm based on label propagation. J. Comput. Sci. **16**, 43–50 (2016). S1877750316300382
9. Du, C., Wang, Z., Xing, Z.: Overlapping community detection algorithm based on improved multi-label propagation. J. Data Acquisition Process. **33**(2), 288–298 (2018)
10. Liu, S., Zhu, F., Gan, L.: A label-propagation-probability-based algorithm for overlapping community detection. Chin. J. Comput. **39**(4), 717–729 (2016)
11. Fei, Y., Ming, Z., Yuwei, T., et al.: Community discovery based on actors' interests and social network structure. J. Comput. Res. Dev. **47**, 357–362 (2010)
12. Yu, X., Jing, Y., Tang, C., et al.: An overlapping semantic community detection algorithm based on local semantic cluster. J. Comput. Res. Dev. **52**(7), 1510–1521 (2015)
13. Nicosia, V., Mangioni, G., Carchiolo, V., et al.: Extending the definition of modularity to directed graphs with overlapping communities. J. Stat. Mech.: Theory Exp. **3**, 3166–3168 (2009)
14. Han, Z., Chen, Y., Liu, W., et al.: Research on node innuence analysis in social networks. J. Softw. **28**(1), 84–104 (2017)
15. Huang, C., Yin, J., Hou, F.: A text similarity measurement combining word semantic information with TF—IDF method. Chin. J. Comput. **34**(5), 856–864 (2011)
16. Luo, J., Wang, Q., Li, Y.: Word clustering based on word2vec and semantic similarity. In: 33rd Chinese Control Conference (CCC), pp. 508–511. IEEE (2014)
17. Dong, Y., Li, W., Yu, H.: Hierarchical relation mining of Chinese text based on mixed cosine similarity. Appl. Res. Comput. **34**(5), 1406–1409 (2017)
18. Shen, H., Cheng, X., Cai, K., et al.: Detect overlapping and hierarchical community structure in networks. Phys.: Stat. Mech. Appl. **388**(8), 1706–1712 (2009)

Micro-blog Retweeting Prediction Based on Combined-Features and Random Forest

Keliang Jia$^{(\boxtimes)}$ and Xiaotian Zhang

Shandong University of Finance and Economics, Jinan, China
sdjiakeliang@qq.com

Abstract. Aiming at the issue of incomplete features selection in micro-blog retweeting prediction researches, a micro-blog retweeting prediction method based on comprehensive features and random forest was proposed. Firstly, macroscopically, we analyzed the distinguishing features between high-retweeted micro-blog and low-retweeted micro-blog, between high-retweeted users and low-retweeted users, between high-retweeting users and low-retweeting users, and extracted nineteen features as the characteristics of micro-blog retweeting prediction. From the micro perspective, we extracted three local features which included the user's activity, the user's interest to the followee, the user's retweeting interest to the micro-blog, as microscopic features to predict micro-blog retweeting. And then we defined the computed methods of the features. Finally, combined with the features, a micro-blog retweeting prediction method based on Random Forest was proposed. The experimental results on Sina micro-blog datasets showed that the proposed method was superior to the classical classification prediction algorithms such as Logistic, BayesNet and Support Vector Machine (SMO), and had good stability on datasets of different scales.

Keywords: Random forest · Combined-features · Micro-blog · Retweeting prediction

1 Introduction and Related Researches

With the development of web2.0, social networks have become the mainstream social media which used by Chinese netizens due to their characteristics of rapid information transmission and strong interactions. It has become an important platform for people to express their opinions, obtain information and share information. According to the statistics of CNNIC (China Internet network information center), as of June 2018, the scale of Sina Micro-blog users in China expanded 337 million, and the usage rate was 42.1% [1]. The Micro-blog platform provides a unique push-retweeting information dissemination mechanism and makes the receiver to be the disseminator of the next information dissemination, makes the micro-blog information to achieve fission propagation, and even affect the Internet public opinion in the short term. Therefore, it is of great significance for topic detection, public opinion monitoring and enterprise product marketing to analyze the features which affect the retweeting behavior and to predict the retweeting behavior.

Y. Sun et al. (Eds.): ChineseCSCW 2019, CCIS 1042, pp. 429–440, 2019.
https://doi.org/10.1007/978-981-15-1377-0_34

Deng et al. [2] considered a case of the conflicts between urban and management officials, extracted the features from the poster and the micro-blog content, and predicted the retweeting amount of micro-blog based on BP neural network, the results showed the fans were more attracted by the content rather than the poster. Li [3] extracted the features of micro-blog, user features and theme features, then used SVM model to predict whether a micro-blog was retweeted by a user. In order to highlight the personality differences of users' retweeting behavior, Tang [4] et al. predicted the retweeting behavior of individuals based on the calculation of user similarity, and improved the prediction accuracy. Wang et al. [5] integrated the social network relationship between users and users' interests into the Bayesian Poisson Factorization model to predict users' retweeting behavior and got good results. Zhang et al. [6] studied the local regional social network, and considered the local social network structure formed by the users directly followed, and then used the traditional logistic regression method to predict the retweeting behavior and got a better result. Wang et al. [7] investigated the interaction between active users and notified users to study the maximization of social network message diffusion, and achieved good results. Chen [8] et al. considered the semantic information and the data form of a micro-blog, the user features and the interaction between users, verified that the receiver's activity, the semantic similarity of micro-blog and the user's interest, the interaction degree between the receiver and the releaser affected significantly on the retweeting behavior. Wang et al. [9] comprehensively analyzed the user features, the micro-blog contents, the user retweeting behavior constraints under the framework of Markov random field, and found that the retweeting behavior depended on these features and the retweeting behavior of the neighbor users. Zhou [10] et al. developed a modified model based on human dynamic models to describe the retweeting behavior, found that the distribution of the time intervals between successive retweeting activities was heavy-tailed. Liu [11] et al. found that the number of fans, the average retweeted number of micro-blog, the intensity of users' interaction and the similarity between micro-blog topics and users' topic interests could significantly influence the retweeting behavior. Guo [12] et al. established a novel retweeting behavior prediction method based on the LDA model to predict the behavior of users. Tang [13] et al. extracted five characteristics: the user's behavior, the user's interaction, the micro-blog, the user's interest, and the emotional divergence, and then used SVM to predict the retweeting behavior.

The above methods extracted the features from different perspectives to predict the retweeting behavior, but those were not comprehensive. Different from the above methods, we comprehensively considered the macroscopic and microcosmic factors which affected the retweeting behavior, analyzed and extracted the features of micro-blog retweeting, and proposed a retweeting prediction method based on combined-features and random forest (RPM-CFRF) to predict the retweeting behavior.

2   Description of the Issue

Sina micro-blog is a social network platform based on users' relationships. Potential dissemination network of micro-blog information is formed by one user following another one on the platform. When a user releases a micro-blog, this information would be pushed to all his fans, the fans could see the micro-blog after they log on to the platform, and could be at a certain probability to retweet the micro-blog, the micro-blog would be pushed to the fans' fans and so on, the micro-blog information could spread along the followed network to implement information explosive dissemination. The paper mainly studied the retweeting behavior based on the followed network. In order to describe the micro-blog user followed network, a micro-blog user network $G = <V, E>$ was constructed with the micro-blog users as the vertexes and the followed relationships between users as the edges, where V was the set of micro-blog users and E was the set of the edges. If a user v_i followed a user v_j, then $Y = F(v_i, v_j, blog)$ was used to indicate whether a micro-blog would be retweeted by v_i when it was released by v_j, $Y = 1$ meant the micro-blog was retweeted and $Y = 0$ meant the micro-blog was not retweeted. In this way, the prediction problem of micro-blog information retweeting was transformed to the binary classification problem of retweeting and non-retweeting.

3 Analysis of Features of Micro-blog Retweeting

The observation indicates that on the micro-blog platform some micro-blogs are retweeted in large quantities, while others are rarely retweeted; some users' micro-blogs are extensively retweeted by their fans, while other users' micro-blogs are rarely retweeted; moreover some users often retweet the receiving micro-blogs, while the others rarely retweet the receiving micro-blogs. In order to analyze and extract the characteristics of micro-blog retweeting, the paper used web crawlers to capture data set from sina micro-blog platform, and collected a total of 100,000 micro-blog users, 4 million micro-blog messages, 100,000 thumbed up messages and 7.5 million comments. Then from the macro perspective, the paper statistically analyzed the features of the micro-blogs which had been retweeted in large quantities, the features of the users whose micro-blogs were retweeted in large quantities, the features of the users who retweeted extensively micro-blogs as the macro features of micro-blog retweeting prediction; from the micro perspective, the paper analyzed the relationship between micro-blog releasers and receivers, the relationship between micro-blog and the interest of the receivers as the micro characteristics that affected the prediction of micro-blog retweeting.

3.1 Macroscopic Features Analysis

Micro-blog Features Analysis. When a micro-blog is widely retweeted, it is indicated that its content is popular among micro-blog users. From the perspective of the popularity of micro-blog content, some contents in the micro-blog are more attractive to the fans. So whether the micro-blog is retweeted is related to the popularity of the micro-blog. The popularity of the micro-blog is higher; it is more likely to attract its fans to retweet it. Therefore, the paper introduces the popularity of micro-blog to predict the retweeting behavior of users. The processing steps were as following:

Step 1 Classification of micro-blog. The obtained micro-blogs was sorted in descending by the retweeted counts. The first 30% of micro-blog was high-retweeted micro-blogs set $Bset_h$, and the last 30% was low-retweeted micro-blogs set $Bset_l$.

Step 2 Lexical analyses. For each micro-blog in $Bset_h$ and $Bset_l$, we used the word segmentation tool "jieba" to process it, and removed the stopped words.

Step 3 Keyword sorting. We counted the frequency of keywords in $Bset_l$ and $Bset_l$ and sorted respectively the keywords in the two sets in descending by the word frequency, and then obtained the vector $Vset_h$ and the vector $Vset_l$.

Step 4 Popular word vectors. We respectively took the first n words in $Vset_h$ and $Vset_l$ and removed the keywords that appeared in the two lists (if the number of the words in the list less than n after the removal of the repeated words, and the lists were filled by the remaining words in the set $Vset_h$ and $Vset_l$), and got the vector V_h of popular micro-blog hot words and the vector V_l of popular micro-blog cold words.

Step 5 Calculation of micro-blog popularity. For predicting micro-blog, after word segmentation and stop words removal, we got keywords vector V_i, and calculated V_i respectively with the vector V_h and V_l and got the similarity sim_h and sim_l. If $sim_h > sim_l$, then this tweet popularity was 1, meat that the fans would retweet the micro-blog. If $sim_h < sim_l$, the tweet popularity was 0, meat that the fans would not retweet it. If $sim_h = sim_l$, the tweet popularity was 2, meat that it was not sure whether the fans would retweet it or not. Jaccard similarity coefficient was used to calculate the similarity.

$$Jaccard(X, Y) = (X \cap Y)/(X \cup Y) \tag{1}$$

From the perspective of micro-blog content, the forms of micro-blog were more likely to attract fans to retweet. To analyze the content features of high-retweeted micro-blog, the paper respectively extracted users mentioned, subject labels, containing a URL or not, micro-blog length, emotional words number, etc. from $Bset_h$ and $Bset_l$. The distribution of the features in the two sets was as showed in Table 1.

Table 1. Comparison of features in $Bset_h$ and $Bset_l$.

Features	$Bset_h$	$Bset_l$
The proportion of users mentioned in micro-blog	0.21221	0.173216
The proportion of micro-blog with subject labels	0.28415	0.177768
The proportion of micro-blog contained the URL	0.055758	0.09889
Average length of micro-blog	21.4531	14.3347
The average number of emotional words	2.5297	1.3393
The proportion of emotional punctuation marks	0.05806	0.049911

Table 1 showed that there was a strong distinction between the set $Bset_h$ and the set $Bset_l$ to the length of micro-blog and the number of emotional words, which could effect on whether micro-blog was retweeted. The features of micro-blog with subject labels, users mentioned in micro-blog or not, micro-blog contained the URL or not, micro-blog with emotional punctuation marks or not also had a certain degree of differentiation between the micro-blogs in set $Bset_l$ and set $Bset_h$. To sum up, the paper selected these 7 features to introduce the feature system of micro-blog retweeting.

Releaser Features Analysis. From the perspective of micro-blog releasers, the observation indicates that some micro-blog releasers receive more attention, and their micro-blog can be quickly retweeted in large quantities. In order to obtain the features of high-retweeted micro-blog users, the paper divided micro-blog users into high-retweeted micro-blog users and low-retweeted micro-blog users: the obtained micro-blog users were sorted in descending according to the average of their retweeted micro-blog. The first 30% users were high-retweeted users and the last 30% were low-retweeted users. Then, we respectively extracted 7 features from the two sets, and calculated the distribution of the features among the two sets, as showed in Table 2.

Table 2. Comparison of features of micro-blog releasers.

Features	High-retweeted user	Low-retweeted user
Average of fans	1970808.8204	5565.3843
Average of micro-blogs	11678.3703	2389.6405
Average of being retweeted	1593.077907439	0.00
Average of followees	805.8388	987.2813
Average number of user labels	5.5285	4.381
Average of being commented	594.021523349	0.220504508
Average of being thumbed micro-blog	2048.435081541	0.315007163

Table 2 showed that the number of fans, the number of micro-blogs, the average of being retweeted micro-blogs, the average of being commented micro-blogs and the average of being thumbed micro-blog had a strong distinction between high-retweeted users and low-retweeted users. So these features could be used for the prediction of micro-blog retweeting. The number of user labels and the number of followees could also distinguish the two sets of users. Therefore, this paper chose the above seven features as the characteristics of the releasers to predict the micro-blog retweeting.

Retweeter Features Analysis. From the perspective of micro-blog retweeters, the observation indicates that some users are used to retweeting the micro-blog released by their followees, which is conducive to the rapid spread of the micro-blog, while other users rarely retweet the micro-blog released by the followees. In order to obtain the features of the users with high retweeting rate, we defined and calculated the retweeting rate of micro-blog users, that was the proportion of micro-blogs retweeted by each user in their released micro-blogs. The obtained micro-blog users were sorted in descending by the retweeting rate, with the first 30% as high-retweeting users and the last 30% as low-retweeting users. After that, four features from the two sets were extracted respectively, and the distribution of the features was as showed in Table 3. The forwarding rate formula was as follows.

$$F_{re} = s_{re}/s \tag{2}$$

Table 3. Comparison of features of micro-blog retweeters.

Features	High-retweeting user	Low-retweeting user
Average of fans	365370.1529	682719.5386
Average of micro-blogs	7209.0016	6785.1785
Average of followees	1140.7357	845.7280
Average number of user labels	1.923610301829623	2.26529542011420

Table 3 showed that the number of fans could better distinguish between high-retweeting users and low-retweeting users. There were some differences in the other three features. So the paper extracted these four features and the retweeting rate as the features of the retweeter.

3.2   Microscopic Features Analysis

From the microcosmic perspective, whether a micro-blog is retweeted is related to the interaction between the micro-blog releaser and the receiver in the micro-blog platform, the activity of the receiver, and the interest of the receiver.

In the platform, the number of followees of each user is different, and the same user shows different attentiveness to different followees. Specifically, the same user might retweet more micro-blogs issued by some followees, while might retweet less micro-blogs issued by the others. Therefore, we defined the user's interest to followees to measure the user's interest in retweeting micro-blogs from different followees, that was, the proportion of the number of micro-blog retweeted from one followee by a user to the total number of micro-blog retweeted from all his followees.

$$FI(v_i, v_j) = sum(v_i, v_j)/s_{re}(v_i) \tag{3}$$

Here, v_i was a fan of v_j, $sum(v_i, v_j)$ represented the total micro blogs retweeted by v_i from v_j, and $s_{re}(v_i)$ represented the total micro blogs retweeted by v_i.

Whether users are active in the micro-blog platform is particularly critical to the retweeting of micro-blog. So the number of posts $s_p(v_i)$, the number of reposts $s_{re}(v_i)$, the number of comments $s_c(v_i)$ and the number of thumbs up $s_z(v_i)$ per user in the recent period of time were used to measure the user's activity $A(v_i)$.

$$A(v_i) = (s_p(v_i) + s_{re}(v_i) + s_c(v_i) + s_z(v_i))/(t_{end} - t_{start}) \tag{4}$$

In the micro-blog platform, users always like to browse and retweet the micro-blog that they are interested in. Therefore, we defined the retweeting interest, that was, the similarity between a micro-blog received by a user and all the micro-blogs of the user. The specific calculation steps were as follows: firstly, word segmentation was performed on the entire user's micro-blogs and the stopped words were removed. The word frequency of keywords was counted and sorted in descending order. The m words with the highest word frequency were selected to form the user interest vector model V_{vi}. Secondly, we conducted word segmentation for the micro-blog to be predicted, and got the micro-blog vector V_{bi} after removing the stopped words. Thirdly, the Jaccard coefficient method was adopted to calculate the similarity between V_{vi} and V_{bi}, and the retweeting interest $RI(v_i, b_i)$ of user v_i on micro-blog b_i was obtained. The formula was as formula (1). To sum up, the paper selected these 3 features as microscopic features to introduce the feature system of micro-blog retweeting.

4 Model Building

4.1 Retweeting Prediction Features

According to the above analysis, the features and its' value range were in Table 4.

Table 4. Retweeting prediction features and its' value range.

The types	Features	Value range
Features of micro-blog	Users mentioned in micro-blog or not	{0, 1}
	Micro-blog with subject labels or not	{0, 1}
	Micro-blog contained the URL or not	{0, 1}
	With emotional punctuation marks or not	{0, 1}
	The length of micro-blog	The numerical
	The number of emotional words	The numerical
	The popularity of micro-blog	{0, 1, 2}
Releaser's features	The number of Fans	The numerical
	The number of micro-blog	The numerical
	The average of being retweeted	The numerical
	The number of followees	The numerical
	The number of User Labels	The numerical
	The average of being commented	The numerical
	The average of being thumbed	The numerical
Retweeter's features	The number of fans	The numerical
	The number of followees	The numerical
	The number of micro-blogs	The numerical
	The number of user labels	The numerical
	The retweeting rate	[0–1]
Microscopic features	The user's activity	[0–1]
	The user's interest to the followee	[0–1]
	The user's interest to the micro-blog	[0–1]

4.2 Predicting Model Based on Random Forest

Random forest is a combinatorial classifier derived from ensemble learning based on multiple decision trees. In prediction, the optimal classification results are obtained by voting multiple decision tree classifiers. The specific construction process was as follows: Step 1 Sample extraction. N sample sets were randomly selected using Bootstrap sampling method from all the sample sets that were put back. Step 2 Feature extractions. The iso probabilistic extraction was adopted to extract m features from all the retweeting features to form a feature subset, and then an optimal classification feature was selected as the node to construct CART decision tree. Step 3 Construct random forests. We repeated the above steps k times and got k decision trees, formed a random forest. Step 4 Retweeting predictions. The micro-blog samples were entered to be predicted, and each decision tree in the random forest would take participate in the classification prediction, and finally voted to form the prediction results.

In the random forest algorithm, the number of predicted feature sampling m affected the effect of random forest prediction. Here, m in the random forest algorithm is usually taken as $|log_2(M)| + 1$, and M is the total number of features. There are 23 features in this paper, so m is taken as 5.

5   Test Results and Analysis

5.1 Sample Data and Evaluation Criteria

The experimental sample data was extracted from the crawled sina micro-blog data set, and the final sample data should include the micro-blog releaser v_j, the forecasted retweeter v_i, the forecasted micro-blog $blog_k$ and its feature set, and the result Y, $Y = F$ $(v_i, v_j, blog_k, featureset)$. For the retweeting samples, micro-blogs retweeted by micro-blog users in the data set were directly selected as the retweeting samples.

For the non-retweeted samples, it was not simple to directly select from the data set. The micro-blog seen by the user but not retweeted should be selected as the non-retweeted samples. It was hard to obtain when the user checked in micro-blog platform, therefore, we judged whether the user logged on the micro-blog platform by whether he/she posted, retweeted, thumbed up, commented and other behaviors at a certain moment, so as to judge whether the user saw a micro-blog and did not retweet it. Finally, 194395 test samples were extracted which included 107304 retweeted positive samples and 87091 non-retweeted negative samples.

The precision, recall and F-Measure [14] were used to evaluate the prediction results. Precision was used to examine the accuracy of retweeting prediction model. Recall was used to examine the comprehensiveness of retweeting prediction model, and F-Measure was a comprehensive measure of precision and recall.

5.2 Comparison with Different Methods

In order to verify the validity of the RPM-CFRF method in the paper, the prediction experiments were compared with the classical classification algorithms such as

Table 5. Comparison with different methods.

Methods	Precision	Recall	F-Measure
J48	0.922	0.920	0.921
Logistic	0.894	0.891	0.892
BayesNet	0.876	0.864	0.870
SMO	0.887	0.885	0.886
RPM-CFRF	0.927	0.926	0.926

Logistic, Decision Tree (J48), BayesNet and Support Vector Machine (SMO). The experimental results were as follows.

In Table 5, the experimental results showed that the features set in this paper had a good prediction performance in different classification prediction algorithms, which showed the validity of the features selected in this paper. Decision tree (J48) algorithm had a good prediction performance. It was superior to the other classical classification algorithms such as Logistic, BayesNet and Support Vector Machine (SMO) in precision, recall and F-Measure. The RPM-CFRF algorithm in this paper was slightly superior to decision tree (J48) algorithm in each index. Because many decision trees were integrated in RPM-CFRF algorithm, so the prediction effect was better than that of J48 algorithm.

5.3 Contribution Rates of Different Types of Features

In order to verify the contribution of the different types of features proposed in this paper to improve the accuracy of micro-blog retweeting prediction, the precision of each type of features relative to all features was compared experimentally. A new feature set was constructed on the basis of the complete features by removing a type of features, and the contribution rate of this kind of features was expressed by the difference between the prediction accuracy of the complete feature and the prediction accuracy of the new feature set. The predicted results under the RPM-CFRF method in different features set were shown in the Table 6.

Table 6. Contribution rates of different types of features.

Features	Precision	Recall	F-Measure	Contribution rate
C-Feature - micro-blog features	0.922	0.919	0.921	0.5%
C-Feature - releaser features	0.830	0.828	0.829	9.7%
C-Feature - retweeter features	0.913	0.910	0.911	1.4%
C-Feature - Microscopic features	0.887	0.885	0.886	5.0%
Combined-Feature	0.927	0.926	0.926	–

The data in Table 6 showed that the releaser features contributed the most to the prediction of micro-blog retweeting, reached 9.7%; secondly, the microscopic features contributed 5.0%; the retweeter features contributed 1.4%; and the micro-blog content contributed the least, only 0.5%. Combined with the data analysis in Tables 1, 2 and 3, the comparison of micro-blog features on high-retweeted micro-blog and low-retweeted micro-blog had a certain degree of differentiation, but it was not as significant as that between high-retweeted users and low-retweeted users. Similarly, the features of retweeter were not significantly different between high-retweeting users and low-retweeting users. Therefore, the releaser features with significant distinction had the greatest impact on micro-blog retweeting prediction.

5.4   Comparison of Data Sets of Different Scales

In order to verify the stability of RPM-CFRF method in the paper, cross-validation method was used to test the stability of prediction results under different scale training data. Cross-validation is a commonly used accuracy test method. The data sets are divided into N parts, among which $N-1$ part is trained in turn, the remaining one is tested, and the average of N results is used as the evaluation criteria of the algorithm. Comparisons from 2-folds to 10-folds were showed in the Table 7.

Table 7. Comparisons from 2-folds to 10-folds for cross-validation accuracy.

n-fold	Precision	Recall	F-Measure
2-fold	0.9235	0.9220	0.9227
3-fold	0.9254	0.9239	0.9246
4-fold	0.9261	0.9246	0.9253
5-fold	0.9267	0.9252	0.9259
6-fold	0.9267	0.9252	0.9259
7-fold	0.9270	0.9256	0.9263
8-fold	0.9270	0.9256	0.9263
9-fold	0.9272	0.9257	0.9264
10-fold	0.9270	0.9256	0.9263

The Table 7 showed the precision, recall and F-Measure of cross validation of RPM-CFRF method from 2-folds to 10-folds. With the increase of training data scale, the performance of RPM-CFRF method had a slight improvement. The accuracy of RPM-CFRF method only increased from 92.35% of 2-folds to 92.70% of 10-folds, the recall rate only increased from 92.20% of 2-folds to 92.56% of 10-folds, and the comprehensive evaluation F-measure only increased from 92.27% of 2-folds to 92.63% of 10-folds. It showed that the RPM-CFRF method in this paper was less affected by the size of training set, and could obtain better prediction performance in the case of smaller training set. It verified that the RPM-CFRF method in this paper could adapt to different training sets and had strong stability.

6   Conclusion

The paper investigated whether a fan would be retweeted a micro-blog released by his followee. From the macro point of view, the features of releaser, retweeter and high-retweeted micro-blog were analyzed respectively. From the micro point of view, the relationship between fans and bloggers, fans' interest to received micro-blog were analyzed. And then we defined the computing methods of these features. Combining the 22 features of micro-blog, releaser, retweeter and micro-features, a retweeting prediction model based on random forest was designed. The precision, recall and F-measures were used to evaluate the prediction results of different experiments. On Sina data set, three groups of experiments were designed in this paper. One was to compare the performance of retweeting prediction between the proposed method and different machine learning classification algorithms. The second one was to verify the contribution of the proposed features to the accuracy of retweeting prediction. The third one was to verify the stability of the method under different data sets with different sizes. The results showed that the proposed method had the best prediction performance, and could adapt to different training sets, showed a better stability.

From the macro and micro view, this paper analyzed the features which affected micro-blog retweeting. The experimental results showed that the releaser features and micro-features significantly affected the prediction results of micro-blog retweeting. The proposed micro-blog retweeting prediction model based on combined-features and random forests could be effectively used in social network public opinion monitoring, hot topic discovery, influence dissemination and other related researches. In the next step, we will further study the calculation method of micro-blog popularity, refine the popularity level, and analyze the propagation law of micro-blog with different popularity levels, so as to further improve the precision of micro-blog retweeting prediction.

Acknowledgments. This work is partially supported by The National Social Science Fund of China (No. 15BGL203) and MOE (Ministry of Education in China) Project of Humanities and Social Sciences (Project No. 14YJC860011).

References

1. 42nd Statistical Report on the Development of Internet in China. http://www.cnnic.net.cn/hlwfzyj/hlwxzbg/hlwtjbg/201808/t20180820_70488.htm. Accessed 21 Nov 2018
2. Deng, Q., Ma, Y., Liu, Y.: Zhang, H: Prediction of retweet counts by a back propagation neural network. J. Tsinghua Univ. (Sci. Technol.) **55**(12), 1342–1347 (2015)
3. Li, Z.: Predicting retweeting behavior based on LDA topic features. J. Intell. **34**(9), 158–162 (2015)
4. Tang, X., Miao, Q., Quan, Y., et al.: Predicting individual retweet behavior by user similarity: a multi-task learning approach. Knowl. Based Syst. **89**, 681–688 (2015)
5. Wang, S.-Q., Li, C.-P., Wang, Z., et al.: Prediction of retweet behavior based on multiple trust relationships. J. Tsinghua Univ. (Sci. Technol.) **59**(04), 270–275 (2019)
6. Zhang, J., Tang, J., Li, J., et al.: Who influenced you? Predicting retweet via social influence locality. ACM Trans. Knowl. Discov. Data **9**(3), 1–26 (2015)

7. Wang, Z., Chen, E., Liu, Q., et al.: Maximizing the coverage of information propagation in social networks. In: Proceedings of 29th International Conference on Artificial Intelligence, pp. 2104–2110. AAAI Press, Austin (2015)
8. Chen, S., Dou, Y., Zhang, Q.: Research on influential factors of the reposting behavior of micro-blog users based on the theory of reasoned behavior. J. Intell. **36**(11), 147–152+160 (2017)
9. Wang, N., Gao, G., Chai, Z.: Predicting micro-blog users forwarding behavior based on markov random fields. J. Chin. Inf. Process. **32**(06), 107–113 (2018)
10. Zhou, C., Zhao, Q., Lu, W.: Modeling of the forwarding behavior in micro-blogging with adaptive interest. J. Tsinghua Univ. (Sci. Technol.) **55**(11), 1163–1170 (2015)
11. Liu, H.-L., Huang, Y.-L., Luo, C.-H., et al.: Modeling information diffusion on micro-blog network based on users' behaviors. Acta Phys. Sin. **65**(15), 158901-1–158901-12 (2016)
12. Guo, Y., Gong, Y., Zhang, Q., et al.: Retweeting behavior prediction using topic model. J. Chin. Inf. Process. **32**(04), 130–136 (2018)
13. Tang, X., Luo, Y.: Integrating emotional divergence and user interests into the prediction of microblog retweeting. Libr. Inf. Serv. **61**(09), 102–110 (2017)
14. Liu, W., He, M., Wang, L., et al.: Research on microblog retweeting prediction based on user behavior features. Chin. J. Comput. **39**(10), 1992–2006 (2016)

AI for CSCW and Social Computing

Cross-Domain Developer Recommendation Algorithm Based on Feature Matching

Xu Yu[1,2], Yadong He[1], Yu Fu[1], Yu Xin[3], Junwei Du[1(✉)],
and Weijian Ni[2]

[1] School of Information Science and Technology,
Qingdao University of Science and Technology,
Qingdao 266061, Shandong, China
djwqd@163.com
[2] Shandong Key Laboratory of Wisdom Mine Information Technology,
Shandong University of Science and Technology,
Qingdao 266590, Shandong, China
[3] Faculty of Electrical Engineering and Computer Science, Ningbo University,
Ningbo 315211, China

Abstract. In recent years, the software crowdsourcing has become a new software development pattern. More and more developers choose to publish, search for software tasks, and solve software problems on software crowdsourcing platform. As such, the platform generates a large amount of developer and development task information every day, which makes it difficult for developers to find appropriate tasks from massive tasks. Therefore, it is significant to deploy developer recommendation system on crowdsourcing platforms. Now, most developer recommendation algorithms can only use single platform data. Since the new software crowdsourcing platforms do not have enough historical behavior information of developers, previous developer recommendation algorithms cannot recommend developers to new tasks effectively. To solve the sparsity problem, this paper proposes a cross-domain developer recommendation algorithm based on feature matching. Firstly, we seek from the auxiliary domain for the most similar tasks to the current target domain task. Then, we retrieved the corresponding developers of these tasks. Finally, we select from the target domain the most similar developer to the developers retrieved to compose the recommendation developer set of the current task. In order to verify the effectiveness of the proposed algorithm, we crawls data from two different software crowdsourcing platforms to conduct experiments and compare the proposed model with various advanced developer recommendation algorithms. The experimental results show that the proposed algorithm has advantages over the previous algorithms on different evaluation metrics.

Keywords: Developer recommendation · Software crowdsourcing platform · Cross-domain recommendation · Feature matching

© Springer Nature Singapore Pte Ltd. 2019
Y. Sun et al. (Eds.): ChineseCSCW 2019, CCIS 1042, pp. 443–457, 2019.
https://doi.org/10.1007/978-981-15-1377-0_35

1 Introduction

In recent years, the software crowdsourcing has become a new software development pattern [1–3]. It can reduce development costs, shorten task cycle, and improve software innovation. Hence, it brings new opportunities for traditional software development methods, and has become the main way for developers to solve problems. However, with the continuous expansion of the platform scale, the platform has accumulated a large number of developers, and it is difficult for task publishers to quickly find suitable candidates among all the developers [4]. Therefore, the platform needs an intelligent recommendation system to match the tasks with appropriate developers.

In order to complete the correct matching of tasks and developers, the researchers studied resource acquisition and developer collaboration in the field of software engineering. Research [5] maps the description of technical language to the problem label of the platform, and then constructs association rules to achieve expert recommendation. Research [6] proposes a social impact-based approach to matching appropriate tasks for active and inactive developers. Research [7] extracts the topic description of development tasks and developers by using the Latent Dirichlet Allocation model (LDA), and then calculates the similarity between tasks and developers by topic. However, most developer recommendation algorithms are designed for mature software platforms, and do not consider new software crowdsourcing platforms with insufficient information resources. Moreover, only the data of a single platform is used and the research on cross-platform is lacking.

Therefore, this paper proposes a cross-domain collaborative filtering developer recommendation algorithm based on feature matching (CDCF-FM), which mainly includes task matching and developer matching. When recommending the suitable developer for a target domain development task, firstly, the development task is matched with the auxiliary domain development task. Hence, the similar development task from the auxiliary domain is obtained, and so the corresponding auxiliary domain developer can be retrieved. Then we calculate the similarity between the obtained developers and the target domain developers, and recommend the most similar developers from the target domain to the development task.

The main contributions of this paper include the following two points:

(1) We calculate the similarities between developers and between development tasks from different angles.
(2) We solve the data sparse problem in the target domain by means of cross-domain recommendation.

The remainder of this paper is organized as follows. Section 2 reviews related work. The proposed CDCF-FM algorithm is detailed in Sect. 3. In Sect. 4, we conduct extensive experiments in ZhuBaJie and Joint Force dataset. Finally, Sect. 5 summarizes the whole paper.

2   Related Work

In recent years, traditional recommendation system algorithms [8, 9] have been widely used in e-commerce websites and online social media to solve the increasingly serious problem of "information overload". Among the recommendation algorithms, collaborative filtering (CF) algorithm has achieved great success with strong versatility, good interpretability, and no need for content information features, but it also faces very serious data sparsity and cold start problem. In order to alleviate this problem, the cross-domain recommendation system [10, 11] combined with CF algorithm and transfer learning has been highly praised. By transferring knowledge from information-rich auxiliary domains to information-deficient target domains, it alleviates data sparsity in target domains and improves recommendation performance. In research [12], in order to decompose multiple matrices simultaneously, a transfer learning model of collective matrix decomposition is proposed, which uses the entities involved in multiple domains to share the feature factors of these entities. Research [13] proposed a collaborative filtering recommendation algorithm for coordinate system migration (CST), which reduces the data sparsity of target domain by transferring knowledge of users and projects in the auxiliary domain. Research [14] designed a rating matrix generation model, and the main idea is to use the potential clustering level rating matrix to establish the relationship between multiple domain rating matrices. Then it shares the clustering matrix to help the target domain rating matrix to fill the missing ratings.

Although the traditional recommendation system has good results in various application platforms, there are few recommendation algorithms for software engineering [15]. In software crowdsourcing platform, the final delivery result of software tasks are influenced by many factors, such as the ability level of developers and the task descriptions. However, there is currently no complete and standard evaluation system in the industry, which makes it difficult to construct the developer capability model and the development task model involved in the recommendation process [16]. The researchers have made a lot of efforts to address the above issues. Researches [17, 18] utilize the static information of the developer to find the similar relationship between the developer and the task to be developed, so as to recommend the suitable developer for the task. However, due to the influence of data sparsity, the accuracy and satisfaction of the recommendation results are not ideal. Research [19] proposes a feature model for recommending developers in software crowdsourcing tasks based on neural network and semantic analysis methods, but the model only focuses on task features and does not utilize the characteristics of developers. Research [20] proposes a multi-feature fusion recommendation method based on developer's ability and behavior, which can alleviate data sparsity problem, but it is difficult for developers to extract dynamic behavior data. So the model is difficult to extend to other platforms. A recent study [21] introduces heterogeneous information network into developer recommendation model to describe the professional knowledge of developers in software crowdsourcing platform. In addition, the activities of developers in different communities are considered, but the model does not focus on other structured data (rating, location, etc.) of the platform.

3 Our Model

3.1 Model Framework

The cross-domain collaborative filtering developer recommendation algorithm based on feature matching, is shown in Fig. 1, which mainly includes task matching and developer matching. When recommend the suitable developer for a target domain development task, firstly, the development task is matched with the auxiliary domain development task. Hence, the completed development tasks similar to the auxiliary domain are obtained, so that the corresponding auxiliary domain developer can be found. Then calculate the similarity between the obtained developers and the target domain developers, and recommend the most similar developers in the target domain to the development task.

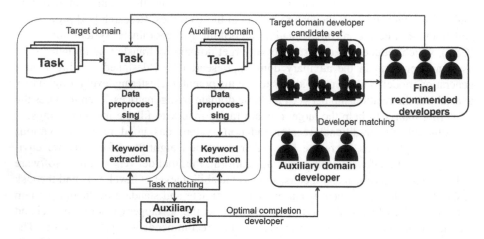

Fig. 1. Flow chart of cross-domain developer recommendation based on feature matching

3.2 Task Matching

The task is first represented in the form of a feature vector, and then the similarity between tasks is calculated. Through the observation of multiple software crowd-sourcing platforms, the description of a development task usually includes several fields, as shown in Table 1.

In order to represent three types of feature in a vector, it is necessary to convert the text type and categorical features into numerical features. The steps of generating task representation vector are in the following.

(1) For the three text features of task title, description and skill requirement, the original text descriptions are converted into a numerical vector by the doc2vec [22] model.
(2) For programming language features, an n-dimensional binary vector is used, and the size of n is determined by the number of possible values of the feature. Each

Table 1. Development task feature information

Feature	Data format	Description
Title	Text	Title of the development task
Task description	Text	Detailed description of development task
Skill requirement	Text	Skill requirement for development task
Programming language	Categorical	The programming language used to complete the development task
Release time	Numerical	Task release time
Completion time	Numerical	Task completion time
Reward	Numerical	Task reward

dimension in the n-dimensional vector represents a programming language. We use 1 to denote that the programming language is required, and 0 otherwise.

(3) For the three numerical features of release time, completion time, and reward, the normalized in formula (1) is used to map the data to [0,1].

$$x' = \frac{x - \min(x)}{\max(x) - \min(x)} \tag{1}$$

(4) The obtained representation vector and the processed numerical vector are combined to obtain a final task representation vector.

The following cosine similarity is used to calculate the similarities between tasks in both domains.

$$similarity = \cos\theta = \frac{W_i \cdot W_j}{|W_i| \times |W_j|} \tag{2}$$

where W represents a task representation vector, and a larger value indicates a large similarity.

3.3 Developer Matching

The similarity between developers is calculated according to their attribute information. Developer's attribute information includes two parts: static features and dynamic features. Static features are the registration information, and dynamic features are the behavior information that is reflected on the platform. The main features of the developer are shown in Table 2.

In this paper, the similarity calculation between developers is divided into three parts: developer skill label matching, developer self-description topic matching and developer dynamic feature vector matching.

(1) Developer skill label matching

Developer skill label matching is based on the similarity calculation between labels. Suppose that the skill labels of developer d_1 and d_2 are $DT_1 =$

Table 2. Developer feature information.

Feature	Data format	Particularity	Description	Example
Skill label	Categorical	Static	Developer skill description	Industry software development, APP development
Self-description	Text	Static	Developer self-description	I am good at excel vba development and have extensive experience in production management and EXCEL applications.
Completion quality	Numerical	Dynamic	Average quality rating for completed tasks by developer	4.89
Work speed	Numerical	Dynamic	Average speed rating for developers to complete development tasks	4.95
Service attitude	Numerical	Dynamic	Average rating of developer service attitude	5

$\{dt_1^1, dt_2^1, \cdots, dt_m^1\}$ and $DT_2 = \{dt_1^2, dt_2^2, \cdots, dt_m^2\}$, respectively. The Jaccard similarity coefficient in formula (3) is used to calculate the label matching degree between developers.

$$Sim_{skill}(d_1, d_2) = \frac{|DT_1 \cap DT_2|}{|DT_1 \cup DT_2|} \tag{3}$$

Sim_{skill} indicates the degree of matching between two developer skill labels. The larger the value Sim_{skill} is, the more similar the two developers are. For example, in ZhuBaJie platform, for two developer skill labels DT_1 = {APP development, public platform development, industry software development, small program development}, and DT_2 = {website custom development, public platform development, industry software development, software localization}, according to formula (3), the matching degree of skill labels of them is 1/3.

(2) Developer self-description topic matching

In order to compare the semantic similarity between two different developers' self-describing, the LDA model is used to process self-description information to obtain the topic-based representation vector. LDA [23] can express text as the probability distribution of topics, which is widely used in natural language processing and other fields. As an unsupervised learning algorithm, LDA does not need to be labeled manually in the training process, but only needs to specify the number of topics. Hence, calculating the similarity between two texts can be converted to calculating the corresponding topic probability distribution obtained by LDA. Let $P(d_1)$ and $P(d_2)$ represent the self-description topic distributions of developers d_1 and d_2, respectively. *KL* distance is usually used to measure the similarity. For two given probability distributions P and Q, *KL* distance is computed in (4):

$$D_{KL}(P\|Q) = \sum_{i=1}^{K} P_i \log \frac{P_i}{Q_i} \tag{4}$$

However, as KL distance is asymmetric, JS distance shown in (5) is used to calculate in this paper. JS distance is the symmetric version of KL distance, and the calculation results are in the range of [0,1].

$$JSD(P\|Q) = \frac{1}{2} D_{KL}(P\|M) + \frac{1}{2} D_{KL}(Q\|M) \tag{5}$$

where $M = \frac{1}{2}(P+Q)$.

Therefore, the self-description topic similarity between developer d_1 and developer d_2 can be calculated by formula (6):

$$sim_{topic}(d_1, d_2) = JSD(P(d_1)\|P(d_2)) \tag{6}$$

(3) Developer dynamic feature vector matching.

For the developer dynamic rating information, three types of rating information are connected to generate a developer dynamic feature vector $R = [r_1, r_2, r_3]$, where r_1, r_2, and r_3 represent the rating of completion quality, work speed and service attitude, respectively. Then, the dynamic feature similarity $Sim_{dynamic feature}(d_1, d_2)$ between developers d_1 and d_2 is obtained by the cosine similarity formula (2).

Finally, formula (7) is used to comprehensively calculate the similarity between developers.

$$sim(d_1, d_2) = sim_{skill}(d_1, d_2) \cdot sim_{topic}(d_1, d_2) \cdot sim_{dynamic feature}(d_1, d_2) \tag{7}$$

3.4   The Proposed Algorithm

Based on the proposed computing methods for similarities between two tasks and between two developers, the cross-domain collaborative filtering developer recommendation algorithm is shown in Table 3.

Table 3. Cross-domain CF developer recommendation algorithm based on feature matching

---

Algorithm: Cross-domain CF developer recommendation algorithm based on feature matching

---

Input: the current development task information of the target domain, developer attribute information of the target domain, development task information of the auxiliary domain, developer attribute information of the auxiliary domain, matching task number S of the auxiliary domain.

Output: developer recommendation list of the target domain.

(1) Obtain the task representation vector with the current task information in the target domain.

(2) Select from the auxiliary domain S tasks that are most similar to the target domain development task, and sort these tasks in descending order of similarity, i.e., $task = \{t_1, t_2, \cdots, t_s\}$.

(3) Retrieve from the auxiliary domain developers $D^a = \{d_1^a, d_2^a, \cdots, d_s^a\}$ corresponding to the tasks, where d_1^a represents the developer of task t_1.

(4) Select from the target domain the most similar developer to d_i^a ($i = 1, 2, \cdots, s$) to compose the recommendation developer set of the current task.

---

4   Experiment

In this section, in order to verify the effectiveness of cross-domain collaborative filtering developer recommendation algorithm based on feature matching, experiments are performed in the crawled datasets. In the experiment, the proposed algorithm in this paper is compared with other developer recommendation algorithms. The computer environment used in the experiment is 2.20 GHz Intel (R) Core (TM) i5-5200U CPU, 8 GB main memory and Windows 10. The related algorithms are implemented with Python 2.7 and jieba, gensim libraries.

4.1   Experimental Dataset

The dataset used in this paper is crawled from ZhuBaJie and Joint Force, two well-known software crowdsourcing platforms. According to the experimental needs, only part of software development related content is selected, mainly including the contents of software development, APP development and website construction. The crawling process mainly extracts information about development tasks and developers, including developer self-description information and skill labels. At the same time, the rating data of transaction evaluation interface are extracted. For the ZhuBaJie, the ratings of completion quality, service attitude and work speed are extracted. For the Joint Force, the completion quality, communication skill and schedule are extracted.

In order to facilitate the experiment, the obtained datasets need to be further processed. Firstly, data with incomplete information is removed. Then, in order to ensure that the information in dataset is rich enough, developers are required to complete at least two development tasks, so developers below this requirement are deleted. Finally, the repetitive developers in three tasks (software development, APP development and website construction) are merged. Table 4 shows the final relevant information of two platform datasets.

Table 4. Statistic of data set from two software crowdsourcing platforms

Platform	Number of development tasks	Number of developers
ZhuBaJie	19139	6855
Joint Force	6123	3899

According to Table 4, the effective data in Joint Force is significantly less than ZhuBaJie, so Joint Force is selected as the target domain and ZhuBaJie is used as the auxiliary domain. In order to simulate the sparsity problem in the real environment, four different training sets of different sparsity levels were constructed by extracting 75%, 60%, 45% and 30% data from the target domain dataset, respectively, denoted as TR75, TR60, TR45 and TR30.

4.2 Evaluation Metrics

Top-N Accuracy

$$Accuracy_{TopN} = \frac{1}{|T|} \sum_{t \in T} correct(N(t)) \tag{8}$$

where $N(t)$ represents the N developers recommended for a task t, and T represents a set of all development tasks. *Accuracy* indicates the percentage of recommended developers to actual developers. For a development task t, if the actual developer is in the recommendation list of length N, the recommendation is successful. Then $correct(N(t)) = 1$, otherwise 0. The higher *Top-N Accuracy* is, the better the recommendation performance is.

Coverage

$$Coverage = \frac{|\cup_{t \in T} N(t)|}{|U|} \tag{9}$$

where U represents the set of all developers.

MRR (Mean Reciprocal Rank)

$$MRR = \frac{1}{|T|} \sum_{i=1}^{T} \frac{1}{R_i} \tag{10}$$

where T represents the set of all development tasks. R_i represents the first ranking of correct developer in the list recommended for the i-th development task. If the correct developer does not appear in the list, the value is 0. The larger the value of MRR is, the better the recommendation performance is.

4.3 Compared Methods

(1) LR: the model proposed in [7] use learning to Rank to realize recommendation. The setting of parameters is consistent with [7]. When the rating is smoothed by Wilson interval algorithm, the parameter α is 5%. When LDA is used to represent task features, the number of topics is set to $T = 50$, and the initial value of hyperparameter is set to $\alpha = 50/T$, $\beta = 0.01$. Gibbs sampling was used in the process, and the iteration is 1000. According to [7], pointwise ranking algorithm is used in experiment.

(2) NN: A feature model for recommending developers in software crowdsourcing tasks is proposed using neural networks and semantic analysis as core methods. The setting of parameters is consistent with [19]: as shown in Table 5.

Table 5. NN parameter setting

Parameter	Size S1	Window k1	Min _count	Size S1	Window k1	Min _count	Similar set m
Value	100	4	1	120	4	5	200

(3) MFD: A multi-feature fusion recommendation method based on developer's ability and behavior is proposed. The setting of parameters is consistent with [20]: as shown in Table 6.

Table 6. MFD parameter setting

Parameter	σ	λ	η_0	λ_1	λ_2	λ_3
Value	0.5	0.025	0.00005	0.4	0.3	0.3

(4) CDCF-FM: The jieba lexicon is used to process word segmentation of text information in the dataset. Using gensim library in python to implement doc2vec method to get a vector representation of text (task title, task description and skill requirement). The related parameter settings in Doc2vec as shown in Table 7. The feature representation vector dimension n of the programming language is set to 80. LDA model is used to process self-description information by controlling variables. The parameter $\beta = 0.01$ and the iteration of Gibbs sampling is 1000. The range of topic number T is set in [10, 15, 20, 25, 30], and the range of initial value α is set in [0.05, 0.1, 0.2, 0.5, 1.0]. The optimal value of parameters is selected in the training process.

Table 7. CDCF-FM parameter setting

Parameter	Min_count	window	Works	vector_size	sample	negative	dm
Value	1	5	4	400	le−3	5	0

4.4   Analysis of Experimental Results

Table 8 records the *Accuracy* of Top-15 under different topic number T and hyper-parameter α when TR75 is selected as training set. The corresponding parameter values under the optimal result are selected as parameter settings when compared with other algorithm experiments. For other training sets, perform the same operation.

Table 8. *Top*-15 results with different value of T and α (%)

α	T				
	10	15	20	25	30
0.05	26	37	38	33	35
0.1	32	27	29	**44**	40
0.2	30	29	24	35	34
0.5	22	31	33	36	38
1.0	27	28	25	29	31

Table 9 and Fig. 2 show the experimental results. By analyzing the experimental results, the following conclusions can be drawn:

Table 9. *MRR* value of the compared models

Algorithm	Dataset			
	TR75	TR60	TR45	TR30
LR	0.455	0.410	0.387	0.355
NN	0.532	0.499	0.462	0.403
MFD	0.612	0.586	0.550	0.468
CDCF-FM	**0.633**	**0.605**	**0.587**	**0.554**

(1) The recommended performance of MFD, NN and CDCF-FM(this paper) is better than LR model. Because LR model has strict rules for task extraction, it requires that both skill requirement and location description should appear in the description of development tasks. However, there are few task descriptions to meet these requirements in actual development tasks, which makes it difficult to extract topic information accurately and achieve ideal recommendation results.

(2) The recommended performance of MFD and CDCF-FM is better than NN model. Because NN model only simply utilizes development task features, and it does not take advantage of the relevant features of developers. At the same time, the

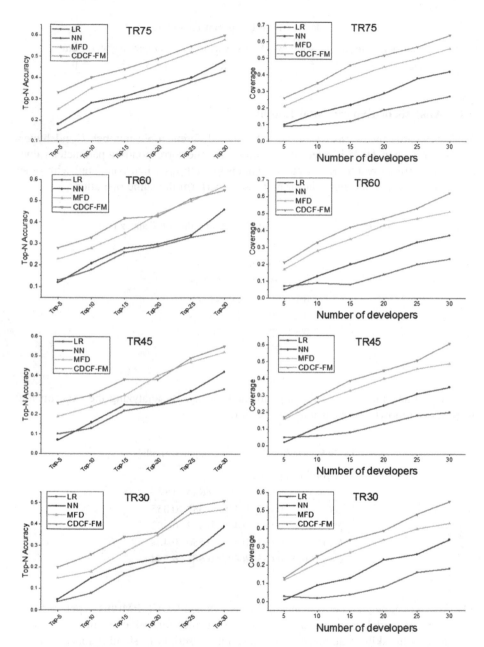

Fig. 2. Accuracy and Coverage results for different algorithms

algorithm needs to train two neural networks separately, so the training complexity is high and the interpretability of the recommended results is not strong.

(3) The recommended performance of CDCF-FM is better than MFD model. MFD only utilizes information from a single data domain. When the target domain data is sparsity, it is difficult to obtain enough information in the learning process. Therefore, the ideal recommendation effect cannot be achieved.

5   Conclusion

At present, most of algorithms for developer recommendation only use a single platform, but not use effective information of the auxiliary domain to help recommendation. For some new software crowdsourcing platforms, it is difficult to accurately learn the data features due to insufficient information in the target domain, so it is impossible to generate suitable recommendation results. In order to solve the influence of data sparsity in the developer recommendation system, this paper proposes a cross-domain collaborative filtering developer recommendation algorithm based on feature matching, which can improve recommendation performance of the target domain by using rich information of the auxiliary domain. The effectiveness of the algorithm is verified by extensive experiments. The experimental results show that the proposed algorithm can solve the problem of data sparsity in the current developer recommendation system, and the recommendation performance is better than other previous recommendation algorithms.

Acknowledgments. This work is jointly sponsored by National Natural Science Foundation of China (Nos. 61402246, 61273180, 61602133, U1806201, 61671261), Natural Science Foundation of Shandong Province (Nos. ZR2019MF014, ZR2018MF007), and key research and development program of Shandong Province (No. 2018GGX101052).

References

1. Li, G.L., Wang, J.N., Zheng, Y.D., Franklin, M.J.: Crowdsourced data management: a survey. IEEE Trans. Knowl. Data Eng. **28**(9), 2296–2319 (2016). https://doi.org/10.1109/TKDE.2016.2535242
2. Fu, Y., Sun, H., Ye, L.: Competition-aware task routing for contest based crowdsourced software development. In: Li, M., Wang, X.Y., Lo, D. (eds.) 6th IEEE International Workshop on Software Mining (SoftwareMining 2017), pp. 32–39. IEEE, Illinois (2017). https://doi.org/10.1109/softwaremining.2017.8100851
3. Begel, A., Bosch, J., Storey, M.A.: Social networking meets software development: perspectives from github, msdn, stack exchange, and topcoder. IEEE Softw. **30**(1), 52–66 (2013). https://doi.org/10.1109/MS.2013.13
4. Mao, K., Capra, L., Harman, M., Jia, Y.: A survey of the use of crowdsourcing in software engineering. J. Syst. Softw. **126**, 57–84 (2017). https://doi.org/10.1016/J.JSS.2016.09.015
5. Mao, K., Yang, Y., Wang, Q., Jia, Y., Harman, M.: Developer recommendation for crowdsourced software development tasks. In: 9th IEEE Symposium on Service-Oriented

System Engineering, pp. 347–356. IEEE, San Francisco (2015). https://doi.org/10.1109/sose.2015.46

6. Li, N., Mo, W., Shen, B.: Task recommendation with developer social network in software crowdsourcing. In: Potanin, A., Murphy, G.C., Reeves, S., Dietrich, J. (eds.) 23rd Asia-Pacific Software Engineering Conference (APSEC), pp. 9–16. IEEE, Hamilton (2016). https://doi.org/10.1109/apsec.2016.013

7. Zhu, J., Shen, B., Hu, F.: A learning to rank framework for developer recommendation in software crowdsourcing. In: Sun, J., Reddy, Y.R., Bahulkar, A., Pasala, A. (eds.) 22nd Asia-Pacific Software Engineering Conference (APSEC), pp. 285–292. IEEE, New Delhi (2015). https://doi.org/10.1109/apsec.2015.50

8. Bouraga, S., Jureta, I., Faulkner, S., Herssens, C.: Knowledge-based recommendation systems: a survey. Int. J. Intell. Inf. Technol. **10**(2), 1–19 (2014). https://doi.org/10.1016/J.KNOSYS.2016.04.020

9. Suganeshwari, G., Syed Ibrahim, S.P.: A survey on collaborative filtering based recommendation system. In: Vijayakumar, V., Neelanarayanan, V. (eds.) Proceedings of the 3rd International Symposium on Big Data and Cloud Computing Challenges (ISBCC – 16'). SIST, vol. 49, pp. 503–518. Springer, Cham (2016). https://doi.org/10.1007/978-3-319-30348-2_42

10. Cremonesi, P., Tripodi, A., Turrin, R.: Cross-domain recommender systems. In: Spiliopoulou, M., et al. (eds.) 11th IEEE International Conference on Data Mining Workshops (ICDM), pp. 496–503. IEEE, Vancouver (2011). https://doi.org/10.1109/icdmw.2011.57

11. Cantador, I., Fernández-Tobías, I., Berkovsky, S., Cremonesi, P.: Cross-domain recommender systems. In: Recommender systems Handbook, pp. 919–959 (2015). https://doi.org/10.1007/978-1-4899-7637-6_27

12. Singh, A.P., Gordon, G.J.: Relational learning via collective matrix factorization. In: 14th ACM SIGKDD International Conference on Knowledge Discovery and Data Mining, pp. 650–658. ACM, Nevada (2008). https://doi.org/10.1145/1401890.1401969

13. Li, B., Yang, Q., Xue, X.Y.: Transfer learning for collaborative filtering via a rating-matrix generative model. In: 26th ACM Annual International Conference on Machine Learning, pp. 617–624. ACM (2009). https://doi.org/10.1145/1553374.1553454

14. Pan, W.K., Xiang, E.W., Liu, N.N., Yang, Q.: Transfer learning in collaborative filtering for sparsity reduction. In: 24th AAAI Conference on Artificial Intelligence, pp. 230–235. AAAI, Atlanta (2010). https://doi.org/10.13328/j.cnki.jos.000000

15. Pan, R., et al.: One-class collaborative filtering. In: Giannotti, F., Gunopulos, D., Turini, F., Zaniolo, C., Ramakrishnan, N., Wu, X.D. (eds.) 8th IEEE International Conference on Data Mining (ICDM), pp. 502–511. IEEE, Pisa (2008). https://doi.org/10.1109/icdm.2008.16

16. Happel, H.J., Maalej, W.: Potentials and challenges of recommendation systems for software development. In: International Workshop on Recommendation Systems for Software Engineering, pp. 11–15. ACM, Atlanta (2008). https://doi.org/10.1145/1454247.1454251

17. Robillard, M., Walker, R., Zimmermann, T.: Recommendation systems for software engineering. IEEE Softw. **27**(4), 80–86 (2009). https://doi.org/10.1109/MS.2009.161

18. Tang, J., Wu, S., Sun, J.M., Su, H.: Cross-domain collaboration recommendation. In: 18th ACM SIGKDD International Conference on Knowledge Discovery and Data Mining, pp. 1285–1293. ACM, Beijing (2012). https://doi.org/10.1145/2339530.2339730

19. Shao, W., Wang, X.N., Jiao, W.P.: A developer recommendation framework in software crowdsourcing development. In: Zhang, L., Xu, C. (eds.) 15th National Software Application Conference (CCIS), pp. 151–164. Springer, Kunming (2016). https://doi.org/10.1007/978-981-10-3482-4_11

20. Xie, X.Q., Yang, X.C., Wang, B., Zhang, X., Ji, Y., Huang, Z.G.: A multi-feature fused software developer recommendation. J. Softw. **29**(8), 2306–2321 (2018). https://doi.org/10.13328/J.CNKI.JOS.005525

21. Yan, J., Sun, H.L., Wang, X., Liu, X.D., Song, X.T.: Profiling developer expertise across software communities with heterogeneous information network analysis. In: 10th Asia-Pacific Symposium on Internetware, p. 2. ACM, Beijing (2018). https://doi.org/10.1145/3275219.3275226

22. Maslova, N., Potapov, V.: Neural network Doc2vec in automated sentiment analysis for short informal texts. In: Karpov, A., Potapova, R., Mporas, I. (eds.) SPECOM 2017. LNCS (LNAI), vol. 10458, pp. 546–554. Springer, Cham (2017). https://doi.org/10.1007/978-3-319-66429-3_54

23. Dong, Y.Y., Chen, J.L., Tang, X.X.: Unsupervised feature selection method based on latent Dirichlet allocation model and mutual information. J. Comput. Appl. **8** (2012). https://doi.org/10.3724/sp.j.1087.2012.02250

Improved Collaborative Filtering Algorithm Incorporating User Information and Using Differential Privacy

Jiahui Ren[✉], Xian Xu[✉], and Huiqun Yu[✉]

Department of Computer Science and Engineering,
East China University of Science and Technology, Shanghai 200237, China
renhui774411@163.com, {xuxian,yhq}@ecust.edu.cn

Abstract. Collaborative filtering algorithm is one of the most popular recommendation algorithms. There is, however, the risk of privacy leakage when making effective recommendation. Differential privacy is a relatively new privacy protection mechanism in the field, and has been used in recommendation systems. To this end, the existing research still has some disadvantages. Particularly, they do not have satisfactory performance and have difficulty in solving the cold start scenarios. In this paper, we propose an improved differential privacy enabled collaborative filtering algorithm incorporating user information. The algorithm improves similarity calculation, and solves the user cold start problem by making effective use of user information (related attributes). Experiments show that with the same privacy guarantee, the proposed algorithm improves the performance of the recommendation system and indeed solves the problem of cold start to some good extent.

Keywords: Differential privacy · Collaborative filtering · Recommendation system · Cold start · User similarity

1 Introduction

Collaborative filtering [1, 2] is one of the most popular recommendation technologies at present. It generates effective personalized recommendation for users by analyzing and mining historical behavior data of users, and is applied in industry widely. However, in the process of producing recommendation results, collaborative filtering inevitably uses the user's private information, so there is a risk of privacy leakage. The information collected during the period of recommendation may be leaked by the service provider, or it may be attacked by hackers, compromising user privacy.

Privacy protection of recommendation systems is an important research direction. Traditional privacy protection techniques, such as data anonymity, perturbation, rely on specific attack assumptions and cannot be proved correct by rigorous mathematical theory. Differential privacy, proposed by Dwork [3], is a robust mechanism to implement privacy protection, with rigorous mathematical proof of correctness. It has become a hot spot of research in the field of both information security and data analysis.

Y. Sun et al. (Eds.): ChineseCSCW 2019, CCIS 1042, pp. 458–471, 2019.
https://doi.org/10.1007/978-981-15-1377-0_36

McSherry [4] first applied differential privacy to the recommendation system in 2009. Specifically, they add Laplace noise to the process of both data preprocessing and construction of user rating covariance matrix, to achieve differential privacy. The noised matrix is then submitted to the recommendation system for recommendation. Friman et al. [5] proposed several noise-adding methods for matrix factorization models, including input disturbances, SGD disturbances, and so on. Xian et al. [6] combined the privacy protection mechanism in the matrix factorization with the SVD+ + algorithm, to obtain a better recommendation outcome. Zhu [7] first combined neighbor-based collaborative filtering with differential privacy. To achieve differential privacy protection in that setting, they add Laplace noise to the similarity calculation and use exponential mechanism to select neighboring users. Chen [8] proposed a nearest neighbor collaborative filtering algorithm based on clustering and improved exponential mechanism, and promotes the recommendation performance. Although there have been quite some works about recommendation algorithm based on differential privacy, some inherent problems in collaborative filtering still need to be solved better, in particular the cold start problem [9], and the performance also has a large promotion potential.

Aiming at these problems, based on the work in the literature, we propose an improved nearest neighbor recommendation algorithm that incorporates user information and meanwhile bases its data processing on differential privacy. Compared with previous works, this paper solves the cold start problem of the recommendation system by adding user information into the similarity calculation process. At the same time, for the performance issue, we designed an improved exponential mechanism to reasonably allocate the privacy budget. This improves the recommendation effect and reduces the complexity of the recommendation process.

The structure of this paper is as follows: Sect. 2 gives the related concepts, algorithms and mechanisms. Section 3 introduces our algorithm, including the main idea and the detailed description. Section 4 presents the experiments that we use to evaluate our algorithm's performance. Section 5 makes a conclusion of the paper.

2 Preliminaries

2.1 Differential Privacy

Differential privacy is fundamentally different from traditional techniques for privacy protection. It is built on a very strict attack model and gives a rigorous, quantitative representation of privacy risk and detailed proof. Differential privacy protection greatly reduces the risk of privacy breaches while greatly ensuring data availability. We now give the definition of differential privacy and other related concepts we use in this paper.

Differential Privacy. Differential privacy conceals the difference of queries between adjacent datasets by adding noise. A differential privacy query ensures that the query result remains unchanged after adding or deleting a piece of data in a data set, so that an attacker cannot infer any information concerning user privacy from the output result. The model of differential privacy still works when the attacker has a very strong background knowledge. Even if the attacker has all the information except the target, it still cannot infer any sensitive information about the target user.

Definition 1 (ϵ-differential Privacy). Given an algorithm M, if for any two neighbor datasets D and D', the output of M satisfies the following Eq. (1), then we can say that the algorithm M is satisfy ϵ-differential privacy (ϵ-DP).

$$Pr[M(D) = O] \leq e^{\epsilon} \times Pr[M(D') = O] \tag{1}$$

The neighbor datasets D and D' are two datasets which only differ in one record, that is $|D - D'| = 1$. The ϵ is privacy budget. It indicates the degree of privacy protection. Generally speaking, the smaller its value is, the stronger the privacy protection becomes.

Sensitivity. The sensitivity $S(f)$ shows the magnitude of the perturbation introduced by the differential privacy to the query function f, i.e., the maximal difference in the output of a query function from two neighbor datasets. Two sensitivity is defined in the differential privacy protection method: global sensitivity [3] and local sensitivity [11].

Definition 2 (Global Sensitivity). Let $f = D \rightarrow R^d$ be a query function and R^d denote a d-dimensional real vector (query result) of the output. Then for any two neighbor datasets D and D', the global sensitivity of function f is as follows:

$$GS(f) = \max_{D,D'} |f(D) - f(D')| \tag{2}$$

The global sensitivity is determined by the way the function is calculated, and different functions have different sensitivities. When the function has large global sensitivity, the introduced noise is often large, resulting in poor data usability, so [11] proposes local sensitivity.

Definition 3 (Local Sensitivity). Let $f = D \rightarrow R^d$ be a query function and R^d denote a d-dimensional real vector (query result) of the output. Then the local sensitivity of function f is:

$$LS(f) = \max_{D'} |f(D) - f(D')| \tag{3}$$

Different from the global sensitivity, the local sensitivity is also determined the dataset D. So it is much smaller than the global sensitivity, and thus greatly reduces the introduced noise scale and improves the efficiency of use.

Laplace Mechanism. The main principle of Laplacian mechanism [3] is adding Laplacian noise to the result of the (query) function to satisfy to definition of differential privacy. Noise is sampling from a Laplace distribution. Its probability density function is as follows:

$$Lap(x) = \frac{1}{2b}\exp(\frac{|x|}{b}) \tag{4}$$

The parameter b represents the scale of the Laplace distribution, which determines the size of Laplace noise and is generally determined by the sensitivity of the function.

Theorem 1 (Laplace Mechanism) [3]. Given a query function $f = D \rightarrow R^d$, dataset D, and algorithm M, if an algorithm M fits the Eq. (5), then the algorithm M can provide ϵ-differential privacy.

$$M(D) = f(D) + Lap\left(\frac{S(f)}{\epsilon}\right) \tag{5}$$

The scale of noise added by the Laplacian mechanism depends on the sensitivity of query function $S(f)$ and privacy budget ϵ.

Exponential Mechanism [12]. Laplacian mechanism is mainly applicable to numerical output. For non-numerical output, Mcsherry et al. proposed exponential mechanism to achieve differential privacy. The exponential mechanism uses the scoring function $q(D, \phi)$ to evaluate the output ϕ and then outputs the result according to certain probability distribution. The main result of the exponential mechanism is as follows.

Theorem 2 (Exponential Mechanism). Given a query function $f = D \rightarrow R^d$, dataset D, and algorithm M, if the algorithm M fits the following Eq. (6), then the algorithm M can provide ϵ-differential privacy.

$$M(D) = \{\text{output } \phi \text{ with probability of } \exp(\frac{\epsilon q(D, \phi)}{2\Delta q})\} \tag{6}$$

Where Δq denotes the sensitivity of the scoring function.

2.2   K-Nearest Neighbor Collaborative Filtering

Collaborative Filtering. As a popular recommendation technique at present, collaborative filtering searches for users who share common characteristics with target users by mining their historical records (such as browsing, favorites setting, scoring, etc.), and chooses items of interest by these similar users to recommend to target users.

The collaborative filtering problem is defined as follows: given the user set $U = \{u_1 \cdots u_n\}$ and the item set $I = \{i_1 \cdots i_m\}$, one obtains the rating matrix R. Each element r_{ui} in R represents the score user u gives to item i. The purpose of the recommendation system is to give the predicted value of the data not rated yet in the matrix.

Figure 1 shows the main process of K-Nearest neighbor collaborative filtering. It mainly includes two parts: neighbor selection and rating prediction. Neighbor selection stage is responsible for selecting K users who are most similar to the target users by calculating the similarity. In the rating prediction stage, the items not scored by the target user are predicted according to the existing ratings of the similar users selected in the selection stage, and the similarity between them and the target user.

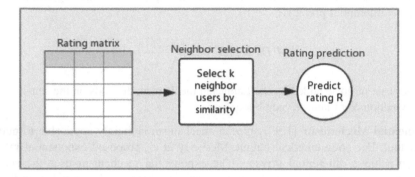

Fig. 1. K-nearest neighbor collaborative filtering

Similarity is a core component of the (K-Nearest Neighbor) collaborative filtering. In the field, there are basically two ways to calculate similarity: Pearson similarity and cosine similarity. We will explain both of them in our algorithm shortly.

2.3   Private Neighbor Collaborative Filtering

The KNN Attack Model. Calandrino [13] proposed a KNN attack model. By observing the changes in the results of the recommendation system and combining with the "auxiliary information", they success fully infer the historical score and behavior of a specific user. Assuming the attacker knows m history record of the target user, then the attacker can construct k targets of "neighbors". These "neighbors" have m same records with the target. Since the rating data set in the actual system is very sparse, a small m would be sufficient to make the target fall into fabricated "near neighbor set". Since the neighbors only have fake users and target user, the attacker can guess with high probability the historical rating record of the target according to the result of the recommendation system.

In order to provide effective privacy protection, Zhu et al. [9] proposed Private Neighbor Collaborative Filtering (PNCF). In their work, differential privacy protection is implemented by adding disturbance in the two steps of the traditional nearest neighbor collaborative filtering. The principle of PNCF is shown in Fig. 2. Firstly, when selecting k-nearest neighbor users, the selection process is disturbed by

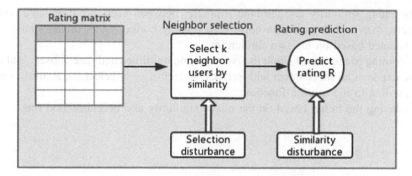

Fig. 2. PNCF (Private Neighbor Collaborative Filtering)

exponential mechanism. Then the similarity calculation is disturbed by Laplacian mechanism in rating prediction.

2.4   Cold Start

The core of collaborative filtering algorithm is to analyze users' rating records and gain their similarity. However, when a user appears in a system for the first time, or if he/she does not have any historical rating data in the system, the traditional algorithm would not be able to calculate the similarity between users, so he cannot be recommended at that moment. This is called the new user cold-start problem [10]. To date, the main idea to solve the cold start problem is to incorporate other contextual information [9] to recommend new users.

3   Collaborative Filtering Algorithm Incorporating User Information and Using Differential Privacy

3.1   Overview of the Algorithm

Although PNCF introduces differential privacy protection, there are still some problems. For example, too much disturbance and inefficient selection of exponential mechanism lead to poor recommendation performance. Besides, some problems of collaborative filtering algorithm itself, such as cold start, are not considered. Facing these problems, we proposes an improved recommendation algorithm based on differential privacy, which incorporates user information as well. By incorporating users' personal information and improving the similarity calculation method, the cold start problem is solved, and the performance is enhanced by improving the exponential mechanism of neighbor selection. The main process of the algorithm consists of three steps and is illustrated in Fig. 3.

1. The rating similarity and attribute similarity between users are calculated respectively, then added Laplacian noise with privacy budget ϵ_1. A mixed similarity is calculated based on the two similarities.
2. According to the mixed similarity, we generate multiple candidate subsets, and then use exponential mechanism with privacy budget of ϵ_2 to select the nearest user set according to the scoring function.
3. Predicting the rating based on the mixed similarity and neighborhood sets.

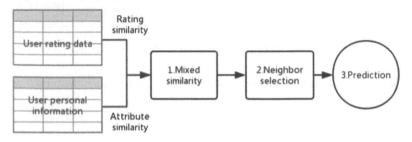

Fig. 3. The improved algorithm

Below we explain in detail how the mixed similarity are calculated and how the neighbors are selected.

3.2   Mixed Similarity

Usually in the database of a recommendation system website, besides the interactive records of users' historical evaluation, there are also user attributes. Common user attributes include gender, age, job and so on. Users with the same or similar attributes are more likely to have the same interests. For new users, the recommendation systems often fail to recognize their interests, but they can recommend new users by (old) users with attributes similar to those of the new users. Based on this start point, this paper adjusts the similarity calculation method and incorporates the users' personal information.

In calculating the similarity of the rating data, we use the traditional Pearson coefficient to calculate the similarity of the rating data. Assume the rating similarity thus obtained is *rsim*. When calculating the similarity of user attributes, because it may be non-discrete data, it is necessary to process user attributes with onehot coding. Onehot coding uses n-dimensional vectors to encode n different attributes, each attribute has its own dimension, and only one is valid at any time. In this paper, the users' occupation may have several values, which are represented by one dimension respectively. If the user has a specific occupation, the corresponding dimension is set to be 1, and the rest are all set to be 0. Attributes other than the occupation can be similarly encoded. Because Pearson coefficients need to use the mean value of attributes, but the mean of onehot vectors is meaningless, this paper uses cosine similarity to calculate the users' attribute similarity. We assume the attribute similarity thus obtained is *asim*.

In order to achieve differential privacy, Laplacian noise is added to rating similarity and attribute similarity respectively. The computational process are shown in Eqs. (7) and (8).

$$rsim^* = rsim + Lap(\frac{\Delta f_r}{\varepsilon_1}) \tag{7}$$

$$asim^* = asim + Lap(\frac{\Delta f_a}{\varepsilon_1}) \tag{8}$$

rsim* and *asim** denote the rating similarity and attribute similarity after adding noise. The sensitivity of rating similarity and attribute similarity in these equation are related to the calculation (function) of similarity. In order to avoid introducing too much noise, this paper uses local sensitivity.

Now we explain how Eqs. (7) and (8) are calculated, and in turn the mixed similarity.

Rating Similarity. For convenience in calculating the rating similarity using Pearson's method, we subtract from each rating value the corresponding mean to simplify calculation. According to the definition of Pearson correlation coefficient, the rating similarity can be changed to Eq. (9):

$$rsim(u,v) = \frac{\sum_{i \in I_{uv}} r_{ui} r_{vi}}{\sqrt{\sum_{i \in I_{uv}} r_{ui}^2 \sum_{i \in I_{uv}} r_{ui}^2}} \tag{9}$$

Then we calculate the local sensitivity, as shown in the following Eq. (10):

$$\Delta f_r = \max_{u,v \in U} \|rsim(u,v) - rsim'(u,v)\|_1$$

$$= \max_{u,v \in U} \left(\frac{\sum_{i \in I_{uv}} r_{ui} r_{vi}}{\sqrt{\sum_{i \in I_{uv}} r_{ui}^2 \sum_{i \in I_{uv}} r_{ui}^2}} - \frac{\sum_{i \in I'_{uv}} r_{ui} r_{vi}}{\sqrt{\sum_{i \in I'_{uv}} r_{ui}^2 \sum_{i \in I'_{uv}} r_{ui}^2}} \right) \tag{10}$$

$$\leq \max_{u,v \in U} \frac{\max_{i \in I'_{uv}} r_{ui} r_{vi}}{\sqrt{\sum_{i \in I'_{uv}} r_{ui}^2 \sum_{i \in I'_{uv}} r_{ui}^2}}$$

$sim'(i,j)$ is the similarity of two users' ratings after deleting an item. I'_{uv} is the collection of items for user u and user v after deleting an item.

Then the rating similarity after noise-adding in (7) can be calculated using Eqs. (9) and (10).

Attribute Similarity. For attribute similarity, according to the definition of cosine similarity, we have:

$$asim(u, v) = \frac{r_u \cdot r_v}{\|r_u\|\|r_v\|} \tag{11}$$

Where r_u, r_v denote the onehot encoding value of the user attributes of user u and user v. For two users, if one attribute of one of them changes, only one value of the encoded vector r_u, r_v will be affected. Thus, the local sensitivity of attribute similarity is:

$$\Delta f_a = \frac{1}{\|r_u\|\|r_v\|} \tag{12}$$

Then the attribute similarity after noise-adding in (8) can be calculated using Eqs. (11) and (12).

Mixed Similarity. Having $asim^*$ $rsim^*$ at hand, the two similarities are weighted and added according to Eq. (13) to get the mixed similarity sim^*:

$$sim^* = (1-\alpha) \bullet rsim^* + \alpha \bullet asim^* \tag{13}$$

Where $\alpha = 1/(1 + n_u/3)$, in which n_u denotes the counting of users' historical rating records.

The setting of the parameter α refers to that in [11]. The advantage of this (mixed) method is that for a user without rating, our algorithm can also make some recommendation according to the user information. For the old users, since the value of the attribute data is not as large as the historical rating records, the system can smoothly adjust the weight of the two similarities to achieve more accurate recommendation.

3.3   Neighbor (User) Selection

In PNCF algorithm, the similarity between users is used as the scoring function, i.e. $q(U, j) = sim(i, j)$, in which $q(U, j)$ is the scoring function and $sim(i, j)$ is the similarity. Then the exponential mechanism outputs the neighbors with the probability of $exp\left(\frac{\epsilon q}{2\Delta q}\right)$. One simply iterates k times to select k neighbor users. In that procedure, each iteration consumes the privacy budget, and each selected privacy budget for one iteration is very small, because the total privacy budget is amortized onto the k iterations. This makes it difficult for effective neighbor users to be selected, resulting in poor recommendation effect in the end.

Unlike previous works, we randomly select from the nearest neighbor user set N, to generate multiple candidate (sub) sets $C_1 \cdots C_n$, each of which contains k users. In order to ensure privacy, we use the exponential mechanism to choose from these candidate sets and output the result ones.

The effect of the exponential mechanism is closely related to the design of the scoring function. Here we use the sum of the absolute similarity values of each candidate set as the scoring function of the exponential mechanism. That is,

$$q(N, C_i) = \sum_{v \in C_i} |sim(u, v)| \tag{14}$$

Where u is the target user and v represents the users in the candidate set.

According to the definition of the scoring function, the sensitivity of the scoring function is:

$$\Delta q = \max_{\|C_1 - C_2\|_1 \le 1} |q(N, C_1) - q(N, C_2)| = 1 \tag{15}$$

Using the exponential mechanism and the scoring function as defined above, a candidate set C is outputted as the neighbor set with probability $exp\left(\frac{\epsilon q}{2\Delta q}\right)$. In this procedure, the algorithm uses the exponential mechanism only once, which reduces the complexity of the algorithm and improves the utility of the privacy budget. Moreover, this also improves the effect of neighbor selection.

3.4   Privacy Analysis of Algorithms

In this paper, the mechanism of differential privacy are used in several places. To measure the level of privacy protection of the whole algorithm, two combinatorial properties of differential privacy are needed.

Lemma 1 (Parallel Combinability) [14]. Assume that there are a set of privacy mechanism algorithms $M_1, M_2, \cdots M_n$ whose privacy protection budgets are $\epsilon_1, \epsilon_2, \cdots \epsilon_n$ respectively. When applied to disjoint datasets, the combinatorial algorithm provides $(\max \epsilon_i)$-differential privacy protection.

Lemma 2 (Sequence Combinability) [14]. Assume that there are a set of privacy mechanism algorithms $M_1, M_2, \cdots M_n$ whose privacy protection budgets are $\epsilon_1, \epsilon_2, \cdots \epsilon_n$ respectively. When applied to the same dataset, the combinatorial algorithm provides $(\sum \epsilon_i)$-differential privacy protection.

According to Theorem 1, the noise added in the process of calculating two similarities satisfies the Laplacian mechanism. At the same time, because the noise is added to different datasets, according to the parallel combinability of Lemma 1, step 1 in Fig. 3 satisfies ϵ_1-differential privacy. According to Theorem 2, the algorithm satisfies the condition of exponential mechanism when selecting neighbor candidate sets, so step 2 in Fig. 3 satisfies ϵ_2-differential privacy. Therefore according to Lemma 2, it can be concluded that our algorithm satisfies $(\epsilon_1 + \epsilon_2)$-differential privacy.

4 Experiment

We now use some public datasets to evaluate the performance of our algorithm, so as to verify its effectiveness. The experiments is based on the system with Intel i5-3230 M CPU 2.60 GHz, 8G RAM, and Windows 10 operating system. Python is used to implement the algorithm. We explain the experimental process and results, and then compare them with the existing algorithms. The comparison results truly show the effectiveness of the improved design in this paper.

Dataset. The experiment uses the open dataset Movielen-1M which is commonly used in recommendation system algorithm experiments. The dataset is provided by Group-Len. It includes a total of more than 1000000 rating records of 6040 users for 3 900 movies and their personal information records. Ratings are ranged from 1 to 5, and the higher the rating is, the greater the degree of interest would be.

Evaluation. According to the evaluation method widely used in related research work, we use the Mean Absolute Deviation (MAE) as the evaluation method of the accuracy of the recommended results. The smaller MAE is, the more accurate the recommended result is, and the better the effect of the algorithm is. The equation for calculating MAE is as follows:

$$MAE = \frac{\sum\limits_{(u,i) \in R} r_{ui} - \hat{r}_{ui}}{n} \tag{16}$$

Where $\hat{r}_{ui}$ denotes the predictive rating, and n denotes the counting of predictive ratings.

To show the effectiveness and performance of our algorithm, we compare our algorithm (we call U-PNCF) with the following two algorithms in the experiments.

- Comparison with KNN cooperative filtering algorithm without privacy protection. This is to verify the performance after adding differential privacy.
- Comparison with the privacy neighbor collaborative filtering algorithm based on differential privacy, i.e., PNCF. This is to verify the performance improvement of our algorithm.

Experiment 1. The performance of the algorithm under different K (i.e., the size of the set of neighbors). The privacy budget ϵ is set to 1, and we set $\epsilon_1 = \epsilon_2 = \epsilon/2$.

The experimental results are shown in Fig. 4 (a). We can observe from the result that with the increase of the number of neighbors, the MAE values of the three recommendation algorithms are declining. Meanwhile, because the privacy budget allocation in the exponential mechanism of our algorithm is different from PNCF algorithm and appears more reasonable in the calculation of similarity), it introduces smaller perturbations. Therefore the algorithm in this paper performs better than that of PNCF algorithm in effect. Besides, when the K is small, the performance of our algorithm is not significantly different from that of PNCF. This is because when k is small, PNCF consumes a relatively larger privacy budget for each selection, leading to a better result. When the number of nearest neighbors is about 30, the decline speed of MAE begins to slow down and the algorithm tends to be stable.

Experiment 2. The performance comparison of the algorithms with different privacy budgets. According to Experiment 1 we set K to 30, and again $\epsilon_1 = \epsilon_2 = \epsilon/2$.

The experimental results are shown in Fig. 4(b). As it can be seen, with the increase of privacy budget, the performance of the two algorithms is increasing. However, the algorithm U-PNCF is better than PNCF on the whole. Similarly, because the privacy budget allocation in the exponential mechanism appears to be more reasonable (related to similarity calculation), under the same privacy budget, a smaller disturbance is imported and thus improves the performance.

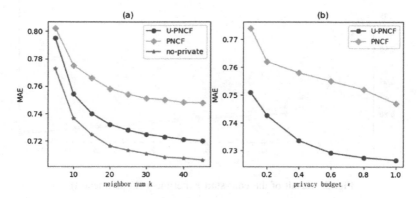

Fig. 4. Results of Experiment 1 and 2

Experiment 3. In this experiment, we design a cold-start scenario to measure our algorithm's performance in the cold start situation. Firstly, we select 500 users as the new users, and add 0, 2, 4, 6, 8, 16 and 32 rating records of the selected new users to the training set. The rest of the dataset are used as the test set. We make a comparison between the version of our algorithm in which the similarity calculating does not use user information at all, and the version in which user information is added to the calculation of similarity. The privacy budget is set as 1, and the size of the set of neighbors is set as 30.

Figure 5 shows the experimental result. We can see that the performance of the algorithm improves with the increase of the number of the new user rating records. For the algorithm without considering user information, it is impossible to predict and recommend without any rating record at the beginning. So when the user rating number is 0, the algorithm produces no result. In the case of fewer user ratings, the performance of our algorithm in which user information is added is better than the algorithm without adding user information. With the increase of user ratings, the coefficient of rating similarity in Eq. (10) increases, and the two algorithms tend to be similar. Another observation is that although user information has been added, the prediction accuracy of users with fewer ratings has not been greatly improved. This may be because the user information provided by Movie-Len dataset is too little and not sufficient. The Movie-Len dataset only provides the attributes of gender, age and job. The corresponding attributes similarity is relatively simple. If there are more attributes of users like in actual applications, the degree of improvement of our algorithm is expected to be more obvious.

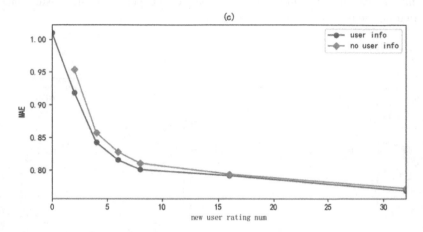

Fig. 5. Result of the cold start experiment (Experiment 3)

5   Conclusion

Based on the available privacy protection enabled nearest neighbor collaborative filtering algorithms, this paper proposes an improved collaborative filtering algorithm based on differential privacy and considering user information. By incorporating user information, the cold start problem is well settled. Also by using the improved differential privacy mechanism, we achieve better privacy protection while enhancing effectiveness and performance of recommendation. Theoretical analysis and experiments show that the algorithm we propose improves the performance of the collaborative filtering algorithm with privacy protection, and indeed has a good accuracy in the cold start situation. In the future, we can explorer how the differential privacy can be combined with other recommendation systems and also how to make a better balance between performance and privacy.

Acknowledgement. The research was supported in part by NSF China (61872142; 61772200; 61702334; 61572318), NSF Shanghai (17ZR140690017ZR1429700), ECUST Research Fund for Education (ZH1726108), Shanghai Pu Jiang (17PJ1401900), Special Funds for Information Developing by SEIC (201602008). We thank Tao Huang for checking the paper.

References

1. Schafer, J.B., Frankowski, D., Herlocker, J., et al.: Collaborative filtering recommender systems. ACM Trans. Inf. Syst. **22**(1), 5–53 (2004)
2. Shi, Y., Larson, M., Hanjalic, A.: Collaborative filtering beyond the user-item matrix: a survey of the state of the art and future challenges. ACM Comput. Surv. **47**(1), 1–45 (2014)
3. Dwork, C.: Differential privacy. Lect. Notes Comput. Sci. **26**(2), 1–12 (2006)
4. Mcsherry F., Mcsherry F., Mironov, I., et al.: Differentially private recommender systems: building privacy into the net. In: ACM SIGKDD International Conference on Knowledge Discovery & Data Mining. ACM (2009)
5. Friedman, A., Berkovsky, S., Kaafar, M.A.: A differential privacy framework for matrix factorization recommender systems. User Model. User-Adap. Inter. **26**(5), 425–458 (2016)
6. Xian, Z., Li, Q., Huang, X., et al.: New SVD-based collaborative filtering algorithms with differential privacy. J. Intell. Fuzzy Syst. **33**(4), 2133–2144 (2017)
7. Zhu, T., Li, G., Ren, Y., et al.: Differential privacy for neighborhood-based collaborative filtering. In: Proceedings of the 2013 IEEE/ACM International Conference on Advances in Social Networks Analysis and Mining, pp. 752–759. ACM (2013)
8. Chen, Z., Wang, Y., Zhang, S., et al.: Differentially private user-based collaborative filtering recommendation based on K-means clustering. arXiv preprint arXiv:1812.01782 2018)
9. Wei, D., Haung, Y.: Incorporating User Attribute Data in Recommendation System. Appl. Electron. Tech. **5**(137–140), 144 (2017)
10. Shao, Y., Xie, Y.: Research on cold-start problem of collaborative filtering algorithm. Comput. Syst. Appl. **28**(2), 246–252 (2019)
11. Nissim, K., Raskhodnikova, S.: Smooth sensitivity and sampling in private data analysis. In: Thirty-Ninth ACM Symposium on Theory of Computing (2007)
12. Mcsherry, F., Talwar, K.: Mechanism design via differential privacy. In: IEEE Symposium on Foundations of Computer Science (2007)
13. Calandrino, J.A., Kilzer, A., Narayanan, A., et al.: "You might also like:" privacy risks of collaborative filtering. In: Security & Privacy. IEEE (2012)
14. Mcsherry, F.: Privacy integrated queries: an extensible platform for privacy-preserving data analysis. Commun. ACM **53**(9), 89–97 (2010)

Friend Recommendation Model Based on Multi-dimensional Academic Feature and Attention Mechanism

Yi He, Liu Wang, Chengjie Mao[(✉)], Ying Li, Saimei Sun, and Yixiang Cai

South China Normal University, Guangzhou 510631, China
{yihe,lwang,smsun,2018022654}@m.scnu.edu.cn,
maochj@qq.com, 2345903@qq.com

Abstract. The academic social network platform is different from the popular social network platform. Most users of the academic social network platform are researchers and they have more academic related information. How to effectively recommend friend for them according to the characteristics of academic social network have become one of the current research directions. Because of traditional recommendation methods only consider the superficial interaction characteristics, it cannot obtain the more complex and diverse nonlinear relationship between the target user and the recommendation item. The method based on deep learning also has problems such as information loss. In this paper, we propose a friend recommendation model based on multi-dimensional academic characteristics and attention mechanism (MLP–AMAF). We capture the attribute features, relation features and text features of the user from academic social networking. Then we combine these academic features and get the key information related to the current recommendation task automatically by attention mechanism. According to the different preferences of the user, we can get more personalized recommendation results. In our experiments, we use friend data of SCHOLAT, a real academic social network platform, to evaluate our model and get better recommendation results than other widely used recommendation algorithms.

Keywords: Academic social networks · Deep learning · Attention mechanism · Personalized recommendation · Multi-dimensional academic feature

1 Introduction

In recent years, with the rapid development of Internet technology, online social networking platforms have been popularized and become an essential part of People's Daily life. Through the platform, users can conveniently exchange and share information. With the increasing number of users, users can get more and more information. However, a large amount of redundant information will affect the normal use. As one of the effective means to solve the information overload problem, recommendation system gets widely research and application, and it can find and recommend content from a

© Springer Nature Singapore Pte Ltd. 2019
Y. Sun et al. (Eds.): ChineseCSCW 2019, CCIS 1042, pp. 472–484, 2019.
https://doi.org/10.1007/978-981-15-1377-0_37

mass of information by user interests. The majority of social networks are users and friend relations between users are the basis of social network. Like most network information, the distribution of friend relationship in social networks also exhibit long tail characteristics [1], that few high-impact users build a large number of social connections and most ordinary users have relatively few friends. Because of the limited number of friends, the majority of users can only obtain and spread information in limited field. Through the social network friend recommendation system, we can mine user interest characteristics and recommend friends for them based on these characteristics better. So, a great social network friend recommendation system can help users to establish high-quality social contact and promote the development of social networks.

Different from popular social networks, the mainly users of the academic social networks are scholars and researchers. By creating the personal academic home pages in academic social networks, users can more comprehensively display their scientific research achievements and conveniently communicate with related scientific research scholars [2]. The academic social network contains more comprehensive and rich academic information, such as user's academic achievements, unit, research interests, etc. This information can reflect the user's academic characteristics better [3]. According to the characteristics of the academic social networks, we propose a friend recommendation model based on multi-dimensional academic feature and attention mechanism (hereinafter referred to as the MLP-AMAF model). We extracted a variety of different types of data such as the attribute data, relational data and text data from academic social networks to extract the corresponding multi-dimensional academic characteristics. Then calculate effects for user's different academic characteristics through the attention mechanism and endow dynamic weight for it. In the friend recommendation process, we can get better recommendation result by using these weights. Compared with the traditional recommendation method based on content, this approach can calculate user preferences for different academic features and get the more personal recommendation results based on these preferences.

The remainder of this paper is organized as follows. Section 2 explain related work. In Sect. 3 we propose our MLP-AMAF model and introduce in detail. Section 4 contains our experiments and results analysis and Sect. 5 we summarize our works.

2 Related Work

At present, there are mainly two types of social network friend recommendation algorithm, one is the method based on relationship in the social network and the other is content-based recommendation method. The method based on relationship through the established social relations to recommend friend, and the theoretical basis is the Triadic Closure principle [4] that the possibility of two people become friends in the future will be increased if they have a common friend. Literature [5] put forward the definition of link prediction and researched the friend recommendation task as the link prediction problem in social network. In the paper, they predicted the potential user interaction in the future based on the existing social network relationship. Yu et al. [6] built a heterogeneous information network by integrating user geographic location information and trajectory information, then applied the random walk process in the heterogeneous network to recommended friends for users. Huang et al. [7] proposed a hybrid method based on time-varying weights. By using extra time information, they obtained better results in the link prediction task of the co-author network. The content-based recommendation methods mine the different preferences of users, calculate the similarity between them according to these preferences and recommend users who have high similarity. Liu et al. [8] extracted user data like browse records and comments in the weibo and generated user tag by semantic analysis. They get better recommend results by using the collaborative filtering in these tags. Sang et al. [9] combined the image information, text information and user information in Flickr to modeling and obtained the user preferences according to the topic model.

Traditional content-based recommendation methods just make recommendations by the simple linear relationships among features and fail to capture more complex relationships. With the continuous development of deep learning, many researchers began to capture the deep interaction relations by the deep neural networks. These methods based on deep learning can greatly improve the performance of recommendation system. He et al. [10] put forward a general framework named NCF, through calculated potential characteristics of the relationship between the user and the project in MLP layer to alleviate the over generalization problem for the traditional matrix decomposition method. Wide&Deep model [11] and DeepMF model [12] combined the traditional method and the deep neural network model to extract the low-order and the higher-order combination features at the same time. Compared with the traditional method, these methods can obtain better recommendation results by the more complicated nonlinear relationship between the acquisition target users and recommend project. This kind of nonlinear relations is associated with all the features. However, in the actual cases, the target users usually do not pay attention to all features. Such as friends recommended tasks in social networks, different users would choose friends according to the different characteristics. Some users would choice friends whose research field is similar to and the other people would like to establish friend relation with his colleagues.

Attention mechanism is a simulation for the visual signal processing mechanism of human brain. For a specific task, more attention should be allocated to the area where is important for this task, and less attention should be allocated to the unimportant part. In this way, we can obtain the key information related to the task. In recent years, the deep learning model based on attention mechanism has been widely applied in various fields. Thang et al. [13] proposed two effective attention mechanisms to improve the performance of NML task. The RA-CNN model [14] recursively extracted necessary features from local information and achieved good results in image recognition task. Lin et al. [15] presented RCNN model to solve the problems of high sparseness and weak feature expression in traditional classification methods. By adding the attention mechanism into the model, we can calculate the attention degree of the different features for each user and get dynamically weight, thereby obtain user preferences better. So, we can improve the performance of recommendation tasks by this way.

Above all, we use the deep learning based on attention mechanism to solve the friends recommend task in the academic social network and puts forward a new friend recommendation model named MLP-AMAF. Experiments using the data extracted from academic social networking platform (www.scholat.com) prove the method is practicability and validity.

3 Methodology

Different from the popular social network, academic social network users have more academic characteristics, including academic achievements such as published papers and patents, degree and research field. These features can represent the characteristics of academic users better for friend recommendation tasks in academic social networks. In the recommendation model we proposed, user characteristics extracted from academic social networks are divided into three categories, namely attribute feature, relationship feature and text feature. The overall description is Table 1 and We will explain each of these in detail.

Table 1. The general description of feature.

Category	Feature group	Dimensionality
Attribute feature (A)	unit	$\sim 10^3$
	degree	$\sim 10^2$
	academic title	$\sim 10^2$
	research field	$\sim 10^3$
Relationship feature (R)	co_academic	1
	co_course	1
	co_team	1
Text feature (T)	achievement	64

3.1 Multi-dimensional Academic Feature Selection

Attribute Feature. There are many attributes of each user in the academic social network and we define all attributes per user as an attribute list. Note that we assure all attributes in the list are single-valued. So, we restructure the attribute list if the attribute category contains more than one value. The process of restructuring is shown in Eq. 1,

$$\begin{bmatrix} unit = scnu \\ degree = D.E. \\ research\,field = \{RS, CV\} \end{bmatrix} = \begin{bmatrix} unit = scnu \\ degree = D.E. \\ research\,field\,1 = RS \\ research\,field\,2 = CV \end{bmatrix} \quad (1)$$

Then we express per attribute in the list as a binary vector by encoding [16]. We defined $attr_{ui}$ as the i-th attribute in the attribute list for u, and V_{ui} as the corresponding binary vector, $V_{ui} \in \mathbb{R}^{K_i}$, and K_i is the dimension for the attribute category i, namely this attribute category include K_i unique attributes. We defined $V_{ui}[j] \in \{0, 1\}$, and set

$V_{ui}[j] = 1$, if u with the j-th attribute in the attribute category i, otherwise, $V_{ui}[j] = 0$. We get the corresponding unique one-hot encoding for each attribute in the list for u. Finial, we combine all one-hot encoding and get the attribute features for u as $A_u = [V_{u_1}, \ldots, V_{u_i}, \ldots, V_{u_n}]$.

Relationship Feature. In addition to the established friend relationship, there are other types of relationships can also reflect the characteristics of users in the academic social networks. In this paper, we extracted the academic cooperation relationship, team relationship and course relationship which are common in the academic social networks to represent the relationship characteristics between users. If there are academic cooperation relationships between u and v, we define co_academic$(u, v) = 1$, otherwise, co_academic$(u, v) = 0$. Similarly, we define co_course$(u, v) = 1$ represent u and v chose the same course, and co_team$(u, v) = 1$ represent u and v join the same team. Then we combine these relationships and convert into a three-dimensional binary-code to represent the relationship characteristics of user pair. For example, if u and v have academic cooperation relationship, but no team relationship or course relationship, we obtain the corresponding relationship characteristics between u and v as $R_{uv} = [1, 0, 0]$.

Text Feature. In academic social network, each academic achievement usually expressed as a text include the type, the title and the main content. For most of users in academic social networking, academic achievements can reflect their research interests well. we use the theme model to mine user research interests. LDA (Latent Dirichlet Allocation) [17] is a kind of unsupervised three layers of bayesian probability graph model, its theoretical basis is the theme of the text and the topic of the words both obeys the Dirichlet prior distribution. We can calculate the posterior theme distribution through the information from the words sample and carry on theme clustering or text classification tasks. Different from the traditional methods based on tf-idf, the LDA takes into account the semantic association of text and achieves better results. We extract all academic achievement for each user and merge into a long text. Then we calculate the interest research distribution and build the corresponding research interests through LDA. Finally, we get the research interest vector for each user and express the text feature of u as T_u.

3.2   Friend Recommendation Model

This paper puts forward the MLP-AMAF model to recommend friend in academic social network and the architecture is shown in Fig. 1. In our model, the MLP module obtain the potential nonlinear relationship between users and the candidates, and the attention-based multi-dimensional academic feature extraction module (hereinafter referred to as the AMAF module) calculate the preferences for different academic features by attention mechanism. The core concepts are that the additional academic feature information would contribute to select friends for user and the attention mechanism could effectively reduce the noise interference caused by this information. So, we can improve the performance of friend recommendation system in academic social network and obtain more personalized recommendation results.

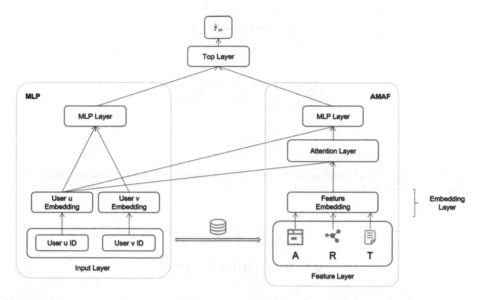

Fig. 1. Model architecture of MLP-AMAF, its left side is the MLP module and the right side is AMAF module.

Input Layer. We combine the ID of u and v as $\mathbf{pair}(u, v)$, then put it into our model to predict the probability that there is friend relationship between them in social network.

Future Layer. We get $\mathbf{pair}(u, v)$ from input layer and extract the corresponding multi-dimensional academic features including attribute feature A_v, relationship feature R_{uv} and text feature T_v. Then, we combine all of these as an academic features list F_{uv}. Because users have different amounts of features, the length of resulting list is also different.

Embedding Layer. The function of the embedding layer is map the high-dimensional sparse binary vector to the corresponding low-dimensional dense vector. For a user u, we get the corresponding binary vector represented by one-hot coding as $C_u = [0, \ldots, 1, \ldots, 0] \in \mathbb{R}^M$, M is the total number of users. Each user has a unique id and corresponds to a unique one-hot coding. There is an embedded matrix $W^U = [W_1^U, \ldots, W_u^U, \ldots, W_M^U] \in \mathbb{R}^{D \times M}$ for all users, D is the dimension of the output embedded vector. Through this matrix, the binary vector of u can be transformed into a low-dimensional dense embedding vector $E_u = W_u^U \in \mathbb{R}^D$. The process is shown as Fig. 1. Similarly, all features in the feature list are also transformed into the embedding vectors through embedding layer (Fig. 2).

MLP Layer. The low-dimensional dense embedded vectors are merged and sent to the full connection layer of the neural network through forward. The model can learn the potential nonlinear relationship between vectors automatically. This structure is called MLP and the calculating process is shown in Eqs. 2 and 3,

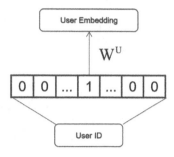

Fig. 2. Through the embedding layer, user id is transformed into the corresponding low-dimensional dense embedding vector.

$$Z^{(1)} = [E_u; E_v] \tag{2}$$

$$Z^{(L+1)} = \sigma\left(W^{(L)}Z^{(L)} + b^{(L)}\right) \tag{3}$$

Where L is the layer number of MLP, the $W^{(L)}$ and $b^{(L)}$ are respectively the weight matrix and bias for the L-th layer. We choose the rectifier linear units (ReLU) as the activation function $\sigma(\cdot)$ in our model, because it is a more appropriate way to deal with the sparse data.

Attention Layer. For the different user pair, the length of the feature list we extract is different. It cannot satisfy the request to MLP layer that the input must be fixed-length vectors. One of the common ways is average pooling [18]. The pooling layer can calculate average for all input vector as Eq. 4 and get a fixed-length vector as the feature vectors for **pair**(u, v).

$$\text{pooling}(F_{uv_1}, \ldots, F_{uv_k}, .., F_{uv_N}) = \frac{1}{N}\sum_{k=1}^{N} F_{uv_k} \tag{4}$$

However, this approach will lead to the losing information and the resulting vector cannot represent the academic characteristics of users well. In our model, the pooling layer is replaced by the attention layer, and the structure is shown in Fig. 3.

Through attention mechanism, this model can capture the attention of multi-dimensional academic feature dynamically. Different impact factors of these academic features on the recommendation task were calculated, and the corresponding vectors were dynamically weighted (as shown in Eq. 5). Academic features obtained in this way can reflect the real preferences of users better.

$$\text{attention}(E_u; F_{uv_1}, \ldots, F_{uv_k}, .., F_{uv_N}) \sum_{k=1}^{N} a(E_u, F_{uv_k})F_{uv_k} = \sum_{k=1}^{N} W^k F_{uv_k} \tag{5}$$

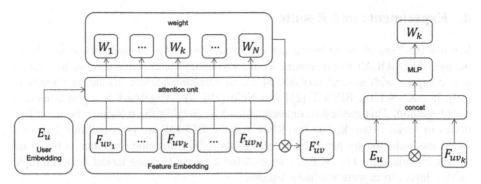

Fig. 3. The structure of attention layer. We get the corresponding weight W_k of each feature F_{uv_k} in the list through this layer. The left part is the overall structure and the right part is detail of the attention unit.

In attention unit (as shown in the right part of Fig. 3) $a(\cdot)$, we calculate the cross product of the user vector E_u and the feature vector F_{uv_k} in the feature individually, then combine them get the vector H_k. Finally, by inputting this vector into neural network full connection layer, we calculate the attention expressed by weight W_k. The relevant calculating processes are shown in Eqs. 6 and 7,

$$h_k = [E_u; E_u \times F_{uv_k}; F_{uv_k}] \tag{6}$$

$$a(E_u, F_{uv_k}) = \sigma(Wh_k + b) \tag{7}$$

Top Layer. In this layer, we combine the output of both sides of model, and the joint prediction function is defined as shown in Eq. 8,

$$\hat{y}_{uv} = \sigma\left(W^T \begin{bmatrix} \hat{y}_{uv}^{MLP} \\ \hat{y}_{uv}^{AMAF} \end{bmatrix}\right) \tag{8}$$

Where $\hat{y}_{uv}^{MLP}$ is output of the MLP module, it captures the potential nonlinear relationship between target users and recommended items by the neural network. And $\hat{y}_{uv}^{AMAF}$ is output of the AMAF module. In this module, we get user preferences for different academic features through an academic preferences network based on attention mechanism. W is the weight learned automatically from the network.

Loss. We regard the friend recommendation task in academic social network as a two-classification problem and define the loss function as shown in Eq. 9,

$$Loss = -\sum_{(u,v)\in S} y_{uv} \log(\hat{y}_{uv}) + (1 - y_{uv}) \log(1 - \hat{y}_{uv}) \tag{9}$$

Where S represents the sets of all training data, $y_{uv} \in \{1, 0\}$ is the real situation between u and v, $\hat{y}_{uv}$ is the predicted value.

4 Experiments and Results

We use the Tensorflow to implement our model MLP-AMAF and choose the friend dataset of SCHOLAT to experiment. In order to verify the effectiveness, we implement and compare with several methods of friend recommendation, including PageRank [19], itemKNN [20], BPRMF [21] and NCF [10]. The PageRank is a graph-features based method. This method recommend friend by judging the relevancy between two nodes in social network. The itemKNN and the BPRMF are both traditional recommend methods, the former is based on collaborative filter and the latter is based on matrix factorization. The NCF is a state-of-the-art deep learning model using multiple hidden layers to capture nonlinear feature.

In parameters setting, we initialize the learning rate as 0.001 and the training epochs as 20. Adam (Adaptive Moment Estimation) method [22] is used as the optimization algorithm in our model.

4.1 Datasets

The model proposed in this paper is the friend recommendation in academic social networks. We choose the academic friend relationship dataset of SCHOLAT, an online academic information service platform, to experiment. After data preprocessing, we obtain 27,758 users and 92,815 one-way friend relationship among them in total.

4.2 Assessment Method

The LOOCV (Leave One Out Cross Validation) [23] is used to evaluate the performance of our model. Specifically, for each user, we choose the last record of his friend relationships as the testing data and take the remaining as the training data to input the model. In the test phase, for a test user, we combine his testing data obtained previously and the 99 users from the training data who have no friend relationship with him. Then we get 100 users as the total friend recommendation list and order it according to the possibility calculated by the model. We evaluate performance of the model by measuring the ranking of the testing data in this ranking list. Specifically, we use HR (Hit Rate) and MRR (Mean Reciprocal Rank) [24] which are commonly used in the related tasks to judge the performance of the ranked list. The definitions are shown in Eqs. 10 and 11,

$$HR@N = \frac{1}{|U|} \sum_{i=1}^{U} \text{hit}_i \tag{10}$$

$$MRR@N = \frac{1}{|U|} \sum_{i=1}^{U} \frac{1}{rank_i}. \tag{11}$$

Where N is the length of current recommendation list and U shows the all test users. The $hit_i \in \{1, 0\}$ indicate whether the testing data in this recommendation and the $rank_i$ is the corresponding rankings. So, by using $HR@N$, we can intuitively observe whether the testing appears in the recommendation list with length of N. And the *MRR@N* is the quality of this list.

4.3   Results and Analysis

Table 2 reports the performance of all the methods. And we have the following observations by these results.

Table 2. Experimental results for academic friend relationship dataset of SCHOLAT.

		PageRank	itemKNN	BPRMF	NCF	MLP-AMAF
Top-1	HR	0.257	0.193	0.147	0.277	**0.353**
	MRR	0.257	0.193	0.147	0.277	**0.353**
Top-2	HR	0.342	0.242	0.194	0.377	**0.459**
	MRR	0.300	0.233	0.170	0.328	**0.392**
Top-5	HR	0.450	0.357	0.269	0.487	**0.700**
	MRR	0.330	0.291	0.191	0.355	**0.507**
Top-10	HR	0.516	0.476	0.321	0.584	**0.784**
	MRR	0.339	0.313	0.193	0.365	**0.516**
Top-20	HR	0.576	0.584	0.413	0.704	**0.871**
	MRR	0.343	0.344	0.205	0.387	**0.517**

For the friend recommendation task in academic social network, the PageRank achieve the better performance than other traditional methods. We hypothesize the uneven distributed data may be the cause of this thing. The users who have high influence in academic social network establish the most of friendships and the general users have few friends. This causes the bias in recommend result of traditional methods.

These methods using deep learning model (such as NCF and MLP-AMAF) are better than general algorithm (PageRank, itemKNN and BPRMF). The deep neural network can effectively capture the potential characteristic relationship between the target user and the recommended item. Methods based on deep learning utilize these to improve recommendation performance.

Compare with the NCF, there is a performance increase in the MLP-AMAF model we proposed. This indicates the attention mechanism can extract the key information related to the current recommendation task more effectively from the multi-dimensional academic features.

We also discuss the effects of the attention mechanism and the multi-dimensional academic feature. In the Fig. 4, the MLP-MAF represents the model with no attention mechanism and the attention layer is replaced by average pooling layer (as shown in Eq. 4). The MLP-MAF(a) only uses attribute features in the feature layer and the MLP-MAF(b) uses all of the academic features were acquired. The results from experiments indicate that the MLP-MAF(b) is worse than MLP-MAF(a). It means using all academic features cause performance degradation, instead. We hypothesized that the extra noise from the multi-dimensional academic feature is the main reasons which affected the effect of recommendation. By using the attention mechanism to assign greater attention to the key information and ignore the unimportant information, MLP-AMAF model can avoid noise interference and improve the recommendation quality.

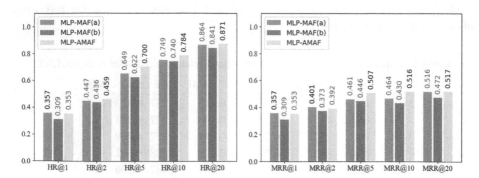

Fig. 4. The effect of attention mechanism on recommendation quality.

5 Conclusion

In this paper, we present a friend recommendation model based on multi-dimensional academic feature and attention mechanism. We enhance the performance of friend recommendation system by multi-dimensional academic feature, including the attribute features, the relationship features and the text features extracted from the academic social network. At the same time, we calculate users' preferences for different academic features by using the attention mechanism. This method can improve the performance bottlenecks in traditional methods caused by using all the feature and provide more personalized recommendations for different users according to their preferences. We use the friend relationship dataset of SCHOLAT to evaluate our model. The results demonstrate that our MLP-AMAF model is a more appropriate approach for friend recommendation tasks in academic social networks.

In future, we will construct a special academic field knowledge graph. By using this auxiliary information in knowledge graph to enhance our model, we can unveil the decision-making process for user in establish friend relationships at academic social networks. As a lot of research [25] shows that these methods added knowledge information can get more effective and interpretable recommend result by this way.

Acknowledgements. Our works were supported by the National Natural Science Foundation of China (No. U1811263, No. 61772211) and Innovation Team in Guangdong Provincial Department of Education (No. 2018-64/8S0177).

References

1. Brynjolfsson, E., Hu, Y., Smith, M.D.: Consumer surplus in the digital economy: estimating the value of increased product variety at online booksellers. Manag. Sci. **49**(11), 1580–1596 (2003)
2. Xia, Q., Li, W., Xue, J., et al.: New social networking platform for the academic fields: research networking systems. J. Intell. **33**(09), 167–172 (2014)
3. Xu, Y., Zhou, D., Ma, J.: Scholar-friend recommendation in online academic communities: an approach based on heterogeneous network. Decis. Support Syst. **119**, 1–13 (2019)
4. Easley, D.: Networks, crowds and markets: reasoning about a highly connected world. J. Roy. Stat. Soc. **175**(4), 1073 (2012)
5. Liben-Nowell, D., Kleinberg, J.: The link prediction problem for social networks. J. Am. Soc. Inf. Sci. Technol. **58**(7), 1019–1031 (2003)
6. Yu, X., Pan, A., Tang, L.A., Li, Z., Han, J.: Geo-friends recommendation in GPS-based cyber-physical social network. In: 2011 3rd International Conference on Advances in Social Networks Analysis and Mining, pp. 25–27. IEEE Computer Society, Kaohsiung (2011)
7. Huang, S., Tang, Y., Tang, F., Li, J.: Link prediction based on time-varied weight in co-authorship network. In: 18th IEEE International Conference on Computer Supported Cooperative Work in Design, pp. 706–709. IEEE, Hsinchu (2014)
8. Liu, L., Yu, S., Wei, X., Ning, Z.: An improved Apriori–based algorithm for friends recommendation in microblog. Int. J. Commun. Syst. **31**, e3453 (2018)
9. Sang, J., Xu, C.: Right buddy makes the difference: an early exploration of social relation analysis in multimedia applications. In: Proceedings of the 20th ACM International Conference on Multimedia, pp. 19–28. ACM, Nara (2012)
10. He, X., Liao, L., Zhang, H., Nie, L., Hu, X., Chua, T.S.: Neural collaborative filtering. In: Proceedings of the 26th International Conference on World Wide Web, pp. 173–182. ACM, Perth (2017)
11. Cheng, H.T., Koc, L., Harmsen, J., Shaked, T., Chandra, T., Aradhye, H., et al.: Wide & deep learning for recommender systems. arXiv preprint arXiv:1606.07792 (2016)
12. Guo, H., Tang, R., Ye, Y., Li, Z., He, X.: DeepFM: a factorization-machine based neural network for CTR prediction. In: Proceedings of the Twenty-Sixth International Joint Conference on Artificial Intelligence, pp. 1725–1731. ijcai.org, Melbourne (2017)
13. Thang, L., Hieu, P., Christopher, D.M.: Effective approaches to attention-based neural machine translation. In: Proceedings of the 2015 Conference on Empirical Methods in Natural Language Processing, pp. 1412–1421. The Association for Computational Linguistics, Lisbon (2015)
14. Jianlong, F., Heliang, Z., Tao, M.: Look closer to see better: recurrent attention convolutional neural network for fine-grained image recognition. In: 2017 IEEE Conference on Computer Vision and Pattern Recognition, pp. 4476–4484. IEEE Computer Society, Honolulu (2017)
15. Lin, R., Fu, C., Mao, C., Wei, J., Li, J.: Academic news text classification model based on attention mechanism and RCNN. In: Sun, Y., Lu, T., Xie, X., Gao, L., Fan, H. (eds.) ChineseCSCW 2018. CCIS, vol. 917, pp. 507–516. Springer, Singapore (2019). https://doi.org/10.1007/978-981-13-3044-5_38
16. Zhou, G., Song, C., Zhu, X., Fan, Y., Zhu, H., Ma, X., et al.: Deep interest network for click-through rate prediction. In: Proceedings of the 24th ACM SIGKDD International Conference on Knowledge Discovery & Data Mining, pp. 1059–1068. ACM, London (2018)
17. Blei, D.M., Ng, A.Y., Jordan, M.I., Lafferty, J.: Latent Dirichlet allocation. J. Mach. Learn. Res. **3**, 993–1022 (2003)

18. Paul, C., Jay, A., Emre, S.: Deep neural networks for YouTube recommendations. In: Proceedings of the 10th ACM Conference on Recommender Systems, pp. 191–198. ACM, Boston (2016)
19. Haveliwala, T.H.: Topic-sensitive PageRank. In: 2002 International Conference on World Wide Web, pp. 517–526. ACM, Honolulu (2018)
20. Ning, X., Karypis, G.: SLIM: Sparse linear methods for top-N recommender systems. In: 11th IEEE International Conference on Data Mining, pp. 497–506. IEEE Computer Society, Vancouver (2011)
21. Rendle, S., Freudenthaler, C., Gantner, Z., Schmidt-Thieme, L.: BPR: Bayesian personalized ranking from implicit feedback. In: Conference on Uncertainty in Artificial Intelligence, pp. 452–461. AUAI Press (2009)
22. Kingma, D.P., Ba, J.: Adam: a method for stochastic optimization. arXiv preprint arXiv: 1412.6980 (2014)
23. Cawley, G.C., Talbot, N.L.C.: Preventing over-fitting during model selection via bayesian regularisation of the hyper-parameters. J. Mach. Learn. Res. **8**(8), 841–861 (2007)
24. Voorhees, E.M.: The TREC question answering track. Natural Language Engineering **7**(04), 361–378 (2001)
25. Wang, X., He, X., Cao, Y., Liu, M., Chua, T.S.: KGAT: knowledge graph attention network for recommendation. arXiv preprint arXiv:1905.07854 (2019)

Research on Telecom Customer Churn Prediction Method Based on Data Mining

Xuechun Liang, Shuqi Chen[✉], Chen Chen, and Taoning Zhang

Nanjing Tech University, Nanjing, China
767271332@qq.com

Abstract. Aiming at overcoming the shortcomings of common telecommunication customer churn prediction models as single model and poor classification performance, a gradient decision tree integration model (GBDT) prediction model is proposed, and the important parameters are searched by harmonic search algorithm (HS). We built a telecom customer churn prediction model based on HS-GBDT algorithm. This model compares the parameter combinations to be optimized in the GBDT algorithm into the synthesized harmony in the HS algorithm, and seeks the optimal parameter combination of the GBDT model through continuous iteration of the harmony. The experimental results show that the combined model has higher classification accuracy than Logistic regression, support vector machine and random forest, and can provide good decision support for major telecom operators in the process of customer churn management.

Keywords: Telecom customer churn prediction · Harmony search algorithm · Gradient Boosting Decision Tree · Combined model

1 Introduction

With the rapid development of China's socialist market economy and the increasing popularity of communication tools, the competition in the telecommunications industry has become increasingly fierce. Nowadays, major telecom operators are paying more and more attention to the management and maintenance of corporate customer relationships. Customer churn prediction helps improve telecom customer relationship management. Data mining modeling technology can effectively predict customers with loss tendency. Adopting the corresponding marketing strategy and trying to retain customers can bring more benefits to telecom operators [1].

Since the 1990s, Researchers in China and other countries have conducted a lot of research on telecom customer churn prediction and produced rich scientific research results. Based on a large number of domestic and foreign scientific research results, this paper concludes that the telecom customer churn prediction model research is mainly divided into two stages [2]: first, the statistical and machine learning prediction model are commonly used, including Decision tree, Bayesian classification, logistic regression, cluster analysis, artificial neural network, support vector machine algorithm, etc. In the second stage, the relatively mature integrated learning prediction model is mainly used, including bagging algorithm and promotion (Boosting) Algorithm, random

© Springer Nature Singapore Pte Ltd. 2019
Y. Sun et al. (Eds.): ChineseCSCW 2019, CCIS 1042, pp. 485–496, 2019.
https://doi.org/10.1007/978-981-15-1377-0_38

forest, Gradient Boosting Decision Tree (GBDT) algorithm [3]. In recent years, the research on telecom customer churn prediction model mainly focuses on intelligent machine learning model, and gradually transforms from a single intelligent algorithm model to a complex integrated algorithm model. The integrated algorithm model has better classification performance than the single algorithm model. Zhang [4] et al. used the mobile communication enterprise customer churn data set to construct a strong classifier model based on classification regression tree and adaptive Boosting algorithm through its training samples. The experiment results show that their model has higher classification accuracy than the single prediction model (C4.5, fuzzy Bayes). Ding et al. [5] proposed an improved random forest algorithm (IRFA) to improve the prediction accuracy of high value lost customers, and the algorithm was applied to a telecom customer churn data set. The results show that IRFA phase has better classification performance than traditional weighted SVM and hybrid neural networks. The decision tree algorithm has good performance in unbalanced data classification. Therefore, Li [6] uses the gradient lifting tree integrated learning model as the telecom customer churn prediction model. By analyzing the "Orange data set", the gradient lifting tree model is more statistical than the statistics. Traditional single-classification models such as regression and decision trees have significantly improved AUC performance and generalization ability.

Based on this research, this paper uses the new gradient decision tree algorithm in Boosting integrated algorithm for telecom customer churn prediction modeling, and optimizes its parameters through the harmony search algorithm [7] to form the final telecom customer churn prediction model. The experimental results show that the improved gradient-lifting decision tree algorithm (HS-GBDT) improves the classification accuracy of unbalanced telecom customer churn data sets.

2   Theoretical Basis

2.1   Harmony Search(HS)

Harmony search (HS) is an emerging optimization algorithm proposed by Geem Z W et al. in 2001. It simulates the process of instrument masters repeatedly adjusting the pitch of each instrument to achieve a perfect harmony. The process can be analogized to the solution process of the optimization problem [8]. The analogy of the harmony is the objective function value of the optimization problem. The harmonics of each instrument's tonal synthesis can be analogized to the solution vector of the optimization problem, and the adjustment of each tone is synthesizing the new sum. The process analog of the sound is the iteration of the solution vector, and the pitch of the instrument corresponds to the specific variable value. The basic steps of the harmony search algorithm are as follows [9]:

Step 1: Initialize the basic parameters of the harmony algorithm. The harmony memory size HMS, The harmony memory value probability $HMCR$, the pitch fine-tuning probability PAR, the pitch fine-tuning bandwidth BM, the maximum number of iterations of the solution vector T_{max}, and the number of individual harmony vectors N.

Step 2: Initialize the harmony memory. Randomly constructing HMS harmony x^1, x^2, ..., x^{HMS}, HMS values in the harmony memory is similar to the number of populations in the genetic algorithm, and the acoustic memory can be expressed as:

$$HM = \begin{bmatrix} x^1 & f(x^1) \\ x^2 & f(x^2) \\ \vdots & \vdots \\ x^{HMS} & f(x^{HMS}) \end{bmatrix} = \begin{bmatrix} x_1^1 & x_2^1 & \cdots & x_N^1 & f(x^1) \\ x_1^2 & x_2^2 & \cdots & x_N^2 & f(x^2) \\ \vdots & \vdots & & \vdots & \vdots \\ x_1^{HMS} & x_2^{HMS} & \cdots & x_N^{HMS} & f(x^{HMS}) \end{bmatrix}. \quad (1)$$

Step 3: Generate a new harmony. Each tone $x_i'(i = 1, 2, \ldots, N)$ in the harmony $x_i' = (x_1', x_2', \ldots, x_N')$ is constructed in three ways, mainly including selecting a tone in the harmony memory, fine-tuning the tone, and randomly selecting the tone. The first pitch variable x_i' in the harmony is used. The probability that the variable has $HMCR$ is randomly selected from HMS in the sound library $(x_1^1 \sim x_1^{HMS})$, and the probability of 1 - HMCR is selected from outside HMS (in the total value range). Similarly, the construction process of other harmony decision variables is

$$x_i' = \begin{cases} x_i' \in (x_i^1, x_i^2, \cdots, x_i^{HMS}), & \text{if } rand < HMCR \\ x_i' \in X_i, \text{ otherwise;} & i = 1, 2, \cdots, N \end{cases}, \quad (2)$$

where $rand$ is a random number on [0, 1]. The last new tone is generated by fine-tuning the tones in the acoustic memory HMS. The specific fine-tuning process is

$$x_i' = \begin{cases} x_i' + rand1 * bw, & \text{if } rand1 < PAR(\text{continuous}), \\ x_i' * (k+m), m \in \{-1, 1\}, & \text{if } rand1 < PAR(\text{discrete}), \\ x_i' & \text{otherwise,} \end{cases} \quad (3)$$

where bw represents the pitch trimming bandwidth, PAR represents the pitch trimming probability, and $rand1$ represents the random number on [0,1].

Step 4: Update the harmony memory. The new harmony obtained in **Step3** compares the harmony with the worst function value in the HMS. If it is better than the worst value, the new harmony is replaced. The specific process is

$$\text{if } f(x') < f(x^{worst}) = \max_{j=1,2,\cdots,HMS} f(x^j), \text{ then } x^{worst} = x'. \quad (4)$$

Step 5: Check the algorithm stop condition. Return **Step3** until the algorithm stops when the number of iterations T_{max} is reached.

2.2 Gradient Boosting Decision Tree (GBDT)

The Gradient Boosting Decision Tree (GBDT) algorithm is a Boosting-based decision tree algorithm proposed by Friedman in 2001. The algorithm uses the Gradient Boosting algorithm to optimize the Boosting Tree. In the GBDT iterative process, assuming that the strong learner of the previous iteration is $F_{m-1}(x)$ and the loss function is $L(y, F_{m-1}(x))$, then the decision tree weak learner of this iteration is $h_m(x)$.

In order to minimize the learning loss function $L(y, F_{m-1}(x)) = L(y, F_{m-1}(x) + h_m(x))$, the GBDT algorithm uses the pre-distribution algorithm and the additive model method find the optimal decision tree parameter $\hat{\Theta}_m$ [10, 11] in the negative gradient direction of the loss function.

This paper applies GBDT algorithm to solve the problem of telecommunication customer churn prediction, so its loss function under the two classification is [12]

$$L(y, F(x)) = \log(1 + \exp(-2yF(x))), \tag{5}$$

where $y \in \{-1, 1\}$, $F(x) = \frac{1}{2} \log[\frac{Pr(y=1|x)}{Pr(y=-1|x)}]$, the negative gradient error is $\frac{2y}{1+\exp(2yF_{m-1}(x))}$, $Pr(y = 1|x)$ is the probability that the classifier predicts that the sample x is 1, and $Pr(y = -1|x)$ is the probability that the classifier predicts that the sample x is -1. The steps of the GBDT algorithm for the two-class problem are as follows:

Step 1: Initialize the weak classifier function to $F_0(x) = \frac{1}{2} \log \frac{1+\bar{y}}{1-\bar{y}}$, its loss function is $L(y_i, F(x_i)), i = (1, 2, \cdots N)$, iteration number is M.

Step 2: Equation $\hat{g}_i = \frac{2y_i}{(1+\exp(2y_iF_{m-1}(x_i)))}, i = (1, 2, \cdots N)$ calculates the negative gradient direction (i.e., the residual) of the model function and obtains a new residual data set $G = \{(x_i, \hat{g}_{mi})\}_1^N$.

Step 3: The residual data set G is used as a training sample of the next tree, and a new classification tree $F_m(x)$ is obtained, and its corresponding leaf node area is $\{R_{jm}\}_1^J = J - final\ node\ tree\ (\{\hat{g}_i, x_i\}_1^N), j = 1, 2 \cdots, J$, where J is the number of leaf nodes of the classification tree.

Step 4: Calculate the corresponding best fit value for $\gamma_{jm} = \frac{\sum_{x_i \in R_{jm}} \hat{g}_i}{\sum_{x_i \in R_{jm}} |\hat{g}_i|(2-|\hat{g}_i|)}, j = (1, 2, \cdots J)$ for $\{R_{jm}\}_1^J$.

Step 5: Update the strong classifier $F_m(x) = F_{m-1}(x) + \sum_{j=1}^{J} \rho_{jm} I(x \in R_{jm})$.

Step 6: Let $m = m + 1$ and go to Step1–Step4.

Step 7: Strong classifier $F_M(x) = F_0(x) + \sum_{m=1}^{M} \sum_{j=1}^{J} \gamma_{jm} I(x \in R_{jm})$.

3   HS-GBDT Model

There are many parameters in the GBDT algorithm to adjust the accuracy of the model and control the degree of model fitting. These parameters are the main decision tree related parameters (*min_samples_split, min_weight_fraction_leaf, min_samples_leaf, max_depth, max_leaf_nodes, max_features*); gradient lifting related parameters (*learning_rate, n_estimators, Subsample*), other parameters (*loss, criterion, init, random_state*).

In the GBDT algorithm, when the maximum depth parameter of the tree in the decision tree related parameters is set too small, the algorithm will be under-fitting, and the problem of setting the over-fitting will be over-set; When the learning rate parameter in the gradient-enhancing parameters is set too much, the algorithm cannot converge. If the setting is too small, the algorithm will overfit. Therefore, the maximum depth (*max_depth*) and learning rate (*learning_rate*) of the tree are two important parameters to be investigated. When other parameters are fixed values, adjusting these two parameter values will directly effect the classification accuracy of the GBDT model. In order to acquire the best model, improve the parameter optimization efficiency, and avoid the randomness, complexity, speculation of the traditional mechanical enumeration optimization method, and the poor performance of genetic algorithm and PSO algorithm, this paper chooses and the acoustic search algorithm (HS) optimizes the *learning_rate* and *max_depth* parameters in the GBDT algorithm, and compares the two parameters to the two instruments of synthesizing harmony, iterating these two parameters, finding the fitness function to the optimal value. The best solution vector.

The HS-GBDT optimization algorithm steps are as follows:

Step1: Set the parameter (*learning_rate*) and (*max_depth*) ranges and HS algorithm related parameters. The number of individual harmony variables N, the harmony memory size *HMS*, the acoustic memory value probability *HMCR*, the pitch fine-tuning probability *PAR*, the pitch fine-tuning bandwidth *BW*, the maximum number of times the solution vector needs to be iterated T_{max}, and the GBDT learning rate (*learning_rate*) and tree The upper and lower limits of the maximum depth (*max_depth*) are x_{max}, x_{min}.

Step2: Initialize and sound library *HMS*. According to the upper and lower limits of the two optimization parameters in GBDT, the *HMS* harmony is generated by the formula $x_{min} + rand(1, N) \times (x_{max} - x_{min})$, and they are combined into one harmony library.

Step3: Calculate the fitness of each harmony in the initial sound library. Each harmony in the initial harmony library is a combination of (*learning_rate, max_depth*), and the GBDT model is constructed with each harmony as a parameter, and the average accuracy of the cross validation set is taken as its fitness value.

Step4: Build new harmony. If *rand* < *HMCR*, a new harmony can be randomly selected from the harmony library by Eqs. (1 and 2), and the selected and tone can be fine-tuned by the probability in Eqs. (1–3); if *rand* > *HMCR*, a new harmony will be reselected within the variable total value field.

Step5: Update the harmony memory. Calculate the fitness value of the new harmonics constructed by Step4, and update the sound library by formula (1–4) to generate a new harmony library.

Step6: Determine if the algorithm is terminated. If the set number of iterations or the fitness value (error requirement) is reached, the algorithm stops and the optimal parameters are output; otherwise, the algorithm proceeds to Step 4 to continue execution.

The specific flow chart of the HS-GBDT algorithm is shown in Fig. 1:

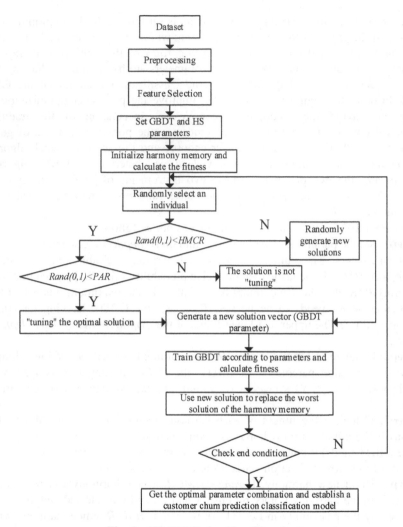

Fig. 1. HS-GBDT algorithm flow chart

4 Experiment Analysis

We verify the validity of the HS-GBDT algorithm, we first selects the Kaggle data contest website US x company telecom customer churn data set for experiments. The number of original samples of the telecom customer churn data set is 4481 and 33 attribute characteristics. After data preprocessing, balancing processing and feature selection, Table 1 gives the basic information of this data set

Table 1. Telecom_churn data set basic information table

Data set status	Actual number	Positive sample	Negative sample	Attributes	Category
Original	3333	2850	483	33	2
After processing	4481	2646	1835	24	2

Initialize the parameters related to HS and GBDT algorithm parameters: $HMS=60$, $HMCR = 0.8$, $PAR=0.5$, $bw = 0.01$, $T_{\max} = 200$, $N = 2$, $learning\_rate \in (0, 1]$, $max\_depth \in [0, 100]$, and use the HS algorithm to optimize the parameters of the GBDT model, and take the optimal iterative process in 20 independent experiments as the final experimental result. Among them, Fig. 2 shows the optimal iterative process for optimizing the GBDT model parameters using the HS algorithm on the Telecom_churn telecom customer churn data set.

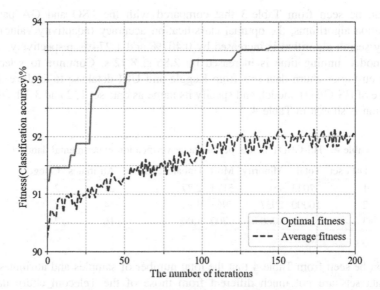

Fig. 2. Iterative convergence graph (Telecom_churn)

It can be seen from Fig. 2 that within 200 iterations of the Telecom_churn data set, the algorithm achieves convergence, and the optimal parameters and optimal fitness values of the model are obtained. The Optimal parameters and other results are shown in Table 2.

Table 2. Optimal parameters and classification accuracy of Telecom_churn

Data set	GBDT accuracy (initial parameters)	Positive sample	Negative sample	Attributes
		Highest accuracy	*learning_rate*	*max_depth*
Telecom_churn	91.231%	93.553%	0.183	6

It can be seen from Table 2 that under the optimization of the harmony search algorithm (HS), the optimal classification accuracy of the GBDT model is 2.78% higher than that of the GBDT model under the initial parameters. In order to verify the optimization performance of the harmony search algorithm, the optimization results of different parameter optimization algorithms on the Telecom_churn data set are compared, as shown in Table 3.

Table 3. Accuracy comparison of the three optimization algorithms

Optimization	Optimal parameter combination	Classification accuracy	Time
HS-GBDT	(0.183, 6)	93.553%	10.41 s
PSO-GBDT	(0.225, 4)	93.146%	13.27 s
GA-GBDT	(0.318, 11)	92.279%	18.53 s

It can be seen from Table 3 that compared with the PSO and GA parameter optimization algorithms, the optimal classification accuracy (adaptivity value) of the harmony search algorithm is increased by 0.407% and 1.274%, respectively, and the single model running time is increased by 2.86 s, 8.12 s. Continue to select three telecom customer churn data sets from Kaggle and UCI database to analyze the performance of HS-GBDT model, and specify its name as data set 1, 2 and 3. The detailed description is shown in Table 4.

Table 4. HS-GBDT model performance verification experimental data set

Data set	Total	Minority	Most	Unbalanced rate	Attributes	Category
1	7043	1869	5174	2.77	21	2
2	10000	2037	7963	3.91	14	2
3	1400	700	700	1.00	16	2

It can be seen from Table 4 that the total number of samples and attributes of the three data sets are not much different from those of the Telecom_churn data set. Selecting these three data sets to check the classification accuracy of the HS-GBDT model has certain rationality. The HS-GBDT model is constructed for the data sets 1, 2, and 3, respectively, and the corresponding optimal parameters and the highest classification accuracy are found. The results are shown in Table 5. The optimal iterative process diagrams of the three data sets are shown in Figs. 3, 4 and 5.

Table 5. Optimal parameters and classification accuracy of different data sets

Data set	GBDT accuracy (initial parameters)	Positive sample	Negative sample	Attributes
		Highest accuracy	learning_rate	max_depth
1	81.232%	83.410%	0.176	2
2	84.240%	86.221%	0.121	7
3	81.571%	84.428%	0.093	1

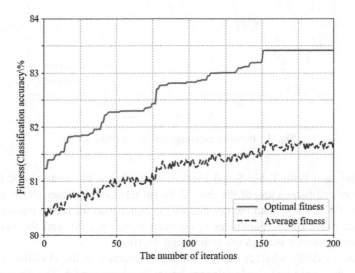

Fig. 3. Iteration convergence graph (data set 1)

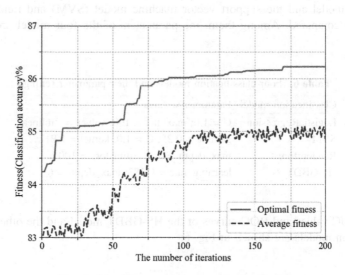

Fig. 4. Iterative convergence graph (data set 2)

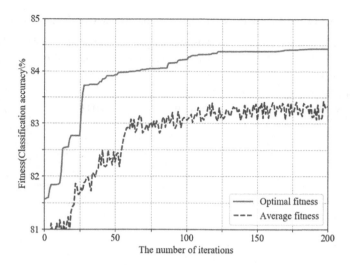

Fig. 5. Iterative convergence graph (data set 3)

It can be seen from Table 5 and Figs. 3, 4 and 5 that the accuracy of the model under the optimal parameters of the three different telecom customer churn data sets is 2.78%, 1.981%, 2.857%, and the optimal rate is better than the initial parameters. The optimization range of the parameter is between *learning_rate*, and the optimization range of the parameter *max_depth* is between $[0, 10]$.

Continue to verify whether the classification accuracy of the HS-GBDT model is higher than the traditional single telecom customer churn model. The Telecom_churn data set after pre-processing and feature selection is taken as input, and the Logistics regression model and the support vector machine model (SVM) and random forest models are compared. Among them, the parameters of the four model are given in Table 6:

Table 6. Four classification models and their parameter settings

Classification model	Parameter settings
LogisticRegression	C = 1.0; max_iter = 100; solver = 'iblinear'
SVM	gamma = 0.25; C = 1.0; kernel = RBF
Randomforest	max_depth = 3; n_estimators = 100
HS-GBDT	learning_rate = 0.183; max_depth = 6

The AUC values and ROC curves of the HS-GBDT model and the other common classification models are shown in Fig. 6:

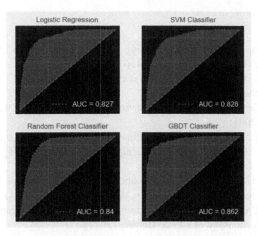

Fig. 6. ROC curve of four classification models

It can be seen from Fig. 6 that the HS-GBDT model has the highest AUC value compared to the other three models on the Telecom_churn dataset, which fully demonstrates the superior classification performance of the HS-GBDT model on the unbalanced telecom customer churn dataset.

5 Conclusion

The construction of telecom customer churn prediction model is a significant link in data mining theory, and it is also the focus of many scholars. The quality of predictive models determines the accuracy of telecom customer management personnel to assist decision-making. In this research, we analyzing the GBDT model and its parameter, determining the model parameter combination (*learning_rate, max_depth*) that needs to be optimized, setting the optimization range of the parameters, and using the harmony search algorithm to find the optimal parameter combination suitable for the model within the two parameters, we build the final predictive model. Finally, through the model evaluation and comparative analysis, it is verified that the classification performance of the combined model is better than other commonly used classification algorithms. Because the harmony search algorithm only optimizes the two parameters *learning_rate and max_depth* in the GBDT algorithm, and does not consider the influence of other important parameters on the classification performance of the model, can it search for multiple important parameters in the GBDT algorithm at the same time? Excellent is the direction of research on subsequent algorithm optimization problems.

Acknowledgment. This paper was supported in part by the Postgraduate Research & Practice Innovation Program of Jiangsu Province under Grant No. KYCX19-0874, National Natural Science Foundation of China under Grant No 11801267, and the Natural Science Foundation of the Jiangsu Higher Education Institutions of China under Grant No. 18KJB520007.

References

1. Qin, H.F.: The application of data mining in telecommunication churn customer. Res. J. Appl. Sci. Eng. Technol. **4**(11), 1504–1507 (2012)
2. Yu, X.B., Cao, J., Gong, Z.W.: Review on customer churn issue. Comput. Integr. Manuf. Syst. **18**(10), 2253–2263 (2016)
3. Friedman, J.H.: Greedy function approximation: a gradient boosting machine. Ann. Stat. **29** (5), 1189–1232 (2001)
4. Zhang, W., Yang, S.L., Liu, T.T.: Customer churn prediction in mobile communication enterprises based on CART and boosting algorithm. Chin. J. Manag. Sci. **22**(10), 90–96 (2014)
5. Ding, J.M., Liu, G.Q., Li, H.: The application of improved random forest in the telecom customer churn prediction. Pattern Recogn. Artif. Intell. **28**(11), 1041–1049 (2015)
6. Li, R.Q.: Churn Prediction Models for Anonymous Telecom Customer Dataset. University of Science and Technology of China, Beijing (2017)
7. Geem, J., Kim, J.H., Loganathan, G.V.: A new heuristic optimization algorithm: harmony search. Simulation **76**(2), 60–68 (2001)
8. Ding, J.L., Li, L.S.: Network traffic predicting based on SVM optimized by harmony search algorithm. Microcomput. Appl. **33**(01), 67–70 (2017)
9. Guo, W.Y.: Research on Selective Ensemble Algorithm Based on Support Vector Machine. Heifei University of Technology, Heifei (2015)
10. Schapire, R.E., Freund, Y., Bartlett, P.: Boosting the margin: a new explanation for the effectiveness of voting methods. Ann. Stat. **26**(5), 1651–1686 (1998)
11. Li, H.: Statistical Learning Method. Tsinghua University Press, Beijing (2012)
12. Zhang, X.: Quantitative Investment Model Based on Improved GBDT. Guangxi University, Nanning (2018)

Temporal Relationship Recognition of Chinese and Vietnamese Bilingual News Events Based on BLCATT

Jidi Wang[1,2], Junjun Guo[1,2], Zhengtao Yu[1,2], Shengxiang Gao[1,2(✉)], and Yuxin Huang[1,2]

[1] Faculty of Information Engineering and Automation,
Kunming University of Science and Technology, Kunming 650500, China
gaoshengxiang.yn@foxmail.com
[2] Yunnan Key Laboratory of Artificial Intelligence,
Kunming University of Science and Technology, Kunming 650500, China

Abstract. The temporal relationship identification of bilingual news events is essentially to find the temporal relationship between news events in different languages under the same topic. At present, the event temporal relationship recognition requires a lot of manpower to design a timeline-based template. The implicit semantic information in the sentence is difficult to obtain, and different language texts are difficult to represent in the same feature space. Therefore, it is difficult to obtain the temporal relationship of cross-language news events. To this end, this paper proposes a Bi-LSTM Cross Attention (BLCATT) model to identify the temporal relationship between two events. Firstly, the bilingual word vector is used to represent the Chinese and Vietnamese bilingual news texts. Secondly, using Bi-LSTM to capture the semantic information of the sentence. Using the attention mechanism combined with the trigger word to obtain the event semantic information of the enhanced trigger word information. The event semantic information including the temporal logic relationship is obtained by using the cross-attention mechanism combined with the trigger word and combines the two parts of the event semantic information as the event coding. Finally, the event coding and the event rule feature are combined to obtain the temporal relationship information, and then the event temporal relationship recognition is transformed into the multi-class problem. To solve. The experimental results show that the proposed model achieves good results.

Keywords: Chinese and Vietnamese bilingual news events · Event temporal relationship identification · Bi-LSTM · Attention mechanism · Cross-attention mechanism

1 Introduction

With the rapid development of the Internet, The mutual influence of economic and political aspects between different countries is deepening, and there are more and more events of common concern. Usually, these events do not exist in isolation, but there are inherent temporal relationships. Therefore, timely and effective situation analysis of

© Springer Nature Singapore Pte Ltd. 2019
Y. Sun et al. (Eds.): ChineseCSCW 2019, CCIS 1042, pp. 497–509, 2019.
https://doi.org/10.1007/978-981-15-1377-0_39

news events in other countries is particularly important. Vietnam is one of the countries with strong economic strength among Southeast Asian countries. Trade and cultural exchanges with China are very frequent. Dealing with international relations with Vietnam plays an important role in regional economic development and political stability. Therefore, this paper takes the Chinese and Vietnamese bilingual news as an example to study the temporal series relationship identification method of bilingual news.

Vietnam is adjacent to China, analyze and process the news texts issued by China and Vietnam, and obtain the temporal relationship between different events under the same topic. This helps the relevant departments to grasp the event dynamics and orientation of the two countries and do the correct response measures are taken.

Most of the current event temporal relationship recognition research is based on time timeline expressions, but not all time in the news text contains time expressions. It is difficult to find the temporal relationship between news events that do not contain time expressions. To this end, this paper explores the sequence of events occurring at the level of sequential logic, not the temporal relationship between time expressions. The temporal logical relationship between events usually depends on the acquisition of implicit semantic information of the event sentence. Therefore, the temporal relationship recognition of events is mainly composed of two sub-tasks: (1) mining the implicit semantic information of event sentences; (2) mining the temporal logic relationship between event pairs to determine the temporal relationship of event pairs.

In addition, the study of event temporal relationship recognition is basically carried out in a monolingual environment. Since Vietnamese is a resource-poor language, there are few studies on event temporal relationship recognition based on Chinese Vietnamese. Traditional monolingual event temporal relationship recognition methods mostly rely on templates [1, 2] and machine learning methods [3–6]. The event relationship templates manually defined by the knowledge and contextual features of linguistics are often subjective, the recall rate is not ideal, and the application value is low. Pantel [3] automatically constructs templates through Espresso algorithm, designs shallow templates with a small amount of data, and iteratively expands shallow templates through machine learning methods, which improves the generalization ability of templates to a certain extent. With the excellent performance of deep learning in various fields, it has significant effects on event temporal relationship recognition tasks [7–11]. Tourille et al. [7] proposed a model for identifying a temporal relationship in medical events and has had the best results to date. Tian et al. [8] proposed a Bi-LSTM model combining the attention mechanism for the Uighur event temporal relationship recognition task, which verified the effectiveness of the attention mechanism in LSTM and completed the event temporal relationship identification task, Vo et al. [9] propose an event networks ECIR to extracting temporal event relations by unsupervised method. Ghorbel et al. [10] proposed a model to deal with the temporal relation between imprecise time intervals. Vashishtha et al. [11] propose a new framework for modeling the semantic to reorganize the temporal relations of events. but the event The temporal relationship is still identified by feature selection.

Compared with the event temporal relationship recognition in the monolingual environment, there are few related types of research in the bilingual environment. The key question is how to cross the language barrier. Currently, it is mainly based on the following two methods. (1) A method based on machine translation. It converts news texts in different languages into the same target language and identifies event time series relationships in a monolingual environment. However, the accuracy of Chinese and Vietnamese machine translation is not good, which greatly reduces the efficiency of event temporal relationship recognition. (2) A method of using a bilingual dictionary. This method translates the entities, trigger words, and event phrases in the text. This method ignores words that have the same meaning but no similar translations, such as "Chunfu Ruan" and "Thủ tướng việt nam (Vietnamese Prime Minister)". There is no mutual translation relationship in the dictionary, but they are express the same person.

Based on the analysis of the above problems, this paper proposes a Bi-LSTM Cross Attention (BLCATT) model to solve the cross-language event temporal relationship recognition task. Since the news is objective, the semantic information of the same news event expressed in different languages is roughly the same. The model uses the bilingual word vector to represent the Chinese and Vietnamese bilingual news texts and maps the Chinese Vietnamese words into the same semantic space. In this space, Words with similar semantics have similar distances, and word vectors with low semantic relevance are far apart. This paper uses Bi-LSTM to obtain the context implicit semantic features of event sentences, then uses the attention mechanism of trigger words obtain the enhanced semantic information of event sentences, uses a cross-attention mechanism that combines different language trigger words obtain Event semantic features of enhanced temporal information. The feature merging of these two parts serves as the semantic information of the current sentence. Combining the bilingual features and the features of the temporal relationship between events, this paper proposes six kinds of event-based rule features and merges them with event semantic features, and classified by soft-max layer. Realize the identification of the temporal series relationship between Chinese and Vietnamese bilingual news events.

2   Related Event Definition

(1) Event [12]: Event composed of event trigger words and event parameters. An event that occurs at a specific time and environment, it is attended by several characters, exhibiting action characteristics.
(2) Event trigger words: The main word that triggers the occurrence of the event, it can clearly express the occurrence of a type of event.
(3) Event parameters: describe the time, place, person and other information about the event.
(4) Event Pair: The event pair in which all the events in the Chinese and Vietnamese news texts are paired according to the matching rules under the same topic.
(5) Event temporal relationship: The temporal relationship between two events in the event pair. Referring to the temporal relationship classification scheme proposed by TimeML [13], this paper divides the event temporal relationship into three types: discontinuous context relationship, continuous context relationship, and parallel relationship.

3   Chinese and Vietnamese Bilingual News Event Temporal Relationship Recognition Model

In order to solve the problem of temporal relationship recognition of Chinese and Vietnamese bilingual news events, this paper adopts the BLCATT model (the model is also applicable to Chinese events and Vietnamese event temporal relationship recognition tasks).

The model is shown in Fig. 1. It consists of three parts: (1) coding layer, (2) bidirectional cross attention layer, (3) the classification layer.

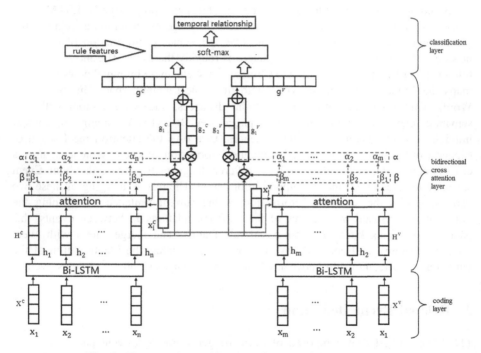

Fig. 1. The overview architecture of proposed BLCATT model

3.1   The Encoding Layer

In order to map Chinese and Vietnamese bilinguals to the same space. This paper uses the mixed training method proposed by Luong [14] to train bilingual word vectors, and the Skip-gram language model is extended to the Chinese-Vietnamese bilingual environment. The target words are used to predict the context information in Chinese, and the target words are used to predict the context information of the aligned words in Vietnamese, so as to obtain the Chinese and Vietnamese bilingual word vectors.

First, the words in the sentence are represented by bilingual word vectors. $X^C = \{x_1, x_2, x_3, \ldots, x_n\}$ and $X^V = \{x_1, x_2, x_3, \ldots, x_m\}$ are the Chinese sentences of length n

and the Vietnamese sentence of length m, which x_j is the jth word in the sentence. The superscript C stands for Chinese and the superscript V stands for Vietnamese.

In natural language processing, words have different semantic information at different positions in the sentence. This model uses bilingual word vectors and position vectors as the neural network's input. The attention mechanism cannot distinguish different locations, and encoding the position of each word facilitates the acquisition of semantic structure information. In this paper, each position is numbered by Vaswani [15]. Each number corresponds to a vector. Through the combination of the position vector and word vector, some position information is introduced for each word, which leads to the attention mechanism to distinguish words at different positions. Therefore, this paper uses the splicing of bilingual word vectors and position vectors as input to the neural network.

3.2 The Bidirectional Cross-Attention Layer

The cross-over attention layer consists of three parts, namely Bi-LSTM, attention mechanism, and cross-attention mechanism. These three parts are used in turn to obtain deep semantic information including sequential logic information.

Bi-LSTM. As a serialization model, Bi-LSTM treats text as an ordered lexical sequence. It solves the long-distance dependence problem of RNN, and can simultaneously use the context information of time series to comprehensively judge the result. The Chinese event sentence and the Vietnamese event sentence of the coding layer are respectively input into Bi-LSTM, and the context implicit semantic information ($H \in R^{d*n}$, d is the dimension of the word, n His the current sentence length) of the two sentences can be respectively obtained.

Attention Mechanism. Attention mechanisms can increase the weight of related content in the global context. The trigger word can clearly express a kind of event. This paper combines the trigger word x_i (i is the i-th word of the current event sentence) to establish the attention mechanism in the current event sentence. About the coding part of the Chinese sentence, this paper combines the trigger word x_i^C of Chinese sentences to establish the attention mechanism of Chinese sentence,so as to obtain the Chinese event semantic feature of enhanced the current Chinese trigger word, the expression is as follows:

$$K = \tanh(W_{XH} \begin{bmatrix} X^{*C} \\ H^C \end{bmatrix}^T) \tag{1}$$

$$\alpha = \text{sotfmax}(W_K K) \tag{2}$$

$$\mathbf{r} = \alpha H^C \tag{3}$$

$$g_1^C = \tanh(r) \tag{4}$$

where $X^{*C} = \{x_i, x_i, x_i, \ldots x_i\}$ is the trigger word matrix of the Chinese event sentence $(X^{*C} \in R^{d*n})$, and $W_{XH}(W_{XH} \in R^{2d*2d})$ is the weight matrix of the combination of X^{*C} and H^C; $\alpha = \{\alpha_1, \alpha_2, \alpha_3, \ldots \alpha_n\}(\alpha \in R^n)$ is the weight vector of the attention mechanism weight vector $W_k(W_k \in R^{2d})$ is the weight matrix of K; $r(r \in R^d)$ is the weight vector of the corresponding event sentence when the event trigger word is given; $g_1^C(g_1^C \in R^d)$ is the Chinese event semantic feature vector obtained by the attention mechanism.

The coding part of the Vietnamese sentence, this paper combines the trigger word x_i^V of Vietnamese sentences to establish the attention mechanism of Vietnamese sentence, so as to obtain the Vietnamese event semantic feature of enhanced the current Vietnamese trigger word, the expression is as follows:

$$K = \tanh(W_{XH}\begin{bmatrix} X^{*V} \\ H^V \end{bmatrix}^T) \tag{5}$$

$$\alpha = \mathrm{softmax}(W_K K) \tag{6}$$

$$r = \alpha H^V \tag{7}$$

$$g_1^V = \tanh(r) \tag{8}$$

Where $X^V = \{x_i, x_i, x_i, \ldots, x_i\}$ is the trigger word matrix of the Vietnamese event sentence $(X^{*V} \in R^{d*m})$, and $W_{XH}(W_{XH} \in R^{2d*2d})$ is the weight matrix of the combination of X^{*V} and H^V; $\alpha = \{\alpha_1, \alpha_2, \alpha_3, \ldots \alpha_n\}$ $(\alpha \in R^m)$ is the weight vector of the attention mechanism weight vector $W_k(W_k \in R^{2d})$ is the weight matrix of K; $r(r \in R^d)$ is the weight vector of the corresponding event sentence when the event trigger word is given; $g_1^V(g_1^V \in R^d)$ is the Chinese event semantic feature vector obtained by the attention mechanism.

Cross-Attention Mechanism. Because the attention mechanism can increase the weight of certain information in the current scope. The cross-attention mechanism is for two ranges of information. Use some information within the scope of the other party to establish an attention mechanism for the information in the current scope.

Since the trigger word is the core of the event sentence, a cross-attention mechanism is established for the trigger word in the Chinese-Vietnamese bilingual event pair. Combine the trigger words of Vietnamese sentences to establish attention mechanisms for Chinese sentences. Combine the trigger words of Chinese sentences to establish an attention mechanism for Vietnamese sentences. The cross-attention mechanism incorporates temporal logic information into the current sentence. This part of the information contains events that may occur continuously with the current event, intermittent events, and simultaneous events.

About the coding part of the Chinese sentence, this paper combines the trigger words of Vietnamese sentences to establish the cross-focus mechanism of Chinese sentences, so as to obtain the semantic features of Chinese events containing the temporal logic relationship of Vietnamese trigger words, the expressions are as follows:

$$K = \tanh\left(W_{XH}\begin{bmatrix} X^{*V} \\ H^C \end{bmatrix}^T\right) \tag{9}$$

$$\alpha = \text{sotfmax}(W_K K) \tag{10}$$

$$\mathrm{r} = \alpha H^C \tag{11}$$

$$g_2^C = \tanh(r) \tag{12}$$

where $X^{*V} = \{x_i, x_i, x_i, \ldots x_i\}$ $(X^{*V} \in R^{d*n})$ is the trigger word matrix of the Vietnamese event sentence, and $W_{XH}(W_{XH} \in R^{2d*2d})$ is the weight matrix of the combination of X^{*V} and H^C; $\alpha = \{\alpha_1, \alpha_2, \alpha_3, \ldots \alpha_n\}$ $(\alpha \in R^n)$ is the weight vector of the attention mechanism, $W_k(W_k \in R^{2d})$ is the weight matrix of K; $r(r \in R^d)$ is the weight vector corresponding to the Chinese event sentence when the event trigger word is given; $g_2^C(g_2^C \in R^d)$ is a Chinese event semantic feature vector containing temporal logic relationships which obtain by cross-attention mechanism.

The coding part of the Vietnamese sentence, this paper combines the trigger words of Chinese sentences to establish the cross-focus mechanism of Vietnamese sentences, so as to obtain the semantic features of Vietnamese events containing the temporal logic relationship of Chinese trigger words, the expressions are as follows:

$$K = \tanh\left(W_{XH}\begin{bmatrix} X^{*C} \\ H^V \end{bmatrix}^T\right) \tag{13}$$

$$\alpha = \text{softmax}(W_K K) \tag{14}$$

$$\mathrm{r} = \alpha H^V \tag{15}$$

$$g_2^V = \tanh(r) \tag{16}$$

where $X^{*C} = \{x_i, x_i, x_i, \ldots, x_i\}$ $(X^{*C} \in R^{d*m})$ is the trigger word matrix of the Chinese event sentence, and $W_{XH}(W_{XH} \in R^{2d*2d})$ is the weight matrix of the combination of X^{*C} and H^V; $\alpha = \{\alpha_1, \alpha_2, \alpha_3, \ldots \alpha_n\}$ $(\alpha \in R^n)$ is the weight vector of the attention mechanism, $W_k(W_k \in R^{2d})$ is the weight matrix of K; $r(r \in R^d)$ is the weight vector corresponding to the Vietnamese event sentence when the event trigger word is given; $g_2^V(g_2^V \in R^d)$ is a Vietnamese event semantic feature vector containing temporal logic relationships which obtain by cross-attention mechanism. Finally, merge the semantic information obtained by the attention mechanism and the cross-attention mechanism as the event semantic information of the event sentence, and then classify it.

3.3   The Classification Layer

Extract Rule Features. Through the study of the linguistic features of Chinese and Vietnamese and the temporal relationship between events, six rules of events were extracted. These features improve the identification of event temporal relationships. The features are as follows:

(1) Event trigger word part of speech: If the word-to-speech of the event trigger word is the same, the feature value is taken as 1, otherwise the feature value is taken as 0.
(2) The semantic role of the trigger word: if the semantic role of the trigger word is the same, the feature value takes 1; otherwise the feature value is 0.
(3) Event type: if the event type is the same, the feature is 1, otherwise it is 0.
(4) Event subtype: If the event subtype is the same, the feature is 1, otherwise the feature is 0.
(5) Event polarity: Whether the described event is a positive event or a negative event. When the polarity of the event pair is the same, the feature is 1, otherwise, it is 0.
(6) Event tense: The tense of the described event is past, present, or future. The three tenses are represented by 0, 1, and 2 respectively.

SOFTMAX. Finally, merge the event semantic features of the event sentence and the six rule features. Then, the soft-max classifier is used to classify the temporal relationship between the Chinese and Vietnamese bilingual news event pairs.

$$y = \text{sotfmax}(W[g^C g^V u] + b) \tag{17}$$

where g^C and g^V are the event semantic features of Chinese and Vietnamese event sentences; u is rule characteristic between events, $W \in R^{(2d+s)*t}$ (s is the number of extracted event rule characteristic, t is the number of classifications) is the weight matrix of the final soft-max layer input variable is the temporal relationship of the model output.

4   Experiment and Analysis

4.1   Corpus

So far, no public data sets have been found on the study of Chinese and Vietnamese bilingual events. The corpus of this experiment is derived from the Southeast Asian public opinion analysis platform, and the corpus is marked with reference to the TimeML annotation system. Pairing the screened corpus, the steps are as follows:

(1) The events extracted from the bilingual news texts under the 10 hot topics are placed in the event list d under the corresponding topic $D_i = \{e_1, e_2, \ldots, e_n\}$, $i = 1, 2, \ldots, 10$, n is the total number of events.

(2) The two events in $<e_j, e_k>$ $(j, k = 1, 2, \ldots, n)$ are combined to form an event pair, where ej, ek can be a Chinese event sentence, Vietnamese event sentence or Chinese-Vietnamese event sentence

(3) The event pairs are labeled with corresponding labels $E<ej, ek, y>$, $y \in \{0, 1, 2\}$ and y represents the three types of temporal relationships between the two events, where 0 represents the discontinuous context relationship, 1 represents the continuous context relationship, and 2 represents the parallel relationship.

(4) Put the event pair into the event pair set Es.

In this paper, a total of 500 Chinese and Vietnamese news texts under 10 hot topics were selected for experimental research. Among them, there were 3,862 pairs of discontinuous context relationship, 2013 pairs of continuous context relationship, and 1927 pairs of parallel relationship. For example:

Vietnamese: Ngày 12 tháng 11, Chủ tịch Trung Quốc Tập Cận Bình đã gặp Ban Chấp hành Trung ương Đảng Cộng sản Việt Nam trong Ban Chấp hành Trung ương Đảng Cộng sản Việt Nam.

(Translation: On November 12th, Chinese President Xi Jinping met with the Central Committee of the Communist Party of Vietnam Ruan Fuzhong in the Central Committee of the Communist Party of Vietnam).

Chinese: At the invitation of General Secretary of the Central Committee of the Communist Party of Vietnam, Ruan Fuzhong and Vietnamese President Chen Daguang, General Secretary of the CPC Central Committee and Chinese President Xi Jinping arrived in the capital Hanoi and began a state visit to Vietnam from November 12th to 13th.

The above two event sentences contain three event types, Invited (e1), Visit (e2), Meet (e3), and the corresponding trigger words are đã gặp (talk) invitation, access. The event pairs formed according to the rules are <e1,e3,0>, <e1,e2,1>, <e2,e3,1>.

4.2 Comparative Experiment

In this paper, the accuracy rate (P), recall rate (R), and F value (F) are used as evaluation indicators.

$$P = \frac{A}{A + B} \tag{18}$$

$$R = \frac{A}{A + C} \tag{19}$$

$$F = \frac{2PR}{P + R} \tag{20}$$

where A is the number of event types that are correctly identified, and B is the number of error recognition event type.

Influence of Position Vector on the Model. In order to explore whether the location information can obtain additional semantic information, thereby improving the

Table 1. Experiment results of whether to add position vector

Whether to add a position vector	P(%)	R(%)	F(%)
Do not add position vector	83.16	80.15	81.54
Add position vector	87.32	84.07	85.98

performance of the model. In this experiment, the model in which the coding layer does not add a position vector is compared with the model in which the coding layer adds a position vector, and the other parts of the model are the same. As shown in Table 1.

From the data in Table 1, it can be found that the return rate, accuracy and F value of the added position vector model are higher than the non-added position vector model. Because the attention mechanism cannot distinguish between different locations, it is important to encode the location of each input word. Therefore, combining the position vector and the word vector, introducing semantic structure information for each word helps to improve the performance of the model.

Influence of the Cross-Attention Mechanism. In order to explore the effectiveness of the cross-attention mechanism in Chinese-Vietnamese bilingual event temporal rela-

Table 2. Experiment results of whether to add attention mechanism

Model	P(%)	R(%)	F(%)
Attention mechanism	85.12	82.43	83.26
Cross-attention mechanism	86.04	83.51	84.19
Model of this paper	87.32	84.07	85.98

tionship recognition task, the model is compared with the model only using the attention mechanism and the model only using the cross- attention mechanism. The experimental results are shown in Table 2.

It can be seen from the experiment that the model only using the cross-attention mechanism is better than the model using only the attention mechanism, because the attention mechanism adds the weight of the trigger word in the current event, and the cross-attention can capture the temporal logic between the two event sentences. In the event, the temporal relationship recognition task, the accuracy of the temporal logic relationship has a more important role. The model adopts these two attention mechanisms at the same time, which makes the Chinese-Vietnamese bilingual event temporal relationship recognition task achieve the best results.

The Influence of Cross-Attention Mechanism and Rule Features. The cross-focus mechanism in this paper plays a major role in the identification of temporal relationships, and the rule features between events play a binding role. To explore that most of

Table 3. Experiment results of different model

Model	P(%)	R(%)	F(%)
Cross-attention	86.04	83.51	84.19
Rule features	82.22	80.40	80.36
Model of this paper	87.32	84.07	85.98

the temporal relationships between events are obtained by cross-attention, the experiment in Table 3 is performed.

It can be seen from the experiment that the model only use the cross-attention mechanism is better than the model only use the rule features between events, indicating that the temporal relationship between events is captured by cross-attention mechanism, and the rule features between events play a supporting role. So this paper uses these two parts to get the temporal relationship.

Model Comparison. RNN, LSTM, and Bi-LSTM are Recurrent neural network which can use context-related information in the mapping process between input and output sequences. To compare their performance, replace the Bi-LSTM of bidirectional cross

Table 4. Comparison of experimental results

Model	P(%)	R(%)	F(%)
RNN	76.21	76.33	77.92
LSTM	79.98	80.13	80.01
Bi-LSTM	87.32	84.07	85.98

attention layer with RNN and LSTM,and the other parts of the model were the same. The experimental results are shown in Table 4.

The experimental results show that the LSTM model is superior to the RNN model in the recognition of the temporal relationship of the Chinese and Vietnamese bilingual events, and the Bi-LSTM model is superior to the LSTM model. The range of historical information that RNN can access is very limited so that the influence of the input of the hidden layer on the network output is degraded with the recurrence of the network loop. LSTM can solve the problem of long-distance dependence and more effectively mine the hidden sentence of the event sentence. In fact, context information can understand semantic information better. Bi-LSTM model can extract information from two directions. Therefore, the text model uses Bi-LSTM to obtain the semantic information of event sentences.

5   Conclusion

In this paper, the deep learning method is used to solve the problem of temporal series relationship recognition of bilingual news events, and the timeline based method is abandoned, and the time logic level is explored. For this reason, this paper proposes a

Bi-LSTM Cross Attention model. The bilingual word vector is used to map the Chinese and Vietnamese bilinguals to the same space, Event semantic information containing time series features is obtained through Bi-LSTM, attention mechanism, cross-over attention mechanism, and inter-event rule features. Experiments in the model coding stage and the feature extraction stage prove that the model achieves the best results in the temporal relationship recognition in the Chinese-Vietnamese bilingual environment. As the quality and quantity of news corpus increase, the model will have a better effect. This paper focuses on the identification of temporal relationships for event sentences. Later, the temporal logic relationship between texts can be realized for the document level.

Acknowledgments. This work was supported by National Key Research and Development Plan (Grant Nos. 2018YFC0830105, 2018YFC0830101, 2018YFC0830100); National Natural Science Foundation of China (Grant Nos. 61732005, 61761026, 61866019, 61672271, 61762056, 61866020, 61972186); Science and Technology Leading Talents in Yunnan, and Yunnan High and New Technology Industry Project (Grant No. 201606); Talent Fund for Kunming University of Science and Technology (Grant No. KKSY201703005, KKSY201703015).

References

1. Ning, Q., Wu, H., Peng, H.: Improving temporal relation extraction with a globally acquired statistical re-source. In: 16th NAACL North American Chapter of the Association for Computational Linguistics, pp. 841–851. IEEE Press, New Orleans (2018). https://doi.org/10.18653/v1/n18-1077
2. Lu, Y.A.: Complex Event Detection Method Based on Normal Tree Pattern Matching. Beijing University of Technology, Beijing (2016)
3. Pantel, P., Pennacchiotti, M.C.: Espresso: leveraging generic patterns for automatically harvesting semantic relations. In: 44th ACL Association for Computational Linguistics, pp. 113–120. IEEE Press, New Orleans (2006). https://doi.org/10.3115/1220175.1220190
4. Zhao, S., Gao, Y., Ding, G.: Real-time multimedia social event detection in microblog. J. IEEE Trans. Cybern. **48**, 3218–3231 (2017). https://doi.org/10.1109/tcyb.2017.2762344
5. Li, Y., Zhang, H., Wang, X.: Temporal relationship recognition method based on news event fragment. J. Netw. Inf. Secur. **3**, 33–41 (2017). https://doi.org/10.1016/j.protcy.2013.12.332
6. Mandal, M., Mukhopadhyay, A.: An improved minimum redundancy maximum relevance approach for feature selection in gene expression data. Proc. Technol. **10**, 20–27 (2013). https://doi.org/10.1016/j.protcy.2013.12.332
7. Tourille, J., Ferret, O., Neveol, A.: Neural architecture for temporal relation extraction: a Bi-LSTM approach for detecting narrative containers. In: 55th ACL Association for Computational Linguistics, pp. 224–230. IEEE Press, Vancouver (2017). https://doi.org/10.18653/v1/p17-2035
8. Tian, W., Hu, W., Yu, L.: Bi-LSTM uyghur event temporal relationship identification based on attention mechanism. J. SE Univ. (Nat. Sci. Ed.) **10**, 393–399 (2018). CNKI:SUN:DNDX.0.2018-03-004
9. Vo, D.-T., Bagheri, E.: Extracting temporal event relations based on event networks. In: Azzopardi, L., et al. (eds.) ECIR 2019. LNCS, vol. 11437, pp. 844–851. Springer, Cham (2019). https://doi.org/10.1007/978-3-030-15712-8_61
10. Ghorbel, F., Hamdi, F., Métais, E.: Temporal relations between imprecise time intervals: representation and reasoning. In: Endres, D., Alam, M., Şotropa, D. (eds.) ICCS 2019.

LNCS (LNAI), vol. 11530, pp. 86–101. Springer, Cham (2019). https://doi.org/10.1007/978-3-030-23182-8_7

11. Vashishtha, S.I., Van Durme B., White, A.S.: Fine-Grained Temporal Relation Extraction. arXiv preprint arXiv:1902.01390 (2019)

12. Fu, J.F.: Event-oriented knowledge processing research. Shanghai University, Shang Hai (2010)

13. Pustejovsky, J., Hanks, P., Sauri, R.: The TIMEBANK corpus. In: 10th CL Corpus linguistics, pp. 40–45. Springer (2003)

14. Luong, T. Pham, H. Manning, C.D.: Bilingual word representations with monolingual quality in mind. In: 1th WVSMNLP Workshop on Vector Space Modeling for Natural Language Processing, pp. 151–159. Springer Press, Atlanta (2015). https://doi.org/10.3115/v1/w15-1521

15. Vaswani, A., Shazeer, N., Parmar, N.C.: Attention is all you need. In: 31th NIPS Advances in Neural Information Processing Systems, pp. 5998–6008. Springer Press, California (2017). https://doi.org/10.2155/978-3-235-21482

Users' Comment Mining for App Software's Quality-in-Use

Ying Jiang[1,2(✉)], Tianyuan Hu[1,2], and Hong Zhao[1,2]

[1] Yunnan Key Lab of Computer Technology Application,
Kunming University of Science and Technology, Kunming 650500, YN, China
jy_910@163.com
[2] Faculty of Information Engineering and Automation,
Kunming University of Science and Technology, Kunming 650500, YN, China

Abstract. Only parts of app software users' comments reflect software quality-in-use. A large number of useless users' comments will affect the analysis of software quality-in-use. In order to mine the users' comments about the quality-in-use from the massive app software users' comments, a mining method of users' comments based on comment seed is proposed. At first, the users' comments reflecting the app software's quality-in-use are mined through initial comment seeds. For the comments that cannot be matched with the comment seed, it will be determined whether they reflect quality-in-use according to the app software's quality-in-use feature words. Then, the candidate comment modes are extracted from the comments that can be matched with the app software's quality-in-use feature words. The new comment seeds are extracted based on the candidate comment mode library to further mine the users' comments related to the quality-in-use. Finally, the experimental results show that the proposed method can effectively mine users' comments reflecting app software quality-in-use with an average mining rate of 79.49%.

Keywords: App software · Quality-in-Use · Comment mining · Comment seed · Comment mode

1 Introduction

At present, app software types are becoming more and more diversified. App software not only provides people with information and simple services but also affects all aspects of people's lives. Therefore, it is critical to evaluate the app software's quality. ISO issued 25010 standards in 2011 which defined software quality as the 'degree to which a software product satisfies stated and implied needs when used under specified conditions' [8]. Software quality includes internal quality, external quality, quality-in-use and process quality. Quality-in-use is the degree to which a specific user uses a product or system to meet the effectiveness, efficiency, safety and satisfaction required to achieve a specific goal in a certain use environment. Quality-in-use is a non-functional quality characteristic that is perceived from the user's point of view in the actual application environment of the software. The measurement of quality-in-use depends on the environment in which the measurements are made and varies with the

© Springer Nature Singapore Pte Ltd. 2019
Y. Sun et al. (Eds.): ChineseCSCW 2019, CCIS 1042, pp. 510–525, 2019.
https://doi.org/10.1007/978-981-15-1377-0_40

evaluators. Some users will comment after using app software. The comments are important sources of real information that reflect the usage of app software and some specific characteristics that users focus on. Moreover, the app software with higher quality-in-use has a higher reputation and can attract more customers. In order to maintain and improve the quality of app software, it is important for developers to evaluate the quality-in-use of app software. After analyzing a large number of app software users' comments, we found that there are many comments irrelevant with quality-in-use because of the randomness of online comments. Less than 70% of users' comments reflect quality-in-use. It is necessary for both potential users and software developers of app software to mine users' comments related to quality-in-use from massive comments of app software.

2   Related Works

There are some researches about the analysis of software quality-in-use. According to Atoum, quality is 'meeting customer needs' and software quality-in-use is comprehension from the user's viewpoint [1]. Leopairote et al. proposes a model that can extract and summarize software reviews in order to predict software quality-in-use. The model builds ontology based on ISO/IEC 9126 model and expands ontology by WordNet 3.0 [11, 12]. Atoum proposes an overarching QinU framework that is able to predict and score QinU from software reviews. The framework utilizes a semantic similarity measure to classify software review-sentences to QinU topics [2–4]. Based on the research of semantic similarity, Atoum proposes a scalable operational framework to learn, predict, and recognize requirements defects [5]. Qian et al. presents a fine-grained topic model to evaluate quality-in-use of FLOSS [16].

Currently, studies of users' comments are becoming more mature in the field of internet goods. Zhang et al. proposes a fine-grained sentiment analysis method based on speech-grammar rules and LDA feature clustering [18]. Qiu et al. proposes a method of commodity opinion target extraction based on speech features, syntactic analysis and syntactic paths [15]. Considering certain features of Chinese comments, Peng et al. designs methods to obtain semantic relationships among words, which using syntactic analysis, word meaning understanding and context relevance [13]. Chen et al. proposes a method of sentiment-topic based on seed words of topic to find topics and related sentiments [7].

According to research on users' comments, we found that: (1) The number of quality-in-use feature words that are extracted on the basis of the definition of quality-in-use in ISO/IEC 9126 is limited. WordNet 3.0 synonyms expansion may add words irrelevant to quality-in-use. (2) Semantic similarity is unsuitable for the short sentence, which ignores the short sentence reflecting quality-in-use. (3) It is limited to use speech rules to expand the sentence pattern structure of the subjective predicate relationship. Many other sentence pattern structures common in users' comments could not be mined. (4) More users' comments could be mined by building and expanding the syntactic path library. However, the noises were introduced because of some kinds of syntactic paths, which reduced the accuracy.

In this paper, we aim at how to mine more app software users' comments related to quality-in-use characteristics in ISO/IEC 25010. Firstly, we try to avoid limitations that are caused by feature words extracted from the definition of quality-in-use and feature words expansion by lexicon. Secondly, we try to solve the problems that the analysis of sentence pattern structure is single and the number of syntactic paths is limited. This paper proposes concepts of comment mode and comment seed that are suitable for short comments. The comments that are the same as or similar with comment seeds on expression content or sentence pattern structure will be mined through comment seeds. New comment seeds are extracted using the method of semi-supervised learning for different users' comments, which is used to mine more users' comments reflecting quality-in-use continuously.

3   Quality-in-Use of App Software

There are five characteristics in the quality-in-use model of ISO/IEC 25010, including effectiveness, efficiency, satisfaction, freedom from risk and context coverage [8]. We found that some words in users' comments are related to app software's quality-in-use. Therefore, the commonly quality-in-use feature words are summarized, and the quality-in-use feature word table is established through analyzing users' comments. After analyzing 56348 users' comments, 15 representative words are extracted to constitute the quality-in-use feature word table, as shown in Table 1:

Table 1. The quality-in-use feature word table

Characteristic	Definition	Feature word/part of speech
Effectiveness	Accuracy and completeness with which users achieve specified goals	失败(fail)/vi, 出错(go awry)/vi, 实用(practical)/a, 功能(function)/n
Efficiency	Resources expended in relation to the accuracy and completeness with which users achieve goals	速度(speed)/n, 效率(efficiency)/n, 内存(memory)/n
Freedom from risk	Degree to which a product or system mitigates the potential risk to economic status, human life, health, or the environment	安全(safe)/a, 收费(charge)/vi, 广告(advertisement)/n
Context coverage	Degree to which user needs are satisfied when a product or system is used in a specified context of use	版本(version)/n, 产品(product)/n, 系统(system)/n
Satisfaction	Degree to which user needs are satisfied when a product or system is used in a specified context of use	不错(good)/a, 差(bad)/a

In Table 1, 'vi' indicates an intransitive verb, 'a' indicates an adjective, 'n' indicates a noun, and 'd' indicates an adverb. It is possible to mine some users' comments related to quality-in-use by determining whether the comment contains feature words in the quality-in-use feature word table. The type of quality-in-use about the comment can be determined.

4 Mining Users' Comments for App Software's Quality-in-Use

4.1 Comment Mode and Comment Seed

In order to analyze app software's quality-in-use, it is important to identify which aspect that user evaluates, such as effectiveness, efficiency, satisfaction, freedom from risk or context coverage. If a user comment reflects app software's quality-in-use, the comment will include the evaluation target and evaluation view or satisfy a certain sentence pattern structure. The evaluation target is also called a sentiment target or opinion target, which refers to the topic discussed in a text [9]. Evaluation view refers to the word with the emotional tendency that can express the user's own point of view, which is the fundamental basis for judging the emotion of the user [14].

By analyzing users' comments of app software, the core of users' comments can be divided into words and parts of speech. Words reflect the content of users' comments. The combination of parts of speech reflects the sentence pattern structure of a user's comment. Studying the part of speech combination for a sentence is a major means to analyzing the sentence pattern. Users' comments including the same evaluation object and evaluation view and reflecting the same quality-in-use are mined through words. Users' comments that have a specific sentence pattern structure and reflect the same quality-in-use are mined through parts of speech without restriction on the evaluation object and the evaluation view. In order to mine users' comments reflecting the quality-in-use characteristics more accurately, this paper proposes the concept of comment mode and comment seed. For the purpose of facilitating the mining of users' comments, the comment mode is defined as the core content of a user's comment.

Definition 1. Comment mode: The word segmentation results of users' comments without stop words. There are r users' comments in the comment library and r comment modes correspondingly:

$$mode = \{mode_1, \ldots, mode_m, \ldots, mode_r\}(1 \leq m \leq r),$$

$$mode_m = <word_{m1} + \ldots + word_{mn} + \ldots + word_{mq}, speech_{m1} + \ldots + speech_{mn} + \ldots + speech_{mq}, quality\_in\_use_m > (1 \leq n \leq q)$$

word represents words. *speech* represents parts of speech, And *q* represents the number of words (parts of speech) in comment mode. $quality\_in\_use_m$ represents the quality-in-use that reflected by $comment_m$. Before judging the quality-in-use of $comment_m$, the value of $quality\_in\_use_m$ is *unknown*.

According to the characteristics of users' comments, such as irregular expression, large quantity and fast update [19], the idea of seed is used in this paper. The concept of seed comes from Bootstrapping [6], a semi-supervised machine learning technique that is widely used in knowledge acquisition. Huang uses a small number of annotated corpus as comment seeds. On this basis, one or more classifiers are used to automatically and iteratively extend the comment seed set from a large number of unlabeled corpus. The comment seed of app software is defined as follows:

Definition 2. Comment seed: A representative expression format of comments reflecting the quality-in-use that includes words, parts of speech, distance and quality-in-use characteristic. There are s comment seeds in comment seed library:

$$seed = \{seed_1, \ldots, seed_i, \ldots, seed_s\}(1 \leq i \leq s),$$

$$seed_i = <word_{i1} + \ldots + word_{ij} + \ldots + word_{ip}, speech_{i1} + \ldots + speech_{ij} + \ldots + speech_{ip},$$
$$dis_i, quality\_in\_use_i > (1 \leq j \leq p)$$

word represents words. *speech* represents parts of speech. p represents the number of words (parts of speech) in comment seed. dis_i represents the distance of $seed_i$, which equals the sum of the maximum distance between the evaluation object and the evaluation view and the number of words (parts of speech) in the comment seed. $quality\_in\_use_i$ represents the quality-in-use characteristic reflected by $seed_i$.

According to the above definition, the expression formats of comment modes are diverse, repetitive and not representative. The comment seeds are extracted based on a number of comment modes, which are more abstract representations of users' comments about quality-in-use characteristics. The words whose parts of speech include 'n', 'v', 'd' or 'a' are important information of a comment.

4.2   Process of Mining Users' Comments

It is an essential issue to mine users' comments related to quality-in-use from enormous comments in this paper. The flow diagram of mining users' comments related to quality-in-use is shown in Fig. 1. The semi-supervised learning is used to improve the

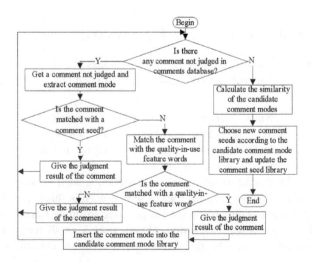

Fig. 1. The flow diagram of mining users' comments related to quality-in-use

ability of mining users' comments related to quality-in-use in the case where manual marking data is limited. The candidate comment mode library will be established according to the user's comments that match with comment seeds unsuccessfully and match with quality-in-use feature word successfully. The comment seed library would be updated dynamically based on the candidate comment mode library.

Matching a User's Comment with Comment Seeds Reflecting Quality-in-Use. In Fig. 1, the first step is to judge whether or not the user's comment matches with a comment seed successfully according to the comment mode of the user's comment. There are 4 attributes in the comment seed, including word, parts of speech, distance and quality-in-use. Keys to judging the matching between the user's comment and comment seed are word, part of speech, and distance.

Word Matching. When word matching happens, $word_{ij}$ of $seed_i$ matches with words of $comment_m$ sequentially. If $word_{mn}$ matches with $word_{ij}$ successfully, the position of n is recorded into a set pos_word. Since the $word_{ij}$ in $seed_i$ matches with the words of $comment_m$ sequentially, numbers in the set pos_word will be an incremental sequence if the sentence has correct grammar. In order to judge whether words that match successfully between the user's comment and comment seed meet the limitation of distance defined in the comment seed, dis_word_{mi}, the maximum distance of words matching successfully between $comment_m$ and $seed_i$, is calculated to determine whether it is smaller than the distance of $seed_i$. dis_word_{mi} is calculated using Eq. (1).

$$dis\_word_{mi} = \text{MAX}\{pos\_word\} - \text{MIN}\{pos\_word\} \tag{1}$$

Part of Speech Matching. When part of speech matching happens, $speech_{ij}$ of $seed_i$ matches with parts of speech of $comment_m$ sequentially. Since the part of speech is general and representative, a part of speech may appear repeatedly in a part of speech combination of a user's comment. In order to ensure the correctness of part of speech matching, the position of the part of speech matching successfully latest must be behind the position of the part of speech already matching successfully. The sequence including positions of part of speech matching successfully should be an incremental sequence. If $speech_{mn}$ matches with $speech_{ij}$ successfully, the position of n is recorded into a set pos_speech. In order to judge whether parts of speech that match successfully between the user's comment and comment seed meet the limitation of distance defined in the comment seed, dis_speech_{mi}, that the maximum distance of part of speech matching successfully between $comment_m$ and $seed_i$, is calculated using Eq. (2).

$$dis\_speech_{mi} = \text{MAX}\{pos\_speech\} - \text{MIN}\{pos\_speech\} \tag{2}$$

Distance Restriction. When the maximum distance of words (parts of speech) matching successfully between $comment_m$ and $seed_i$ is bigger than the distance of $seed_i$, it is indicated that words (parts of speech) matching successfully between the user's comment and comment seed are outside the position scope defined by the maximum distance between evaluation object and evaluation view. At this time, the user's comment does not conform to the grammar expression rules defined by the comment seed. In this situation, the comprehensive matching value is 0 regardless of the number of words (parts of speech) matching successfully between the user's comment and comment seed. If the maximum distance of words (parts of speech) matching successfully is smaller than the distance of $seed_i$, it is shown that words (parts of speech) matching successfully between the user's comment and comment seed conform to the grammar expression rules defined by the comment seed. The word-matching (part-of-speech-matching) value is calculated using Eqs. (3) and (4).

$$commentWord\_sim_{mi} = \begin{cases} NUM(pos\_word)/p & (dis\_word_{mi} \leq dis_i) \\ 0 & (dis\_word_{mi} > dis_i) \end{cases} \quad (3)$$

$$commentSpeech\_sim_{mi} = \begin{cases} NUM(pos\_speech)/p & (dis\_speech_{mi} \leq dis_i) \\ 0 & (dis\_speech_{mi} > dis_i) \end{cases} \quad (4)$$

$commentWord\_sim_{mi}$ represents the word-matching value between $comment_m$ and $seed_i$. $commentSpeech\_sim_{mi}$ represents the part-of-speech-matching value between $comment_m$ and $seed_i$. NUM(pos) represents the number of items in pos, which equals the number of words (parts of speech) matching successfully between user's comment and comment seed. p represents the number of words (parts of speech) in $seed_i$.

Comprehensive Judgment. Comprehensive judgment must be conducted after word matching, part of speech matching and distance restriction. Through these experiments, several cases as follows have been found when a user's comment matches with a comment seed.

a. When words and parts of speech both match exactly, the matching is successful.
b. When parts of speech match exactly, the part-of-speech-matching value is 1;

- When words match partially, the word-matching value is bigger than 0. If the user's comment matches with the comment seed successfully, the user's comment reflecting the same quality-in-use as the comment seed will be more possible.
- When words do not match at all, the word-matching value is 0. If the user's comment matches with the comment seed successfully, the user's comment reflecting the same quality-in-use as the comment seed will be less possible.

Whether the user's comment matches with the comment seed successfully is judged based on the comprehensive matching value $commentBoth\_sim_{mi}$. $commentBoth\_sim_{mi}$ is calculated using Eq. (5).

$$\begin{aligned} &commentBoth\_sim_{mi} \\ &= 0.5 * commentWord\_sim_{mi} + 0.5 * commentSpeech\_sim_{mi} (1 \leq i \leq s) \end{aligned} \quad (5)$$

After $comment_m$ matches with all comment seeds sequentially, max_com-$mentBoth\_sim_{mg}$ extracts the maximum of the comprehensive matching values between $comment_m$ and $seed_g$. The maximum of the comprehensive matching values will be used to judge whether $comment_m$ matches with one comment seed successfully. $max\_commentBoth\_sim_{mg}$ is extracted using Eq. (6).

$$max\_commentBoth\_sim_{mg} = \text{MAX}\{commentBoth\_sim_{m1},$$
$$..., commentBoth\_sim_{mg}, ..., commentBoth\_sim_{ms}\}(1 \le g \le s) \tag{6}$$

The higher the comprehensive matching value between $comment_m$ and $seed_g$, the higher the possibility that $comment_m$ matches with a comment seed successfully. In addition, a threshold needs to be set here. Only when $max\_commentBoth\_sim_{mg}$ is greater than the threshold can the user's comment match with the comment seed successfully. After a comprehensive analysis for the possible cases when a user's comment matches with a comment seed, the threshold is set to 0.5.

Matching a User's Comment that Matches with a Comment Seed Unsuccessfully with Quality-in-Use Feature Words. Every user's comment matching with comment seeds unsuccessfully needs to match with words in quality-in-user feature word table successively. If a user's comment matches with a quality-in-use feature word successfully, it will reflect the same quality-in-use as the quality-in-use feature word.

Matching a User's Comment that Matches with a Comment Seed Unsuccessfully with New Comment Seeds. The user's comment that matched with a quality-in-use feature word successfully contain the contents that are not included in the comment seed library. Thus, new comment seeds should be extracted based on the user's comments matching with a quality-in-use feature word successfully, which will facilitate subsequent mining.

Extracting Comment Modes. In the process of Fig. 1, $comment_m$ failing to match with comment seeds will match with a quality-in-use feature word table. It is used to judge whether $comment_m$ reflects quality-in-use. If $comment_m$ contains a quality-in-use word, modem of $comment_m$ will be inserted into the candidate comment mode library. Comment modes reflecting the same quality-in-use characteristic have more similarity in expression content and sentence pattern structure.

Extracting New Comment Seeds. In order to ensure that the newly extracted comment seeds reflecting the quality-in-use are representative and abstract, text similarity is calculated for every comment mode in the candidate comment mode library by the category of the quality-in-use characteristic. New representative comment seeds for different quality-in-use characteristics will be extracted based on comprehensive text similarity. The flow diagram of extracting new comment seeds is shown in Fig. 2.

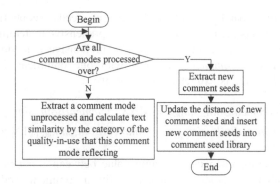

Fig. 2. The flow diagram of extracting new comment seeds

modeWord_sim$_{mm'}$ and *modeSpeech_sim$_{mm'}$* are calculated according to the *Levenshtein* [10], a calculation method of similarity. *modeWord_sim$_{mm'}$* is word-text similarity between *mode$_m$* and *mode$_{m'}$*. *modeSpeech_sim$_{mm'}$* is part-of-speech-text similarity between *mode$_m$* and *mode$_{m'}$*. The comment modes reflecting the same quality-in-use is defined as *mode$_{m'}$*. The effect of text similarity calculation is mainly embodied in two aspects:

(1) **Extracting New Comment Seeds**

As the part of speech is representative and universal, it is possible that there is no correlation between two Chinese expressions even though part-of-speech-text similarity is high. The Part-of-speech-text similarity is proportional to word-text similarity and vice versa. Thus, new comment seeds are extracted through word-text similarity. *sum_modeWord_sim$_m$*, the sum of word-text similarity between *mode$_m$* and *mode$_{m'}$*, is calculated to ensure that the new comment seeds are from the representative comment modes among the candidate comment mode library. A higher value of *sum_modeWord_sim$_m$* means that the similarity between *mode$_m$* and other comment modes in the candidate comment mode library is higher and *mode$_m$* is more representative. The sum of word-text similarity is calculated using Eq. (7). The largest sum of word-text similarity, *mode$_{max}$*, will be obtained for every quality-in-use characteristic using Eq. (8).

$$sum\_modeWord\_sim_m = \text{SUM}\{modeWord\_sim_{mm'}\}(m'! = m) \qquad (7)$$

$$sum\_modeWord\_sim_{max} = \text{MAX}\{sum\_modeWord\_sim_m\}(1 \leq max \leq m) \qquad (8)$$

After all calculations of text similarity for comment modes reflecting quality-in-use, 5 comment modes are obtained at most. Each comment mode of these 5 comment modes represents different quality-in-use characteristics. To avoid the number of comment seeds being too massive and the overall quality of comment seeds being low, the first 3 comment modes with maximum word-text similarity value are extracted as new comment seeds *new_seed$_{max}$*. *dis$_{max}$* of *new_seed$_{max}$* equals the number of 'n', 'v', 'd', or 'a'. *dis$_{max}$* can only represent the maximum distance of *mode$_{max}$*. It cannot represent the max distance of the comment mode that is similar with $mode_{max}$. Therefore, dis_{max} must be updated according to the comment mode similar with $mode_{max}$ in the following step.

(2) **Updating the Distance of New Comment Seed**

The max extensible distance between evaluation object and evaluation opinion is a kind of grammatical distance. Because the grammatical distance is a limitation on sentence pattern structure, the update of distance relies on comprehensive text similarity. Even though word-text similarity between two sentences is low, part of speech combinations of these two sentences may be similar. For two comment modes having similar part of speech combination, the larger words distance of one comment mode could represent the grammatical distance of other comment modes. Therefore, we choose comprehensive text similarity as the standard to update the distance of new comment seed. $modeBoth\_sim_{maxmax'}$ is comprehensive text similarity between $mode_{max}$ of new_seed_{max} and $mode_{max'}$ ($1 \leq max' \leq m$, $max'! = max$). $mode_{max'}$ is the other comment modes having the same quality-in-use characteristic. $modeBoth\_sim_{maxmax'}$ is calculated using Eq. (9).

$$modeBoth\_sim_{maxmax'} = 0.5 * modeWord\_sim_{maxmax'} + 0.5 * modeSpeech\_sim_{maxmax'}$$
$$(1 \leq max' \leq m, max'! = max)$$

$$(9)$$

We select all the comment modes whose comprehensive text similarity with comment mode of the new comment seed is greater than 0.5. The maximum distance of words among these comment modes will be used to update the distance of the new comment. Experiences show that it is most reasonable to update the distance of the new comment when the threshold is 0.5. At the same time, it is effective to enlarge the scope of users' comment mining and ensure that the comment mined through comment seeds could reflect quality-in-use. After all calculations of comprehensive text similarity are finished, $dis_{max'}$ of $mode_{max'}$ whose comprehensive text similarity is greater than 0.5 are gathered. Maximum distance $new\_seed\_dis_{max}$ is extracted using Eq. (10).

$$new\_seed\_dis_{max} = MAX\{dis_{max'}\}$$

$$(10)$$

If $new\_seed\_dis_{max}$ is greater than the distance of new_seed_{max}, the distance of new_seed_{max} is updated as $new\_seed\_dis_{max}$. Finally, new_seed_{max} is updated in the comment seed library. In the process of users' comment mining for app software's quality-in-use in this paper, users' comments are mined through initial comment seeds and app software's quality-in-use feature word table. The comment seed library is expanded dynamically through semi-supervised learning for specific users' comments in every instance of mining.

(3) **Matching a User's Comment with New Comment Seeds**

Users' comments matching with a comment seed unsuccessfully will be matched with new comment seeds using the method described in Sect. 4.2.1. The semi-supervised learning is used while cyclic mining. Since new comment seeds are extracted from different users' comments data, cyclic mining with new comment seeds can mine many users' comments reflecting quality-in-use.

5   Experimental Results and Analysis

In order to verify the effectiveness of the above method, various types of users' comments of app software are picked up randomly from the Android electronic market (http://apk.hiapk.com/apps). The users' comments database is established. ICTCLAS 2018 is used as a data processing tool in this paper, which completes comment word segmentation and part-of-speech marking. Users' comments that contain no evaluation object (whose parts of speech contain no 'n' or 'v') are filtered [17]. Based on the above method, a mining prototype tool is developed.

5.1   Users' Comments Mining Effect for App Software's Quality-in-Use

45466 users' comments of audio, video, social and application software are selected. There are a total of 37965 users' comments related to quality-in-use after manual marking. We set 7 initial comment seeds and use 15 feature words in Table 1. Several minings were conducted for these experimental data. In every instance of mining, comments that matched with a comment seed successfully would not be in the next mining. After three instances of mining, the number of comment seeds increased to 16. There were 23273 comments matched with comment seeds and 9123 comments matched with feature words. Analysis of the experimental results is shown in Table 2.

Table 2. Results of users' comment mining for app software's quality-in-use

Quality-in use characteristic	NoUCMM	Proportion of total users' comments	NoUCMwCS	NoUCMwQFWT	Mining rate*
Effectiveness	7344	19.34%	4480	1534	81.89%
Efficiency	3319	0.08%	1201	1189	72.01%
Satisfaction	24655	64.94%	17159	4784	89.00%
Context coverage	1808	0.04%	317	1094	78.04%
Freedom from risk	839	0.02%	116	522	76.04%

*NoUCMM: The Number of Users' Comments Marked Manually; *NoUCMwCS: The Number of Users' Comments Mined with Comment Seeds; *NoUCMwQFWT: The Number of Users' Comments Mined with Quality-in-use Feature Word Table;

*Mining Rate = (The Number of Users' Comments Mined with Comment Seeds + The Number of Users' Comments Mined with Quality-in-use Feature Word Table)/The Number of Users' Comments Marked Manually.

In Table 2, it is shown that the proportion of the users' comments indicating satisfaction is the biggest (64.94%). The reason is that most of app software users' comments are about satisfaction. Expressions of app software users' comments about satisfaction are usually short and frequent. The average mining rate is 79.49% for 5 quality-in-use characteristics. It is demonstrated that the proposed method is effective.

5.2   Application of Comment Seeds

Mining Users' Comments Using Comment Seeds. Partial results are shown in Table 3 where users' comments are matched with comment seeds. Users' comments 1–4 are comments about Camera Software. Users' comments 5–8 are comments about Audio and Video Software. In this mining time, the total number of users' comments is 10275, and the number of comment seeds is 7.

Table 3. Users' comments matched with comment seeds successfully (some example)

Serial number	User's comment	Quality-in-use comment reflects	Comment seed	Is the user's comment matched with comment seed?
1	软件/n 非常/d 好/a (software/n very/d good/a)	Satisfaction	<很 + 好(very + good), d + a,3, satisfaction >	Yes
2	不错/a(good/a)	Satisfaction		No
3	图/n 效果/n 也/d 不错/a (picture/n effect/n also/d good/a)	Satisfaction		No
4	也/d 说/v 不错/a (also/d said/v good/a)	Satisfaction		No
5	安装/v 速度/n 真/d 太/d 慢/a (install/v really/d very/d slow/a)	Efficiency	<下载 + 速度 + 慢(download +speed + slow), v + n+a, 7, efficiency >	Yes
6	占/v 内存/n (occupy/v memory/n)	Efficiency		No
7	占/v 空间/n 太/d 大/a (occupy/v space/n too/d large/a)	Efficiency		No
8	占用/v 内存/n 大/a (occupy/v memory/n large/a)	Efficiency		No

It is indicated in Table 3 that sentence pattern structures of users' comments reflecting the same quality-in-use are similar. Using the above method, parts of users' comments are matched with comment seeds. Users' comments that fail to match with comment seeds will be used to extract new comment seeds.

Extracting New Comment Seeds. According to the above method, comments in Table 3 that failed to match with comment seeds will match with quality-in-use feature words. On the basis of comments matching with feature words successfully, candidate comment modes are extracted and the candidate comment mode library is established. The candidate comment mode library is shown in Table 4.

Table 4. The candidate comment mode library (some example)

Serial number	Words of comment modes	TPoSCoCM	TQFWMwUC	Quality-in-use characteristic	ItUCRQ
1	不错(good)	a	不错(good)/a	Satisfaction	Yes
2	图 + 效果 + 也+不错 (picture + effect + also + good)	n + n + d + a			Yes
3	也 + 说+不错 (also + said + good)	d + v + a			Yes
4	占 + 内存(occupy + memory)	v + n	内存 (memory)/n	Efficiency	Yes
5	占 + 空间 + 太+大 (occupy + space + too + large)	v + n + d + a			Yes
6	占用 + 内存 + 大 (occupy + memory + large)	v + n + a			Yes

*TPoSCoCM: The Part of Speech Combination of Comment Mcodes;
*TQFWMwUC: The Quality-in-Use Feature Word Matched with User's Comment;
*ItUCRQ: Is the User's Comment Reflect Quality-in-Use (Manual Marking).

According to the candidate comment modes shown in Table 4, the comment seed extracted from comment modes 1–3 is '<不错(good), a, 1, satisfaction>', and the comment seed extracted from comment modes 4–6 is '<占 + 内存(occupy + memory), v + n, 3, efficiency>'. If the app software's quality-in-use feature word table does not contain quality-in-use reflected by a user's comment, the matching of user's comment will fail. The reason is limited number of quality-in-use feature words.

The Mining Effect of Different Comment Seeds. The method of semi-supervised learning is adopted in this paper for mining users' comments reflecting app software's quality-in-use. It is a process that the number of comment seeds gradually increases and the effects of mining are constantly improved. For different comments, dynamically enlarging the comment seed library effectively improves the mining rate of users' comments about quality-in-use.

In order to verify the effectiveness of cyclic mining, we designed three different experiments of cyclic mining using 10275 users' comments data. There is a total of 6065 users' comments related to quality-in-use after manual marking. In every experiment, initial comment seeds or mining times are different. The numbers of initial comment seeds in experiments 1 and 2 are the same, but initial comment seeds of these two experiments are different. The number of initial comment seeds in experiment 3 is less than those in experiments 1 and 2, and the initial comment seeds of experiment 3 are different from those of experiments 1 and 2. The results of three experiments are shown in Table 5.

Table 5. Comment seeds of three experiments

ExpN	TNoICS	Initial comment seeds (some example)	TToCM	TNoNCS	New comment seeds (some example)	TNoUCMC	TAoM
1	7	<好+用 (easy+use),a+v,3,satisfaction>, <界面+简洁(interface+ consice), n+a,6, satisfaction >, <不+能+使用(can+not +use), d+v+v, 5, effectiveness>,<安装+不+了(can+not+ install), v+d+y, 3, effectiveness>,<验证+失败(validate +fail), v+vi, 4, effectiveness>	3	9	<功能+很+强大(function+very+powerful), n+d+a,6, effectiveness>,<占+内存 (occupy+memory),v+n,3, efficiency>, <广告+多(advertisement+many), n+a,4, freedom from risk>,<新+版本+不错(new+ version+good),a+n+a,7, context coverage>, <不错(good),a,1,satisfaction>,<垃圾+垃圾 (garbage +garbage),n+n,7, satisfaction>	3808	62.79%
2	7	<很+好(very+good),d+a,3, satisfaction>,<不+能+看(can+not+ watch), d+v+v,5, effectiveness >, <广告+多 (advertisement+ many), n+a, 6, freedom from risk >, <占+内存(occupy+ memory),v+n,4, efficiency>,<功能+很+强大(function+very+powerful), n+d+a,6, effectiveness >	3	9	<新+版本+不错(new+ version+ good), a+n+a,7,context coverage >, <内存+太+大 (memory+too+ large), n+d+a,5, effectiveness>,<下载+失败(download+fail),v+vi,2, effectiveness >,<好+喜欢(very+like),a+vi, 3, satisfaction >,<垃圾+垃圾(garbage +garbage),n+n,7, satisfaction>	4375	72.14%
3	6	<内存+太+大(memory+too+ large), n+d+a, 6, effectiveness>, <好+用 (easy+use), a+v,3, satisfaction>,<安装+不+了(can+ not+install), v+d+y,3, effectiveness>,<新+版本(new+version), a+n,3, context coverage>,<很+好 (very+good), d+a, 3, satisfaction>	5	15	<没+广告(not have+advertisement),v+n,4, freedom from risk>,<不错(good),a, 1, satisfaction>,<怎么+下载+失败(why+ download +fail), ryv+v+vi, 4, effectiveness>, <好+喜欢(very+like), a+vi,3, satisfaction>, <垃圾+垃圾(garbage +garbage), n+n, satisfaction>,<占+内存(occupy+memory), v+n,3,efficiency>	4784	78.88%

*ExpN: Experiment Number; *TNoICS: The Number of Initial Comment Seeds;
*TToCM: The Times of Cyclic Mining; *TNoNCS: The Number of New Comment Seeds;
*TNoUCMC: The Number of Users' Comments Mined Correctly; *TAoM: The Accuracy of Mining.

As shown in Table 5, the new comment seeds differing from the initial comment seeds can be extracted in every experiment, and 3 new comment seeds were extracted in every mining time. We found that comment seeds of three experiments are similar. Even though the initial comment seeds of three experiments are different, new comment seeds that are representative of the 10,275 users' comment data were found using our method. Furthermore, different initial comment seeds will influence the mining effect. Due to different initial comment seeds and times of cyclic mining, the mining result of experiment 3 was better than those of experiments 1 and 2, while experiment 3 has less initial comment seeds. The primary reason is that initial comment seeds of experiment 3 include many high-frequency expressions, which are representative of the 10,275 users' comments data. The average accuracy of mining in three experiments is 71.27%, which means that most of the users' comments reflecting quality-in-use can be mined using the method proposed in this paper. Mining effects are determined by the quality of comment seeds. The experimental results above show that the proposed method of users' comment mining for app software's quality-in-use is effective.

Mining effects are determined by the quality of comment seeds. In our previous research, users' comments were classified using Bayesian network. The experimental results show that the imbalance of training data leads to the poor performance in mining users' comments. In addition, due to the limitation of the evaluation objects and the evaluation views in the training data, the Bayesian network classifier can not mine users' comments with similar content or sentence structure. Therefore, the proposed method of users' comment mining for app software's quality-in-use in this paper is effective.

6 Conclusion

This paper proposes a method of users' comment mining for app software's quality-in-use. Users' comments that are the same as or similar to comment seeds could be mined through matching with comment seeds. For the comments that cannot be matched with the comment seed, it will be determined whether they reflect quality-in-use through matching with app software's quality-in-use feature words. Then, the candidate comment modes are extracted from the comments matching with the app software's quality-in-use feature word successfully. The new comment seeds are extracted based on the candidate comment mode library to further mine the users' comments related to the quality-in-use.

The experimental results show that our method is effective. It can mine the users' comments that are same as or similar to comment seeds and reflect the same quality-in-use with comment seeds more accurately. It can also extract new representative comment seeds with high quality based on users' comments that fail to match with comment seeds. Due to the limitation of the number of app software's quality-in-use feature words and the abstraction of comment seeds, the accuracy of mining is affected. Future works would update app software's quality-in-use feature word table dynamically and improve the description of comment seed.

Acknowledgements. This research is sponsored by the National Science Foundation of China No. 61462049, 60703116, and 61063006, Key Project of Yunnan Applied Basic Research No. 2017FA033 and the Scientific Research Fund Project of the Yunnan Education Department No. 2018Y016.

References

1. Atoum, I., Bong, C.H.: A framework to predict software "quality in use" from software reviews. In: Herawan, T., Deris, M.M., Abawajy, J. (eds.) Proceedings of the First International Conference on Advanced Data and Information Engineering (DaEng-2013). LNEE, vol. 285, pp. 429–436. Springer, Singapore (2014). https://doi.org/10.1007/978-981-4585-18-7_48
2. Atoum, I., Bong, C.H., Kulathuramaiyer, N.: Towards Resolving Software Quality-in-Use Measurement Challenges. arXiv preprint arXiv:1501.07676 (2015)
3. Atoum, I., Otoom, A.: Mining software quality from software reviews: research trends and open issues. Int. J. Emerg. Trends Technol. Comput. Sci. **31**(2), 74–83 (2016)
4. Atoum, I.: A novel framework for measuring software quality-in-use based on semantic similarity and sentiment analysis of software reviews. J. King Saud Univ.-Comput. Inf. Sci. (2018)

5. Atoum, I.: A scalable operational framework for requirements validation using semantic and functional models. In: Proceedings of the 2nd International Conference on Software Engineering and Information Management, pp. 1–6. ACM (2019)
6. Abney, S.: Bootstrapping. In: Proceedings of the 40th Annual Meeting on Association for Computational Linguistics, pp. 360–367 (2002)
7. Chen, Y., Zuo, W., Lin, Y.: Sentiment-aspect analysis method based on seed words. J. Comput. Appl. **35**(9), 2560–2564 (2015)
8. International Organization for Standardization.: Systems and Software Engineering – Systems and Software Quality Requirements and Evaluation (SQuaRE) – System and Software Quality Models, International Organization for Standardization, pp. 34–35 (2011)
9. Li, J.Y., Zhang, Y.S., Jiang, Y.R.: Opinion target extraction method based on multi-features in Chinese micro blog. Appl. Res. Comput. **33**(2), 378-3 (2016)
10. Levenshtein, V.I.: Binary codes capable of correcting deletions. Insertions Reversals Sov. Phys. Dokl. **10**(8), 707–710 (1966)
11. Leopairote, W., Surarerks, A., Prompoon, N.: Evaluating software quality in use using user reviews mining. In: International Joint Conference on Computer Science and Software Engineering IEEE, pp. 257–262 (2013)
12. Leopairote, W., Surarerks, A., Prompoon, N.: Software quality in use characteristic mining from customer reviews. In: 2012 Second International Conference on Digital Information and Communication Technology and it's Applications (DICTAP), pp. 434–439 (2012)
13. Peng, Y., Wan, C.X., Jiang, T.J., Liu, D.X., Liu, X.P., Liao, G.Q.: Extracting product aspects and user opinions based on semantic constrained LDA model. J. Softw. **28**(3), 676–693 (2017)
14. Qiu, G., Liu, B., Bu, J., Chen, C.: Opinion word expansion and target extraction through double propagation. Comput. Linguist **37**(1), 9–27 (2011)
15. Qiu, Y.F., Chen, Y.F., Wang, W., Shao, L.: Commodity opinion target extraction based on part of speech feature and syntactic analysis. Comput. Eng. **42**(7), 173–180 (2016)
16. Qian, Z.Z., Wan, C.C., Chen, Y.T.: Evaluating quality-in-use of FLOSS through analyzing user reviews. In: International Conference on Software Engineering, Artificial Intelligence, Network and Parallel/Distributed Computing, pp. 547–552 (2016)
17. Ran, M., Jiang, Y., Xiang, Q., Ding, J., Wang, H.: Method of consistency judgment for app software's user comments. In: Che, W., et al. (eds.) ICYCSEE 2016. CCIS, vol. 623, pp. 470–483. Springer, Singapore (2016). https://doi.org/10.1007/978-981-10-2053-7_42
18. Zhang, J.: The Sentiment Analysis about User Reviews based on LDA and Speech-Grammar Rules, Guangxi University, Guangxi (2014)
19. Zhang, L., Qian, G.Q., Fan, W.G., Hua, K., Zhang, L.: Sentiment analysis based on light reviews. J. Softw. **25**(12), 2790–2807 (2014)
20. Zhang, Y.S.: Study on the Adverbs in Modern Chinese, pp. 180–181. Xue Lin Publishing House, Beijing (2009)

Chinese-Vietnamese News Documents Summarization Based on Feature-related Attention Mechanism

Jinjuan Wu[1,2], Zhengtao Yu[1,2(✉)], Shengxiang Gao[1,2], Junjun Guo[1,2], and Ran Song[1,2]

[1] Faculty of Information Engineering and Automation,
Kunming University of Science and Technology, Kunming 650500, China
ztyu@hotmail.com
[2] Yunnan Key Laboratory of Artificial Intelligence,
Kunming University of Science and Technology, Kunming 650500, China

Abstract. The association analysis of cross-linguistic information is still a challenging problem in the task of multi-language summarization. To address this issue, we propose an LSTM framework based on feature-related attention mechanism to extract the summarization of Chinese-Vietnamese bilingual news. Firstly, the word embedding with multi-features is used as input to the model such as word frequency, sentence position and relevance. Then, the degree of elements co-occurrence in bilingual documents is analyzed, and the attention mechanism based on bilingual features is proposed to calculate the importance scores of sentences. Finally, the sentence with high score is selected and the redundant information is deleted according to the similarity analysis to generate summary. The results of comparison experiments show that the method has achieved good results.

Keywords: Chinese-Vietnamese · Documents summarization · Feature-related attention mechanism

1 Introduction

With the explosive increase of information in the new era, quantities of hot topic news are published in various linguistic forms. Therefore, much attention has been paid to quickly grasp hot topics news with multiple languages on the internet. In order to solve this problem, it is necessary to summarize the document information from various sources and provide the user with a short version of the summary. Consequently, the multilingual text summarization system that uses a multi-language document set as input to produce a concise summary reflecting the original meaning of the original document collection with refined text has gained fast development. As the thriving exchanges between China and Vietnam, more and more relevant reports have been published through different languages. Actually, it is not only time-consuming but also difficult for readers to extract the key content from the vast texts because of the existence of language barriers. To overcome the drawback, this work aims to summarize the bilingual news documents describing relevant events to obtain the main

© Springer Nature Singapore Pte Ltd. 2019
Y. Sun et al. (Eds.): ChineseCSCW 2019, CCIS 1042, pp. 526–539, 2019.
https://doi.org/10.1007/978-981-15-1377-0_41

content of Chinese and Vietnamese bilingual news to help people understand events quickly and comprehensively.

For the sake of analyzing multilingual documents, the difference between two languages should be solved firstly. In recent years, most of the research has employed the machine translation technology [1] to map bilingual texts into the same semantic space, or to build pairs of bilingual translations by means of bilingual dictionaries [2]. With the development of deep learning, the expression of bilingual word vector [3] provides a new way of thinking for cross-linguistic analysis. Generally, the generation of multiple documents summarization is defined as extraction [4], that is to say, some of the significant sentences in the article are directly selected to stitch together, in which the significance of sentence is assessed through the combination of heuristic based statistics characteristic [5] and the characteristics of languages. Current researches [6] mostly transforms the summarization extraction problem into sentence classification or regression modeling where the attention mechanism is considered to be one of the key factors to improve performance. Besides, the fusion of text feature information and attention mechanism are involved in many fields. For example, Narayan et al. [7] integrated the auxiliary information such as title and picture description into the abstract extraction process under the guidance of attention mechanism. Wang et al. [8] proposed to realize the task of emotion classification based on the attention mechanism of aspect-level.

Our work firstly focus on constructing a word vector representation learning that combines word frequency, correlation of sentences and location information. Then, constructing regression function to assess the scores of different sentences according to the neural network's ability to acquire the semantic information of sentences and feature learning, through which the task of abstract extraction is transformed into a sorting problem based on the significance of sentence. Further, an attention-based mechanism based on the co-occurrence of bilingual features is employed to score the sentences. Finally, the summarization is obtained by removing the redundant information according to the similarity of sentences. The contributions of the proposed method is summarized as follows: (1) We propose an LSTM-based regression model for the task of cross-linguistic summarization, which combines the feature information, i.e. bilingual factor co-occurrence, word frequency, sentence position and relevance to achieve the summarization extraction of bilingual text. (2) Considering the bilingual news that describe the same event often contain same elements, we propose to build the score function to evaluate sentences by incorporating the bilingual documents elements features into the attention mechanism to predict the significant scores of the sentences.

2 Related Work

Multilingual summarization system [9–11] has attracted much attention in recent years. In general, the differences among different languages should be eliminated firstly to construct the multilingual text abstract system. For example, grammatical and lexical differences [12] between two languages affect the text processing, which involves abstract extraction process, the scoring mechanism and the overall performance of the system. In order to alleviate those differences, the information of bilingual text is usually mapped to the same vector space through machine translation [13] as the input of the system and the feature of sentence extraction. Then, adding some related vocabulary concept knowledge to improve the language quality and information content of the translation system. However, the performance of this method will be limited when dealing with minority languages since the translation quality is ignored. Therefore, it is necessary to take the translation quality into consideration and carefully weighting the sentences. Moreover, multi-language distributed word representation learning system [14] gives new idea for cross-linguistic analysis.

The text summarization system that extracts representative sentences from the original text to form summaries has devoted to the study of the extraction method, which can be categorized into two parts: unsupervised method and supervised method. The unsupervised method employs the feature rules designed by manual work to estimate the significance of sentences. Generally, according to the writing character-istics of news, the first sentence of the document plays a role of outline. Therefore, the key information can be extracted based on the position or length [15] of a sentence in the document. Besides, the word frequency [16] information also has a guiding role for news abstracts. Some of the key conceptions or words appear in the document at a higher frequency, so the more high-frequency words contained in the sentence, the more likely they become summarization sentences. Furthermore, the distribution of news topics contributes [17] to the abstract extraction of documents or news. In abstract extraction, the sentences can be sorted according to the degree of overlap between sentences and features. Additionally, the graph-based summarization algorithms [18] have grown into a research focus, whose core idea is to conduct the abstract extraction through comparing the similarities among sentences, but cross-language analysis steps should be added based on monolingual summarization.

In order to enhance the ability of the graph-sorting algorithm to recognize the important sentences, the semantic role information [19] can be integrated to define the similarity criterion of the edges in the graph. Also, using the relationship between topic and sentence to integrate the topic model [20] into graph sorting to improve the improve the recognition ability of important sentences. The supervised methods mainly rely on labeled training data to learn model parameters. Quantities of supervised methods have been developed to extract summarization, such as using neural network to construct the framework for sentences ordering [21] and define the task of sentences ordering as hierarchical regression process. Cao [21] use the recursive neural network to automatically learn the sorting features on the parsing tree and set the importance scores for each sentence in the document. However, the performance of this method lies in the structure of the syntax tree largely. Zhang [22] developed an summarization extraction model based on convolutional neural network to learn the features of sen-tences and execute the procedure of sentences ordering. Nallapati [23] predicted the sentences' scores through considering the significance and redundancy of information simultaneously. Actually, sequential labeling is helpful for the abstract extraction of document, that is, using the potential labels to decide whether the sentence should be extracted as part of abstract. It is noticeable that, in this method, "0" denotes the corresponding sentence is not involved in the abstract, while "1" indicates the corre-sponded sentence is extracted to construct the abstract. Cheng et al. [24] and Nallapati [25] employed the recurrent neural networks to obtain the sentences order and the semantic vector representations of whole document, and then predict each sentence through sequential labeling.

3 Chinese-Vietnamese News Documents Summarization Based on Multi-feature Fusion

The goal of Chinese-Vietnamese news documents summarization system lies in extract the key sentences involved in the cross-linguistic related news document sets, through a sequence of operations to generate summarization, such as sentences partition, significance recognition, eliminate redundancy, etc. In this work, we propose to integrate features of bilingual news text into neural network and build the scoring model to evaluate sentences. The framework of the proposed method is shown in Fig. 1.

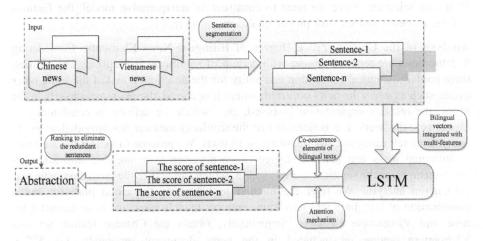

Fig. 1. Chinese and Vietnamese bilingual news summarization based on LSTM

In our proposed framework, the problem of generating document summarization is firstly transformed into the sentence regression. Then, integrate the text features into LSTM model to predict the corresponding scores. Generally, news reports are often built around the key concepts and named-entity elements. Thus, one of the challenges is to identify the key factors information. To achieve this goal, the features of news text are integrated into word embedding, such as name the statistical information of entity and TF-IDF to be the external knowledge joint words to embed into the input LSTM network. In the stage of predict sentence score by regression, connect the bilingual news elements and the concurrence weights to the hidden layer representation which is used to calculate the attention weights. And then, construct the function to predict the regression scores of sentences. Finally, generate the Chinese-Vietnamese news documents summarization though extracting the sentences with high scores and filtering the redundant information.

3.1 Multi-feature Word Embedded Representation Learning

Conventionally, whether a sentence in bilingual news text is selected as a summarization sentence lies in whether the sentence has the ability to characterize the key content of the document. For the news events concerned by China and Vietnam, although the language forms are different, they usually have similar characteristics. For example, the time and place of occurrence, the participants and other information are consistent. The first sentence of the article or paragraph has the role of outline and important. Further, important information is repeatedly emphasized during the writing process, and these features contribute to the joint analysis of Chinese and Vietnamese bilingual news. Therefore, we propose to incorporate the heuristic statistical rules such as the degree of co-occurrence of bilingual news elements, sentence position, and word frequency into neural networks. Besides, the feature vector and the semantic information of a sentence are taken into consideration, which are play a guidance role for sentences selection. Since we need to construct an interpretative model, the features will be modeled and set with proper qualification score.

Analysis of the Co-occurrence Degree of Bilingual News Elements. Considering that the news are reported around with certain of significant conceptions and entities, there contains identical conception and entity for the description text of the same news event, such as when the news occurred, where it occurred, the main participants in the event, the relevant organization involved, etc., which are defined as consistency of elements in this work. It is noticeable that the similarity measure has provided access to analyze the relevance of cross-linguistic news texts. We propose to extract the elements of bilingual news text and the degree of co-occurrence to integrate into attention mechanism to set the sentence score function, which can realize the cross-linguistic text association analysis. We refer the method in [3], using the cloud platform and a combination of templates and maximum entropy models respectively to extract Chinese and Vietnamese elements. Sequentially, obtain the Chinese feature set and Vietnamese feature set included in the news document separately, i.e. $E^{cn} = \{e_1^{cn}, e_2^{cn}, \ldots, e_n^{cn}\}$ and $E^{vi} = \{e_1^{vi}, e_2^{vi}, \ldots, e_n^{vi}\}$. To evaluate the degree of co-occurrence of news elements between bilinguals, it is necessary to use the Chinese and Vietnamese bilingual dictionary [3]. We align bilingual news elements with inter-translation relationships and attain the aligned set of Chinese-Vietnamese news $E^{cv} = \{\{e_1^{cn}, e_1^{vi}\}, \{e_2^{cn}, e_2^{vi}\}, \ldots, \{e_k^{cn}, e_k^{vi}\}\}$. The Chinese sentences containing news elements are denoted as $s_i = \{e_1^{cn}, e_2^{cn}, \ldots, e_p^{cn}\}$. If there exists an intersection between the Chinese elements and the set of Vietnamese news elements, then there is a co-occurrence relationship between the two languages. The degree of co-occurrence can be calculated by:

$$R_e\left(s_i^{cn}\right) = \frac{Count\left(s_i^{cn} \cap D^{ve}\right)}{Count\left(s_i^{cn}\right)} \tag{1}$$

where $Count\left(s_i^{cn} \cap D^{vi}\right)$ stands for the quantity of common news elements between Chinese sentence s_i and Vietnamese sentence D^{ve}. $Count\left(s_i^{cn}\right)$ denotes the number of news elements contained in Chinese sentences. Especially, if there exists no element in the sentence, it is set to zero. Analogy calculation can be implemented to calculate the degree of co-occurrence of Vietnamese elements.

Moreover, the degree of elements co-occurrence is integrated into the coding vector to assign the weight of attention. We can not only achieve multi-linguistics association analysis through the consistency of factors, but also capture the strength of the relationship of elements through the principle of probability distribution.

Position of the Sentence. Generally, the sentences that deliver the core idea of the full text may locate at the beginning or end of the article for news documents with strong normalization. That is, if the position of the sentence is closer to the beginning and end of the news, the sentence is more likely to be extracted as a summary. It is noticeable that the sentence position score in the multiple documents is not the one in the merged document, while the sentence position score is only calculated as that in the original single document to which the sentence belongs. Therefore, there exist some identical scores of sentence position among the sentences. If two documents have the same number of sentences, the sentences in the same position will have the same sentence position score, which can be formulated as:

$$P(s_i) = \frac{1}{\min(i, N - i + 1)} \tag{2}$$

where N denotes the total number of sentences in a single document to which the sentence belongs. s_i is the i-th sentence in the document.

Features of Words Frequency. In general, the key words will be emphasized repeatedly during composing a news document. Thus, in the abstract system, it is considered that the pivotal conceptions or vocabularies possess high document frequency. The more vocabularies with high document frequency in a sentence, the more likely it is to be selected as abstract. In this work, the sentences are assigned with proper weights in accordance with vocabulary information contained in the sentences to conduct the selection process. The word frequency is calculated through TF-IDF, and integrate the probability of word frequency into the pre-trained word vector as the input of encoder to assess whether each sentence is subordinate to the abstract qualification score. The weights of vocabulary is formulated as:

$$W_{i,j} = tf_{i,j} \times \log \frac{N}{n_j + 1} \tag{3}$$

where $W_{i,j}$ is the weight of vocabulary, $tf_{i,j}$ denotes the frequency that the word t_i appearing in document d_j. N is the number of all texts in the text collection. n_j represents the number of texts that contain word t_i in the text collection.

Features of Sentences Relevance. It is considered that the similarity between sentences can be employed as an measurement index of significance in the abstract extraction method based on graph association analysis [18]. In the abstractive extraction method based on graph association analysis [18], it is considered that the similarity between sentences can be used as an evaluation index of its importance. Further, the relevance score can be calculated through the PageRank formula. Since the cosine similarity measure is involved in the algorithm, the embedded vector representation of a sentence can be attained by average the embedded of each word to obtain the LSTM encoder.

3.2   LSTM Based on Bilingual Feature Attention Mechanism

In the task of abstraction summarization of texts, the recurrent neural networks can be employed to acquire the deep semantic information through the process of word vector coding. In this work, we adopt the LSTM structure to improve the problem of gradient vanish during training the long sequence. At time t, the status of hidden layer is updated as below:

$$I_t = \sigma(X_t W_{xi} + h_{t-1} W_{hi} + + b_i) \tag{4}$$

$$F_t = \sigma(X_t W_{xf} + h_{t-1} W_{hf} + + b_f) \tag{5}$$

$$O_t = \sigma(X_t W_{xo} + h_{t-1} W_{ho} + + b_o) \tag{6}$$

$$\tilde{C}_t = \tanh(X_t W_{xc} + h_{t-1} W_{hc} + + b_c) \tag{7}$$

$$C_t = F_t \odot C_{t-1} + I_t \odot \tilde{C}_t \tag{8}$$

$$H_t = O_t \odot \tanh(C_t) \tag{9}$$

where W stands for the weight matrix that connect the two layers. b is the bias vector. σ and tanh denote the activation function. C_t and $\tilde{C}_t$ are memory cells. The operator $\odot$ denote to multiply by element.

The pivotal problem in generating bilingual abstract lies in how to build the relevant relationship between bilingual texts. To tackle this problem, we integrate the degree of co-occurrence of bilingual elements into the attention mechanism to learn the relationship between the bilingual news texts and the sentence significant scores. For those sentences that can be selected as abstract, the representative information should be contained in the sentences. Scores of sentences are obtained via synthetically considering the semantic information, elements co-occurrence, word frequency and the relevance of the sentences. In this session, the detailed description of sentence scoring model based on bilingual features related attention mechanism will be presented. The basic idea is illustrated in Fig. 2.

There are two input parts in this framework, i.e., the pre-trained word vector and news feature vector. Those two parts are spliced to form a multi-features embedding representation with the same dimension. It is worthy note that, different from conventional input vector of word in RNN, the feature vector is involved in the input of text rather than word embedding representation of automatically learning. It is mainly because in the task of summarization extraction, words often have different importance in different documents, and the addition of features can help to identify their

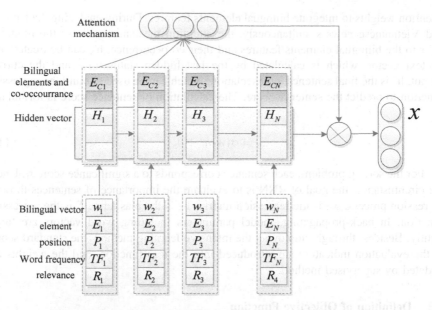

Fig. 2. Attention-based LSTM score prediction regression model

importance. In the forward propagation, the bilingual word vectors integrated with multi-features are mapped into the hidden layer $H \in R^{d \times N}$ through the LTSM network, where d and N are the size of hidden layer and the number of word in a sentence, respectively. Those vectors in hidden layer are used as sentences representation and fused with bilingual elements features to calculate the attention weight α. Then, set the regression score after multiple transformations.

$$M = \tanh \left(\begin{bmatrix} W_1 H \\ W_2 E_c \end{bmatrix} \right) \tag{10}$$

$$\alpha = soft\max \left(w^T M \right) \tag{11}$$

$$x = H\alpha^T \tag{12}$$

where E_c, x, W_1, W_2, w are projection parameters. α is the attention mechanism. x is a vector expression that is obtained through the combination of attention and bilingual elements features. The bilingual information has been applied to assign the weight of attention and the sentence expression. Besides, the changes based on attention weight can capture the most important features in the sentence of bilingual texts and generate the feature vector with fixed length.

$$h^* = \tanh \left(W_3 x + W_4 h_N \right) \tag{13}$$

where W_3 and W_4 are transformation vectors. tanh denotes the activation function. x and h_N are outputs of the encoder. x is the vector expression obtained by encoding the attention weights to integrate bilingual elements features. During analyzing the Chinese and Vietnamese news simultaneously, the attention mechanism allows the model to refer to the bilingual elements features and their co-occurrence. h_N can be treated as a context vector, which is calculated by the last hidden status h_{N-1} and the current output. h^* is the final sentence representation, which can be used to build the regression function to predict the sentence score. The calculation of sentence score is formulated as:

$$\hat{s}_i = \sigma(w_5 h^* + b_5) \tag{14}$$

For the sorting problem, each sentence corresponds to a significance score $\hat{s}_i$. Under the circumstance, the goal of RNN is to evaluate the importance of sentences through regression process. w_5 is the regression matrix. b_5 is the bias term. Σ is the regression function. In back-propagating, model parameters including word vectors are tuned lightly. Besides, through employing the intensive learning method, the standard scores in the evaluation indicators are introduced into the loss function, and the weights are updated by supervised method.

3.3   Definition of Objective Function

In order to train the RNN model, the definition of loss function is referred to [21]; that is, the cross entropy of the predicted score based on the sentence and the standard score. The document is firstly processed to obtain the significance score of each sentence, which is defined as norm score s_i. The norm score is calculated through the general automatic criteria ROUGE-1 (R1) and ROUGE-2 (R2). The formula is as follows:

$$s_i = R_1 + R_2 \tag{15}$$

The objective function is minimizing the cross entropy $CE(s_i, \hat{s}_i)$:

$$CE(s_i, \hat{s}_i) = -(s_i \ln \hat{s}_i + (1 - s_i) \ln(1 - \hat{s}_i)) \tag{16}$$

3.4   Sentence Selection

In general, there exist many duplicate descriptions in multi-document news describing the same event, thus, the tradeoff between significance and redundancy should be solved properly in the document abstract system. The widely used sentence selection methods in the task of abstract extraction are greedy algorithms and integer linear programming. Considering the computational complexity, we use the greedy algorithms to filter the redundant information. It is noticeable that the status of hidden layer H is utilized as the vector expression of sentence in the filter process according to the cosine similarity measure of sentences.

4 Experiment

4.1 Dataset

Since there is no large-scale related open source corpus, we adopt the technology of Internet crawler to automatically obtain online news text information as the dataset. Those news documents are derived mainly from some domestic media such as Xinhuanet, International Online Chinese and Sina Weibo, the Vietnamese news websites of Vietnam Daily News, Vietnam Economic Daily, Vietnam News Agency, etc. The obtained data contains news headline, news details, release time, media source. Then, the Chinese and Vietnamese bilingual text-processing method based on graph clustering [4] is used to classify the cross-language texts describing identical events into a class, and manually verify the results of the collation. Especially, each couple of news documents includes at least two documents, i.e. Chinese document and Vietnamese document, respectively. Considering that we need to train the neural networks, thus, the abstract the sentences of documents should be labeled. In this work, 20,000 document abstract datasets are constructed, where 12,000 are Chinese news documents and 8,000 are Vietnamese. It is noticeable that those documents involve the hot topics that are jointly concerned by China and Vietnam in recent years, such as the policy topics, the tourism, study abroad, etc. For each collection of events, the selection of the referenced summary is to extract four sentences from each language as a norm.

4.2 Parameter Setting

Some parameters are required to determine before constructing the model. Since there are two parts in the input of the model, i.e. the Chinese-Vietnamese bilingual word vectors and the news feature vectors. The two categories of vectors are optimized during the training process and the hidden status is fixed as 200. To avoid overfitting, we use the L2 regularization. Besides, the learning rate is set as 0.005, the regularization factor is 0.1 and the batch size is set as 32.

4.3 Evaluation

In this work, the ROUGE score [27] is used as evaluation metric, which is a universal evaluation metric for abstract evaluating and used in many international conferences such as DUC and TAC. The basic idea is to analyze the proximity between the model output summary and the standard one. Specifically, the quantization calculation is performed according to the co-occurrence unit in the two linguistics. The formula is shown as below:

$$ROUGE - N = \frac{\sum_{S \in \{Ref\,Sum\}} \sum_{n-gram \in s} count_{match}(n - gram)}{\sum_{S \in \{Ref\,Sum\}} \sum_{n-gram \in s} count(n - gram)} \qquad (17)$$

where n denotes the length of $n - gram$. $count_{match}(n - gram)$ denotes the number of $n - gram$ that occurs frequently in the generation summary and manually written standard summary. In detail, $ROUGE - 1$, $ROUGE - 2$ and the co-occurrence statistics of the longest common subsequence in a sentence ROUGE – L are chose to the experiment evaluation metrics.

4.4 Results and Analysis

To validate the effectiveness of the proposed method, we design two sets of experiments to evaluate the performance on different datasets. The detailed description of the compared method is as follows:

TextRank. Firstly, calculate the similarity between sentences according to the word co-occurrence. Then, employ the TextRank algorithm to calculate the significance scores of sentences. However, the TextRank algorithm used in [28] is limited to single document abstract extraction. To apply this algorithm for solving the multi-linguistics documents, we add a comparison for bilingual dictionary in the pre-processing stage. Besides, considering there is a large number of duplicate descriptions of multi-document news, the greedy algorithm is introduced to filter redundancy in the sentence selection phase as a follow-up step.

LReg. Construct the classifiers for sentence extraction using logistic regression and news text features. Intuitively, the feature selection contains sentence position, word frequency, element information and degree of association between sentences. To clarify the effectiveness of this method, in this paper, the processing steps such as word embedding, feature selection, and training data are consistent with the model in the comparison experiment of the regression model.

LSTM-Feature. The proposed method in this work, the goal of the proposed method is to evaluate the importance of the sentences through calculating the regression scores. Therefore, the LTSM algorithm is adopted to construct a sentence vector representation of multi-feature fusion. Furthermore, integrate bilingual elements association analysis into neural networks to deploy the attention weight. The output of the model is the prediction score of sentences.

Table 1 gives the compared results of the different schemes of extractive summary generation. Table 2 shows the impact of adding feature association analysis on summary performance.

Table 1. Comparison results of different methods

	ROUGE–1	ROUGE–2	ROUGE–3
TextRank	0.3166	0.1352	0.2519
LReg	0.3227	0.1470	0.2650
LSTM-Feature	0.3824	0.2270	0.3169

From Table 1 we can see that LSTM-Feature achieves significant best performance. In detail, although TextRank is a strong benchmark model for extractive summary, when it comes to bilingual document analysis, it is necessary to consider the degree of overlap of words in different language sentences. Therefore, in this paper, we propose

Table 2. The comparison experiment of whether to add the feature association analysis

	ROUGE–1	ROUGE–2	ROUGE–3
LSTM-Att	0.3618	0.2159	0.3041
LSTM-Feature	0.3824	0.2270	0.3169

to pre-process the documents with the bilingual dictionary, which may help to improve the model performance. LREG utilizes multiple features related to abstract to perform the sentence classification, that is, to determine whether a sentence in a text collection belongs to the abstract. Besides, the classifier is trained on the same data set as the neural network model, and the acquisition of the feature vector is consistent with the proposed method. The experiment result of LSTM-Feature is related to the learning ability of the model for the text semantic structure information and abstract features. Further, construct the model to reflect the relationship between bilingual texts through the attention mechanism based on feature association, which enables the model to give higher significance scores on representative sentences than ordinary sentences in the documents. To further understand the impact of feature association analysis on the performance of abstracts, we develop a group of comparative experiments additionally.

LSTM-Attention. To verify the effectiveness of relevance attention mechanism based on bilingual elements features, the traditional attention mechanism will be used for score prediction, that is, the hidden layer status $\mathbf{H} \in \mathbf{R}^{d \times N}$ is directly used for linear transformation to obtain the significant score of the sentence in this method.

The comparison results of LSTM-Att and LSTM-Feature disclose that it is more conducive for the identification of abstract sentences through merging the bilingual attention-based attentional mechanism and the weight distribution method. There are two main reasons: (1) the addition of news features can effectively improve the performance of abstract extraction; (2) there is a large amount of consistent information in the Chinese and Vietnamese bilingual news documents describing the same news event, and the co-occurrence of news information promotes the recognition of important sentences in the original document collection.

5 Conclusion

In this paper, a new LSTM regression method based on the elements co-occurrence and feature-embedded strategies is proposed for the task of abstraction extraction. In this method, some statistics characters such as news elements, word frequency and the relevance of sentences are firstly integrated into the neural networks. Then, the attention mechanism based on bilingual elements features is proposed to calculate the importance of sentences to obtain corresponding scores. Finally, pick out the sentences with high scores and eliminate the redundant sentences according to the similarity measure to obtain the document abstraction. Although effective results are obtained by the proposed method, it still depends on large scale labeled data to improve the model performance. Therefore, we are going to introduce the semi-supervised learning or transfer learning to expand the proposed framework in the future exploitation.

Acknowledgments. This work was supported by National Key Research and Development Plan (Grant Nos. 2018YFC0830105, 2018YFC0830101, 2018YFC0830100); National Natural Science Foundation of China (Grant Nos. 61732005,61761026, 61866019,61672271,61762056, 61866020); Science and Technology Leading Talents in Yunnan, and Yunnan High and New Technology Industry Project (Grant No.201606); Talent Fund for Kunming University of Science and Technology (Grant No. KKSY201703005, KKSY201703015).

References

1. Wan, X., Li, H., Xiao, J.: Cross-language document summarization based on machine translation quality prediction. In: 48th ACL Meeting of the Association for Computational Linguistics, pp. 917–926. IEEE Press, Uppsala (2010). https://doi.org/10.17562/pb-43-16
2. Mathieu, B., Besançon, R., Fluhr, C.: Multilingual document clusters discovery. In: 7th CAIR Computer-Assisted Information Retrieval, pp. 116–125. IEEE Press, Naples (2004)
3. Lei, Y.: The comparative summarization of Chinese and Vietnamese bilingual news. Kunming University of Science and Technology (2018)
4. Wang, Y.S.: Detecting hot news topics and generating summarization from bilingual news texts. Kunming University of Science and Technology (2018)
5. Ko, Y., Seo, J.: An effective sentence-extraction technique using contextual information and statistical approaches for text summarization. Pattern Recogn. Lett. **29**, 1366–1371 (2008). https://doi.org/10.1016/j.patrec.2008.02.008
6. Yao, J.G., Wan, X., Xiao, J.: Recent advances in document summarization. Knowl. Inf. Syst. **53**, 1–40 (2017). https://doi.org/10.1007/s10115-017-1042-4
7. Narayan, S., Papasarantopoulos, N., Cohen, S.B., Lapata, M.: Neural extractive summarization with side information. In: 31th AAAI Conference on Artificial Intelligence, pp. 116–125. IEEE Press, San Francisco (2017). https://doi.org/10.11606/d.55.2018.tde-24102018-155954
8. Wang, Y., Huang, M., Zhu, X.: Attention-based LSTM for aspect-level sentiment classification. In: 34th EMNLP Empirical Methods in Natural Language Processing, pp. 606–615. IEEE Press, Austin (2016). https://doi.org/10.1109/access.2019.2893806
9. Huang, T., Li, L., Zhang, Y.: Multilingual multi-document summarization with enhanced hLDA features. In: Sun, M., Huang, X., Lin, H., Liu, Z., Liu, Y. (eds.) CCL/NLP-NABD - 2016. LNCS (LNAI), vol. 10035, pp. 299–312. Springer, Cham (2016). https://doi.org/10.1007/978-3-319-47674-2_25
10. Wan, X., Jia, H., Huang, S.: Summarizing the differences in multilingual news. In: 34th SIGIR Conference on Research and Development in Information Retrieval, pp. 735–744. Springer, Beijing (2011). https://doi.org/10.1145/2009916.2010015
11. Singh, S.P., Kumar, A., Mangal, A., Singhal, S.: Bilingual automatic text summarization using unsupervised deep learning. In: 12th ICEEOT International Conference on Electrical, Electronics, and Optimization Techniques, pp. 1195–1200. IEEE Press, Chennai (2016). https://doi.org/10.1109/iceeot.2016.7754874
12. Wang, F.L., Yang, C.C.: The impact analysis of language differences on an automatic multilingual text summarization system. J. Assoc. Inf. Sci. Technol. **57**, 684–696 (2014). https://doi.org/10.1002/asi.20330
13. Di Felippo, A., Tosta, Fabrício E.S., Pardo, Thiago A.S.: Applying lexical-conceptual knowledge for multilingual multi-document summarization. In: Silva, J., Ribeiro, R., Quaresma, P., Adami, A., Branco, A. (eds.) PROPOR 2016. LNCS (LNAI), vol. 9727, pp. 38–49. Springer, Cham (2016). https://doi.org/10.1007/978-3-319-41552-9_4

14. Oufaida, H., Blache, P., Nouali, O.: Using distributed word representations and mRMR discriminant analysis for multilingual text summarization. In: Biemann, C., Handschuh, S., Freitas, A., Meziane, F., Métais, E. (eds.) NLDB 2015. LNCS, vol. 9103, pp. 51–63. Springer, Cham (2015). https://doi.org/10.1007/978-3-319-19581-0_4

15. Cruz, C.M., Urrea, A.M.: Extractive summarization based on word information and sentence position. In: Gelbukh, A. (ed.) 6th CLCLing International Conference on Computational Linguistics and Intelligent Text Processing, pp. 653–656. Springer, Berlin (2005). https://doi.org/10.1007/978-3-540-30586-6_73

16. Vanderwende, L., Suzuki, H., Brockett, C., Nenkova, A.: Beyond sumbasic: task-focused summarization with sentence simplification and lexical expansion. Inf. Process. Manag. **43**, 1606–1618 (2007). https://doi.org/10.1016/j.ipm.2007.01.023

17. Meng, X., Wei, F., Liu, X.: Graph-based lexical centrality as salience in text summarization entity-centric topic-oriented opinion summarization in twitter. In: 18th SIGKDD Proceedings of the ACM International Conference on Knowledge Discovery and Data Mining, vol. 10, pp. 93–102 (2015). https://doi.org/10.4018/ijirr.2015070102

18. Radev, D.R.: LexRank: graph-based lexical centrality as salience in text summarization. J. Artif. Intell. Res. **22**, 457–479 (2004). https://doi.org/10.1613/jair.1523

19. Yan, S., Wan, X.: SRRank: leveraging semantic roles for extractive multi-document summarization. In: 22th TASLP Transactions on Audio Speech and Language Processing, pp. 2048–2058. IEEE Press, Piscataway (2012). https://doi.org/10.1109/taslp.2014.2360461

20. Li, J., Li, S.: Query-focused multi-document summarization: combining a novel topic model with graph-based semi-supervised learning. In: 25th COLING International Conference on Computational Linguistics, pp. 1197–1207. IEEE Press, Dublin (2014). https://doi.org/10.1305/coling.2014.232131

21. Cao, Z., Dong, L.: Ranking with recursive neural networks and its application to multi-document summarization. In: 29th AAAI Conference on Artificial Intelligence, pp. 114–120. IEEE Press, Washington (2013). https://doi.org/10.3455/aaai.2013.1249131

22. Yong, Z., Meng, J.E., Ning, W., Pratama, M.: Extractive document summarization based on convolutional neural networks. In: 42th IECON Conference of the IEEE Industrial Electronics Society, pp. 918–922. IEEE Press, Beijing (2016). https://doi.org/10.1109/iecon.2016.7793761

23. Nallapati, R., Zhou, B., Ma, M.: Classify or select: neural architectures for extractive document summarization. In: 5th ICLR International Conference on Learning Representations Conference Submission, pp. 928–936. IEEE Press, Toulon (2017). https://doi.org/10.1109/iclr.2017.7793761

24. Cheng, J., Lapata, M.: Neural summarization by extracting sentences and words. In: 54th ACL Annual Meeting of the Association for Computational Linguistics, pp. 1138–1146. IEEE Press, Berlin (2016). https://doi.org/10.18653/v1/p16-1046

25. Nallapati, R., Zhai, F., Zhou, B.: SummaRuNNer: a recurrent neural network based sequence model for extractive summarization of documents. In: 31th AAAI Conference on Artificial Intelligence, pp. 1318–1329. IEEE Press, San Francisco (2017)

26. Tang, P.L.: The method of the discovery and evolution of bilingual news topics in the Chinese and Vietnamese. Kunming University of Science and Technology (2018)

27. Lin, C.Y.: ROUGE: a package for automatic evaluation of summaries. In: Proceedings of the Workshop on Text Summarization Branches Out, pp. 74–81. IEEE Press, San Francisco (2004)

28. Mihalcea, R., Tarau, P.: TextRank: Bringing order into texts. In: Emnlp Proceedings Conference on Empirical Methods in Natural Language Processing, pp. 404–411. IEEE Press, Barcelona (2004)

A Simple and Convex Formulation for Multi-label Feature Selection

Peng Lin, Zhenqiang Sun, Jia Zhang, Zhiming Luo, and Shaozi Li[✉]

Department of Cognitive Science,
Xiamen University, Xiamen 361005, People's Republic of China
szlig@xmu.edu.cn

Abstract. In recent years, multi-label study has received extensive attention and research in many fields. The feature dimensions of a multi-label data set are high but contain a large amount of noise as well as irrelevant and redundant features. This not only leads to huge storage and time overhead, but also brings serious dimensional disaster problems, making multi-label learning tasks very difficult. Therefore, how to effectively select multi-label features is an important research content in multi-label learning. However, most of the current methods are converted from the methods of single-label feature selection, and feature selection is easy to fall into the local optimal heuristic search strategy. Time complexity has always been the biggest problem of such methods. Based on those considerations, our paper proposes a fast and effective multi-label feature selection method, which uses the optimization strategy to replace the previous search strategy for multi-label feature selection, and transforms the search problem into convex optimization problem. Therefore, the time performance of the traditional method is improved by two to three orders of magnitude. Finally, the experimental results of five evaluation indicators on the four data sets show that our method is superior to many popular methods in feature selection field.

Keywords: Convex optimization · Multi-label learning · Feature selection

1 Introduction

In common supervised learning tasks, one sample is defaulted to have only one type of semantic information, that is, only one classification label. However, such assumptions often do not match the real situation in the real world. For example, in the image classification task, a picture of a beach landscape often contains scenes such as "the sea", "ship", "sunset", etc. A label cannot fully express its semantic i nformation. Thus, researchers began to pay attention to multi-label study (multi-label learning) [1], and retrieve information [2], the biometric information [3], the pharmaceutical research [4] and medicine diagnosis [5,6], and other fields achieved excellent effect.

© Springer Nature Singapore Pte Ltd. 2019
Y. Sun et al. (Eds.): ChineseCSCW 2019, CCIS 1042, pp. 540–553, 2019.
https://doi.org/10.1007/978-981-15-1377-0_42

Compared with traditional supervised learning, multi-label learning has a larger input-output spatial dimension. In most cases, the characteristics of multi-label data tend to be more redundant and sparse in order to support multi-label learning tasks than single-label data sets. However, Excessive feature dimensions will lead to dimensional disasters, which will make multi-label learning tasks inefficient and difficult. Therefore, it is very important to effectively solve the dimensional disaster problem in multi-label learning tasks.

Since the feature subset directly select from the feature space, it retains the realistic meaning of the original feature, and is highly interpretable and easy to operate, especially in high-dimensional data sets and data processing of limited data sets. The mission has a very important position. The feature selection method selects a set of optimal feature subsets from the feature space according to some evaluation criteria, thereby reducing feature dimensions and improving classification performance. Feature selection is an indispensable part of machine learning, optimization problems and deep learning. It has established an effective screening mechanism from the data source to ensure that the subsequent work is light and effective. At present, the most common evaluation criteria are: dependency metrics, distance metrics, and information metrics.

Among the many methods for measuring feature correlation and label correlation, mutual information based methods are adopted by many researchers because of their reversibility and robustness. Therefore, this paper will calculate the redundancy between features and features, and the correlation between features and labels by using mutual information. However, the time complexity of traditional mutual information-based search methods increases exponentially with the increase of the labelup dimension. This method is difficult to solve high-dimensional data. Therefore, this paper attempts to use the optimized method instead of the search method by analyzing the process of search strategy to reduce the time complexity, thus greatly improving the efficiency of multi-label feature selection. Because this method is about mutual information (Mutual Information) and convex optimization (Convex Optimization), while the most rapid analytical method (Analytic) find an optimum solution, so the proposed method is named MICO_Ana. The specific algorithm details are described in Sect. 4.

2   Related Work

With the more and more attention of multi-label task, a lot of outstanding works has emerged [7–9]. Some workers have carefully compared and classified existing methods. There are two widely accepted classification methods. In the first classification method, feature selection method is divided into encapsulation method [10], filtering method [11] and embedding method [12],which considered it from the strategy. The second classification method divides the feature selection method into supervised [10,12], unsupervised [13], and semi-supervised [14] feature selection methods from the perspective of label utilization.

Since the work of this paper mainly uses mutual information and label correlation, the following will briefly introduce the feature selection work related to the above two. Among the many methods of using feature theory for feature selection, the two most classic works are called MIFS [15]and mRMR [16]. In the MIFS method, The researchers uses mutual information to estimate the amount of information of a one-dimensional feature, and uses the "greedy" search method to find the subset of features; But the mRMR method, The author believes that the selected feature subsets should have the characteristics of "maximum correlation, minimum redundancy" and propose related algorithms. In addition, there are many excellent work using information theory for feature selection. For example, Lin et al. [17] proposed the Maximum Independent Minimum Redundancy Algorithm(MDMR); Lee et al. [18]proposed a method (PMU) by computing multivariate mutual information to maximize the correlation between categories and markers; Song et al. [11] combines the principle of maximum entropy and Z-test technology to achieve rapid selection and optimization of features(MEFS); Li et al. [19] proposed an algorithm that considers the text frequency of the document and the word frequency to classify and optimize the selection of mutual information features.

In addition to the above-mentioned information-related work, the method of using feature information for label selection has also emerged in recent years. Brown et al. [20] proposed a joint feature selection framework for information, and summarized the feature selection methods based on information theory; Sun et al. [21] revealed a new method about information of mutual labels. This method constrains convex optimization to obtain optimal solution in less time by considering label correlation to generate generalized model; Zhang et al. [22] designed a new feature weighting method to measure the association between features.; Wang et al. [23] solves the redundancy problem by maximizing the non-independent classification information; Braytee et al. [14] proposed a method which using the non-negative matrix factorization to solve this problem; Spolar et al. [24] proposed a method for reconstructing the label space based on the original label correlation, considering that the reconstruction space contains high-order correlation information. Liu et al. [25] presented for us with a weighted method of selecting features by means of label correlation, and if can improve the robustness of the results.

3   Related Information

In 1948, in the paper [26] published by Shannon, the famous information theory was first mentioned, and this method became the theoretical basis for many later work. In this paper, mutual information is an effective means of measuring correlation, which describes the degree of information sharing between two sets of variables. The mutual information can be expressed as follows:

$$I(A;B) = \sum_{a \in A} \sum_{b \in B} p(a,b) log(\frac{p(a,b)}{p(a)p(b)}) \tag{1}$$

As a classic algorithm for feature selection, mRMR has evolved many variants. Our article is also inspired by this outstanding work, so this article will briefly introduce the core idea of mRMR so that readers can better understand the method. The authors of mRMR believe that "The best m features may not be the top m best features", because there may largely correlated, that is, carrying a lot of similar information, so the authors took out the mRMR framework, and Formally describe it as:

$$max\Phi(D, R) = D - R \tag{2}$$

In this formula, D represent Dependence and R represent Redundancy, while D and R are calculated as follows:

$$D(S, c) = \frac{1}{|S|} \sum_{x_i \in S} I(x_i; c) \tag{3}$$

$$R(S) = \frac{1}{|S|^2} \sum_{x_i, y_j} I(x_i; y_j) \tag{4}$$

where S is the subset feature and c is the corresponding label.

3.1   Evaluation Index

This paper selects five basic evaluation indicators [27] as a standard for comparing the methods of this paper with other methods, and briefly introduces them. In the following, this article will uniformly use $y_i \in L$ for the true label, y_i' for the predicted label of the feature vector x_i, N for the number of samples, and m for the label dimension.

Hamming loss:

$$HL = \frac{1}{N} \sum_{i=1}^{N} \frac{y_i' \oplus y_i}{m} \tag{5}$$

where $\oplus y_i'$ is an exclusive OR operation, this formula measures the proportion of mistakes made in all N * m prediction label. Its value is between 0 and 1, which is inversely proportional to its performance.

One-error:

$$OE = \frac{1}{N} \sum_{i=1}^{N} [(argmax_{y_i \in L} f(x_i, y_i)) \notin y_i'] \tag{6}$$

This formula measures the most likely marker prediction error ratio in the prediction. When its value is close to 0, it proves that the model is better.

Coverage:

$$CV = \frac{1}{N} \sum_{i=1}^{N} \max_{\lambda \in y_i} rank(\lambda) - 1 \tag{7}$$

In the predicted marker sequence, when all markers are covered, the maximum search depth at this time is represented by calculating the coverage. It is negatively related to the ranking of the relevant labels.

Ranking loss:

$$RL = \frac{1}{N} \sum_{i=1}^{N} \frac{1}{|y_i| \, |\overline{y_i}|} \, |(\lambda_1 \lambda_2) \lambda_1 \leqslant \lambda_2, (\lambda_1, \lambda_2) \in y_i \times \overline{y_i}| \qquad (8)$$

In this formula, y_i represents the set of related labels for xi, $\hat{y}_i$ represents its set of extraneous labels, and then calculates the error rate for each pair of related and unrelated labels. Obviously, the smaller the index, the better the performance of the algorithm.

Average precision:

$$AP = \frac{1}{N} \sum_{i=1}^{1} \frac{1}{|y_i|} \sum_{\gamma \in y_i} \frac{|\gamma \in y_i : r_i(\gamma \leqslant r_i(\gamma))|}{r_i(\gamma)} \qquad (9)$$

This formula indicates one case that if the labels are sorted according to the predicted values,those markers that are placed before the relevant markers are the relevant markers in the sequence.

4   The Proposed Approach

In this part, we will elaborate the motivation of this method and the related formulation, and then briefly introduce the solution method used in this article in the second section, and provide a description of the corresponding pseudo code.

4.1   Method Motivation and Formulation

In order to make the mRMR method migrate to multi-label problem, it can have better generalization performance. This paper attempts to add the second-order label correlation information to the mRMR computing framework. First, a calculation method for the most simple second-order label correlation is introduced:

$$Cor(l_j; l_k) = Jaccard(l_j; l_k) = \frac{|l_j \bigcap l_k|}{|l_j \bigcup l_k|} \qquad (10)$$

Jaccard(;) is a distance measurement method that can also be replaced by different methods such as Euclidean distance or Hamming distance. The mRMR algorithm incorporating label correlation can be expressed as:

$$\max_{S} J = \frac{\sum\limits_{f_i \in S} \sum\limits_{l_j, l_k \in L} I(f_i; l_j) Cor(l_j; l_k) + \sum\limits_{f_i \in S} \sum\limits_{l_j \in L} I(f_i; l_j)}{\sum\limits_{f_i, f_j \in S} I(f_i; f_j)} \qquad (11)$$

The above expression adds the label correlation information to the evaluation index of the feature selection. The method considers that the different feature label correlations should have different weights, so the correlation between the label is used to weight them, for example, a feature and When a label is related and the label is associated with other label, the feature is preferred.

However, when using this method for feature selection, if the first k features are selected as the feature subset, the analysis shows the time complexity $O(C_n^k(nm^3 + n^2 + nm))$, so when the data dimension is high, The actual problem can still not be solved with a simple search strategy.

By analyzing the above calculation process, it is known that the mutual information I(;) and the label correlation Cor(;) between the features are repeatedly calculated multiple times in the process of searching for the optimal subset. Therefore, pre-calculating all the data that needs to be used and saving it to the matrix will greatly reduce the repetitive calculations generated during the search process. Therefore, we use matrix D and vectors c and e to save the required data. The calculation details are as follows:

$$D_{i,j} = I(f_i; f_j) = \sum_{a \in f_i} \sum_{b \in f_j} p(a,b) log(\frac{p(a,b)}{p(a)p(b)}) \tag{12}$$

where $D_{i,j}$ is the i-th row and the j-th column element in the matrix D, representing the mutual information between the i-th eigenvector and the j-th eigenvector. $a \in f_i$, $b \in f_j$ are the values of different elements appearing in the feature vector (the same below).

$$c_i = \sum_{l_j \in L} I(f_i; l_j) = \sum_{l_j \in L} \sum_{a \in f_i} \sum_{b \in l_j} p(a,b) log(\frac{p(a,b)}{p(a)p(b)}) \tag{13}$$

In formula, c_i is the ith element of vector c, representing the sum of the mutual information between the ith feature and all label.

$$e_i = \sum_{l_j \in L} I(f_i; l_j) \sum_{l_k \in L} Cor(l_j; l_k) \tag{14}$$

E_i is the i-th element of the vector e, which represents the sum of the mutual information of the i-th feature and all the label weighted by the label distance.

By calculating D, c, and e in advance, the above formula can be rewritten as:

$$\max_s J = \sum_{f_i \in S} e_i + \sum_{f_i \in S} c_i - \sum_{f_i \in S} \sum_{f_j \in S} D_{i,j}, \tag{15}$$

Thereby, the time complexity can be reduced to $O(C_n^k)$.

However, in most cases, feature dimensions tend to be higher and the number of optimal feature subsets is unknown, and the commonly adopted strategy is to iterate through all possible feature subsets. Resulting in an increase in the total number of searches to $C_n^1 + C_n^2 + ... + C_n^n = 2^n$. The time complexity of searching for the best feature subset is still high.

In order to further simplify the calculation, a strategy similar to the sparse learning method is adopted in this paper to assign corresponding weights to each dimension feature. The greater the weight, the more important the feature. This paper defines the n-dimensional vector x to represent the feature weight vector, where x_1 represents the ith feature. For the convenience of solving, this paper converts the maximization problem into a minimization problem, then the above formula is converted into:

$$\max_x J = c^T x + \alpha e^T x - \beta x^T D x \qquad (16)$$

$$s.t. \ x_1, x_2, x_3, ..., x_n \geq 0, \ \sum_{i=1}^{n} x_i = 1. \qquad (17)$$

The constraint of the above formula is to ensure the obtained weight distribution has practical significance. By the above formula, the search problem becomes a constrained convex optimization problem, and the optimal solution obtained by global optimization has better robustness.

It is worth noting that the main reason for this paper is to convert the traditional search strategy into an optimization strategy. The main reason is that the matrix D is found to be a symmetric matrix in the process of observing the label redundancy matrix D, so it is mostly loosely defined. Under the condition is a semi-positive definite matrix. In other words, the binomial of the vector x formed by the D matrix as a parameter must be convex, and the binomial must have the highest value under the condition of x, regardless of the position of the most value at the edge of the x-defined domain or some point inside. Under the condition that this analysis is established, the corresponding global optimization strategy can be designed for the above framework.

4.2   Solution Process

There are many ways to solve the convex optimization problem, including the classical Newton method and the gradient descent method. However, many methods based on approximation thought have natural defects, and the obtained results can only approximate the optimal solution infinitely, and the time complexity is very high. Therefore, this paper attempts to solve the objective function under the strict mathematical constraints, so as to get the global optimal solution quickly. Since the analytical solution cannot satisfy the obtained solution in the defined domain, this summary will not temporarily limit the solution of the objective function, that is, the solution that is not required to have practical significance in numerical value, the two types of objective function are transformed as follows:

$$F(x) = -c^T x - \alpha e^T x + \beta x^T D x \qquad (18)$$

Then according to the matrix vector relationship derivation rule, the first derivative of the above objective function can be obtained as follows:

$$F'(x) = -c - \alpha e + \beta (D + D^T) x \qquad (19)$$

When the above reciprocal is zero, the obtained solution is the analytical solution of the global minimum value sought as follows:

$$x = \frac{(c + \alpha e)}{\beta}(D + D^T)^{-1} \tag{20}$$

A fatal shortcoming of the analytical solution is that the inverse matrix of D is required to exist. This condition cannot be guaranteed to be theoretically feasible, but in most cases, a reversible D matrix can be obtained when solving practical problems (such as all data sets used in this paper). Therefore, the analytical solution as a solution method would be an ideal choice.

5 Experimental Setup and Results Comparison

This paper will compare the other three feature selection algorithms on the six real data sets, using MLkNN (default neighbor number k=10)[27] as the classification algorithm to select the data set after feature selection. Finally, we will give a brief analysis of the experimental results.

5.1 Experimental Data and Settings

The experimental data in this paper is mainly composed of the data of the website MULAN (http://mulan.sourceforge.net/datasets.htm), in which the four web datasets of Arts, Business, Health and Recreation belong to Yahoo dataset, each data The set contains 5000 samples, and the label represents the category information of the text; The TCM dataset was provided by Fujian University of Traditional Chinese Medicine and contained 1146 samples. Each feature represents quantitative information on the patient's symptoms. The label indicates the patient's condition. The Yeast dataset describe the basic gene function classification of the yeast, including 2,417. sample. Table 1 lists the details of the data set used. At the same time, this paper will compare with PMU [18], MDDMproj [28], MDMR [17], three multi-label feature selection algorithms to compare the performance of then. The comparison algorithm parameter settings used in this paper use the default parameter settings in the original paper.

5.2 Experimental Results and Analysis

In order to compare the best classification effects that can be achieved by each algorithm, this paper selects the feature subsets of different scales from 1 to the maximum to conduct experiments and draw score curve. The maximum number of features of the MLNB algorithm is different from other algorithms, but it is still comparable.

Figures 1, 2, 3, 4 and 5 show the results of the feature selection of the six methods on different data sets. In the figure, we use the value of the x-axis to indicate the number of selected features, and the value of the y-axis to represent

Table 1. Common multi-label datasets and details

Data	Samples	Classes	Features	Training	Test	Domains
Arts	5000	26	462	2000	3000	Text
Health	5000	32	612	2000	3000	Text
Education	5000	33	550	2000	3000	Text
Yeast	2417	14	103	918	1499	Biology

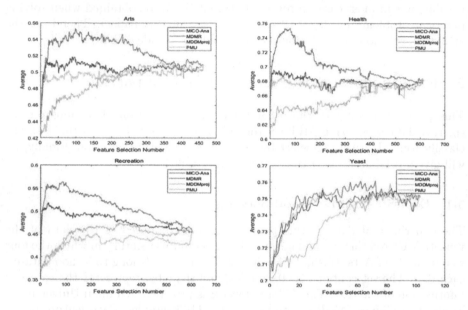

Fig. 1. Experimental result of each method for Average Precision

the scores of different indicators. Among them, the smaller the Haiming loss, coverage, sorting loss, and one-error rate, the better the classification performance, and the higher the accuracy index, the better the classification performance.

At the same time, the experiment was carried out on one computer (Intel Core i7, 4G CPU 8 GB RAM)to verify the efficiency of this method. The three methods were compared time. According to the results of Table 2, we find that Compared with MDMR and PMU, this method takes better consideration of time performance and algorithm accuracy. The specific performance is as follows: while the running time is significantly improved, the result of feature selection is also very robust and effective.

According to Figs. 1, 2, 3, 4, 5 and Table 2:

(1) The experimental curve shows a significant increase or decrease trend in the first few features, and it remains stable or shows the opposite trend after reaching the optimal value, indicating that the proposed algorithm is effective for most data sets and can be selected to be significantly better than the use. A subset of features for which all features are classified.

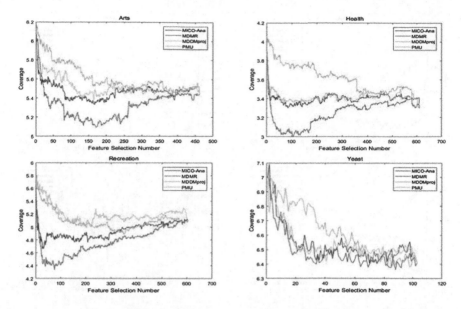

Fig. 2. Experimental result of each method for Coverage

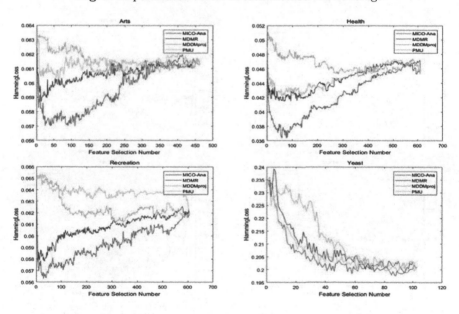

Fig. 3. Experimental result of each method for Hamming loss

(2) The five evaluation indexes of the method in Arts, Health, Recreation, Yeast and other four data sets can achieve the best results, especially on the Arts and Health and Recreation data sets, which shows that this method has obvious advantages over other methods. Because the Yeast dataset has a

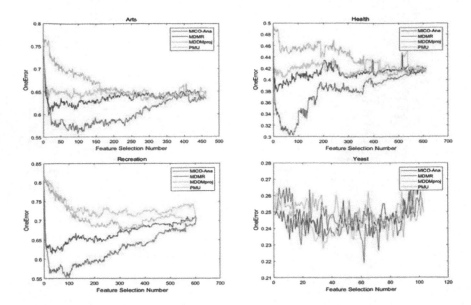

Fig. 4. Experimental result of each method for One Error

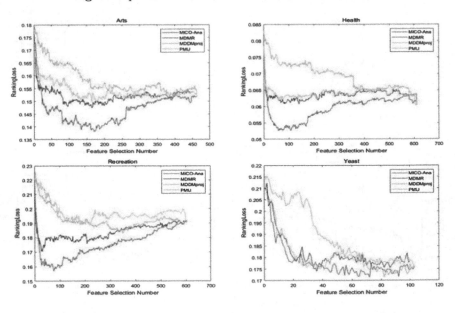

Fig. 5. Experimental result of each method for Ranking loss

small feature dimension, the four feature selection methods are not ideal. Therefore, this paper temporarily believes that the existence of multi-label datasets is not suitable for feature selection methods for dimensionality reduction, because the correlation between features may be small. There is no need to make feature selections.

(3) Since the MDDM method is a matrix-based decomposition method, it has obvious advanlabeles in time performance. This paper has three orders of magnitude improvement in time performance, when it compares to MDMR and PMU.

(4) Our proposed method has better performance in the same time, which means we can not sacrifice accuracy, but at the same time greatly improve the efficiency of the algorithm.

Table 2. Running time of different method (unit:s)

Data	MICO_Ana	MDMR	MDDMproj	PMU
Arts	7.23	8630	0.97	9373
Health	12.63	18628	1.92	18739
Recreation	11.99	12738	1.68	13153
Yes	0.37	180	0.19	196

In general, the MICO_Ana algorithm has a good performance under the above comparison indicators, especially when ensuring the feature selection result, and greatly improving the time efficiency, so that the method can be applied to some data flow scenarios with better performance.

6 Conclusion

Based on the traditional mRMR, In our paper, we proposes a more fast feature selection method to transform the search problem into an optimization problem, which greatly improves the efficiency of feature selection. This paper integrates the label correlation into the optimization framework, and uses the label correlation to weight the features to make the algorithm performance more robust.

Since the MICO_Ana algorithm proposed in this paper only utilizes the simplest second-order label correlation, ignoring the influence of higher order and local label correlation, how to find more reliable label correlation and integrate it into the feature selection method. It will become the focus of the next step of this paper to further improve the feature selection effect and classifier performance. In addition, the use of analytical methods does not use the most stringent mathematical constraints, so finding a solution that is both fast, accurate, and robust will be the focus of the next step.

Acknowledgements. This work is supported by the National Nature Science Foundation of China (No. 61876159, No. 61806172, No. 61572409, No. U1705286 & 61571188), the National Key Research and Development Program of China (No.2018YFC0831402),

Fujian Province 2011Collaborative Innovation Center of TCM Health Management, Collaborative Innovation Center of Chinese Oolong Tea Industry-Collaborative Innovation Center (2011) of Fujian Province.

References

1. Bucak, S.S., Jin, R., Jain, A.K.: Multi-label learning with incomplete class assignments. In: CVPR 2011, pp. 2801–2808 (2011)
2. Schapire, R.E., Singer, Y.: Boostexter: a boosting-based systemfor text categorization. Mach. Learn. **39**(2), 135–168 (2000). https://doi.org/10.1023/A:1007649029923
3. Diplaris, S., Tsoumakas, G., Mitkas, P.A., Vlahavas, I.: Protein classification with multiple algorithms. In: Bozanis, P., Houstis, E.N. (eds.) PCI 2005. LNCS, vol. 3746, pp. 448–456. Springer, Heidelberg (2005). https://doi.org/10.1007/11573036_42
4. Chen, Z., Chen, M., Weinberger, K.Q., Zhang, W.: Marginalized denoising for link prediction and multi-label learning. In: Twenty-Ninth AAAI Conference on Artificial Intelligence, pp. 1707–1713 (2015)
5. Liu, G., Li, G., Wang, Y., Wang, Y.: Modelling of inquiry diagnosis for coronary heart disease in traditional Chinese medicine by using multi-label learning. BMC Complement. Altern. Med. **10**(1), 37–37 (2010)
6. Gu, Q., Li, Z., Han, J.: Correlated multi-label feature selection. In: Proceedings of the 20th ACM International Conference on Information and knowledge management, pp. 1087–1096 (2011)
7. Sun, Z., Zhang, J., Luo, Z., Cao, D., Li, S.: A fast feature selection method based on mutual information in multi-label learning. In: Sun, Y., Lu, T., Xie, X., Gao, L., Fan, H. (eds.) ChineseCSCW 2018. CCIS, vol. 917, pp. 424–437. Springer, Singapore (2019). https://doi.org/10.1007/978-981-13-3044-5_31
8. Zhang, J., Li, C., Sun, Z., Luo, Z., Zhou, C., Li, S.: Towards a unified multi-source-based optimization framework for multi-label learning. Appl. Soft Comput. **76**, 425–435 (2019)
9. Lin, Y., Hu, Q., Jia, Z., Wu, X.: Multi-label feature selection with streaming labels. Inf. Sci. **372**, 256–275 (2016)
10. Guyon, I., Elisseeff, A.: An introduction to variable and feature selection. J. Mach. Learn. Res. **3**, 1157–1182 (2003)
11. Zhu, S., Wu, Y.N., Mumford, D.: Minimax entropy principle and its application to texture modeling. Neural Comput. **9**(9), 1627–1660 (1997)
12. Guyon, I., Weston, J., Barnhill, S., Vapnik, V.: Gene selection for cancer classification using support vector machines. Mach. Learn. **46**(1), 389–422 (2002). https://doi.org/10.1023/A:1012487302797
13. Dy, J.G., Brodley, C.E., Kak, A.C., Broderick, L.S., Aisen, A.M.: Unsupervised feature selection applied to content-based retrieval of lung images. IEEE Trans. Pattern Anal. Mach. Intell. **25**(3), 373–378 (2003)
14. Braytee, A., Liu, W., Catchpoole, D.R., Kennedy, P.J.: Multi-label feature selection using correlation information. In: Proceedings of the 2017 ACM on Conference on Information and Knowledge Management, pp. 1649–1656 (2017)
15. Battiti, R.: Using mutual information for selecting features in supervised neural net learning. IEEE Trans. Neural Netw. **5**(4), 537–550 (1994)

16. Peng, H., Long, F., Ding, C.H.Q.: Feature selection based on mutual information criteria of max-dependency, max-relevance, and min-redundancy. IEEE Trans. Pattern Anal. Mach. Intell. **27**(8), 1226–1238 (2005)
17. Lin, Y., Hu, Q., Liu, J., Duan, J.: Multi-label feature selection based on max-dependency and min-redundancy. Neurocomputing **168**, 92–103 (2015)
18. Lee, J., Kim, D.: Feature selection for multi-label classification using multivariate mutual information. Pattern Recogn. Lett. **34**(3), 349–357 (2013)
19. Li, H.: Optimized mutual information feature selection method. Comput. Eng. Appl. **46**(26), 122–124 (2010)
20. Brown, G., Pocock, A.C., Zhao, M., Lujan, M.: Conditional likelihood maximisation: a unifying framework for information theoretic feature selection. J. Mach. Learn. Res. **13**(1), 27–66 (2012)
21. Sun, Z., et al.: Mutual information based multi-label feature selection via constrained convex optimization. Neurocomputing **329**, 447–456 (2019)
22. Zhang, J., et al.: Multi-label learning with label-specific features by resolving label correlations. Knowl. Based Syst. **159**, 148–157 (2018)
23. Wang, J., Wei, J., Yang, Z., Wang, S.: Feature selection by maximizing independent classification information. IEEE Trans. Knowl. Data Eng. **29**(4), 828–841 (2017)
24. Spolaor, N., Monard, M.C., Tsoumakas, G., Lee, H.D.: A systematic review of multi-label feature selection and a new method based on label construction. Neurocomputing **180**, 3–15 (2016)
25. Liu, L., Zhang, J., Li, P., Zhang, Y., Hu, X.: A label correlation based weighting feature selection approach for multi-label data. In: Cui, B., Zhang, N., Xu, J., Lian, X., Liu, D. (eds.) WAIM 2016. LNCS, vol. 9659, pp. 369–379. Springer, Cham (2016). https://doi.org/10.1007/978-3-319-39958-4_29
26. Shannon, C.E.: A mathematical theory of communication. Bell Syst. Tech. J. **27**(3), 379–423 (1948)
27. Wang, J., Zucker, J.: Solving multiple-instance problem: A lazy learning approach, pp. 1119–1126 (2000)
28. Zhang, Y., Zhou, Z.: Multi-label dimensionality reduction via dependence maximization, pp. 1503–1505 (2008)

Efficiently Evolutionary Computation on the Weak Structural Imbalance of Large Scale Signed Networks

Weijin Jiang[1,2], Yirong Jiang[3]([⊠]), Jiahui Chen[1], Yang Wang[1], and Yuhui Xu[1]

[1] Institute of Big Data and Internet Innovation, Mobile E-Business Collaborative Innovation Center of Hunan Province, Hunan University of Technology and Business, Changsha 410205, China
jlwxjh@163.com, 810663304@qq.com, 18508488203@163.com, 363168449@qq.com
[2] School of Computer Science and Technology, Wuhan University of Technology, Wuhan 430073, China
[3] Tonghua Normal University, Tonghua 134002, China
307553803@qq.com

Abstract. With The symbolic network adds the emotional information of the relationship, that is, the "+" and "−" information of the edge, which greatly enhances the modeling ability and has wide application in many fields. Weak unbalance is an important indicator to measure the network tension. This paper starts from the weak structural equilibrium theorem, and integrates the work of predecessors, and proposes the weak unbalanced algorithm EAWSB based on evolutionary algorithm. Experiments on the large symbolic networks Epinions, Slashdot and WikiElections show the effectiveness and efficiency of the proposed method. In EAWSB, this paper proposes a compression-based indirect representation method, which effectively reduces the size of the genotype space, thus making the algorithm search more complete and easier to get better solutions.

Keywords: Weak structural balance signed networks · Evolutionary algorithms incremental computation · Compressed representation

Network is a general model of many complex systems. It represents the things in the system with nodes and the relations between things with edges. Starting from the emotional attributes of the side, the network can be divided into symbolic networks [1–3] and unsigned networks. It is widely used in politics [4–6], society [7–9], biology [10], e-commerce [11], cyberspace [12], etc. applications.

Structural balance theory is the basic theory in symbolic networks. It was first proposed by Heider [13] from the perspective of social psychology in the 1940s. Cartwright and Harary [14] then redefined and expanded the theory in graph theory in the 1950s.

Arahona [15] pointed out that solving the structural imbalance problem is an NP-hard problem. Terzi et al. [16] proposed a spectral method for solving the imbalance.

© Springer Nature Singapore Pte Ltd. 2019
Y. Sun et al. (Eds.): ChineseCSCW 2019, CCIS 1042, pp. 554–566, 2019.
https://doi.org/10.1007/978-981-15-1377-0_43

Facchetti et al. [17] used the canonical transformation to give an efficient greedy algorithm for solving the imbalance. Chiang et al. [18] used the Katz metric to find the number of negative loops and used this to measure the imbalance of the symbol network. Sun et al. [19] proposed a dense-mother algorithm for solving structural imbalances by using the characteristics of evolutionary algorithm global optimization.

In the 1960s, Davis [20] improved the structural balance theory. He believed that "the enemy of the enemy is a friend" is not necessarily correct in many occasions. His theory is called weak structural balance theory. Leskovec et al. [21, 22] have shown through experiments that weak structural equilibrium is more common than structural equilibrium in a large number of actual symbolic networks. However, it is usually not feasible to simply extend the method of solving structural imbalances such as [16–19] to solve the weak structural imbalance. Earlier research on this problem was Doreian and Mrvar [23, 24]. In view of the good performance of evolutionary algorithms [26, 27] in solving many NP-hard problems, and also inspired by the literature [19], this paper proposes an evolutionary algorithm EAWSB for solving the weak imbalance of symbol networks. Experiments on large symbolic networks Epinions, Slashdot and WikiElections show that this method is effective and efficient.

1 Problem Definition

1.1 Structural Balance and Weak Structural Balance

A symbolic network can be defined as a graph $G(V, E, \sigma)$, where V and E are node sets and edge sets, respectively. The mapping $\sigma: E \rightarrow \{+, -\}$ defines the symbol properties of each edge. Figure 1 is the four basic paradigms of the symbolic network. In the case of structural equilibrium, (a) (b) is balanced, (c) (d) is unbalanced. In the case of weak structural equilibrium, (a) (b) (d) is balanced and (c) is unbalanced.

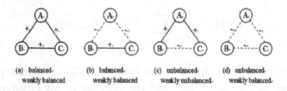

(a) balanced-
weakly balanced

(b) balanced-
weakly balanced

(c) unbalanced-
weakly unbalanced

(d) unbalanced-
weakly balanced

Fig. 1. Four basic paradigms of signed networks

For a complete graph of symbols with edges connected between any two nodes, the balance and weak balance can be measured by counting the occurrence of the above four paradigms. However, for the general symbolic network, the statistical method is no longer valid. At this time, its balance and weak balance are given by the following Theorem 1 [14] and Theorem 2 [20].

Theorem 1. A symbolic network is structurally balanced if and only if its node set can be divided into two classes and satisfy the following conditions: the edges in the same class are all positive, and the edges between different classes are all negative.

Theorem 2. A symbolic network is weakly structurally balanced if and only if its node set can be divided into multiple classes and satisfy the following conditions: the edges in the same class are all positive, and the edges between different classes are all negative.

1.2 Calculation of Structural Balance and Weak Structural Balance

By using Theorems 1 and 2, we can give another definition of (weak) unbalance. The nodes of a symbolic network are divided into several classes.

At present, the methods of seeking (weak) imbalance are mainly spectral methods [16], canonical transformation [17], Katz metric [18], evolutionary algorithm [19] and block model [23, 24]. The literature [19] is also based on evolutionary algorithms, but it is only solved and discussed in a relatively small scale symbolic network and structural equilibrium case.

2 The Algorithm EAWSB

Considering the complexity of large-scale symbolic networks and the global optimization of evolutionary algorithms, combined with Theorem 2, this paper proposes an EAWSB (Evolutionary Algorithms for Weak Structural Balance) algorithm for solving the weak imbalance of symbol networks. The details are as follows.

2.1 Energy Function and Fitness Function

According to Theorem 2, the energy function reflecting the weak imbalance can be defined as follows.

$$E(s) = \frac{1}{2} \sum_{(V_i,V_j) \in E} (1 - a_{ij}\delta(s_i, s_j)) \tag{1}$$

Where $\delta(s_i, s_j) = 1$, if $s_i = s_j$, otherwise take -1. E(S) is the sum of the number of all negative edges in the same class and the number of all positive edges between different classes. The minimum value is the weak imbalance of the symbol network G.

The algorithm for defining the EAWSB is:

$$F(s) = \sum_{(V_i,V_j) \in E} a_{ij}\delta(s_i, s_j) \tag{2}$$

Since E(S) = (m − F(S))/2, minimizing E(S) is equivalent to maximizing F(S). In this case, if the maximum number of categories k is specified in advance, the optimization problem to be solved by the algorithm EAWSB is transformed into

$$\max F(s) = \max_{\substack{s_i \in \{0,1,\dots,k-1\}, \\ i=1,2,\dots,n}} \sum_{(V_i,V_j) \in E} a_{ij}\delta(s_i, s_j) \tag{3}$$

2.2 The Natural Representation and Compression Representation of the Individual

The general individual representation is divided into direct representation and indirect representation [26, 27].

For large symbolic networks, the value of n is too large, often tens of thousands, which seriously affects the performance of genetic operations and the overall algorithm.

Theorem 3. Given the symbolic network G (V, E, _), suppose that A is the optimal solution of the optimization problem represented by (3), then for any I {1, ... N}, there are all

$$s_i^* = \arg\max_{s_i \in \{0,1,\dots,k-1\}} \sum_{v_j \in N(v_i)} a_{ij}\delta(s_i, s_j) \tag{4}$$

The set $N(v_i) = \{v_k | (v_i, v_k) \in E\}$ is the neighborhood of v_i. It is proved that if suppose the existence of $h \in (1,\dots,n)$ does not satisfy the condition in the theorem, that is, $s_h^* \neq \arg\max_{s_h \in \{0,\dots,k-1\}} \sum_{v_j \in N(v_h)} a_{hj}\delta(s_h, s_j)..$

Because $\displaystyle\sum_{(v_i,v_j) \in E} a_{ij}\delta(s_i, s_j) = \sum_{(v_h,v_j) \in E} a_{hj}\delta(s_h, s_j) + \sum_{(v_i,v_j) \in E \wedge i \neq h} a_{ij}\delta(s_i, s_j)$

$$= \sum_{v_j \in N(v_h)} a_{hj}\delta(s_h, s_j) + \sum_{(v_i,v_j) \in E \wedge i \neq h} a_{ij}\delta(s_i, s_j)$$

Note that h only appears in the first summation in the above formula, and does not appear in the second summation, so we can define $s_h^\# \neq \arg\max_{s_h \in \{0,\dots,k-1\}} \sum_{v_j \in N(v_h)} a_{hj}\delta(s_h, s_j)$,

$s_j^\# = s_j^*$ for $j \neq h, j \in \{1,\dots,n\}$. This gives us a better solution than S*. Because

$$\sum_{(v_i,v_j) \in E} a_{ij}\delta(s_i^\#, s_j^\#) = \sum_{v_j \in N(v_h)} a_{hj}\delta(s_h^\#, s_j^\#) + \sum_{(v_i,v_j) \in E \wedge i \neq h} a_{ij}\delta(s_i^\#, s_j^\#)$$

$$> \sum_{v_j \in N(v_h)} a_{hj}\delta(s_h^*, s_j^*) + \sum_{(v_i,v_j) \in E \wedge i \neq h} a_{ij}\delta(s_i^\#, s_j^\#)$$

$$= \sum_{v_j \in N(v_h)} a_{hj}\delta(s_h^*, s_j^*) + \sum_{(v_i,v_j) \in E \wedge i \neq h} a_{ij}\delta(s_i^*, s_j^*)$$

$$= \sum_{(v_i,v_j) \in E} a_{ij}\delta(s_i^*, s_j^*)$$

Theorem 3 tells us the state of a node, that is, the optimal class value to which it belongs can be found by the state of its neighbor node using Eq. (4). So how do you find the dominating set U? Algorithm 1 gives a solution.

Algorithm 1. A compressed representation

1.Input: Adjacency matrix of symbol network G: $A = (a_{ij})_{n \times n}$
2. Calculate node degree array deg[0..n-1]
3. ori_deg[0..n-1]= deg[0.. n-1]
4. The value of the initialization array selNode[0..n-1] is 0.
5: for each i with ori_deg[i]=1 and deg[i]>0 do
6:      j=the neighbor of i
7:      selNode[j]=1, deg[j]=0
8:      for each j's neighbor p do
9:          if(deg[p]>0)deg[p]= deg[p]-1
10:    end for
11: endfor
12: repeat
13. select a node with degree > 0 in roulette mode Randomly
14:     selNode[j]=1, deg[j]=0
15:     for each j's neighbor p do
16:         if(deg[p]>0)deg[p]= deg[p]-1
17:     endfor
18: until deg[i]=0 for alli∈{0.. n-1}
19: Output: All nodes i satisfying selNode[i]=1

Algorithm 1 consists of 3 parts. Part 1 (lines 2–4) defines three arrays, ori_deg and deg, which hold the degree information of nodes exactly the same at the beginning. Part 2 (lines 5–11) handles leaf nodes (i.e., nodes with degree 1). The leaf node has a unique neighbor node. Part 3 (lines 12–18) uses a degree ratio selection strategy to select a node, i.e. the probability that a node is selected is the sum of the degrees of a node divided by the degrees of all nodes.

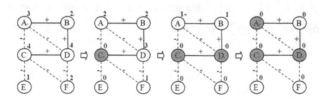

Fig. 2. Illustration of Algorithm 1 (The number in the upper right corner of vertex i indicatesdeg[i])

Figure 2 is an example of Algorithm 1 generating compression coding. E is a leaf node that is generally not selected, but its neighbor nodes must be elected to the dominating set. The last three nodes A, C, and D are selected to dominate the set U. The individual compression code is ind_c = sAsCsD, the natural code is ind = sAsBsCsDsEsF, and the compression ratio is 50%.

2.3 Population Initialization

The theory of homogeneity tells us that we will become more similar to our friends. The above selection-assignment process is repeated iniK times. Where iniK is a positive integer representing the initialization strength. The time complexity of population initialization is O(iniK*davg).

Genetic Operator. (1) cross This paper uses the one-way crossover operator proposed by Tasgin et al. [28]. The main idea is as follows. Find all the nodes in ind1 whose category value is s, change the category values of these nodes to s in ind2, and return the modified ind2. (2) variation. In this paper, a single point mutation is used to randomly select a node on the individual to be mutated and assign it a new category value. The time complexity of the mutation is O(1). (3) Choice. This paper adopts the league selection of league size 2 [26, 27], and adopts the elite retention strategy [26, 27]. The time complexity of the selection is O(1). (4) Rotation. The value of each gene of each individual is {0. In the evolutionary process, each individual rotates with a small probability (generally 0.05 in this paper), i.e. the class value is $0 \rightarrow , 1 \rightarrow 2, \ldots, k - 1 \rightarrow 0$.

Local Search. Starting from Theorem 3, the local search can be designed as follows: For a given individual ind, a node vi on it is randomly selected, and the state of the node is modified.

$$s_i = \underset{s_i \in \{0,1,\ldots,k-1\}}{\arg \max} \sum_{v_j \in N(v_j)} a_{ij} \delta(s_i, s_j)$$

Incremental Calculation of Fitness Values. Equation (2) can be used to directly calculate the individual's fitness value, but for large networks, the amount of calculation is large because the length of the individual is the number of network nodes.

Algorithm 2 Incremental calculation of fitness values after individual variation

1. Input: Current individual: ind, mutation position: h, the value of the hth gene sh before mutation: clsOld, the value of sh after mutation: clsNew
2. delta = 0 //The fitness value increment is initialized to 0
3. For each neighbor j of vertex at h
4. if(J is in the dominating set)
5. clsNbr = class label at position j in ind
6. if (clsNbr = clsOld) delta = delta - 2*a_{hj}
7. else if(clsNbr = clsNew) delta = delta + 2*a_{hj}
8.   }
9.   else if (j in the degradation concentration){
10.   maxEnergyNew =Before mutation maxEnergy(v_j)
11.   maxEnergyOld=After mutation maxEnergy(v_j)
12.   delta = maxEnergyNew - maxEnergyOld
13. }
14. Endfor
15. Output: Adaptation value of ind before mutation + delta
$$\max Energy(v_h) = \max_{s_h \in \{0,...k-1\}} \sum_{v_j \in N(v_h)} a_{hj}\delta(s_h, s_j)$$

EAWSB Algorithm Framework. Algorithm 3 is the overall framework of the EAWSB algorithm.

Algorithm 3 EAWSB algorithm framework

1. Input: G(V, E) adjacency matrix: A=(aij)n×n, population size: popSize, initialization strength: iniK, local search strength: locK, maximum evolution algebra: maxGen, league size: tourSize, Crossover probability: pc, probability of variation: pm, rotation probability: pr
2. Use natural representation or generate a compressed representation (A)
3. P← population initialization (popSize, iniK)
4. repeat
5. Pparent← select (P, tourSize)
6. Pchild ← cross (Pparent,p_c
7. Pchild ← variation (Pchild, p_m)
8. Pchild ← rotation (Pchild, p_r)
9. P← local search (Pchild, locK)
10. until the evolution termination condition (maxGen) is met
11. Output: the best individual in P

3 Experiments Analysis

3.1 EAWSB Algorithm Composition

To be exact, EAWSB is a cluster of algorithms, which consists of EAWSB_N, EAWSB_I, EAWSB_C and EAWSB_IC. They have the same function, but have different performance in different occasions. The difference between them is shown in Table 1.

Table 1. Four constituents of eawsb

	EAWSB_N	EAWSB_I	EAWSB_C	EAWSB_IC
Incremental calculation	×	√	×	√
Compressed representation	×	×	√	√

3.2 Experimental Environment

Table 2 is the Experimental Environment of Algorithm EAWSB in this paper.

Table 2. Experimental environment of eawsb

Hardware environment	Lenovo laptop savior e520,Quad-core processor, logical eight core, 16G memory
Operating system	Microsoft Windows [version 10.0.15063]
Development environment	java version "1.7.0_15" Java(TM) SE Runtime Environment (build 1.7.0_15-b03)

3.3 Data Set

This article was conducted on three large symbolic network datasets, Epinions, Slashdot, and WikiElections. Epinions (epinions.com) is a product review website [8]. Slashdot (slashdot.com) is a technology news site [29] that allows users to mark authors as "friends" or "enemies" for other users' articles, forming a network of friends/enemies. WikiElections [21] is a dataset for Wikipedia users voting for elections. It is a support or objection network. Table 3 is the original case of the three data sets. The experiment is mainly carried out on the large undirected symbolic network shown in Table 4.

Table 3. Original datasets

Raw data set	Number of nodes	Number of sides	Description
soc-sign-epini ons	131,828	841,372	Epinions Symbolic network
soc-sign-Slash dot090221	82,144	549,202	Slashdot Zoo Symbolic network February 21, 2009, Snapshot
wiki-Elec	8,297	103,591	Wikipedia Administrator election symbol network

Table 4. Preprocessed datasets

Experimental data set	Number of nodes	Number of sides
Epinions	131,513	708,507
Slashdot	82,062	498,532
WikiElections	7,114	99892

3.4   Operation Results and Running Time

Parameter Setting. For all algorithms, all data sets, EAWSB parameter settings. The results are as follows: population size popSize = 500, initialization strength iniK = 500, local search strength locK = 500, maxGen = 500, tourSize = 2, crossover probability PC = 0.8, mutation probability PM = 0.1, rotation probability PR = 0.05.

Operation Results. Figures 3, 4 and 5 shows the results of the four algorithms EAWSB_N, EAWSB_I, EAWSB_C, and EAWSB_IC on Epinions, Slashdot, and WikiElections. The operation is performed in five cases according to the number of categories k = 2, 3, 4, 5, and 6, where k = 2 is a structural equilibrium situation, which can be regarded as a special case of weak structural balance.

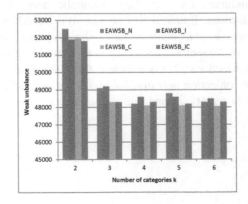

Fig. 3. Results of the four algorithms on epinions

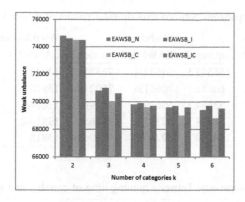

Fig. 4. Results of the four algorithms on slashdot

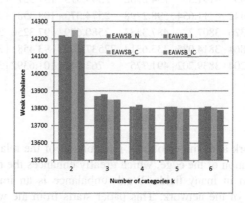

Fig. 5. Results of the four algorithms on WikiElections

3.5 Performance Comparison with Similar Algorithms

Meme-sb [19] is a structural unbalanced algorithm based on the timid algorithm. Tables 5 and 6 show the experimental results and running time of EAWSB_I and meme-sb u shows under three large-scale symbolic network datasets in the number of categories k = 2~6. Experiments show that EAWSB_I is significantly better than meme-sb on the two large datasets of Epinions and Slashdot. In addition, meme-sb is slightly better than EAWSB_I on WikiElections. Because meme-sb has a large "tearing" negative impact, the one-way crossover used by EAWSB_I is easier to maintain the integrity of the building block than the 2-point crossover used by meme-sb.

Table 5. Comparisons between experimental results of eawsb_i and meme-sb

	Epinions		Slashdot		WikiElections	
k	EAWSB_I	meme-sb	EAWSB_I	meme-sb	EAWSB_I	meme-sb
2	51867	56544.5	74634.4	76334.33	14220.6	14204.67
3	49288.8	60851	70661.6	75021.67	13870.8	13858
4	48637.8	58854.33	69699.2	75299.67	13824.6	13809
5	48625	58628	69382.4	75465.67	13814.2	13793
6	48535.6	56799	69260.2	77196.67	13815	13805.33

Table 6. Comparisons between running time of eawsb_i and meme_sb (s)

	Epinions		Slashdot		Wikielections	
k	eawsb_i	meme-sb	eawsb_i	meme-sb	eawsb_i	meme-sb
2	1282.295	4191.39	754.2902	2549.973	131.527	306.6153
3	996.9442	4211.654	598.2798	2574.373	108.358	302.8893
4	811.7326	3807.2	487.3583	2580.679	97.3256	305.7383
5	706.6864	3814.085	433.6314	22537.29	93.458	344.1983
6	681.7204	3839.202	491.775	763.142	101.6392	342.976

4 Conclusion

The symbolic network adds the emotional information of the relationship, that is, the "+" and "−" information of the edge, which greatly enhances the modeling ability and has wide application in many fields. Weak unbalance is an important indicator to measure the tension of the network. This paper starts from the weak structural equilibrium theorem, and integrates the work of predecessors, and proposes the weak unbalanced algorithm EAWSB based on evolutionary algorithm. Experiments on large symbolic networks Epinions, Slashdot, and WikiElections demonstrate the effectiveness and efficiency of this approach. In EAWSB, this paper proposes a compression-based individual indirect representation method, which effectively reduces the size of the genotype space, thus making the algorithm search more complete and easier to get a better solution. In this paper, an incremental fitness calculation method is proposed, which reduces the time complexity of fitness calculation from O (n) to O (davg), and greatly improves the efficiency of the algorithm. In order to maintain the diversity of the population and avoid premature convergence, this paper proposes a round-scale converter that can improve the diversity of the population without affecting the individual's fitness value. Compared with the similar algorithms in meme-sb and the famous network analysis software Pajek on the above three large data sets, the algorithm is not only more efficient but also has a better solution. In the future, we will continue the work of this article from two directions. First, study the weak imbalance of symbolic networks in distributed computing environment, upgrade EAWSB to distributed evolutionary algorithm version. Second, in the field of symbolic network, there are only two emotional attributes of positive and negative sides in the current symbolic network. We want to study the multi-symbolic network, that is, each edge may have more than two emotional tags except "+" and "−". Such symbolic networks are more common in real life, and the corresponding algorithms are more practical.

References

1. Easley, D., Kleinberg, J.: Networks, Crowds, and Markets: Reasoning About a Highly Connected World, pp. 119–152. Cambridge University Press, New York (2010)
2. Lan, M., Li, C., et al.: Survey of sign prediction algorithms in signed social networks. J. Comput. Res. Dev. **52**(02), 410–422 (2015)
3. Zheng, X., Zeng, D., Wang, F.Y.: Social balance in signed networks. Inf. Syst. Front. **17**(5), 1077–1095 (2015)
4. Harary, F.: A structural analysis of the situation in the Middle Eastin 1956. J. Conflict Resolut. **5**, 167–178 (1961)
5. Moore, M.: An international application of Heider's balance theory. Eur. J. Soc. Psychol. **8**, 401–405 (1978)
6. Ghosn, F., Palmer, G., Bremier, S.A.: The MID3 data set 1993-2001: procedures, coding rules, and description. Confl. Manag. Peace Sci. **21**(2), 133–154 (2004)
7. Wasserman, S., Faust, K.: Social Networks Analysis: Methods and Applications. Cambridge University Press, Cambridge (1994)
8. Guha, R., Kumar, R., Raghavan, P., et al.: Propagation of trust and distrust. In: International Conference on World Wide Web, pp. 403–412 (2004)
9. Kunegis, J., Preusse, J., Schwagereit, F.: What is the added value of negative links in online social networks. In: Proceedings of the 22nd International Conference on World Wide Web, pp. 727–736 (2013)
10. Parisien, C., Anderson, C.H., Eliasmith, C.: Solving the problem of negative synaptic weights in cortical models. Neural Comput. **20**(6), 1473–1494 (2008)
11. Zolfaghar, K., Aghaie, A.: Mining trust and distrust relationships in social Web applications. In: IEEE International Conference on Intelligent Computer Communication and Processing, pp. 73–80. IEEE (2010)
12. Burke, M., Kraut, R.: Mopping up: modeling wikipedia promotion decisions. In: ACM Conference on Computer Supported Cooperative Work, pp. 27–36. ACM (2008)
13. Heider, F.: Attitudes and cognitive organization. J. Psychol. **21**(1), 107–112 (1946)
14. Cartwright, D., Harary, F.: Structural balance: a generalization of Heider's theory. Soc. Netw. **63**(5), 277–293 (1956)
15. Barahona, F.: On the computational complexity of Ising spin glass models. J. Phys. A Gen. Phys. **15**(10), 3241 (1999)
16. Terzi, E., Winkler, M.: A spectral algorithm for computing social balance. In: Frieze, A., Horn, P., Prałat, P. (eds.) WAW 2011. LNCS, vol. 6732, pp. 1–13. Springer, Heidelberg (2011). https://doi.org/10.1007/978-3-642-21286-4_1
17. Facchetti, G., Iacono, G., Altafini, C.: Computing global structural balance in large-scale signed social networks. Proc. Natl. Acad. Sci. U.S.A. **108**(52), 20953–20958 (2011)
18. Chiang, K.Y., Hsieh, C.J., Natarajan, N., et al.: Prediction and clustering in signed networks: a local to global perspective. J. Mach. Learn. Res. **15**(1), 1177–1213 (2013)
19. Sun, Y., Du, H., Gong, M., et al.: Fast computing global structural balance in signed networks based on memetic algorithm. Phys. A Stat. Mech. Appl. **415**(415), 261–272 (2014)
20. Davis, J.A.: Clustering and structural balance in graphs. Soc. Netw. **20**(2), 27–33 (1977)

21. Leskovec, J., Huttenlocher, D., Kleinberg, J.: Signed networks in social media. In: Sigchi Conference on Human Factors in Computing Systems, pp. 1361–1370. ACM (2010)
22. Leskovec, J., Huttenlocher, D., Kleinberg, J.: Predicting positive and negative links in online social networks. In: International Conference on World Wide Web, pp. 641–65. ACM (2010)
23. Doreian, P., Mrvar, A.: A partitioning approach to structural balance. Soc. Netw. **18**(2), 149–168 (1996)
24. Doreian, P., Mrvar, A.: Partitioning signed social networks. Soc. Netw. **31**(1), 1–11 (2009)
25. De Jong, K.A.: Evolutionary Computation: A Unified Approach. MIT Press, Cambridge (2016)
26. Li, M., Kou, J., Lin, D., et al.: Genetic Algorithms, Theory and Applications. Science Press, Beijing (2002)
27. Tasgin, M., Herdagdelen, A., Bingol, H.: Community detection in complex networks using genetic algorithms. Corr **2005**(3120), 1067–1068 (2006)
28. Kunegis, J., Lommatzsch, A., Bauckhage, C.: The slashdot zoo: mining a social network with negative edges. In: Proceedings of the International World Wide Web Conference, pp. 741–750 (2009)
29. Milo, R., Shen-Orr, S., Itzkovitz, S., et al.: Network motifs: simple building blocks of complex networks. Science **298**(5594), 824–827 (2002)

Text Generation from Triple
via Generative Adversarial Nets

Xiangyan Chen, Dazhen Lin[✉], and Donglin Cao

Cognitive Science Department, Xiamen University, Xiamen, China
xychen@stu.xmu.edu.cn, {dzlin,another}@xmu.edu.cn

Abstract. Text generation plays an influential role in NLP (Natural Language Processing), but this task is still challenging. In this paper, we focus on generating text from a triple (entity, relation, entity), and we propose a new sequence to sequence model via GAN (Generative Adversarial Networks) rather than MLE (Maximum Likelihood Estimate) to avoid exposure bias. In this model, the generator is a Transformer and the discriminator is a Transformer based binary classifier, both of which use encoder-decoder structure. With regard to generator, the input sequence of encoder is a triple, then the decoder generates sentence in sequence. The input of discriminator consists of a triple and its corresponding sentence, and the output denotes the probability of being real sample. In this experiment, we use different metrics including Bleu score, Rouge-L and Perplexity to evaluate similarity, sufficiency and fluency of the text generated by three models on test set. The experimental results prove our model has achieved the best performance.

Keywords: Generative adversarial nets · Text generation · Triple

1 Introduction

Text generation is an important branch of NLP. However, text generation with non-input is challenging due to the information asymmetry. Inspired by this, we consider incorporating more information into the text generation. One method is to take the fact of the knowledge graph as input to our generation task.

Knowledge graph has been applied to various fields and drawn much attention of scholars in recent years. The fact of knowledge graph is described as triple composed by two entities and a relation. For example, Freebase [1] is a famous knowledge graph containing a number of facts.

We regard our generation task as text to text generation, such as machine translation [2–5], abstractive text summarization [6] and question answering [7], which has attracted the attention of many scholars. Compared to other natural language generation task, there are two research significance for this paper. Firstly, the model we proposed can be combined with other knowledge based generation task, for instance, it allows task like question answering considering triple to improve the quality of answer. Secondly, extracting numerous specified

© Springer Nature Singapore Pte Ltd. 2019
Y. Sun et al. (Eds.): ChineseCSCW 2019, CCIS 1042, pp. 567–578, 2019.
https://doi.org/10.1007/978-981-15-1377-0_44

triples from the knowledge graph for text generation can solve the problem of insufficient text corpus for specific requirements, which lays the root for following studies.

For this task, the text generation is accomplished by a sequence to sequence model [2]. Sequence to sequence model is always trained based on MLE, which suffers from exposure bias [8]. Hence, GAN [9] was selected to train our model. However, native generator in adversarial network updates parameters from the gradient of discriminator where the output of generator must be continuous value. It is hard to applied to natural language generation directly due to the discrete representation of words.

In this paper, gradient policy method of reinforcement learning is used to update the generator. The generator of our model is a standard Transformer [5] model. The input sequence of encoder is a triple and decoder generates tokens one after another. Discriminator is also an encoder-decoder architecture, unlike generator, the output of that is a probability of being true.

In the experiment, we compared our model to baselines using automatic metrics including Bleu score [10], Rouge-L [11] and Perplexity. Results show that the model we proposed achieve the best performance.

The contribution of this paper can be listed as follows:

1. We proposed a text generation model from triple. The model can be combined with other text generation task to enhance the quality of generated text and expands the text corpus under special requirements.
2. For generating text from triple, we proposed a Transformer trained by GAN. The experimental results declare that our model has achieved the best result.

2   Related Work

Knowledge graph is proposed by Google in 2012 and widely applied in web search, medical research and so on. The construction of knowledge graph can be divided to three parts incorporating information acquisition like NER [12] (Named Entity Recognition) and relation extraction [13], knowledge fusion and knowledge reasoning [14].

Bengio et al. [15] firstly proposed NNLM (Neural Network Language model). But using this model to generate text is hard to capture the dependency of words. To address the problem, Mikolov et al. [16] proposed the RNN (Recurrent Neural Network) [17] based model RNNLM (Recurrent Neural Network Language Model), which is used to learn long term dependency successfully while RNNLM is hard to map sequence to sequence.

Hence, Sutskever et al. [2] proposed an encoder-decoder based model to map input sequence to output sequence. In this model, encoder encodes input sequence to a fixed-sized vector and decoder decodes the vector as a textual sequence. However, the drawback of this model is that fixed-sized vector is difficult to deliver enough information for decoding.

To solve the problem above, attention mechanism [3,4] is applied in sequence to sequence model in machine translation. In this work, every step of decoding creates a context vector instead of fixed-length vector and it makes decoder to obtain more information to decode.

Transformer [5] proposed by Google is an encoder-decoder structure language model. Transformer uses the self-attention mechanism and the positional encoding to learn long range dependency instead of RNN architecture. It is widely used in some sequence modeling tasks like machine translation and achieves impressive success. Experimental results show that this method is far better than RNN based language models.

Google proposed a powerful pre-training model called BERT [18] (Bidirectional Encoder Representations from Transformers) based on MLM (Masked Language Model) and NSP (Next Sentence Predication). BERT has the ability of acquiring context information through cross-sentence modeling and has achieved state-of-the-art in 11 fields such as text classification, reading comprehension and question answering. The one of successful reason can be attributed to Transformer.

The development mentioned above laid a solid foundation for text to text generation. However, such language models usually use the method of MLE, which easily suffers from exposure bias, that is, in the process of training, words are trained using real corpus. In the inference, the generated words are based on previously generated words which may be inaccuracy, so this will lead to inconsistency between the two processes. One possible solution is to make use of the combination of GAN and reinforcement learning for training.

Goodfellow et al. [9] proposed the Minimax based generative adversarial network framework. In this framework, generator is responsible for generating text and discriminator discriminator outputs the sentences probability of being true and feedback the probability to generator. Through the adversarial process, the Nash balance is achieved. Then the discriminator may output a probability close to 0.5, namely the discriminator can not determine whether the sentence is from corpus or generated by our model, which means our model succeed in "cheating" the discriminator. Native GAN is usually applied in computer vision since the output of generator is continuous value which can be updated by gradient. By contrary, words are treated as discrete values. The updated of generated words can not be accomplished by gradient of discriminator.

Proposing SeqGAN, Yu et al. [19] obtain parameters of generator by applying the method of reinforcement learning, which make GAN available in NLP tasks. In their model, incomplete sentences can be complete by Monte Carlo Tree Search. Then the reward of each sequence of state is obtained and the generator is updated.

Much work has been done based on SeqGAN. Guo el al. [20] proposed Leak-GAN to obtain more guide information from discriminator. In their model, the interior feature of discriminator deliver to generator, which enhances the effects of long sentence generating. TreeGAN [21] is tree-form LSTM (Long Short-Term Memory)[17], which incorporate context-free grammar into generator to obtain sentence confirm to syntax. Instead of using the feedback probability of discriminator as reward of state sequence, DP-GAN [22] map the probability of discriminant by using cross entropy, to promoting diversity of generated sentence. Wang et al. [23], aiming to generating emotional sentence, propose a framework with multi-generator and multi-class discriminator. They use an objective function based on penalty to organize sentimental information into model. Park et al. [24] proposed a generative model combining standard sequence to sequence model with GAN from keyword.

3   Model

Our model contains a generator and a discriminator.We define C as vocabulary and define input sequence as $X = (x_1, x_2, x_3, ..., x_M)$, where $x \in C$. Then our generator will map input sequence to output sequence $Y = (y_1, y_2, y_3, ..., y_N)$, $y \in C$, where M and N are the length of input sequence and output sequence respectively.

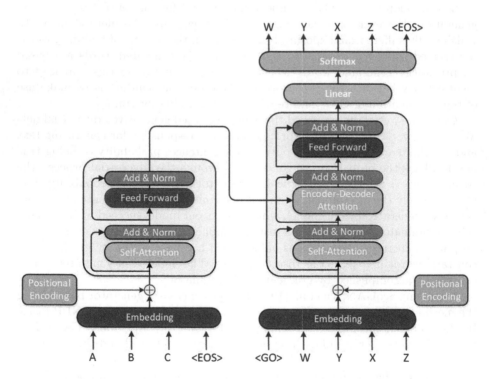

Fig. 1. The structure of our generator. Our generator consists of multiple encoder layers and decoder layers. The input (A, B, C, <EOS>) is a triple and decoder generates tokens one by one.

3.1   Generator

Generator is a Transformer language model in encoder-decoder structure. Self-attention mechanism and positional encoding are used to learn long-range dependencies instead of RNN architecture. Meanwhile, attention mechanism is also applied to encoder and decoder and it makes decoder have enough information to decode. We regard the generating process as sequential decision making processes [25]. In our generator, the input sequence X is a triple and $G_\theta(s_n|s_{n-1}, X)$ outputs the probability of next state s_n given X and the s_{n-1} where s_{n-1} denotes the generated sequence $y_{1:n-1}$. After encoding is finished, the first input for decoder is the start token <GO>, then the token generated will be the next input until the end token <EOS> is obtained. $V_{D_\varphi}^{G_\theta}(s_n, X)$ is the reward calculated by discriminator. The objective function of generator is to maximize the reward of generated sentence.

$$J(\theta) = E_{s_n \sim G_\theta} \left[-\log\left(G_\theta(s_n|s_{n-1}, X)\right) V_{D_\varphi}^{G_\theta}(s_n, X) \right] \tag{1}$$

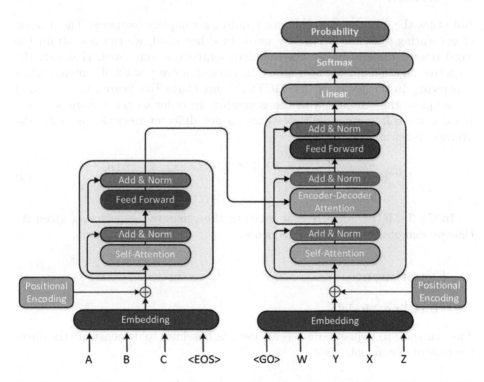

Fig. 2. The architecture of discriminator. The input for encoder is a triple (A, B, C, <EOS>) and for decoder is generated text (<GO>, W, Y, X, Z). The output probability measures the reality of the sentence.

3.2   Discriminator

The discriminator has similar structure as Transformer based architecture, but differ in that the input of discriminator is a triple for encoder and generated text for decoder. The output $D_\varphi(Y, X)$ is value denoting the sentence's probability of being true. Y denotes the generated text and X is the triple. We use this discriminator for the following reasons. First of all, self-attention mechanism is powerful to process text which prompts the discriminant ability. Besides, the parallel computing of Transformer reduces time spent for training. Lastly, our model takes input sequences into account so that the output and input sequences are closely related.

Discriminator aims at being able to distinguish between real and fake samples. The objective function is as followed:

$$\min_{\varphi} - \mathbb{E}_{Y \sim p_{data}} [\log D_\varphi (Y, X)] - \mathbb{E}_{Y \sim G_\theta} [\log (1 - D_\varphi (Y, X))] \tag{2}$$

3.3   Reward

Note that the input of discriminator requires a complete sequence. The process of generating sentence is word by word. In other word, we can not attain the word reward directly before the complete sentence is generated. However, the objective function of generator needs the reward of every word of sentence when generating. In this paper, we use MCTS (Monte Carlo Tree Search Tree) method to complete the sentence with the generator. In order to get a more accurate reward, we will sample multiple times to get different rewards and take the average as reward.

$$V_{D_\varphi}^{G_\theta}(s_n, X) = \begin{cases} D_\varphi (Y, X), Y \in MCTS(s_n, X) & , n < N \\ D_\varphi (Y, X) & , n = N \end{cases} \tag{3}$$

In Eq. 3, MCTS adds $|N - n|$ words to the generated sequence s_n given X, thus we can obtain a fixed size sequence.

4   Experiment

4.1   Baseline Model

Two sequence to sequence models are used as baselines to demonstrate the effectiveness of our model.

Seq2seq-Attention. Seq2seq-Attention is a standard sequence to sequence model with LuongAttention [4]. This model is trained by MLE.

Transformer. A standard Transformer trained with by MLE.

Transformer-GAN. The model we proposed. Generator is a standard Transformer language model and discriminator is a binary classifier. We train the model via GAN.

4.2 Training Details

Every word needs to be embedded into vector space, we used skip-gram [26] model to pre-train the word embedding. The dimension of word embedding is 512 for three models. Seq2seq-Attention has 2 layers of LSTM with residual networks [27], and the hidden size of each layer is 1024. Moreover, six encoder layers and six decoder layers are applied to transformer.

Furthermore, pre-training of generator and discriminator is necessary in GAN. In our model, The generator is pre-trained by MLE, and the discriminator is pre-trained with a cross entropy loss function.

4.3 Evaluation Metrics

Bleu. Bleu [10] score is automatic metrics for generative model. It compared the count of n-gram between generated sentence and reference sentence. Bleu score has proven to be effective for measuring the similarity between generated sentence and reference sentences.

$$BLEU = BP \left(\sum_n^N w_n \log(p_n) \right) \tag{4}$$

The calculating of Bleu is shown as Eq. 4, where BP is a penalty term. When the length of generated sentence is less than the reference sentence, it will be multiplied by a penalty value to reduce the Bleu score because the short sentences are more likely to get higher score than long sentence in the equation.

Rouge-L. Rouge [11] is a series of evaluation metrics including Rouge-N, Rouge-L, Rouge-W and Rouge-S. Rouge-L is an F metric based on LCS (Longest Common Subsequence) of generated sentence and reference sentence. These metrics embodies the sufficiency of sentence.

Perplexity. Perplexity can well reflect the fluency of generated text. In Eq. 5, p denotes the probabilities of w_i given the generated sequence $(w_1 w_2 ... w_{i-1})$, which is in an inverse ratio to perplexity. Therefore, the lower perplexity denotes higher fluency of a sentence.

$$PP(S) = \sqrt[N]{\prod_{i=1}^N \frac{1}{p(w_i | w_1 w_2 ... w_{i-1})}} \tag{5}$$

4.4 Dataset

The data used in our experiment is extracted from Sina Weibo and filtered from noise such as titles, emoticons and some other meaningless tokens. Then we extract triples of sentences based on dependency parsing. The dataset has a total of 248584 examples. Length of example is less than 20 because the longer sentence will increase information asymmetry. Then we randomly selected 90% data as train set and 10% as test set. In addition, we used the Bpe [28] to makeup vocabulary with a size of 41803 to address the rare word problem.

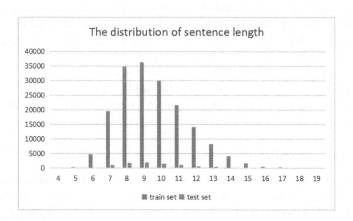

Fig. 3. This picture show that the distribution of sentence length of train set and test set.

4.5 Analysis

We trained our model on train set of Sina Weibo dataset mentioned above. The triple of test set is used as input sequence, and the models generates the sentences. The generated sentences are used to evaluate the effectiveness of three models with Bleu, Rouge-L and Perplexity respectively.

Table 1. Bleu score of test set

Models	Bleu
Seq2seq-Attention	18.66
Transformer	24.54
Transformer-GAN	**25.35**

Referring to Table 1, we can see that the Bleu score of Transformer trained with MLE is far higher than Seq2seq-Attention, which declares that Transformer is useful for our task.

What is notable is that our model achieves a Bleu score of 25.35, while other two models achieve Bleu score of 18.66 and 24.54, which means the sentences generated by our model are more similar with reference sentence. We own the improvement to different choice of training method.

Then, the three models were evaluated with Rouge-L on nlg-eval [29] platform. Table 2 shows that our model achieves the highest score than other baseline models. This result indicates the model we proposed is more sufficient than other two models.

Table 2. Rouge-L of test set

Models	Rouge-L
Seq2seq-Attention	0.4980
Transformer	0.5575
Transformer-GAN	**0.5664**

Besides, we want to measure the fluency of the text generated by three models. We computed the perplexity using SRILM [30] tool with generated text from three models and ground truth. The result is shown as Fig. 4. The perplexity of our model is closest to the real sample. Therefore, the text generated Transformer-GAN is the most fluent.

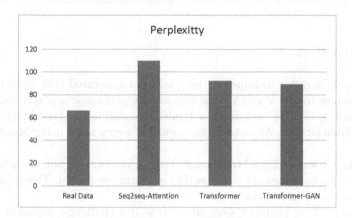

Fig. 4. The perplexity of real data and text generated by three models. The smaller perplexity means higher fluency.

Measured by the 3 metrics, our model outperforms the other two models. Results prove that our model improves the generating ability.

In the end of experiment, we show examples in Table 3 to illustrate our generation task. It is noticeable that the table is expressed in Chinese because our data set is Chinese, and we have translated it into English sentences. GBKsong

Table 3. The triple and reference sentence was selected from test set. The sentence was generated by Seq2seq-Attention, Transformer and Transformer-GAN.

Triple	1. (生活, 关, 一扇门) (life, close, a door) 2. (女人, 取悦, 自己) (women, please, oneself)
Real Data	1. 生活给我们关上了一扇门。 Life closes a door for us. 2. 取悦自己的女人必定倾倒世人。 Women who please themselves must be beautiful.
Seq2seq-Attention	1. 当生活关上一扇门的时候。 When life closes a door. 2. 女人要用自己的去取悦自己。 Women want to use their own to please themselves.
Transformer	1. 生活关上了一扇门也不会让你觉得幸福。 Closing a door in life will not make you feel happy. 2. 女人要取悦自己才能真正珍惜自己。 Women cherish themselves by pleasing themselves.
Transformer-GAN	1. 生活为你关上了一扇门。 Life closes a door for you. 2. 女人要取悦自己才能有更好的魅力。 Women who please themselves have a better charm.

5   Conclusion

In this paper, in order to improve the quality of generated text from triple, we proposed a new model via GAN. In this model, generator is a standard Transformer which generates text from a triple. Discriminator is a binary classifier based on Transformer. We obtain the reward of every word using Monte Search Tree.

Experiment was designed to evaluate the performance of our model on generating text. Two baselines are composed to evaluate our model. The experimental results show its effectiveness.

At present, long sentence generation is still a challenge, because when we only use one triple to generate sentence, the longer the sentence is, the larger the information asymmetry will be. In the future work, we will focus on how to better integrate the knowledge graph to get more information for text generation.

Acknowledgments. This work is supported by the National Key Research and Development Program of China (No. 2018YFC0831402), the Nature Science Foundation of China (No. 61402386, No. 61502105, No. 61572409, No. 81230087 and No. 61571188), Open Fund Project of Fujian Provincial Key Laboratory of Information Processing and Intelligent Control (Minjiang University) (No. MJUKF201743), Education and scientific research projects of young and middle-aged teachers in Fujian Province under Grand No. JA15075. Fujian Province 2011 Collaborative Innovation Center of TCM Health Management and Collaborative Innovation Center of Chinese Oolong Tea Industry-Collaborative Innovation Center (2011) of Fujian Province.

References

1. Bollacker, K., Evans, C., Paritosh, P., Sturge, T., Taylor, J.: Freebase: a collaboratively created graph database for structuring human knowledge. In: Proceedings of the 2008 ACM SIGMOD International Conference on Management of Data, pp. 1247–1250. ACM (2008)
2. Sutskever, I., Vinyals, O., Le, Q.V.: Sequence to sequence learning with neural networks. In: Advances in Neural Information Processing Systems, pp. 3104–3112 (2014)
3. Bahdanau, D., Cho, K., Bengio, Y.: Neural machine translation by jointly learning to align and translate. arXiv preprint arXiv:1409.0473 (2014)
4. Luong, M.T., Pham, H., Manning, C.D.: Effective approaches to attention-based neural machine translation. arXiv preprint arXiv:1508.04025 (2015)
5. Vaswani, A., et al.: Attention is all you need. In: Advances in Neural Information Processing Systems, pp. 5998–6008 (2017)
6. Nallapati, R., Zhou, B., Gulcehre, C., Xiang, B., et al.: Abstractive text summarization using sequence-to-sequence RNNs and beyond. arXiv preprint arXiv:1602.06023 (2016)
7. Reddy, S., Raghu, D., Khapra, M.M., Joshi, S.: Generating natural language question-answer pairs from a knowledge graph using a RNN based question generation model. In: Proceedings of the 15th Conference of the European Chapter of the Association for Computational Linguistics: Volume 1, Long Papers, pp. 376–385 (2017)
8. Bengio, S., Vinyals, O., Jaitly, N., Shazeer, N.: Scheduled sampling for sequence prediction with recurrent neural networks. In: Advances in Neural Information Processing Systems, pp. 1171–1179 (2015)
9. Goodfellow, I., et al.: Generative adversarial nets. In: Advances in Neural Information Processing Systems, pp. 2672–2680 (2014)
10. Papineni, K., Roukos, S., Ward, T., Zhu, W.J.: BLEU: a method for automatic evaluation of machine translation. In: Proceedings of the 40th Annual Meeting on Association for Computational Linguistics, pp. 311–318. Association for Computational Linguistics (2002)
11. Lin, C.Y.: Rouge: A package for automatic evaluation of summaries. Text Summarization Branches Out (2004)
12. Huang, Z., Xu, W., Yu, K.: Bidirectional LSTM-CRF models for sequence tagging. arXiv preprint arXiv:1508.01991 (2015)

13. Mintz, M., Bills, S., Snow, R., Jurafsky, D.: Distant supervision for relation extraction without labeled data. In: Proceedings of the Joint Conference of the 47th Annual Meeting of the ACL and the 4th International Joint Conference on Natural Language Processing of the AFNLP: Volume 2-Volume 2, pp. 1003–1011. Association for Computational Linguistics (2009)
14. Levesque, H.J.: Knowledge representation and reasoning. Annu. Rev. Comput. Sci. 1(1), 255–287 (1986)
15. Bengio, Y., Ducharme, R., Vincent, P., Jauvin, C.: A neural probabilistic language model. J. Mach. Learn. Res. 3(Feb), 1137–1155 (2003)
16. Mikolov, T., Karafiát, M., Burget, L., Černocký, J., Khudanpur, S.: Recurrent neural network based language model. In: Eleventh Annual Conference of the International Speech Communication Association (2010)
17. Hochreiter, S., Schmidhuber, J.: Long short-term memory. Neural Comput. 9(8), 1735–1780 (1997)
18. Devlin, J., Chang, M.W., Lee, K., Toutanova, K.: BERT: pre-training of deep bidirectional transformers for language understanding. arXiv preprint arXiv:1810.04805 (2018)
19. Yu, L., Zhang, W., Wang, J., Yu, Y.: SeqGAN: sequence generative adversarial nets with policy gradient. In: Thirty-First AAAI Conference on Artificial Intelligence (2017)
20. Guo, J., Lu, S., Cai, H., Zhang, W., Yu, Y., Wang, J.: Long text generation via adversarial training with leaked information. In: Thirty-Second AAAI Conference on Artificial Intelligence (2018)
21. Liu, X., Kong, X., Liu, L., Chiang, K.: TreeGAN: syntax-aware sequence generation with generative adversarial networks. In: 2018 IEEE International Conference on Data Mining (ICDM), pp. 1140–1145. IEEE (2018)
22. Xu, J., Ren, X., Lin, J., Sun, X.: Diversity-promoting GAN: a cross-entropy based generative adversarial network for diversified text generation. In: Proceedings of the 2018 Conference on Empirical Methods in Natural Language Processing, pp. 3940–3949 (2018)
23. Wang, K., Wan, X.: SentiGAN: generating sentimental texts via mixture adversarial networks. In: IJCAI, pp. 4446–4452 (2018)
24. Park, D., Ahn, C.W.: LSTM encoder-decoder with adversarial network for text generation from keyword. In: Qiao, J., et al. (eds.) BIC-TA 2018. CCIS, vol. 952, pp. 388–396. Springer, Singapore (2018). https://doi.org/10.1007/978-981-13-2829-9_35
25. Bachman, P., Precup, D.: Data generation as sequential decision making. In: Advances in Neural Information Processing Systems, pp. 3249–3257 (2015)
26. Mikolov, T., Chen, K., Corrado, G., Dean, J.: Efficient estimation of word representations in vector space. arXiv preprint arXiv:1301.3781 (2013)
27. He, K., Zhang, X., Ren, S., Sun, J.: Deep residual learning for image recognition. In: Proceedings of the IEEE Conference on Computer Vision and Pattern Recognition, pp. 770–778 (2016)
28. Sennrich, R., Haddow, B., Birch, A.: Neural machine translation of rare words with subword units. arXiv preprint arXiv:1508.07909 (2015)
29. Sharma, S., Asri, L.E., Schulz, H., Zumer, J.: Relevance of unsupervised metrics in task-oriented dialogue for evaluating natural language generation. arXiv preprint arXiv:1706.09799 (2017)
30. Stolcke, A.: SRILM-an extensible language modeling toolkit. In: Seventh International Conference on Spoken Language Processing (2002)

Measuring Node Similarity for the Collective Attention Flow Network

Manfu Ma[1,2], Zhangyun Gong[1,2], Yong Li[1,2(✉)], Huifang Li[1],
Qiang Zhang[1,2], Xiaokang Zhang[3], and Changqing Wang[4]

[1] College of Computer Science and Engineering, Northwest Normal University,
Lanzhou, China
facingworld@126.com
[2] Gansu IOT Research Center, Lanzhou 730000, China
[3] Lanzhou Qidu Data Technology Co., Ltd., Lanzhou 730070, China
[4] DNSLAB, China Internet Network Information Center, Beijing 100190, China

Abstract. Quantifying the similarity of nodes in collective attention flow network has an important theoretical and practical value. In this paper, we defined the generation time Rt, the influence radius Sr and the representation Vs (Rt, Sr) of the nodes in the collective attention flow network based on the optimization of Spatial Preferred Attachment (SPA) model. NID algorithm, based on the influence distance Sd that was calculated by the spatial norm, to measure the similarity of the nodes in the collective attention flow network was proposed. Experiments show that our algorithm not only accurately quantify the similarity of nodes in the collective attention flow network, but has a higher universality.

Keywords: Spatial Preferred Attachment (SPA) · Collective attention flow network · Node similarity algorithm

1 Introduction

In 1990s, the discovery of small-world networks [1] and scale-free networks [2] has set off an upsurge in the study of the CN (Complex Networks). Because the concept of CN has been controversial by scholars of social sciences, computer science and other disciplines. In 2012, Nature published a paper officially proposing the concept of "network science" [3]. Modeling with network science theory is an effective method and means to describe complex systems. Many systems in reality, such as biological, social and natural systems, can be described and quantified by the relevant characteristics of network science.

The attention flow is a sorted sequence about websites clicked by a consumer [10]. The CAFN (collective attention flow network), node represents website and edge represents the switch of websites clicked by users, is a weighted directed graph that represents users click and jump between different websites [6]. The CAFN, as an important branch of network science, has attracted the attention of many researchers in recent years. Many important universal laws, such as dissipation law, gravity law, Heap's law, Kleiber's law [4, 5] have been found. As one of the most important directions, measuring the node's similarity in CAFN has an important theoretical and

© Springer Nature Singapore Pte Ltd. 2019
Y. Sun et al. (Eds.): ChineseCSCW 2019, CCIS 1042, pp. 579–590, 2019.
https://doi.org/10.1007/978-981-15-1377-0_45

practical value. For example, network community detecting, online user clicks behavior prediction, link prediction, website impact analysis, website classification, website ranking and accurate delivery of ads are based on the node's similarity in CAFN.

Although the application of the CAFN is very extensive, there are no scholars to study the similarity of nodes in the online CAFN. This paper attempts to fill this gap. This paper, based on network science theory and online user collective behavior data, is the first to study the similarity of nodes in CAFN. We based on the optimization of the SPA model proposed a new method to measure the similarity of nodes in CAFN. The spatial norm was used to calculate the influence distance between nodes in the virtual space. The closer influence distance between nodes are, the more similar the nodes are. Finally, we obtain the main factors affecting the similarity of nodes in the attention flow network, and which was verified by experiments.

Our contributions are as follows:

(1) Based on the theoretical deduction and optimization of the SPA model, the generation time (Rt), the influence radius (Sr), and the representation Vs (Rt, Sr) of the nodes in virtual network space are defined.
(2) Based on the definition in (1), a new theoretical method for measuring the similarity of nodes in the collective attention flow network was proposed. The spatial norm was used to calculate the influence distance (Sd) between nodes in the virtual network space. The closer influence distance between nodes are, the more similar the nodes are.
(3) Experiments show that our method have the advantage of the stronger theoretical and higher universality for measuring node similarity in collective attention flow networks. It also can be applied to large-scale networks formed by massive data.

The rest of this paper is organized as follows. Section 2 presents the related work that introduce the research production about the similarity of the collective attention flow networks and network nodes in recent years. Section 3 introduces the theory and method. The experimental analysis is showed in Sect. 4. We can get conclusion and outlook in Sect. 5.

2   Related Work

2.1   Collective Attention Flow Network

As an important branch of network science, collective attention flow network has attracted wide attention of researchers. The collective attention flow network is a weighted directed graph that represents users click and jump between different websites. Where node represents website and edge represents the switch of websites clicked by users. In recent years, most of the researches on collective attention flow network are based on the structure features, aiming at revealing the evolution law of network, solving the problems of the allometric growth and dissipation, the decentralized flow structure and geometric representation.

Along with the rapid development of Internet, "Information overload" has become an urgent problem. In [7], a geometric representation method is proposed to represent the collective attention distribution among different websites. They embed abundant websites into the high-dimensional Euclidean space based on the new definition of flow distance. For the geometric representation, they embedded all the websites into a 20-dimensional ball. The study found that 20% popular sites attracting 75% attention flows. In [9] studied the decentralized flow structure of clickstreams on the web. They constructed three clickstreams networks, node represents website and edge represents the switch of websites clicked by users. C_i represents the influence of website i and is measured by the click flow controlled by the site in the clickstreams cycle. In order to accurately quantify the effect of sites in CAFN. In [10] based on the metabolism theory, they regard the Internet as a virtual "living things" and the websites want to grow and develop have to absorb energy. The collective attention is the necessary energy for the website.

2.2   Network Node Similarity

In recent years, scholars have paid more attention to the application and algorithm research about the node similarity. The existing research work mainly focuses on the following two aspects:

Implement Community Detection and Network Reconfiguration Based on Network Node Similarity. Community detection, to divide social network nodes into different communities, is momentous topic in the field of network science. In the past, a variety of community detection algorithms based on topology have been proposed. However, these algorithms only consider the structure of the network and ignore the attributes of the network nodes. In [11], the similarity of node attributes based on topology is studied. They speculated whether the nodes derived from the topology-based community detection algorithm also had interest similarities. The results show that about 50% of the nodes in all communities have similar interests. To group nodes with similar interests into a single community, [12] proposes a new community clustering algorithm based on the nodes' location and the similarity of interest in social networks. This algorithm has high accuracy of community detection. In [13, 14] studied the network completion with node similarity. Based on the auxiliary similarity information about nodes an algorithm was proposed to complete the network.

Research the Node Similarity Algorithms for Specific Networks. A simple adjustable measurement method is proposed in [15] to analyze the similarity between nodes according to the different link weights. They collected the information from 659 freshmen students at a large university through different channels, and constructed a weighted multi-layer network. This paper showed that in terms of basic personality traits, even highly correlated individuals are not more similar than randomly selected pairs of individuals. In [16] studied the node similarity with q-grams for real-world labeled networks, presented randomized algorithms and data structures for sketching node similarity. In order to measure nodes' similarity paper [17] proposed a new effective method based on relative entropy. They treated the structural feature as quantitative information of each node, and the difference of quantitative information is calculated by relative entropy.

3 Theory and Methods

3.1 SPA Theoretical Model

The SPA model is a random network space graph model first gave in paper [8]. In this model, S is used to represent a network metric space, V is used to represent nodes in S, and each node has an influence range S (v) in S. Supposing that a new node x can link to the existing node v in S, it means that the new node x falls within the effect range of node v, where p indicates the probability that u links to v. In this model, the influence range of each node in metric space S mainly depends on the in-degree of node. The higher the popularity of the node, the larger the impact range.

The step of the SPA model generates stochastic sequences of graphs as follows:

step1: Time-step t = 0, G_0 is an empty graph;
step2: Time-step t $\geq$ 1, Replace Gt $-$ 1with Gt;
step3: In order to create Vt, where the beginning of each time step t, a new node vt is chosen uniformly at random from S to Vt-1.

The parameters of the model include two parts: link probability p $\in$ [0, 1] and positive constants (a_1, a_2, a_3), so we can get $pa_1 \leq 1$.

In this model, Gt = (Vt, Et), and Vt $\in$ S, where I-degree (v, t) express the in-degree of node, O-degree (v, t) express the out-degree of node. And when time t $\geq$ 1, the influence range S (v) of node v is defined as the volume S (v, t) of a sphere centered on v. as follows:

$$S(v,t) = \frac{a_1 I - degree(v,t) + a_2}{t + a_3} \tag{1}$$

Definition 1. [18] if e $\gg$ 1 and t $\gg$ e, when t $\to$ ∞, the expected in-degree at time t of a node born at time e, as follows:

$$EI - degree(v, t) = (1 + o(1)) \frac{a_2}{a_1} \left(\frac{t}{e}\right)^{Pa_1} \tag{2}$$

3.2 Theoretical Model Optimization

For formula (1), the influence range of node is a random function rather than an accurate value. However, we want to get a certain value. In order to characterize the structural features of CAFN, reveal the similarity of nodes in CAFN, and improve the practical application value of the model. This paper optimizes the model. The node generation time (Rt), the node influence radius (Sr), the new representation of the nodes (Vs) and the influence distance (Sd) between the nodes are defined.

Definition 2. The generation time (Rt) of the nodes in the collective attention flow network.

$$R_t = \begin{cases} nk^{-(1/p)}, & k > 1 \\ \infty, & else \end{cases} \tag{3}$$

Where n is the total number of nodes in the attention flow network, k indicates the number of in-degrees of the nodes, p indicates the probability of generating a link in the network. when $k \leq 1$, $R_t = \infty$, explain that the node is generated at the latest in the collective attention network.

In order to quantify the influence radius of nodes more accurately, this paper optimizes it based on formula (1) and formula (2). The specific operation is as follows: Formula (1) and formula (2) are combined into a system of equations, and it can be obtained:

$$\begin{cases} S(v,t) = \dfrac{a_1 I\text{-}degree\ (v,\ t) + a_2}{t + a_3} & (4) \\[2em] EI\text{-}degree\ (v,\ t) = (1 + o(1)) \dfrac{a_2}{a_1} \left(\dfrac{t}{e}\right)^{pa_1} & (5) \end{cases}$$

Bring (5) into (4) and get it:

$$S(v,t) = \frac{a_1 (1 + o(1)) \frac{a_2}{a_1} \left(\frac{t}{e}\right)^{pa_1} + a_2}{t + a_3} \tag{6}$$

Let $a_1 = a_2 = a_3 = 1$, we can get:

$$S(v,t) = \left(\frac{t}{e}\right)^{p} / t \tag{7}$$

The Euclidean torus metric is used to measure the influence range of the node, this influence range is a circle with radius R and $S\ (v,\ t) = \pi R^2$, as follows:

$$S(v,t) = \left(\frac{t}{e}\right)^{p} / t = \pi R^2 \tag{8}$$

We can get the R, as follows:

$$R = t^{(p-1)/2} \cdot e^{-(p/2)} \cdot \pi^{-(1/2)} \tag{9}$$

Definition 3. The influence radius (Sr) of the nodes in the collective attention flow network, as follows:

$$Sr = t^{(p-1)/2} \cdot e^{-(p/2)} \cdot \pi^{-(1/2)} \tag{10}$$

Definition 4. The quantized form of the nodes in the collective attention flow network, as follows:

$$Vs(Rt, Sr) \tag{11}$$

Definition 5. The influence distance between nodes in collective attention flow network.

The distance between each Vs (Rt, Sr) calculated by space 2 norm, represented by Sd.

3.3   Collective Attention Flow Network Node Similarity Algorithm NID (Nodes Influence Distance)

Based on the optimization of Sect. 3.2 theoretical model, this paper proposes an algorithm to quantify the similarity of nodes in CAFN. The algorithm consists of the following four parts:

Generation Time Algorithm of Nodes in Collective Attention Flow Network (Algorithm1). In Algorithm 1, the Rt value is used to measure the time of the node generation time, and the smaller the Rt value, the earlier the time of the node in CAFN is, which attracts a large number of users' attention. Conversely, the larger the Rt value, the later the time generated by the node. When k ≤ 1, it can be concluded that the node generates the latest time, and its ability to attract the user's attention is far less than the node with a small Rt value, and thus the user's attention to the node is relatively low.

Influence Radius Algorithm of Nodes in Collective Attention Flow Network (Algorithm 2). According to the generation time Rt of nodes and the deduction of the theoretical model, the influence radius Sr of nodes in collective attention flow network is defined. In Algorithm 2, the Sr value is used to measure the influence radius Sr of the node. The larger the Sr value, the larger the influence radius of the node, that is, the larger the influence range, indicating that the node attracts more online users in the collective attention flow network. Attention to it, there are a great many number of neighbor nodes, and the nodes also play a central role in the collective attention flow network.

Node Representation in the Collective Attention Flow Network. According to the Rt and Sr of the collective attention flow network, its node representation Vs (Rt, Sr) is defined in a completely new way.

Influence Distance Algorithm between Nodes in Collective Attention Flow Network. In Algorithm 3, Vs (Rt, Sr) is used to calculate the influence distance (Sd) between nodes in the attention flow network, and Sd is used to measure the distance between nodes. The smaller the Sd is, the higher the similarity between nodes is, and vice versa. Finally, the similarity between nodes in the collective attention flow network is quantified according to the Sd matrix between nodes.

Algorithm 1: The generation time (Rt) of the node algorithm

Input: n, k, p

Output: Rt

1 Define Function Rt (k)

2   for i in k:

3   Input: n and p

4   formula: Rt = n * math.pow(i, -1/p)

5   Input the value of k

6 Output Rt

Algorithm 2: The influence radius (Sr) of the node algorithm

Input: π, p

Output: Rt

1 Define Function Sr (i , t)

2   for i, j in zip (k, t):

3   Input: π and p

4   Formula: Sr = math.pow(π, -1/2)*math.pow(e,-p/2)*math.pow(j,(-(1-p)/2))

5   Input the values of I and Rt

6 Output Sr

Algorithm 3: Influence distance algorithm between nodes in the collective attention flow network

---

Input: List of node pairs in the attention flow network

Output: Sd

1 Define Function Sd (argument1, argument2):

2   return sqrt ((a[i] - b[i]) **2 + (b[j] - b[j]) **2)

3   for index, item1 in enumerate(list):

4   for item 2 in A [index + 1:]:

5   Run Sd

6   Sort (Sd)

7 Output (Sd)

4   Experiment Analysis

4.1   Data Source

Data is provided by CNNIC (China Internet Network Information Center) online user click behavior log data, this paper uses a large number of user's online behavior log data to conduct research. Through data cleaning, a collective attention flow network is constructed, which includes 1107 network nodes and 4351 linked lists, as shown in Fig. 1.

Table 1 shows the node in-degree k and generation time rank the top 10 node information in the collective attention flow network. Figure 2 shows the relationship between the node's in-degree k and the generation time Rt. From Table 1 and Fig. 2, when the in-degree k value of the node is larger, the smaller the value of Rt, indicating that the earlier the node is generated. For example, the earlier the node qq.com is generated, the earlier it is clicked by the user, attracting a large number of users' attention, followed by baidu.com, hao123.com…, which also proves that the high degree is more attractive.

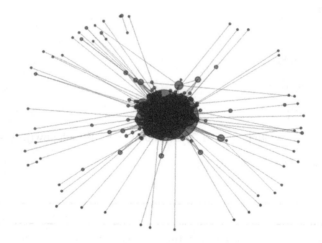

Fig. 1.   Collective attention flow network

4.2   Experimental Analysis

Table 2 shows the top 10 nodes of Rt and Sr in the collective attention flow network, and Fig. 3 shows the relationship between Rt and Sr of the node. From the Table 2 and Fig. 3, it can be seen that the smaller the Rt value of the node, the larger the Sr value, indicating that the earlier the node is generated, the larger the radius of the node and the larger the scope of the impact. In the collective attention flow network, the early generation nodes are qq.com, baidu.com and hao123.com, which have a larger impact range, and have the greatest impact on the click-and-browse behavior of online users. These nodes have attracted the attention of abundant online consumers for a long time.

Table 1. The k and Rt of the nodes top 10

Nodes	k	R_t
qq.com	253	3.10
baidu.com	211	3.76
hao123.com	140	5.79
taobao.com	132	6.16
360.cn	104	7.91
weibo.com	98	8.43
sogou.com	88	9.44
sohu.com	68	12.38
sina.com	59	14.37
163.com	59	14.37

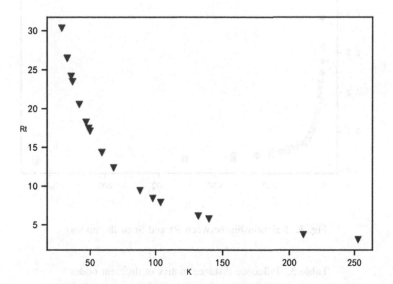

Fig. 2. Relationship between the k and Rt of the nodes

Table 2. The Rt and Sr of the nodes top 10

Nodes	Rt	Sr
qq.com	3.10	0.55
baidu.com	3.76	0.39
hao123.com	5.79	0.32
taobao.com	6.16	0.28
360.cn	7.91	0.25
weibo.com	8.43	0.23
sogou.com	9.44	0.21
sohu.com	12.38	0.20
sina.com	14.37	0.19
163.com	14.37	0.18

From Table 3, the most similar nodes to qq.com are baidu.com, taobao.com, etc. It is found that these nodes have the same characteristics of an early generation and large influence radius Sr. Thus, in the collective attention flow network, when the nodes with large radius and earlier generation time, the influence distance in the virtual space is small, and the similarity between them is also the highest. Experiments show that there are three factors affecting the similarity of nodes in a collective attention flow network, namely, the generation time of nodes, the influence radius Sr and based on Vs (Rt, Sr) calculation of the influence distance Sd between nodes.

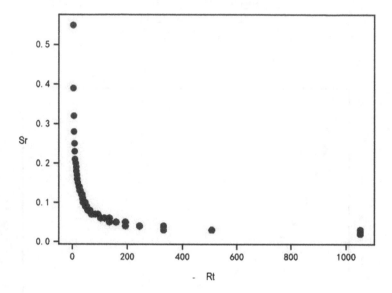

Fig. 3. Relationship between Rt and Sr of the nodes

Table 3. Influence distance matrix of different nodes

	qq.com	baidu.com	hao123.com	taobao.com	360.cn	weibo.com
qq.com	0	0.68	2.7	3.07	4.82	5.34
baidu.com	0.68	0	2.03	2.4	4.15	4.67
hao123.com	2.7	2.03	0	0.37	2.12	2.64
taobao.com	3.07	2.4	0.37	0	1.75	2.27
360.cn	4.82	4.15	2.12	1.75	0	0.52
weibo.com	5.34	4.67	2.64	2.27	0.52	0
sogou.com	6.35	5.68	3.65	3.28	1.53	1.01
sohu.com	9.29	8.62	6.59	6.22	4.47	3.95
......	......	......	......	......	......	......

5 Summary and Outlook

In order to make the model more universal, show the practical application value of the model, and reveal the similarity of nodes in the collective attention flow network, we optimized the SPA model. We defined the generation time, influence radius, representation form of the nodes and influence distance between nodes in CAFN, and proposed an Algorithm NID with high universality to quantify the similarity of nodes in CAFN. Three factors affecting node similarity in CAFN are found through research. It is also explained from a theoretical point of view that the high degree is more attractive, which has been verified by experiments. Experiments show that our algorithm NID can more objectively represent the structural characteristics of CAFN when the probability of links generated by nodes is p = 0.95 in the SPA model with uniform random selection. Through this study, not only can we clearly show the similarity between the nodes in CAFN, but also can evaluate the nodes of the website more comprehensively, and provide reasonable suggestions for online advertising and users' online behavior prediction. The shortcoming of this paper is that the generation time and influence radius of the nodes depend on the in- degree k of the node, which inevitably ignores the nodes with small in-degree of penetration.

Acknowledgements. This paper was supported by the Natural Science Foundation of China (No. 71764025, 61863032, 61662070); the Research Project on Educational Science Planning of Gansu, China (Grant No. GS[2018]GHBBKZ021, GS[2018]GHBBKW007); the Scientific Research Foundation of the Higher Education Department of Gansu, China (Grant No. 2018A-001). Author contributions: Manfu Ma and Zhangyun Gong are co-first authors who jointly designed the research. Correspondence and requests for materials should be addressed to Yong Li.

References

1. Watts, D.J., Strogatz, S.H.: Collective dynamics of 'small-world' networks. Nature **393** (6684), 440 (1998)
2. Barabási, A.L., Albert, R.: Emergence of scaling in random networks. Science **286**(5439), 509–512 (1999)
3. Barabási, A.L.: Network science: luck or reason. Nature **489**(7417), 507 (2012)
4. Li, Y., Meng, X.F., Zhang, Q., Zhang, J., Wang, C.Q.: Common patterns of online collective attention flow. Sci. China Inf. Sci. **60**(5), 59102 (2017)
5. Wu, F., Huberman, B.A.: Novelty and collective attention. Proc. Natl. Acad. Sci. **104**(45), 17599–17601 (2007)
6. Lou, X., Li, Y., Gu, W., Zhang, J.: The atlas of Chinese world wide web ecosystem shaped by the collective attention flows. PLoS ONE **11**(11), e0165240 (2016)
7. Shi, P., Huang, X., Wang, J., Zhang, J., Deng, S., Wu, Y.: A geometric representation of collective attention flows. PLoS ONE **10**(9), e0136243 (2015)
8. Aiello, W., Bonato, A., Cooper, C., Janssen, J., Prałat, P.: A spatial web graph model with local influence regions. Internet Math. **5**(1–2), 175–196 (2008)
9. Wu, L., Zhang, J.: The decentralized flow structure of clickstreams on the web. Eur. Phys. J. B **86**(6), 266 (2013)

10. Li, Y., Zhang, J., Meng, X.F., Wang, C.Q.: Quantifying the influence of websites based on online collective attention flow. J. Comput. Sci. Technol. **30**(6), 1175–1187 (2015)
11. Sharma, R., Montesi, D.: Investigating similarity of nodes' attributes in topological based communities. In: The Web Conference, pp. 1253–126 (2018)
12. Hutair, M.B., Aghbari, Z.A., Kamel, I.: Social community detection based on node distance and interest. In: 3rd IEEE/ACM International Conference on Big Data Computing, Applications and Technologies, pp. 274–289. ACM (2016)
13. Forsati, R., Barjasteh, I., Ross, D., Esfahanian, A.H., Radha, H.: Network completion by leveraging similarity of nodes. Soc. Netw. Anal. Min. **6**(1), 102 (2016)
14. Masrour, F., Barjesteh, I., Forsati, R., Esfahanian, A.H., Radha, H.: Network completion with node similarity: a matrix completion approach with provable guarantees. In: IEEE/ACM International Conference on Advances in Social Networks Analysis and Mining, pp. 302–307. ACM (2015)
15. Mollgaard, A., Zettler, I., Dammeyer, J., Jensen, M.H., Lehmann, S., Mathiesen, J.: Measure of node similarity in multilayer networks. PLoS ONE **11**(6), e0157436 (2016)
16. Conte, A., Ferraro, G., Grossi, R., Marino, A., Sadakane, K., Uno, T.: Node similarity with q-Grams for real-world labeled networks. In: 24th ACM SIGKDD International Conference on Knowledge Discovery & Data Mining, pp. 1282–1291. ACM (2018)
17. Zhang, Q., Li, M., Deng, Y.: Measure the structure similarity of nodes in complex networks based on relative entropy. Phys. A: Stat. Mech. Appl. **491**, 749–763 (2018)
18. Janssen, J., Prałat, P., Wilson, R.: Estimating node similarity from co-citation in a spatial graph model. In: ACM Symposium on Applied Computing, pp. 1329–1333. ACM (2010)

A Graph Representation Learning Algorithm Based on Attention Mechanism and Node Similarity

Kun Guo[1,2,3], Deqin Wang[1,2,3], Jiangsheng Huang[4],
Yuzhong Chen[1,2,3(✉)], Zhihao Zhu[1], and Jianning Zheng[4]

[1] College of Mathematics and Computer Sciences, Fuzhou University,
Fuzhou 350116, China
gukn123@163.com, dqwang_fzu@163.com,
yzchen@fzu.edu.cn, zhzhu_fzu@163.com
[2] Fujian Provincial Key Laboratory of Network Computing and Intelligence
Information Processing, Fuzhou, China
[3] Key Laboratory of Spatial Data Mining and Information Sharing,
Ministry of Education, Fuzhou 350116, China
[4] Power Science and Technology Corporation State Grid
Information and Telecommunication Group, Xiamen 351008, China
huangjiangsheng@sgitg.sgcc.com.cn, 12720740@qq.com

Abstract. Recently graph representation learning has attracted much attention of researchers, aiming to capture and preserve the graph structure by encoding it into low-dimensional vectors. Attention mechanism is a recent research hotspot in learning the representation of graph. In this paper, a graph representation learning algorithm based on Attention Mechanism and Node Similarity (AMNS for short) is proposed. Firstly, the similarity neighborhood is generated for each node in graph. Secondly, attention mechanism is used to learn weight coefficients for each node and its similarity neighborhood. Thirdly, the node vectors are generated by aggregating its similarity neighborhood with weight coefficients. Finally, node vectors are applied to many tasks, e.g., node classification and clustering. The experiments on real-world network datasets prove that the AMNS algorithm achieves excellent results.

Keywords: Graph representation learning · Attention mechanism · Node similarity

1 Introduction

Graphs, also known as Networks, widely exist in real life scenarios including citation graph, social network and information network. Graph analytics is conducive to capture useful information hidden in graphs. Analyzing graphs plays an important part in many applications, e.g., node classification [1], node clustering [2], recommender system [3], community discovery [4].

Recently, graph representation learning [5] has aroused intense scholarly interest. It is proposed to embed graph into vector space while preserving graph structure

© Springer Nature Singapore Pte Ltd. 2019
Y. Sun et al. (Eds.): ChineseCSCW 2019, CCIS 1042, pp. 591–604, 2019.
https://doi.org/10.1007/978-981-15-1377-0_46

information. As the output of graph representation learning methods, low dimensional vectors can be easily applied to many machine learning algorithms. Lots of useful and efficient algorithms are presented, such as DeepWalk [6], node2vec [7], LINE [8] and SDNE [9], etc. Among them, Graph Neural Networks (GNNs) [10] is proved to be a useful graph representation method based on deep learning, such as GCN [11]. A recent research hotspot in deep learning is attention mechanism [12] which concentrates on the most relevant sections of input for making decision. For instance, a framework named GAT [13] learns the representations of graph via using attention mechanism. Many other methods based on attention mechanism have achieved excellent results in graph analysis tasks, too.

Besides, measuring similarity [14] between nodes is useful and beneficial. Lots of algorithms with respect to node similarity have been put forward. However, most of them are unable to fit for machine learning applications directly. Thus, it is a good way to take advantage of node similarity by embedding them into vectors.

In this paper, the primary contributions are as follows: (1) A graph representation learning algorithm combines Attention Mechanism with Node Similarity (AMNS) is proposed. Firstly, the similarity neighborhood is generated for each node. Secondly, the attention coefficients of each node and its neighborhood are calculated by an attention model. Thirdly, the node vectors are reconstructed as a weighted sum of its neighborhood's features. (2) A new node similarity coefficient is proposed so as to generate node's highly similar neighborhood which can be applied to attention model as prior knowledge.

2 Related Work

2.1 Attention Mechanism

Attention mechanism is becoming so popular in field of deep learning. It aims to take advantage of the most relevant input for decision making. Recently, many algorithms based on attention models for graphs have been proposed. For instance, methods like GAT [13] and AGNN [15] incorporate an explicit attention mechanism to GCNs. AttentionWalk [16] uses a set of graph attention models that allows learning arbitrary context distributions. In addition, there are many other methods using attention mechanism to improve their algorithmic performance.

3 The AMNS Algorithm

3.1 Generating Neighborhood via Node Similarity

It is acknowledged that each node has its own neighborhood in a complex network, such as direct neighbors and second-order neighbors:

Definition 1. (Neighbors of Node u) Neighbors of node u, N (u)'s definition:

$$N(u) = \{v \in V \mid (u, v) \in E\} \tag{1}$$

Definition 2. (Second-order Neighbors of Node u) Second-order Neighbors of node u, SN (u)'s definition:

$$SN(u) = \{v \in V | w \in N(u) \wedge (w, v) \in E\} \tag{2}$$

Definition 3. (Jaccard coefficient of Node u and v) Jaccard coefficient of node u and v, J(u, v)'s definition:

$$J(u, v) = \frac{|N(u) \cap N(v)|}{|N(u) \cup N(v)|} \tag{3}$$

Jaccard coefficient [18] is defined to estimate the similarity between two nodes. There is a positive relation between the value and similarity of two nodes.

It is well known that the node's degree is an important attribute of a node. Here we extend the definition of node's degree, which is based on node's neighbors:

Definition 4. (Total Degree of Node u's Neighbors) Total degree of node u's Neighbors, TD(u)'s definition:

$$TD(u) = \sum_{v \in N(u)} d(v) \tag{4}$$

Many algorithms use node's local structure information to embed nodes into vectors. But some nodes have little local structure information because they are short of neighbors. Therefore, it is essential for such nodes to expand its neighborhoods. Meanwhile, some methods are able to generate nodes' neighborhoods, such as BFS and DFS. However, most of these methods cannot ensure the most relevant nodes to be selected. To overcome this problem, a neighborhood generating algorithm is proposed based on node similarity. A similarity coefficient is needed. First, we define an attraction coefficient as follows:

Definition 5. (The Attraction Coefficient of Node u to v) The attraction coefficient of node u to v, Attr(u, v)'s definition:

$$Attr(u, v) = log_2\left(\frac{TD(u)}{TD(v)} + 1\right) \tag{5}$$

Here we employ log function to lessen the ratio of TD(u) and TD(v). Finally, we can get the similarity coefficient:

Definition 6. (The Similarity Coefficient of Node u to v) The similarity coefficient between node u and v, Sim(u, v)'s definition:

$$Sim(u, v) = Attr(u, v) * J(u, v) \tag{6}$$

The similarity neighborhood called N_u^{sim} of node u can be generated by selecting highly relevant and important nodes to it. Moreover, the values of Sim(u, v) and Sim(v, u) are different.

Algorithm 1. The Neighborhood Generating algorithm based on Node Similarity (NGNS) is designed as follows:

Algorithm 1. NGNS

Input: G, $N(V)$, $SN(V)$
        The parameter a;
Output: a set of similar neighborhood $\{N_u^{sim} \mid u \in V\}$
1.   SIM=$\emptyset$, $N_u^{sim}=\emptyset$
2.   **for each** $u \in V$ **do**
3.   **if** (d(u) < 0.5*a) **then**
4.   **if** (v $\in$ SN(u) $\cup$ N(u)) **then**
5. $N_u^{sim}= N_u^{sim} \cup \{v\}$;
6.   **end if**
7.   **end if**
8.   **if** (d(u) > =0.5*a and d(u) <a) **then**   //d(u) means degree of node u
9. $N_u^{sim}=N_u^{sim} \cup$ N(u);
10.   **for each** $w \in SN(u)$ **do**
11.   calculate Sim(u, w) according to equation(3), (5), (6);
12.   SIM=SIM $\cup$ $\{Sim(u, w)\}$;
13.   **end for**
14.   Top-k-Sim = heap(k, SIM);
15.   //heap is a function to find the k-th largest number of input
16. $N_u^{sim}= N_u^{sim} \cup$ $\{w\mid Sim(u, w) \in$ Top-k-Sim$\}$;
17.   **end if**
18.   **if** (d(u) >=a and d(u) < 2*a) **then**
19. $N_u^{sim}=N_u^{sim} \cup$ N(u);
20.   **end if**
21.   **if**(d(u) > 2*a) **then**
22.   **for each** v $\in$ N(u)
23.   calculate $Sim(u, v)$ according to equation(3), (5), (6);
24.   SIM=SIM $\cup$ $\{Sim(u, v)\}$;
25.   **end for**
26.   Top-k-Sim = heap(k, SIM);
27. $N_u^{sim}= N_u^{sim} \cup$ $\{v \mid Sim(u, v) \in$Top-k-Sim$\}$;
28.   **end if**
29.   **end for**
30.   **return** $\{ N_u^{sim} \mid u \in V \}$

In this algorithm, the parameter a is of great importance, because it can divide nodes into different groups. If a node's degree is less than a, we need to expand its neighborhood. If not, we should limit the size of node's neighborhood. The value of a can be set to the average number or the mode of nodes' degree. In this algorithm, we select the average value of nodes' degree because we cannot ignore nodes with large degrees. Moreover, a graph G, N(V) and SN(V) are also the input. N(V) represents each node's direct neighbors, while SN(V) represents each node's second-order neighbors, and V is a node-set of G. The output is node u's similarity neighborhood N_u^{sim}.

3.2 Attention Model

The graph attention layer [13] is used as model's basic layer. Its input is a set of node features from a graph, Its definition is:

Definition 7. (Node Features H) Node features H from a graph, H's definition:

$$H = \{h_v \in R^F | v \in V\} \tag{7}$$

Where R^F means F-dimensional vectors that represent nodes features. They are initially originated as one-hot encoding or bag-of-words encoding which implies semantic information of node. The output is $Z = \{ z_v \in R^{F'} | v \in V \}$ and its definition is similar to H.

In the attention model, we use a linear transformation (a trainable weight matrix, essentially) $W \in R^{F'F}$ that applies to every node feature h.. It is defined as:

Definition 8. (A Linear Transformation) A linear transformation (i.e., a trainable weight matrix W), is defined as:

$$h' = Wh \tag{8}$$

Attention coefficients are calculated by attention mechanism, with N_u^{sim} as its priori knowledge. The attention mechanism a is the key part of attention model. Its details are as follows:

Definition 9. (An Attention Mechanism a) An attention mechanism a that computes attention coefficients e_{uv} which denotes the importance of node v to u, where $v \in N_u^{sim}$, the definition is:

$$e_{uv} = a\left(h'_u, h'_v; N_u^{sim}\right) \tag{9}$$

The attention mechanism a can be illustrated as follows:

$$e_{uv} = LeakReLU(a^T[h'_u \| h'_v] \ominus|_{v \in N_u^{sim}} \tag{10}$$

where LeakReLU denotes the activation function, while $\|$ denotes concatenation. The weight vector a^T is trainable. In addition, to make coefficients easily comparable, it is vital to normalize them, with the help of the softmax function:

$$\alpha_{uv} = softmax_u(e_{uv}) = \frac{\exp(e_{uv})}{\sum_{k \in N_u^{sim}} \exp(e_{uk})} \tag{11}$$

A sketch map is drawn to understand attention model clearly, as presented in Fig. 1.

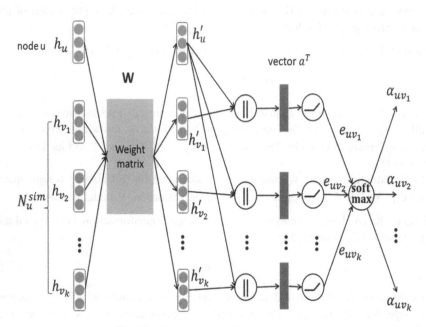

Fig. 1. Graph attention layer's structure

Then, the vector of node u can be aggregated by its similarity neighbor's features with the attention coefficients.

Definition 10. (The vector of Node u) The vector of node u, z_u's definition:

$$z_u = \sigma\left(\sum_{v \in N_u^{sim}} \alpha_{uv} h_v^{'}\right) \tag{12}$$

Specifically, the attention mechanism is repeated for K times and the learned vectors are concatenated:

$$z_u = \|_{k=1}^{K} \sigma \sum_{v \in N_u^{sim}} \alpha_{uv}^k h_v^{'} \tag{13}$$

where ‖ represents concatenation, through the k-th attention mechanism, weight attention coefficients α_{uv}^k are calculated. The concept that describes the process of generating node vectors is shown in Fig. 2.

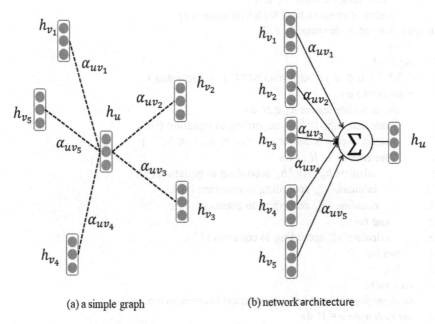

(a) a simple graph   (b) network architecture

Fig. 2. (a) a simple graph (b) a graph neural network architecture

3.3   Algorithm Details

The AMNS algorithm mainly falls into two parts. For the first part, the similarity neighborhood is generated via node similarity. For the second part, an attention model based on node similarity is adopted to obtain node vectors. The AMNS algorithm can be illustrated as follows:

Algorithm 2.

Algorithm 2. AMNS
Input: $G=(U, V)$, $N(V)$, $SN(V)$,
A set of nodes features $h_u \in H$,
Number of attention head K (default value is 8)
Output: A set of node vector $Z_u \in Z$
1. $Z = \emptyset$
2. $set\ k = K$
3. $\{ N_u^{sim} \mid u \in V \}$ = Algorithm NGNS ;   // Algorithm 1
4.   **while**($k!=0$) **do**
5.   **for** each node feature $h_u \in H$ **do**
6.   calculate $h_u'^k = W^k h_u$ according to equation (9);
7.   select a set of node features $H_{sim} = \{ h_v \mid v \in N_u^{sim} \}$;
8.   **for** each $h_v \in H_{sim}$ **do**
9.   calculate $h_v'^k = W^k h_v$ according to equation (9);
10.   calculate e_{uv}^k according to equation (10);
11.   calculate α_{uv}^k according to equation (11);
12.   **end for**
13.   calculate z_u^k according to equation (12);
14.   **end for**
15. k--;
16.   **end while**
17.   Back propagation and update parameters in attention model;
18.   **for** each node $u \in V$ **do**
19.   calculate z_u according to equation (13);
20. $Z=Z \cup \{z_u\}$
21.   **end for**
22.   **return** Z

3.4   Algorithm Complexity Analysis

For NGNS algorithm, supposed the average number of SN(V) is d, then the time complexity is O(|V|*d), where |V| is nodes' number.

For attention model of a single head, its time complexity is O(|V|FF' + N*F'), where F is the dimension of input features, F' is the dimension of output features. N is the average number of each node's similarity neighborhood. In summary, AMNS algorithm's time complexity is O(|V|*d + K(|V|FF' + N*F')), where K is attention head's number.

4   Experiments

The experiments adopted three real network datasets to evaluate the algorithms' performances. For College Football network, each node has its value indicating which clubs it belongs to. The Cora and Pubmed [18] datasets are both citation networks. The details of them are shown in Table 1.

Table 1. Three real-world network datasets.

Datasets	Nodes	Edges	Class
College football	115	616	12
Cora	2708	5429	7
Pubmed	19717	44338	3

4.1   Evaluation Metrics

The tasks of multi-class classification, node clustering and visualization are performed. Different evaluation metrics are employed to measure their performances.

For multi-class node classification task, macro-F1 score is adopted as many other works do [19]. Macro-F1 is defined as:

$$macro - F1 = \frac{\sum_{L \in C} F1(L)}{|C|} \tag{15}$$

where C is label set, $F1(L)$ is the $F1$-measure for label L.

For node clustering task, the normalized mutual information (NMI) [20] is employed. The equation is:

$$\text{NMI} = \frac{-2 \sum_{i=1}^{C_A} \sum_{j=1}^{C_B} C_{ij} \cdot \log\left(\frac{C_{ij} \cdot N}{C_i \cdot C_j}\right)}{\sum_{i=1}^{C_A} C_i \cdot \log\left(\frac{C_i}{N}\right) + \sum_{j=1}^{C_B} C_{ij} \cdot \log\left(\frac{C_j}{N}\right)} \tag{16}$$

where N is nodes' number, C is a confusion matrix. C_A is the number of real clusters, C_B is the number of clusters formed by algorithm.

For visualization task, a tool called t-SNE [21] is used to visualize the node vectors so that its performance can be intuitively observed.

4.2   The Compared Methods

The algorithm AMNS will be compared with several recent graph representation learning methods:

(1) DeepWalk/Node2vec: They are both random walk methods using skip-gram model to obtain node vectors.

(2) LINE: It uses loss functions to maintain the first and second order proximity respectively. The representations are connected after optimizing loss functions.
(3) GraRep: It is a method using matrix factorization and training by the SVD.
(4) GCN: A graph convolution network framework that employs the convolutional operation to graph.
(5) GAT: It is a graph neural networks using attention Mechanism.

4.3 Parameter Settings

For algorithms GCN, GAT and AWNS, the learning rate is set to 0.005, the number of attention head K to 8. Then these models are all trained 200 epochs. In random walk based methods, the walk length is set to 80, the window size to 10, Negative sample's number to 5. The dimension of vector is set to 64 for the sake of a fair comparison.

4.4 Experimental Results

Multi-class Classification. For the classification task, third-party labels are adopted to determine the class of each node. A logistic regression classifier is employed to perform node classification. The training set's ratio varies from 10% to 90%. Macro-F1 score assists in evaluating the performance. Then each classification experiment is repeated five times. At last, the average performance is described at Table 2.

Table 2. Multi-class node classification results in three datasets

Datasets	Methods	10%	30%	50%	70%	90%
College football	Deepwalk/node2vec	0.3649	0.5468	0.8199	0.7448	0.5767
	LINE	0.3714	0.7043	0.8398	**0.8399**	**0.6333**
	GraRep	0.3714	0.6878	**0.8589**	0.778	0.5325
	GCN	**0.3841**	0.7132	0.8364	0.8191	0.6060
	GAT	0.3810	0.7224	0.8340	0.8191	0.6060
	AMNS (ours)	0.3789	**0.7427**	0.8401	**0.8399**	**0.6333**
Cora	Deepwalk/node2vec	0.7458	0.7821	0.8077	0.8015	0.8150
	LINE	0.6549	0.7129	0.7425	0.7626	0.7712
	GraRep	0.7451	0.7704	0.7732	0.7790	0.7938
	GCN	0.7908	0.7951	0.8118	0.8179	0.8215
	GAT	0.8013	0.8037	0.8166	**0.8295**	0.8315
	AMNS (ours)	**0.8227**	**0.8399**	**0.8336**	0.8242	**0.8639**
Pubmed	Deepwalk/node2vec	0.5895	0.5998	0.6039	0.6085	0.6104
	LINE	0.6092	0.6313	0.6379	0.6317	0.6338
	GraRep	0.7837	0.7847	0.7823	0.7819	0.7837
	GCN	0.7832	0.7925	0.7949	0.7980	0.8014
	GAT	0.7898	**0.7998**	0.8012	0.7998	0.8092
	AMNS (ours)	**0.7927**	0.7936	**0.8126**	**0.8144**	**0.8228**

For football dataset, methods above receive excellent results and some of them even achieve 90% accuracy. Especially, our algorithm achieves or matches the best performance. For the two other datasets, methods in deep learning perform better than others because they have a stronger ability to extract features than shallow model. They get 2%–6% promoted compared with methods in shallow model. Meanwhile, by expanding the neighborhood of nodes, our algorithm can capture more information than GCN and GAT, so that it gets promoted nearly 1%–2% over them.

Node Clustering. For the node clustering task, the learned vectors by each method are inputted to a clustering model. K-means algorithm is leveraged to this task. The clustering results are evaluated by NMI [20]. All clustering experiments are repeated five times then average score is taken down. The result is given in Fig. 3.

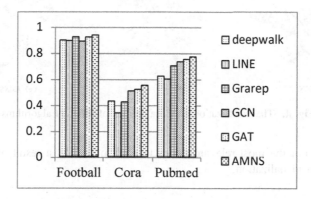

Fig. 3. The NMI score of above algorithms on three real-world networks.

Figure 3 shows that AMNS algorithm's performance is better than others. This is mainly due to the fact that our algorithm aggregates node features by its highly similar and relevant neighbors so that the similar nodes will be divided into one cluster easily, while the others cannot ensure that.

Visualization. Visualization of a graph is also an important application in graph representation learning. The node vectors learned by above methods are taken as the input to the t-SNE tool. Then each vector is visualized as a point on a two dimensional space. Obviously, the same categories of nodes should be close to each other if the node representations achieve excellent performance. The node presentations of Cora dataset are chosen to generate a visualization result, which is presented in Fig. 4.

As is illustrated in Fig. 4, The node representations of GCN, GAT and our algorithm AMNS get excellent result while the others cannot divide node representations into different classes clearly. The reason why different classes can be distinguished by vectors is that a neighborhood aggregation strategy is used so that connected points are also similar in vector representation. Other methods like DeepWalk use a random walker to generate node sequences so as to embed them. However, these node

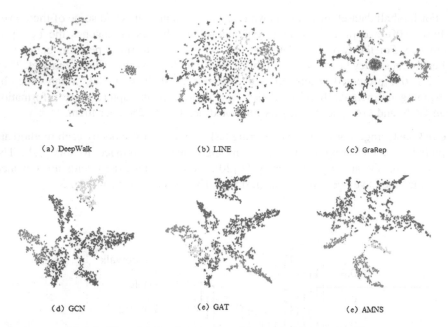

(a) DeepWalk   (b) LINE   (c) GraRep

(d) GCN   (e) GAT   (e) AMNS

Fig. 4. The visualization on Cora dataset of different algorithms

sequences are not the most relevant nodes to each other. As a result, it causes poor performance in visualization.

5   Conclusion

This paper proposes a graph representation learning algorithm called AMNS, where structural information of the graph is obtained by attention mechanism and node similarity. First, in this algorithm, the similarity neighborhood is generated for each node. Second, attention mechanism is applied to learn weight coefficients for each node pair related by similarity neighborhood. The node vectors are generated by aggregating its similarity neighborhood. Finally, node vectors, i.e. node embedding are used to complete different graph mining tasks. For future work, we can enhance this algorithm to heterogeneous graph or reduce its time consumption, or make some relevant improvements with respect to attention mechanism.

Acknowledgements. This work is partly supported by the National Natural Science Foundation of China under Grant No. 61300104, No. 61300103 and No. 61672159, the Fujian Province High School Science Fund for Distinguished Young Scholars under Grant No. JA12016, the Fujian Natural Science Funds for Distinguished Young Scholar under Grant No. 2015J06014, the Fujian Industry-Academy Cooperation Project under Grant No. 2017H6008 and No. 2018H6010, and Haixi Government Big Data Application Cooperative Innovation Center.

References

1. Bhagat, S., Cormode, G., Muthukrishnan, S.: Node classification in social networks. In: Aggarwal, C. (ed.) Social Network Data Analytics, pp. 115–148. Springer, Heidelberg (2011). https://doi.org/10.1007/978-1-4419-8462-3_5
2. Malliaros, F.D., Vazirgiannis, M.: Clustering and community detection in directed networks: a survey. Phys. Rep. **533**(4), 95–142 (2013). https://doi.org/10.1016/j.physrep.2013.08.002
3. Resnick, P., Varian, H.: Recommender systems. Commun. ACM **40**(3), 56–59 (1997). https://doi.org/10.1145/245108.245121
4. Parthasarathy, S., Ruan, Y., Satuluri, V.: Community discovery in social networks: applications, methods and emerging trends. In: Aggarwal, C. (ed.) Social network data analytics, pp. 79–113. Springer, Heidelberg (2011). https://doi.org/10.1007/978-1-4419-8462-3_4
5. Hamilton, W.L., Ying, R., Leskovec, J.: Representation learning on graphs: methods and applications. arXiv preprint arXiv:1709.05584 (2017)
6. Perozzi, B., Al-Rfou, R., Skiena, S: DeepWalk: online learning of social representations. In: Proceedings of the 20th ACM SIGKDD International Conference on Knowledge Discovery and Data Mining, pp. 701–710. ACM (2014). https://doi.org/10.1145/2623330.2623732
7. Grover, A., Leskovec, J.: node2vec: scalable feature learning for networks. In: Proceedings of the 22nd ACM SIGKDD International Conference on Knowledge Discovery and Data Mining, pp. 855–864. ACM (2016). https://doi.org/10.1145/2939672.2939754
8. Tang, J., Qu, M., Wang, M., Zhang, M., Yan, J., Mei, Q.: Line: large-scale information network embedding. In: Proceedings of the 24th International Conference on World Wide Web, pp. 1067–1077. International World Wide Web Conferences Steering Committee (2015). https://doi.org/10.1145/2736277.2741093
9. Wang, D., Cui, P., Zhu, W.: Structural deep network embedding. In: Proceedings of the 22nd ACM SIGKDD International Conference on Knowledge Discovery and Data Mining, pp. 1225–1234. ACM (2016). https://doi.org/10.1145/2939672.2939753
10. Scarselli, F., Gori, M., Tsoi, A.C., Hagenbuchner, M., Monfardini, G.: The graph neural network model. IEEE Trans. Neural Netw. **20**(1), 61–80 (2009). https://doi.org/10.1109/TNN.2008.2005605
11. Kipf, T.N., Welling, M.: Semi-supervised classification with graph convolutional networks. arXiv preprint arXiv:1609.02907 (2016)
12. Vaswani, A., et al.: Attention is all you need. In: Advances in Neural Information Processing Systems, pp. 5998–6008 (2017)
13. Veličković, P., Cucurull, G., Casanova, A., Romero, A., Lio, P., Bengio, Y.: Graph attention networks. arXiv preprint arXiv:1710.10903 (2017)
14. Lu, W., Janssen, J., Milios, E., Japkowicz, N., Zhang, Y.: Node similarity in the citation graph. Knowl. Inf. Syst. **11**(1), 105–129 (2007). https://doi.org/10.1007/s10115-006-0023-9
15. Thekumparampil, K.K., et al.: Attention-based graph neural network for semi-supervised learning. arXiv preprint arXiv:1803.03735 (2018)
16. Abu-El-Haija, S., et al.: Watch your step: learning node embeddings via graph attention. In: Advances in Neural Information Processing Systems, pp. 9180–9190 (2018)
17. Niwattanakul, S., Singthongchai, J., Naenudorn, E., Wanapu, S.: Using of Jaccard coefficient for keywords similarity. In: Proceedings of the International Multiconference of Engineers and Computer Scientists, pp. 380–384 (2013)
18. Sen, P., Namata, G., Bilgic, M., Getoor, L., Galligher, B., Eliassi-Rad, T.: Collective classification in network data. AI Mag. **29**(3), 93 (2008)

19. Tang, L., Liu, H.: Relational learning via latent social dimensions. In: Proceedings of the 15th ACM SIGKDD International Conference on Knowledge Discovery and Data Mining, pp. 817–826. ACM (2009). https://doi.org/10.1145/1557019.1557109
20. Danon, L., Diaz-Guilera, A., Duch, J., Arenas, A.: Comparing community structure identification. J. Stat. Mech: Theory Exp. **2005**(09), P09008 (2005)
21. Maaten, L.V.D., Hinton, G.: Visualizing data using t-SNE. J. Mach. Learn. Res. **9**(Nov), 2579–2605 (2008)

A Method of Analysis on Consumer Behavior Characteristics Based on Self-supervised Learning

Bin Liu[1], Wei Guo[1,2(✉)], Xudong Lu[1,2], Meng Xu[2,3],
and Lizhen Cui[1,2]

[1] School of Software, Shandong University, Jinan 250100, China
guowei@sdu.edu.cn
[2] Key Laboratory of Shandong Software Engineering, Jinan 250100, China
[3] School of Computer Science and Technology, Shandong Technology
and Business University, Yantai 264005, China

Abstract. Consumer shopping decision-making style is a kind of mental orientation, which represents the way consumers make decisions and has cognitive and emotional characteristics. It determines the behaviors of consumers and is relatively stable for a long time, as a result of which can be used as the basis for market segmentation. The traditional way to determine shopping decision-making style is mostly through questionnaire, which is time-consuming and laborious. This paper uses e-commerce data, combines the different commodities purchased by consumers, and comprehensively measures the mental characteristics of consumers when they make purchasing decisions based on their behaviors. By clustering analysis, the shopping decision-making behaviors of consumers are divided into two categories. Considering that clustering is an unsupervised method, its results are often not perfect, which is especially reflected in the data at the junction of two categories. In view of this, we divide the data at the junction of two categories into unclassified data and then train naive Bayes with classified data to classify unclassified data. Ultimately, by synthesizing all shopping decision-making behaviors of each consumer, the decision-making styles of consumers are divided into three categories: direct style, cautious style and neutral style. The experimental results show that the model proposed in this paper makes the classification of consumer decision-making behaviors more intuitive and is obviously superior to the comparison model. This model can effectively determine the consumer decision-making style.

Keywords: Naive Bayes · Self-supervised learning · Clustering analysis ·
Consumer shopping decision-making · Style analysis on consumer behavior

1 Introduction

Consumer shopping decision-making style is a kind of mental orientation, which represents the way consumers make decisions, with cognitive and emotional characteristics. It is relatively stable for a long time and determines the behavior of consumers, so it can be used as a basis for market segmentation [1].

© Springer Nature Singapore Pte Ltd. 2019
Y. Sun et al. (Eds.): ChineseCSCW 2019, CCIS 1042, pp. 605–617, 2019.
https://doi.org/10.1007/978-981-15-1377-0_47

However, the current determination of consumer shopping decision-making styles is mostly carried out through questionnaires, and the manpower issuance and collection of questionnaires is time-consuming and laborious. At present, the rise of e-commerce, the background of e-commerce platform can easily record, statistics, and analyze consumer behavior data, such as browsing products, collecting goods, purchasing goods, browsing speed, buying frequency, etc. Adequate consumer behavior data makes it possible to determine consumer shopping decision-making style through data mining technology.

Careful analysis reveals that consumers' purchasing behavior characteristics are affected by both the purchased goods and the consumer's mental characteristics. In general, cautious consumers like to carefully compare the products of different merchants before purchasing goods, and collect the products they like, paying great attention to the price/performance ratio of the products. While direct consumers see their favorite, novelty products, they prefer to buy directly.

There are many data mining methods to analyze consumer behavior. For example, RFM model [2, 3], Tanimoto similarity [4], Pareto/NBD model [5], etc. The RFM model is just a consumer value model, the focus is on assessing the value of consumers, while Tanimoto similarity measures the similarity between consumers based on consumer buying behavior. Neither of these has assessed the mental characteristics of consumer shopping decision-making styles.

To this end, the consumer behavior analysis model we want to construct must have the following characteristics: (1) The data used by the model comes from the background of the e-commerce platform system and does not need to be obtained through questionnaires, etc. (2) The model is combined with the information of the goods purchased by the consumer for comprehensive analysis. (3) The model should be analyzed in combination with behaviors that reflect the mental characteristics of consumers (such as browsing behavior, collection behavior, etc.).

The consumer behavior analysis method proposed in this paper satisfies the above requirements. It consists of five steps: the determination of commodity ratings, the breakdown of consumer shopping decision behavior, the use of naive Bayes, the calculation of scores for different purchase behaviors, and the determination of consumer shopping decision styles. It first assigns different scores to the merchandise according to the comprehensive behavior of all consumers. The higher the score, the more inclined the consumers are to purchase cautiously. The lower the score, the more consumers prefer to buy directly. Then use the clustering method [7, 8] to divide the different decision-making behaviors of consumers into direct purchase behavior and cautious purchase behavior. Considering this unsupervised model, the results are often not perfect, especially in the data at the junction of the two categories. In view of this, we divide the data of the two types of junctions into unclassified data. Then use the classification data to train naive Bayes and classify the unclassified data. The use of this step represents the model to explore the environment that has never been seen based on its own good decisions. It is a self-supervised model. Finally, the different decision-making behaviors of all consumers are scored, and consumers are divided into direct style, cautious style and neutral style according to each consumer's decision-making behaviors.

The structure of this paper is as follows: Sect. 2 introduces the relevant research work on consumer behavior analysis, Sect. 3 details the consumer shopping decision-making style determination method, Sect. 4 gives the experimental results and comparative analysis, the last section of this paper Summarize and give a future research plan.

2 Related Work

Research on consumer decision-making styles can be roughly divided into two categories: non-data mining methods and data mining methods.

Non-data mining methods mostly use case surveys or questionnaires to obtain data, which is time-consuming and laborious. Sproles and Kendall [9] proposed the Consumer Styles Inventory to determine the consumer shopping decision style, which has been recognized by many scholars. Although the method classifies consumers' shopping decision-making styles in detail and specific, the data used by them is obtained through questionnaires, and it is necessary to fill in the questionnaires in a special way, which is time-consuming and labor-intensive.

Zhang [10] ordered 377 middle school students to complete the middle school students' consumption values questionnaire, the family financial education questionnaire and the middle school students' consumption decision style questionnaire, and concluded that the middle school students' consumption values play a mediating role between the family financial education mode and the consumption decision style. This method is also analyzed by questionnaire, which is time consuming and laborious.

Many scholars have proposed their methods for measuring consumer behavior. Bao [11] proposed an improved RFM model based on the perspective of customer consumption behavior. The method uses the analytic hierarchy process to determine the weight of each variable in the model, and on this basis, the K-Means clustering algorithm is used for customer segmentation to calculate and determine the customer's personal value to the merchant. Sun [12] used the RFM model to reflect the customer's purchasing preferences and customer value, and combined the RFM model with the original collaborative filtering mechanism to improve the traditional collaborative filtering algorithm. And developed a differentiated e-commerce recommendation strategy to make the recommendation method more in line with the individualization of different customers, thus achieving full personalization of recommended content and recommended forms. But the RFM model is just a consumer value model and does not measure consumer shopping decision styles.

Zhang [13] used Tanimoto similarity to measure the purchase behavior among customers, and designed a genetic clustering algorithm to divide the customer groups and aggregate customers with similar purchase behaviors into one category. However, this method only measures the similarity of purchase behavior between consumers, and does not apply to the measurement of consumer shopping decision style.

There are also many ways to predict the purchasing behavior of consumers. Using the behavior data of customers using RFID in Japanese supermarkets, Yi [14] proposed to use the support vector machine to classify consumers and predict the behavior of each type of consumers, and compare them with linear regression and naive Bayes. It shows that the support vector machine has the highest prediction accuracy. Yada [15] uses user narrative behavior as a research point to model consumer behavior using Naive Bayes and neural networks respectively. Experiments show that the neural network can learn more about the relationship between consumer behavior characteristics than Naive Bayes. Hui [16] aims to extract the relevant characteristics of the user brand based on the user's shopping behavior database on the Tmall website, and combine the random forest and the initial Gradient Boost Decision Tree model to predict the user's purchase behavior in the next month. However, the above method only predicts the consumer's purchase behavior, and does not apply to the determination of the consumer's purchase decision style.

Wang [17] proposed an individual consumption behavior prediction method based on quadratic clustering and hidden Markov chain theory. This method also only predicts consumer behavior and does not apply to the determination of the consumer's decision-making style.

In summary, the non-data mining method classifies consumers' shopping decision-making styles in detail and specific, but most of them use the questionnaires, case surveys, etc. However, the existing data mining methods are difficult to measure the mental characteristics of consumers' shopping decisions, and they cannot accurately determine the consumer shopping decision-making style. Therefore, it is necessary to propose a new method, which can comprehensively measure the mental characteristics of consumers' shopping decisions, and use data mining methods to determine the consumer shopping decision style.

3   Detailed Method

3.1   Determination of Product Ratings

Each type of product has the following characteristics (Fig. 1):

$$Feature_{category} = \{Cpv\_count, Cfav\_count, Ccart\_count\} \tag{1}$$

Among them, Cpv_count, $Cfav\_count$, and $Ccart\_count$ respectively represent the number of times of browsing, the number of collections, and the number of purchases of the goods purchased by the consumers. And all the features of the product make up a collection, named it *featureSet*, the minimum point to define this collection is:

$$point_{min} = \{min(Cpv\_count), min(Cfav\_count), min(Ccart\_count)\} \tag{2}$$

$min(Cpv\_count)$, $min(Cfav\_count)$ and $min(Ccart\_count)$ respectively represent minimum number of items viewed, minimum number of items stored, and minimum

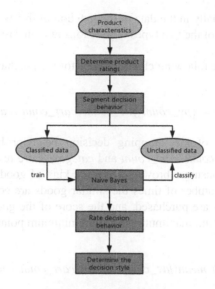

Fig. 1. Overall flow chart.

number of items purchased in set *featureSet*. *DIS(A,B)* represents the Euclidean distance between the *A* and *B* vectors, and the distance between the *i*-th point *point$_i$* and *point$_{min}$* is *d$_i$*:

$$d_i = DIS(point_i, point_{min}) \tag{3}$$

The set of distances of all points in *featureSet* from *point$_{min}$* is denoted as *d*:

$$d = \{d_1, d_2, \ldots \ldots d_n\} \tag{4}$$

Normalize *d*, *normD* represents the normalized set, and *normD$_i$* is the normalized result of *d$_i$*:

$$normD = \{normD_1, normD_2, \ldots \ldots normD_n\} \tag{5}$$

Then *normD* is the rating of all kinds of products, and *normD$_i$* is the rating of the goods corresponding to the *i*-th element in *featureSet*.

3.2   Segmentation of Consumer Shopping Decision Behavior

Considering the interpretability and convenience of subsequent applications and algorithm performance factors, we hope to divide consumers' shopping decision behavior into two opposite behaviors: direct purchase behavior and cautious purchase behavior. So, we directly specify the number of clusters to 2, and manually specify two initial points. Here, we divide consumers' decision-making behavior into two categories through *k-means*, but for the unsupervised model of *k-means*, the results are often not perfect, especially in the data of the junction of the two categories. In view of this, we divide the data of the two types of junctions into unclassified data. The detailed method is as follows.

Consumers have the following characteristics for the purchase decision behavior of certain types of goods:

$$Feature_{decision} = \{pv\_count, fav\_count, cart\_count, categoryScore\} \qquad (6)$$

The characteristics of the shopping decision behavior have four dimensions, wherein pv_count, fav_count, $cart\_count$ and $categoryScore$ respectively represent the number of times the consumer browses the same kind of goods when purchasing the product decision, the number of times the similar goods are collected, the number of times the similar goods are purchased, and the score of the goods.

Define the center point, maximum point, and minimum point in the data as P, P_{max}, and P_{min}:

$$P = \{mean(pv\_count), mean(fav\_count), mean(cart\_count), mean(categoryScore)\} \qquad (7)$$

$$P_{max} = \{max(pv\_count), max(fav\_count), max(cart\_count), max(categoryScore)\} \qquad (8)$$

$$P_{min} = \{min(pv\_count), min(fav\_count), min(cart\_count), min(categoryScore)\} \qquad (9)$$

Then calculate CP_1 and CP_2 as follows:

$$CP_1 = (P_{max} - P)/2 \qquad (10)$$

$$CP_2 = (P - P_{min})/2 \qquad (11)$$

Use CP_1 and CP_2 as the initial point of the k-means algorithm, use k-means algorithm to cluster all commodities. Specifying two initial points can reduce the impact of the k-means algorithm's random selection of initial points on the clustering results. The Euclidean distance is used as a measure of similarity between two consumer behaviors to obtain clustering results.

Calculate the center point of the two clusters in the clustering result, and use $CP_0 = \{0, 0, 0, 0\}$ as the origin to calculate the Euclidean distance between the center point of the two clusters and CP_0. The cluster corresponding to the larger one is marked as a cautious purchase behavior, and the cluster corresponding to the small one is marked as a direct purchase behavior.

For cautious purchase behavior class $clusterA$, the distance between the i-th point $clusterA_i$ and P_{max} is dA_i:

$$dA_i = DIS(clusterA_i, P_{max}) \qquad (12)$$

The set of distances between all points in *clusterA* and P_{max} is denoted as *dA*. When *dA* is sorted in ascending order, the first 61.8%(Golden ratio) of the data is classified as cautious purchase behavior, and the other data is classified as unclassified data.

For direct purchase class behavior *clusterB*, the distance between the *i*-th point *clusterB_i* and P_{min} is dB_i:

$$dB_i = DIS(clusterB_i, P_{min}) \tag{13}$$

The set of distances between all points in *clusterB* and P_{min} is denoted as *dB*. When *dB* is sorted in ascending order, the first 61.8% of the data is direct purchase behavior, and the other data is classified as unclassified data.

3.3 Use of Naive Bayes

In the previous section, we divided consumer decision-making behavior into direct purchase behavior and cautious purchase behavior, and we also got a special collection —unclassified data sets. We need to classify unclassified data sets using a supervised model. Naive Bayes is an excellent supervised classification model that has been successfully applied in many ways [18, 19]. We use categorical data to train naive Bayes and divide unclassified data sets into direct purchase behavior and cautious purchase behavior.

As we all know, if the accuracy of a tag data is very low or only a small part is tagged data, then no matter how perfect this supervised model is, its performance is definitely very poor. For an unsupervised model like the *k-means* algorithm, the results may not be very accurate. But here, our training data is chosen to be away from the data of the two types of junctions. For data in clustering results that are far from the two types of junctions, the results are often credible. At the same time, we also specify two initial points to reduce the impact of the *k-means* algorithm randomly selecting the initial point on the clustering result.

3.4 Calculate Scores for Different Purchase Behaviors

We hope that the cautious purchase behavior score is in the range [0, 1], while the direct purchase behavior score is in the [−1, 0] interval, which is achieved by the following method.

For the cautious purchase behavior class, define P_{max} as a reference. Calculate the distance between the vector corresponding to each point in each of the cautious purchase behavior classes and P_{max}. The *i*-th point is denoted by *cPoint_i*, then the distance between *cPoint_i* and P_{max} is recorded as *cDi*:

$$cD_i = DIS(cPoint_i, P_{max}) \tag{14}$$

The set of distances between all points and P_{max} is denoted as *cDSet*, and the reciprocal is denoted as *cDSetR:*

$$cDsetR = \{1/cD_1, 1/cD_2\ldots\ldots1/cD_n\} \tag{15}$$

Normalizing $cDSetR$, the normalized $cDSetR$ represents a collection of ratings for all consumer behavior.

The treatment of the direct purchase behavior class is like the treatment of the cautious purchase behavior class. However, there are some differences. The direct purchase behavior class is referenced to P_{min}, and the final score needs to be mapped to the $[-1, 0]$ interval by taking the opposite number.

3.5 Determination of Consumer Shopping Decision-Making Style

We divide consumer shopping decision-making styles into three categories: neutral style, direct style, and cautious style. The specific processing procedure is as shown in Algorithm 1. The *behaviorScore* in Algorithm 1 represents the behavioral score we calculated through Sect. 3.4.

Algorithm 1: Consumer Shopping Decision Style Determination Algorithm

---

Input: The score of n items purchased by consumers

Output: The style of consumers

1. For (The score of n items purchased by consumers):
2. $scoreList=\{behaviorScore_1, behaviorScore_2 \cdots\cdots behaviorScore_n\}$
3.   If $sum(scoreList) \in [-\infty, -0.05]$
4.   Classify the consumer as a direct style;
5.   Else if $sum(scoreList) \in [-0.05, 0.05]$
6.   Classify the consumer as a neutral style;
7.   Else
8.   Classify consumers as cautious style;
9. End For

---

4   Experiment Analysis

4.1   Experimental Setup and Experimental Parameters

This section will conduct an experimental analysis of the method proposed in this paper. The experimental environment is: INTEL Corei5 CPU, 2.80 GHz; 4G memory. The data used is user behavior data between November 25, 2017 and December 3, 2017, with approximately 1 million behavioral users. Behaviors include browsing merchandise, collecting merchandise, adding merchandise, and purchasing merchandise (Table 1).

Table 1. Data set example

USER_ID	GOOD_ID	CATEGORY_ID	BEHAVIOR	TIME
100002	2452885	4806751	pv	1511854685
100002	1585267	4806751	pv	1511861332
100002	2452885	4806751	pv	1511861477
100002	1585267	4806751	pv	1511861485
100002	1585267	4806751	pv	1511861880
100002	2968552	4806751	pv	1511862934
100002	2452885	4806751	pv	1511864359
100002	5159088	4806751	pv	1511865232
100002	2452885	4806751	pv	1511866043

The experiment randomly selected three groups of users and set up two comparison models. The three groups of users were named as group A, group B and group C. By default, these users and user purchase behavior are conditionally independent and are not affected by other unrelated factors. The three sets of user data are:

(1) Group A: 10021 users, 26,554 purchases, an average of 2.64 items per user
(2) Group B: 9881 users, 25,836 purchases, an average of 2.61 items per user
(3) Group C: 9,981 users, 26,068 purchases, an average of 2.61 items per user

Two comparison models:

(1) The complete model proposed in this paper, named it *CompleteModel*
(2) On the basis of *CompleteModel*, the naive Bayes is removed, and only *k-means* is used to divide the consumer shopping decision behavior into direct purchase behavior and cautious purchase behavior, which is named *BasicModel*.

4.2   Experimental Results and Analysis

Figures 2, 3 and 4 show the experimental results of two comparison models on three groups of users. The x, y, and z axes represent the number of views, the number of favorites, and the number of purchases. The size of the bubble represents the product rating. The larger the bubble, the higher the product rating. Bubble color represents category, red bubble represents direct purchase behavior, and yellow bubble represents cautious purchase behavior. The figure on the left shows the result of A, and the graph on the right shows the result of B. The comparison of the figure can be used to visually see the effect of naive Bayes.

Figure 2 shows the results of the two models on the group A. There are a lot of scattered red bubbles in the result of *BasicModel*. These scattered red bubbles are intuitively a cautious purchase, and *BasicModel* divides them into direct purchases. In the result of *CompleteModel*, the red bubbles are more concentrated, and the division of the two categories is in line with the intuitive understanding.

Figures 3 and 4 show the results of the two models on the group B and the group C. There are many red bubbles in the circled part. In visual understanding, these red bubbles should not be classified as direct purchases. In contrast, the red bubble is more concentrated in the result of *CompleteModel*. Compared with *BasicModel*, *CompleteModel* has fewer error classification points. In summary, *CompleteModel*'s effect on three user groups is better than *BasicModel*, and its classification of consumer behavior categories is more in line with human intuitive understanding.

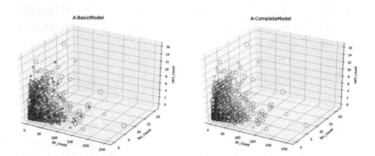

Fig. 2. Group A: Visual representation of consumer decision behavior breakdown results for BasicModel (left) and Complete Model (right) (Color figure online)

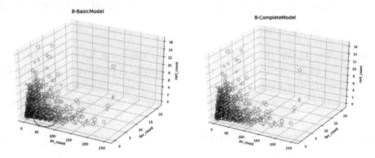

Fig. 3. Group B: Visual representation of consumer decision behavior breakdown results for BasicModel (left) and Complete Model (right) (Color figure online)

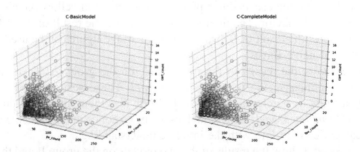

Fig. 4. Group C: Visual representation of consumer decision behavior breakdown results for BasicModel (left) and Complete Model (right) (Color figure online)

In Table 2, the central point of the direct purchase behavior class for the two models on three user groups is shown. Compared with the central point of the direct purchase behavior class in the result of BasicModel, the center point of the direct purchase behavior class in CompleteModel results is closer to the origin. Because CompleteModel correctly divides BasicModel's misclassification point into cautious purchase behavior class. This makes the direct purchase behavior class more convergent, and closer to the origin. Therefore, in the result of CompleteModel, the central point of the direct purchase behavior class is closer to the origin, which is an inevitable result. Table 2 verifies that CompleteModel is better than BasicModel with accurate data.

Table 2. The central point of direct purchase behavior.

User group	Model	pv_count	Fav_count	Cart_count	categoryScore
Group A	BasicModel	4.833	0.132	0.339	0.009
Group A	CompleteModel	3.817	0.083	0.233	0.009
Group B	BasicModel	4.156	0.112	0.239	0.010
Group B	CompleteModel	3.623	0.060	0.209	0.009
Group C	BasicModel	4.392	0.115	0.321	0.009
Group C	CompleteModel	3.598	0.066	0.235	0.009

In the end, we combine all the behaviors of each consumer to divide the consumer's shopping decision style into direct style and cautious style. And we use a neutral style to define those consumers who don't have obvious characteristics (Table 3).

Table 3. Number of people in each category

User group	Model	Cautious style	Neutral style	Direct style
Group A	BasicModel	4705	2503	2813
Group A	CompleteModel	5953	2208	1881
Group B	BasicModel	2945	1326	5609
Group B	CompleteModel	3983	2123	3774
Group C	BasicModel	3547	1077	5357
Group C	CompleteModel	4684	1543	3754

5 Summary and Research Outlook

This paper proposes a self-supervised data mining method, using e-commerce data to derive the style of consumer shopping decisions. Specifically, it comprehensively represents the mental characteristics of consumers' shopping decisions based on the differences in the products purchased by consumers and the behaviors of consumers when purchasing decisions (for example, browsing products, collecting goods, etc.). It explores an environment that has never been seen based on its good past decisions. Consumer behavior is first divided into direct purchase behavior, cautious purchase behavior, and unclassified data. Then use the categorical data to train naive Bayes and classify the categorical data. And comprehensive consumer behavior has derived the consumer's shopping decision style. The experimental results show that the proposed model makes the classification of consumer decision-making behavior more in line with the intuitive understanding, which is obviously better than the comparison model. This model can effectively determine the consumer decision style.

This article has some areas for improvement. Consumers' decision-making styles are complex, not just the three styles that are available in the text: cautious buying style, neutral style, direct buying style. For example, some consumers pay more attention to brands, and they are more inclined to buy well-known brands with high prices. Other consumers value new fashion, these consumers pursue trends and pursue diversity. How to extract more consumer decision-making styles from data through e-commerce back-end data utilization data mining is the next research goal.

References

1. Xue, H.: Review and prospect of research on consumer shopping decision style scale. Consum. Econ. (05), 92–96 (2007)
2. Ching, C., You, C.: Classifying the segmentation of customer value via RFM model and RS theory. Expert Syst. Appl. **36**(3), 4176–4184 (2009)
3. Xu, X., Wang, J., Tu, H., Mu, M.: Customer classification of E-commerce based on improved RFM model. J. Comput. Appl. **32**(05), 1439–1442 (2012)
4. Andreas, M., Thomas, R.: An improved collaborative filtering approach for predicting cross-category purchases based on binary market basket data. J. Retail. Consum. Serv. **10**(3), 123–133 (2003)
5. David, C.S., Donald, G.M., Richard, C.: Counting your customers: who are they and what will they do next? Manag. Sci. **33**(1), 1–24 (1987)
6. Ma, S.: The Pareto/NBD model extensions. Syst. Eng. **26**(8), 123–126 (2008)
7. Sun, J., Liu, J., Zhao, L.: Clustering algorithms research. J. Softw. **19**(01), 48–61 (2008)
8. Zgou, T., Lu, H.: Clustering algorithm research advances on data mining. Comput. Eng. Appl. **48**(12), 100–111 (2012)
9. Sproles, G.B., Kendall, E.L.: A methodology for profiling consumers' decision-marking styles. J. Consum. Affairs **20**(2), 267–279 (1986)
10. Zhang, J., Zou, Y.: The mediating role of adolescents' consumption values between family financial education and consumer decision-making styles. J. Psychol. Sci. **35**(02), 376–383 (2012)
11. Bao, Z., Zhao, Y., Zhao, Y., Hu, X., Gao, F.: Segmentation of Baidu takeaway customer based on RFA model and cluster analysis. Comput. Sci. **45**(S2), 436–438 (2018)
12. Sun, L., Zhang, W.: Electronic recommendation mechanism based on RFM model and collaborative filtering. J. Jiangsu Univ. Sci. Technol. (Nat. Sci. Ed.) **24**(03), 285–289 (2010)
13. Zhang, Z., Yan, J., Chen, F., Li, M.: Feature extraction of customer purchase behavior based on genetic algorithm. Pattern Recognit. Artif. Intell. **23**(02), 256–266 (2010)
14. Yi, Z., Shawkat, A., Katsutoshi, Y.: Consumer purchasing behavior extraction using statistical learning theory. Procedia Comput. Sci. **35**, 1464–1473 (2014)

15. Yada, K.: String analysis technique for shopping path in a supermarket. J. Intell. Inf. Syst. **36** (3), 385–402 (2011)
16. Hui, S.K., Bradlow, E.T., Fader, P.S.: Testing behavioral hypotheses using an integrated model of grocery store shopping path and purchase behavior. J. Consum. Res. **36**(3), 478–493 (2009)
17. Song, T., Wang, X.: Customer behavior prediction for card consumption based on two-step clustering and hidden Markov chain. J. Comput. Appl. **36**(07), 1904–1908 (2016)
18. Zhang, P., Tang, S.: Privacy preserving Naive Bayes classification. Chin. J. Comput. (08), 1267–1276 (2007)
19. He, J., Meng, Z., Chen, X., Wang, Z., Fan, X.: Semi-supervised ensemble learning approach for cross-project defect prediction. J. Softw. **28**(06), 1455–1473 (2017)

DeepLoc: A Location Preference Prediction System for Online Lodging Platforms

Yihan Ma[1,2], Hua Sun[1,2], Yang Chen[1,2(✉)], Jiayun Zhang[1,2], Yang Xu[1],
Xin Wang[1,2], and Pan Hui[3,4]

[1] School of Computer Science, Fudan University, Shanghai, China
{mayh18,hsun15,chenyang,jiayunzhang15,xuy,xinw}@fudan.edu.cn
[2] Shanghai Key Lab of Intelligent Information Processing, Fudan University,
Shanghai, China
[3] Department of Computer Science, University of Helsinki, Helsinki, Finland
panhui@cs.helsinki.fi
[4] Hong Kong University of Science and Technology, Hong Kong SAR, China

Abstract. Online lodging platforms have become very popular around
the world. To make a booking, a user normally needs to select a city first,
then browses among prospective options. To improve the user experience,
understanding the location preference of a user's booking behavior will
be useful. In this paper, we propose DeepLoc, a location preference prediction system, adopting deep learning technologies to predict the location preference of a user's next booking, based on both the descriptive
features and the user's historical booking records. Using the real data
collected from Airbnb, we can see that DeepLoc can achieve an F1-score
of 0.885 for booking apartments in the city of London.

Keywords: Online lodging systems · Location preference ·
Prediction · Deep learning

1 Introduction

Traveling has been an important part in people's daily life. A satisfactory
accommodation in an unfamiliar city can greatly promote people's travel experience. There are a number of online lodging services, such as Airbnb [1–3],
Booking.com [4] and Homestay[1], offering both traditional hotels and residential
accommodations for visitors.

When booking accommodations online, users usually have a specific destination, such as London, Paris and New York. In addition, online lodging services
such as Airbnb, Booking.com and Homestay offer a set of filters. Users can
choose accommodations which meet their requirements by setting these filters.
In particular, many users have location preferences for accommodations. When

---

[1] https://www.homestay.com/, accessed on May 1, 2019.

Y. Sun et al. (Eds.): ChineseCSCW 2019, CCIS 1042, pp. 618–630, 2019.
https://doi.org/10.1007/978-981-15-1377-0_48

traveling to a new city, some people like to live in prosperous places, for example, the downtown of the city. These places are normally well connected, and have attractions for tourism and shopping. On the contrary, some people may tend to live in less prosperous places, where they can get rid of the traffic jam and the noisy environment of the downtown. However, users cannot select accommodations based on their location preferences directly on most online lodging platforms.

Given the importance of a user's location preference, we aim to understand a user's location preference and recommend her accommodations accordingly for her next trip. In this paper, we propose a location preference prediction system named DeepLoc. In our system, we utilize users' history booking records to predict the desired location of users' next booking in a given city. Our methodology combines the long short-term memory (LSTM) neural networks [5,6] and some conventional supervised machine learning algorithms to make use of both dynamic and descriptive characteristics of each user. LSTM is well-known for its ability to process time sequence data and analyze the dynamic patterns.

In this paper, we use Airbnb as a case study. Founded in 2008, Airbnb has quickly grown to be one of the world's most popular online lodging platforms with about 200 million registered users. It has attracted millions of hosts to rent out their apartments. In Airbnb, users can find over 6 million accommodations in more than 81,000 cities in 191 countries[2]. In particular, we choose London as a sample city for our study. London is one of the world's leading tourism destinations, which attracted over 20 million international visitors in 2018 and it has over 80 thousand accommodations in Airbnb by Dec. 2018. By dividing the city of London into central London and non-central London, our DeepLoc system can predict which part a traveler will choose to live. The results show that our system can achieve an F1-score of 0.885, indicating that DeepLoc performs very well when predicting a user's location preference. The contributions of this paper are summarized as follows.

First, we formulate the location preference prediction problem in online lodging platforms and design a system which can obtain the location preference for booking and give users better recommendations.

In addition, we propose a location preference prediction system named DeepLoc to predict the desired location of a user's next booking in a selected city. Our system combines the advantages of LSTM and traditional supervised machine learning algorithms to deal with both dynamic and descriptive features of the dataset.

Last but not least, we evaluate our system using a real dataset collected from Airbnb. The results show that our system can predict the location of user's next booking in London with an F1-score of 0.885.

---

[2] https://press.airbnb.com/about-us/, accessed on May 1, 2019.

2   Data Collection and Feature Extraction

In this section, we give an overall introduction of our datasets. We delineate the preprocessing part of our datasets and describe the features we used in the experimental part.

2.1   Datasets and Preprocessing

In this paper, we want to design a system which can use the history booking records of a user to predict which part she will live in the next trip in a given city. Also, every individual is special, the personal information of the user might be useful for us to do the prediction task. Thus, our datasets consist of 2 parts, InsideAirbnb dataset from which we obtain the detailed information of the history booking record and user profile dataset where we can get the personal information of each user.

InsideAirbnb Dataset. First, we obtain the accommodation data and review data of all Airbnb accommodations in 84 cities from InsideAirbnb[3]. InsideAirbnb dataset is widely used in studies about Airbnb [2,3]. For each city, we get two .csv files which store the accommodation data and review data of the city, respectively. The accommodation data includes price, longitude, latitude, type

Table 1. Description of the InsideAirbnb dataset

Categories	Features	Description
Accommodation data	Price	The price of accommodation
	Location	The longitude and latitude of accommodation
	Type	The type of accommodation, including villa, apartment...
	Amenities	The amenities of accommodation, including hair dryer, kitchen...
Review data	ID	The ID of reviewed recommendation
	Time	The time when the review is given
	Guest_ID	The ID of the guest
	Guest_name	The name of the guest
	Comment	The detailed comment of the review

---

[3] InsideAirbnb (http://insideairbnb.com/, accessed on May 1, 2019) is a website which offers open sourced dataset contains the detailed information of the accommodations and reviews in 84 cities in Airbnb.

of Airbnb accommodations, amenities of accommodations, demographical information of hosts and so on. The review data is the collection of reviews of all
the accommodations in the city. Each line of review data represents an actual
visit including the ID of the accommodation, time stamp of the review, ID and
name of the guest who lived in this accommodation and wrote this review. The
detailed information of InsideAirbnb dataset is summarized in Table 1.

Since we want to utilize the history records of a particular user to predict
the desired location of her next booking, we extract all the reviews of the same
user and build a user-review related database based on MongoDB, which is a
cross-platform document-oriented database program. Unlike NoSQL, MongoDB
uses JSON-like documents to store data. Because the lengths of history booking
records of different users are not the same, it is convenient to store the history
booking records with MongoDB. Finally we get a database consisting of more
than 20 million users. To make sure we can get enough information from previous
booking records, we select 15,442 users who have at least 7 history records.
As introduced in previous sections, users usually select city first when booking
accommodations in online lodging platforms. We choose London as our target
city at first, and get a dataset containing the detailed information of 2,045 users
whose latest booking records are in London.

User Profile Dataset. Second, we build a crawler to get the profiles of selected
users. A user profile includes name, location, registration time, self description,
the total number of reviews and other verification items, such as work, language,
credit card, government ID and so on. Note that when crawling the profiles of
users, some of them might not be available[4]. Finally, we get 2,004 user profiles.
All the profile data was crawled between 10 Mar. 2019 and 11 Mar. 2019.

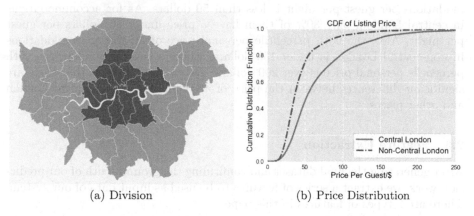

(a) Division   (b) Price Distribution

Fig. 1. The division and Airbnb accommodation price distribution of London

---

[4] If an account is deleted by the corresponding user or by the Airbnb platform, the
profile page of this user will be unavailable.

2.2   Obtaining Ground Truth

In this paper, our main target is to predict which part of a given city a user will choose for her next booking, so that we can obtain the preferences of users for location and improve the quality of users' booking experiences. To evaluate the performance of our system, we use the latest booking record as our ground truth. With our processed InsideAirbnb dataset, we get 2,045 selected users. In our DeepLoc system, we utilize the detailed information of 6 booking records before the latest one as input data, and the output data is the location of the last booking record.

Since we want to predict whether a user will choose to live in prosperous downtown or quiet outskirts, we need to divide London into two different parts. In this paper, the division criterion is according to *London Plan*[5] which defines the *Central Activities Zone* as a set of 10 Boroughs. This area is described as "a unique cluster of vitally import activities". In the light of this description, we divide London into two parts, central London which includes the 10 boroughs and the other boroughs in Greater London, as shown in Fig. 1(a). Users whose latest booking records are located in central London are labeled as the first class. Users with latest booking record in non-central London are labeled as the second class. The ratio of the number of samples in first class to second class is 1,071:974.

To validate our classification criterion of London, we compare the booking price of accommodations in central London and non-central London in Airbnb. Figure 1(b) shows the comparison of price distribution of accommodations in central London and non-central London. We can see from the figure that the price of accommodations in central London are higher than that in non-central London on average. In non-central places of London, the price of over 60% accommodations per guest per night is less than 50 dollars. As for accommodations in central London, only 30% of them have a price under 50 dollars per guest per night. Also, the 90th percentile personal price of all the accommodations in non-central London is under 150 dollars, while in central London, the 90th percentile personal price is over 200 dollars. This figure indicates that there are significant differences between the price of accommodations in central London and other places.

2.3   Feature Extraction

After generating the final dataset and confirming the ground truth of our prediction work, we extract a series of features to be used as input data of our system. There are 2 types of features in this paper.

Historical Booking Features. In our final InsideAirbnb user-review related dataset, we have 2,045 users who have at least 7 history records. As shown before, we utilize the 6 history records each user before the latest one to predict

---

[5] https://www.london.gov.uk/what-we-do/planning/london-plan, accessed on May 1, 2019.

which part she will live in London. And we use the location of the latest booking record as our ground truth. At first, we need to extract features of the previous 6 booking records.

For each record, the review data contains time, ID and city of this accommodation and the comments to it. First, we acquire the sentiment of each comment by VADER Sentiment [7]. VADER (Valence Aware Dictionary and sEntiment Reasoner) is a lexicon and rule-based sentiment analysis tool and it is specifically used to detect sentiments expressed in social media. The input data of VADER is sentences and the results of VADER usually contain four items, i.e. the ratio of positive words, negative words, neural words and a compound score. The compound score is a general measure of sentiment of a given sentence. Then we obtain the detailed information of accommodations through the accommodation csv file of each city. Note that in Airbnb, each accommodation has a fixed ID, so we can look up the detailed information of this accommodation by its ID. From the accommodation file, we can get some information of this accommodation itself, such as the price, longitude and latitude, property type[6], room type[7], accommodates[8], amenities, cleaning fee. Also, we can get some information about the host, including demographic characteristics, the response rate of the host, the total number of accommodations owned by the host, some verified information the host offered.

Also, for each history booking record, we can get the longitude and latitude of the corresponding accommodation. To enrich our feature set, we obtain the POI information within 5 miles of each accommodation via Google Places API[9]. Google Places API is a service that returns information about places. When sending HTTP requests with the longitude and latitude of accommodations to Google Places service, users can choose to return information of certain kinds of Points of Interest (POI), such as shopping malls and museums. It will return a JSON-formatted file which contains the information of all the required categories of POIs. Since Google Places API has very strict access speed restrictions, we choose to get the information of only 5 categories, including subway station, bus station, train station, airport and shopping mall.

Finally, we use the extracted historical booking features to formulate a Historical Matrix $H^{6 \times 263}$, which stores the extracted features of 6 history booking records, where each booking record consists of a 263-dimensional feature set.

User Profile Features. The user profile features are extracted from our user profile dataset, which contains the city where the user lives in, the created time of Airbnb account, the total number of reviews from guests[10], the total number of reviews from hosts, the verified information and so on. The feature set of users whose profiles are no longer available are filled by -1.

---

[6] The types of Airbnb accommodations, including apartments, villas, tree houses.

[7] There are three types of rooms in Airbnb: Private room, Shared room and Entire home.

[8] It refers to the number of people that this accommodations can host at one time.

[9] https://developers.google.com/places/web-service/intro, accessed on May 1, 2019.

[10] The reviews user received from her guests.

Finally, we get a dataset including 2,045 samples. Each sample has a 1594-dimensional feature set consisting of information of 6 history booking and profile data of the user. Table 2 is the summarized feature set.

Table 2. Feature set of each user

Category	Features
Historical booking features	ID of accommodation in each booking record
	City of accommodation in each booking record
	Time of each booking record
	Sentiment of comments in each booking record
	Amenities of accommodation in each booking record
	Demographic information of host in each booking record
	Geographical information of accommodation in each booking record
User profile features	ID of user
	City of user
	Created time of user's Airbnb account
	Number of reviews from hosts
	Number of reviews from guests
	Verified information

3   System Design

Online lodging platforms usually offer filters like price, amenities to help users choose accommodations, but the filters cannot help users select accommodations based on their location preferences directly. Thus, we propose a location preference prediction system names DeepLoc to predict the location of next booking in a given city for users. DeepLoc is based on a model named DeepScan which was proposed by Gong et al. [8]. To utilize the history booking records, we need to involve an algorithm which is capable to process time series information. There have been a lot of techniques which can process time sequence data. LSTM networks [5,6] have shown its power in recent studies. So we involve LSTM in our system to acquire the dynamic features of sequential booking records. In this section, we will first give an overall description of our system, and then specifically introduce its workflow.

Our online lodging recommendation system is mainly made of 2 parts: a bidirectional LSTM (BLSTM) module which processes time sequence data and a Decision Maker which utilizes combined conventional features and dynamic features as input data and outputs the classification results of each sample. The architecture of DeepLoc is shown in Fig. 2.

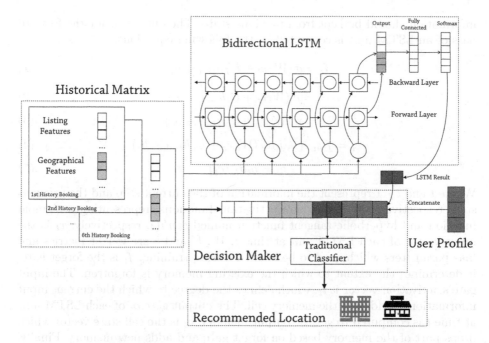

Fig. 2. The architecture of DeepLoc

As shown in Fig. 2, we first input extracted features of booking records into a BLSTM layer to get the dynamic information of each user. The historical matrix represents the features of 6 history booking records. After putting the last output of BLSTM into a softmax layer, we can get 2 normalized probabilities. Then we concatenate the probability features, the historical booking features and the user profile features to get the final feature set. In the Decision Maker module, we train the classification model with the final feature set to get classification results of each user.

4   Implementation and Evaluation

In this section, we show the implementation details of our system, and evaluate its prediction performance with real data collected from Airbnb. In the next subsections, we will introduce the implementation of BLSTM, Decision Maker and the evaluation results in detail.

4.1   Bidirectional LSTM

LSTM units were proposed by Hochreiter et al. [5]. They are designed to process long-term dependency information and commonly used to overcome gradient vanishing problem. Compared to traditional recurrent unit, the main improvement of LSTM is the introduction of *forget gate*, which determines how much information should be kept from previous state. The output h_t for the forward pass of an LSTM unit is computed by the following equations:

$$f_t = \sigma \left(W_f x_t + U_f h_{t-1} + b_f \right) \tag{1}$$

$$i_t = \sigma \left(W_i x_t + U_i h_{t-1} + b_i \right) \tag{2}$$

$$o_t = \sigma \left(W_o x_t + U_o h_{t-1} + b_o \right) \tag{3}$$

$$c_t = f_t c_{t-1} + i_t \sigma_h \left(W_c x_t + U_c h_{t-1} + b_c \right) \tag{4}$$

$$h_t = o_t \sigma_h \left(c_t \right) \tag{5}$$

Where t and $t-1$ represent the information of the current step and the previous step, respectively. σ_g and σ_h are activation functions, representing a sigmoid function and hyperbolic tangent function named $tanh()$, respectively. x_t is the input vector of each LSTM unit at time t. W, U and b are weight metrics and bias parameters which need to be learned during training. f_t is the forget gate, it determines the extent to which the existing memory is forgotten. The input gate's activation vector is i_t, which defines the degree to which the current input information is added to the memory cell. The output gate o_t of each LSTM unit at time t is computed to get the output memory. c_t is the cell state vector which drops part of the memory based on forget gate and adds new memory. Finally, the output h_t is computed based on the output gate. h_t is the hidden state vector and also known as output vector of the LSTM unit at time t.

The current cell state and the output of an LSTM unit are generated by previous and current input vectors. However, for some sequence modeling tasks, future information can improve the performance of LSTM model a lot. Hence, we introduce BLSTM network, which is an extension to unidirectional LSTM network by adding a backward LSTM layer. BLSTM has a capability to utilize both previous and future input vectors. And finally, the output h_t of a BLSTM unit of the current step t is computed as follows:

$$h_t = [\overrightarrow{h_t} \oplus \overleftarrow{h_t}] \tag{6}$$

The loss function we use is called binary cross-entropy loss, which is commonly used in binary prediction tasks. The equation of binary cross entropy loss is given:

$$loss = -y_i \log \left(\sigma_g \left(s_i \right) \right) - (1 - y_i) \log \left(1 - \sigma_g \left(s_i \right) \right) \tag{7}$$

Where y_i is the ground truth label and s_i represents the probability to be the first class. σ_g represents a sigmoid function.

In this paper, our LSTM model is constructed by Keras[11], a high-level neural networks API which is capable of running on top of some deep learning platforms, such as TensorFlow[12]. The LSTM we use is a fully connected Bidirectional LSTM. The learning rate is set as 0.01. We utilize a dropout layer to prevent

---

[11] https://keras.io/, accessed on May 1, 2019.
[12] https://www.tensorflow.org/, accessed on May 1, 2019.

over-fitting problem, and the dropout ratio is set to 0.1. Also, we use the Adam optimizer to optimize our model in the training process. In the learning process, the last output of the BLSTM unit will be sent to a 3-layer fully connected layer (FC layer). Then, the output of FC layer will go through a softmax layer to compute the probabilities to be samples in the first class and in the second class. Thus, the ultimate output of our BLSTM model is a two-dimensional probability vector.

4.2   Decision Maker

The Decision Maker is made of conventional supervised machine learning algorithms. In this paper, we choose several frequently-used machine learning algorithms, including RandomForest [9], Decision Tree [10], XGBoost [11], Light-GBM [12] and Catboost [13]. In the training process, GridSearchCV [14] is applied to get the optimal parameters of each model automatically. Given a set of values of each parameter which needs to be tuned, GridSearchCV iterates through each parameter combination and records the parameters which lead to the best F1-score. For all the machine learning algorithms, we use a 5-fold cross-validation to avoid over-fitting.

4.3   Evaluation

In this work, we first randomly select 90% of samples as training and validation set, and use the other 10% as test set. Note that we use the same training and validation set and test set in BLSTM and Decision Maker. The evaluation metrics we use for Decision Maker are F1-score and AUC [15]. F1-score is a combination and balance of precision and recall. Precision reflects the performance of a model when identifying a sample as positive one. And recall is introduced to measure the ratio of positive samples that have been correctly predicted. AUC is the area under the ROC (receiver operating characteristic) curve, which tells how much the model is capable of distinguishing between classes. All these metrics range from 0 to 1. The larger the value is, the better the performance is.

To better evaluate the performance of our system, we introduce some baselines. We utilize basic BLSTM and several representative supervised machine learning algorithms as baselines. For LSTM algorithm, the implementation details are the same as that in our online lodging recommendation system. The input feature of LSTM is the history booking features of each user. The algorithms we use for machine learning task coincide with our system. For each machine learning algorithm, the input feature includes the historical matrix and user profile features. The classification results of our system and all baselines are summarized in Table 3. In general, we can see that among all the experiments, our system with a Decision Maker of LightGBM performs the best, with an F1-score of 0.885 and an AUC-ROC score of 0.877. Also, when comparing the classification results of our system and machine learning algorithms which coincide with the Decision Maker in our system, a better result can be noticed except for XGBoost. It indicates that in general, our system can utilize the dynamic

Table 3. The performance of our system and baselines

Algorithms	Precision	Recall	F1-score	AUC-ROC
DeepLoc (RF)	0.862	0.859	0.858	0.857
DeepLoc (DT)	0.832	0.824	0.823	0.832
DeepLoc (XGBoost)	0.858	0.854	0.853	0.852
DeepLoc (LightGBM)	0.857	0.914	**0.885**	**0.877**
DeepLoc (Catboost)	0.854	0.849	0.863	0.862
RF	0.852	0.849	0.848	0.847
DT	0.820	0.820	0.819	0.819
XGBoost	0.870	0.868	0.868	0.848
LightGBM	0.870	0.868	0.868	0.867
Catboost	0.857	0.854	0.853	0.852
BLSTM	0.772	0.905	0.824	0.830

history information of users better. It can also be noticed that the result of LSTM is almost the worst. The results suggest that for a given city, our system can successfully predict which area a user will live in the city.

5   Related Work

Research on Airbnb. On account of the rapid development of Airbnb, there have been some researches about the profiles of Airbnb users, the accommodation in Airbnb and the comparison of Airbnb accommodations and traditional hotels. Fradkin et al. [1] did a field experiment on reviews from Airbnb and found that reviews were typically informative but negative experiences were under-reported. Ma et al. [2] studied the profile of Airbnb users and got a conclusion that Airbnb hosts who disclosed more information on the profile could gain more trust from guests. Lee et al. [21] analyzed the social features associated with accommodations and found the most significant features for room sale in Airbnb. Quattrone et al. [3] did an cross-ref analysis of Airbnb economy with Foursquare data, census data and hotel data in London. Also, Grbovic et al. [16] gave real-time recommendations for users in Airbnb based on their click data and search history. Zhou et al. [17] presented a comprehensive and evolutionary study of Airbnb, using the information of 43.8 million users. However, none of previous work studied the location preferences of users when booking in Airbnb.

Research on Hotel Recommendation System. Zhang et al. [18] combined collaboration filtering (CF) with content-based (CBF) method to overcome sparsity issue in hotel recommendation. Lin et al. [19] utilized users' browsing information when reading hotel reviews on mobile devices to obtain users' preference to make personal recommendations. Raul et al. [20] tried to acquire features to capture the user's price sensitivity, and then constructed a recommendation system which was price sensitive. Most of these approaches were based on the search history or click history. As for our work, we use the history records to obtain the location preferences of users to give them better recommendations.

6 Conclusion and Future Work

In this paper, we study the users' location preferences in booking accommodations on online lodging platforms. To improve the user experience, we propose a deep learning-based location preference prediction system, called DeepLoc, for online lodging platforms. Our system combines BLSTM and traditional machine learning algorithms. It can utilize a user's fine-grained historical booking records and descriptive characteristics. We implement our system with a real dataset collected from Airbnb using London as the target city. Our evaluation results show that DeepLoc can predict the location preference of a user's next booking in London with an F1-score of 0.885.

In the future, we will give more fine-grained prediction by dividing a given city into multiple parts. We will use the data of other cities to further validate its performance on Airbnb. Also, experiments with datasets from other online lodging platforms like Booking.com will be conducted to evaluate the compatibility of DeepLoc.

Acknowledgment. This work is sponsored by National Natural Science Foundation of China (No. 61602122, No. 71731004), the Research Grants Council of Hong Kong (No. 16214817) and the 5GEAR project from the Academy of Finland.

References

1. Fradkin, A., Grewal, E., Holtz, D., Pearson, M.: Bias and reciprocity in online reviews: evidence from field experiments on Airbnb. In: Proceedings of EC (2015)
2. Ma, X., Hancock, J.T., Mingjie, K.L., Naaman, M.: Self-disclosure and perceived trustworthiness of Airbnb host profiles. In: Proceedings of ACM CSCW (2017)
3. Quattrone, G., Proserpio, D., Quercia, D., Capra, L., Musolesi, M.: Who benefits from the "sharing" economy of airbnb? In: Proceedings of WWW (2016)
4. Mellinas, J.P., María-Dolores, S.M.M., García, J.J.B.: Booking.com: the unexpected scoring system. Tourism Manag. **49**, 72–74 (2015)
5. Hochreiter, S., Schmidhuber, J.: Long short-term memory. Neural Comput. **9**(8), 1735–1780 (1997)
6. Graves, A., Mohamed, A., Hinton, G.E.: Speech recognition with deep recurrent neural networks. In: Proceedings of ICASSP (2013)
7. Hutto, C.J., Gilbert, E.: VADER: a parsimonious rule-based model for sentiment analysis of social media text. In: Proceedings of ICWSM (2014)
8. Gong, Q., et al.: DeepScan: exploiting deep learning for malicious account detection in location-based social networks. IEEE Commun. Mag. **56**(11), 21–27 (2018)
9. Breiman, L.: Random forests. Mach. Learn. **45**(1), 5–32 (2001)
10. Quinlan, J.R.: C4.5: Programs for Machine Learning. Morgan Kaufmann, Burlington (1993)

11. Chen, T., Guestrin, C.: XGBoost: a scalable tree boosting system. In: Proceedings of ACM SIGKDD (2016)
12. Ke, G., et al.: LightGBM: a highly efficient gradient boosting decision tree. In: Proceedings of NIPS (2017)
13. Prokhorenkova, L.O., Gusev, G., Vorobev, A., Dorogush, A.V., Gulin, A.: Cat-Boost: unbiased boosting with categorical features. In: Proceedings of NeurIPS (2018)
14. Fabian, P., et al.: Scikit-learn: machine learning in Python. J. Mach. Learn. Res. **12**, 2825–2830 (2011)
15. Fawcett, T.: An introduction to ROC analysis. Pattern Recogn. Lett. **27**(8), 861–874 (2006)
16. Grbovic, M., Cheng, H.: Real-time personalization using embeddings for search ranking at Airbnb. In: Proceedings of ACM SIGKDD (2018)
17. Zhou, Q., et al.: Measurement and analysis of the reviews in Airbnb. In: Proceedings of IFIP Networking (2018)
18. Zhang, K., Wang, K., Wang, X., Jin, C., Zhou, A.: Hotel recommendation based on user preference analysis. In: Proceedings of ICDE Workshops (2015)
19. Lin, K., Lai, C., Chen, P., Hwang, S.: Personalized hotel recommendation using text mining and mobile browsing tracking. In: Proceedings of IEEE SMC (2015)
20. Raul, S., Jordan, S., Rodrygo, L.T.S.: Exploiting socio-economic models for lodging recommendation in the sharing economy. In: Proceedings of ACM RecSys (2017)
21. Lee, D., Hyun, W., Ryu, J., Lee, W., Rhee, W., Suh, B.: An analysis of social features associated with room sales of Airbnb. In: Proceedings of CSCW Companion (2015)

A Multi-domain Named Entity Recognition Method Based on Part-of-Speech Attention Mechanism

Shun Zhang[1], Ying Sheng[1], Jiangfan Gao[1], Jianhui Chen[1,3(✉)],
Jiajin Huang[1,3], and Shaofu Lin[1,2]

[1] Faculty of Information Technology, Beijing University of Technology,
Beijing 100024, China
{zhanshun, shengying, S201761411,
hjj}@emails.bjut.edu.cn,
{chenjianhui, linshaofu}@bjut.edu.cn
[2] Beijing Institute of Smart City, Beijing University of Technology,
Beijing 100024, China
[3] Beijing Key Laboratory of MRI and Brain Informatics, Beijing, China

Abstract. Named entity recognition is an important and basic work in text mining. To overcome the shortcomings of existing multi-domain named entity recognition methods, a multi-domain named entity recognition method based on the part-of-speech attention mechanism, called BiLSTM-ATTENTION-CRF, was proposed in this paper. The domain dictionary was constructed to represent multi-domain semantic information and the BiLSTM network was used to capture the grammatical and syntactic features, as well as multi-domain semantic features in context information. A part-of-speech attention mechanism was designed to obtain the contribution weight of part-of-speech for entity recognition. Finally, a group of experiments were performed on the multi-domain dataset to compare various fusion strategies of multi-level entity information. The experimental results show that BiLSTM-ATTENTION-CRF has a high precision and recall rate, and can effectively recognizes the multi-domain named entities.

Keywords: Multi-domain entity recognition · Attention mechanism · BiLSTM · CRF

1 Introduction

Named Entity Recognition (NER) [1] refers to the recognition of entities with specific special meanings in texts, which is the basis of information extraction, information filtering, information retrieval, question-and-answer system, machine translation and other researches. From the early methods based on the dictionary and rule [2], the methods based on statistics [3], to the recent methods based on machine learning, various NER methods have been developed. At present, the NER methods based on deep learning are a research hotspot. Main deep learning models include the Recurrent Neural Network (RNN), Convolutional Neural Network (CNN), as well as their

© Springer Nature Singapore Pte Ltd. 2019
Y. Sun et al. (Eds.): ChineseCSCW 2019, CCIS 1042, pp. 631–644, 2019.
https://doi.org/10.1007/978-981-15-1377-0_49

combinations and variants. Some researchers also combined the deep learning models with traditional machine learning methods for improving the precision of NER [4].

Multi-domain named entity recognition is a specific NER task, which recognizes entities belonging to different domains. It is a key technology in machine translation, topic discovery, social network analysis, and spoken language understanding [5]. Compared with other NER tasks, the multi-domain named entity recognition needs to capture the richer semantic information from different domains, design the more complex annotation system, and solve the polysemous phenomenon among different domains. Therefore, it is a valuable research topic.

Based on the above observations, this paper presents a multi-domain named entity recognition method based on the part-of-speech (POS) attention mechanism. The structure of this paper is as follows. The related researches is introduced in Sect. 2. Section 3 gives a description of the proposed method and Sect. 4 discusses experimental results. Finally, Sect. 5 gives concluding remarks.

2   Relevant Work

The rule-based, dictionary-based and statistics-based NER methods require a large number of manual rules and annotations, which lead to the high time cost and poor portability. Therefore, NER researches mainly focus on the methods based on machine learning, including shallow model-based methods and deep learning-based methods. The NER methods based on shallow models, such as Conditional Random Fields (CRFs) [6], Support Vector Machine (SVM) [7], Hidden Markov Model (HMM) [8] is a "feature engineering" which relies on hand-crafted features and task-specific resources [9]. This often leads to the high time cost and poor portability. Aiming at this shortcoming, deep learning-based NER methods become the research focuses in recent years. CNN is one kind of important deep network model in NER. It was proposed by Lipenkova [10] in 1996, and has been used in various natural language processing (NLP) tasks. RNN is another kind of deep network model. Because of the problem of gradient dissipation in the training process, traditional RNN cannot capture long-distance text features, Joty et al. [11] proposed a variant of RNN, i.e., Long Short Term Memory (LSTM). Furthermore, Graves et al. proposed a bidirectional long-term and short-term memory network (Bi-LSTM). To a certain extent, Bi-LSTM solves the problem of long-distance information loss and can fully capture semantic features of contexts by using its forward and backward networks. Aiming at the output of Bi-LSTM, CRF can restrict the forward and backward dependencies between adjacent labels. Therefore, Bi-LSTM-CRF becomes the deep learning model widely used in domain-specific entity recognition tasks. Liu et al. [12] firstly applied Bi-LSTM-CRF to sequence annotation tasks in NLP. Yang et al. [13] applied Bi-LSTM-CRF to the domain-specific NER task and achieved good results. A large number of researches have proved that such a hybrid model can effectively improve the effect of NER.

After various deep learning-based methods effectively solve the problems of feature extraction and selection, the representation of entity information has become a research hotspot, which involves not only the spelling, POS and other grammatical or syntactic information of entities, but also external domain semantic information, such as the domain dictionary [14]. However, most of the current researches concatenated a series of information representations, such as words, POS, external dictionaries, phonologies and locations, as the input of deep learning models [15–17]. They did not fully consider the organizational hierarchy and weight of different types of information representations, especially in multi-domain entity recognition tasks [18, 19].

Based on above-mention analysis, this paper proposes a new multi-domain named entity recognition method. By using domain dictionary vectors and the POS attention mechanism, domain semantic information and POS information are utilized to improve the performance of multi-domain named entity recognition. The technical details are covered in the following section.

3 The Proposed Method

This paper proposes a multi-domain named entity recognition method, called BiLSTM-ATTENTION-CRF, based on the part-of-speech attention mechanism. Figure 1 gives the overall architecture of BiLSTM-ATTENTION-CRF, including three layers, text embedding, feature learning, and entity recognition.

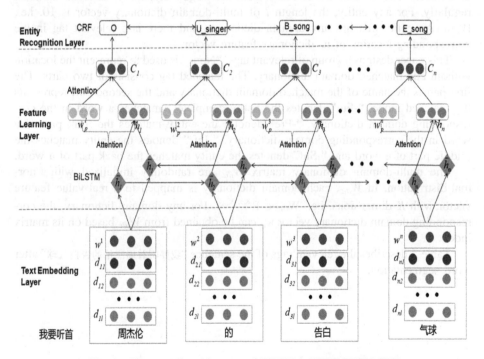

Fig. 1. The overall architecture of BiLSTM-ATTENTION-CRF

3.1   Text Embedding

The text embedding layer is used to encode multi-level of text information, including lexical units, parts of speech and the multi-domain dictionary by using fixed-length vectors. These vectors are concatenated to form text vectors, which are the input of the next layer.

Word Vectors: The word vector w^n is a mixed vector and obtained by merging two word vectors pre-trained by Glove [20] and Word2vec [21], respectively.

POS Vector: The POS matrix W_{pos} is randomly initialized with a normal distribution. In W_{pos}, each POS is mapped to a real-value vector with M dimensions, i.e., $W_{pos} \in R^{M \times |V_{pos}|}$, where V_{pos} is the POS set. Therefore, for any POS, the corresponding POS vector w_p can be obtained from W_{pos} based on its matrix index.

Multi-domain Dictionary Vector: The main task of this paper is to identify multi-domain Chinese entities. In order to effectively solve the polysemous phenomenon among different domains, a multi-domain dictionary set V_{dic} were constructed according to the domain distribution of datasets. In this paper, V_{dic} includes 10 domain dictionaries. All domain words come from the Sogou dictionary, which is updated regularly. For any entity, the length l of multi-domain dictionary vector is 10, i.e., $|V_{dic}|$. An entity matches any domain dictionary, and there is a relevant tag in the corresponding location of its dictionary feature vector.

This paper designs a group of relevant tags. "None" is used to represent the location without any matched domain dictionary. The matched tag consists of two parts. The first part is the name of the matched domain dictionary and the second part represents the matched mode. "-Ful" denotes the entity completely matches a word in the corresponding domain dictionary. "-Pre" denotes the entity matches the front part of a word in the corresponding domain dictionary. "-Inf " denotes the entity matches the middle part of a word and "-Suf" denotes the entity matches the back part of a word.

The multi-domain dictionary matrix W_{dic} are randomly initialized with a normal distribution. In W_{dic}, each domain dictionary is mapped to a real-value feature vector with K dimensions, i.e., $W_{dic} \in R^{K \times |V_{dic}|}$. For any domain dictionary, the corresponding domain dictionary vector w_d can be obtained from W_{dic} based on its matrix index.

Table 1 describes the relevant tags of the sentence "我要听首周杰伦的告白气球" after word segmentation.

Table 1. Sample matching tags for multi-domain dictionary

	Flight	Stock	...	Singer	Telephone	Song
我	None	None	...	None	None	None
要	None	None	...	None	None	None
听	None	None	...	None	None	None
首	None	None	...	None	None	None
周杰伦	None	None	...	singer-Ful	None	None
的	None	None	...	None	None	None
告白	None	None	...	None	None	song-Pre
气球	None	None	...	None	None	song-Suf

3.2 Feature Learning

The feature learning layer mainly combines BiLSTM and the attention mechanism to extract various features from texts. As shown in Fig. 1, it includes two parts, the BiLSTM layer and the attention layer. The BiLSTM layer is used to extract textual features and multi-domain dictionary features. The attention layer is used to learn the contribution weight of POS of candidate entities for multi-domain named entity recognition.

BiLSTM Unit. In order to solve the problem of gradient disappearance and gradient explosion of traditional RNN, LSTM was proposed [22]. It includes three gate mechanisms. The forgetting gate realizes the selective forgetting of cell state information. The input gate realizes the selective recording of new cell state information. The output gate realizes the transmission of cell state information to the outside world [23]. Their calculation formulas are described as follows [24].

$$i_t = \sigma(W_{ii}x_t + b_{ii} + W_{hi}h_{(t-1)} + b_{hi}) \tag{1}$$

$$f_t = \sigma(W_{if}x_t + b_{if} + W_{hf}h_{(t-1)} + b_{hf}) \tag{2}$$

$$g_t = \tanh(W_{ig}x_t + b_{ig} + W_{hg}h_{(t-1)} + b_{hg}) \tag{3}$$

$$o_t = \sigma(W_{io}x_t + b_{io} + W_{ho}h_{(t-1)} + b_{ho}) \tag{4}$$

$$C_t = f_t \otimes C_{(t-1)} + i_t \otimes g_t \tag{5}$$

$$h_t = o_t \otimes tanh(C_t) \tag{6}$$

where $\otimes$ denotes the tensor product operation, h_t denotes the hidden state at t time, C_t denotes the cell state at t time, $h_{(t-1)}$ denotes the hidden state at $t-1$ time, i_t, f_t, o_t

denote the input gate, the forgetting gate and the output gate at t time respectively, σ denotes the sigmoid activation function, W and b denotes the corresponding weight matrix and bias vector.

LSTM only considers one direction of context information and ignores another direction [25]. Based on LSTM, BiLSTM [26, 27] adopts the forward and backward parallel layers to capture two directions of context information. It can be realized by concatenate two LSTM layers.

Fig. 2. A mixed vector w'_d

Every word contained in the sentence is transformed into a mixed vector w'_d.as the input of BiLSTM. As shown in Fig. 2, w'_d is constructed by merging the word vector w with the corresponding multi-domain dictionary vector w_d. After w'_d is inputted into BiLSTM, the output of forward and backward hidden layer is shown as follows:

$$\overrightarrow{h}_j = \text{LS}\overrightarrow{T}\text{M}\left(w'_d\right), j \in [1, n] \tag{7}$$

$$\overleftarrow{h}_j = \text{LS}\overleftarrow{T}\text{M}\left(w'_d\right), j \in [1, n] \tag{8}$$

Finally, the output is shown as follows:

$$h_j = [\overrightarrow{h}_j, \overleftarrow{h}_j] \tag{9}$$

Attention Mechanism. A POS attention mechanism is designed to utilize the POS information and identify their contribution weight for multi-domain named entity recognition. This paper adopts the global attention mechanism [28]. The relevance between the jth word in a text and its POS vector w_p can be calculated as follows:

$$\alpha_j = \frac{\exp(e(h_j, w_p))}{\sum_{k=1}^{n} \exp(e(h_j, w_p))} \tag{10}$$

where

$$e(h_j, w_p) = v^T \tanh(W_H h_j + W_U w_p + b) \tag{11}$$

formula $e(h_j, w_p)$ is used to calculate the importance of h_j and w_p for multi-domain entity recognition. Parameter matrices W_H, W_U, V, b are initialized randomly and optimized continuously during training. Then, the output of the attention layer of the *j*th word *Attention$_j$* is shown as follows:

$$C_j = \alpha_j * h_j \tag{12}$$

$$Attention_j = tanh[C_j, h_j] \tag{13}$$

3.3   Multi-domain Entity Recognition

The last layer of the method is the CRF layer, which considers the tag relationship between the current word and the adjacent words for decoding the global optimal tag sequence [29]. It has a state transition matrix A as parameters and its input is *Attention$_j$*, i.e., the output of the feature learning layer. Then, the predicted annotation output of the observation sequence X is shown as follows:

$$L(X, y) = \sum_{i=1}^{n} \left(A_{y_i, y_{i+1}} + P_{i, y_i} \right) \tag{14}$$

where $A_{i,j}$ denotes the transition probability of time series from the *i*th state to the *j*th state, and $P_{i,j}$ denotes the probability that the *i*th word in the input sequence is the *j*th tag.

4   Experimental Results and Analysis

4.1   Experimental Settings

This study adopted a commercial dataset from the task-based online QA system, involved with 10 domains. Figure 3 gives a group sample data. The ATIS data set [30], which composed of audio recordings from flight reservation personnel and is widely applied into NLP, was also translated into Chinese and added into the experimental dataset. The final experimental dataset includes 242,000 recordings involved with 234 entity types from 10 domains. As shown in Table 2, this mixed dataset was split into the training and test datasets according to the ratio of 9:1.

Data Preprocessing. This study used the popular Chinese NLP tool HanLP [31] to perform word segmentation and POS tagging. Different from our previous studies [32, 33], other data preprocessing, such as stop word removing, weren't performed. After word segmentation, all of words were directly inputted into the proposed method for text embedding.

Annotation Scheme. In this paper, "BMEUO" tagging system was adopted to annotate different components and boundary of named entities [34]. "B" denotes the first word of the entity, "M" denotes inner words of the entity, "E" denotes the last word of the entity, "U" denotes a single entity, and "O" denotes a word that does not belong to any entity. Tags also include the information of entity types. For examples,

Music：喂喂喂请帮我播放一首王菲的著名歌曲红豆吧

Flight：你好啊帮我预定一张12月15号北京飞往哈尔滨的机票好吗

Stock：嘿帮我查一查今天象屿股份股票的收盘价格吧

Weather：听懂的话就帮我播放明天北京的天气吧谢谢你

Movies：喂喂喂我要看刘德华和成龙出演的武打片电影快帮我播放一下啊

Navigation：帮我打开高德地图找到从郑州到吉林长春的路线

Education：喂喂喂请帮我读出来三个水加起来是什么字吧

News：嗨嗨快帮我找一下昨天杨幂和冯绍峰的热点新闻

Talking：喂我想听十万个为什么里面关于小熊的冷笑话

Telephone：现在就请帮我拨打电话133███████给刘毅雷好吧

Fig. 3. Sample data

Table 2. The experimental data set

Domain	Train	Test
Music	20,500	2,500
Flight	22,500	2,500
Stock	21,500	2,500
Weather	20,500	2,500
Movies	23,500	2,500
Navigation	21,500	2,500
Education	20,500	2,500
News	22,500	2,500
Talking	20,500	2,500
Telephone	23,500	2,500

"_name" denotes the name of people, "_geo" denotes the place name, "_stock_name" denotes the stock name, "_time" denotes the time, and "_song" denotes the song name. Table 3 gives some examples of multi-domain entity annotation.

Training Parameters. In order to explore the best way of word embedding, both Glove and word2vec were used to train word vectors with different dimensions. The training corpora adopted collected news and the data of commercial QA system (different from the experimental data in this paper). As shown in Fig. 4, the mixed word vector *mutivec*, which was constructed by merging the Glove-based word vector with the word2vec-based word vector, can obtained the best results. The optimal dimensional number of word vector is 128.

In the experiment, the dimensional number of POS vector and multi-domain dictionary vector were set to 50. The parameters were set to update during model training. The dimensional number of LSTM hidden layer was set to 300, the mini-batch size was set to 64, and dropout was set to 0.5. Adam algorithm [35] was used to optimize the parameters and the learning rate was set to 1e−5. Other optimization algorithms were also tried in this experiment, but the final performance is not as good as Adam algorithm.

Table 3. Examples of multi-domain entity annotation

Annotation Sample 1					
来	首	周杰伦	的	告白	气球
O	O	U_singer	O	B_song	E_song

Annotation Sample 2					
送	我	到	石家庄	人民	医院
O	O	O	B_geo	M_geo	E_geo

Annotation Sample 3					
预定	后天	飞往	北京	的	航班
O	U_time	O	U_geo	O	O

Annotation Sample 4					
明天	石家庄	的	天气	情况	如何
U_ti	U_geo	O	O	O	O

Annotation Sample 5					
看	今日	通润	装备	股票	价格
O	U_time	B_stockname	E_stockname	O	O

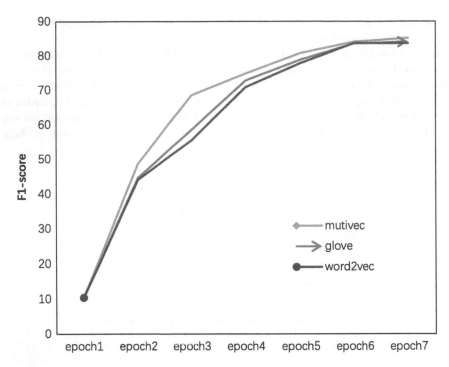

Fig. 4. The comparison of three types of word vectors.

4.2 Experimental Results

The model was used on the test dataset to recognize multi-domain named entities. Repeat 10 times to obtain the average value of F_1. Table 4 gives a comparison between BiLSTM-ATTENTION-CRF and contrast methods.

Table 4. A comparison between BiLSTM-ATTENTION-CRF and contrast methods

ID	Method	F_1 score
01	CRF	0.740
02	LSTM	0.762
03	BiLSTM	0.783
04	BiLSTM-CRF	0.800
05	BiLSTM-CRF+POS	0.818
06	BiLSTM-CRF+Multi-Dic	0.827
07	BiLSTM-CRF+POS+Multi-Dic	0.830
08	BiLSTM-Attention-CRF+POS	0.846
09	BiLSTM-Attention-CRF+Multi-Dic	0.833
10	**BiLSTM-Attention-CRF+POS+Multi-Dic**	**0.851**

In Table 4, "POS" denotes POS vectors and "Multi-Dic" denotes multi-domain dictionary vectors. Method 01-04 only use word vectors. Method 05-07 use POS vectors, multi-domain dictionary vectors and their mixed vector in the text embedding layer respectively. Method 08 and 09 add attention mechanism. Method 10 is just the BiLSTM-ATTENTION-CRF.

This paper uses F_1 value to evaluate the experimental results. Its formula is shown as follows:

$$\text{Pr}ecision = \frac{|TP|}{|TP| + |FP|} \tag{15}$$

$$\text{Re}call = \frac{|TP|}{|TP| + |FN|} \tag{16}$$

$$F_1 = \frac{2 \times \text{Pr}ecision \times \text{Re}call}{\text{Pr}ecision + \text{Re}call} \tag{17}$$

where $|TP|$ represents the number of target entities in recognized entities, $|FP|$ represents the number of no-target entities in recognized entities, and $|FN|$ represents the number of target entities which aren't recognized.

As shown in Table 4, BiLSTM-ATTENTION-CRF has the best performance than other methods. Furthermore, all methods can be compared and analysis as follows:

- By comparing the methods 01, 02, 03 and 04, it can be found that the F_1 value of method 04, which adopts a hybrid model, is nearly 5% higher than those of methods 01, 02 and 03, which adopt a single model. This shows that the hybrid model integrates the advantages of BiLSTM and CRF, and can achieve the better performance in the sequence annotation task, such as multi-domain named entity recognition.
- The F_1 value of methods 05, 06 and 07 is higher than that of method 04. This shows that the POS information and multi-domain dictionary information are useful for multi-domain named entity recognition. By comparing the F_1 values of methods 05, 06 and 07, it can be found that rich inner and external textual information can effectively improve the performance of multi-domain named entity recognition.
- By comparing the F_1 value of methods 08, 09 and 10, it can be found that the attention mechanism is useful for improving multi-domain named entity recognition. This shows that word embedding, POS and multi-domain dictionary have different contribution weights for multi-domain named entity recognition. Finally, BiLSTM-ATTENTION-CRF proposed by this paper has the best performance.

5   Conclusion

This paper introduce a multi-domain named entity recognition method based on the POS attention mechanism, called BiLSTM-ATTENTION-CRF. Its effectiveness can be demonstrated by experimental results. The main contributions of this paper include:

- Aiming at the characteristics of multi-domain named entity recognition, a multi-domain dictionary was designed and added into text embedding for modeling multi-domain semantic information. Furthermore, a POS attention mechanism was introduced into BiLSTM-CRF for not only utilizing the POS information but also identifying its contribution weight.
- This paper compared various fusion strategies of multi-level entity information and found the most effective hybrid method for multi-domain named entity recognition. The experimental results show that the proposed method has a high precision and recall rate, and can effectively recognizes the multi-domain named entities.

Acknowledgment. The work received support from Science and Technology Project of Beijing Municipal Commission of Education (No. KM201710005026), National Basic Research Program of China (No. 2014CB744600), Open Foundation of Beijing Key Laboratory of MRI and Brain Informatics, Open Foundation of Beijing Key Laboratory of Multimedia and Intelligent Software (Beijing University of Technology).

References

1. Mikheev, A., Moens, M., Grover, C.: Named entity recognition without gazetteers. In: Ninth Conference on European Chapter of the Association for Computational Linguistics (EACL 1999), pp. 1–8. ACM Press (1999)
2. Chandel, A., Nagesh, P.C., Sarawagi, S.: Efficient batch top-k search for dictionary-based entity recognition. In: 22nd International Conference on Data Engineering (ICDE 2006), p. 28. ACM Press (2006)
3. Abacha, A.B., Zweigenbaum, P.: Medical entity recognition: a comparison of semantic and statistical methods. In: BioNLP 2011 Workshop (BioNLP 2011), pp. 56–64. ACM Press (2011)
4. Eftimov, T., Seljak, B.K., Korošec, P.: A rule-based named-entity recognition method for knowledge extraction of evidence-based dietary recommendations. PLoS One **12**(6), e0179488 (2017)
5. Gandhe, A., Rastrow, A., Hoffmeister, B.: Scalable language model adaptation for spoken dialogue systems. In: 2018 IEEE Spoken Language Technology Workshop (SLT 2018) (2018)
6. Teixeira, J., Sarmento, L., Oliveira, E.: A bootstrapping approach for training a NER with conditional random fields. In: Antunes, L., Pinto, H.S. (eds.) EPIA 2011. LNCS (LNAI), vol. 7026, pp. 664–678. Springer, Heidelberg (2011). https://doi.org/10.1007/978-3-642-24769-9_48
7. Ju, Z.F., Wang, J., Zhu, F.: Named entity recognition from biomedical text using SVM. In: 5th International Conference on Bioinformatics and Biomedical Engineering. IEEE Press (2011). https://doi.org/10.1109/icbbe.2011.5779984
8. Morwal, S., Jahan, N., Chopra, D.: Named entity recognition using hidden markov model (HMM). Int. J. Nat. Lang. Comput. (IJNLC) **1**(4), 15–23 (2012)
9. Ding, P., Zhou, X.B., Zhang X.J., Wang, J., Lei, Z.F.: An attentive neural sequence labeling model for adverse drug reactions mentions extraction. IEEE Access **6** (2018). https://doi.org/10.1109/access.2018.2882443

10. Lipenkova, J.: A system for fine-grained aspect-based sentiment analysis of Chinese. In: 53rd Annual Meeting of the Association for Computational Linguistics and the 7th International Joint Conference on Natural Language Processing (ACL-IJCNLP 2015), pp. 55–60 (2015)

11. Liu, P.F., Joty, S., Meng, H.: Fine-grained opinion mining with recurrent neural networks and word embeddings. In: 2015 Conference on Empirical Methods in Natural Language Processing (EMNLP 2015), pp. 1433–1443. ACL (2015)

12. Liu, Q., Liu, B., Zhang, Y., Kim, D.S., Gao, Z.: Improving opinion aspect extraction using semantic similarity and aspect associations. ACM SIGARCH Comput. Archit. News **44**(3), 506–518 (2016)

13. Silver, D.L., Yang, Q., Li, L.: Lifelong machine learning systems: beyond learning algorithms. In: AAAI 2013 Spring Symposium on Lifelong Machine Learning (2013)

14. Liu, P.F., Qiu, X.P., Chen, X.C, Wu, S.Y.: Multi-timescale long short-term memory neural network for modelling sentences and documents. In: 2015 Conference on Empirical Methods in Natural Language Processing (EMNLP 2015), pp. 2326–2335 (2015)

15. Jakob, N., Gurevych, I.: Extracting opinion targets in a single- and cross-domain setting with conditional random fields. In: 2010 Conference on Empirical Methods in Natural Language Processing (EMNLP 2010), pp. 1035–1045 (2010)

16. Zhao, Y.Y., Che, W.X., Guo, H.L., Qin, B, Su, Z., Liu, T.: Sentence compression for target-polarity word collocation extraction. In: 25th International Conference on Computational Linguistics (COLING 2014), pp. 1360–1369 (2014)

17. Schmidhuber, J.: Deep learning in neural networks: an overview. Neural Netw. **61**, 85–117 (2015)

18. Mikolov, T., Chen, K., Corrado, G.: Efficient estimation of word representations in vector space. Comput. Sci. (2013)

19. Beck, D., Cohn, T., Hardmeier, C., Specia, L.: Learning structural kernels for natural language processing. Trans. Assoc. Comput. Linguist. **3**, 461–473 (2015)

20. Shalaby, W., Zadrozny, W.: Mined semantic analysis: a new concept space model for semantic representation of textual data. In: 2017 IEEE International Conference on Big Data (Big Data 2017). IEEE Press (2017)

21. Ustun, V., Rosenbloom, P.S., Sagae, K., Demski, A.: Distributed vector representations of words in the sigma cognitive architecture. In: Goertzel, B., Orseau, L., Snaider, J. (eds.) AGI 2014. LNCS (LNAI), vol. 8598, pp. 196–207. Springer, Cham (2014). https://doi.org/10.1007/978-3-319-09274-4_19

22. Zilly, J.G., Srivastava, R.K., Koutník, J., Schmidhuber, J.: Recurrent highway networks. In: 34th International Conference on Machine Learning, pp. 4189–4198 (2017)

23. Strobelt, H., Gehrmann, S., Huber, B., Pfister, H.: LSTMVis: a tool for visual analysis of hidden state dynamics in recurrent neural networks. IEEE Trans. Visual. Comput. Graphics (2016)

24. Yu, Z., et al.: Using bidirectional LSTM recurrent neural networks to learn high-level abstractions of sequential features for automated scoring of non-native spontaneous speech. In: 2015 IEEE Workshop on Automatic Speech Recognition and Understanding (ASRU), pp. 338–345. IEEE Press (2015)

25. Long, D., Zhang, R., Mao, Y.Y.: Prototypical recurrent unit. Neurocomputing **311**, 146–154 (2018)

26. Graves, A., Fernández, S., Schmidhuber, J.: Bidirectional LSTM networks for improved phoneme classification and recognition. In: Duch, W., Kacprzyk, J., Oja, E., Zadrożny, S. (eds.) ICANN 2005. LNCS, vol. 3697, pp. 799–804. Springer, Heidelberg (2005). https://doi.org/10.1007/11550907_126

27. Chen, Y., et al.: Named entity recognition from Chinese adverse drug event reports with lexical feature based BiLSTM-CRF and tri-training. J. Biomed. Inform. **96** (2019). https://doi.org/10.1016/j.jbi.2019.103252
28. Sasaki, Y., et al.: Local and global attention are mapped retinotopically in human occipital cortex. Natl. Acad. Sci. U.S.A. **98**(4), 2077–2082 (2001)
29. Quan, C.Q., Ren, F.J.: Target based review classification for fine-grained sentiment analysis. Int. J. Innov. Comput. Inf. Control **10**(1), 257–268 (2016)
30. Huang, H.J., Li, Z.-C.: A multiclass, multicriteria logit-based traffic equilibrium assignment model under ATIS. Eur. J. Oper. Res. **176**(3), 1464–1477 (2007)
31. HanLP Tool. https://github.com/hankcs/HanLP
32. Zhang, S., Lin, S.F., Gao, J.F., Chen, J.H.: Recognizing small-sample biomedical named entity based on contextual domain relevance. In: 2019 IEEE 3rd Information Technology, Networking, Electronic and Automation Control Conference (ITNEC). IEEE Press (2019)
33. Dong, G.C., Chen, J.H., Wang, H.Y., Zhong, N.: A narrow-domain entity recognition method based on domain relevance measurement and context information. In: 2017 IEEE/WIC/ACM International Conference on Web Intelligence (WI 2017), pp. 623–628. ACM Press (2017)
34. Jagannatha, A.N., Yu, H.: Structured prediction models for RNN based sequence labeling in clinical text. In: 2016 Conference on Empirical Methods in Natural Language Processing, pp. 856–865 (2016)
35. Miwa, M., Bansal, M.: End-to-end relation extraction using LSTMs on sequences and tree structures. In: 54th Annual Meeting of the Association for Computational Linguistics, pp. 1105–1116 (2016)

Understanding Lexical Features for Chinese Essay Grading

Yifei Guan[1,2], Yi Xie[1,2], Xiaoyue Liu[1], Yuqing Sun[1,3(✉)],
and Bin Gong[1,3(✉)]

[1] School of Software, Ministry of Education, Shandong University, Jinan, China
poppy@mail.sdu.edu.cn, heilongjiangxieyi@163.com,
suiqiyue@163.com, {sun_yuqing,gb}@sdu.edu.cn
[2] School of Computer Science and Technology, Ministry of Education,
Shandong University, Jinan, China
[3] Engineering Research Center of Digital Media Technology,
Ministry of Education, Shandong University, Jinan, China

Abstract. Essay grading is an important and difficult task in natural language processing. Most of the existing works focus on grading non-native English essays, such as essays in TOEFL. However, these works are not applicable for Chinese essays due to word segmentation and different syntax features. Considering lexical features are important for essay grading, in this paper, we study the expert evaluation standard and propose an interpretable lexical grading method for essays. We first study different levels of vocabulary provided by experts and introduce a quantitative evaluation framework on lexical features. Based on these standards, we quantify the Chinese essay dataset of 12 education grades in primary and middle schools and propose a set of interpretable features. Then a Bi-LSTM network model is proposed for semantically grading essay, which accepts a sequence of word vectors as input and integrates attention mechanism in terms of lexical richness. We evaluate our method on real datasets and the experimental results show that it outperforms other methods on the task of lexically Chinese essay grading. Besides, our method gives interpretable results, which are helpful for practical applications.

Keywords: Essay grading · LSTM · Lexical richness · Interpretable

1 Introduction

It is an important and difficult task to automatically grade essays in natural language processing. Most existing works focus on non-native English essay grading. For example, E-Rater [1], a rating system developed by ETS, has been applied in major official examinations, such as TOFEL and GMAT since 2001, with an accuracy rate of over 97%. The *juku* [2] is a website that provides services on automatic correction of English essay, on which students can submit their essays and get feedback on corrections. However, the English essay grading methods cannot be applied to Chinese tasks due to the differences between the two languages, such as lexical separator and

Y. Sun et al. (Eds.): ChineseCSCW 2019, CCIS 1042, pp. 645–657, 2019.
https://doi.org/10.1007/978-981-15-1377-0_50

tense. To the best of our knowledge, there is not any publicly available work on Chinese essay grading.

Lexical richness is an important indicator to evaluate a student's linguistic level, which reflects his vocabulary and the ability to use the words. Therefore, it is reasonable to select the lexical richness as features to grade the essays. This paper has the following contribution:

(1) We propose a lexical grading framework that integrates expert evaluation. By studying the different levels of vocabulary, idioms and advanced verbs provided by experts, we analyze the lexical features on the Chinese essays of 12 education grades in primary and middle schools and introduce interpretable metrics on the lexical richness of essay.

(2) We propose the Bi-LSTM network [3, 4] with attention mechanism method to extract the semantic features of essay. The model combines two layers of Bi-LSTMs to generate the semantic vector of an essay that considers both the sentence and text aspects.

(3) We adopt the multilayer perceptron network for essay grading with the attention mechanism on the lexical aspect. Based on the lexical features extracted by the experts, the grading results are interpretable, which are much helpful for practical applications.

(4) The method is verified against real datasets and the experimental results show that it outperforms other methods on the task of lexically grading Chinese essay.

The rest of this paper is organized as follows. Section 2 presents the related work. In Sect. 3, we introduce expert review rules on essays and the data sets, and discuss how to extract the lexical features. In Sect. 4, we present the grading model on Chinese essays. Section 5 evaluates our model on real datasets. We conclude the paper in Sect. 6.

2 Related Work

In this section, we present the influential approaches on essay grading. Existing essay grading models include two categories: traditional machine learning and deep learning.

Classical regression and classification algorithms often use the features extracted by experts in automatic essay grading tasks. Project Essay Grade (PEG) [5, 6] is one of the earliest essay grading systems, using linear regression over vectors of lexical features to predict an essay level. PEG relies on the analysis of the latent semantic features of the essays without understanding the semantic content of the essays, such that it cannot give feedback to students. Intelligent Essay Assessor (IEA) [7] adopts Latent Semantic Analysis (LSA) [8] to calculate the semantic similarity between essays without considering the language expression. The E-rater system [1], developed by the Educational Testing Service, has been deployed in the English language test, such as Test of English as a Foreign Language (TOEFL) and Graduate Record Examination (GRE). The system uses a number of different features, including different aspects of vocabulary and grammar. BETSY [9] is a program, funded by the United States Department of Education, which is based on the probability theory and the statistics on a training corpus to classify texts. In 2012, the Hewlett Foundation sponsored a competition on Kaggle[1] called the Automated Student Assessment Prize (ASAP) [10], aiming to find efficient automated essay grading methods. The dataset released has been widely used for automatic essay grading tasks [11, 12].

In recent years, motivated by the success of deep learning in different domains, many deep neural networks have been proposed for essay grading. Cozma et al. [13] proposed a method combining word vector and SVM with the string kernel function. Alikaniotis et al. [11] employed an LSTM model to learn features for the essay grading task, which learns score-specific word embeddings (SSWEs) for word representation. Taghipour et al. [14] combined LSTM and CNN for automatic essay grading, which outperforms many methods that require handcrafted features. Dong et al. [12] introduced the attention mechanism on the basis of CNN and RNN, and found that the attention mechanism on keywords and sentences helps to judge the quality of essays. Jin et al. [15] proposed a two-stage neural network model to automatically grade prompt-independent essays, and built three stacked Bi-LSTMs to extract the semantic, part-of-speech and syntactic features of essays. Based on LSTM, a new SKIPFLOW mechanism was proposed by Tay et al. [16], which incorporated semantic and logical information of essays.

However, the English essay grading methods cannot be directly applied to Chinese tasks due to the differences between the two languages, such as lexical separator and tense. Although, Fu et al. [17] analyze the gracefulness of sentences in Chinese essays by the combination of CNN and LSTM, but their model cannot grade a complete essay. To the best of our knowledge, there is not any publicly available work on Chinese essay grading. Moreover, the automatic grading tasks require the interpretable results, especially the deep neural network model. To this end, we propose an interpretable Chinese essay grading model, which gives a reasonable explanation for essay grading.

3   Understanding Expert Rules for Essay Lexical Features

In this section, we first introduce a set of expert essay grading standards. Then we present the experimental dataset and define the lexical features of essays. Based on the essay grading rules, we introduce a quantitative evaluation framework on lexical features.

3.1   Expert Review Rules

Since essay grading is the somewhat subjective task, to have the normalized rules on Chinese essay grading, the Ministry of Education asks experts to set up the *Essay Scoring Standard for National New Curriculum Standards College Entrance Examination (the standard* for short). This standard evaluates the essays into four levels according to the expressions and the characteristics, the details are given in Table 1. We can see that the lexical features play an important role in essay grading, such as the lexical richness and the usage of advanced words. It is reasonable to use lexical features for Chinese essay grading.

---

[1] http://www.kaggle.com/c/asap-aes/.

We adopt the *Outline of Chinese Proficiency Vocabulary and Chinese Characters* [18] (*the outline* for short) to extract the measurable lexical features from essays. The outline was officially released by the *Examination Center of the Office of the National HSK Examination Committee* to grade Chinese words. The Chinese vocabulary are graded into four levels, from advanced to simple: A-level, B-level, C-level and D-level. Specifically, A-level and B-level always contain advanced words, such as "哀悼". The common used words are classified to C-level or D-level, such as "帮助" and "发生". In this paper, we adopt the word levels in the outline as rules to generate the lexical features of Chinese essays.

Table 1. Some rules on essay grading in the *Essay Scoring Standard for National New Curriculum Standards College Entrance Examination*.

	First level	Second level	Third level	Fourth level
Expression	Precise content structure Quite fluency verbs	Complete content structure Fluency verbs	Almost complete content structure Fairly fluency verbs	Confusing content structure Not fluency verbs
Characteristic	Quite rich content Quite literary writing	Rich content Literary writing	Fairly rich content Fairly literary writing	Not rich content Not literary writing

3.2 Dataset

We adopt a Chinese essay dataset from primary and middle schools that are provided by our partner. It contains 59,142 student essays covering from Primary Grade Two (P2 for short) to Senior Grade Three (S3 for short). Table 2 shows the statistics of the dataset on each education grade, including the number of essays, the average essay length and the average number of idioms. We count the number of advanced verbs according to an *Advanced Chinese Verb List* (*the list* for short), which includes 199 advanced verbs such as "觊觎" and "斟酌". The average numbers of advanced verbs are listed on the second row from the bottom in Table 2.

Table 2. Statistics on the dataset.

	Primary school					Junior high school			Senior high school		
Education grade	P2	P3	P4	P5	P6	J1	J2	J3	S1	S2	S3
#essay	4867	11636	13194	12028	10566	2045	1969	1492	494	469	382
Avg. #character	178	252	323	354	379	408	425	578	789	904	867
Avg. #idiom	0.96	2.15	2.96	3.88	4.34	4.71	5.10	5.62	7.08	7.54	8.03
Avg. #advanced verb	0.010	0.019	0.053	0.071	0.092	0.13	0.19	0.23	0.31	0.34	0.41
Essay grade	2	3	4	5	6	7	8	9	10	11	12

Since the education grades reflect the average ability of writing skills of students, we adopt the education grades as the essay grade in the learning process. The higher the education grade that the essay is selected from, the higher the corresponding essay grade.

3.3 Interpretable Lexical Features of Essays

In this section, we propose the interpretable features on lexical richness of essay by understating the statistics on the essay dataset from Chinese primary and middle schools with the help of the word levels extracted from the outline.

Vocabulary is one of the basic elements of essays. An essay is more likely to have a higher grade if it contains many high-level words. To understand the correlations between the usage of words and the essay grade, we calculate the number of words in different word levels against the grades of students. As shown in Fig. 1, there are obviously positive correlations between the student grade and the number of high-level words used in each essay. Therefore, it is reasonable to adopt the metric of lexical richness as an indicator in essay grading.

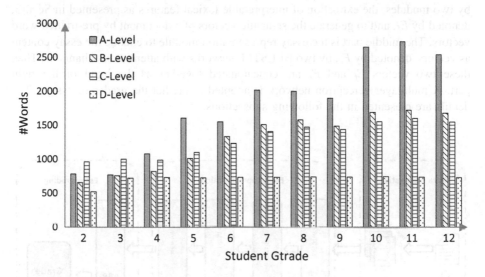

Fig. 1. The correlation between the student grades and the lexical richness.

We also consider other measurable lexical features to represent an essay, including the length of essay, the number of idioms and the number of advanced verbs used in an essay. We also quantify the importance on how much a word contributes for the judgement of the grade of essay by information gain and select 44 words with high information gains against student grades. These words are adopted as the lexical features as well. The interpretable lexical features are summarized in Table 3 for grading Chinese essay. This lexical feature vector for each essay is denoted by E_f, which would be used in the following grading process.

Table 3. Interpretable lexical features of an essay.

1	#A-level word
2	#B-level word
3	#C-level word
4	#advanced verb
5	#character
6	#idiom
7	#high information gain word

4   Chinese Essay Grading Based on Lexical Features

In this section, we discuss how to grade Chinese essays based on the lexical features and the content of essay. There are three parts in our model, as illustrated in a left-right view in Fig. 2.

The left part of data processing is the extraction of lexical features and mapping a document to a sequence of word vectors. The original content of an essay is processed by two modules, the extraction of interpretable lexical features as presented in Sect. 3, denoted by E_f, and to generate the semantic vectors of a document by pre-trained word vectors. The middle part is the essay representation module to encode the essay content as vectors, denoted by E_e, by two Bi-LSTM networks with attention mechanisms. Then these two vectors E_f and E_e are concatenated together as the input of the right part. A multilayer perceptron network is adopted to predict the grades of essays. The details are presented in the following subsections.

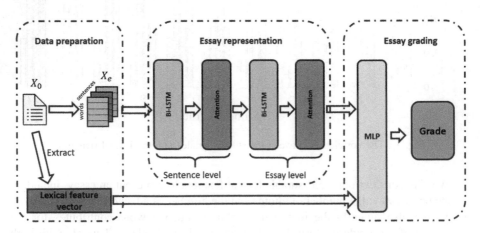

Fig. 2. The interpretable essay grading model.

4.1   Learning the Semantic Representation of Chinese Essay

Given a Chinese essay, the semantic representation is learned by a deep network, denoted by E_e. Let the sequence of sentences $s_1, s_2, \cdots\cdots, s_L$ denotes the contents of an essay, where L is the length of essay, and each sentence s_i contains a sequence of words, represented by $w_1^i, w_2^i, \cdots\cdots, w_{T_i}^i$, where T_i is the length of sentence. The word w_t^i represents the t-th word in the i-th sentence, and is embedded to a word vector x_t^i by *Word2vec* [19] or *Glove* [20]. Then the sequence of word vectors is fed to a Bi-LSTM network, which contains a forward LSTM network reading the sentence s_i from w_1^i to $w_{T_i}^i$, and a backward LSTM network reading the words from $w_{T_i}^i$ to w_1^i:

$$\vec{h}_t^i = \overrightarrow{LSTM}(x_t^i), \, t \in [1, \, T_i] \tag{1}$$

$$\overleftarrow{h}_t^i = \overleftarrow{LSTM}(x_t^i), \, t \in [T_i, \, 1] \tag{2}$$

$$h_t^i = \vec{h}_t^i \oplus \overleftarrow{h}_t^i \tag{3}$$

where $\vec{h}_t^i$ and $\overleftarrow{h}_t^i$ represent the hidden states of t-th cell in the forward LSTM and the backward LSTM, respectively. The symbol $\oplus$ denotes the vector concatenation.

To have the semantic representation of a sentence, we adopt the attention mechanism to learn the different contribution α_t^i of word w_t^i in sentence s_i. The h_t^i is fed to a one-layer perception network to extract the hidden state u_t^i. Then the normalized weight α_t^i is learned through a *softmax* function. The context vector u_w is introduced as the combination weights on the outputs of the network, which are randomly initialized and jointly learned during the training process. Finally, the sentence vector s_i is the sum of h_t^i against weights α_t^i:

$$u_t^i = \tanh\left(W_w h_t^i + b_w\right) \tag{4}$$

$$\alpha_t^i = \frac{exp\left(u_t^{iT} u_w\right)}{\sum_j exp\left(u_j^{iT} u_w\right)} \tag{5}$$

$$s_i = \sum_t \alpha_t^i h_t^i \tag{6}$$

Similarly, we further learn the essay representation vector by a Bi-LSTM based on the sentence vectors. The attention mechanism here is used to analyze the importance of each sentence in an essay. The context vector u_s is randomly initialized and jointly learned during the training process. The semantic representation of essay E_e is learned by the following functions:

$$\vec{h}_i = \overrightarrow{LSTM}(s_i), \, i \in [1, \, L_i] \tag{7}$$

$$\overleftarrow{h}_i = \overleftarrow{LSTM}(s_i), \, i \in [L_i, \, 1] \tag{8}$$

$$h_i = \vec{h}_i \oplus \overleftarrow{h}_i \tag{9}$$

$$u_i = \tanh(W_s h_i + b_s) \tag{10}$$

$$\alpha_i = \frac{exp\left(u_i^{\mathrm{T}} u_s\right)}{\sum_j exp\left(u_j^{\mathrm{T}} u_s\right)} \tag{11}$$

$$E_e = \sum_t \alpha_i h_i \tag{12}$$

4.2   Chinese Essay Grading

Considering the lexical features and the contents are both important elements of Chinese essays, we concatenate the lexical feature vector E_f and the semantic representation E_e together, and feed it into a multilayer perceptron. The sigmoid function is adopted as the activation function to predict the grade $\hat{y}$ of essay:

$$\hat{y} = sigmoid\left\{W_c\left[E_e \oplus E_f\right] + b_c\right\} \tag{13}$$

The MSE is adopted as the loss function to measure the variance between the predicted grade and the ground-truth y:

$$\mathcal{L} = mse\left(y, \hat{y}\right) = \frac{1}{n}\sum_{i=1}^{n}\left(y_i - \hat{y}_i\right)^2 \tag{14}$$

5   Experiments

5.1   Experimental Setup

Each essay is segmented into sentences and each sentence is segmented into words. We adopt the 300-dimensional embeddings provided by Beijing Language and Culture University [21] who preform *Word2vec* [19] on the 22.6 G corpus from Wikipedia and other Chinese corpus. Then, we use the word embeddings to initialize the embedding matrix W_e.

In our experiments, the maximum number of words per sentence is limited to 100, and the maximum number of sentences per document to 50. Padding is used to maintain the length of word sequences and sentence. We fix the LSTM hidden state size at 64, and the dimension of both sentence and essay representations obtained by

Bi-LSTM are then 128. The context vectors in the attention layer also have a dimension of 128.

For training, the batch size is 16. We use the ADAM [22] optimizer with learning rate $= 0.001$, $\beta_1 = 0.9$, $\beta_2 = 0.999$ as parameters. We use 80% of the data for training and 20% for testing.

5.2 Evaluation Metrics

The Quadratic Weighted Kappa (QWK), the Pearson Correlation Coefficient (PCC) and the Spearman Correlation Coefficient (SCC) are adopted as the evaluation metrics in this paper, which are widely applied to measure essay grading models.

The Kappa coefficient is an evaluation metric used for consistency testing or measuring classification accuracy. In this paper, the Kappa is used to measure the consistency between the predicted essay grade and the ground-truth. QWK is improved from Kappa by adding quadratic weights. QWK is calculated as follow:

$$\kappa = 1 - \frac{\sum W_{i,j} O_{i,j}}{\sum W_{i,j} E_{i,j}} \tag{15}$$

$$W_{i,j} = \frac{(i-j)^2}{(R-1)^2} \tag{16}$$

Where $W_{i,j}$ denotes the square weight matrix. j is the predicted essay grade based on our model and i is the ground truth, formally $\hat{y} = j$, $y = i$. R represents the number of essay grades, R = 11. The element $O_{i,j}$ in the observation matrix O denotes the number of essays that satisfy $\hat{y} = j \cap y = i$. The expectation matrix E is calculated from the outer product of the true histogram vector and the predicted histogram vector, and is normalized.

5.3 Comparison Methods

- **SVM.** Support-vector machines are supervised learning models that analyze data used for classification and regression analysis with the lexical features. We use this method as a baseline in the comparison method.
- **2L-LSTM-word2vec** [11]. A two-layer Bi-LSTM model is used to generate a representation vector of the essays, and then the vector is used to obtain the essay grade.
- **CNN-LSTM** [14]. The essay vector is generated by CNN and LSTM, and then the vector is used to obtain the essay grade.
- **CNN-LSTM-ATT** [12]. A CNN layer is employed to encode word sequences into sentences, followed by an LSTM layer to generate the essay representation. An attention mechanism is added to model the influence of each sentence on the final essay representation.
- **TDNN** [15]. This model employs three two-layer Bi-LSTMs to extract the features of the essays in terms of semantics, part-of-speech and syntax, and finally grade the

essays. Since the syntactic tree extracted by this method is not suitable for Chinese, we use the semantic and part-of-speech features in the experiment only.

- **2L-Bi-LSTM-ATT-lexical.** This is our proposed model, using word vectors and lexical features as input. We next compare three variances of our model.
- **2L-Bi-LSTM-ATT.** This model only uses the word vector as the input, which is similar to 2L-Bi-LSTM-ATT-lexical but without using the lexical feature.
- **2L-Bi-GRU-ATT-lexical.** This model replaces the LSTM unit with GRU, using the word vector and combining the interpretable features as input.
- **2L-Bi-GRU-ATT.** Similarly, this model replaces the LSTM unit with GRU and only uses the word vector as the input.

5.4   Results and Analyzes

In this section, different components of our model are compared and analyzed using three correlation metrics. The performance results of each variance on different evaluation metrics is shown in Fig. 3. Then, we compare our model with other state-of-art methods, where the best result for each metric is highlighted in bold in Table 4.

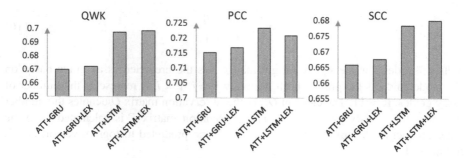

Fig. 3. Variances comparison on different metrics.

By comparing the variances of the methods proposed in this paper, we can see that 2L-Bi-LSTM-ATT-lexical performs better than 2L-Bi-LSTM-ATT in QWK and SCC, indicating that the lexical features are helpful to improve the performance of the model on the essay grading task. Meanwhile, by comparing our model with the method of grading the essays directly using SVM, we find that the performance of using only the lexical features on the essay grading task is not very satisfactory. This indicates that, apart from the lexical features, semantic representations of essays are also essential in essay grading task.

In our experiments, our method with GRU is not as effective as the LSTM method in consistency with ground truth, but the GRU takes less time in training. In the first few epochs, the convergence rate of the GRU method is fast, while in the next few epochs, the convergence rate is slowed down. Since LSTM outperforms GRU, we choose 2L-Bi-LSTM-ATT-lexical instead of 2L-Bi-GRU-ATT-lexical.

Table 4. The QWK, PCC and SCC scores of different models.

Method	QWK	PCC	SCC
2L-Bi-LSTM-ATT-lexical	**0.6977**	**0.7208**	**0.6789**
SVM	0.443	0.506	0.471
2L-LSTM-word2vec	0.6395	0.6615	0.6356
CNN-LSTM	0.6793	0.6803	0.6492
CNN-LSTM-ATT	0.6659	0.6924	0.6658
TDNN	0.6952	0.7191	0.6781

The experimental results show that our model performs better than other methods on QWK, PCC and SCC. The performance results of each model on different evaluation metrics is shown in Table 4. In terms of QWK, 2L-Bi-LSTM-ATT-lexical performs the best among different comparison models. More precisely, 2L-Bi-LSTM-ATT-lexical outperforms 2L-LSTM-word2vec by 10%, demonstrating that the proposed model has a higher consistency with the real essay grading. However, in terms of PCC, 2L-Bi-LSTM-ATT-lexical performs worse than the model without lexical features, but still performs better than other methods. Our model is obviously superior to other comparison models in PCC score except TDNN. Similar to PCC, our model has the best performance in terms of SCC, demonstrating that the proposed model monotonically correlates better with the real essay grading.

At the same time, TDNN has the best performance in comparison models, which is close to our proposed model. However, this model is more complicated and less interpretable and it does not incorporate expert knowledge. Due to the interpretable features extracted by experts, our model is easier to understand and has higher interpretability than the model using only deep neural networks.

6 Conclusion

In this paper, we studied the expert evaluation standard, and proposed an interpretable lexical grading method for essays. Our model accepted a sequence of word vectors as input and integrated attention mechanism in terms of lexical richness. Experimental results show that our model outperforms state-of art models for Chinese essay grading task. Besides, our method gives interpretable results, which are helpful for practical applications.

For future works, we are planning to study the syntactic characteristics and use them together for the essay grading task. One promising solution is to introduce the features on syntactic complexity and elegant sentences of essays. Another important direction is essay grading for the students in the same exam. Since their writing abilities are very close, the essay grading task is more challenging. We will also explore the prompt based Chinese essay grading task and provide useful feedback to authors.

Acknowledgments. This work was supported by the National Key Research and Development Program of China under Grant No. 2018YFC0831401, the National Natural Science Foundation of China under Grant No. 91646119, the Major Project of NSF Shandong Province under Grant No. ZR2018ZB0420, and the Key Research and Development Program of Shandong province under Grant No. 2017GGX10114. The scientific calculations in this paper have been done on the HPC Cloud Platform of Shandong University.

References

1. Attali, Y., Burstein, J.: Automated essay scoring with e-rater® V. 2. J. Technol. Learn. Assess. **4**(3), 1–30 (2006)
2. Juku Correction Website. https://www.pigai.org/
3. Graves, A.: Supervised sequence labelling with recurrent neural networks. Stud. Comput. Intell. **385**, 1–131 (2012)
4. Hochreiter, S., Schmidhuber, J.: Long short-term memory. Neural Comput. **9**(8), 1735–1780 (1997)
5. Page, E.B.: Grading essays by computer: progress report. In: Proceedings of the Invitational Conference on Testing Problems, pp. 87–100 (1967)
6. Daigon, A.: Computer grading of English essays. Engl. J. **55**(1), 46–52 (1966)
7. Foltz, P.W., Laham, D., Landauer, T.K.: The intelligent essay assessor: applications to educational technology. Interact. Multimedia Electron. J. Comput.-Enhanc. Learn. **1**(2), 939–944 (1999)
8. Landauer, T.K., Foltz, P.W., Laham, D.: An introduction to latent semantic analysis. Discourse Process. **25**(2–3), 259–284 (1998)
9. Rudner, L.: Computer grading using Bayesian networks-overview. Wayback Machine (2012)
10. Automated Student Assessment Prize (ASAP). https://www.kaggle.com/c/asap-aes
11. Alikaniotis, D., Yannakoudakis, H., Rei, M.: Automatic text scoring using neural networks. arXiv preprint. arXiv:1606.04289 (2016)
12. Dong, F., Zhang, Y., Yang, J.: Attention-based recurrent convolutional neural network for automatic essay scoring. In: Proceedings of the 21st Conference on Computational Natural Language Learning, pp. 153–162. ACL, Vancouver (2017)
13. Cozma, M., Butnaru, A.M., Ionescu, R.T.: Automated essay scoring with string kernels and word embeddings. In: Proceedings of the 56th Annual Meeting of the Association for Computational Linguistics, pp. 503–509. ACL, Melbourne (2018)
14. Taghipour, K., Ng, H.T.: A neural approach to automated essay scoring. In: Proceedings of the 2016 Conference on Empirical Methods in Natural Language Processing, pp. 1882–1891. ACL, Austin (2016)
15. Jin, C., He, B., Hui, K., et al.: TDNN: a two-stage deep neural network for prompt-independent automated essay scoring. In: Proceedings of the 56th Annual Meeting of the Association for Computational Linguistics, pp. 1088–1097. ACL, Melbourne (2018)
16. Tay, Y., Phan, M.C., Tuan, L.A., et al.: SkipFlow: incorporating neural coherence features for end-to-end automatic text scoring. In: Thirty-Second AAAI Conference on Artificial Intelligence, pp. 5948–5955. AAAI, New Orleans (2018)
17. Ruiji, F., Dong, W., Shijin, W., Guoping, H., Ting, L.: Elegart sentence recognition for automated essay scoring. J. Chin. Inf. Process. **32**(6), 88–97 (2018)

18. Examination Center of the Office of the National HSK Examination Committee: Outline of Chinese Proficiency Vocabulary and Chinese Characters. Economic Science Press, Beijing (2001)
19. Le, Q.V., Mikolov, T.: Distributed representations of sentences and documents. In: International Conference on Machine Learning, pp. 1188–1196. IMLS, Beijing (2014)
20. Pennington, J., Socher, R., Manning, C.: Glove: global vectors for word representation. In: Proceedings of the 2014 Conference on Empirical Methods in Natural Language Processing, pp. 1532–1543. ACL, Doha (2014)
21. Shen, L., Zhe, Z., Renfen, H., Wensi, L., Tao, L., Xiaoyong, D.: Analogical reasoning on Chinese morphological and semantic relations. In: Proceedings of the 56th Annual Meeting of the Association for Computational Linguistics, pp. 138–143. ACL, Melbourne (2018)
22. Kingma, D.P., Ba, J.: Adam: a method for stochastic optimization. In: International Conference on Learning Representations, Microtome, San Diego (2015)

An Email Visualization System Based on Event Analysis

Qiang Lu[1,2(✉)], Qingyu Zhang[1], Xun Luo[1], and Fang Fang[3]

[1] School of Computer and Information, Hefei University of Technology,
Hefei, Anhui, China
luqiang@hfut.edu.cn
[2] Anhui Province Key Laboratory of Industry Safety and Emergency
Technology (Hefei University of Technology), Hefei 230009, Anhui, China
[3] School of Management, Hefei University of Technology, Hefei, Anhui, China

Abstract. E-mail has a wealth of information, including work topics, interactions between people, and the evolution of events over time. The emails will give users a better understanding that how things have changed and evolved in the past. Much of the effort to visualize email has focused on three areas of email archiving: exploring the relationship between email volumes, mining the evolution of topics and events in emails, or the relationship of email owners to their counterparts. But there are currently fewer systems for analyzing their background stories through mail dataset. In this paper, we present the Mail event, which is an email visualization system. Its main purpose is to help users analyze the main information in the mail data set, such as keywords, topics, and event contents of the mail. Firstly, it helps users understand the keywords and themes of the mail through a variety of different attempts. Secondly, the way the email is matched into an event allows the user to understand the story of the email corresponding to the email at a certain point in time so that users can deeply understand the story behind the email. In this system, through rich visual elements, users can understand the e-mail dataset and have a further understanding of the development of events and their anomalies, so as to better coordinate or improve future work. Finally, the effectiveness of the system is verified by case studies and user evaluation experiments.

Keywords: Email visualization · Email event analyze · Cooperative information

1 Introduction

In daily collaborative tasks, e-mail is one of the communication tools, which runs through every step of the project. It contains rich information such as time, events, and people-to-person interactions. If we can deeply understand the news of the mail, we can better understand how things have evolved and developed (Fig. 1).

Event data refers to a series of ordered events that occur over a period of time. By analyzing the sequence of events, we can derive the regularity or progression of events. For example, event sequence analysis may reveal a clinical path for hospitals to improve outcomes [1], or an understanding of the career trajectory of certain

© Springer Nature Singapore Pte Ltd. 2019
Y. Sun et al. (Eds.): ChineseCSCW 2019, CCIS 1042, pp. 658–669, 2019.
https://doi.org/10.1007/978-981-15-1377-0_51

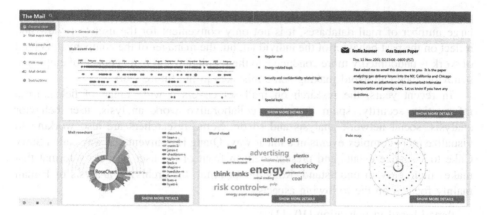

Fig. 1. The Mail event main interface

workers [2], and a service model that provides better user experience. It can effectively help managers to complete task planning, resource allocation, project scheduling, communication and other aspects to promote teamwork. The existing methods focus on the use of statistical analysis to clearly create the stage of progress of events over time [3–5].

Information visualization is an effort to show and help people understand and analyze more abstract data in an intuitive way by using techniques and methods in graphic images. In recent years, collaborative visualization [6] has gradually been recognized and has formed a new field. A large number of scholars have proved that they have a positive impact on collaborative work in the direction of information visualization technology [7–9].

Email is a valuable record of interaction between people. Different from face-to-face communication, this communication method is mainly recorded by means of text, photos, pictures, and supplements. As a more intuitive way, it's in the back to see when we need to add more energy, we hope that we can have a better way to describe the story in the email.

If we can convert the message information into a sequence of events and present it in front of people through visualization, then we can more clearly understanding the role of certain users in the work and the relationship among groups in a certain period of time. Interacting behaviors, etc., can even find out the causes of certain abnormal situations in the work, thus more effectively helping managers to manage the internal communication and potential security risks of the company.

2 Related Work

Over time, many people's mail databases will become very large. Among them, there are a lot of unread mails, cc mails, and long-lasting mails. Their themes, quantity, and importance are different, which makes it difficult for even e-mail users to meet past events through their own e-mail addresses. In short, the interaction information of a person is not obvious in the email. If we can classify and extract the information in a large number of mail databases, it is not only convenient for the user to review and reflect on the past behavior of the individual, but the manager of the company, this kind of work is undoubtedly more conducive to the regulation and management of employee behavior, they can avoid making some mistakes in the future work.

In recent years, the research on mail set covers a wide range of fields, from information security, spam detection, collaborative work analysis, user behavior analysis, social network analysis and so on. Some researchers are also working to visualize more complex events in a linear way, Qiang [12] invented a way called Story Cake to visualize events and present them in front of the user in a story format that makes them easy to understand. Roughly speaking, the archiving analysis of E-mail mainly focuses on the following aspects:

- thread-based visualization [10, 11]
- social network visualization [13, 14, 17]
- time visualization [15, 16].

In terms of mail threads, Luo [10] and Kerr [11] described in detail how to visualize the mail sending and receiving relationship and how to visualize events by visualizing the mail sending and receiving relationship. Fernanda [17] focused on the extraction of email keywords for users' social networks, and predicted what might happen to users in a period of time through the change of users' email keywords over time. Shneiderman [15], Joorabchi [16] put their focus on the correspondence between mail and time, trying to get some mail sending and receiving relationships in this way, and showing what might happen in the future. In the process of visualization, we hope to present the content to users mainly on the following two issues:

- What information, or backstory I can get about the email data owner from the email dataset?
  The information includes the sender's behavior, direction of work, intensity, position, or the relationship between the above elements over time.
- What stories I can find out from the mail dataset about what happened to the mail owner?
  That is, at an important point in time, what happened, and whether you can get from the mail to some of the behavior of the sender?

3 Structure View

As mentioned above, the Mail event hopes to complete a visual view. On the one hand, it can display the subject of emails through text analysis of word frequency, word meaning and so on. On the other hand, this view can also reflect the interaction between people through the screening of recipients.

In the mail dataset we obtained, there were about 500 thousand emails from 150 users in total. The email dataset shows some of the company's email traffic from 1999 to 2001.

Firstly, we should deal with the mail dataset. Through the screening of various dimensions, we can intuitively understand the main business direction of the company and the changing relationship between its businesses over time. At the same time, we can better understand the important information such as the business proportion of each member in the mail dataset in the company.

Generally speaking, the importance of an employee's position is often related to a variety of factors, among which the more important factors are the number of emails and the number of people associated with emails. If you have a high proportion of both dimensions, it is likely that the member is a higher level. In a rose petal diagram, its radian represents the area across which its message topic spans, and its radius length represents the number of messages. As shown in Fig. 2, we can clearly find the users with the highest importance by the weighted sum of the two dimensions. Dasovich has the most email exchange information, while Kean is involved in the most extensive email information field, which indicates that both of them have relatively high positions in the company.

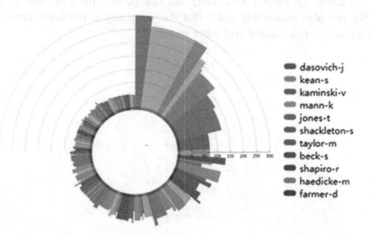

- dasovich-j
- kean-s
- kaminski-v
- mann-k
- jones-t
- shackleton-s
- taylor-m
- beck-s
- shapiro-r
- haedicke-m
- farmer-d

Fig. 2. Shows the order of importance of users in two dimensions

For the business areas mainly involved in the mail data set, the key words involved in the core members are very valuable. In order to determine which members are "core members", we consider the three dimensions, in addition to the number of people emails, the human email relates to the field described above, the third dimension is the length of time that the person's email is sent and received. Through the first two dimensions, we can help us screen out a large number of ordinary employees, because the number of mails of managers is usually much higher than that of ordinary employees, but there are certain exceptions. For example, in the customer service position, they usually have a large number of mails, but their business is often solved in a short period of time, and the manager's role often has a long-term mail relationship. Through these aspects, we can quickly locate core members.

We categorized the mail dataset and counted the subject frequency based on the content of the email. We use the nltk natural language package under python for word segmentation, deleting useless words and frequency statistics on words. Then use lda for analysis in Python to get the 30 areas covered by the full mail data set.

For each member x, we first count the domain scope f of his email subject, the total number of mails n, and the longest mail delivery time d, where d is the number of days, we calculate the importance of the member I:

$$I(x) = f(x) * N(x) * t(x). \tag{1}$$

Where $N(x) = -\log(1/n(x))$, this is done to reduce the weight of the number of messages, so that the importance of the corresponding personnel can be more reasonable.

As shown in Fig. 3, we screened the top 30 topics covered by the email content, and generate the corresponding word cloud, the company in the business sectors is involved in the energy, such as wind, water treatment, kerosene, gas, etc. Meanwhile, within the company, media, advertising and risk control have become its main daily work. We can also reasonably guess that this company is relatively large, so it pays more attention to risk control and other projects.

Fig. 3. Business areas of high importance users

Another aspect to measure the work intensity of a company is the change relationship between the number of emails sent and the time. Figure 4 shows the change relationship among the total amount of emails sent in three years with the hours. In the polar coordinate scatter plot, the time represents the different years from the inside to the outside, and the counterclockwise in each ring represents from 0 to 24 o'clock. The small circle in the circle indicates the total amount of mail sent and received at the corresponding time. The more the total amount of sending and receiving, the longer the radius corresponding to the small circle. As shown in Fig. 4, it can be seen that the main working hours of the company started at 8 am and lasted until nearly 10 PM, with a long coverage time.

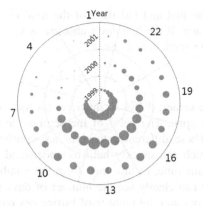

Fig. 4. The relationship between the number of mail delivery and the number of hours in three years.

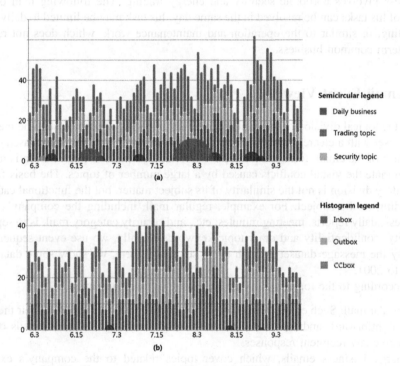

Fig. 5. Demonstrate two mail owners for different business types

Figure 5 mainly shows us the number of emails sent and received by employees in two different positions over time, and the time between emails. The horizontal coordinate in the figure indicates the date the mail was sent and gradually increased. The ordinate represents the total number of messages, and the histogram represents different types of messages in different colors. When the mailings of a two contact have spanned for days, we will use the first and last days of the mail to be the diameter d in the horizontal axis of the graph. If the first day's abscissa is x_1 and the last day is x_2. We define the abscissa of the semicircle x_3 as:

$$x_3 = (x_1 + x_2)/2. \tag{2}$$

With $(x_3, d/2)$ as the center, $d/2$ is a semicircle for the radius, and a semicircle is used to reflect the time span. The color of the semi-circle corresponds to different categories of themes. These different categories are consistent with the categories expressed in Fig. 7, which can quickly help us understand what the main topic of communication during this time. In this way, from the number and size of the white circles in the figure, we can clearly see the number of days involved in the different business of the mail owner and the number of businesses communicating frequently. As shown in Fig. 5, we believe that the above mail owner clearly has a higher status in the company than the mail owner below, because his business spans a long time and his business involves corporate security and energy trading. The following mail owner, most of his tasks can be resolved in the same day, his task may be limited to daily work reporting, or similar to the operation and maintenance work, which does not require long-term common business.

4 Email Event View

After the topical division of the mail dataset, we hope to be able to present the message to the user with a clearer look. We set up our own keyword database and, based on the response of the email, to divide the topic twice. The purpose of this division is mainly to eliminate the visual conflicts caused by a large number of topics. The basis for the secondary division is not the similarity of its subject matter, but the functional category according to the subject. For example, regular mail, including the company's main business, daily reports, meeting minutes, etc., and security category mail, is to separate security, confidentiality and other topic keywords. Finally, we use event sequence to display the message dataset through event sequence. Here, we show some data from 2000 to 2001.

According to the topic, we divide the mail into five categories:

- Regular mail. Such emails are usually daily reports and work reports. Their theme is straightforward, and most emails are one-way, meaning that most emails do not receive any recipient responses.
- Energy business emails, which cover topics related to the company's external energy business, including natural gas, energy supply strategy and other topics;
- Trade emails, including business sales, overseas business development, commodity contract signing, conference and other topics;
- Security and confidentiality related emails, including internal information confidentiality, personnel security protection and other related topics;
- Special events, when the main content of the email is too long or there are too many exchanges with the same email, it will be classified into the same category.

Although this kind of email does not have a clear theme, it can be used to draw important events that may happen at a certain time.

After filtering, display it in the mail sequence, as shown in Fig. 6. You can view the details of this message by hovering over the corresponding message.

	2000	February	March	April	May	June	July	August	September	October	November	December	2001	February
Regular mail(11273)														
Energy related topics(1761)														
Security and confidentiality related topics(366)														
Foreign trade mail topic(1139)														
Special topic(7)														

> rosalee.fleming On Wed, 25 Jul 2001 09:16:13 -0700 (PDT)
> **Climate News from Bonn**

Fig. 6. The horizontal axis shows the number of messages changed over time, while the vertical axis shows different topics.

5   User Interaction

The mail system provides some means of user interaction to help users understand mail datasets. On The mail's home screen, it displays an overview view, including the mail event view, pole, rose, word cloud, and mail content view. The 'show more details' button allows users to view the details of the corresponding data. In the rose view, by clicking on different characters, you can display their corresponding column charts to show the three dimensions described above that measure the importance of user mail.

Figure 7 shows the mail event view, users can click on the E-mail with brief information on the right side to see the mail, it can let users with email roughly the content of the subject matter of a simple understanding, including the theme, the sender and content of the brief, choose mail to highlight. If required, the user can also check in

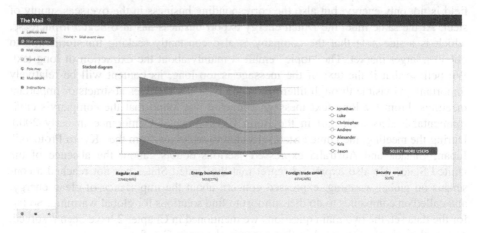

Fig. 7. Change of mail quantity over time

the mail the details of the complete email. If the user simultaneously tracks the relationship between the subject and number of messages, the detailed view also lists the number of messages sent and received by different users over time, which is displayed in the form of a stacked chart.

6 The Results Analyze

By analyzing the results of the mail data set, from 2000 to 2001, the number of ordinary mails occupied the main part of the entire article, which indicates that the company's business can be carried out in a stable and orderly manner. In terms of energy mail, it has been running through the company's main business. The key word for its main energy business is "natural gas", followed by "electricity", "coal", "energy supply", "energy management" and "equipment management", which explains the company's The main business is around energy and related industries, and is a company that provides a variety of energy sources, and it is also consistent with our previous word cloud. In the management of company services, the emergence of keywords such as "risk control", "advertising" and "brainstorming" indicates that the company's security guarantees for the business and external publicity work are in place. It is worth noting that in terms of email security, it mainly starts at the beginning of each quarter, which means that the company conducts stricter screening and implementation at the beginning of each quarter. In early 2000, we noticed a significant increase in the number of emails related to security directions. By inquiring about relevant e-mails, we can analyze that in the early 2000s, overseas business may start to have violent incidents and the impact is bad. "gunners", "vehicle damage", "violence" as keywords have begun to appear in e-mails, and hope that through the assistance and support of the government through e-mail, the call for promoting security while also actively seeking to resolve this A new way of class-like events; "overseas business" is mostly mail for foreign trade. According to the screening, the company began to appear in the beginning of 2001, and the words "steel" and "the Netherlands" began to appear. We can reasonably speculate that this company The company's involvement in the field is not only energy, but also the corresponding business in the overseas supply of steel. At the same time, the Dutch energy export business has also begun to increase, which is a side note that the company is also constantly seeking transformation to obtain a larger market. The "topic" email is mainly about the extraction of long text. We believe that if the text of the message is too long, its content will be relatively important and visible through filtering, which typically involves abstracts or important meetings. From the long text message extracted, we found that the company's environmental leaders took part in the Bonn Environmental Conference in early 2000. During the meeting, the United States decided to withdraw from the "Kyoto Protocol". Japan, Canada and Australia expressed "serious doubts" about the absence of the United States. He also expressed regret that the United States has not reached a consensus on global warming, expressed concern about the importance of clean energy, and called on companies to do their utmost to find solutions for global warming. So far, for the data set, the two main questions we mentioned in Chapter 2 have been resolved. The analytical structure used in this paper is shown in Fig. 8.

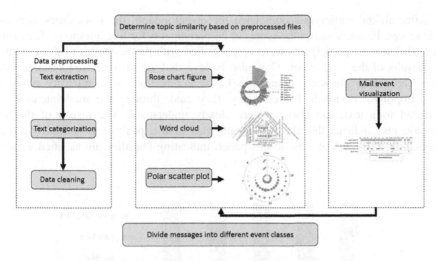

Fig. 8. The basic flow of mail sequence generation used in this article.

7   User Study

In order to further test whether the system can correctly guide users needed to come to the practical information, and solve two major problems of Sect. 2, verified the effectiveness and practicability, this paper designed a user survey assessment by 36 participants evaluate visualization system, among them, there are 20 professional IT engineers, we want the system to email owners behavior analysis can help for their work, the other 13 professional no relationship with the IT direction, We also want to see if users with less knowledge of the IT business can come up with useful information about the system. The whole investigation process includes the introduction of basic information and the completion of the questionnaire. For each participant, the researchers first explained the basic meaning of each subgraph and familiarized the participants with the purpose of the sequence of mail events shown in Fig. 9. Participants were then asked to explore freely and complete an evaluation of the questionnaire.

In this paper, users were asked to use a five-point likert scale to score five questions in the questionnaire to express their satisfaction. Five is very satisfied, four is satisfied, three is neutral, two is dissatisfied, and one is very dissatisfied. The problem is described as follows:

Q1: is a view aesthetically readable?
Q2: can you learn about the company's main business direction from the view?
Q3: can you analyze the regularity of work intensity and working hours of the company through the view?
Q4: can you see through the view that important events happened in the company at a certain time?
Q5: do you think the system is practical?
Q6: can you describe the views of the two visualization systems?

After all the participants completed the questionnaire, the researchers measured their scores. Figure 9 shows the scores of 36 participants for five questions. The results showed that most participants were very satisfied with the aesthetics of the visualization results of the Mail event. They also believe that the visualization system in this paper can clearly show the company's business direction, work intensity and work time. Important events in the company, they said, through the mail intensity and extracted long text, can also be more clearly understood. The results of the fifth question also indicate that the participants also hold a positive attitude towards the practicality of the visual view in this paper, indicating that they are satisfied with it.

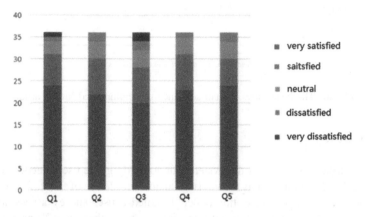

Fig. 9. User satisfaction with the system

8 Conclusion and Future Work

E-mail plays an important role in people's communication and cooperation. It contains rich information for recalling and understanding the past. The content, quantity, and time of email are often intertwined in our lives, but most tools designed to present email archives provide only a limited understanding of an email, making it difficult for people to understand the content of an email in general. In this article, we have analyzed the E-mail set in multiple views so that users can understand the information and data contained in the message through supplementary contextual information. This new way of understanding email archives can better assist users in archiving email information. At the same time, for company managers, mailing is also a summary of past events so that in the future work can learn lessons and avoid repeating the same mistakes.

Acknowledgments. This work was supported in part by the Natural Science Foundation of Anhui Province of China under Grant 1708085MF158, in part by the Visiting Scholar Researcher Program at North Texas University through the State Scholarship Fund of the China Scholarship Council under Grant 201806695039, and in part by the Key Project of Transformation and Industrialization of Scientific and Technological Achievements of Intelligent Manufacturing Technology Research Institute of Hefei University of Technology under Grant IMICZ2017010.

References

1. Shunan, G., Zhuochen, J., David, G., et al.: Visual progression analysis of event sequence data. IEEE Trans. Visual. Comput. Graph. **25**(1), 418–426 (2018)
2. Guo, S., Jin, Z., Gotz, D., et al.: Visual progression analysis of event sequence data, 1–6 (2019)
3. Hong, S., Wu, M., Li, H., Wu, Z.: Event2vec: learning representations of events on temporal sequences. In: Chen, L., Jensen, C.S., Shahabi, C., Yang, X., Lian, X. (eds.) APWeb-WAIM 2017. LNCS, vol. 10367, pp. 33–47. Springer, Cham (2017). https://doi.org/10.1007/978-3-319-63564-4_3
4. Du, F., Plaisant, C., Spring, N., et al.: EventAction: visual analytics for temporal event sequence recommendation. In: 2016 IEEE Conference on Visual Analytics Science and Technology (VAST), pp. 2–4. IEEE (2016)
5. Gotz, D., Stavropoulos, H.: DecisionFlow: visual analytics for high-dimensional temporal event sequence data. IEEE Trans. Visual. Comput. Graph. **20**(12), 1783–1792 (2014)
6. Jianu, R., Rusu, A., Hu, Y., et al.: How to display group information on node-link diagrams: an evaluation. IEEE Trans. Visual. Comput. Graph. **20**(11), 1530 (2014)
7. Bach, B., Pietriga, E., Fekete, J.D.: Visualizing dynamic networks with matrix cubes. In: Proceedings of the SIGCHI Conference on Human Factors in Computer Systems, pp. 877–886. ACM Press, New York (2014)
8. Ogawa, M., Ma, K.L.: Software evolution storylines. In: ACM 2010 Symposium on Software Visualization, Salt Lake City, Ut, USA. DBLP, pp. 35–42 (October 2010)
9. Liu, S., Wu, Y., Wei, E., et al.: StoryFlow: tracking the evolution of stories. IEEE Trans. Visual. Comput. Graph. **19**(12), 2436–2445 (2013)
10. Luo, S.J., Huang, L.T., Chen, B.Y., et al.: EmailMap: visualizing event evolution and contact interaction within email archives. In: 2014 IEEE Pacific Visualization Symposium (PacificVis), pp. 320–324. IEEE Computer Society (2014)
11. Kerr, B.: Thread arcs: an email thread visualization. IBM Corporation, pp. 2–4 (2003)
12. Qiang, L., Bingjie, C., Haibo, Z.: Storytelling by the storycake visualization. Visual Comput. **3**(10), 1241–1252 (2017)
13. Li, W.J., Hershkop, S., Stolfo, S.: Email archive analysis through graphical visualization. In: Workshop on Visualization and Data Mining for Computer Security. DBLP, pp. 2–4 (2004)
14. Biukaghai, R.P.: Visualization of interactions in an online collaboration environment. In: International Conference on Collaborative Technologies and Systems. IEEE Computer Society, pp. 1–5 (2005)
15. Perer, A., Shneiderman, B., Oard, D.W.: Using rhythms of relationships to understand e-mail archives. J. Assoc. Inf. Sci. Technol. **57**(14), 1936–1948 (2010)
16. Joorabchi, M.E., Yim, J.D., Shaw, C.D.: EmailTime: visual analytics of emails. In: Visual Analytics Science and Technology, pp. 233–234. IEEE (2010)
17. Viégas, F.B., Golder, S.A., Donath, J.S.: Visualizing email content: portraying relationships from conversational histories. In: SIGCHI Conference on Human Factors in Computing Systems, CHI 06, pp. 2–4 (2006)

Grading Chinese Answers on Specialty Subjective Questions

Dongjin Li[1,2], Tianyuan Liu[1,2], Wei Pan[1], Xiaoyue Liu[1],
Yuqing Sun[1(✉)], and Feng Yuan[3]

[1] School of Software, Shandong University, Jinan, China
`lidongjin1994@163`, `zodiacg@foxmail.com`,
`panwei_sdu@163`, `suiqiyue@163.com`,
`sun_yuqing@sdu.edu.cn`
[2] School of Computer Science and Technology,
Shandong University, Jinan, China
[3] Shandong University Ouma Software Co., Ltd., Jinan, China
`sdyuanf@sina.com`

Abstract. It is an important task to grade answers on specialty subjective questions, which is helpful for the supervision of human review and improving the efficiency and quality of review process. Since this grading process should be performed at the same time with human review, there are only a few samples available for each question that can be provided by specialty experts before review process. We investigate the problem of grading Chinese answers on specialty subjective questions with a reference answer in this paper by proposing a grading model that combines two Bi-LSTM networks with attention mechanism. The first part is a sequence to sequence Bi-LSTM network that adopts the pre-trained word embeddings as input. Since there is no embedding for some specialty words, we instead use the fine-grained word embeddings. After the max-pooling on each sentence, we adopt the mutual attention mechanism to learn the matching degree on specialty knowledge between each pair of sentences of answer and reference. Then we adopt another Bi-LSTM with max-pooling to have an overall vector. By concatenating these two vectors from answer and reference, a multilayer perceptron is adopted to predicate the scores. We adopt the real datasets on a national specialty examination to thoroughly verify the model performance against different amount of training data, network structures, pooling strategies and attention mechanisms. The experimental results show the effectiveness of our method.

Keywords: Grading Chinese answer · Specialty subjective questions · Attention mechanism

1 Introduction

It is an important task to grade answers on specialty subjective questions. We investigate the problem of grading Chinese answers on specialty subjective questions with a reference answer in this paper. Although there are quite a few works on grading English essays, they are not applicable for our problem due to the following challenges.

Y. Sun et al. (Eds.): ChineseCSCW 2019, CCIS 1042, pp. 670–682, 2019.
https://doi.org/10.1007/978-981-15-1377-0_52

One is the reference answer. In English essay scoring, there is not any reference. For example, the E-rater system developed by Burstein [1] scored English essays from the perspectives of syntactic analysis, subject analysis and other semantic aspects. Instead, in the subjective question problem, we are given the standard answer as reference for each question. When evaluating the student answers, we need to exam how much they match on knowledge points. For a specificity subjective question, the content precisely defines the direction and scope of answer. Some answers hit the key words in reference and seem similar in phrase level, but they might be logically wrong. Many student answers contain the same specialty words such that the evaluation on lexical or even syntax feature does not work.

The second is the insufficient amount of training data. Text classification methods based on deep learning generally require a large number of training samples. In our scoring scene, the model needs to learn based on a small number of labeled samples. Since the exam questions change every year, the data of previous years are not suitable for the grading task of this year.

The third is the discrete scores. Generally, experts examine how many knowledge points are targeted by a student answer, and assign different discrete scores. It is not suitable to directly adopt the classification or regression methods for this scoring process.

There is also another challenge on the specialty word embeddings. The pre-trained universal word embeddings do not exactly contain all specialty words. Since there are not enough specialty corpus, it is difficult to learn stable embeddings for specialty words.

To tackle these challenges, we propose a grading model based on mutual attention mechanism. When a specialty word has no embedding, we use its fine-grained words embeddings to represent the word. We combine bidirectional Long Short-Term Memory (Bi-LSTM) network and mutual attention mechanism to grade student answers, taking into account the semantic information of student answer and its matching degree to the reference answer. We adopt the real datasets on a national specialty examination to thoroughly examine the performance against different amount of training data, network structures, pooling strategies and attention mechanisms. The experimental results show the effectiveness of our proposed method.

The rest of this paper is organized as follows. Section 2 introduces related works. Section 3 introduces our grading model. In Sect. 4, we validate our model on real datasets and analyze the experimental results. Section 5 summaries this paper and presents future work.

2 Related Work

So far, to the best of our knowledge, there is not any publicly available works on the task of grading Chinese answer of specialty questions that are exactly related with our work. In this section, we present some works that are technically related. At present, the Recurrent Neural Network (RNN) and Convolutional Neural Network (CNN) are often adopted to extract semantic features from text. Sutskever et al. [2] performed the language translation task using a RNN model based on Long Short-Term Memory

(LSTM) to obtain sentence vector. Colleber et al. [3] and Kim et al. [4] extracted features using CNN, and achieved good results in tasks such as part-of-speech tagging, sentiment classification, and named entity recognition. Zhang et al. [5] used CNN to model sentences at the character level and applied the obtained sentence vector to text classification task. Kalchbrenner et al. [6] proposed the Dynamic Convolutional Neural Network which used dynamic k-max pooling, and the model achieved good performance on multi-class sentiment prediction tasks. Schwenk et al. [7] and Johnson et al. [8] extracted deeper semantic features through multi-level convolution and performed well on text classification.

The combination of CNN and RNN are also adopted to extract the semantic features. Tang et al. [9] generated a sentence vector by extracting features through CNN at lexical level, and then generated a text vector by extracting sentences sequence features based on a Gated Recurrent Unit (GRU) network. Lai et al. [10] used RNN to encode a sentence, and then obtained the sentence vector via pooling operations. Shi et al. [11] replaced the convolution kernel of CNN with LSTM to encode sentence and used the generated vector for text classification. Xiao et al. [12] used CNN and RNN to process sentences respectively, and then concatenated the generated vectors as the sentence vector which was applied to text classification. By using CNN and RNN together, the local features and context-sensitive features of the text are extracted separately.

In the subjective review task with reference, the final score of a student answer is not only determined by features of answer text, but also by the matching degree between the answer and reference. The introduction of attention mechanism has enabled the model to capture the points of focus on each answer. Bahdanau et al. [13] first introduced attention mechanism into natural language processing field in 2014. In machine translation, the authors calculated the related information between current word and each word of the sentence to be translated, and dynamically searched the information related to current word during decoding. The attention mechanism can dynamically acquire the key information focused by current word or sentence, and was later applied to multiple natural language processing tasks such as question answering, text entailment and text classification. In question answering task, Tan et al. [14] generated a representation for a specific question by calculating attention weights of the candidate answers and the question. In the reading comprehension task, Chaturvedi et al. [15] concatenated the question with each candidate answer, and calculated attention weights on each sentence in the context. Yang et al. [16] introduced the attention mechanism into the GRU network on text classification tasks. In the Chinese cloze-style reading comprehension task, Cui et al. [17] proposed a consensus attention mechanism to calculate the attention weights between each words in the query and the document. In the English cloze-style reading comprehension task, Cui et al. [18] proposed a mutual attention mechanism by calculating the text-based attention and the question-based attention respectively, and combing two attention weights as the probability of each word in the text to be the standard answer. We adopt this idea into our model to calculate the matching degree of answer and reference.

3 The Grading Model on Specialty Subjective Questions

3.1 The Grading Model on Specialty Subjective Questions

The grading task on specialty subjective questions with reference answer is defined as follows. For the subjective question Q, the student answer text X_0 and the reference answer text A_0, the problem is to predict student answer's score $c \in C$, where $C = \{c_1, c_2, \ldots, c_r\}$ is a set of categories according to the score range of Q. We propose a grading model based on Bi-LSTM and mutual attention mechanism, which is shown in Fig. 1. Details of our model are given below.

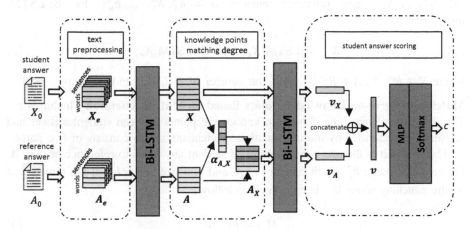

Fig. 1. The grading model on specialty subjective questions.

Text Preprocessing. First, the student answer X_0 and the reference answer A_0 are segmented into sentences according to commas, periods, semicolons and colons. Then each sentence is segmented into words. We adopt pre-trained Chinese word embeddings [19] as the word embedding. Since this is a specialty exam, the text may contain some specialty words which have no embedding. In order to retain the semantic information of the word, we combine the fine-grained word vectors to obtain the embedding of the specialty word. The specialty word without embedding is segmented into subwords. If a subword still has no embedding, the segmentation is performed to the subword again until it has embedding or is split to characters.

After the text preprocessing, we have the embedding form of student answer X_e and the reference answer A_e, where $X_e \in R^{m \times l \times d_0}$, $A_e \in R^{n \times l \times d_0}$, m is the number of answer sentences, n is the number of reference sentences, l is the number of words segmented by each sentence after padding, and d_0 is the dimension of word embedding.

The Bi-LSTM Network for Semantic Feature Extraction. We adopt a sequence to sequence Bi-LSTM model to extract the semantic features of both student answer X_e and reference answer A_e with the max-pooling on each sentence. For sentence $s = [w_1, w_2, \ldots, w_l]$, where w_l is word embedding of l th word in s. The forward LSTM encodes words sequence along the direction of from the first word to the last word, and the backward LSTM encodes words sequence along the reverse direction.

We adopt the max-pooling on the hidden state vectors of all timesteps of forward LSTM and backward LSTM, respectively, to obtain the forward vector $\vec{h}$ and backward vector $\overleftarrow{h}$. $\vec{h}$ and $\overleftarrow{h}$ are concatenated as the final sentence vector h.

$$\vec{h} = \text{LSTM}_f(s), \ \overleftarrow{h} = \text{LSTM}_b(s), \ h = \vec{h} \oplus \overleftarrow{h} \tag{1}$$

For each pair of X_e and A_e, we can get the encoded student answer $X = [h_1^X, h_2^X, \ldots, h_m^X]$ and reference answer $A = [h_1^A, h_2^A, \ldots, h_n^A]$ by Bi-LSTM, respectively.

$$X = \text{BiLSTM}(X_e), \ A = \text{BiLSTM}(A_e) \tag{2}$$

where $X \in R^{m \times 2d_1}$, $A \in R^{n \times 2d_1}$, d_1 is the number of Bi-LSTM hidden units.

Matching Degree on Knowledge Points Based on Mutual Attention Mechanism. The detection of matching degree on knowledge points between student answer and reference is performed by mutual attention mechanism, which consists of two parts.

The first part is the one-way attention of student answer X to reference answer A. For sentence vector h_i^A of i th sentence of A, and sentence vector h_j^X of j th sentence of X, the matching score $M_{i,j}$ is calculated as follows.

$$M_{i,j} = h_i^A \cdot h_j^{X^T} \tag{3}$$

The matching score for the whole student answer X and reference A is calculated pairwise as the matching score matrix $M \in R^{n \times m}$. The column-wise softmax function is applied to M to obtain the one-way attention matrix $\alpha \in R^{n \times m}$ of the student answer to reference answer. For sentence h_p^X in student answer X, let $\alpha(p)$ represent the distributions of matching degree between h_p^X and each sentence of reference answer A.

$$\alpha(p) = softmax(M_{1,p}, \ldots, M_{n,p})$$
$$\alpha = [\alpha(1), \ldots, \alpha(m)] \tag{4}$$

The second part is the mutual attention of student answer and reference answer. In the general attention mechanism, each row of the one-way attention matrix α is simply added or averaged as the final attention weights. In this grading task, for the pth sentence h_p^X of student answer X, even if the content of the sentence is completely irrelevant to reference answer A, after column-wise softmax of the matching score matrix M, the sum of probabilities of matching degree of h_p^X on reference answer A is still 1, so that the model with the general attention mechanism cannot effectively distinguish the invalid sentences in the student answer. We utilize mutual attention mechanism to solve this problem.

Row-wise softmax function is applied to M to obtain the one-way attention matrix β of the reference answer to the student answer, where $\beta \in R^{n \times m}$. For sentence h_q^A in reference answer A, $\beta(q)$ represents the probability distributions of matching degree between h_q^A and each sentence in student answer X.

$$\beta(q) = softmax\left(M_{q,1}, \ldots, M_{q,m}\right)$$

$$\beta = [\beta(1), \ldots, \beta(n)] \tag{5}$$

β is averaged on each column direction, and we get a weight vector β_{ave}.

$$\beta_{ave} = \frac{1}{n} \sum_{q=1}^{n} \beta(q) \tag{6}$$

$\beta_{ave} = \left(\beta_{ave}^1, \beta_{ave}^2, \ldots, \beta_{ave}^m\right)$, where β_{ave}^m represents the matching weight of m th sentence in student answer to the whole reference answer. Next, we calculate the mutual attention-based weight vector α_{A_X} between the student answer and reference answer.

$$\alpha_{A\_X} = \alpha \cdot \beta_{ave}^T \tag{7}$$

$\alpha_{A\_X} = \left(\alpha_{A\_X}^1, \alpha_{A\_X}^2, \ldots, \alpha_{A\_X}^n\right)$, where $\alpha_{A\_X}^n$ is the matching score of the overall student answer to n th sentence in reference answer. According to the mutual attention-based weight vector α_{A_X}, the reference representation $A_X = \left[h_1^{A_X}, h_2^{A_X}, \ldots, h_n^{A_X}\right]$ is calculated specifically for student answer X, which indicates the matching degree between student answer X and reference answer A, where $A_X \in R^{n \times 2d_1}$.

$$A_X = A \times \alpha_{A\_X} \tag{8}$$

The Bi-LSTM Network for Text Feature Extraction. By using the semantic feature extraction network and mutual attention network, we obtain student answer X and reference answer A_X. Sentences in X and A_X are respectively encoded by Bi-LSTM to capture the dependency between sentences. After the max-pooling over the hidden state vectors of all timesteps, we can get the encoded student answer vector v_X and reference answer vector v_A, respectively.

$$v_X = \text{BiLSTM}(X), \quad v_A = \text{BiLSTM}(A_X) \tag{9}$$

where $v_X \in R^{2d_2}$, $v_A \in R^{2d_2}$, d_2 is the number of Bi-LSTM hidden units.

Student Answer Scoring. The student answer vector v_X and the reference answer vector v_A are concatenated as the overall vector v. Then v is fed into a two-layer feedforward neural network, and we can get the category c as the final score of the student answer through a softmax function.

$$v = v_X \oplus v_A$$

$$v_1 = relu(W_1 \cdot v + b_1), v_2 = relu(W_2 \cdot v_1 + b_2)$$

$$c = softmax(v_2) \tag{10}$$

We minimize the following cross entropy loss function when training the model.

$$L(\Theta) = - \sum_{i=1}^{r} c_i \log p_{c_i} \tag{11}$$

Where r is the number of categories, $c_i \in \{0, 1\}$ is the real category of the sample, p_{c_i} is the probability that the sample is predicted to be category c_i, and Θ is the set of all parameters in the model.

4   Experiments and Analysis

4.1   Datasets

We adopt the real datasets on a national specialty examination provided by our partner, which include student answers and expert reviews, as well as the reference answers. The dataset I contains 45,000 answers and scores range from 0, 1.5 and 3. The dataset II contains 40,000 student answers and scores range from 0, 1 and 1.5.

Each question is associated with a reading material on a specialty case. It requires the student to make a judgement according to the question and present his reasons. For example, the question "李某是否有权拒绝张某的赔偿请求?请简要说明理由 (Is Li's right to refuse Zhang's claim for compensation? Briefly explain the reason)". If a student makes a wrong judgement, he gets 0 point. If his judgement is correct but the reason is wrong, he gets 1.5 points. Only both his judgement and reason are correct, he gets 3 points. The statistics of datasets are shown in Table 1.

Table 1. The statistics of datasets.

Datasets	Full score	Number of student answers	Score categories and counts	
I	3	45000	0	8545
			1.5	10928
			3	25527
II	1.5	40000	0	5590
			1	18607
			1.5	15803

Each dataset is divided into training set, validation set and test set with the proportions 60%, 20%, and 20%, respectively. Taking into account the practical requirement on a small amount of samples, we also select the proportion 0.5%, 1%, 5%, 10% and 30% as training set, respectively. For comparison purpose, the verification set and test set remain 20%.

4.2 Comparison Models

Conv-GRNN. Conv-GRNN was proposed by Tang et al. [9]. The model first used CNN to encode sentence at lexical level, then generated a text vector through GRU at sentence level, and finally classified the text according to the text vector.

LSTM-GRNN. LSTM-GRNN was proposed by Tang et al. [9]. The model first used LSTM to encode sentences at lexical level, then generated a text vector through GRU at sentence level, and finally classified the text according to the text vector.

HN-AVE. HN-AVE was proposed by Yang et al. [16]. The model first used bidirectional GRU to encode sentence at lexical level, taking the mean of hidden state vectors of all timesteps as the sentence vector. Then the sentence vector sequence was input into another bidirectional GRU, taking the mean of hidden state vectors of all timesteps as the text vector and finally classified the text according to the text vector.

HN-MAX. HN-MAX was proposed by Yang et al. [16]. The model first encoded the sentence at lexical level using bidirectional GRU, taking the max-pooling result of the hidden state vectors of all timesteps as the sentence vector. Then the sentence vector sequence was input into another bidirectional GRU, taking the max-pooling result of hidden state vectors of all timesteps as the text vector and finally classified the text according to the text vector.
Our model and variants are as follows.

Bi-LSTM-CA-MAX. Bi-LSTM-CA-MAX is our model.

Bi-LSTM-CA-AVE. Bi-LSTM-CA-AVE is a variant of our model. The average of hidden state vectors of all timesteps of Bi-LSTM is taken as the output, and the other parts are the same as Bi-LSTM-MAX.

Bi-LSTM-CA. Bi-LSTM-CA is a variant of our model. The hidden state of the last timestep of Bi-LSTM is taken as the output, and the other parts are the same as Bi-LSTM-MAX.

Bi-LSTM-A. Bi-LSTM-A is a variant of our model. The general attention mechanism is used to calculate attention weights between student answer and reference answer. Each row of the one-way attention matrix α is summed to obtain the final attention weights. The other parts are the same as Bi-LSTM-CA.

4.3 Experiment Setting and Metrics

Each student answer and reference answer are segmented into 20 sentences and 10 sentences, respectively, with zero vectors padded when the number of sentences was insufficient. Each sentence is segmented into 20 words, with zero vectors padded when the number of words is insufficient. The dimension of word embedding is 300. The number of Bi-LSTM hidden units is set as 100. In Conv-GRNN, LSTM-GRNN, HN-AVE and HN-MAX, the student answer and the reference answer are input into the model, respectively. In Conv-GRNN and LSTM-GRNN, the outputs of GRU layer of the student answer and reference answer are concatenated as the input of next layer. In HN-AVE and HN-MAX, the outputs of the second GRU layer of the student answer and reference answer are concatenated as the input of next layer. The other parts were consistent with the original model.

We adopt the accuracy as the overall evaluation metric, with the precision P, recall R and F1 score on each category as metrics. The precision P_{c_i} on category c_i is the proportion of the number of samples classified to c_i whose real category is c_i to the total number of samples classified to c_i by the model. The recall R_{c_i} on c_i is the proportion of the number of samples classified to c_i by the model whose real category is c_i to the total number of samples whose real category is c_i.

4.4 Experimental Results Analysis

Accuracy Against Different Amount of Training Data. We first verify how much the amount of samples influence the performance. The comparison results are shown in Table 2. The Bi-LSTM-CA-MAX model outperforms other methods in most cases. The attention mechanism, such as in the models Bi-LSTM-A, Bi-LSTM-CA and Bi-LSTM-CA-MAX, contributes a lot on improving the performance, especially in the cases with less training data. For example, for training set 0.5%, the accuracy of our model increases by 2.2% and 2.5% on dataset I and dataset II comparing with non-attention models, respectively. For training set 1%, compared with models without attention mechanism, the accuracy of our models increases by 1.3% and 1.4% on dataset I and dataset II, respectively. We think the reason is that models without attention mechanism cannot capture the matching degree between the student answer and reference, so that the lack of training data has a greater limitation on the learning ability of these models. With the attention mechanism, we can use the matching information between the student answer and reference answer, and improve model performance when having less training data.

Table 2. Model comparison and evaluation against the training data ratio.

Models	DatasetI					DatasetII				
	0.5%	1%	5%	10%	30%	0.5%	1%	5%	10%	30%
Conv-GRNN	0.780	0.820	0.843	0.864	0.871	0.636	0.685	0.701	0.726	0.742
LSTM-GRNN	0.794	0.830	0.858	0.871	0.876	0.621	0.674	0.716	0.735	0.748
HN-AVE	0.826	0.841	0.869	0.878	0.882	0.675	0.715	0.731	0.749	0.751
HN-MAX	0.829	0.843	0.872	0.878	0.880	0.681	0.718	0.735	0.749	0.755
Bi-LSTM-A	0.831	0.848	0.876	0.876	0.881	0.680	0.719	0.739	0.742	0.754
Bi-LSTM-CA	0.838	0.849	0.878	0.876	**0.884**	0.695	0.725	0.743	0.75	0.756
Bi-LSTM-CA-MAX	**0.851**	**0.856**	**0.880**	**0.882**	0.883	**0.706**	**0.732**	**0.743**	**0.754**	0.756
Bi-LSTM-CA-AVE	0.845	0.850	0.879	0.879	0.881	0.700	0.725	0.741	0.751	**0.758**

As the amount of training data increases, the accuracy of each model grows lower. The gap of different models gradually becomes small, but our models are always at a leading position. Because the scope of answers to the question is relatively fixed, the changes of student answers are relatively small. Although comparison models cannot effectively utilize the matching information between the student answer and the reference answer, the results get better with the increasing size of data.

Analysis on Different Components of Models. Then we verify the performance of different network structures, pooling strategies and attention mechanisms, and get three conclusions according to the experimental results in Table 2.

The mutual attention mechanism is superior to the general attention mechanism. The performance of models with mutual attention mechanism, such as Bi-LSTM-CA, Bi-LSTM-CA-MAX and Bi-LSTM-CA-AVE, is better than Bi-LSTM-A with general attention mechanism. For training set 0.5%, Bi-LSTM-CA-MAX has an accuracy of 2.0% and 2.6% higher than Bi-LSTM-A on dataset I and dataset II, respectively. For training set 1%, the accuracy of Bi-LSTM-CA-MAX is 0.8% and 1.3% higher than Bi-LSTM-A on dataset I and dataset II respectively. The reason is that the mutual attention mechanism can capture the matching degree between student answer and reference answer more effectively than the general attention mechanism.

Models with the max-pooling perform better than models with the average-pooling. Compared with the models using average-pooling, the models with max-pooling achieved a better performance in terms of the overall accuracy. For training set 0.5% of dataset I, Bi-LSTM-CA-MAX and HN-MAX improved the accuracy by 0.6% and 0.3% than Bi-LSTM-CA-AVE and HN-AVE respectively. This might be caused by that the max-pooling strategy weakens the interference caused by the padding of zero vector compared with the average-pooling.

LSTM is superior to CNN. Based on the overall experimental results, LSTM-GRNN achieved a better performance than Conv-GRNN. This may be caused by that CNN structure in Conv-GRNN ignores the long-distance dependence features between words when encoding sentence vectors, and loses lots of semantic information, which cannot handle the problem of misjudgment caused by similarity in lexical and phrase level between student answer and reference answer. LSTM-GRNN captured the long-distance dependency between words through the first LSTM layer, and retained more semantic information than CNN, which can overcome the above errors to some extent.

Analysis of Model Performance on Different Categories. In order to further analyze the performance difference of each model on each category, we calculate the recall and F1 score of each categories on dataset I, as shown in Fig. 2. Figures (a) and (b) show F1 score of each model with 0.5% and 30% of training data, respectively. Figures (c) and (d) show the recall of each model with 0.5% and 30% of training data, respectively.

For training set 0.5%, the overall performance of our model is significantly better than comparison models in terms of F1 score and recall, which indicates the outstanding ability of our model in the situation of less training data. As the amount of training data increases to 30%, the gap between models gradually narrows, but the overall performance of our model is still in a leading position.

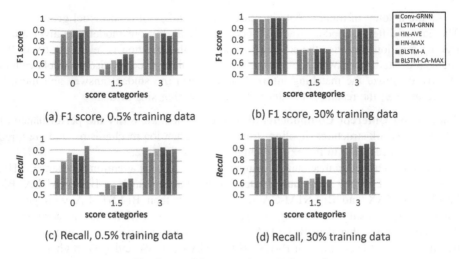

(a) F1 score, 0.5% training data

(b) F1 score, 30% training data

(c) Recall, 0.5% training data

(d) Recall, 30% training data

Fig. 2. Models comparison on dataset I.

The recall and F1 score of our model are significantly high than comparison models on the category of 0-point, such as Conv-GRNN, when training set is 0.5%. This indicates our model is good at capturing the student's judgement to the question, especially when the training data is insufficient.

It is notable that the recall and F1 score of 1.5-points category are significantly lower than the other two categories. The improvement of models on 1.5-points category is less than other categories as the training data increases. This is caused by the characteristics of datasets mentioned in previous section. The student answers might hit key words or phrases in reference answer, but are with wrong reason. The model cannot distinguish them and classify the student answer to 3-points category wrongly, which resulting in the low recall and F1 score of 1.5-points category.

5   Conclusion

For the task of grading Chinese answers on specialty subjective questions with reference answers, we propose a grading model which captures the matching degree between student answer and reference through Bi-LSTM network and mutual attention mechanism. We verify our model on real datasets of a national specialty examination against different amount of training samples, and analyze the performance of different network structures, pooling strategies and attention mechanisms. The experimental results demonstrate the effectiveness of our method. In the future, we are planning to investigate how to extract the knowledge points related to the specialty question from the textbook, then utilize the specialty knowledge to solve this grading task and answer the specialty question automatically.

Acknowledgments. This work was supported by the National Key R&D Program of China (Grant No. 2018YFC0831401), the National Natural Science Foundation of China (Grant No. 91646119), the Major Project of NSF Shandong Province (Grant No. ZR2018ZB0420), and the Key Research and Development Program of Shandong province (Grant No. 2017GGX10114). The scientific calculations in this paper have been done on the HPC Cloud Platform of Shandong University.

References

1. Burstein, J.: The E-rater® scoring engine: automated essay scoring with natural language processing. Shermis, M.D., Burstein, J.C. (eds.), pp. 113–121 (2003)
2. Sutskever, I., Vinyals, O., Le, Q.V.: Sequence to sequence learning with neural networks. In: Advances in Neural Information Processing Systems 27, Annual Conference on Neural Information Processing Systems, pp. 3104–3112. MIT Press, Montreal (2014)
3. Collobert, R., Weston, J., Bottou, L., Karlen, M., Kavukcuoglu, K., Kuksa, P.: Natural language processing (almost) from scratch. J. Mach. Learn. Res. **12**, 2493–2537 (2011)
4. Kim, Y.: Convolutional neural networks for sentence classification. In: Proceedings of the 2014 Conference on Empirical Methods in Natural Language Processing, pp. 1746–1751. ACL, Doha (2014)
5. Zhang, X., Zhao, J., LeCun, Y.: Character-level convolutional networks for text classification. In: Advances in Neural Information Processing Systems 28, Annual Conference on Neural Information Processing Systems, pp. 649–657. ACL, Montreal (2015)
6. Kalchbrenner, N., Grefenstette, E., Blunsom, P.: A convolutional neural network for modelling sentences. In: Proceedings of the 52nd Annual Meeting of the Association for Computational Linguistics, pp. 655–665. ACL, Baltimore (2014)
7. Schwenk, H., Barrault, L., Conneau, A., LeCun, Y.: Very deep convolutional networks for text classification. In: Proceedings of the 15th Conference of the European Chapter of the Association for Computational Linguistics, pp. 1107–1116. ACL, Valencia (2017)
8. Johnson, R., Zhang, T.: Deep pyramid convolutional neural networks for text categorization. In: Proceedings of the 55th Annual Meeting of the Association for Computational Linguistics, pp. 562–570. ACL, Vancouver (2017)
9. Tang, D., Qin, B., Liu, T.: Document modeling with gated recurrent neural network for sentiment classification. In: Proceedings of the 2015 Conference on Empirical Methods in Natural Language Processing, pp. 1422–1432. ACL, Lisbon (2015)
10. Lai, S., Xu, L., Liu, K., Zhao, J.: Recurrent convolutional neural networks for text classification. In: Proceedings of the Twenty-Ninth AAAI Conference on Artificial Intelligence, pp. 2267–2273. AAAI, Austin (2015)
11. Shi, Y., Yao, K., Tian, L., Jiang, D.: Deep LSTM based feature mapping for query classification. In: The 2016 Conference of the North American Chapter of the Association for Computational Linguistics, pp. 1501–1511. NAACL, San Diego (2016)
12. Xiao, Y., Cho, K.: Efficient character-level text classification by combining convolution and recurrent layers. arXiv:1602.00367 (2016)
13. Bahdanau, D., Cho, K., Bengio, Y.: Neural machine translation by jointly learning to align and translate. In: 3rd International Conference on Learning Representations. San Diego (2015)
14. Tan, M., Xiang, B., Zhou, B.: LSTM-based deep learning models for non-factoid answer selection. arXiv:1511.04108 (2015)

15. Chaturvedi, A., Pandit, O.A., Garain, U.: CNN for text-based multiple choice question answering. In: Proceedings of the 56th Annual Meeting of the Association for Computational Linguistics, pp. 272–277. ACL, Melbourne (2018)
16. Yang, Z., Yang, D., Dyer, C., He, X., Smola, A.J., Hovy, E.H.: Hierarchical attention networks for text classification. In: The 2016 Conference of the North American Chapter of the Association for Computational Linguistics, pp. 1480–1489. NAACL, San Diego (2016)
17. Cui, Y., Liu, T., Chen, Z., Wang, S., Hu, G.: Consensus attention-based neural networks for Chinese reading comprehension. In: 26th International Conference on Computational Linguistics, pp. 1777–1786. ACM, Osaka (2016)
18. Cui, Y., Chen, Z., Wei, S., Wang, S., Liu, T., Hu, G.: Attention-over-attention neural networks for reading comprehension. In: Proceedings of the 55th Annual Meeting of the Association for Computational Linguistics, pp. 593–602. ACL, Vancouver (2017)
19. Li, S., Zhao, Z., Hu, R., Li, W., Liu, T., Du, X.: Analogical reasoning on Chinese morphological and semantic relations. In: Proceedings of the 56th Annual Meeting of the Association for Computational Linguistics, pp. 138–143. ACL, Melbourne (2018)

A Mobile Application Classification Method with Enhanced Topic Attention Mechanism

Junjie Chen, Buqing Cao$^{(\boxtimes)}$, Yingcheng Cao, Jianxun Liu, Rong Hu, and Yiping Wen

School of Computer Science and Engineering,
Hunan University of Science and Technology, Xiangtan, China
hnust_cjj@163.com, buqingcao@gmail.com,
caoyingcheng12138@gmail.com, ljx529@gmail.com,
ronghu@126.com, ypwen81@gmail.com

Abstract. Faced with the explosive growth of mobile applications, how to classify mobile applications correctly and efficiently is more helpful for users to choose their own mobile applications, which has become a challenging issue. For this reason, we propose a mobile application classification method with enhanced topic attention mechanism. Firstly, our approach uses LSA to obtain the global topic of mobile application description text. Then, the local hidden representations of mobile application are trained by BiLSTM model. Secondly, for mobile application content representation text rich in global topic information and local semantic information, attention mechanism is introduced to distinguish the contribution degree of different words and calculate their weight values. Thirdly, the classification and prediction of mobile application can be completed by using the softmax activation function through a full connection layer. Finally, we evaluate our method on a real and open dataset Mobile App Store. On the whole, the experimental results illustrate that the performance of our approach is better than other comparison methods, and the classification accuracy of mobile applications is indeed improved. Particularly, compared with the standard LSTM model, the method proposed in this paper increased more than 12.7% in F1 score.

Keywords: Mobile application · BiLSTM · LSA model · Attention mechanism

1 Introduction

With the continuous development of mobile Internet, mobile devices play an increasingly significant role in people's daily life. Up to December 2018, according to the 43rd Statistical Report on the Development of Internet in China, mobile phone users has reached 817 million, and the proportion of Internet users using mobile phones has reached 98.6%, which is far higher than that using desktop computers and laptops. Mobile phones are gradually replacing traditional internet devices.

With the increasing use of mobile devices such as smartphones, the number of mobile applications has shown explosive growth [1]. It is difficult for users to find their own mobile applications [2], facing a large number of mobile applications with rich

© Springer Nature Singapore Pte Ltd. 2019
Y. Sun et al. (Eds.): ChineseCSCW 2019, CCIS 1042, pp. 683–695, 2019.
https://doi.org/10.1007/978-981-15-1377-0_53

content. In order to manage these mobile applications and facilitate users to download and use them, there are various kinds of mobile application stores on the network, such as domestic Pea Pods and 360 Mobile Assistants, foreign Google Play, App Store, and so on. These mobile application stores provide mobile applications for users to download and employ in two ways: (1) The application store searches for mobile applications based on keywords and returns the corresponding mobile applications according to the user's input. (2) Recommend similar mobile applications to users according to their historical records. The research shows that introducing text categorization technology to classify mobile applications in advance will greatly improve the search ability of mobile applications, and provide convenience for finding mobile applications from massive data [3]. Therefore, it can effectively reduce the space and scope for searching if we locate user requirements in a specific application cluster and select the required mobile applications. Meanwhile, this process could improve the efficiency and accuracy for application search.

At present, there have been some research results on mobile application classification [4–6]. They regard mobile application classification as a text categorization problem. The mobile application classification based on functional semantic features can be performed by modeling mobile application content text (including name, description, label and other functions) as a vector through vector space model, topic model and other technologies. These approaches improve the efficiency and accuracy of mobile application classification to some extent, but some issues have not yet been considered: (1) Generally, the description information of mobile applications is long. The existing text representation technologies (such as LDA topic modeling [7]) cannot represent the content text of mobile applications accurately. (2) Not all words in mobile application content contribute equally for mobile application classification. (3) The word order between feature words and context information in mobile application content are not considered. To address the above problems, we propose a BiLSTM model based on enhanced topic attention mechanism (called LSA-BiLSTM) for mobile application classification. In this model, firstly, we use LSA (Latent Semantic Analysis) [8] to model the global topic of mobile application content text, and adopt the truncated SVD (singular value decomposition) method to mine the latent semantic information of text. It can obtain more accurate global modeling capability for a large amount of text information and vocabulary. Next, BiLSTM (Bidirectional Long Short-Term Memory Networks) model is applied to mine the word order between feature words and context information in mobile application text, and obtain the local hidden representation of mobile application content text. Then, attention mechanism is introduced to distinguish the contribution of different words in mobile application description text and calculate their weight values. Finally, combining the local hidden vectors and global topic vectors of mobile application content text, the classification and prediction of mobile applications are completed by using the softmax activation function.

The remainder of this paper is organized as follows: Sect. 2 shows the related work. In Sect. 3, we describe the proposed methods in details. Section 4 carries out the concrete experiment process and analysis. And the last Sect. 5 is the summary of this work and the introduction of follow-up research.

2 Related Work

With the explosive growth of mobile applications in recent years, the problem of information overload has been brought to users. How to quickly find mobile applications that you are interested in from a large number of mobile applications with rich content has become a challenging problem. Therefore, relevant scholars have carried out research on the classification and recommendation of mobile applications in order to improve the efficiency and accuracy of mobile application search.

Mobile application categorization can be regarded as a specific text categorization problem, which refers to functional categorization of content text information for mobile applications. Text classification is roughly divided into two major categories: One is utilizing traditional machine learning algorithms (such as decision tree, naive Bayes, etc.) to learn the representation of content text and achieve text classification. For example, Zhang et al. [3] proposed using SVM (support vector machine) to classify text. McCallum et al. [4] compared the different event models of Naive Bayesian text classification, found that the multi-variable Bernoulli model performed well in small vocabulary, but polynomial model usually performed better in large vocabulary. Considering the conditional distribution of class variables in a given document, Nigam et al. [5] used maximum entropy to classify text, but the experimental results using maximum entropy are sometimes significantly better, but also sometimes worse. When the number of text and words is large, NMF (non-negative matrix factorization) and SVD technology can be used to reduce the dimension of text features. But it will cause incomplete text information to some extent. Bakeryz et al. [6] grouped words by using the distribution of class labels associated with each word. This approach could compress the feature space much more aggressively, while still maintaining the accuracy of document classification.

The other is using the deep neural network technology (such as LSTM, Bi-LSTM, etc.) to mine and characterize the deep semantic information of text, and to achieve text classification and prediction. A classic method is to utilize LSTM (long-short memory network) to learn representation vector of mobile application description [9], and then classify the text according to the representation. Ye et al. [10] proposed a Web service classification approach based on Wide & Bi-LSTM Model, which effectively improved accuracy of Web service classification. It has been observed that the description information of mobile applications is usually long. How to accurately characterize mobile applications become an extremely important problem. LSA topic modeling, using truncated SVD decomposition dimension reduction method to mine the latent semantics of text, has advantages in modeling and representing long text. For this reason, we introduce LSA to model and represent mobile application content text. In recent years, attention mechanism has been successfully applied in the natural language processing and computer vision field [11]. Inspired by the research of Yang et al. [12], we introduce the attention mechanism into our model. To achieve more accurate text representation and classification, we calculate the weight of words in mobile application content text and quantify their contribution.

The essence of mobile application recommendation is that the system recommends mobile applications that are closest to users' preferences. And the main methods of it include content-based mobile application recommendation, context-based mobile application recommendation and hybrid mobile application recommendation. Chen et al. [13] analyzed the content similarity of mobile applications according to the semantic relationship and category between mobile applications, and then performed content-oriented mobile application recommendation. Based on the context information of mobile applications, Woerndl et al. [14] proposed a context-based mobile application recommendation model. Wang et al. [15] introduced both social network information and context information into mobile service recommendation system, and proposed a hybrid mobile service recommendation method. Xie et al. [16] measured the semantic similarity between users and applications by weighted meta-graphs through heterogeneous information networks, using complex structure and semantic information, and accomplish hybrid mobile application recommendation.

3   The LSA-BiLSTM Method

The basic idea of LSA-BiLSTM is to represent mobile application description text, which combines local hidden vectors of mobile application description text with its global topic vectors through attention mechanism. And then, the classification and prediction of mobile applications is accomplished by using the softmax activation function in a full connection layer. The framework of our model is shown in Fig. 1. It consists of four parts: description text representation of mobile applications, sequence coding based on BiLSTM, attention mechanism based on LSA topic modeling and classification of mobile applications.

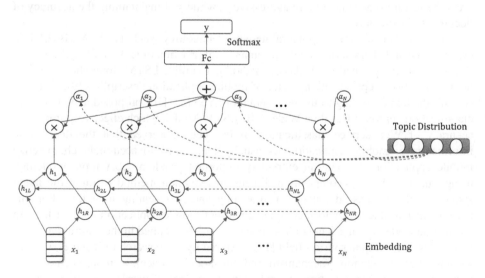

Fig. 1. BiLSTM model for enhanced topic attention mechanism

3.1 Description Text Representation of Mobile Applications

Each word in mobile application description text can be represented as a continuous, low-dimensional and real-valued vector, also known as word embedding. Word embedding learning algorithms such as Word2vec [17] can be used to pre-train word vectors in mobile application description corpus, which can make better use of the semantic and grammatical association of words. Given an input of mobile application description s, we get the word embedding $x_t \in \mathbb{R}^{dim}$ for each word in s. Hence, as shown in Fig. 1, a mobile application description text with length N can be represented as $X = (x_1, x_2, \ldots, x_N)$.

3.2 Sequence Coding Based on BiLSTM

BiLSTM is a combination of forward LSTM and backward LSTM. LSTM is a special form of recurrent neural network. Also, LSTM is suitable for temporal data modeling such as text data. We can memorize more important information and forget less important information to capture longer-distance semantic dependencies better by training LSTM model. But, using LSTM model is unable to encode back-to-front information of the sentence. As for finer-grained classification, BiLSTM model can capture bidirectional semantic dependencies better.

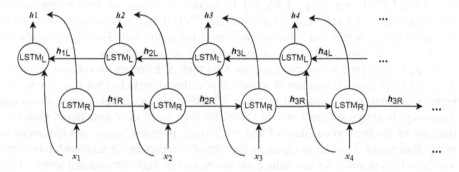

Fig. 2. BiLSTM model encodes single mobile application description

Given a mobile application description text $X = (x_1, x_2, \ldots, x_N)$, forward LSTM and backward LSTM will process the same mobile application description text sequentially. For the input word x_t at time t, given the previous hidden state h_{t-1} and cell state c_{t-1}, LSTM model can generate the hidden state h_t and cell state c_t at the next moment through the forget gate f_t, the input gate i_t and the output gate o_t. The calculation formula of LSTM model is defined as follows:

$$i_t = \sigma(W_i x_t + U_i h_{t-1} + b_i) \tag{1}$$

$$f_t = \sigma(W_f x_t + U_f h_{t-1} + b_f) \tag{2}$$

$$o_t = \sigma(W_o x_t + U_o h_{t-1} + b_o) \tag{3}$$

$$c_t = f_t \odot c_{t-1} + i_t \odot \tanh(W_c x_t + U_c h_{t-1} + b_c) \tag{4}$$

$$h_t = o_t \odot \tanh(c_t) \tag{5}$$

Where $\sigma(\cdot)$ is a logistic function and its output interval is $(0, 1)$. $\odot$ represents element-wise multiplication. All W, U and b are grid parameters.

From Fig. 2, we can see that the forward and backward hidden vectors are joined together to obtain the hidden state sequence $\{h_1, h_2, \ldots, h_N\}$ which has the same length as X.

3.3   Attention Mechanism Based on LSA Topic Modeling

In LSA-BiLSTM, we adopt LSA [8] to model the topic of mobile application description text. LSA topic model uses truncated SVD decomposition to mine the latent semantics of text, which can solve the problem of polysemy well. For a large number of text information and vocabulary, we can get more accurate results.

In particular, for a set of mobile application texts S, a document-vocabulary matrix $A_{m \times n}$ (m represents the number of mobile application description texts, n denotes the number of words) is constructed by using TF-IDF (term frequency–inverse document frequency) to assign different weight values for each word of mobile description text. Because of the large vocabulary of text set S, $A_{m \times n}$ is sparse, noisy and redundant in many dimensions. In order to capture a few potential topics in the relationship between words and documents, we can reduce the dimension of high-dimensional matrix $A_{m \times n}$ by truncating SVD decomposition. The truncated SVD decomposition formula is as follows:

$$A_{m \times n} = U_{m \times m} S_{m \times n} V_{n \times n}^T \approx U_{m \times t} S_{t \times t} V_{t \times n}^T \tag{6}$$

As shown in Fig. 3, the singular values are arranged from large to small after SVD decomposition, and the top t singular values are taken as an approximate representation of the original matrix $A_{m \times n}$. $U_{m \times t}$ corresponds to the document-topic matrix of mobile application description text. Each mobile application description text has a t-dimension topic distribution θ_s.

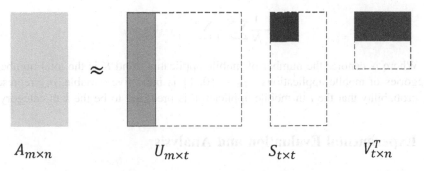

$$A_{m \times n} \qquad U_{m \times t} \qquad S_{t \times t} \qquad V_{t \times n}^{T}$$

Fig. 3. Truncated SVD decomposition

In fact, in mobile application description text, not every word contributes equally to mobile application classification, so it is necessary to acquire the weight of different words α_i. We obtain the weights by the hidden state sequence and the external topic vectors. The formulas are as follows:

$$g_i = v_a^T \tanh(W^a \theta_s + U^a h_i) \qquad (7)$$

Where v_a, W^a and U^a are training weight matrices.

After obtaining $[g_1, g_2, \ldots, g_N]$, we can generate the final weight scores by the softmax function. Next, our model will output a continuous context vector $vec \in \mathbb{R}^d$ for each mobile application text. The output vector is calculated by a weighted sum of hidden state h_i:

$$vec = \sum_{i=1}^{N} \alpha_i h_i \qquad (8)$$

Where $\alpha_i \in [0, 1]$ stands for the attention weight of each hidden state h_i, and $\sum_i \alpha_i = 1$.

3.4 Classification of Mobile Applications

We use the output vector vec as an input of the full connection layer whose output length is the same as the total number of mobile application categories. And then a softmax activation function is employed to obtain the probability distributions of all candidate mobile application categories. The softmax function is calculated as follows:

$$\text{softmax}(m_j) = \frac{\exp(m_j)}{\sum_{i=1}^{M} \exp(m_i)} \qquad (9)$$

By minimizing the cross-entropy error of mobile application classification, Our model is trained by supervised manner. The loss function is as follows:

$$\mathcal{L} = -\frac{1}{N}\sum\nolimits_{i=1}^{N}\sum\nolimits_{k=1}^{K} y_{i,k}\log p_{i,k} \tag{10}$$

Where N denotes the number of mobile applications and K is the total number of categories of mobile applications. $y_{i,k} \in \{0, 1\}$ is indicative variable, $p_{i,k}$ represents the probability that the i-th mobile application is predicted to be the k-th category.

4　Experimental Evaluation and Analysis

4.1　Dataset

We use the open dataset Mobile App Store on Kaggle website as the experimental dataset for mobile application classification. The dataset contains 23 categories, 7,197 IOS mobile applications from Apple Store. And the detailed distribution of the top 20 categories are shown in Table 1. For the fairness and accuracy of the experimental results, we remove as many mobile applications as possible that are not described in English. In addition, the sample distribution of dataset after cleaning is not uniform. Among them, there are 3,381 mobile applications categorized as 'Games', while only 82 mobile applications categorized as 'Shopping'. we randomly selected 480 subsets of 'Games' as experimental data because of uneven distribution of dataset, or it will greatly affect the experimental results.

Table 1. Category number statistics of top 20 mobile applications.

Category	Number	Category	Number
Games	3381	Lifestyle	98
Entertainment	456	Shopping	82
Education	408	Weather	69
Photo & video	339	Book	59
Utilities	202	Travel	58
Health & fitness	166	News	55
Productivity	164	Business	54
Music	136	Reference	53
Social networking	113	Finance	47
Sports	103	Food & drink	45

4.2　Pre-processing

Before using mobile application description text as topic modeling input, it is necessary to pre-process mobile application description text and extract meaningful words. The specific operations are as follows:

(1) Use regular expressions to match mobile application description text, remove data containing Chinese characters, only focus on mobile application data described in English. All words in mobile application text description should be lowercase so that mobile application description text contains only lowercase words. Remove punctuation and special symbols because they are meaningless.

(2) Word Segmentation, cut each sentence into a list of words. Remove stop words in mobile application text, such as "she", "I", "can", "must", and so on.

(3) Obtain POS Tagging for each word in mobile application description text, remove the words whose parts of speech is useless such as numerals, conjunctions, interjections, and so on.

(4) Get the stem for every word in text, because usually the same stem has the same meaning. For example, the three words "agreed", "agreeing" and "agreeable" have the same root.

4.3 Baseline Methods

For comparison, we select the following methods as baseline methods:

- **LSTM:** The mobile description text is processed with fixed length, then the hidden vectors are obtained by LSTM training, and the mobile applications are classified by softmax function.

- **LSA-SVM:** Firstly, the global topic distribution of mobile application text is learned by LSA model, and then the text is trained and predicted by SVM [18]. We use the document-topic vector trained by LSA model as the input of SVM, and use grid search to set different C parameters and kernel parameters to improve the accuracy of classification.

- **LDA-SVM:** Compared with LSA model, we use LDA [7] topic model to learn document-topic distribution of mobile application text, and SVM is used to classify vectors. Similarly, the grid search is used to learn the optimal hyper-parameters and optimize accuracy rating of the classifier.

- **LAB-BiLSTM** [19]: We use LDA topic model instead of LSA model to learn document-topic vector offline, and train BiLSTM model to get hidden vector representation of mobile application description text. After the feature representation of mobile applications is enhanced by topic attention, the classification and prediction of mobile applications are finished by using the softmax activation function.

4.4 Evaluation Metrics

Generally, for binary classification problems, we apply recall (R), precision (P) and their comprehensive evaluation F1-score as classification indicators. For multi-category tasks, there will be multiple confusion matrices. So, we adopt macro-Recall, macro-Precision and macro-F1-score as evaluation indicators, that is to calculate the precision and recall of each category for mobile applications separately and then calculate their arithmetic average value:

$$macro - P = \frac{1}{N} \sum_1^N P_i \tag{11}$$

$$macro - R = \frac{1}{N} \sum_1^N R_i \tag{12}$$

$$marco - F1 = \frac{1}{N} \sum_1^N \frac{2 \times P_i \times R_i}{p_i + R_i} \tag{13}$$

Where N represents the total number of mobile applications, R_i and P_i are the recall and precision of the i-th category respectively.

4.5   Experimental Result

Experimental Setup: In this experiment, 70% of the dataset were selected as training set and 30% as test set. Considering that LSTM model and BiLSTM model can only handle fixed length text, while mobile application description text average length is 165.9, the text sequence length N is set to 170. In addition, Adam [20] method is used as the optimizer of the model. The hyperparameter β_1 is set to 0.9 and β_2 set to 0.9999. The learning rate is set at 0.001, and the batch size is 25. For topic model LSA, the implicit semantic dimension which also called topic number t can be set artificially. Because the total number of mobile application categories is 23, t is set to 20 firstly. Then, in Sect. 4.5, we test and analyze the influence of topic number t on our experimental results to select its best value. In the contrast methods, set the same number of topics for the LDA model.

Classification Performance: We test mobile application data of 5 categories, 10 categories, 15 categories and 20 categories respectively. The experimental results are shown in Figs. 4, 5 and 6. On the whole, our model LSA-BiLSTM is superior to the other four comparison methods in different indicators. Especially, when the data of mobile applications are 10 categories, the F1-scores of LSA-BiLSTM achieve an improvement of 29.2%, 29.3%, 12.8% and 3.9% compared to the LSTM, LDA-SVM, LSA-SVM and LAB-BiLSTM, respectively.

In the light of this experimental results, we can find that: (1) LSA-SVM is better than LDA-SVM. Because of the large vocabulary of mobile application description text, LSA can make full use of redundant data and denoise them to get better topic vector representation. (2) LSA-BiLSTM and LAB-BiLSTM models are much better than ordinary LSTM model. It can be seen that the F1-score of LSA-BiLSTM has a certain improvement under the same parameter setting, which shows that the incorporating attention mechanism into the model is indeed conducive to the classification for mobile applications. (3) LSA-BiLSTM is slightly better than LAB-BiLSTM in F1-score. The results again show that LDA topic model is affected by text length, resulting in inaccurate topics.

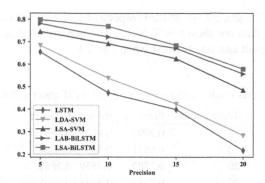

Fig. 4. Precision changes from different categories of data

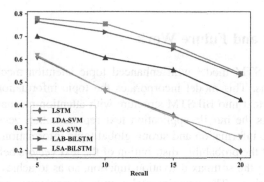

Fig. 5. Recall changes from different categories of data

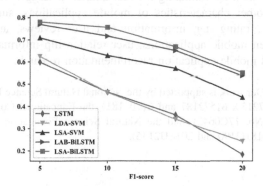

Fig. 6. F1-score changes from different categories of data

The Effect of Topic Number on Classification Results: Furthermore, we consider the effect of the topic number t on the classification results of mobile applications in the LSA-BiLSTM model. We change the values of the topic number t while fixing the other parameters. When the number of categories for experimental data is 10 and the topic number is 5, 10, 20, 30, 40 and 50 respectively, the classification test results of mobile application data are shown in Table 2. The experimental results illustrate that the best possible result can be obtained when $t = 20$.

Table 2. Experimental result comparison in LSA-BiLSTM under different topic numbers

Topic number	Precision	Recall	F1-score
5	0.7097	0.6825	0.6831
10	0.6871	0.6661	0.6688
20	**0.7702**	**0.7559**	**0.7562**
30	0.7213	0.7348	0.7248
40	0.7226	0.7012	0.7019
50	0.7122	0.6926	0.6917

5 Conclusion and Future Work

We present a BiLSTM model with enhanced topic attention mechanism to classify mobile applications. This model incorporates the topic information of mobile application description text into BiLSTM structure with attention mechanism through LSA model, and obtains the mobile application text representation vector which contains rich local semantic information and strong global topic information. A full connection layer is used to get the probability distribution of the text vectors belonging to different categories by using the softmax activation function, so as to achieve the classification of mobile applications. The experimental results demonstrate that our approach is superior to other comparison methods, and it can improve the accuracy of mobile application classification by validating the real dataset in Kaggle. In our future work, we will consider other characteristics of mobile applications, such as downloads, number of scorers, rating, tag information and user reviews, and even structure information between mobile applications, user relationship information, to carry out classification-based mobile application recommendation research.

Acknowledgment. Our work is supported by the National Natural Science Foundation of China (No. 61873316, 61872139, 61572187 and 61702181), the Educational Commission of Hunan Province of China (No. 17C0642), and the Natural Science Foundation of Hunan Province (No. 2017JJ2098, 2018JJ3190 and 2018JJ2136).

References

1. Gao, S., Zang, Z., Gopalakrishnan, S.: A study on distribution methods of mobile applications in China. In: Seventh International Conference on Digital Information Management ICDIM 2012, pp. 375–380. IEEE (2012)
2. Deng, S., et al.: Toward mobile service computing: opportunities and challenges. IEEE Cloud Comput. **3**(4), 32–41 (2016)

3. Zhang, D., Lee, W.S.: Question classification using support vector machines. In: Proceedings of the 26th annual international ACM SIGIR Conference on Research and Development in Information Retrieval 2003, pp. 26–32. ACM (2003)
4. McCallum, A., Nigam, K.: A comparison of event models for naive bayes text classification. In: AAAI 1998 Workshop on Learning for Text Categorization 1998, vol. 752, no. 1, pp. 41–48 (1998)
5. Nigam, K., Lafferty, J., McCallum, A.: Using maximum entropy for text classification. In: IJCAI 1999 Workshop on Machine Learning for Information Filtering 1999, vol. 1, no. 1, pp. 61–67 (1999)
6. Baker, L.D., McCallum, A.K.: Distributional clustering of words for text classification. In: Proceedings of the 21st Annual International ACM SIGIR Conference on Research and Development in Information Retrieval, pp. 96–103. ACM (1998)
7. Blei, D.M., Ng, A.Y., Jordan, M.I.: Latent Dirichlet allocation. Mach. Learn. Res. Arch. 3, 993–1022 (2003)
8. Deerwester, S., Dumais, S.T., Furnas, G.W., et al.: Indexing by latent semantic analysis. J. Assoc. Inf. Sci. Technol. 41(6), 391–407 (2010)
9. Hochreiter, S., Schmidhuber, J.: Long short-term memory. Neural Comput. 9, 1735–1780 (1997)
10. Ye, H., Cao, B., Peng, Z., Chen, T., Wen, Y., Liu, J.: Web services classification based on wide & Bi-LSTM model. IEEE Access 7, 43697–43706 (2019)
11. Pappas, N., Popescu-Belis, A.: Multilingual hierarchical attention networks for document classification (2017). https://arXiv.org/abs/1707.00896
12. Li, Y., Liu, T., Jiang, J., Zhang, L.: Hashtag recommendation with topical attention-based LSTM. In: COLING 2016, 26th International Conference on Computational Linguistics, Proceedings of the Conference on Technical Papers 2016, Osaka, Japan, pp. 3019–3029 (2016)
13. Chen N, Hoiy S, Li S, Xiao X.: SimApp: a framework for detecting similar mobile applications by online kernel learning. In: WSDM—Proceedings of the 8th ACM International Conference on Web Search and Data Mining 2015, pp. 305–314 (2015)
14. Woerndl, W., Schueller, C., Wojtech, R.: A hybrid recommender system for context-aware recommendations of mobile applications. In: ICDE 2007, pp. 871–878 (2007)
15. Wang, L.C., Meng, X.W., Zhang, Y.J.: A heuristic approach to social network-based and context-aware mobile services recommendation. J. Convergence Inf. Technol. 6(10), 339–346 (2011)
16. Xie, F., Chen, L., Ye, Y., Liu, Y., Zheng, Z., Lin, X.: A weighted meta-graph based approach for mobile application recommendation on heterogeneous information networks. In: Pahl, C., Vukovic, M., Yin, J., Yu, Q. (eds.) ICSOC 2018. LNCS, vol. 11236, pp. 404–420. Springer, Cham (2018). https://doi.org/10.1007/978-3-030-03596-9_29
17. Mikolov, T., Chen, K., Corrado, G., Dean, J.: Efficient estimation of word representations in vector space. arXiv:1301.3781 (2013)
18. Hearst, M.A., Dumais, S.T., Osman, E., Platt, J., Scholkopf, B.: Support vector machines. IEEE Intell. Syst. Appl. 13(4), 18–28 (1998)
19. Cao, Y, Liu, J, Cao, B, et al.: Web services classification with topical attention based Bi-LSTM. In: 15th EAI International Conference on Collaborative Computing: Networking, Applications and Worksharing 2019, (2019, in press)
20. Kingma, D., Ba, J.: Adam: a method for stochastic optimization. In: ICLR (2015)

Water Level Prediction of Taocha Based on CCS-GBDT Model

Yibin Wang[1], Tao Sun[1], Jiapei Su[2(✉)], and Daibin Pan[2]

[1] The Eastern Route of South-to-North Water Diversion Project Jiangsu Water
Source Co., Ltd., Nanjing 210000, China
[2] Nanjing Tech University, Nanjing 211816, China
1455884001@qq.com

Abstract. The aim is to forecast water level accurately and to provide scientific basis for decision-making of water regulation of South-to-North Water Transfer Project. Firstly, the cuckoo algorithm is improved. Secondly, the improved chaotic cuckoo algorithm is used to optimize the gradient boosted decision tree. Taking the water level at the head of Taocha canal as an example, the model is used to predict the water level. The analysis shows that the relative error between the improved gradient boosted decision tree and the measured water level is reduced by 2.70%. Compared with BP neural network, RBF network and SVR show better, accuracy.

Keywords: South-to-North Water Transfer · Taocha canal head · Water level prediction · Gradient boosted decision tree · Cuckoo algorithms

1 Introduction

The traditional mathematical statistics model can only realize the stationary water level data under time series. However, machine learning method breaks the limitation of traditional methods in the prediction of non-stationary time series data. Machine learning has better generalization ability and improves the accuracy of water level prediction. Coppola Jr. et al. [1] proposed to use artificial neural networks (ANNs) to predict well water level. The research shows that ANNs can provide good prediction results and valuable sensitivity analysis, which is more suitable for decision-making of groundwater management. Zhao et al. [2] used correlation vector machine to simulate the dynamic change of groundwater. Experiments show that this method can significantly reduce the relative average error and mean square error in predicting Groundwater level. Manzione et al. [3] established a coupling model based on multivariate time series analysis and geostatistics, and applied it to groundwater level prediction in the Guarani aquifer system outcropped area in southeastern Brazil. The practicability of machine learning method in water level prediction is verified.

The first phase of the Middle Route of the South-to-North Water Transfer Project was completed on December 12, 2014 [4]. The Middle Route of South-to-North Water Transfer Project has important strategic significance for improving the water environment and promoting the sustainable development of regional economy and society [5, 6]. Taocha Project is the head of the main water diversion channel of the middle

© Springer Nature Singapore Pte Ltd. 2019
Y. Sun et al. (Eds.): ChineseCSCW 2019, CCIS 1042, pp. 696–708, 2019.
https://doi.org/10.1007/978-981-15-1377-0_54

route of South-to-North Water Transfer Project. As a major part of the first phase of the middle line project, water level prediction at the head of Taocha is an important basis for water resources management. Accurate prediction of water level at the head of Taocha is very significant for reservoir dispatching, flood control, power generation, irrigation, etc. [7]. However, due to the complexity of river water level change, it has the characteristics of time and space change, multi-dimensional, dynamic and uncertain, which brings difficulties to river water level prediction [8]. In view of the nonlinear characteristics of the water level prediction of the head of Taocha, this paper applies the GBDT of the machine learning model to the water level prediction, and optimizes the model parameters to ensure the applicability of the model. It can provide some scientific basis and reference for the following research.

2 Theoretical Basis

2.1 Gradient Boosting Decision Tree (GBDT)

Gradient Boosting Decision Tree model was proposed by Jerome Friedman [9] in 1999, which is a combination of decision tree and Boosting method. The GBDT model can better realize the classification and regression tasks, and it is not easy to overfit [10, 11]. Moreover, GBDT establishes a new decision tree in the gradient direction of residual error reduction in the previous model, so as to continuously reduce the residual error. Finally, the conclusions of all trees are added up as the final classifier, which can be expressed as the following model [12]

$$\widehat{y}_i = \sum_{k=1}^{n} f_k(x_i), f_k \in F. \tag{1}$$

In Eq. (1), $f_k(x_i)$ represents the first decision tree, and $\widehat{y}_i$ represents the strong classifier formed by the linear addition of weak classifiers. That is to say, a new decision tree function $f_k(x_i)$ is added to the last prediction value to minimize the residual error from the real value. The objective function of GBDT is

$$L^{(t)} = \sum_{i=1}^{n} l(y_i, \widehat{y}_i) + \sum_{k=1}^{K} \Omega(f_k), \tag{2}$$

In Eq. (2), l is a differentiable loss function representing the difference between the predicted value $\widehat{y}_i$ and the true value y_i, $\sum_{k=1}^{K} \Omega(f_k)$ is the added regularization [13]. Ω represents the complexity of the decision tree, which can constrain the number of nodes in the decision tree, the depth of the tree or the norm of L_2 corresponding to the leaf node, mainly to prevent overfitting of the model.

$$L^{(t)} = \sum_{i=1}^{n} l(y_i, \hat{y}_i^{(t-1)} + f_t(x_i)) + \Omega(f_t) + C \qquad (3)$$

Equation (3) is the objective function of the $t-$ th iteration, where C is a constant. According to Taylor's formula, the above equation is expanded to take the second order form as the approximate value of the objective function. The formula is as follows:

$$L^{(t)} \approx \sum_{i=1}^{n} [l(y_i, \hat{y}_i^{(t-1)}) + g_i f_t(x_i) + \frac{1}{2} h_i f_t^2(x_i)] + \Omega(f_t) + C. \qquad (4)$$

In Eq. (4), $g_i = \partial_{\hat{y}_i^{(t-1)}} l(y_i, \hat{y}_i^{(t-1)})$, $h_i = \partial^2_{\hat{y}_i^{(t-1)}} l(y_i, \hat{y}_i^{(t-1)})$ respectively represent the first and second derivatives of the loss function with respect to $\hat{y}_i^{(t-1)}$ [14]. The objective function of $t-$ iteration can be simplified to Eq. (5) by removing the constant term, and the tree complexity function used in this paper can be reduced to Eq. (6):

$$L^{(t)} \approx \sum_{i=1}^{n} [g_i f_t(x_i) + \frac{1}{2} h_i f_t^2(x_i)] + \Omega(f_t), \qquad (5)$$

$$\Omega(f) = \gamma T + \frac{1}{2} \lambda \sum_{j=1}^{T} \omega_j^2. \qquad (6)$$

In Eq. (6), γ is leaf node coefficient, T is leaf node number [15]. λ as the square modulus coefficient of L_2, also plays a role in preventing overfitting, and ω represents the leaf weight. The decision tree function f is redefined as $f_t(x) = \omega_{q(x)}$ [16], that is, the tree is divided into structural function q and leaf weight part ω, where q maps the input to the index of the leaf, i.e. $q : R^d \rightarrow \{1, 2, 3, \cdots, T\}$, and defines the sample set of each leaf as $I_j = \{i | q(x_i) = j\}$, thus rewriting the objective function as

$$L^{(t)} \approx \sum_{i=1}^{n} [g_i \omega_{q(x_i)} + \frac{1}{2} h_i \omega_{q(x_i)}^2] + \gamma T + \frac{1}{2} \lambda \sum_{j=1}^{T} \omega_j^2$$

$$= \sum_{j=1}^{T} [(\sum_{i \in I_j} g_i) \omega_j + \frac{1}{2} (\sum_{i \in I_j} h_i + \lambda) \omega_j^2] + \gamma T \qquad (7)$$

$$= \sum_{j=1}^{T} [G_j \omega_j + \frac{1}{2} (H_j + \lambda) \omega_j^2] + \gamma T.$$

In (7), $G_j = \sum_{i \in I_j} g_i$ and $H_j = \sum_{i \in I_j} h_i$, use the minimum value of the quadratic equation to get the optimal solution ω_j^* and the optimal solution L^* of the objective function

$$\omega_j^* = -\frac{G_j}{H_j + \lambda},$$ (8)

$$L^* = -\frac{1}{2} \sum_{j=1}^{T} \frac{G_j^2}{H_j + \lambda} + \lambda T.$$ (9)

From the above, when the structural function q of the decision tree is obtained, the objective function can be obtained according to the calculation of the above equation. The final problem is to find the optimal tree structure q^*, so that the objective function has the minimum value.

2.2 Cuckoo Search (CS)

Cuckoo search algorithm was proposed by Deb and Yang in 2009, which is a heuristic search algorithm based on bionics [17, 18]. The algorithm mimics the cuckoo's random flight to find its own nest, which is used for iterative search and optimization. The location of the nest represents the solution, and the location is updated as follows in the iterative process

$$x_i^{(t+1)} = x_i^t + \alpha \oplus L(s, \lambda).$$ (10)

In Eq. (10), x_i^t represents the position of the t th nest in the i iteration, α is the scaling factor of step size, $\oplus$ represents the dot product of the vector, s is the step size, λ is the Levy flight index. The function L imitates the typical characteristics of bird's Levy flight [19]. Its function is

$$L(s, \lambda) = \frac{\lambda \Gamma(\lambda) \sin(\lambda \pi / 2)}{\pi s^{1 + \lambda}}$$ (11)

The implementation process of cuckoo search algorithm [20] is as follows:

(1) Setting the initial state, the number of nests n, the search space dimension d^T of the solution and the initial position $X_0 = [x_1^0, x_2^0, \cdots, x_n^0]^T$ of the nest.
(2) At the initial position, the optimal solution x_b^0 and the optimal solution f_{min} of the objective function are obtained by solving the objective function.
(3) After position iteration updating, the optimal position in the last iteration is retained while the position of other nest utilization formula (10) is updated, and the new nest location $X_{t-1} = [x_1^{t-1}, x_2^{t-1}, \cdots, x_n^{t-1}]^T$ is obtained. The fitness of each nest is recalculated and compared with the previous one, and the former position $X_t = [x_1^t, x_2^t, \cdots, x_n^t]^T$ is replaced by the fitness position.
(4) Each nest in g_t will get a random number $r \in (0, 1)$, which is compared with the probability p_a of exotic eggs being found, and the nest will be randomly changed. Then the fitness of the objective function is tested, and the nest with high fitness is replaced by the nest position. The bird's nest with better location was obtained. Then the fitness of the objective function is tested, and the nest position with high

fitness is used to replace the nest in g_t. The bird's nest $X_t = [x_1^t, x_2^t, \cdots, x_n^t]^T$ with better location is obtained.

(5) Update the optimal solution f_{min} of the objective function and the location x_b^t of the optimal solution bird's nest to determine whether the optimal solution satisfies the set conditions and whether it reaches the maximum number of iterations. If not, turn to step (3) for iteration. If it satisfies, the optimal solution f_{min} of the output function and the location x_b^t of the optimal solution bird's nest are determined.

2.3   Improved Cuckoo Search Algorithm

The cuckoo algorithm has the characteristics of easy implementation and high stability [21]. Since the algorithm has few parameters, it is not necessary to configure too many parameters when solving the problem. Research shows that the algorithm is more effective than genetic algorithms, PSO and other algorithms [22]. However, the cuckoo algorithm has poor performance in local search, resulting in slow search speed and low precision [23]. This paper improves it through the following two aspects.

(1) Introducing Inertia Weights

In the process of solving the problem, it is generally expected that the optimization algorithm will show good global search ability in the early stage and fine local development ability in the later stage. The location update of cuckoo algorithm is random. In order to improve the performance of cuckoo algorithm, the inertia weight [22] is introduced into Eq. (10), such as Eq. (12).

$$x_i^{(t+1)} = \omega x_i^t + \alpha \oplus L(s, \lambda) \tag{12}$$

$$\omega = \left(\frac{2}{t}\right)^{0.3} \tag{13}$$

In Eq. (13), ω decreases with the number of $t-$ iterations, which ensure that the cuckoo algorithm has a good search space. The large ω value in the early stage is beneficial to jump out of the local optimal solution and ensure the global search ability of the algorithm. The later ω value is small, which ensures the algorithm's local search ability and speeds up the search speed of the algorithm later.

(2) Adding Chaotic Mutation System

In order to improve the shortcomings of the algorithm's optimization accuracy, the chaotic system traversal characteristics are used to make the partial solution jump out of the local optimum. As shown in Eq. (8) is a chaotic system of logistic mapping [25]:

$$X_{n+1} = uX_n(1 - X_n) \quad n = 0, 1, 2 \cdots \tag{14}$$

In Eq. (14), u is the control parameter, generally take u as 4 and the system is completely in chaos. Given any initial value $X_0 \in [0, 1]$, Logistic is completely chaotic, ensuring the global uniformity and uniformity of the nest. When the optimal solution x_b^t remains unchanged in the $h-$ iteration process, the mapping condition is triggered. The chaotic system is used to make the optimal solution x_b^t iterate to produce an optimal particle sequence, thereby replacing some of the stopping individuals, making the algorithm not easy to fall into the local optimum and ensuring the accuracy of the algorithm in the later period.

3 Implementation Process

In this paper, in order to accurately predict the Taoxuan head data, the implementation process is mainly divided into the following steps, as shown in Fig. 1:

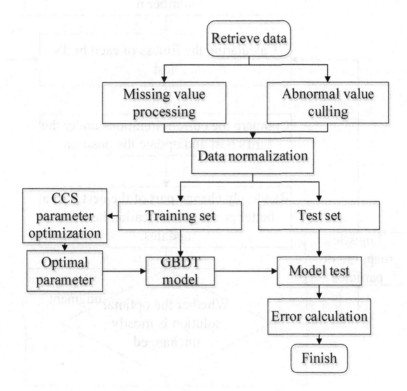

Fig. 1. Model implementation process

Data Processing:

Step 1: Obtain the data and perform preliminary processing, including outlier elimination, missing value processing, and data normalization.

Step 2: Divide the processed data into training sets D_{train} and test sets D_{test}.

Parameter Tuning:

The choice of parameters directly determines the accuracy of the machine learning model. The commonly used methods of tuning are expert experience and grid search. The former is too dependent on human subjective judgment, and the latter has the disadvantage that the parameter optimization range is too narrow and it is difficult to find the optimal parameters. In view of the above problems, this paper proposes to use the CCS algorithm to optimize the c parameters of the GBDT model as shown in flow Fig. 2

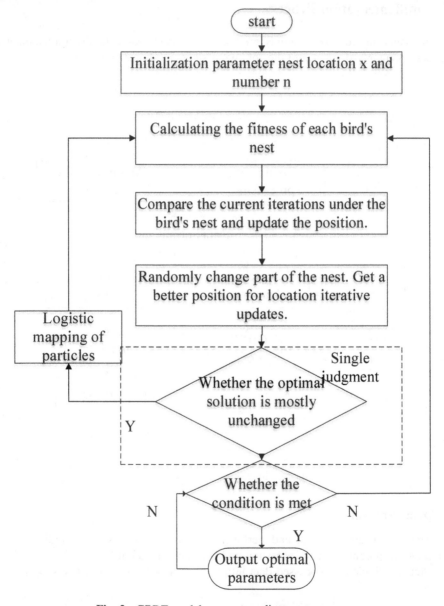

Fig. 2. GBDT model parameter adjustment process

Step 3: Randomly generate n groups of solutions $X_0 = [x_1^0, x_2^0, \cdots, x_n^0]^T$. Each group contains three parameters of GBDT: learning_rate, max_depth of decision tree and max_leaf_nodes. The error between the predicted value and the actual value of the training set is taken as the fitness function $f(x)$ of CCS.

Step 4: Using the optimized cuckoo algorithm to improve the parameters of GBDT, and iteratively obtain the optimal parameters.

Model Construction and Verification:

Verify the performance of the model before and after optimization, and test the performance of the model.

Step 5: Perform training on the model using the training set D_{train} to obtain the GBDT model under the optimal parameters.

Step 6: Combine the test set D_{test} with the model test, calculate the error from the actual value, and verify that the model is excellent.

4 Application Cases

4.1 Data Situation

This paper collects the data of Taocha Channel Head from November 3, 2017 to June 30, 2018 with an interval of 1 h. The data includes time stamps, instantaneous flow, cumulative water volume, flow rate and water level for a total of 5,727 data sets. Since the input indicators in the data set are too small, this paper uses the time series method to introduce the water level data of the first ten time nodes into the input indicators. With the time stamp, instantaneous flow, cumulative water volume and flow rate as the 14-dimensional input index data, the water level data as the prediction index for an example analysis.

The experiment used Core(TM) i5-7500, memory 8G, graphics card GeForce GTX 1050Ti computer. Run in python 3.6 environment, use pycharm IDE to write experimental content.

4.2 Analysis of Results

The raw data set is preprocessed, the missing and outliers in the data are processed and the processed data set is normalized. 5154 sets of data were randomly extracted from the processed data set as a training set, and the remaining 573 was used as a test set. First, use the training set for parameter tuning. In the process of model iteration, the error of actual value is used as fitness function to optimize three sets of parameters: step size, max_depth of decision tree and max_leaf_nodes.

Figure 3 is an iterative curve for CCS and CS parameter tuning. The initial number of optimized population is 30. In the 100 iterations, the CCS has better local optimization ability. Compared with the CS algorithm, the average error in the previous period is relatively small. After the chaotic mapping, the CCS algorithm successfully jumps out of the local optimum. When the parameters learning_rate = 0.025, max_depth = 8, max_leaf_nodes = 11, the average training error is 0.049. The CCS algorithm in the example verification shows good global optimization ability. The average error of the GBDT model test set under parameter optimization is 0.063, and the average error of the unoptimized GBDT model test is 0.096. The feasibility of CCS for parameter tuning of GBDT algorithm is verified.

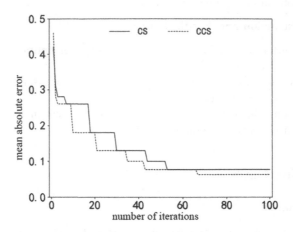

Fig. 3. Iteration curve for CCS and CS parameter tuning

BP neural network, RBF network and SVR are common water level prediction models. To verify the advanced nature of the methods used in this paper, the above common methods are used for comparison. The important parameters of the three common methods are set as shown in Table 1. The data training set of this paper is selected separately for model construction. The data test set is used to test the performance of the model, and 100 sets of sample sets of test sets are randomly selected for display, as shown in Figs. 4, 5 6 and 7.

Table 1. Common model parameter setting table

	BP neural network	RBF network	SVR
Programming language	Matlab	Matlab	Python
Important	hiddennum = 8	GOAL = 0.0	Kernel = 'rbf '
parameter	net.trainParam.epochs = 1000	SPREAD = 1.0	C = 2.0
setting	net.trainParam.lr = 0.01	DF = 25	Gamma = 'auto'

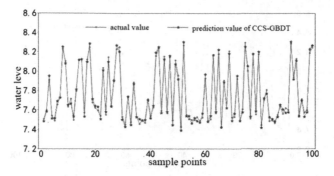

Fig. 4. Fitting of CCS-GBDT prediction value

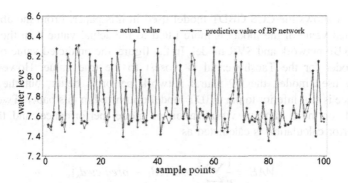

Fig. 5. Fitting of BP network prediction value

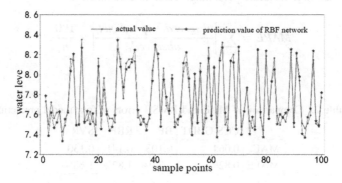

Fig. 6. Fitting of RBF network prediction value

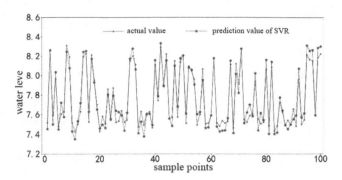

Fig. 7. Fitting of SVR prediction value

Figure 4 shows the CCS-GBDT model used in this paper. From the above figure, the predicted value of the model is more fitted to the actual value than the BP neural network, RBF network and SVR model. In the figure, the predicted value of the CCS-GBDT model for the Taoshu canal is closest to the true value. However, in the commonly used model, the BP neural network performs better, but the prediction performance is still inferior to the GBDT model optimized in this paper. Two errors are introduced for more intuitive visualization of model performance, and the average absolute error calculation is calculated as

$$MAE = \frac{1}{n} \sum_{k=1}^{n} |observed_k - predicted_k|, \tag{15}$$

And the average absolute percentage error is calculated as

$$MAPE = \sum_{k=1}^{n} \left| \frac{observed_t - predictet_t}{observed_t} \right| \times \frac{100}{n}. \tag{16}$$

Table 2. Comparison of MAE and RMSE errors of four model predictions

	CCS-GBDT	BP	RBF	SVR
MAE	0.064	0.103	0.141	0.130
MAPE	0.82%	1.33%	1.82%	1.68%

In summary, compared with other models in the CCS-GBDT model, the average absolute value error and the average absolute percentage error of the test set are smaller, and the prediction accuracy is more accurate. The GBDT model has higher applicability for the prediction of water level data, and the reasonable optimization of parameters can ensure that the model has better performance for different nonlinear data prediction. Therefore, the CCS-GBDT model is more suitable for the practical application of Taoshu water level prediction (Table 2).

5   Conclusion

Taocha Channel Head is a water intake project of the first phase of the Middle Route of the South-to-North Water Transfer Project, which has important significance for the accurate prediction of its water level. In this paper, CCS-GBDT model is used to predict the head water level of Taocha canal. Experiments show that the GBDT model with CCS algorithm parameter tuning is more accurate than the original model. Compared with BP neural network, RBF network and SVR method CCS-GBDT are more suitable for water level prediction of Taocha canal head and more practical. Therefore, the CCS-GBDT method proposed in this paper is an effective water level prediction model.

Acknowledgments. This work was supported by the Postgraduate Research & Practice Innovation Program of Jiangsu Province under Grant No. KYCX19-0874, National Natural Science Foundation of China under Grant No. 11801267, and the Natural Science Foundation of the Jiangsu Higher Education Institutions of China under Grant No. 18KJB520007.

References

1. Porter, D.W., Gibbs, B.P., Jones, W.F.: Data fusion modeling for groundwater systems. J. Contam. Hydrol. **42**, 303–335 (2000)
2. Zhao, W., Gao, Y., Li, C.: RVM based on PSO for groundwater level forecasting. J. Comput. **5**, 1073–1079 (2012)
3. Manzione, R.L., Wendland, E., Tanikawa, D.H.: Stochastic simulation of time-series models combined with geo-statistics to predict water-table scenarios in a Guarani aquifer system outcrop area Brazil. Hydrogeol. J. **20**, 1239–1249 (2000)
4. Zhong, Z.Y., Liu, G.Q., Wu, Z.Y.: Analysis and practices of water regulation in the middle route of south-to-north water transfer project. South-to-North Water Transf. Water Sci. Technol. **1**, 95–99 (2018)
5. Li, P., Wang, S.Q., Li, Y.Y.: Eutrophication evaluation for Taocha water quality in Danjiangkou reservoir based on fuzzy comprehensive evaluation method. J. Nanyang Normal Univ. **16**(09), 21–24 (2017)
6. Cao, Y.S., Chang, J.X., Huang, Q.: Real-time control strategy for water conveyance of middle route project of south-to-north water diversion in China. Adv. Water Sci. **28**(1), 133–139 (2017)
7. Li, F.L., Huang, G.T., Han, S.J.: Analysis of major geological disasters in the main canal of middle route project (Taoca-Zhanghe section) of south-to-north water transfer. Ecol. Environ. **15**(4), 889–891 (2006)
8. Wang, C., Zhao, H.C.: River water level forecast based on spatio-temporal series model and RBF neural network. Urban Geotech. Invest. Surv. **5**, 34–39 (2016)
9. Zhang, W.W., Li, R.M., Xie, Z.J.: Multi-step urban road travel time prediction based on PCA-GBDT. Highw. Eng. **6**, 6–11 (2017)

10. Ma, X., Ding, C., Luan, S.: Prioritizing influential factors for freeway incident clearance time prediction using the gradient boosting decision trees method. IEEE Trans. Intell. Transp. Syst. **18**(9), 2303–2310 (2017)

11. Liu, L., Ji, M., Buchroithner, M.: Combining partial least squares and the gradient-boosting method for soil property retrieval using visible near-infrared shortwave infrared spectra. Remote Sens. **9**(12), 1299 (2017)

12. Cheng, Q.W., Wang, W., Ma, D.: Class-imbalance credit scoring using Ext-GBDT ensemble. Appl. Res. Comput. **2**, 421–427 (2018)

13. Chen, T., Guestrin, C.: XGBoost: a scalable tree boosting system. In: Proceedings of the 22nd ACM SIGKDD International Conference on Knowledge Discovery and Data Mining, California, pp. 785–794 (2016)

14. Li, G.: Merging model in freeway weaving section based on gradient boosting decision tree. J. SE Univ. (Nat. Sci. Ed.) **48**(3), 563–567 (2018)

15. Feng, H.M., Li, M.W., Hou, X.L.: Study of network intrusion detection method based on SMOTE and GBDT. Appl. Res. Comput. **34**(12), 3745–3748 (2017)

16. Xia, Y., Liu, C., Li, Y.Y.: A boosted decision tree approach using Bayesian hyper-parameter optimization for credit scoring. Expert Syst. Appl. **78**, 225–241 (2017)

17. Deb, S., Yang, X.S.: Cuckoo search via levy flights. In: Mathematics, pp. 210–214 (2010)

18. Liu, D.J., Liang, B., Yuan, X.Y.: Color image multi-threshold segmentation based on improved CS algorithm. Comput. Eng. Design **37**(12), 3322–3326 (2016)

19. Zhang, D.Y., Wang, P.T., Yuan, Y.B.: An improved cuckoo search algorithm for optimal power flow problem. Water Res. Power **1**, 200–204 (2017)

20. Tao, T., Zhang, J., Xin, K.: Optimal valve control in water distribution systems based on cuckoo search. J. Tongji Univ. **44**(04), 600–604+631 (2016)

21. Oltean, M., Grosan, C.: Multi-objective optimization using adaptive Pareto archived evolution strategy. In: Proceeding of the 5th International Conference on Intelligent Systems Design and Applications, pp. 558–563 (2005)

22. Yang, X.S., Deb, S.: Cuckoo search via lévy flights. In: World Congress on Nature and Biologically Inspired Computing (NaBIC) (2009)

23. Ming, B., Huang, Q., Wang, Y.M.: Cascade reservoir operation optimization based-on improved cuckoo search. J. Hydraul. Eng. **46**(3), 341–349 (2015)

24. An, W.G.: Research on multi-objective optimization method and its application in Engineering. Northwestern Polytechnical University, Xi'an (2005)

25. Han, C.Y., Yu, S.M.: Modified logistic map and its dynamic performances. Periodica. Ocean Univ. China **45**(5), 120–125 (2015)

A High Accuracy Nonlinear Dimensionality Reduction Optimization Method

Zhitong Zhao, Jiantao Zhou[✉], and Haifeng Xing

Inner Mongolia Engineering Lab of Cloud Computing and Service Software,
College of Computer Science, Inner Mongolia University, Hohhot, China
31709038@mail.imu.edu.cn, cszhoujiantao@qq.com, 1323823094@qq.com

Abstract. In the analysis and processing of image recognition, extracting useful and valuable data from the original dataset has become a problem. Since the data to be processed often presents high dimensional and nonlinear feature, reasonable dimensionality reduction is an necessary method for improving the accuracy of data analysis. One of the dimensionality reduction methods Kernel Principal Component Analysis (KPCA) has certain advantages in dealing with nonlinear data, but it also has defects when facing the dataset in which data owe highly complex relationship. The other linear dimensional reduction method Linear Discriminant Analysis (LDA) has supervisory characteristics, which can reduce the dimensionality of dataset in which data owe highly complex relationship. However it can only handle linear data. So we propose a hybrid method which is the combination of the above two methods called KPCA-LDA. By it the new dataset obtained by the step of dimensionality reduction is beneficial to be processed in next step for classification. We combine KPCA-LDA with the Back Propagation Neural Network (BPNN) method to achieve the classification of handwritten numbers. The experimental results show that the classification accuracy of the proposed KPCA-LDA-BPNN model can reach 98.67%, which is about 3%-5% higher than the original method using K Nearest Neighbor (KNN) and Support Vector Machine (SVM).

Keywords: Dimensionality reduction model · Model solving algorithm · Classification framework

1 Introduction

Dimensionality reduction essentially maps one dimension space to another in pattern recognition and machine learning. In practical application scenarios, data tends to exhibit massive, high-dimensional, nonlinear, and other characteristic, which bring many problems to the further processing of data, such as low computational performance caused by massive characteristic, dimensionality disaster caused by high-dimensional characteristic, and linear model failure

© Springer Nature Singapore Pte Ltd. 2019
Y. Sun et al. (Eds.): ChineseCSCW 2019, CCIS 1042, pp. 709–722, 2019.
https://doi.org/10.1007/978-981-15-1377-0_55

problems caused by nonlinear characteristic. Especially, nonlinear dimensionality reduction has become a hot issue at present.

The purpose of dimensionality reduction is mainly in the following aspects. One is to remove redundant features and facilitate data analysis. The other goal is to simplify the training and prediction of machine learning models, reduce the amount of storage space required, and speed up calculations. Common models cause overfitting due to excessive complexity of parameters. Therefore, for complex high-dimensional dataset, we need to extract effective features from the original feature dataset.

At present, the dimensionality reduction of high-dimensional dataset is a hot and inevitable research work. The paper [1,2] proposes the KPCA algorithm to study the dimensionality reduction of nonlinear dataset. The paper [3,4] mainly introduces the supervised dimensionality reduction method LDA algorithm. The paper [5,6] uses SVM and KNN to classify the dataset after feature extraction. The KPCA model can reduce the dimensionality of the nonlinear model, but it faces a complex dataset in the dimension reduction process, and the dimensionality reduction capability is limited. The LDA algorithm is a supervised dimensionality reduction method that makes dataset easier to distinguish. But it can only handle linear dataset. KNN, SVM is a relatively common classifier in machine learning, but the classification ability is limited in the face of multi-class and high-dimensional nonlinear dataset.

Aiming to the shortcomings of common dimensionality reduction methods, this paper designs a high-precision nonlinear dimensionality reduction model, namely KPCA-LDA model. This method solves the problem of nonlinear feature relationship of dataset through KPCA, and supervises the complexity of dataset through LDA, which facilitates the classification of later dataset. At the same time, this paper uses the more popular classifier BPNN to verify the performance of the model.

This paper is mainly composed of the following sections: The second section mainly introduces related work, the third section is optimized models, the fourth section introduces experiments and discussions, the fifth gives the conclusion.

2   Related Work

Two important goals of dimensionality reduction are feature extraction and dataset classification. Therefore, we need to find a suitable feature extraction model and the appropriate classifier. Some feature extractions are given in [1–4,7,8]. Most of the common features are linear dataset dimensionality reduction and nonlinear dataset dimensionality reduction. Some dataset classifications are presented in [6,9,10], which is mainly based on dataset classification after feature extraction.

In terms of linear dimension reduction. In the literature [3,4], it mainly introduces the supervised dimensionality reduction method LDA algorithm, and the paper [4] mainly studies text classification. The article helps to classify annotations by clustering topics and extracting topic keywords. The LDA model is then used to more comprehensively predict document categories. In terms of nonlinear dimensionality reduction. In the paper [7], it mainly introduces a face recognition method combining error detection and KPCA algorithm. KPCA is used to extract small areas that are specifically divided by facial images. The paper [8] identifies compound odors by analyzing chromatograms to determine compound information. The raw dataset is extracted by KPCA to search for better features.

In terms of dataset classification. Paper [9] KPCA solves the linear correlation of the input dataset when it encounters nonlinear characterization dataset. The accuracy of the dataset is predicted by extracting new features of the multi-core SVM and the optimal parameters. The paper [10] is mainly about the diagnosis of diseases. It returns the selected subset of features to the KPCA module to project the data to a higher kernel spatial dimension, thereby reducing the principal component coefficients to achieve linear separability. The KPCA coefficients are then mapped by LDA into a more favorable linear discriminant space. Finally, the multi-core SVM classifier is used to classify the newly projected data. The paper [6] in the field of tumor classification, the KPCA method is mainly used to reduce the feature dimension and feed back the minimum feature value to the KNN classifier.

According to our research, there is still space for improvement in current feature extraction. In terms of nonlinear dimensionality reduction in feature extraction, KPCA has a good effect on dataset dimensionality reduction of non-linear feature relationships. However, the high complexity of dataset often leads to problems such as limited dimensionality reduction, which is detrimental to the classification of dataset. Traditional linear dimension reduction method LDA has supervised characteristics, which makes the dataset points after dimensionality reduction as easy as possible to be distinguished. But it can only handle linear dataset. Therefore, this paper proposes a new feature model KPCA-LDA, which solves the problem of nonlinear feature relationship of dataset through KPCA. It supervises the complexity of datasets through LDA and facilitates the classification of later dataset. At the same time, the popular classifier BPNN in machine learning and the common classification model SVM and KNN are selected to verify the validity of the feature extraction model.

3 Optimized Model

Aiming at the high complexity and limited performance of the KPCA algorithm in the nonlinear dimensionality reduction process, a new dimension reduction algorithm LDA is introduced to generate a new feature extraction model named KPCA-LDA. For the case that the common classifier has limited classification ability for high-dimensional dataset and multi-classification problems. In this paper, BPNN is used as a classifier. It not only has self-learning ability, but also can easily implement relatively complex nonlinear mapping functions when the dataset is large. Currently, BPNN is one of the more popular network models in the field of pattern recognition [11,12]. The following flowchart describes the entire process of feature extraction and classification in model optimization (KPCA-LDA-BPNN).

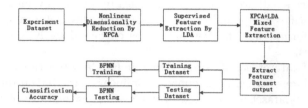

Fig. 1. The flowchart of the KPCA-LDA-BPNN method

3.1   KPCA and Its Improved Algorithm.

Definition 1. *n samples is the training set, and the other part is the testing set, arrange each sample as an N dimensional column vector, the matrix of the training sample $X = [x_1, x_2, \cdots, x_n].\varphi : x \to \varphi(x) \in F$ is a nonlinear mapping implemented on it, φ is a nuclear map, and F is a kernel space.*

Randomly select a sample set from the selected dataset, select n samples as the training set, and the other part is the testing set, and arrange each sample from top to bottom and then from left to right into an N dimensional column vector, so that the matrix of the training sample set is $X = [x_1, x_2, \cdots, x_n].\varphi : x \to \varphi(x) \in F$ is a nonlinear mapping implemented on it, φ is a nuclear map, and F is a kernel space, known as subspace dimensionality reduction in feature space [13,14]

$$Y = W^T \varphi(x). \tag{1}$$

Solving $\varphi(x)$ and arranging the obtained eigenvalues, and selecting the d largest eigenvalues λ_i and their corresponding eigenvectors μ_i, According to formula (1), the dimensionally reduced projection matrix $W = (r_1, r_2, \cdots, r_d)$ can be obtained.

Although the traditional KPCA algorithm includes the nonlinear information of handwritten numbers in the kernel matrix, it makes the originally high sample dimension higher. Therefore, the traditional KPCA algorithm has problems such as computational difficulty and long time. In response to the above problems, this paper introduces an improved KPCA algorithm. Its specific operation is to replace the original sample vector with the best mean vector, so that the sample dimension mapped to the kernel space is also reduced. The details are as follows [15].

Select one sample set from the selected handwritten numeral library, select m numbers from the sample set, each number c pictures, a total of n samples as the training set, and press each sample from top to bottom, then from left to the order of the right is arranged into an N-dimensional column vector, so that the matrix of the training sample set is $X = [x_1, x_2, \cdots, x_n]$.

1. Suppose the sample vector of the i-th class is

$$X_i = [x_1^i, x_2^i, \cdots, x_c^i](0 < x < n). \tag{2}$$

2. Find the sample mean of x_b^i in the i-th class is

$$u_b^i = \sum_{j=1}^{n} x_{bj}^i. \tag{3}$$

3. The sample mean of the i-th sample

$$\mu_i = \frac{1}{n} \sum_{j=1}^{n} x_{ij}. \tag{4}$$

4.

$$X_i' = [\mu_1^i - \mu_i, \mu_2^i - \mu_i, \cdots, \mu_c^i - \mu_i]. \tag{5}$$

Selecting the appropriate mean vector coefficients to estimate the sample mean vector, so that the dimension mapped to the kernel space is reduced, which is equivalent to performing a dimensionality reduction, reducing the amount of calculation, and setting the mean valued coefficient matrix to $u(c \times c)$.

Multiply the mean vector coefficient by Eq. (5) to find the best mean vector.

$$\widehat{X}_i = X_i'^T \cdot \mu_i, \tag{6}$$

$$\widehat{X} = [\widehat{X}_1^T, \widehat{X}_2^T, \cdots, \widehat{X}_c^T]. \tag{7}$$

In order to find the best mean estimation coefficient, the author first constructs an α_i function, where μ_i is the independent variable of the function α_i.

$$\alpha_i = \sqrt{\frac{\widehat{X}^T - X_i'}{X_i'}}. \tag{8}$$

Solving μ_i of $\arg\max(\alpha_i^T \alpha)$ is the best mean estimate coefficient matrix.

$\widehat{X}$ is a $c \times m$ dimensional matrix whose dimension is significantly smaller than the original sample vector set X.

3.2   LDA and Its Improved Algorithm

The traditional LDA method defines the intra-class dispersion matrix S_W and the inter-class dispersion matrix S_B as follows [16,17]:

$$S_w = \sum_{i=1}^{m} \sum_{j=1}^{c} p_i(X_j^i - \mu_i)(X_j^i - \mu_i)^T, \tag{9}$$

$$S_B = \sum_{i=1}^{m} p_i(\mu_i - \mu)(\mu_i - \mu)^T, \tag{10}$$

$$J_F(W) = \arg\max \frac{|W^T S_B^T|}{|W^T S_W W|}, \tag{11}$$

In the formula: W satisfies $S_B W = \lambda S_W W$.

As we all know, the traditional LDA has two prominent shortcomings: one is that it is easy to produce "small sample problem" when facing high-dimensional images, and the other is edge data classification. In view of the two outstanding problems above, this paper makes the following two improvements to the traditional algorithm.

Make the following changes to S_B.

In order to better distinguish the relationship between classes and classes, and to distinguish the indistinguishable samples, the same samples are closely linked. When it find the inter-class divergence matrix, it perform a certain weighting process. The distinction between classes is more obvious, and the information within the class is more compact.

$$S_B' = \sum_{i=1}^{c-1} \sum_{j=i+1}^{c} p_i p_j w_{ij}(d_{ij})(\mu_i - \mu_j)(\mu_i - \mu_j)^T, \tag{12}$$

$$d_{ij} = (\mu_i - \mu_j)^T(\mu_i - \mu_j), \tag{13}$$

$$w(d_{ij}) = \exp(-a d_{ij}^2). \tag{14}$$

Extract the zero space of S_W and remove the zero space of S_B'.

It is known from the criterion that the most discriminative feature is the intersection space of the non-zero space of S_B and the zero space of S_w.

1. Extract the zero space U of S_w

$$U^T S_W U = 0, (U^T U = I). \tag{15}$$

2. Project S_B' onto the zero space U of S_w

$$\widehat{S}_B = U^T S_B' U. \tag{16}$$

3. The best projection direction for $\widehat{S}_B$ is V_B

$$V_B = \arg\max |V^T S_B' V|. \tag{17}$$

Through the above two-step projection transformation, the optimal projection transformation matrix of zero-space linear discriminant analysis is: $\widehat{W} = U V_B$.

3.3   Fusion Improvement Method of KPCA and LDA

Firstly, the improved KPCA is used to obtain the dimensionality reduction Y', then the S_w and $\widehat{S}_B$ are obtained according to the improved LDA, and then the optimal projection matrix $\widehat{W}$ is obtained, and finally the characteristic data Y of the KPCA and LDA fusion improvement method is obtained. $\widehat{Y} = \widehat{W}^T X$.

The specific theory for both KPCA and LDA algorithms can be found in [18]. For the combination of KPCA and LDA, the first m KPCA-LDA principal component sample extracted from the dataset of the preprocessing operation is used as the input of the classifier, which reduces the complexity of the dataset and facilitates the later classification. In this paper, the mixed feature extraction method KPCA-LDA is compared with the KPCA feature extraction.

3.4   BPNN-Based Classifier

At present, pattern recognition is a hot research field, and it needs new ways to solve today's increasingly difficult problems [19]. Since thousands of functions are available in many pattern recognition and machine learning applications, feature selection and feature extraction are still important tasks for finding raw data. At the recognition stage, it is important to choose the appropriate pattern recognition method to build the classifier. The main method of feature selection is to use some mathematical tools to reduce the pattern dimension, in order to find the most effective, lower-dimensional features to form the pattern vector for pattern recognition.

As a relatively common machine learning method, neural networks are relatively easy to implement complex nonlinear mapping functions (mathematical theory verification) to approximate any nonlinear function and obtain ideal results [20]. For the classification of datset, it is necessary to accurately obtain the recognition results. BPNN is a "supervised" training algorithm for multi-layer perceptron networks that can derive the weights and thresholds that correctly classify the network based on a given training sample. At the same time, it can completely approach complex nonlinear relationships and has high classification accuracy.

Specific steps of the BPNN algorithm:

1. Initialization weight: the weight in the network is generally initialized to a relatively small random number (for example: any random number from -1.0 to 1.0); the threshold (offset) of each unit is also often initialized to a small random number.
2. Forward Propagation Input: In this step, each sample X computes the net input and output of each unit in the network hidden and output layers. Firstly, the training samples are provided to the input layer of the network. Note that for unit j of the input layer, its output is equal to its input, for unit j, $O_j = I_j$.

$$I_j = \sum_i W_{ij} O_i + \theta_j \tag{18}$$

Where W_{ij} is the connection weight from the upper unit i to the local unit j; O_i is the output of the upper unit i; and θ_j is the offset of the current unit j. Finally, the net input I_j of each unit in the hidden layer and the output layer is sent to its respective activation function (assumed to be the Sigmoid function), and its output O_j is calculated:

$$O_j = \frac{1}{1 + e^{-I_j}} \tag{19}$$

3. Back Propagation Error: Propagation backwards through the update and offset (threshold) of the weight of the network prediction error.
For the output layer unit j, the error Err_j is calculated by:

$$Err_j = O_j(1 - O_j)(T_j - O_j) \tag{20}$$

Where O_j is the actual output of unit j and T_j is the expected output of j. From the back to the front, the error Err_j of each hidden layer unit j is calculated in order:

$$Err_j = O_j(1 - O_j) \sum_k Err_k w_{kj} \tag{21}$$

Where w_{kj} is the connection weight from unit k to unit j in the next higher layer, and Err_k is the error of unit k. Calculate the correction amount of the weight and update the weight:

$$\Delta w_{ij} = (l)Err_j O_i \longrightarrow w_{ij} = w_{ij} + \Delta w_{ij} \tag{22}$$

Where: l is the learning rate, usually taking a value between 0 and 1. Calculate the amount of correction for the threshold and update the threshold:

$$\Delta \theta_j = (l)Err_j \longrightarrow \theta_j = \theta_j + \Delta \theta_j \tag{23}$$

Where: l is the learning rate, usually taking a value between 0 and 1.
4. Termination condition:
All Δw_{ij} in the previous cycle are too small, less than a certain threshold;
The percentage of samples not correctly classified in the previous cycle is less than a certain threshold;
Exceeding the number of pre-specified cycles (the termination conditions commonly used in practice).

4   Experiments and Discussions

In this paper, handwritten numbers are used as experimental dataset for the dimensionality improvement model. Numeral recognition is one of the most interesting and challenging areas of research in the field of image processing. Due to the variety of shapes, scales and formats in handwritten characters, the recognition rate of handwritten numbers is still limited [21].

Because the diversification of handwritten numbers greatly increases the difficulty of recognition, in order to improve the recognition ability, it is necessary to extract the features of handwritten numbers. Feature extraction and selection are the key factors to improve the recognition rate of handwritten numeral characters. The main idea is to extract some of its structural features, so that the numeral displacement, size change, glyph distortion and other interference are relatively reduced. Moreover, after feature extraction, the amount of data is also greatly reduced, which is convenient for later handwritten numeral classification and recognition.

4.1 Data Preprocessing

In the field of character recognition, image preprocessing is an essential part of the field, which converts the original image into a binary form acceptable to the recognizer.

To recognize image datset, firstly you must preprocess their character images. The main purpose of preprocessing is to remove noise from the image, eliminate redundant information, and obtain a normalized matrix to prepare for feature extraction.

In this paper, the popular handwritten numbers in the field of image recognition are used as experimental dataset to verify the advantages of the model. This dataset has a total of 1797 handwritten numbers. Each number consists of a matrix of 8×8 size. It is stored in numpy.nparray in the third-party module Sklearn commonly used in machine learning. This ndarray has a total of 1797 lines and 64 columns. A line is a number. The display of any handwritten number 3 is shown in Fig. 1. The recognition effect of the feature input depends on the completeness of the feature dataset, which requires the preprocessing to maintain the character characteristics of the original image as much as possible in eliminating the factors unrelated to the recognition in the image. In the preprocessing stage, for the characteristics of handwritten numbers, we need to perform proper preprocessing operations on the characters, including binarization, smooth denoising, segmentation, tilt adjustment, small normalization, and refinement. Figure 2 shows the effect of handwritten number 3 preprocessing standardization.

4.2 Feature Extraction and Proper Classifier

In our study, since handwritten numbers are 64-dimensional feature vectors, PCA can reduce high-dimensional dataset to 3 dimensions in order to visualize dataset to facilitate display of dataset. The handwritten number is displayed after the size is reduced in Fig. 3. We introduce the kernel function RBF, namely KPCA algorithm, on the basis of PCA. KPCA focuses on the dimensionality reduction of nonlinear dataset. However, due to the complexity and high dimensionality reduction capability of KPCA in the dimension reduction process, this paper introduces LDA algorithm to improve feature extraction ability. As a supervised algorithm, LDA is able to extract the most discriminative information

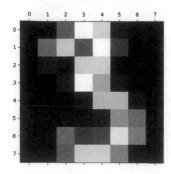

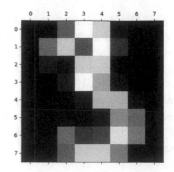

Fig. 2. The number 3 original display   **Fig. 3.** The number 3 preprocessing display

from the dataset, which is beneficial to classification. Therefore, the combined model KPCA-LDA is used to delete the redundant dataset features, and then the feature extracted dataset is used as the input of the classifier (Fig. 4).

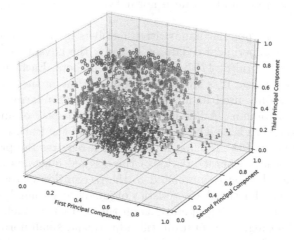

Fig. 4. 3D display by PCA

In this paper, due to the common classifier SVM, KNN has limited ability to classify high-dimensional and multi-class dataset, so we choose BPNN as the classifier. Because it not only has a self-learning ability compared to a larger dataset but also it can handle any complex nonlinear model. After the 64-dimensional handwritten numeral dataset is preprocessed and feature extracted, we will obtain a 20-dimensional feature vector. Because our output is divided into 10 categories 0–9. Therefore, the number of input/output neurons in the network is 20/10. We only need to predict the index with the largest output value, which is our final prediction. Secondly, the activation function of BPNN is Sigmoid, the learning rate of the threshold is chosen to be 0.2, and the number of iterations is 100.

4.3   Results and Discussions

This article is primarily a combination of hybrid algorithms. It mainly combines feature extraction and classifiers to analyze handwritten numbers. The feature space dimension after KPCA-LDA extraction is not determined by itself. In the experiment, this paper selects the best feature number according to the classification rate.

BPNN is built by installing the pybrain module in a python environment. This module is an open source neural network toolkit. The above data analysis method is executed by Pycharm 2018 (Commnunity Edition) software. During the execution of KPCA, the RBF function is selected as the core function.

In the experiment, the number of features of the dataset remained the same, the classification accuracy increased and eventually remained at a steady state level. In the face of each class, it does not properly characterize some features, which can lead to misclassification results. Therefore, the accuracy of a few nodes with a small number of features and hidden layers is relatively low. As shown in the experimental results of the six combined models shown in Fig. 5, it can be found from the experimental results that the number of features at the beginning is relatively small, so the accuracy is also low, and then the number of features is continuously increased, and the accuracy is also improve and eventually remain at a stable level.

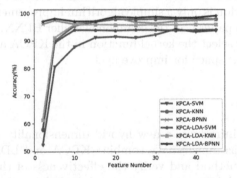

Fig. 5. The classification rate of six kinds of combination models

In Fig. 5, the comparison results of KPCA-LDA and KPCA feature extraction are clearly shown. The experimental data shows that the extraction effect using KPCA-LDA is mostly better than KPCA, mainly in terms of time and accuracy. Especially for the classifier BPNN. When KPCA-LDA is selected as the feature extraction method of the experimental dataset, it can obtain better and more satisfactory classification results. When the number of features is 20, the experimental results are the best, and the highest accuracy of KPCA-LDA-BPNN is 98.67%.

The accuracy of the runtime and cross-validation classifications is documented in Table 1. For different six models, the number of features and the number of hidden layer nodes are guaranteed to be constant. It can be seen from the table that when the number of feature sets of the data set is 20, the performance of the six models is good and the precision is high.

Table 1. Recognition accuracy using six different methods

Types of models	Feature number	Running time(s)	Accuracy (%)
KPCA-SVM	20	38.37	91.32
KPCA-KNN	20	13.32	96.33
KPCA-BPNN	20	26.55	97.77
KPCA-LDA-SVM	20	3.24	93.67
KPCA-LDA-KNN	20	0.13	95.83
KPCA-LDA-BPNN	20	15.05	98.67

In the table, it can be seen that KPCA-LDA and BPNN are an ideal combination. This hybrid model has significant advantages in terms of accuracy and time. We also compare BPNN with the previously proposed multilinear classifier, which has lower classification performance than BPNN.

The kernel function in the KPCA algorithm is selected by empirical value, and the LDA algorithm is only a simple algorithm improvement. For the selection of the classifier, this paper uses the common classifier BPNN. In the future work, we can scientifically select the kernel function in the KPCA algorithm. Secondly, the classifier still has space for improvement.

5   Conclusion

This paper mainly introduces a new hybrid dimensionality reduction model to classify and recognize images. We combine KPCA and LDA as a new hybrid feature extraction method and verify the effectiveness of the method by using handwritten numeral dataset as experimental data. According to the experimental results, KPCA-LDA showed better feature extraction performance. Among the classification effects, BPNN has higher classification performance than ordinary classifier SVM and KNN. Therefore, the combination of KPCA-LDA and BPNN is a high accuracy nonlinear dimensionality reduction method.

Acknowledgement. The research is supported by Natural Science Foundation of China under Grant No. 61662054, 61262082, Inner Mongolia Science and Technology Innovation Team of Cloud Computing and Software Engineering and Inner Mongolia Application Technology Research and Development Funding Project "Mutual Creation Service Platform Research and Development Based on Service Optimizing and Operation Integrating" under Grant 201702168, Inner Mongolia Engineering Lab of Cloud Computing and Service Software and Inner Mongolia Engineering Lab of Big Data Analysis Technology.

References

1. Huang, D., Zhang, D., Liu, Y., Zhang, S., Zhu, W.: A KPCA based fault detection approach for feed water treatment process of coal-fired power plant. In: Proceeding of the 11th World Congress on Intelligent Control and Automation, pp. 3222–3227. IEEE, Shenyang (2014). https://doi.org/10.1109/WCICA.2014.7053247
2. Wang, X., Huang, L., Zhang, Y.: Modeling and monitoring of nonlinear multi-mode processes based on similarity measure-KPCA. J. Central S. Univ. **24**(3), 665–674 (2017). https://doi.org/10.1007/s11771-017-3467-z
3. Ghassabeh, Y.A., Rudzicz, F., Moghaddam, H.A.: Fast incremental LDA feature extraction. Pattern Recogn. **48**(6), 1999–2012 (2015). https://doi.org/10.1016/j.patcog.2014.12.012
4. Fu, R., Qin, B., Liu, T.: Open-categorical text classification based on multi-LDA models. Soft. Comput. **19**(1), 29–38 (2015)
5. Yin, S., Jing, C., Hou, J., Kaynak, O., Gao, H.: PCA and KPCA integrated support vector machine for multi-fault classification. In: IECON 2016–42nd Annual Conference of the IEEE Industrial Electronics Society, pp. 7215–7220. IEEE, Florence (2016). https://doi.org/10.1109/IECON.2016.7793188
6. Ibrahim, A.M., Baharudin, B.: Classification of mammogram images using shearlet transform and kernel principal component analysis. In: 2016 3rd International Conference on Computer and Information Sciences (ICCOINS), pp. 340–344. IEEE, Kuala Lumpur (2016). https://doi.org/10.1109/ICCOINS.2016.7783238
7. Chen, X., Wang, S., Ruan, X.: Recognition of partially occluded face by error detection with logarithmic operator and KPCA. In: 2016 9th International Congress on Image and Signal Processing, BioMedical Engineering and Informatics (CISP-BMEI)), pp. 460–464. IEEEE, Datong (2016). https://doi.org/10.1109/CISP-BMEI.2016.7852755
8. Jha, S.K., Josheski, F., Marina, N., Hayashi, K.: GC-MS characterization of body odour for identification using artificial neural network classifiers fusion. Int. J. Mass Spectrom. **406**, 35–47 (2016). https://doi.org/10.1016/j.ijms.2016.06.002
9. Qian, X., Chen, J.-P., Xiang, L.-J., Zhang, W., Niu, C.-C.: A novel hybrid KPCA and SVM with PSO model for identifying debris flow hazard degree: a case study in Southwest China. Environ. Earth Sci. **75**(11), 991 (2016). https://doi.org/10.1007/s12665-016-5774-3
10. Alam, S., Kwon, G.-R.: The Alzheimer's disease neuroimaging initiative: Alzheimer disease classification using KPCA, LDA, and multi-kernel learning SVM. Int. J. Imaging Syst. Technol. **27**(2), 133–143 (2017). https://doi.org/10.1002/ima.22217
11. Mei, C., Yang, M., Shu, D., Hui, J., Liu, G.: Monitoring wheat straw fermentation process using electronic nose with pattern recognition methods. Anal. Methods **7**(13), 6006–6011 (2015)
12. Urmila, K., Chen, Q., Li, H., Zhao, J., Hui, Z.: Quantifying of total volatile basic nitrogen (TVB-N) content in chicken using a colorimetric sensor array and nonlinear regression tool. Anal. Methods **7**, 5682–5688 (2015)
13. Mika, S., Ratsch, G., Weston, J., Scholkopf, B., Mullers, K.R.: Fisher discriminant analysis with kernels. In: Neural Networks for Signal Processing IX: Proceedings of the 1999 IEEE Signal Processing Society Workshop (Cat. No. 98TH8468), pp. 41–48. IEEE, Madison (1999). https://doi.org/10.1109/NNSP.1999.788121
14. Deng, M., Chen, X., Chen, T.X., Wang, H.R., Lu, H.-X.: Improved kernel principal component analysis based on a clustering algorithm. CAAI Trans. Intell. Syst. (2010)

15. Han, Z.S., Li, Y., Zhang, Y.N.: A comparative study on face recognition using LDA-based algorithm. Microelectron. Comput. (2005)
16. Shermina, J.: Illumination invariant face recognition using discrete cosine transform and principal component analysis. In: 2011 International Conference on Emerging Trends in Electrical and Computer Technology, Nagercoil, pp. 826–830. IEEE, Nagercoil (2011). https://doi.org/10.1109/ICETECT.2011.5760233
17. Xie, Y.L.: LDA algorithm and its application to face recognition. Comput. Eng. Appl. (2010)
18. Jenssen, R.: Kernel entropy component analysis. IEEE Trans. Pattern Anal. Mach. Intell. 32(5), 847–860 (2010). https://doi.org/10.1109/TPAMI.2009.100
19. El Haimoudi, K., Issati, I., Daanoun, A.: The particularities of the counter propagation neural network application in pattern recognition tasks. In: Ezziyyani, M., Bahaj, M., Khoukhi, F. (eds.) AIT2S 2017. LNNS, vol. 25, pp. 474–487. Springer, Cham (2018). https://doi.org/10.1007/978-3-319-69137-4_42
20. Jiang, P., Ge, Y., Wang, C.: Research and application of a hybrid forecasting model based on simulated annealing algorithm: a case study of wind speed forecasting. J. Renew. Sustain. Energy 8(1), 015501 (2016). https://doi.org/10.1063/1.4940408
21. Ashiquzzaman, A., Tushar, A.K.: Handwritten Arabic numeral recognition using deep learning neural networks. In: 2017 IEEE International Conference on Imaging, Vision Pattern Recognition (icIVPR), pp. 1–4. IEEE, Dhaka (2017). https://doi.org/10.1109/ICIVPR.2017.7890866

The Metrics to Evaluate the Health Status of OSS Projects Based on Factor Analysis

Sha Jiang[1], Jian Cao[1(✉)], and Mukesh Prasad[2]

[1] Department of Computer Science and Engineering,
Shanghai Jiaotong University, Shanghai, China
{jiangsha1007,cao-jian}@sjtu.edu.cn
[2] Centre for Artificial Intelligence,
Faculty of Engineering and Information Technology,
University of Technology Sydney, Sydney, Australia
mukesh.prasad@uts.edu.au

Abstract. As open-source software (OSS) development is becoming a trend, an increasing number of businesses and developers are joining OSS projects. For project managers, developers and users, understanding the current health status of a project is very important to manage a development process, select the open-source projects to development or to adopt the software packages developed by projects. Therefore, an efficient approach to evaluate the health status of the open-source project is needed. Unfortunately, although many approaches including metrics have been proposed, they are designed in arbitrary ways. In this paper, a mathematical tool, i.e., factor analysis, is used to build a health evaluation model for OSS projects. As far as we know, this is the first time that factor analysis has been applied to evaluate OSS projects. This model is based on GitHub data and uses the basic indexes that are closely related to the health status of the projects as the input. Then, six new synthetic metrics, namely community activity, project popularity, development activity, completeness, responsiveness and persistence are obtained through factor analysis, which can be used to calculate the overall health score of a project. Moreover, in order to verify the effectiveness of this model, it is applied to some real projects and the results show that the overall scores achieved by this model can reflect the health status of the projects.

Keywords: Open source software project · Health status · Factor analysis

1 Introduction

Open-source software (OSS) hosting platforms, such as GitHub and SourceForge, are becoming increasingly popular among developers because of their intelligent

© Springer Nature Singapore Pte Ltd. 2019
Y. Sun et al. (Eds.): ChineseCSCW 2019, CCIS 1042, pp. 723–737, 2019.
https://doi.org/10.1007/978-981-15-1377-0_56

and ecological development models. GitHub, for instance, is currently used by over 31 million users and 21 million organizations, including well-known companies such as Google, Microsoft, Facebook and Ali. It hosts more than 9,600 projects which generated 2 billion pull requests over the past 12 months[1]. Many well-known projects, such as TensorFlow, jQuery, and Python, are hosted on GitHub.

On the OSS hosting platform, users can easily search and evaluate projects, actively participate in the development processes of projects as developers, or interact with other developers as reviewers [8]. Therefore, on these platforms, users have formed a development community and project operations can be promoted effectively. However, with the rapid development of the OSS hosting platform, some issues have emerged:

1. As OSS hosting platforms become increasingly popular, a large number of external developers use them because of their openness and low entry requirements. Inevitably, some developers may provide low-quality code, annotations and issues, which reduce the quality of open-source projects.
2. An increasing number of developers use OSS hosting platforms for code hosting, hence the quality of projects and code fragments ranges from good to bad, which causes difficulties for the development, maintenance and code reuse of the project. It is difficult to manage and evaluate these software projects using traditional approaches.

After engaging in discussions with project leaders and other stakeholders in the OSS community, Wahyudin formalized the notion of the "health" of a community, namely its state, as it pertains to its survivability [25]. Health can be defined as a comprehensive indicator of an OSS project in terms of various metrics such as popularity, number of defects, number of iterations, and developer engagement. Assessing the health of OSS projects has become an important topic in software engineering research. By understanding the health status of OSS projects, companies can assess the long-term survival probabilities of their projects [14]; developers can choose those active and potential projects to which to contribute; and managers can adjust the project development process in a timely manner. Compared with traditional software projects in which development data is not publicly available, the OSS hosting platforms and repositories save a wealth of data, including code changes, interactions between users, milestones, and so on, which can be accessed and used at any time in a variety of ways [10]. More specifically, GitHub provides a rich set of APIs through which most of the development data, including commit data, issue data and comment data can be accessed. Other platforms such as GHtorrent [7] also provide comprehensive data. Based on these data, evaluation models can be established to evaluate the health of OSS projects.

Some researchers have studied the health evaluation models of OSS projects, but they are often based on subjective analysis or a small set of arbitrary factors, lacking an objective and quantitative basis. In some papers, the health status

---

[1] https://octoverse.GitHub.com/.

of some OSS projects are evaluated based on different models and metrics [3]. Moreover, most evaluations only provide detailed measurements on the indicators that are supposed to be relevant to the health status of the project. They do not provide an overall health score that can reflect a project's health status in an explicit and intuitive way.

In this paper, we derive a comprehensive evaluation model for the health status of OSS projects. Firstly, the relevant indicators'data is obtained through APIs provided by GitHub. These influential indicators include problem-solving speed, commit speed, pull request speed, project duration, user active time and so on. Based on the factor analysis method, a new health assessment model is established which can output an overall health score based on six derived comprehensive indexes. Several projects are evaluated using this model and the health scores and their consequent performances are compared. In addition, a health evaluation tool has been developed by which users can easily query the health scores and other data charts of OSS projects.

Our contributions are as follows:

1. We select a set of indicators of OSS projects to evaluate their health statuses in GitHub.
2. We find some comprehensive metrics that are more related to health status so we can derive the health scores of projects.
3. We evaluate the validity of our model for health evaluation.

The paper is organized as follows. The related work is presented in Sect. 2. The basic indicators used in the evaluation are introduced in Sect. 3. The factor analysis together with how to apply factor analysis to derive the evaluation model is explained in Sect. 4. The validity of the evaluation model is evaluated in Sect. 4. Threats to validity are discussed in Sect. 5. Finally, Sect. 6 concludes the whole paper.

2 Related Work

2.1 Health of OSS Development

For traditional software development projects, factors including development progress, funding and staffing are the biggest concerns. For OSS projects, people often pay more attention to the overall status of the projects, which is often referred to as "health".

The concept of health has been defined by many researchers with different understandings. Wahyudin et al. [25] study the concept of health in OSS projects. They define health as "survivability", i.e., the ability of the project to survive throughout time. Moreover, they divide the OSS model into three parts, i.e., developer community, user community and software products, and each have different implications for health. In [23], an analogy between the health of an ecosystem and human health is made and an approach is proposed to measure the software ecosystem's health based on the intensity productivity, robustness, and niche creation (PRN) measures. Manikas et al. propose a conceptual framework for defining and measuring the health of software ecosystems, which includes three main components, i.e., the actors, the software and the orchestration [13]. Crowston describes the onion model of the open source community, illustrating the characteristics and relationships of developers, leaders, and active users in a healthy OSS community and how to maintain a healthy open source community [16].

Some researchers are particularly interested in evaluating the quality or health of some OSS projects. For example, Jae Yun Moon and others have tracked the Linux kernel for a long time and analyzed the success of Linux from three levels: individual, group and community [18]. Mockus and others [17] qualitatively and quantitatively analyzed the development process of the Apache Web Server, and compared it with four commercial software projects and seven conclusions about the characteristics of the OSS development process were put forward. This study pioneered the use of OSS project development process data (including code submission history, problem reports and repair records, e-mail records) to conduct quantitative research on the activities of open-source participants, establish quantitative indicators, and use these indicators to analyze community size, contributor distribution, core team size, defect density, problem solving speed and so on. In this study, the typical characteristics of OSS development process are also summarized.

At the same time, many researchers have analyzed some of the influencing factors of software projects'health from different aspects:

Gamalielsson pioneered the use of the mailing list response time and quality as measures to assess ecosystem health [6]. Ray et al., after studying 729 open source projects, revealed the impact of programming languages on software quality [20]. Bird et al. delved into the relationship between software development and the ownership of certain codes and software quality [1]. Borges et al. studied metrics related to software popularity, including language and application areas [3].

It can be observed by now that the metrics proposed by different researchers are designed in arbitrary ways. The health assessment model is built based on a solid mathematical tool in this paper.

2.2 Data Acquisition for Health Evaluation

To evaluate the health of an OSS project, data needs to be collected.

Gousios et al. from the software engineering group of the Delft University of Technology developed the GHtorrent tool, which regularly collects data on GitHub through GitHub API and packages it into offline data packets for researchers to download, which effectively resolves many limitations in the use of GitHub API [7].

Almost 1,113,640 open-source projects are collected by Openhub, which synchronized about 475,060 source codes of open source projects and ranked the popularity and activity of OSS according to the number of users and submissions [5]. In addition, code quality analysis, code language composition, monthly submissions, monthly additions of contributors and other dimensions of data are also listed.

3   Indicators for Health Evaluation of OSS Projects

GitHub provides many interfaces for researchers to access the development data of OSS. Based on the GitHub APIs, this study obtains the quantitative data of OSS projects (defined as a repository on GitHub) from multiple dimensions and establishes a health status measurement framework, in which data can be obtained, processed and saved in three steps:

1. Using the Python Scrapy library, a crawler can periodically call the GitHub APIs to obtain the specified data. According to the cluster sampling principle, the list is divided into groups larger than 10000, 5000–10000, 1000–5000, and less than 1000 according to the number of stars, and the group is sampled according to the principle of random sampling. Finally, 658 projects are extracted from the GitHub. The extraction principle follows GitHub's principle of long tail effect, which can better reflect the overall characteristics and also meet the needs of factor analysis for sample capacity [15].
2. Pre-processing the acquired data. The data obtained directly through the APIs incorporates some worthless information and some indicators which cannot directly reflect the characteristics of the data itself, so it is necessary to clean and pre-process the data. For example, some data of issues and pull requests overlap, and issues themselves have duplicated contents, which need to be removed. In addition, due to the different dates of the project establishment, the values of some indicators cannot reflect the health of the project, which should be divided by the life span in order to obtain the average value.
3. Storing the processed data in the database to facilitate visualizing and subsequent processing. After analyzing the stages of OSS projects which includes development, operation and maintenance, the characteristics of open-source communities and the metrics suggested by the related research, the metrics are divided into three classes, i.e., basic metrics, health metrics based on development activities and health metrics based on user activities, which involves 14 metrics in all.

3.1   Basic Metrics

- **Number of Stars (ST)**: The number of stars reflects the popularity of the project. The higher the number of stars, the more effective the evaluation of a project.
- **Number of Subscribers (SU)**: This metric indicates the attention degree of a project.
- **Number of Forks (FO)**: Users can copy a project's code under his own branch through the fork, then edit the branch, submit the requests and merge

them into the main branch of the project.

The above metrics are not the data which are generated in the development process, but the subjective evaluation of users in the open-source community. They are also the most intuitive evaluation metrics.

- **Number of Contributors (CO):** This metric captures the number of different contributors to the project (such as those who have conducted commit behaviors). The more people involved in development, the more intelligence they provide.

3.2 Health Metrics Based on Development Activities

- **Number of Commits (COI):** After a developer changes the code, he can commit the code to his private repository or public repository.
- **Number of Pull Requests (PR):** After a developer commits the code, he can launch a pull request to merge his own changes into the main branch.
- **Number of Pull Merged (PRM):** The codes are merged if there is no problem after the code is carefully reviewed. If most of the submitted code is reviewed, the better the quality of the project.

The above metrics all reflect the activity of project development. The more, the better.

- **Issue close/open Ratio (ICOR):** After an issue is resolved, it will be closed. This ratio indicates whether every issue has been well resolved.
- **Issue Close Time (ICT):** This metric measures the average close time needed for each issue. The shorter the time from open to close, the more efficient the problem solving.
- **Developer Involvement Time (DIT):** This metric captures how long a developer participates in the development of the project, and it is calculated in terms of the first commit time and the latest commit time. The longer it takes, the more attractive and stable the project.

3.3 Health Metrics Based on User Activities

- **Number of Issues (IS):** People submit issues and point out project defects, which can help improve the project.
- **Number of Issue Comments (ISC)** and **Number of Comments (COE):** These two metrics indicate the level of activation in community discussions. In the study of OSS, it has been observed that the more issues and comments, the better the project will be.

4   The Health Evaluation Model Based on Factor Analysis

4.1   Introduction to Factor Analysis

Factor analysis is a statistical technique for extracting common factors from variable groups. It was first proposed by Spearman [21], a British psychologist.

The basic purpose of factor analysis is to use a few factors to describe the relationship between many indicators or factors.

$$x_1 - \mu_1 = a_{11} * F_1 + a_{12} * F_2 + \cdots + a_{1m} * F_m + \varepsilon_1$$
$$x_2 - \mu_2 = a_{21} * F_1 + a_{22} * F_2 + \cdots + a_{2m} * F_m + \varepsilon_2$$
$$\cdots$$
$$x_p - \mu_p = a_{p1} * F_1 + a_{p2} * F_2 + \cdots + a_{pm} * F_m + \varepsilon_p$$

(1)

Equation 1 is called a factor model.

4.2 Analysis of Data by Factor Analysis

IBM's SPSS software is a professional tool for data statistics and analysis [26]. SPSS is used to analyze the obtained data by applying factor analysis to measure the health status of the project. The specific steps are as follows:

1. Exporting tables that need to be analyzed from the database:
2. In order to facilitate factor analysis, data is normalized. In SPSS, data is standardized by Z-score by default.

$$z_{ij} = \frac{(x_{ij} - \overline{x_i})}{s_i}$$

(2)

Of these, x_{ij} is the original data, $\overline{x_i}$ is the average value of class i's data, s_i is the standard deviation of class i's data and z_{ij} is the data after standardization.
3. Factor analysis of standardized data, in which extraction method is principal component analysis method, rotation method is maximum variance method and convergence iteration is 25 times.
4. After factor analysis is completed, we can look at the chart to analyze the results.

Table 1. Test of KMO and Bartlett

Kaiser-Meyer-Olkin measure of sampling sufficiency		0.749
Bartlett's sphericity test	Approximate chi square	7153.436
	Df	78
	Sig.	0

From Table 1, the Kaiser-Meyer-Olkin measure of sampling adequacy is 0.749. A minimum Kaiser-Meyer-Olkin score of 0.50 is considered necessary to reliably use factor analysis for data analysis. When the score is larger than 0.80, the results are considered to be very good. Similarly, the Bartlett test of sphericity (where the higher, the better) is 7153.436 with significance levels of $p = 0.00$ (p is considered to be suitable) [11, 22].

Table 2. Total variance of interpretation

S/N	Extract sum of square loadings			Rotation sums of squared loadings		
	Total	Variance%	Accumulate%	Total	Variance%	Accumulate%
1	5.619	43.226	43.226	3.140	24.157	24.157
2	1.820	14.000	57.226	2.959	22.762	46.919
3	1.178	9.058	66.284	2.296	17.663	64.582
4	1.031	7.928	74.212	1.012	7.784	72.366
5	.950	7.307	81.519	1.011	7.777	80.143
6	.829	6.380	87.899	1.008	7.756	87.899

In relation to how to determine the number of factors, there are several different criteria including eigenvalues [12], the cumulative variance contribution rate and scree plot [4]. A reasonable suggestion is we can use multiple approaches at the same time. In this study, the cumulative variance contribution rate and the scree plot are used at the same time. It can be observed from Table 2 that the total cumulative variance contribution rate of the first six factors reaches 87.961%. Generally, when their total cumulative variance contribution rate is more than 85%, it can be considered that these factors have described most of information in the original data. Therefore, in this study, six common factors are extracted.

4.3 Establishment of a Health Assessment Model

It can be seen from the rotation component matrix that factor 1 is strongly correlated with issues such as Number of Issues, Number of Comments, Number of Issue Comments, Number of Pull Requests, and Number of Pull Merged. Factor 1 can be defined as the **Community Activity** of the project. Community activity indicates the interactions between users and developers, such as raising bugs, improvement opinions and commenting are frequent. Since developers often update versions based on comments, more interactions in a community can promote the progress of the project and better discover the potential problems of the project.

Factor 2 has a strong correlation with indicators such as Number of Stars, Number of Forks, and Number of Subscribers. Factor 2 can be defined as **Project Popularity**. This is an indicator which inherently reflects the quality of the project. Projects in better health tend to receive more attention. Hu et al. also used two factors, i.e., stars and forks, as important indicators to measure the influence of OSS [9].

Factor 3 is strongly correlated with indicators such as the number of Number of Contributors, Number of Commits, Number of Issues, Number of Pull Requests, and Number of Pull Merged. Factor 3 can be defined as the **Development Activity** of the project. As a healthy project, it attracts more contributors [24]. When the project has more development activities, such as continuous code development and version iteration, continuous problem solving and function evolution, the project is in good health.

Factor 4 is strongly correlated with the Issue close/open Ratio indicator, and it can be regarded as the **Completeness** of the project. When this value is large, it indicates that most of the problems raised have been solved. From one side, the developers have strong capabilities to solve problems or have strong feedback capabilities.

Factor 5 is strongly correlated with the Issue Close Time indicator. It can be defined as **Responsiveness** of the project. Oriol et al. also define the problem processing time as timeliness and regard it as an important indicator for evaluating project health [19].

Factor 6 is strongly correlated with the Developer Involvement Time indicator, It can be defined as the Persistence of the project. It is important to maintain a relatively stable development team. Other studies have also shown that when different developers work on the same file at different times, more defects are introduced compared to when a stable team works on it [2].

We can use the scoring matrix to combine the metric data obtained by each project and use matrix multiplication to calculate the scores of each factor for a project.

After obtaining the scores, we need to calculate the overall health score. The function is as follows:

$$score = F_1 * w_1 + F_2 * w_2 + \cdots + F_6 * w_6 \tag{3}$$

The weight of the factor represents the variance contribution rate of the factor, which corresponds to the percent of variance value under the rotation sums of squared loadings in Table 3. The weights corresponding to the six factors are 24.157, 22.762, 17.663, 7.784, 7.777 and 7.756 respectively. Then the six weights are normalized into the range $(0, 1)$.

This is because the factor score fluctuates around 0 and we use the sigmoid function to convert it into a percentage value.

$$y = \frac{1}{1+e^{-x}} \tag{4}$$

Using the above steps, we extract project data, analyze the data by applying the factor analysis method, and establish a health evaluation model based on six synthetic factors. A final health score can be calculated by this model for a project.

4.4   The Validation of the Health Evaluation Model

The health score we obtain is a relative value. When the score of Project A is higher than the score of Project B, we can think that the weighted value of Project A on the six factors participating in the evaluation is higher than that of Project B. According to our definition of health status, we believe that the health status of Project A is better than Project B.

We calculate the health scores of the sample projects using our model. In terms of their health scores, the top 5 and bottom 5 projects are shown in Tables 3 and 4 respectively.

Table 3. Top 5 healthy project

Name	Score
kubernetes/kubernetes	98.961
tensorflow/tensorflow	98.593
freeCodeCamp/freeCodeCamp	95.649
Homebrew/homebrew-core	93.973
apple/swift	92.535

Table 4. Bottom 5 projects in bad health

Name	Score
mmozeiko/RcloneBrowser	37.915
travis-ci/apt-source-safelist	35.763
ipfs/archives	35.479
imagej/imagej-common	35.252
ionic-team/ionic-v1	34.526

From Tables 3 and 4, it can be seen that the model results are consistent with our empirical judgments. The projects with better reputations or high popularity have higher health scores, mainly because of the large number of participants, more attention, faster product iterations and problem-solving speed. These factors make the project more likely to develop in a healthy direction. Projects with low health scores are relatively less well-known. More specifically, these projects have low community activity and high developer mobility.

Let us take TensorFlow/tensorflow and ipfs/archives as two examples. Tensorflow [28] is currently the most popular machine learning project, which is developed and maintained by the Google artificial intelligence team 'Google Brain' [2]. ipfs (Inter Planetary File System) is an open-source project to create a content-addressable, peer-to-peer method of storing and sharing hypermedia in a distributed file system [27]. We compare 13 indicators of TensorFlow and ipfs as shown in Tables 5 and 6 Raw data).

Tensorflow has a score of 100 on Factor 2 (i.e., Project Popularity), while ipfs has a score of 47.83. This factor is related to Star, Fork, Watch, Contributor and other metrics. We can see that Tensorflow is far superior in terms of these metrics. If improvements could be made to these metrics, ipfs would narrow the gap with TensorFlow on this factor. However, in factor 6 (i.e., Persistence) of the project, TensorFlow has a score of 45.79, while ipfs has a score of 50.18. On the factor developer involvement time, TensorFlow has a score of 44.018, less than that of ipfs, i.e., 103.40. This shows that although there are many developers working on TensorFlow, these developers are unstable.

Table 5. Comparison of tensorflow and ipfs (1)

Name	ST	FO	SU	CO	COI	IS	ICOR
tensorflows	56774	12547	2981	8963	33696	25048	0.970
ipfs/archives	122	17	36	15	55	184	0.261

Table 6. Comparison of tensorflow and ipfs (2)

Name	DIT	ICT	COE	ISC	PR	PRM
tensorflows	44.018	36.572	1279	209648	15845	1673
ipfs/archives	103.4	59.438	4	884	18	15

Obviously, in addition to developer involvement time, TensorFlow is far better than ipfs in terms of other metrics, which is well reflected in the final health score and rankings.

At the same time, in order to verify the validity of the model, we collect data from the same project in November and December for comparison purposes. Table 7 shows the health indicators for several projects (data time 2018.11) and Table 8 shows the health indicators for several projects (data time 2018.12).

Table 7. Data in November

Name	Score	F_1	F_2	F_3	F_4	F_5	F_6
nodejs/node	84.046	83.921	96.016	88.395	61.308	42.367	39.121
pytorch/pytorch	75.258	79.105	83.954	83.487	39.445	60.600	47.800
spring-projects/spring-boot	62.639	43.811	90.690	44.711	70.773	40.560	48.075
ReactTraining/react-router	58.954	51.927	78.146	38.267	71.093	61.499	45.929
spyder-ide/spyder	51.996	56.767	48.908	54.164	56.784	40.399	47.915
kubernetes/dashboard	51.746	52.858	50.360	48.572	68.253	47.094	46.673
travis-ci/packer-templates	47.227	43.923	43.895	45.305	64.056	53.130	48.615
quintel/etsource	46.924	44.504	43.041	46.445	68.997	43.013	47.799
vaadin/framework	46.825	51.242	42.673	62.979	19.148	39.087	48.885

Table 8. Data in December

Name	Score	F_1	F_2	F_3	F_4	F_5	F_6
nodejs/node	86.362	82.649	95.892	87.332	61.682	42.504	39.421
pytorch/pytorch	80.601	72.285	87.492	90.933	38.61	61.212	44.864
spring-projects/spring-boot	63.941	43.216	91.077	44.209	70.847	40.677	48.147
ReactTraining/react-router	59.246	51.506	77.68	38.257	71.119	61.465	45.911
spyder-ide/spyder	51.968	56.129	48.986	53.69	56.851	40.349	47.802
kubernetes/dashboard	51.661	52.209	50.668	48.182	68.284	46.955	46.601
travis-ci/packer-templates	46.804	43.852	43.905	45.236	64.064	53.031	48.49
vaadin/framework	46.647	51.046	42.671	62.577	19.166	39.176	48.773
quintel/etsource	46.497	44.446	43.073	46.372	69.007	42.935	47.753

Comparing the health statuses between these two months, the change is not so significant. But the health score of the projects with a high health score in November tended to increase. On the contrary, the health score of the projects with a low health score in November tends to decline. This reflects the health score is an indicator to predict how well this project will develop in the coming future.

To display the health status of OSS projects more intuitively, we developed a tool which presents the health score together with the detailed information on a project (as shown in Figs. 1 and 2). The tool is based on Python and is presented as a website. Users enter the project name in the search box and the tool shows the project's health score and the six factor scores, while showing basic information on the project, owner information and charts such as project code changing, developer participation, commit data and so on.

5 Threats to Validity

This study is based on data extracted from GitHub according to certain rules. Limited by sample size or extraction rules, the final model may change if other projects are introduced. Moreover, since we select 13 metrics as base indicators to build the model, our evaluation model will inevitably be affected by introducing more metrics. However, the approach is general and if new metrics are collected, our method, i.e., factor analysis, can still be used to build the model.

Health is a dynamic metric which means we can obtain the health status at different times. Therefore, for different stages of a project, we may need different health evaluation models. For example, for an early stage and a stable stage of the project, the factors that should be included in the model may be different. Our model is based on stable projects since the projects we collected are stable ones.

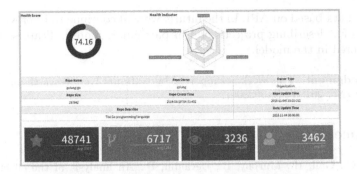

Fig. 1. The Health score show in the tool

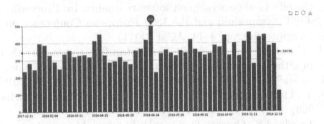

Fig. 2. Number of weekly commits in the past year shown in the tool

6   Conclusion

In this paper, a health status evaluation model is constructed for OSS projects based on factor analysis. Specifically, 13 basic metrics are selected, such as the number of stars, number of forks and number of comments, from which six synthetic indexes are derived using factor analysis, i.e., Community Activity, Project Popularity, Development Activity, Completeness, Responsiveness, and Persistence. This work then shows how the overall health score can be calculated directly through a simple combination of scores on these new synthetic indexes. This model is applied to several repositories, hence providing empirical evidence of its validity.

As a result of our work, a tool is also developed to evaluate the health status of a GitHub repository. Users enter a project's name and the project information into the tool and the overall health score of the project, the scores of the six synthetic factors, and the scores of other synthetic factors which are less important are shown.

This study also has the following limitations:

1. The proportion of projects included in the dataset is relatively small compared with the total number of projects hosted by GitHub. We will expand the dataset by including more projects.
2. The basic indicators adopted may not adequately reflect the health status of projects. At the same time, GitHub has some limitations as to the way to

obtain data based on API. In the future, we will continue to improve the data models for describing projects, and in particular, more indicators should be considered in the model.

Acknowledgment. This work is supported by National Key Research and Development Plan (No. 2018YFB1003800).

References

1. Bird, C., Gall, H., Murphy, B., Devanbu, P.: An analysis of the effect of code ownership on software quality across windows, eclipse, and firefox (2010)
2. Bird, C., Nagappan, N., Murphy, B., Gall, H., Devanbu, P.: Don't touch my code!: examining the effects of ownership on software quality. In: Proceedings of the 19th ACM SIGSOFT Symposium and the 13th European Conference on Foundations of Software Engineering, pp. 4–14. ACM (2011)
3. Borges, H., Hora, A., Valente, M.T.: Understanding the factors that impact the popularity of github repositories. In: 2016 IEEE International Conference on Software Maintenance and Evolution (ICSME), pp. 334–344. IEEE (2016)
4. Cattell, R.B.: The scree test for the number of factors. Multivar. Behav. Res. 1(2), 245–276 (1966)
5. Farah, G., Tejada, J.S., Correal, D.: OpenHub: a scalable architecture for the analysis of software quality attributes. In: Proceedings of the 11th Working Conference on Mining Software Repositories, pp. 420–423. ACM (2014)
6. Gamalielsson, J., Lundell, B., Lings, B.: Responsiveness as a measure for assessing the health of OSS ecosystems. In: Proceedings of the 2nd International Workshop on Building Sustainable Open Source Communities (OSCOMM 2010), pp. 1–8. Tampere University of Technology, Tampere (2010)
7. Gousios, G., Spinellis, D.: GHTorrent: GitHub's data from a firehose. In: 2012 9th IEEE Working Conference on Mining Software Repositories (MSR), pp. 12–21. IEEE (2012)
8. Hippel, E.V., Krogh, G.V.: Open source software and the "private-collective" innovation model: issues for organization science. Organ. Sci. 14(2), 209–223 (2003)
9. Hu, Y., Zhang, J., Bai, X., Yu, S., Yang, Z.: Influence analysis of github repositories. SpringerPlus 5(1), 1268 (2016)
10. Jensen, C., Scacchi, W.: Data mining for software process discovery in open source software development communities. In: Proceedings of Workshop on Mining Software Repositories, pp. 96–100. IET (2004)
11. Junior, J.H., Joseph, F., Anderson, R.E., TATHAM, R.L., et al.: Multivariate Data Analysis with Readings. Macmillan London (1992)
12. Kaiser, H.F.: The application of electronic computers to factor analysis. Educ. Psychol. Meas. 20(1), 141–151 (1960)
13. Manikas, K., Hansen, K.M.: Reviewing the health of software ecosystems - a conceptual framework proposal (2013)
14. Van der Linden, F., Lundell, B., Marttiin, P.: Commodification of industrial software: a case for open source. IEEE Softw. 26(4), 77–83 (2009)
15. MacCallum, R.C., Widaman, K.F., Zhang, S., Hong, S.: Sample size in factor analysis. Psychol. Methods 4(1), 84 (1999)
16. Manikas, K., Hansen, K.M.: Software ecosystems-a systematic literature review. J. Syst. Softw. 86(5), 1294–1306 (2013)

17. Mockus, A., Fielding, R.T., Herbsleb, J.: A case study of open source software development: the apache server. In: Proceedings of the 22nd International Conference on Software Engineering, pp. 263–272. ACM (2000)
18. Moon, J., Sproull, L.: Essence of Distributed Work. Online Communication and Collaboration: A Reader, p. 125 (2010)
19. Oriol, M., Franco-Bedoya, O., Franch, X., Marco, J.: Assessing open source communities' health using service oriented computing concepts. In: 2014 IEEE Eighth International Conference on Research Challenges in Information Science (RCIS), pp. 1–6. IEEE (2014)
20. Ray, B., Posnett, D., Filkov, V., Devanbu, P.: A large scale study of programming languages and code quality in github. In: Proceedings of the 22nd ACM SIGSOFT International Symposium on Foundations of Software Engineering, pp. 155–165. ACM (2014)
21. Spearman, C.: "General intelligence," objectively determined and measured. Am. J. Psychol. 15(2), 201–292 (1904)
22. Tabachnick, B.G., Fidell, L.S.: Using Multivariate Statistics, 5th edn. Allyn & Bacon, Needham Height (2007)
23. Van Den Berk, I., Jansen, S., Luinenburg, L.: Software ecosystems: a software ecosystem strategy assessment model. In: Proceedings of the Fourth European Conference on Software Architecture, pp. 127–134. ACM (2010)
24. Van Maanen, J.E., Schein, E.H.: Toward a theory of organizational socialization (1977)
25. Wahyudin, D., Mustofa, K., Schatten, A., Biffl, S., Min Tjoa, A.: Monitoring the health status of open source web-engineering projects. Int. J. Web Inf. Syst. 3(1/2), 116–139 (2007)
26. Wikipedia contributors: Spss – Wikipedia, the free encyclopedia (2018). https://en.wikipedia.org/w/index.php?title=SPSS&oldid=870276612. Accessed 16 Jan 2019
27. Wikipedia contributors: Interplanetary file system – Wikipedia, the free encyclopedia (2019). https://en.wikipedia.org/w/index.php?title=InterPlanetary_File_System. Accessed 18 Jan 2019
28. Wikipedia contributors: Tensorflow – Wikipedia, the free encyclopedia (2019). https://en.wikipedia.org/w/index.php?title=TensorFlow&oldid=878912059. Accessed 18 Jan 2019

Detection of Algorithmically Generated Domain Names Using SMOTE and Hybrid Neural Network

Yudong Zhang[1,2], Yuzhong Chen[1,2(✉)], Yangyang Lin[1,2],
and Yankun Zhang[1,2]

[1] College of Mathematics and Computer Sciences, Fuzhou University,
Fuzhou 350116, China
ydzhangch@163.com, yzchen@fzu.edu.cn,
grow_up@foxmail.com, yyl.bugloser@gmail.com
[2] Fujian Provincial Key Laboratory of Network Computing and Intelligent
Information Processing, Fuzhou 350116, China

Abstract. Domain generation algorithms (DGA) provide methods that use specific parameters as random seeds to generate a large number of random domain names for preventing malicious domain name detection, which greatly increases the difficulty of detecting and defending botnets and malware. State-of-the-art models for detecting algorithmically generated domain names are generally based on the principle of analyzing the statistical characteristics of the domain name and building a classifier to locate the algorithmically generated ones. However, most current models have problems of requiring the manual construction of feature sets for classification, as they are sensitive to the imbalance of the sample distribution in the domain name dataset and are difficult to adapt to frequent changes of the domain-name algorithm. To address this issue, we propose a hybrid model that combines a convolutional neural network (CNN) and a bidirectional long-term memory network (BLSTM). First, to solve the problem of the number of domain names generated by DGAs being relatively small and the sample distribution being unbalanced, which consequently decreases detection accuracy, the borderline synthetic minority over-sampling technique is employed to optimize the sample balance of the domain name dataset. Second, a hybrid deep neural network that combines CNN and BLSTM is introduced to extract the semantic and context-dependency features from the domain names. The experimental results from different domain-name datasets demonstrate that the proposed model achieves significant improvement over state-of-the-art models with regard to precision and robustness.

Keywords: Domain name generation · SMOTE · LSTM · CNN · Malicious domain name detection

1 Introduction

Domain generation algorithms (DGA) provide methods of generating a large number of pseudo-random domain names using specific parameters, such as date, time, text, etc., as seeds for random initialization. DGAs are often associated with malicious network

© Springer Nature Singapore Pte Ltd. 2019
Y. Sun et al. (Eds.): ChineseCSCW 2019, CCIS 1042, pp. 738–751, 2019.
https://doi.org/10.1007/978-981-15-1377-0_57

behavior. For example, recent botnets (e.g., Conficker, Kraken, and Torpig) used DGAs to quickly generate candidate remote command-and-control server domain lists [1, 2]. Subsequently, they redirected the normal domain name services (DNS) request to the botnet [3] for the purpose of conducting malicious activities, such as distributed denial-of-service attacks, spamming, phishing, click frauds, etc., [4–6] by establishing communication with the infected host through a valid-appearing domain name. Therefore, effectively detecting algorithmically generated domain names is crucial for preventing malicious cyber activities.

In recent years, researchers have proposed several types of models to detect algorithmically generated domain names. Among them, the traditional model requires manual reverse-engineering of the DGA, which is time consuming and laborious. The malware can easily escape detection by changing its DGA during examination. Therefore, reverse engineering models cannot meet the accuracy and timeliness requirements. Models based on black-list filtering have a problem in which the coverage of malicious domain names is limited and cannot adapt to the size of domain-name growth. Models based on DNS request analyses must rely on third-party credit systems, and the detection costs are high while the detection results are unsatisfactory. Generating a domain-name detection model based on traditional machine-learning techniques requires many feature-engineering tasks, such as selection and extraction. When the DGA produces variants, it is necessary to reconstruct the feature set, which is difficult for application scenarios in which the DGAs are large and frequently change. The neural-network-based algorithm generates a domain-name detection model. By constructing a neural network with multiple hidden layers, the domain-name category feature is automatically extracted, which can effectively detect the algorithmically generated domain name. However, it is necessary to rely on large-scale domain-name datasets for training, and the problem of uneven distribution of samples in the dataset is more pronounced.

To address the above-mentioned issues, this paper proposes a domain-name detection model that combines the convolutional neural network (CNN) and bidirectional long- and short-term memory network (BiLSTM).

(1) To solve the data-sample imbalance, an improved borderline synthetic minority over-sampling technique (SMOTE) oversampling algorithm is introduced to optimize the positive and negative sample balances in the domain-name datasets.

(2) A hybrid DNN model that combines CNN and BiLSTM (HCNN-BLSTM) is proposed. HCNN-BLSTM is employed to extract semantic features and context dependencies from the domain names. Thereafter, the domain-name category feature is obtained from feature fusion. Finally, a logistic regression (LR) classifier is constructed to generate domain-name discrimination.

(3) On several domain name datasets, the existing algorithmically generated domain-name model is compared and analyzed. Experimental results show that the proposed model for algorithmically generated domain-name detection has the advantages of detection accuracy and robustness.

2 Related Work

Existing models for detecting algorithmically generated domain names are primarily based on methods, such as blacklist filtering, DNS request analysis, traditional machine learning, and neural networks.

Kuhrer et al. [7] conducted a comprehensive analysis and evaluation of the validity of the domain-name blacklist, finding that blacklists provided by different systems had different coverage rates for algorithmically generated domain names, ranging from 0 to 99.5%. They found that the blacklists require extra assistance to provide adequate protection. Furthermore, an attacker can evade blacklist detection by continuously generating different domain names for its connection attempts.

Wang et al. [8] proposed a model, DBod, designed to detect botnets using DGAs. Dbod clusters hosts according to the differences of DNS traffic between an infected host and a normal host in the query domain-name set, DNS query-time distribution, and number of DNS queries needed to detect the infected host. Truong et al. [9] proposed a model based on DNS traffic analysis to detect infected hosts and botnets. The model analyzes the difference between the periodic characteristics of a botnet-infected host and a normal host in the DNS query interval, extracts relevant features, and establishes a classifier to identify the infected host. The above model is based on DNS traffic analysis, the detection accuracy is not ideal, and the domain-name length, registration, registration time, and registration information are obtained through a DNS request to determine the DGA. This requires a third-party credit system and is more limited in practical applications.

Because the domain name generated by the algorithm and the normal domain name differ in the distribution characteristics of characters, words, word length, and number of words, the domain name can be detected by traditional machine-learning methods. Yadav et al. [10] proposed a random-forest (RF) -based algorithm to generate a domain-name detection model. The extraction algorithm generates statistical properties of the domain name and builds a domain name based on the RF-based classifier detection algorithm. Yadav et al. also analyzed and compared the performance of various similarity measures, such as K–L distance, edit distance, and the Jaccard index. Zang et al. [11] proposed a domain-name detection model based on spectral clustering and the support vector machine (SVM) algorithm. The model first identifies the same kind of DGA or a domain name generated by its variant via clustering association based on a binary graph. Thereafter, it extracts the domain-name feature set of each domain-name cluster, including time to live, parsing internet-protocol distribution, attribution, "whois" update, integrity and activity history characteristics, etc. Thereafter, it uses SVM for discrimination. Antonakakis et al. [12] proposed a domain-name detection model based on the hidden Markov model and an RF classifier. The hidden Markov model is used to extract the characteristics of the domain name in terms of part-of-speech and lexicon, and the RF classifier is used to classify the domain name.

These detection models, based on machine learning, can obtain better detection effects when detecting a domain name generated by an existing DGA. However, it remains necessary to manually construct a domain-name feature sets, requiring users to have a rich feature-engineering experience. When a new DGA or a variant appears, satisfactory detection effects cannot be achieved based on the original domain-name feature set.

Kejun et al. [13] proposed a detection model based on a deep neural network (DNN). The model uses word-hashing coding to map the domain name into the high-dimensional vector space, extracts the text features of the domain name, and uses a five-layer DNN for training classification. Woodbridge et al. [14] proposed a model-based LSTM to algorithmically accomplish domain-name detection, implicitly extracting domain-name character sequence features via LSTM, and subsequently using a logistic regression classifier for the algorithmically generated domain name. Feng et al. [15] analyzed and compared the detection performance of CNN and LSTM, based on a real domain-name dataset. In most experimental scenarios, LSTM performance is better than CNN.

The neural models for detecting algorithmically generated domain names can extract domain-name features automatically using multiple layers of hidden-layer nodes and can solve the problem of feature extraction and feature selection faced by traditional machine learning techniques. However, the neural model must rely on large-scale domain-name datasets for network training and remain sensitive to the sample imbalance problem in the domain-name datasets, limiting their capability. Additionally, the current neural models are based on a single CNN or a recurrent neural network (RNN). Considering the diversity and complexity of the DGA, it is difficult for these neural models to fully reflect the statistical characteristics difference between the DGA generated and the normal domain name.

3 Algorithmically Generated Domain-Name Detection Model

3.1 Model Framework

The framework of the proposed model is shown in Fig. 1. The framework includes data preprocessing, domain-name encoding, domain-name sample equalization, domain-name category feature extraction, and domain-name category classification. Data pre-processing performs string truncation and padding on the input domain name. The domain-name sample code includes domain-name dictionary creation and character sequence encoding. Domain-name sample equalization is based on the improved Borderline-SMOTE oversampling method, used to optimize the balance of positive and negative samples in the domain-name sample set and to improve the training accuracy of the model. Domain-name category feature extraction is based on CNN and BLSTM, extracting the contextual relationship between the semantic information of the domain name and the character sequence and generating the category characteristics of the domain name. The classifier uses the category feature vector as input to discriminate algorithmically generated domain names.

3.2 Domain-Name Sample

For the domain-name sample in the datasets, a domain-name character dictionary is created based on the frequency at which characters appear in the sample. According to the character dictionary, the domain name is encoded and converted into a fixed-length sequence as a hybrid neural network input for domain-name category feature extraction. The specific steps of domain name encoding are as follows:

Step 1: Create a domain-name character dictionary and initialize it to null. Each element in the dictionary is in the form $<c, f>$, where c is a character, and f is the frequency of occurrence of the character.

Step 2: Traversing all domain names in the domain name sample, for each character, c, contained in the domain name, if c exists in the dictionary, its frequency in the domain-name character dictionary is updated. Otherwise, the $<c, 1>$ is added to the character dictionary.

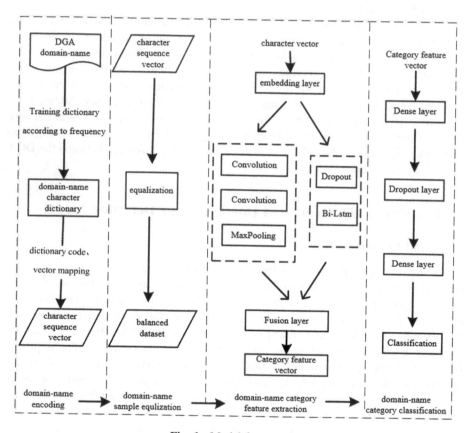

Fig. 1. Model framework

Step 3: Traverse the character dictionary, first assigning each unique character a unique number based on the frequency of occurrence of the characters. If the frequency is the same, different numbers are assigned according to the order of traversal.

Step 4: For the domain name, m, in the domain-name sample, where $m = \{c_i, i = 1, 2, \ldots, d\}$, c_i is the ith character in the domain name. According to the character dictionary, the domain name is converted into a vector matrix and input to a subsequent hybrid neural network. The calculation formula is as follows:

$$v = W * v' \qquad (1)$$

Each character in the domain name is initialized to a real number according to the character dictionary created in Step 3, and v' is a real number vector obtained by embedding each real number. W is a randomly initialized mapping matrix, $W \in R^{d*d'}$, which is used to map the d' dimensional real vector to a d-dimensional vector. $v \in R^d$, and d is the fixed length of the character vector, representing the mapped vector.

3.3 Domain-Name Sample Equalization

There are many types of DGAs. In real domain-name datasets, the sample size of a certain DGA often only accounts to a small part of the domain-name sample set. The imbalance of sample distribution is more serious and has a greater impact on the detection accuracy of neural networks.

The SMOTE algorithm is a random oversampling algorithm. The Borderline-SMOTE algorithm [17] further solves the problem of the SMOTE algorithm randomly selecting samples. It can select the sample points located in the class boundary and the neighborhood to perform interpolation to generate the boundary-blur problem. Therefore for this paper, the Borderline-SMOTE algorithm is used to optimize the impact of domain-name sample imbalance on the accuracy of randomly generated domain-name detection. The specific steps of the domain name sample equalization algorithm are as follows:

Step 1. Traverse the class sample set minority of the malicious DGA for each domain name sample, $p_i (i = 1, 2, \ldots, N)$, in a few sample sets. The m most recent domain name samples of p_i are obtained via the k-nearest neighbors algorithm, and the number of samples of the majority of the m nearest-neighbor samples is expressed as m', and a few of the classes here are DGAs used to generate domain names. Most are normal domain names.

Step 2. If $m = m'$, m nearest-neighbor samples of p_i are majority classes. Subsequently, p_i is considered to be a noise sample and is skipped. If $0 \leq m' < m/2$, thereafter p_i is considered to be far from the sample distribution boundary. Hence, the sample is skipped. If $m/2 \leq m' < m$, the number of majority samples in the m nearest neighbor samples of p_i is larger than the number of samples in a few classes. Therefore, p_i is easily misclassified, and Step 3 is executed.

Step 3. Select s nearest-neighbor samples from p_i's m nearest-neighbor domain samples ($1 < s < m$) and synthesize s minority samples by interpolation between p_i and nearest-neighbor samples. The specific formula is as follows:

$$synthetic_j = p_i + r_j * diff_j (j = 1, 2, \ldots, s) \tag{2}$$

where s is a random integer between 1 and m, indicating the number of samples synthesized by p_i, and synthetic$_j$ is the synthesized j-th domain name sample. $diff_j$ is the difference between p_i and the j-th sample in the selected s nearest-neighbor samples, and r_j is a random number between 0 and 1, reflecting the degree of influence of the difference on the synthesized sample.

Step 4. The synthesized minority sample set is added to the DGA domain training set for balancing.

3.4 Domain-Name Feature Extraction

It is difficult for a single CNN or RNN to comprehensively or accurately extract the category features reflecting the differences between a DGA-generated domain name and a normal domain name in terms of semantic information and character sequences. Therefore, this paper proposes a fusion deep network, HCNN-BLSTM, which extracts and fuses the domain-name classification features and improves the detection accuracy of the algorithm. The structure of the domain-name category feature extraction network is shown in Fig. 2.

3.5 Semantic Feature Extraction

HCNN-BLSTM uses the CNN to extract the semantic features of domain names. First, the character is encoded into the corresponding character vector, as explained in Sect. 3.2. The domain name is subsequently mapped to the corresponding domain-name vector matrix, $A = \{v_1, v_2, \ldots, v_s\}$, where v_s represents the corresponding character vector in the domain-name vector matrix of the s-th character in the domain name.

As the vector matrix of the embedded layer of the CNN model, the domain-name matrix, A, is input to the CNN, and the convolution kernel of size $h*d$ is used for local convolution calculation of the vector matrix corresponding to the domain name. d is the dimension of the character vector, h is the window size of the convolution kernel, and h takes values 4 and 5 to extract different ranges of context semantic features. The calculation formula of the convolutional layer is as follows:

$$o_i = F(w \cdot A[i, \ldots, i + h - 1]) \tag{3}$$

$$c_i = f(o_i + b)(i = 1, 2, \ldots, s - h + 1) \tag{4}$$

$$c = [c_1, c_2, \ldots, c_{s-h+1}] \tag{5}$$

where F represents a filter of size $h * d$, A is the input domain-name matrix, $A \in R^{s \times d}$, and s is the number of characters in the domain name. $A[i, \ldots, i + h - 1]$ indicates the part of the domain-name vector matrix, A, located in the sliding window: the i-th row to the $i + h$-1th row of the domain-name vector matrix. w is the weight matrix of the convolution kernel, $\mathbf{w} \in R^{h \times d}$, o_i is the output of the convolution operation, b is the bias term, $b \in R$, f is the rectified linear unit activation function for nonlinear operation, and c_i is the i-th row to the $i + h - 1$ row of the domain-name vector matrix extracted by the convolution kernel. The local context semantic relationship feature, c_i, constitutes feature map c.

The pooling layer of HCNN-BLSTM adopts the maximum pooling strategy to maximize the pooling of the feature map c generated by the convolution kernel window, perform feature dimensionality reduction and extract key domain name semantic features. The calculation formula is as follows:

$$v_C = max(c_i)(i = 1, 2, \ldots, s - h + 1) \tag{6}$$

Context Feature Extraction

BLSTM is mainly used to process sequence data and has succeeded for several natural-language processing tasks. HCNN-BLSTM uses a BLSTM network to extract context-dependent features of a character sequence of a domain name.

According to the order of appearance of the characters in the domain name, the character vector after character conversion is sequentially input into the BLSTM network. For a positive BLSTM network, enter $v_1, v_2 \ldots, v_L$. For a reverse BLSTM network, enter $v_L, v_{L-1}, \ldots, v_1$. Calculate the forward and reverse hidden-layer state vectors of each character vector, calculate the average value of the hidden-layer state values, multiply by the weight matrix, and use the activation function to get the context vector of the domain name, m. The calculation is as follows:

$$\overrightarrow{h_i} = f(v_i, \overrightarrow{h_{i-1}}) \tag{7}$$

$$\overrightarrow{h_i} = f(v_i, \overrightarrow{h_{i+1}}) \tag{8}$$

$$h_i = \left[\overrightarrow{h_i}, \overrightarrow{\overrightarrow{h_i}} \right] \tag{9}$$

$$o = \frac{1}{n} \sum_{i=1}^{n} h_i \tag{10}$$

$$v_B = Sigmoid\left(W_c o^T\right) \tag{11}$$

where $\overrightarrow{h_i}$ represents the hidden-layer state vector of the i-th character in the domain name, X, in the forward LSTM network ; $\overrightarrow{\overrightarrow{h_i}}$ represents the hidden-layer state vector of the i-th character in the domain name, X, in the reverse LSTM network; and f is the calculation function of the hidden-layer state in the LSTM. h_i is the cascade of the output of the last network element in the forward direction and the output of the last unit in the reverse direction, whereas o is the mean vector of the hidden layer state. o^T is the transpose vector of the vector o; W_c is the weight parameter matrix; the sigmoid is the activation function; and v_B is the context vector of the domain-name sequence.

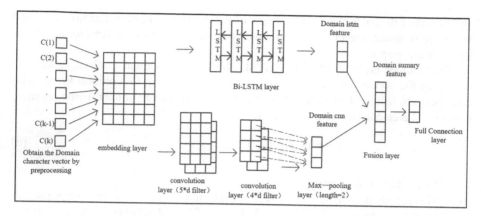

Fig. 2. HCNN-BLSTM for discriminative feature extraction

Feature Fusion

The domain-name semantic feature vector, v_C, obtained with the CNN, is concatenated with the context-dependent feature vector, v_B, of the character sequence of the domain name obtained by the BLSTM network to obtain the category feature vector, v_T, of the domain name. The calculation is as follows:

$$v_T = [v_B, v_C] \tag{12}$$

3.6   Classification and Model Training

The classification layer classifies the domain-name category features extracted by the category feature extraction network as input. It introduces a dropout mechanism between the first fully connected layer and the classification layer and abandons some of the trained parameters for each iteration, so that the weight update does not depend on some inherent features. This prevents overfitting, and the classifier can use the fully connected layer, logistic regression, SVM, etc.

In the HCNN-BLSTM integrated network training process, the gradient of each weight in the neural network is calculated with the corresponding error term. The cross-entropy of the real and prediction categories is used as the loss function. The root mean square (RMS) back propagation method, RMSprop, is used as the optimization. The direction propagation is used to iteratively update the weights and model parameters, and the model is trained to minimize the loss function. The loss function is as follows:

$$L(\widehat{y}, y) = -\frac{1}{N} \sum_{i}^{N} [y_i \log \widehat{y}_i + (1 - y_i) \log(1 - \widehat{y}_i) \tag{13}$$

where $\widehat{y}_i$ is a possible vector for predicting domain names; y is the true tag value corresponding to all domain names; benign domain name is 0; and malicious domain name is 1.

4 Experiment and Result Analysis

4.1 Experimental Environment and Datasets

The experimental environment was built as follows: the operating system was Ubuntu16.04, the CPU was Intel Core i5-5200U, the GPU was GeForce GTX 1050Ti, the memory was 32 GB, and the network was based on Tensorflow 1.10.1 and Keras 2.2.2 [18]. The source-code sample used in the experiment was as follows:

(1) The domain name of the Alexa [19] website was selected as the data source for the normal domain name.
(2) The seed of OSINT DGA [20] from Bambenek Consulting was used to generate a domain name, including the domain-name sample of 30 domain-name generation algorithm families.
(3) The DGA public datasets [21] from the Netlab OpenData project of 360, contained more than 40 kinds of DGA families. A total of about 800,000 algorithms were generated domain-name samples.

With the performance analysis of model generation, this paper selected Alexa's domain name as a positive sample, and the random domain name generated by the OSINT DGA seed from Bambenek Consulting was used as a negative sample. The positive and negative sample combination constituted the domain-name dataset.

4.2 Experimental Parameter Setting and Evaluation Index

The parameters of the network model in the experiment were as follows. The CNN comprised two convolutional layers and one maximum-pooling layer. According to the experimental results, different parameters were tried and the best parameters were selected. To adjust the parameters, the character-vector dimension, d, of the input character selected in this paper was 128. The number of feature maps of the two convolutional layers was set to 64 and 128, and their convolution window sizes were set to 5×128, 4×128, respectively, and the dropout ratio was set to 0.5. The BLSTM comprises two LSTM units with a hidden-layer dimension of 128 and an output dimension of 128. The *RMSProp* gradient optimization algorithm was used to train the process network parameters. The experiment used Precision, Recall, F1, and ROC as performance evaluation indicators, and we took the average of $100 \times$ experimental results.

4.3 Analysis of Experimental Results

The performance of the model proposed in this paper to detect algorithmically generated domain names was verified by multiple experiments.

(1) We verified the ability of the difference between different network models for generating domain-name feature extraction via the algorithm. Through comparison experiments with control variables, we compared algorithmically generated domain-name detection performance.
(2) We verified that the HCNN-BLSTM model algorithmically generated domain-name detection performance. Through an ablation experiment, we analyzed the

effects of SMOTE-based sample equalization on CNN extracted domain-name categories and BLSTM-extracted domain-name categories for the performance of algorithmically generated domain name detection.

(3) By controlling the size of the sample set of domain names used for model training and the proportion of sample sizes of different categories of domain names, the impact of the domain sample size and the sample class imbalance on model performance was analyzed.

(4) The HCNN-BLSTM model was used to test the ability of the unknown DGA algorithm family to verify the generalization ability of the HCNN-BLSTM model.

Performance Analysis and Comparison

It can be seen from the experimental results in Fig. 3 that the network prediction performance of the HCNN-BLSTM model was significantly better than other models using the same data. Because the CNN and the BLSTM network extracted the category features of the domain name from different angles, it better reflected the context statistics and semantic features of the DGA-generated domain name and improved the detection effect.

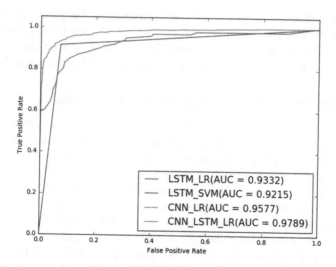

Fig. 3. ROC of different models

Table 1. Impact of data equalization on detection performance

Model\evaluation indicator	Precision	Recall	f1-score	auc
LSTM+LR	0.85	0.85	0.85	91.32%
BSMOTE+LSTM+LR	0.90	0.88	0.89	93.16%
Bi LSTM	0.86	0.84	0.84	92.25%
BSMOTE+BLSTM+LR	0.92	0.91	0.91	94.77%
LSTM+SVM	0.87	0.88	0.87	88.13%
BSMOTE+LSTM+SVM	0.89	0.90	0.88	91.14%
CNN-BLSTM	0.92	0.92	0.92	94.08%
HCNN-BLSTM	0.93	0.93	0.93	96.81%

Table 2. Accuracy results under different imbalance ratios

Model\unbalance ratio	5%	10%	20%	30%	40%
LSTM+LR	90.52%	90.96%	91.32%	92.88%	93.69%
BSMOTE+LSTM+LR	92.76%	92.99%	93.16%	94.22%	94.17%
BLSTM+LR	81.00%	83.78%	89.13%	87.74%	90.23%
BSMOTE+BLSTM+LR	92.38%	92.64%	93.22%	94.76%	94.89%
LSTM+SVM	82.67%	87.19%	92.14%	91.07%	91.72%
BSMOTE+LSTM+SVM	95.42%	96.08%	96. 81%	96.97%	97.01%

Impact of Sample Equalization

It can be seen from the experimental results in Table 1 that using Borderline-SMOTE for equalization improved the model-to-generated domain-name detection performance. Each model improved the detection performance by about 2% via dataset processing following sample equalization under the same imbalance ratio of 20%. This is because the number of domain-name samples of some algorithmically generated domain names was too small, and the domain-name category features could not be effectively extracted during the model-training process. It is easy to misjudge the normal domain name. Notwithstanding the data equalization process, the number of samples generated by the minority domain-name algorithm in the training set increased, which helped the domain-name category feature learning, improving the detection rate.

Impact of Sample Imbalance

In this section, the size of the random domain-name dataset generated by the OSINT DGA seed was 10,000, which changed the proportion of the domain name in the sample set and the imbalance degree to test the detection performance of each model under different balance degrees. LSTM and LR, LSTM and SVM, CNN and LR, and SMOTE were used as comparison models. Repeated experiments were obtained on average, as listed in Table 2.

It can be seen from Table 2 that, as the proportion of algorithmically generated domain names in the sample set increased, the accuracy of model detection gradually increased. This is because, as the algorithmically generated domain name increased, the network could learn better algorithmically generated domain-name feature information. Furthermore, the accuracy of the HCNN-BLSTM, proposed in this paper, was better

Table 3. Accuracy results under different sample sizes

Model\ sample sizes	1,000	5,000	10,000	20,000	50,000
LSTM+LR	91.32%	92.49%	94.33%	96.78%	98.46%
SMOTE+LSTM+LR	93.16%	94.91%	96.49%	97.95%	98.56%
LSTM+SVM	89.13%	91.40%	92.94%	95.51%	97.12%
CNN+LR	93.22%	95.11%	95.98%	97.68%	98.02%
SMOTE+LSTM+SVM	92.14%	93.19%	95.10%	95.85%	97.21%
HCNN-BLSTM	96. 81%	98.24%	98.27%	99.27%	99.64%

than the comparison model under the imbalance ratio of each sample. This is caused by the fusion deep network used by HCNN-BLSTM, which can extract domain-name sample category features from different angles using different types of neural networks.

Table 4. Accuracy results under different domain names

Model\DGAs	chinad	dyre	emotet	fobber	virut	gspy	locky
LSTM+LR	83.69%	83.74%	84.33%	84.80%	71.35%	86.63%	84.08%
CNN+LR	92.62%	92.38%	93.72%	93.10%	78.85%	93.43%	92.64%
CNN+LSTM+LR	93.38%	94.73%	94.43%	95.98%	84.13%	94.42%	93.84%
SMOTE+LSTM+LR	91.66%	92.22%	92.69%	89.57%	75.05%	89.68%	84.25%
SMOTE+CNN+LR	96.15%	96.72%	96.92%	97.36%	81.42%	97.37%	95.83%
HCNN-BLSTM	96.84%	98.67%	98.39%	98.42%	92.66%	97.94%	98.99%

Impact of Sample Size
It can be seen from Table 3 that, as the number of samples increased, the detection accuracy of each model increased correspondingly under the same imbalance ratio of 20%. The results of the HCNN-BLSTM model were relatively superior than other models. This is because the data size increased, and the neural network learned more information about the random domain name, improving the detection performance of the model. In addition to the fusion model presented by this paper, via the combination of different types of neural networks, different angle extraction features retained more information than other single-network models. Therefore, the detection performance was optimal.

Model Robustness Performance Analysis
This part of the experiment added 360 other algorithms to generate domain-name sample data, such as "chinad," "dyre," "emetet," "fobber," "gspy," "locky," and "virut" as an unknown domain name set. The detection ability of each model for the domain name generated by the unknown DGA was analyzed. The experimental results are listed in Table 4. It can be seen from the experimental results that HCNN-BLSTM showed better detection ability than the other comparison models when facing a domain name generated by an unknown DGA. This supports the notion that HCNN-BLSTM has better generalization ability and is of great significance for practical applications. HCNN-BLSTM can extract more information and is more discernible to algorithmically generated domain names and normal domain name categories.

5   Conclusion

This paper proposed a domain-name detection model based on CNN and BLSTM, which improved the algorithmically generated domain-name detection capability. We optimized sample equalization using the improved Borderline-SMOTE oversampling method. Additionally, a DNN model that combined CNN and BLSTM was introduced to extract the semantic features of domain names and the context-dependent features of domain-name character sequences to obtain richer domain-name category features and superior domain name detection performance over existing models.

References

1. Bilge, L., Sen, S., Balzarotti, D., Kirda, E., Kruegel, C.: Exposure: a passive DNS analysis service to detect and report malicious domains. ACM Trans. Inf. Syst. Secur. (TISSEC) **16**(4), 14 (2014)
2. Schiavoni, S., Maggi, F., Cavallaro, L., Zanero, S.: Phoenix: DGA-based botnet tracking and intelligence. In: Dietrich, S. (ed.) DIMVA 2014. LNCS, vol. 8550, pp. 192–211. Springer, Cham (2014). https://doi.org/10.1007/978-3-319-08509-8_11
3. Choi, H., Lee, H., Lee, H., Kim, H.: Botnet detection by monitoring group activities in DNS traffic. In: 7th IEEE International Conference on Computer and Information Technology, pp. 715–720. IEEE, CIT, USA (2007)
4. Qu, Y.Z., Lu, Q.K.: Effectively mining network traffic intelligence to detect malicious stealthy port scanning to cloud servers. J. Internet Technol. **15**(5), 841–852, (2014). https://doi.org/10.6138/jit.2014.15.5.14
5. Jiang, J., Zhuge, J.W., Duan, H.X., Wu, J.P.: Research on botnet mechanisms and defenses. J. Softw. **23**(1), 82–96 (2012)
6. Zhou, H., Guo, W., Feng, Y.: An automatic extraction approach of worm signatures based on behavioral footprint analysis. J. Internet Technol. **15**(3), 405–412 (2014)
7. Kührer, M., Rossow, C., Holz, T.: Paint it black: evaluating the effectiveness of malware blacklists. In: Stavrou, A., Bos, H., Portokalidis, G. (eds.) RAID 2014. LNCS, vol. 8688, pp. 1–21. Springer, Cham (2014). https://doi.org/10.1007/978-3-319-11379-1_1
8. Wang, T.S., Lin, H.T., Cheng, W.T., Chen, C.Y.: DBod: clustering and detecting DGA-based botnets using DNS traffic analysis. Comput. Secur. **64**, 1–15 (2017)
9. Truong, D.T., Cheng, G., Jakalan, A.: Detecting DGA-based botnet with DNS traffic analysis in monitored network. J. Internet Technol. **17**(2), 217–230 (2016)
10. Yadav, S., Reddy, A.K.K., Reddy, A.L., Ranjan, S.: Detecting algorithmically generated malicious domain names. In: Proceedings of the 10th ACM SIGCOMM Conference on Internet Measurement, pp. 48–61. ACM, USA (2010)
11. Xiaodong, Z., Jian, G., Xiaoyan, H.: Detecting malicious domain names based on AGD. J. Commun. **39**(7), 1000–1436 (2018)
12. Antonakakis, M., et al.: From throw-away traffic to bots: detecting the rise of DGA-based malware. Presented as part of the 21st Security Symposium, pp. 491–506, Bellevue, WA (2012)
13. Kejun, Z., Liansheng, G., Fenglin, Q., Xiaoguang, H.: Deep model for DGA botnet detection based on word-hashing. J. Southeast Univ. **373**(07), 19–29 (2017)
14. Woodbridge, J., Anderson, H.S., Ahuja, A.: Predicting domain generation algorithms with long short-term memory networks. arXiv preprint arXiv:1611.00791 (2016)
15. Feng, Z., Shuo, C., Xiaochuan, W.: Classification for DGA-based malicious domain names with deep learning architectures. In: 2017 Second International Conference on Applied Mathematics and Information Technology, vol. 6, no. 6, pp. 67–71 (2017)
16. Han, H., Wang, W.-Y., Mao, B.-H.: Borderline-SMOTE: a new over-sampling method in imbalanced data sets learning. In: Huang, D.-S., Zhang, X.-P., Huang, G.-B. (eds.) ICIC 2005. LNCS, vol. 3644, pp. 878–887. Springer, Heidelberg (2005). https://doi.org/10.1007/11538059_91
17. Chollet, F.: Keras. https://github.com/fchollet/keras. Accessed 2016
18. Does Alexa have a list of its top ranked webites?. https://support.alexa.com/hc/enus/articles/200449834Does-Alexa-have-a-list-of-its-top-ranked-websites. Accessed 2019
19. Bambenek consulting master feeds. http://osint.bambenekconsultin.com/feeds/. Accessed 06 Apr 2016
20. DGA Page. https://data.netlab.360.com/dga. Accessed 2018

Deep Q-Learning with Phased Experience Cooperation

Hongbo Wang[✉], Fanbing Zeng, and Xuyan Tu

School of Computer and Communication Engineering,
University of Science and Technology Beijing, Beijing 100083, China
foreverwhb@ustb.edu.cn

Abstract. The value-based reinforcement learning algorithms train agents by storing previous experience rewards, however, this simply sampling at the same probability results in a slow learning rate. In reality, the importance of each sample is not exactly the same. The use of prioritized experience replay greatly improves the learning rate of reinforcement learning, but good experiences and more effective strategies may be ignored or missed. In order to overcome two shortcomings, a Deep Q-learning with phased experience replay (MixDQN) is put forward in this article, where the priority is used to improve the training rate in the early stage of training and the random sampling in the later stage to make good use of good experience. Experiments with three classic control problems are based on OpenAI Gym. The experimental results prove that the MixDQN can enable an agent's learning more stably, quickly and efficiently.

Keywords: Reinforcement learning · Experience replay · Value-based

1 Introduction

Reinforcement learning (RL) focuses on a satisfied strategy for an agent in a complex decisive space, namely, which action is its better choice in a dilemma. There are a lot of amazing ideas have been introduced for solving RL problem, such as Q-learning [1], SARSA [2], Policy gradient [3] and so on. In a practical application, as its complexity increases, more and more actions need to be judged and selected, which inevitably requires more computing memory and has greatly reduced the performance of RL. Recent years, with the maturity of deep learning technology, many algorithms based on policy gradient [4–6] have achieved vigorous development. The DDPG [7], as a combination of deterministic Actor-critic algorithms [8] with deep learning, rapidly improves the learning speed of the agent. Mnih et al. [9] attempted to use neural networks instead of huge table storage, which solved the problem that traditional RL requires

Supported in part by the National Natural Science Foundation of China under Grant 61572074 and in part by the China Scholarship Council under Grant 201706465028.

Y. Sun et al. (Eds.): ChineseCSCW 2019, CCIS 1042, pp. 752–765, 2019.
https://doi.org/10.1007/978-981-15-1377-0_58

huge space to store data and long time to find information. Narasimhan [10] and Zelinka et al. [11] applied the excellent LSTM in text problems to text games, which achieved amazing results. Deep reinforcement learning (DRL) combines the decision-making ability of RL with the excellent fitting ability of neural networks, and the development of RL has entered a new stage. Especially the most famous DQN algorithm [9], which uses a neural network for its value estimation in Q-learning, and has achieved better results than real players in the Atari games. Many improvements in DQN mainly update learning rates and reduce over-fitting. The double Q-learning [12] uses some different models to complete its selection of optimal actions for keeping a tiny fluctuation during its value estimation. Dueling DQN [13] splits $Q(s, a)$ (which represents the value of action a in state s) into two parts, namely, $v(s) + adv(s, a)$, for some special states to modify their advantage function. The distributional DQN [14]represents those examples in a histogram-like form. Noisy DQN [15] enhances its exploration ability of the model by increasing the randomness of the parameters. The above mentioned algorithms illustrate the Q-learning algorithm combined with deep learning to bloom in all aspects.

The combination of deep learning and RL brings new vitality to RL, but inevitably brings some new challenges. One problem is, in RL, training samples are usually highly data-dependent state sequences. But for a machine learning model based on maximum likelihood, there is an important assumption, namely, the training samples are independent and from the same distribution. Once this assumption is not established, the effect of the model will be greatly weakened. The relevance of data in reinforcement learning may break the assumption of independent and identical distribution. Another problem is that the RL model is discarded after the calculations in the training process, so it needs a lot of time to interact with the environment to collect samples. To deal with these problems, Mnih et al. [16] proposed the data structure of Experience Replay Buffer (ER) to store sample information, and make random samples in this structure. In this way, DQN cuts the sequence correlation of data and makes training more stable. One of the promising methods used to improve the learning efficiency of agents is to prioritize experience replay. In the field of neurology, scientists have found the role of experience replay in the hippocampus [17]. Similarly, RL improve data utilization by replaying learning experiences during the agent learning process. Researchers often replay more reward-related sequences [17] or larger TD-error erroneous experiences [18] to forward the efficiency of agent learning. The Researcher in [10] takes a form of re-sampling and divides the experience into two parts, the positive and the negative, respectively. This method is suitable for the case where the positive and negative samples are uniform but in most cases our samples are non-uniform. Hinton et al. [19] proposed an error-based sampling format for deep learning and achieved higher accuracy. Schaul et al. [20] used this error-based prioritized sampling format on the basis of Deep RL, and adopted the data structure of SUM-Tree to make the experience replay process take as little time as possible. Adam et al. [21] demonstrated the applicability of the RL algorithm of experience replay in real-time control problems. In the RL algorithm based on strategy for continuous problems, the method of replay is also useful. Wawrzynski [22] shows that adding experience replay does not affect the convergence of the RL algorithm. Hou [23] also used the priority experience replay in the DDPG algorithm [7]to achieve better performance. The DeepMind team [24] applied the idea of distributed learning to prioritized experience replays.

There are many experiments show that experience replay is a very good method to improve efficiency of intelligent learning, but it also has some problems which can not be ignored. For example, what indicators are used to measure the importance of these samples? Some people think that setting a higher priority to a poorly performing sample can help the agent learn the lesson, but others think that a good sample is more worthy of priority learning. If the sample is prioritized according to a single indicator, it is inevitable to ignore many sample information that may be useful, what more, the determination of the priority will take a lot of time. Therefore, we propose to carry out the entire learning process in stages, in order to learn quickly, we use the priority strategy to extract the sample with the highest priority at beginning, after that we use the random strategy to extract the sample to effectively use the other samples. From the experimental results it is clearly that the phased strategy has better performance than using a single strategy in practical problems.

The remainder of this paper is organized as follows. Section 1 introduces the development of reinforcement learning algorithms and analyzes related algorithms simply. Section 2 describes the related concepts in reinforcement learning and our proposed phased experience prioritization algorithms. Section 3 demonstrates the effectiveness of our algorithm by comparing it with DQN and PrDQN, and analyzes the influence of some hyperparameters settings on the results. Finally, Sect. 4 concludes the paper with a discussion of future work.

2   Reinforcement Learning

RL is a useful framework in controlling strategy problems, it is generally based on the Markov decision process. The model of RL generally includes two parts in Fig. 1. Agent gets current status $s_t \in S$, and takes an action $a_t \in A$ in state s by consulting a state-action value function $Q(s; a)$, which is This is an indicator of the long-term return of the forecast action. After the action is executed, an evaluation is made on it, and the reward value $r_t \in R$ is given. The agent acquires a new state and reciprocates until the task is completed, so the environment can be represented as a state transition sequence $(s_t, a_t, r_t, s_t + 1)$ of the Markov Decision Process (MDP). The agent aims to learn an optimal strategy $\pi : S \rightarrow A$, which can maximize long-term reward efforts, so the strategy can be equated as: $\max \sum_{t=0}^{T} \gamma^t r_t$, where T is the termination time step, $\gamma \in [0, 1]$ is the discount factor.

Reinforcement Learning algorithms are mainly divided into two major types, oriented value and strategy-based. The value-based approach uses the value func-

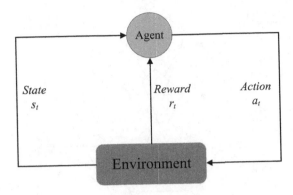

Fig. 1. The interaction between an agent and its environment in RL

tion to determine the optimal strategy. The value function is defined as:

$$Q^\pi(s,a) = E^\pi[\sum_{i=0}^{T} \gamma^t r_{t|}|s_0 = s, a_0 = a].$$ (1)

The Q value calculation under strategy as follows:

$$Q^*(s,a) = \max_\pi Q^\pi(s,a).$$ (2)

and the optimal strategy π^* can be easily derived by

$$Q^*(s) \in \arg\max_a Q^*(s,a).$$ (3)

2.1 The Neural Networks

Traditional RL algorithms, such as Q-learning, store each state in a table, namely, the Q-values that each action has in this table. However, the problems in reality are mostly complicated and there may be countless states. If you use a table to store it, you need a very large amount of memory, and it is a very time-consuming period to search for the corresponding state in such a large table. Fortunately, the neural network in machine learning can solve this problem very well. We will be using a network, which takes the state encoded in a one-hot vector, and produces a vector of all actions' Q-values. We will be using back propagation and a loss function to update the network parameter, we use the sum-of-squares loss as loss function, that is, the interpolation of the predicted Q and the actually calculated Q value, as specified shown in Eq. (4).

$$Loss = \sum (Q_{target} - Q_{value})^2.$$ (4)

The state vector becomes the input of the neural network, then we can obtain the Q-value of all actions and directly decide the action with the largest Q-value as the next step behavior according to the principle of Q-learning in Fig. 2.

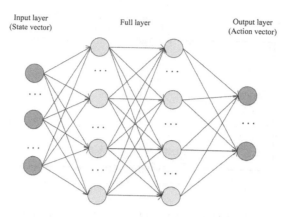

Fig. 2. Neural network structure.

2.2 DQN with Experience Replay

In the traditional reinforcement learning, the value of each action in Q-learning [25], that is, its value function $Q(S, A)$ is calculated in a table. The action-value function in the table is updated by the Eq. (5).

$$Q(s,a) \leftarrow Q(s,a) + \alpha(r + \gamma \max_{a'} Q(s',a') - Q(s,a)) \tag{5}$$

where s' is the next state after the agent selects action a in state s, r is the reward obtained by taking action a under state s, γ is the attenuation rate, and α is the learning rate. It is unrealistic to save all state information in a table when there are so many state. An effective solution is to approximate $Q(s,a)$ using a function approximation parametrized by θ, in this way, a very good algorithm is Deep Q-Network (DQN). It uses a deep neural network to simultaneously predict the Q of all possible actions in state s. The nonlinear fitting ability of deep neural networks has been shown to be better than linear approximation. The update of the parameter θ in the neural network is by minimizing the loss function, which is defined in Eq. (6).

$$L_i(\theta_i) = E_{s' \sim \varepsilon}(y_i - Q(s,a;\theta_i))^2 \tag{6}$$

Where $y_i = E_{s' \sim \varepsilon}(r + \gamma \max_{a'} Q(s',a';\theta_{i-1}))$. In order to make the model more stable, another Target-Network, which is exactly the same structure as the original Eval-Network is inserted. Parameters of the Target-Network writes down those results running a given interval ago. The model for calculating value through Target-net is fixed for a period of time, which is helpful to reduce its volatility. During each iteration, the agent selects the operation with the highest $Q(s,a)$ to maximize its expected future rewards, which often limits its ability to explore more or less. In practice, we use the ϵ-greedy policy to balance exploration and exploitation follows, that is, the agent has a certain probability of selecting its behavior, rather than habits with the largest Q-value. DQN completes as Algorithm 1:

Algorithm 1. Deep Q-learning with experience replay

1: Initialize the replay memory size N and the minibatch size
2: Initialize Q with random weights θ
3: Initialize target $\hat{Q}$ with weights $\theta^- = \theta$
4: **for** episode $= 1, M$ **do**
5:   Initialize state s
6:   **for** t=1,T **do**
7:    Random selection of action a_t with ε Probability
8:    otherwise select $a_t = \arg\max_a Q(s_t, a; \theta)$
9:    Execute action a_t in emulate and observe reward r_t
10:    set $s_{t+1} = s_t$
11:    Store transition (s_t, a_t, r_t, s_{t+1}) in D
12:    Random selection of minibatch size samples from memory D
13:    Set $y_j = r_j + \lambda\max_{a'} \hat{Q}(s_{j+1}, a'; \vartheta^-)$
14:    Update parameters $y_j = r_j + \lambda\max_{a'} \hat{Q}(\varphi_{j+1}, a'; \vartheta^-)$
15:    Every C steps reset $\hat{Q} = Q$
16:   **end for**
17: **end for**

2.3 Prioritized Experience Replay

The main function of experience replay is to eliminate the relevance of information in Memory, where the specific approach is to store each interaction-sample between agent and environment in the Experience replay buffer.

During learning, some samples (minibatch size) are randomly extracted from the memory for training. We break the correlation between training samples at random, but these related experiences replay may cause some problems, especially when the effective reward is very sparse and the learning rate will be particularly slow. In order to effectively solve this problem, the priority experience replay approach was proposed. The priority of prioritized experience replay is that the batch sampling is not randomly sampled, but is sampled according to the priority of the samples in the buffer Memory. The priority is determined by TD-error, which is the difference between the Q-values output of each action by the Target-net and Eval-net. If the TD-error is larger, there is still much room for improvement in the prediction accuracy, the more the sample needs to learn, the higher the priority P. The priority is calculated as follows in Eq. (7).

$$j \sim P(j) = p_j^\alpha / \sum_i P_i^\alpha \tag{7}$$

where P_i is the TD-error we calculated.

2.4 Phased Experience Replay

The prioritized experience replay has some issues that cannot be ignored, one problem is the priority method requires a lot of computational cost. In addition, the priority method will lead to the relatively good experience of TD-error

---

Algorithm 2. Prioritized experience replay

   Initialize replay memory capacity of N and the minibatch size.
2: **for** each step **do**
      Store (s_t, a_t, r_t, s_{t+1}) in D with maximal priority $P_t = \max\limits_{i<t} P_i$
4:    sample transition j $\sim$ P(j) $= p_j^\alpha / \sum_i P_i^\alpha$
      Computer TD-error $\delta_j = r_j + \lambda * \max_a Q(s_j, a) - Q(s_{j-1}, a_{j-1})$
6:    Update transition priority $P_j \leftarrow |\delta_j|$
   end for
8: Select the size of minibatch memory with the highest P value and put it into the neural network for learning.

---

is neglected. Therefore, we propose a Deep Q-learning with phased experience replay (MixDQN), where using the priority strategy in the early stage of learning, so that an agent can select the larger sample of TD-error and improve the learning efficiency. But selecting samples at random in the later stage of learning, in this way, the agent may run away from those local sub-optimal trap. The algorithm of DQN with phased experience replay (MixDQN) is described in Algorithm 3.

Algorithm 3. Deep Q-learning with phased experience replay (MixDQN)

   Initialize the replay memory capacity of N and the minibatch, total episode M, Stage of the conversion strategy t(such as 1/3,1/2 etc).
   Initialize Q function with random weights θ
3: Initialize function$\hat{Q}$ with weights $\theta^- = \theta$
   for episode $= 1, t*M$ **do**
      Learning process as Deep Q-learning with experience replay (Algorithm 1), except choose minibatch Sample Prioritized experience replay (Algorithm 2) from memory D.
6: **end for**
   for episode $= t*M, M$ **do** Deep Q-learning with experience replay (Algorithm1)
   end for

Priority P can be determined by calculating TD-error, but if each sample needs to be sorted for all samples, it will be very computationally intensive, so SumTree's data structure is used. Due to no longer needs to sort the samples, it is apt to reduce the complexity of computing.

3   Simulated Experiments

3.1   Test Environment

We are going to use the OpenAI gym to verify the validity of MixDQN, a collection of reinforcement learning environments. Some classic tasks (such as the Pendulum, MountainCar, Acrobot) will validate its effectiveness through these typical control problems.

Pendulum. The first well-known control problem we used was inverted pendulum, whose goal is to keep a pendulum upright for longer. The action here defines as its horizontal force of the pendulum. An agent writes its state in a three-dimensional vector, which includes its position, current angle of its bar θ and its speed v, namely, $S = \{sin(\theta), cos(\theta), v\}$, where $cos(\theta) \in [-1.0, 1.0]$, $sin(\theta) \in [-1.0, 1.0]$, $v \in [-8.0, 8.0]$. Its action space denotes as $A = \{-2.0, -1.6, ..., 1.6, 2.0\}$, where 11 action values illustrate. We transform angle from $-\pi$ to π with an initial velocity between -1 and 1. In the below experiment, a reward of 1000 is offered when the angle between the pole and the vertical line is less than a certain value. Otherwise, it will fail with a 0 bonus. The maximum step size 2000 is a default in each round. The average rewards of several algorithms in each round will verify three comparators. The mathematical model of an inverted pendulum task is shown in Fig. 3(a).

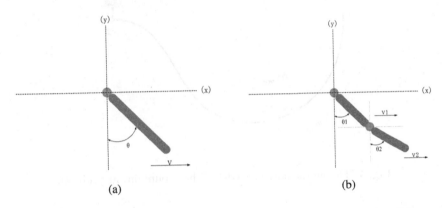

Fig. 3. The mathematical model of the Pendulum (a) and Acrobot (b)

MountainCar. The problem of MountainCar can be described as a car locating on a track between two mountains, the goal is to made the vehicle achieve the big tree on the right. However, the engine strength of the car is not enough to climb the peak in one pass, the only solution to succeed is to back and down slope boost to enhance power and arrival the top of the mountain (top = 0.5 position). An agent's state include position and velocity, where $position \in [-1.2, 0.6]$, $velocity \in [-0.07, 0.07]$. The action space denotes as $A = \{0, 1, 2\}$, which represents push left side, no push and push right side. It's position varies from -0.6 to -0.4 without initial velocity of each episode. In the experiment below, we set the reward -1 for each step, until the goal position of 0.5 is reached, and there is no penalty for climbing the left hill, which upon reached acts as a wall. The goal of this problem is to spend as few steps as possible to climb the top of right side mountain, so we compare the step size of a car climbed to the right side mountain in each episode. The model of MountainCar test is shown in Fig. 4.

Acrobot. The Acrobot system includes two poles and two joints. The two poles move through the joints. Initially, the pole hangs down and the goal is to swing the lower rod to a certain height. The status of this system includes the angle and speed of the two links, which expressed as $s = \{\sin(\theta_1), \cos(\theta_1), v_1, \sin(\theta_2), \cos(\theta_2), v_2\}$, The sine and cosine values of the two angles are between -1 and 1, $v1 \in [-12.75, 12.75]$, $v2 \in [-28.75, 28.75]$. The action is either applying $+1$, 0 or -1 torque on the joint between the two pendulum links. Same as MountainCar, we compare the step size of each algorithm to arrival the goal in each episode. The mathematical model of the Acrobot test is shown in Fig. 3(b).

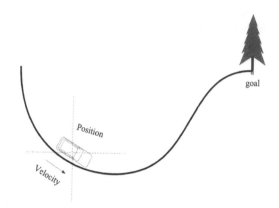

Fig. 4. The mathematical model of the MountainCar problem.

3.2   Experimental Settings

The input of those test problems are only a simple vector with low dimensions, therefore, we choose the simplest low-dimensional neural network, which use two layers of fully connected neural networks. We set up two neural networks with exactly the same structure, called Target-net and Eval-net, the Target-net will copy the Eval-net network's parameters after every 500 steps. The initial parameters of the two neural networks are set exactly the same, the learning rate α here is 0.005, and the attenuation γ is 0.9. The initial weight of the neural network is generated by a random uniform distribution, which is a random uniform distribution with a mean of 0 and a standard deviation of 0.3. The minibatch size is 32 and the capacity N of the memory D is set 2000. We set the greedy degree $\varepsilon = 0.9$ to made some possibility for an agent to choose other actions to improve exploration, that is, the 90% probability to select the largest Q value action, and the other 10% to randomly select the action. All hyperparameter settings are determined by experimental comparisons.

There is a table that stores memory size samples as a replay buffer, we use SumTree's structure to store the priority of each sample in the early stages, this stage is used to find the top minibatch samples as the input to the fully linked neural network. During the process of environmental interaction, the sample information in an agent memory is used as the observation. This approach guarantees the independence of the learning samples to a certain extent.

In the pendulum problems, we use the value of average reward as a criterion for evaluating the performance of three algorithms, and the average reward is obtained by dividing the total reward for each episode by the maximum step size. The goal of other two test questions is to reach the terminal point with as few steps as possible, so we can record the step size used for each episode.

3.3 Experimental Results

Results of Pendulum. The pendulum problem is our main experimental problem,under this problem, we compare the effects of each parameter on the results in detail, and determine the appropriate parameter values.

In Fig. 5, the bold lines are averages rewards over ten independent episode and the shaded area presents one standard deviation σ. The parameters of the three algorithms are exactly the same, as described in the Experimental Settings section. Our algorithm changes the strategy at the 1/2 stage, that is, the first half uses the priority experience replay, and the second half uses the stochastic strategy, this parameter is determined by the following experiment. From Fig. 5 and Table 1, we can see that the nature DQN, which does not use priority experience replay, performs worst. PrDQN, which use only priority experience replay, learns faster, but ultimately does not achieve better results. MixDQN reached the highest average reward and taken less time than PrDQN. The main reason is that if the neural network only samples poorly performing experience likes PrDQN, some samples especially the well-performing samples may never be collected to optimize the parameters. The algorithm MixDQN adopts the method of randomly selecting samples at a certain stage. After accumulating a certain negative sample, all the samples are given the same weight, which increased the likelihood that a good sample will be selected. In addition to increasing the average reward, the Mix DQN algorithm also reduces training time because the sample does not need to be sorted in importance in the second half. In Fig. 6(a), in order to determine when it is more appropriate to change the strategy, we compare the five sets of stages to change strategy and finally determine the conversion strategy when t = 1/2.

Table 1. Average steps, stand deviation and spending time to get the goal on Pendulum test with three algorithms.

Value	Mean	Std	Running time (s)
DQN	54.76	**26.69**	**2.2428e+04**
PrDQN	85.06	107.66	3.7640e+04
MixDQN	**127.11**	165.4	3.0383e+04

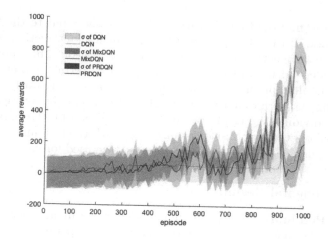

Fig. 5. The average rewards of DQN, PrDQN and MixDQN.

In order to understand the influence of other hyperparameters on the algorithm, we changed some parameter values. For example, the minibatch k (Fig. 6(b)) is set to 16, 32, 64. In figures, the lines are averages rewards over fifty independent episode. The size of the replay memory buffer ϕ (Fig. 6(c)) is set to 1000, 2000, 10000, and the Maximum step size H (Fig. 6(d)) is set to 1500, 2000, 2500. We can see that the setting of these parameters has a great influence of average reward and it is very difficult to find the best parameters, but we can find the corresponding parameter values through several experiments. From the stability of the average reward growth and the final score, we think that when minibatch size = 32, memory size = 2000, and the max steps = 2000 are relatively suitable choices.

Table 2. Average steps, stand deviation and spending time to get the goal on MountainCar test with three algorithms.

Value	Mean	Std	Running time (s)
DQN	1.0425e+03	5.4549e+03	**1.4349e+04**
PrDQN	3.8532e+02	1.4047e+03	2.5033e+04
MixDQN	**3.5835e+02**	**1.2661e+03**	2.2893e+04

Results of MountainCar. The purpose of the mountaincar problem is to reach the target as quickly as possible, so we no longer set the maximum step size, and the other parameter settings are consistent with Pendulum test, smaller the steps of each episode, better the algorithm works.

From the Fig. 7(a), ours algorithm MixDQN, which perform similar to PrDQN, but only half of us use priority to experience replay, reducing the time it takes to calculate each memeory priority. Therefore, our algorithm combines the advantages of both DQN and PrDQN algorithms, which can achieve faster convergence and relatively less time. From the Table 2, to compar the mean and standard deviation of the three algorithms' step in each episode, it's no doubt that the algorithm of MixDQN has best performance.

Results of Acrobot. The goal of the Acrobot is to made the lower pole achieve a certain position, when the pole is placed at the target position this episode ended. The reward for each step is -1 until a acrobot reach the goal, so the smaller the steps of each episode, the better the algorithm works.

Table 3. Average steps, stand deviation and spending time to get the goal on Acrobot test with three algorithms.

Value	Mean	Std	Running time (s)
DQN	1.0113e+03	1.1589e+03	**8.7572e+03**
PrDQN	**3.5390e+02**	4.8496e+02	1.2994e+04
MixDQN	3.6180e+02	**4.3992e+02**	1.0431e+05

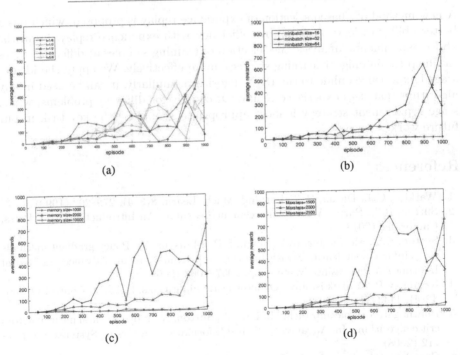

Fig. 6. (a) Average rewards of five time phase to change strategies. (b) Average reward of three mini-batch. (c) Average reward of three memory sizes. (d) Average reward of three max-steps.

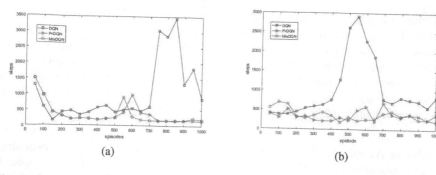

Fig. 7. Average number of steps when reach the goal for every 50 episodes on MountainCar (a) and Acrobot (b) test with DQN, PrDQN and MixDQN.

The Fig. 7(b) and the Table 3 show that the algorithm DQN which spent so many steps in every episode. The performance of other two algorithms basically the same, but the MixDQN effect is more stable, and only half of the time to prioritizes the memory, so it takes much less time than PrDQN.

4   Conclusion

A new method of phased selection of experience replay is proposed, which has a better ability to learn and improve efficiency with experience replay. From the above experimental analysis, the selection of training samples at different stages can help the intelligent learning strategy more effectively. We apply the idea of staged experience online to the DQN algorithm, similarly, it can be used in any algorithm that uses experience replay. Of course, for different problems, which stage replacement strategy is most appropriate should be the key task in our future work.

References

1. Watkins, C.J., Dayan, P.: Q-learning. Mach. Learn. **8**(3–4), 279–292 (1992)
2. Sutton, R.S., Barto, A.G.: Reinforcement Learning: An Introduction. MIT Press, Cambridge (2018)
3. Sutton, R.S., McAllester, D.A., Singh, S.P., Mansour, Y.: Policy gradient methods for reinforcement learning with function approximation. In: Advances in Neural Information Processing Systems, pp. 1057–1063 (2000)
4. Konda, V.R., Tsitsiklis, J.N.: Onactor-critic algorithms. SIAM J. Control Optim. **42**(4), 1143–1166 (2003)
5. Bhatnagar, S., Ghavamzadeh, M., Lee, M., Sutton, R.S.: Incremental natural actor-critic algorithms. In: Advances in Neural Information Processing Systems, pp. 105–112 (2008)
6. Grondman, I., Busoniu, L., Lopes, G.A., Babuska, R.: A survey of actor-critic reinforcement learning: standard and natural policy gradients. IEEE Trans. Syst. Man Cybern. Part C (Appl. Rev.) **42**(6), 1291–1307 (2012)

7. Lillicrap, T.P., et al.: Continuous control with deep reinforcement learning. arXiv preprint arXiv:1509.02971 (2015)
8. Silver, D., Lever, G., Heess, N., Degris, T., Wierstra, D., Riedmiller, M.: Deterministic policy gradient algorithms (2014)
9. Mnih, V., et al.: Playing atari with deep reinforcement learning. arXiv preprint arXiv:1312.5602 (2013)
10. Narasimhan, K., Kulkarni, T., Barzilay, R.: Language understanding for text-based games using deep reinforcement learning. arXiv preprint arXiv:1506.08941 (2015)
11. Zelinka, M.: Using reinforcement learning to learn how to play text-based games. arXiv preprint arXiv:1801.01999 (2018)
12. Van Hasselt, H., Guez, A., Silver, D.: Deep reinforcement learning with double q-learning. In: Thirtieth AAAI Conference on Artificial Intelligence (2016)
13. Wang, Z., Schaul, T., Hessel, M., Van Hasselt, H., Lanctot, M., De Freitas, N.: Dueling network architectures for deep reinforcement learning. arXiv preprint arXiv:1511.06581 (2015)
14. Bellemare, M.G., Dabney, W., Munos, R.: A distributional perspective on reinforcement learning. In: Proceedings of the 34th International Conference on Machine Learning, vol. 70. pp. 449–458. JMLR.org (2017)
15. Fortunato, M., et al.: Noisy networks for exploration. arXiv preprint arXiv:1706.10295 (2017)
16. Mnih, V., et al.: Human-level control through deep reinforcement learning. Nature **518**(7540), 529 (2015)
17. Atherton, L.A., Dupret, D., Mellor, J.R.: Memory trace replay: the shaping of memory consolidation by neuromodulation. Trends Neurosci. **38**(9), 560–570 (2015)
18. McNamara, C.G., Tejero-Cantero, Á., Trouche, S., Campo-Urriza, N., Dupret, D.: Dopaminergic neurons promote hippocampal reactivation and spatial memory persistence. Nat. Neurosci. **17**(12), 1658 (2014)
19. Hinton, G.E.: To recognize shapes, first learn to generate images. Progr. Brain Res. **165**, 535–547 (2007)
20. Schaul, T., Quan, J., Antonoglou, I., Silver, D.: Prioritized experience replay. arXiv preprint arXiv:1511.05952 (2015)
21. Adam, S., Busoniu, L., Babuska, R.: Experience replay for real-time reinforcement learning control. IEEE Trans. Syst. Man Cybern. Part C (Appl. Rev.) **42**(2), 201–212 (2011)
22. Wawrzyński, P.: Real-time reinforcement learning by sequential actor-critics and experience replay. Neural Netw. **22**(10), 1484–1497 (2009)
23. Hou, Y., Liu, L., Wei, Q., Xu, X., Chen, C.: A novel DDPG method with prioritized experience replay. In: 2017 IEEE International Conference on Systems, Man, and Cybernetics (SMC), pp. 316–321. IEEE (2017)
24. Horgan, D., et al.: Distributed prioritized experience replay. arXiv preprint arXiv:1803.00933 (2018)
25. Sauthoff, G., Mhl, M., Janssen, S., Giegerich, R.: Bellmans GAP a language and compiler for dynamic programming in sequence analysis. Bioinformatics **29**(5), 551–560 (2013)

Charge Prediction for Multi-defendant Cases with Multi-scale Attention

Sicheng Pan[1,2,3], Tun Lu[1,2,3,5(✉)], Ning Gu[1,2,3], Huajuan Zhang[4,5],
and Chunlin Xu[6]

[1] School of Computer Science, Fudan University, Shanghai, China
lutun@fudan.edu.cn
[2] Shanghai Key Laboratory of Data Science, Fudan University, Shanghai, China
[3] Shanghai Institute of Intelligent Electronics and Systems, Shanghai, China
[4] Division of Procuratorial Technology,
Guangdong Provincial People's Procuratorate, Guangzhou, Guangdong, China
[5] Guangdong Provincial Joint Laboratory of Natural Language Processing and
Machine Learning, Guangzhou, Guangdong, China
[6] TongFang SaiWeiXun Information Technology Co., Ltd., Chengdu, Sichuan, China

Abstract. The charge prediction task for multi-defendant cases is to determine appropriate charges for a specific defendant according to its name and its fact description. This task is not trivial since it is hard to recognize fact descriptions for different defendants. Therefore, we propose a multi-scale attention model for this problem. We employ local attention, which is highly related to the position of the specific defendant's name appear in the fact description, to restrict our model to the description for a specific defendant and employ global attention, which is calculated by a charge prediction model for single-defendant cases, to supplement the model with global information of the case. We collect about 160,000 indictments for experiments. After data preprocessing, we choose the two most common charge pairs which are Theft with Concealment of Crime-related Income, and Open Casinos with Gamble for experiments. Experimental results show the effectiveness of our model, the multi-scale attention model does benefit from the global information from the complete case compared to the local attention model.

Keywords: Legal intelligence · Charge prediction · Attention

1 Introduction

The charge prediction task is a classical legal intelligence task, aims to determine appropriate charges for each defendant according to its fact description. However, it is not trivial to determine charges for a specific defendant. Since there is only one textual fact description in a case and it contains all the fact descriptions for all the defendants in this case. And there are also researchers [4,10] pointed out that because of the complexity of cases involving multiple defendants and the difficulty of recognition of descriptions for different defendants, previous works

© Springer Nature Singapore Pte Ltd. 2019
Y. Sun et al. (Eds.): ChineseCSCW 2019, CCIS 1042, pp. 766–777, 2019.
https://doi.org/10.1007/978-981-15-1377-0_59

usually deal with cases involving only one defendant, few studies can handle multi-defendant cases.

The charge prediction task for single-defendant cases can be formulated as a classification problem, the classifier assigns a set of predefined charges to a case according to its fact description. Early works [6,7,11] usually concentrate on using mathematical and statistical methods to build systems for legal judgment prediction. With the development of machine learning techniques, researchers [1,8] propose novel methods for feature engineering and employ existing machine learning model to improve the performance on the charge prediction task. Recently, with the successful usage of deep learning techniques on natural language processing tasks, researchers [4,5,10,14,15] employ neural models to solve the charge prediction task. Such approaches not only improve the performance of the charge prediction model but also improve the interpretability of it.

Even though there are lots of studies on the charge prediction task, few works focus on cases involving multiple defendants. Cases involving multiple defendants are more complicated than those involving only one defendant since it is necessary to recognize descriptions for different defendants but the complex relationship between people involved in such cases makes the recognition even harder.

Most of the previous works study cases involving only one defendant, using the fact description of a case as its input. This is not suitable for cases involving multiple defendants since the fact description of a case involving multiple defendants describe all the defendants involved in it, and different defendants might be charged with different crimes, simply using the fact description as a prediction system's input will lead to data conflicts.

"During the period from June 19 to 25, 2018, the defendant Lan rode an unlicensed green electric bike to an industrial zone in Jinli Zhaoqing, stole about 200 kg aluminum products (valued at 6,207.76 CNY) from a workshop, and sold the stolen goods for 1,912 CNY".

The case above only involves one defendant and it will be classified to class "Theft" while the case below involves four defendants that are charged differently. The defendant Shen and the defendant Zhang Jia are charged with theft, the defendant Liang and the defendant Zhang Yi are charged with concealment of crime-related income. Therefore, we cannot simply use the fact description of a case as the input of a charge prediction model for multi-defendant cases.

"The defendant Shen and the defendant Zhang Jia stole the victim Zhu's electric bike in front of an internet cafe in a college town in Xinxiang, and sold the electric bike to the defendant Liang and the defendant Zhang Yi for 600 CNY, according to appraisal center in Jinghui, the electric bike valued 2,534 CNY".

It is necessary to recognize the description for a specific defendant when determining appropriate charges for it, an attention mechanism [2] should be included. We use a two-scale attention network for fact description embedding. The local attention restricts our model to the description for a specific defendant. Using a prior distribution describing the probability of a word that might describe the defendant, the local attention calculated by the model will mainly concentrate on the description for the defendant. The global attention supplements our model with global information of a case. Calculating by a pre-trained charge prediction model for single-defendant, which means treating the fact description of a case as the input, the global attention can capture all the informative words for the case instead of words might describe specific defendant. Based on this two-scale attention network, when predicting the charges of a specific defendant, our model can focus on the fact description of the defendant while considering the global information of the case.

We experiment with our model on indictments involving multiple defendants. Experimental results show that the multi-scale attention network can effectively recognize descriptions for different defendants with the local attention and determine appropriate charges for a specific defendant with the supplements from the global attention.

2 Related Work

Charge prediction task has been studied for several years. In the early years, studies focus on how to utilize mathematical methods to solve the problems. Kort [7] proposes a quantitive method to predict supreme court decisions. Nagel [11] applies correlation analysis to make predictions for reapportioning cases. And Keown [6] experiments with some mathematical models on charge prediction tasks. These attempts show the possibility of legal intelligence.

Since machine learning has shown its effectiveness in many areas, researchers formalize the charge prediction task as a text classification task and employ existing machine learning techniques to improve the performance. Liu and Hsieh [8] using phrases as features and their model performs better than other related work. Aletras et al. [1] use N-grams and topics to represent textual information, experimental results show that their models can predict the court's decisions with strong accuracy. However, limited by traditional machine learning techniques, the performance of the models is highly related to the methods for feature engineering.

Nowadays, with the successful usage of neural network methods on artificial intelligence tasks, researchers begin to employ neural network models for the charge prediction task. Luo et al. [10] using a hierarchical attention network to predict charges with fact description and the legal basis. Hu et al. [4] come up with several charge attributes to predict few-shot charges. Zhong et al. [15] propose a topological model to solve multiple legal tasks simultaneously.

Researchers also study on tasks related to the charge prediction task to improve the interpretability of the models. Ye and Jiang et al. [14] focus on court views generation tasks, they formulate this task as a text-to-text natural language generation problem and design models to improve the interpretability of charge prediction systems. Jiang and Ye et al. [5] employ deep reinforcement learning to extract rationales from the fact description and using rationales for prediction.

However, works mentioned above mainly handle cases involving only one defendant. Our work focus on cases involving multiple defendants. The key point of the charge prediction task for multi-defendant cases is to recognize the description for a specific defendant, which is similar to the object detection task in computer vision. Algorithms [3,9] that solve object detection problems contain the thought of multi-scale feature extraction, which we share similar spirits with. However, our multi-scale attention is for information supplement, while multi-scale in object detection tasks is mainly for finding the appropriate size of the object. Moreover, the relations between defendants involved in a case is more complicated than objects in an image.

3   Method

In the following part, we first give the definition of the charge prediction task for multi-defendant cases. Afterward, we give an overview of our model. Finally, we describe the details of our multi-scale attention network.

3.1   Charge Prediction for Multi-defendant Cases

The fact description of a case can be seen as a word sequence $X = \{x_1, x_2, ..., x_T\}$, where T represents the length of the sequence, $x_t \in V$, V is a dictionary with the index of each word. The name of the specific defendant is C, $C \in V$. The charge prediction task for multi-defendant cases aims to predict charges y for the defendant C according to the fact description X of a case, $y \in Y$, where Y is the set of all charges.

3.2   Multi-scale Attention

As depicted in Fig. 1, our approach contains the following steps: (1) The fact description is fed to the embedding layer to generate the fact embedding, then pass the fact embedding to the charge prediction model for single-defendant cases to calculate the global attention, meanwhile, pass it to the module for the local attention calculation. (2) The charge prediction model for single-defendant cases calculate the global attention distribution and pass it to the aggregation module. (3) The module for the local attention calculates the local attention distribution according to fact embedding and the defendant's name. (4) The aggregation module combining both the local and the global attention distributions, the model uses the two-scale attention distribution to extract features from the fact embedding, and pass the features to the softmax classifier to predict the most possible charge.

3.3   Global Attention

The global attention is calculated by charge prediction model for single-defendant cases. Inspired by Luo et al. [10] and Yang et al. [13], we use a global

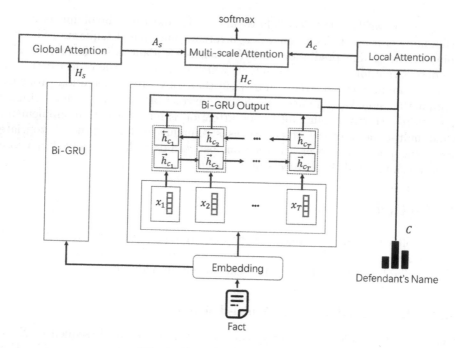

Fig. 1. Overview of our model.

context vector U_s to distinguish informative words from non-informative words. Assume that the output of the Bi-GRU layer is $H_s = \{h_{s_1}, h_{s_2}, ..., h_{s_T}\}$, the state of Bi-GRU at position t is $h_{s_t} = [\overrightarrow{h}_{s_t}, \overleftarrow{h}_{s_t}]$, where $\overrightarrow{h}_{s_t}$ and $\overleftarrow{h}_{s_t}$ are the states of the forward and backward GRU at position t. Given the output of the Bi-GRU state sequence H_s, the model calculates a sequence of global attention values $A_s = \{\alpha_{s_1}, \alpha_{s_2}, ..., \alpha_{s_T}\}$, where $\alpha_{s_t} \in [0, 1]$ and $\sum_t \alpha_{s_t} = 1$. The α_{s_t} is calculated by:

$$u_{s_t} = \tanh(W_{s_{att}} h_{s_t} + b_{s_{att}}) \tag{1}$$

$$a_{s_t} = U_s^\top u_{s_t} \tag{2}$$

$$\alpha_{s_t} = \frac{\exp(a_{s_t})}{\sum_i \exp(a_{s_i})} \tag{3}$$

where $W_{s_{att}}$ is a trainable weight matrix and $b_{s_{stt}}$ is a trainable bias.

3.4 Local Attention

The local attention restricts the model to the description for a specific defendant. Analysis of indictments shows that: (1) Sentences are grammatically well-formed, few pronouns appear in the fact description. (2) Almost all the sentences are simple sentences with subject-predicate structure, phrases that modify subjects or verbs are usually short. Based on these observations, we argue that words close to the name of the specific defendant are more likely to describe the defendant.

We employ the Gaussian distribution as the prior distribution which, hopefully, can help us to find words that describe the specific defendant. The expectation of the Gaussian distribution is the position of the name in the fact description, and the variance is σ. Since the name of the specific defendant may appear several times in fact description, the prior distribution we used in our model is the sum of all the Gaussian distribution whose expectations are the position of the defendant's name in the fact description and are different from each other. Assume that the prior distribution is $D_C = \{d_{C_1}, d_{C_2}, ..., d_{C_T}\}$, where $d_{C_t} = \sum_{\mu \in M_C} f_\mu(t)$, M_C is the set contains the positions that the name of the specific defendant appears in the fact description, and f_μ is the probability density of the gaussian distribution where μ is the expectation and σ is the variance. We use this prior distribution to activate the words close to the name. The output of the Bi-GRU layer is $H_c = \{h_{c_1}, h_{c_2}, ..., h_{c_T}\}$, and the output after activation is $H'_c = H_c \operatorname{diag}(D_C)$, where diag is the function converts a vector into a diagonal matrix.

Pass the output H'_c to the module for attention calculation, same as the method in the model for single-defendant cases. Since the prior distribution has increased the value of the word that might describe the defendant, when calculates the attention value using the context vector, words close to the name of the defendant will gain more attention. And we get the local attention $A_c = \{\alpha_{c_1}, \alpha_{c_2}, ..., \alpha_{c_T}\}$.

3.5   Aggregator

Assign the global attention with a trainable weight β to restrict its influence to the local attention. Add the weighted global attention to the local attention and then normalize the sum, we get the multi-scale attention value suitable for multi-defendant cases.

Assume that the multi-scale attention value is $A_m = \{\alpha_{m_1}, \alpha_{m_2}, ..., \alpha_{m_T}\}$, it is calculated by:

$$a_{m_t} = \alpha_{s_t} \times \beta + \alpha_{c_t} \tag{4}$$

$$\alpha_{m_t} = \frac{a_{m_t}}{\sum_i a_{m_i}} \tag{5}$$

Use the multi-scale attention value to encode the output of the Bi-GRU layer, we get the features g for classification, g is calculated by:

$$g = H'_c A_m^\top \tag{6}$$

4   Experiments

We experiment with our model on indictments and judgment documents. Experimental results show that the local attention can restrict the model to the description for the specific defendant, and global attention does supplement the model with global information from the entire case.

4.1   Data Preparation

We construct dataset from the published legal documents in Case Information Disclosure[1], an example indictment document is shown in Fig. 2. We can extract the fact description and charges from each paragraph. Paragraphs start with "The defendant" are the information of the defendants, which also contains the gender, the birthday, the identification number, and the charges of a defendant. The text between "The court hold that" and "Evidence" is the fact description, which contains the name of the defendants involved in the case. Moreover, we can use regex expressions to extract information. "The defendant(.+?)[,]" can be used to extract the name of the defendant and "(<=is charged with).+?(?=,)" can be used to extract charges. And we can simply recognize the fact description by finding the text between "The court hold that" and "Evidence".

We use the character "*" to replace the numbers of dates and times in the corpus. We also mask all the charges appear in the fact description, since although rare, charges' names may appear in the fact description and highly influence the model when training on it.

Fig. 2. An example indictment in our dataset.

We get 164,997 records after preprocessing of raw data, each record contains one charge for the defendant and the factual description of the case the defendant involved. Cases involving only one defendant count of 114,975 and cases involving multiple defendants count of 50,022. In cases involving multiple defendants, 45,128 records are from cases that all defendants are charged the same, while 4,894 records are from cases that defendants are charged differently. Considering the requirements of the dataset, we first choose cases involving charges counted more than 1,000 in the dataset. Then we find the two most common pairs of charges which usually appear simultaneously in a case, Theft with Concealment of Crime-related Income, and Open Casinos with Gamble, as shown in Fig. 3. We choose records who contains these charges to build the final dataset, and our experiments are based on the dataset constructed with charges in these two pairs.

---

[1] http://www.ajxxgk.jcy.gov.cn/html/index.html.

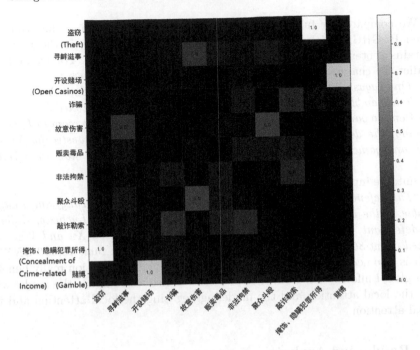

Fig. 3. Charge pairs.

Limited by the size of the dataset, after preprocessing, records from cases that involves defendants that are charged differently are less, we randomly choose 80 records for training, 20 for testing and 20 for validation.

We use the dataset from CAIL2018 [12] to train a model for single-defendant cases for the global attention calculation. The accuracy of the charge prediction task of the model is 0.92.

4.2 Experimental Setup

We use tokens that contains four continuous characters, such as "AAAA", to replace the defendant's name to help the Chinese word segmentation. Use jieba[2] to segment Chinese words. Word embeddings are trained using word2vec on legal documents. The resulting word embeddings contain 100,000 words, with 200 dimensions, length threshold of fact description of 500 is set up.

We set the variance of the probability density of the Gaussian distribution at 10. The optimizer we used during training is rmsprop and the batch size is 128, stop the training after there is no improvement of the accuracy during 10 epochs.

---

[2] https://github.com/fxsjy/jieba.

We compare our full model with the Text-CNN model and the attention-based Bi-GRU model. The input text for these two models is the text only contains sentences with the name of the specific defendant. For example, to predict the charges for the defendant Yuan in the following cases:

"On August 30, 2017, the defendant Wei and the defendant Xu organized people to gamble. The defendant Wei and the defendant Xu invited Yuan, Tang, and Peng to gamble in the way of Pai Gow in an entertainment room in Lucheng, Lujiang, the defendant Yu provided money for gamblers requested by Xu and Wei, and gained about 800 CNY, the total gambling money is more than 70,000 CNY."

Suitable input text for the two models is shown below:

"The defendant Wei and the defendant Xu invited Yuan, Tang, and Peng to gamble in the way of Pai Gow in an entertainment room in Lucheng, Lujiang, the defendant Yu provided money for gamblers requested by Xu and Wei, and gained about 800 CNY, the total gambling money is more than 70,000 CNY".

It is also necessary to verify if the model with global attention really benefits from global information from the total case, hence we compare the model with only the local attention and the model aggregating the local attention and the global attention.

4.3   Results and Analysis

We refer to the Text-CNN model as CNN, the attention-based RNN model as $Bi - GRU_{att}$, the local attention model as $Micro_{att}$ and the multi-scale attention model as $Aggregate_{att}$.

Table 1. Results of charge prediction for multi-defendant cases.

Model	Theft	Concealment of criminal-related income	Gamble	Open casinos
CNN	85	65	65	70
$Bi - GRU_{att}$	80	70	**80**	60
$Micro_{att}$	85	**90**	65	**75**
$Aggregate_{att}$	**90**	**90**	**80**	**75**

As shown in Table 1, the gap of the accuracy between charges in a pair are different from each other, where models with the local attention obtain a smaller gap than other models, which means that the local attention model indeed recognizes description for different defendants and performs better than the CNN model and the $Bi - GRU_{att}$ model. It is comprehensible since the preprocessing method for constructing the input text for the two compared model is not robust enough, it cannot handle the situation when all the sentences in the fact description containing all the defendants' names involved in a specific case.

"During the period from October 2017 to February 2018, the defendant Li Jia stole the victim Qiao's aluminum products, which Qiao placed on the billboard and placed in the warehouse in a manufacturing base in Dongsheng, six times and sold them to the defendant Ye Jia at a price of 24,000 CNY. The defendant Ye Jia still bought them at a lower price compared to the market after knowing that they were stolen by the defendant Li Jia".

The fact description above contains defendants who are charged with Theft and Concealment of Criminal-related Income. However, because of all the sentences in the fact description containing all the defendants' names involved in the case, the CNN model and the $Bi-GRU_{att}$ model may face to data conflicts since the input texts are the same while the labels are different, which results in the accuracy gap in the prediction task for charges pairs. When these two models perform well in predicting a charge, like Theft, their performances are poor in the other charge in the same pair, which means Concealment of Criminal-related Income.

The results also show that the multi-scale attention model performs better than the local attention model, which is the same as our expectation. To comprehend that, we find the text below:

"The defendant Wang opened a gaming room in November 2016 in Qingshan, Baotou, and employed the defendant Zhang Jia to manage it. On November 15, 2016, Zhang Yi came to the gaming room and applied for a job, Wang employed Zhang Yi to clean the gaming room. On the morning of November 21, 2016, Guo came to the gaming room and applied for a job, Wang employed Guo to do clean work, too. At 15 pm, November 21, 2016, the police found 16 gambling machines when inspection".

The fact description does not mention gambling at first, and the descriptions of the defendants do not relate to gambling either. Even though the model might infer that the defendants should be charged with Gamble because of the gaming room since most casinos are opened in or covered by gaming rooms, it is not for sure because the charge can also be Theft or Disturbing The Peace since such crimes are also usually happened in a gaming room. The local attention model which are restricted to the descriptions for defendants needs extra global information to make a prediction, and the multi-scale attention model overcomes such restriction. Figure 4 shows the local and global attention value and their weighted sum. As we can see, the local attention mainly focuses on the phrase "gaming room" while the global attention concentrates on the word "gambling", and aggregation of these two attention values can supplement each other with the information they have missed.

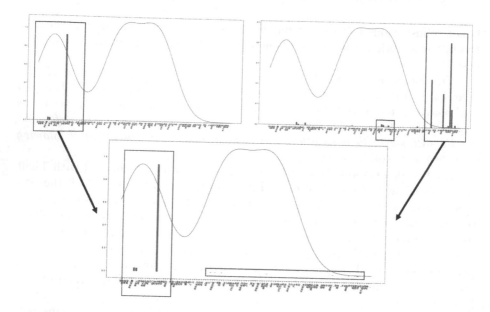

Fig. 4. Multi-scale attention.

5   Conclusion

In this paper, we propose a multi-scale attention model that can consider both the fact description of a specific defendant and global information of the complete case when doing the charge prediction task for multi-defendant cases. Experimental results show the effectiveness of our multi-scale model as the local attention does concentrate on description for a specific defendant and the global attention model does benefit from the global information from the complete case. However, there are also limitations to our work. Since the charge distribution is imbalanced, our dataset for the charge prediction task for multi-defendant cases is small, and due to the lack of data, our model still cannot distinguish confusing charges. We will leave these challenges for future work.

Acknowledgement. This work was supported by the National Key Research and Development Program of China under Grant No. 2018YFC0381402 and the project of Guangdong Provincial Joint Laboratory of Natural Language Processing and Machine Learning.

References

1. Aletras, N., Tsarapatsanis, D., Preoţiuc-Pietro, D., Lampos, V.: Predicting judicial decisions of the european court of human rights: a natural language processing perspective. PeerJ Comput. Sci. **2**, e93 (2016)
2. Bahdanau, D., Cho, K., Bengio, Y.: Neural machine translation by jointly learning to align and translate. arXiv preprint arXiv:1409.0473 (2014)

3. Girshick, R., Donahue, J., Darrell, T., Malik, J.: Rich feature hierarchies for accurate object detection and semantic segmentation. In: Proceedings of the IEEE Conference on Computer Vision and Pattern Recognition, pp. 580–587 (2014)
4. Hu, Z., Li, X., Tu, C., Liu, Z., Sun, M.: Few-shot charge prediction with discriminative legal attributes. In: Proceedings of the 27th International Conference on Computational Linguistics, pp. 487–498 (2018)
5. Jiang, X., Ye, H., Luo, Z., Chao, W., Ma, W.: Interpretable rationale augmented charge prediction system. In: Proceedings of the 27th International Conference on Computational Linguistics: System Demonstrations, pp. 146–151 (2018)
6. Keown, R.: Mathematical models for legal prediction. Computer/lj **2**, 829 (1980)
7. Kort, F.: Predicting supreme court decisions mathematically: a quantitative analysis of the "right to counsel" cases. Am. Polit. Sci. Rev. **51**(1), 1–12 (1957)
8. Liu, C.-L., Hsieh, C.-D.: Exploring phrase-based classification of judicial documents for criminal charges in Chinese. In: Esposito, F., Raś, Z.W., Malerba, D., Semeraro, G. (eds.) ISMIS 2006. LNCS (LNAI), vol. 4203, pp. 681–690. Springer, Heidelberg (2006). https://doi.org/10.1007/11875604_75
9. Liu, W., et al.: SSD: single shot multibox detector. In: Leibe, B., Matas, J., Sebe, N., Welling, M. (eds.) ECCV 2016. LNCS, vol. 9905, pp. 21–37. Springer, Cham (2016). https://doi.org/10.1007/978-3-319-46448-0_2
10. Luo, B., Feng, Y., Xu, J., Zhang, X., Zhao, D.: Learning to predict charges for criminal cases with legal basis. arXiv preprint arXiv:1707.09168 (2017)
11. Nagel, S.S.: Applying correlation analysis to case prediction. Tex. L. Rev. **42**, 1006 (1963)
12. Xiao, C., et al.: Cail 2018: a large-scale legal dataset for judgment prediction. arXiv preprint arXiv:1807.02478 (2018)
13. Yang, Z., Yang, D., Dyer, C., He, X., Smola, A., Hovy, E.: Hierarchical attention networks for document classification. In: Proceedings of the 2016 Conference of the North American Chapter of the Association for Computational Linguistics: Human Language Technologies, pp. 1480–1489 (2016)
14. Ye, H., Jiang, X., Luo, Z., Chao, W.: Interpretable charge predictions for criminal cases: learning to generate court views from fact descriptions. arXiv preprint arXiv:1802.08504 (2018)
15. Zhong, H., Zhipeng, G., Tu, C., Xiao, C., Liu, Z., Sun, M.: Legal judgment prediction via topological learning. In: Proceedings of the 2018 Conference on Empirical Methods in Natural Language Processing, pp. 3540–3549 (2018)

Classification of Subliminal Affective Priming Effect Based on AE and SVM

Yongqiang Yin[1,2], Bin Hu[1,2(✉)], Tiantian Li[3], and Xiangwei Zheng[1,2]

[1] School of Information Science and Engineering, Shandong Normal University,
Jinan 250014, China
binhu@sdnu.edu.cn
[2] Shandong Provincial Key Laboratory for Distributed Computer Software
Novel Technology, Jinan 250014, China
[3] Faculty of Education, Shandong Normal University, Jinan 250014, China

Abstract. The study of the Subliminal Affective Priming Effect (SAPE) mainly uses event-related potential technology and mapping method. Many researches are only for the study of emotional classification, but there are few researches on the classification of the SAPE. That is, the SAPE is directly judged by the psychologist in most experiment. So, this paper designs a classifier based on Automatic Encoder (AE) and Support Vector Machine (SVM) for automatic recognition of SPAE. Initially, this paper collects EEG signal, and then extracts statistical features from EEG signal to form a data set. After that, the data set is dimension reduction by AE and then divided into training set and test set randomly. At last, the already designed model is trained with the training set and validated with the test set. In the experiment, we find that the designed classifier has the best performance compared with the classifiers based on BP neural network, Principal Component Analysis (PCA) and SVM. The experimental results show that the average classification accuracy is 95.31%. The classification results further indicate that the SAPE's judgment is hopeful to reduce the labor with the machine.

Keywords: Subliminal Affective Priming Effect · BP neural network · Automatic encoder · Support vector machine

1 Introduction

Emotional recognition is widely used in many fields, such as rehabilitation therapy, driver status assessment and so on [1]. Meanwhile, current emotional recognition researches focus on speech-based emotional recognition, emotional recognition based on facial expressions, and emotional recognition based on human physiological parameters.

At present, many researches are only for the study of emotional classification, but there are few researches on the classification of the Subliminal Affective Priming Effect (SAPE). Because the presentation time of the prime stimulus is very ephemeral, the subjects are not conscious of it who participate in the experiment; thus, the emotional effect is called SAPE [2]. In other words, the SAPE is unconscious, and objectively

© Springer Nature Singapore Pte Ltd. 2019
Y. Sun et al. (Eds.): ChineseCSCW 2019, CCIS 1042, pp. 778–788, 2019.
https://doi.org/10.1007/978-981-15-1377-0_60

reflects the corresponding physiological response that people produce after being stimulated by psychological emotions.

But the study of the SAPE mainly uses event-related potential technique, repeated measurement ANOVA analysis [3] and mapping method. Namely, they judge whether there is SAPE by artificial method. Manual judgment of SAPE requires a lot of resources, such as taking up manpower, material resources and time, etc. So, this paper wants to reduce manual work with machines. It is a brand-new field to study the classification of SAPE, and there are few related studies in domestic and overseas at present.

In the classification of the SAPE, BP neural network, AE and SVM are applied to the design of the SAPE classifier. And we design a classifier based on AE and SVM. Meanwhile, the classifier is compared with the other two classifiers respectively.

2 Related Work

In the SAPE studies, Murphy and Zajonc [23] first studied the SAPE phenomenon, they found that when the stimulating stimulus is a positive facial expression picture, the subject is more inclined to judge the target stimuli as "good". On the contrary, the subject judged the target stimulus as "bad".

Dannlowski et al. [21] found that the affective priming paradigm provided evidence for differential group effects regarding subliminal affective priming info processing. Jiang et al. [24] found that when the priming stimulus was a picture of crying emotions, the subject judged the target stimulus as a crying expression, but when the priming stimulus was a laughing emotional picture, no priming effect occurred. Li et al. [2] found that the negative priming group had a priming effect, but the positive priming group had no priming effect.

In classification studies, BP neural network and SVM are suitable for classification problems. They have been widely used in medicine, biology, climate, image processing and other fields.

BP neural network and SVM were widely used in medical research. Modai et al. [19] pointed that the Computerized Suicide Risk Scale based on BP neural networks was capable in the discovery of records of patients who had medically serious suicide attempts. Jin et al. [6] used BP neural network to classify the EEG signal of motor imagination, and their recognition rate was more than 87.14%. Wang et al. [7] used BP neural network to classify the EEG signal of motor imagination, and their classification accuracy reached 86%. Yang et al. [8] used F-score feature selection and SVM to identify P300, and their recognition result could reach 100%.

BP neural network and SVM were also widely used in biology. Zhang et al. [9] pointed out that the BP neural network predicting model had a better fit with actual value of the density of Oscillatoria. Guo et al. [10] used SVM to predict biological protein type, and their results showed that SVM had obvious superiority to membrane protein type. Prediction of protein sequences using SVM [20] was "easy to crystallize" or "resistance to crystallization", and the classification accuracy reached 89.53%.

All these studies show that BP neural network and SVM are more reliable for predicting results. Therefore, this paper introduces the idea of BP neural network and SVM into the task of the SAPE recognition.

3   Method

The main purpose of this paper is to judge whether the SAPE exists and how to identify it, that is, it can be regarded as a binary classification problem. In order to improve the accuracy of judging whether the priming effect exists, we superimpose it after pre-processing raw EEG data, and then extract seven statistical features from the super-imposed EEG data. At the same time, a classifier based on AE and SVM is designed in this paper. And flow chart of our proposed method is shown in Fig. 1.

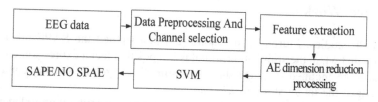

Fig. 1. Flow chart of the proposed method

3.1   Feature Extraction

The classification of this paper is mainly based on the distinction between LPC wave and N400 wave, so we have to extract the potential features of data [4]. The following six statistics and amplitude energy are considered in each bio-potential signal:

Standard deviation of Event-related Potentials (ERP) data is defined as in Eq. (1).

$$\sigma_S = \sqrt{\frac{1}{T}\sum_{t=1}^{T}(S(t) - \mu_S)^2} \tag{1}$$

The means of the absolute values of the first difference is defined as in Eq. (2).

$$\delta_S = \frac{1}{T-1}\sum_{t=1}^{T-1}|S(t+1) - S(t)| \tag{2}$$

The normalized process of the means of the absolute values of the first difference is defined as in Eq. (3).

$$\overline{\delta}_S = \frac{1}{T-1}\sum_{t=1}^{T-1}|\overline{S}(t+1) - \overline{S}(t)| = \frac{\delta_S}{\sigma_S} \tag{3}$$

The means of the absolute values of the second differences is defined as in Eq. (4).

$$\gamma_S = \frac{1}{T-2} \sum_{t=1}^{T-2} |S(t+2) - S(t)| \tag{4}$$

The normalized process of the means of the absolute values is defined as in Eq. (5).

$$\overline{\gamma}_S = \frac{1}{T-2} \sum_{t=1}^{T-2} |\overline{S}(t+2) - \overline{S}(t)| = \frac{\gamma_S}{\sigma_S} \tag{5}$$

The signal energy is defined as in Eq. (6).

$$P = \sum_{t=1}^{T} S^2(t) \tag{6}$$

Where t is the point in time, T is the length of time, and S is the signal corresponding to each time point. And the μ_S is the average level of EEG data. By using these characteristic values, the statistical feature vector FVs is defined as follows:

$$FVs = (\mu_S, \sigma_S, \delta_S, \overline{\delta}_S, \gamma_S, \overline{\gamma}_S, P) \tag{7}$$

3.2 Automatic Encoder Based on BP Neural Network

Inspired by the principal component analysis (PCA) algorithm [18], the automatic encoder (AE) is applied to the dimension reduction of data. AE is an unsupervised prediction algorithm based on BP neural network [11, 12]. The encoding part can compress multiple attributes of a sample into fewer attributes.

From the input layer to the hidden layer is the encoding process, from the hidden layer to the output layer is the decoding process. $f(\bullet)$ and $g(\bullet)$ are the encoding and decoding functions respectively, then [17]:

$$h = f(x) := s_f(wx + b_1) \tag{8}$$

$$y = g(h) := s_g(\dot{w}h + b_2) \tag{9}$$

Where s_f is encoder activation function, s_g is decoder activation function, b_1 and b_2 are bias. The parameters of the AE are $\theta = \{w, b_1, \dot{w}, b_2\}$.

The parameters of AE are trained to make Y and X as close as possible, and the degree of proximity between Y and X is characterized by the reconstruction error function $L(X, Y)$.

$$L(X, Y) = \|X - Y\|^2 \tag{10}$$

The loss function of AE can be minimized by gradient descent algorithm [13], and the parameter θ of AE neural network can be solved.

3.3 Support Vector Machine

The goal of the SVM based on classification boundary is to find the boundary between these classifications through training. For multidimensional data, the linear classifier uses the boundary of the type of hyperplane, and the nonlinear classifier uses the boundary of the type of hypersurface [14]. Because SVM can solve the practical problems of small samples, nonlinearity, high dimensionality, local minimum, and has strong generalization ability [22], we use the SVM to judge the existence of SAPE.

In our study, we choose radial basis function, that is:

$$K(x, x_i) = e^{-\frac{\|x-x_i\|^2}{2g^2}} \tag{11}$$

For Gaussian kernel support vector machines, the parameters c and g directly affect the classification performance of support vector machines, which must be carefully selected. The simple and practical method is to determine the optimal c and g by means of grid-search and cross validation.

4   Experiments and Result Analysis

4.1   The Background of Experimental Data Set

17 students from two universities in Tianjin were selected in the experiment as paid participants. There were 8 males and 9 females in the students. And their age were 20 to 26 years. Their average age was 22.6 years. Each of participants completed a positive priming group and a negative priming group, respectively, and balanced the completion sequence. All participants were right-sided, with normal or corrected visual acuity, no history of brain trauma and physical and mental health problems, and did not participate in similar psychological experiments.

Firstly, ten positive emotional expression face images with low-arousing and ten negative emotional expression face images with low-arousing were selected as the prime stimulus. In these images, there were half male and half female. Secondly, 80 negative emotional expression face pictures with medium-arousing were selected as the probe stimulus. And in these images, there were half man and half woman. In order to reduce effect of the kinds of features of the probe face with the initial emotional priming effect, all probe faces images were converted into mosaic images. Mask stimulation were neutral facial images which were converted into a strong mosaic image. And these pictures were selected from the Chinese Emotional Materials Library.

The experimental procedure was compiled by E-Prime software. In each trial, a black cross gaze point (1000 ms) was first presented on a white screen, then a stimulus (low or high arousal, 12 ms) was presented, followed by a masking stimulus (200 ms) and followed by a blank screen (300 ms), the final target stimulus (middle arousal), the target stimulus disappear after the response, and the next test was taken after the interval of 1200 ms. 20 photos of each of them were presented 8 times each. Each of the 80 target stimulation images was presented twice, each under high arousal and low arousal conditions, with a total of 160 trials. The order of stimulation of high and low arousal was random.

We collected EEG using a 64-channel Neuro Scan EEG acquisition device. And the reference electrode we selected was the left mastoid. The upper limit of all electrode impedances was less than 5 kΩ.We set the sampling rate and filter band of the device to 1000 Hz, 0.05–100 Hz respectively. We recorded vertical eye electrooculogram (VEOG) and horizontal eye electrooculogram (VEOG) to remove eye electrooculogram for data. We put the one VEOG electrode at up left eye. We put the other at down the left eye. At the same time, we put the HEOG electrodes at one centimeter from the outer canthus of right eye and left eye [2].

4.2   Data Preprocessing and Channel Selection

Data Preprocessing. In the negative stimulation group, there are 2 subjects who could clearly see the expression of the stimulating picture, and 2 subjects' EEG data has too many artifacts. So, the data of the 4 subjects is eliminated, and there are 13 participants who are involved in the subsequent analysis. In the positive stimulation group, there are 5 subjects who could clearly see the expression of the stimulating picture, and 5 subjects' EEG data has too many artifacts. So, the data of the 10 subjects is eliminated, and 7 participants are involved in the subsequent analysis.

Channel Selection. According to the study of Li [2], we found that the negative priming group had a priming effect, but the positive priming group had no priming effect. We also observed that the waveforms produced in 15 channels (*F3, Fz, F4, FC3, FCz, FC4, C3, Cz, C4, CP3, CPz, CP4, P3, Pz, P4*) were obvious. Therefore, the same channels are also selected in this paper.

4.3   The Construction of Classifier and Training

Classifier Training Based on BP Neural Network. In this paper, the parameters of each layer of BP neural network constructed by MATLAB [5] are shown in Table 1. In the training process, the adaptive learning rate gradient descending backpropagation algorithm is used to greatly improve the learning speed.

Table 1. Parameter table of BP neural network construction

Network layer number	5
Activation function	Sigmoid
Number of neurons in each layer	105, 52, 26, 14, 2
Training algorithm	Traingdx
Target error	0.001
Learning efficiency	0.01
Maximum training epoch	30000

The average accuracy of the classifier is calculated by multiple random sampling, and the stability of the classification is observed.

Classifier Based on PCA and SVM. Based on the principal component analysis algorithm, the acceptable threshold is set to 99%, and then the six statistical features and one signal energy are extracted from 15 channels respectively. The number of total features is 105(7 × 15). Based on the experimental results, the cumulative contribution rate of variance is shown in Fig. 2. And Fig. 2. shows that the dataset becomes a new data set with 17 new features.

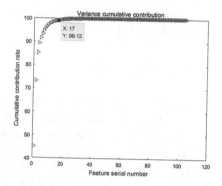

Fig. 2. Differential contribution rate of PCA algorithm

After that, g and c are calculated by cross-validation in SVM [15] and grid search algorithm [16] for the processed data. The details are shown in Fig. 3.

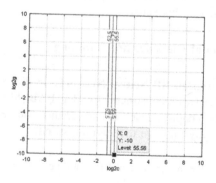

Fig. 3. The results of penalty factor g and cost c

Classifier Based on Combination of AE and SVM. The specific parameters of the AE are designed as shown in Table 2.

Table 2. AE construction parameter list

Network layer number	7
Activation function	Tansig
Training algorithm	Traingdx
Number of neurons in each layer	105, 64, 32, 17, 32, 64, 105
Target error	0.008
Learning efficiency	0.001
Maximum training epoch	500000

The error curve of the AE training is shown in Fig. 4. In Fig. 4, we see that the error between input and output converges to the target error after 87697 training epochs. The aim of this method is to minimize the loss of data information by minimizing the total error of about 0.32.

Fig. 4. Automatic encoder training error curve

And then, we set the grid search range for c and g in the kernel function to [0, 1024]. Figure 5 is a contour map of the accuracy of the grid search algorithm and cross-validation of the parameters of the SVM classifier, and the accuracy is obtained from the SVM cross-validation. From Fig. 5, we can get the maximum cross-validation accuracy at the point (0, 3). Because the data on both the x-axis and y-axis is logarithm based on 2, the g and c are 1 and 8 respectively.

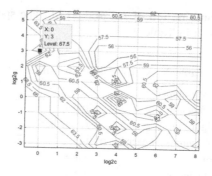

Fig. 5. The results of penalty factor g and cost c

Finally, the processed data is classified by SVM. Because the data set is relatively small, the average accuracy and the stationarity of the classifier are calculated by means of multi-random sampling in this paper.

4.4 Experimental Results

The total number of samples is 40. In this paper, 32 samples are randomly selected as the training set, and 32 repeated sampling are conducted. The final test results are as follows: The average recognition accuracies of the classifiers based on BP neural network, PCA and SVM, AE and SVM are 71.25%, 65.93%, 95.31% respectively. And their variances are 0.0118, 0.0037, 0.0049 respectively. The concrete results are shown in Table 3.

Table 3. Performance comparison of three classifiers

Classifier type	Average accuracy (%)	Variance
BP neural network	71.25	0.0118
PCA+SVM	65.93	0.0037
AE+SVM	95.31	0.0049

In Table 3, it is not difficult to see that the classification accuracy of PCA and SVM is more stable than the classifiers based on based on BP neural network, AE and SVM. And its variance is less than 0.05. But the average classification accuracy of PCA and SVM is lowest in these classifiers. However, we designed the classifier is relatively stable, and its accuracy is highest in these three classifiers.

5 Conclusion

Because there are few researches on the classification of SAPE, the main work of this paper uses machine learning method to help the psychologist achieve automatic identification of SAPE. Firstly, we collect EEG signal data, remove noises of EEG signal and then extract statistical features to form a data set. Secondly, the data set is dimension reduction by AE and then divided into training set and test set randomly. Thirdly, our model is trained with the training set. Finally, use the test set to validate the model. The experimental results indicate that the average classification accuracy of the classifier designed is 95.31%, and higher than the classifiers based on BP neural network, PCA and SVM. Namely, the classifier performs well on experimental data, indicating that it is hopeful to reduce manual detection with the machine.

The classifier in this paper has a good average classification effect, but it takes longer time to find the optimal parameters (penalty factor g and cost c). Therefore, we need to further optimize the algorithm.

References

1. Nie, D., Wang, X.W., Duan, R.N., Lv, B.L.: A survey on EEG based on emotion recognition. J. Chin. J. Biomed. Eng. **31**(4), 595–606 (2012). https://doi.org/10.3969/j.issn.0258-8021.2012.04.018
2. Li, T.T., Lu, Y.: The subliminal affective priming effects of faces displaying various level of arousal: an ERP study. Neuroscience Lett. **583**, 148–153 (2014). https://doi.org/10.1016/j.neulet.2014.09.027
3. Yi, J.Y., Zhong, M.T., Luo, Y.Z., Ling, Y., Yao, S.Q.: A study of subliminal affective priming with affective pictures. Chin. J. Clin. Psychol. **15**(3), 304 (2007). https://doi.org/10.16128/j.cnki.1005-3611.2007.03.030
4. Takahashi, K.: Remarks on SVM-based emotion recognition from multi-modal bio-potential signals. In: 13th IEEE International Workshop on Robot and Human Interactive Communication (IEEE Catalog No.04TH8759) (2004). https://doi.org/10.1109/roman.2004.1374736
5. Liu, L., Chen, J., Xu, L.: Realization and application research of BP neural network based on MATLAB. J. Int. Seminar Future BioMed. Inform. Eng. **2008**, 130–133 (2008). https://doi.org/10.1109/fbie.2008.92
6. Jin, H.L., Zhang, Z.H.: Research of movement imagery EEG based on hilbert-huang transform and BP neural network. Chin. J. Biomed. Eng. **2013**(2), 249–253 (2013). https://doi.org/10.7507/1001-5515.20130047
7. Wang, J.: Analysis of motor imagery EEG based on the BP network. Tianjin Normal University (2012)
8. Yang, L.C., Li, J.L., Yao, Y.B., Wu, X.Q.: A P300 detection algorithm based on f-score feature selection and support vector machines. J. Biomed. Eng. **2008**(1), 23–26 (2008). https://doi.org/10.1142/S0217595908001626
9. Zhang, K.X., Lu, K.H., Zhu, J.Y., Liu, X.S., Xie, L.F.: Predicting model of algal blooms based on BP neural network. Environ. Monit. China **28**(3), 53–57 (2012). https://doi.org/10.19316/j.issn.1002-6002.2012.03.012
10. Guo, Z.M., Zhang, Z.Z., Pan, Y.X., Hua, Z.D., Feng, G.Y., He L.: Prediction of membrane protein types by using support vector machine. J. Shanghai Jiao tong Univ. **38**(5) (2004). https://doi.org/10.1016/s0960-0779(03)00420-x
11. Hagan, M.T., Beale, M., Beale, M.: Neural Network Design, pp. 8–10, PWS Publishing Co., Boston (1996). ISBN:0-534-94332-2
12. Cotter, N.E.: The stone-weierstrass theorem and its application to neural networks. IEEE Trans. Neural Netw. **1**(4), 290–295 (1990). https://doi.org/10.1109/72.80265

13. Ran, W.L.: Soft measring technique researoh appliedin wastewater BOD based on neural computing. Beijing University of Technology (2004)
14. Wei, A.: Design of a classifier based on the SVM algorithm. Electron. Sci. Technol. **28**(4), 23 (2015). https://doi.org/10.16180/j.cnki.issn1007-7820.2015.04.007
15. Chang, C.-C., Lin, C.-J.: LIBSVM: a library for support vector machines (2011).https://doi. org/10.1145/1961189.1961199. http://www.csie.ntu.edu.tw/~cjlin/libsvm
16. Li, K., Liu, P., Li, Y.J., Zhang, G.P., Huang, Y.H.: The parallel algorithms for LIBSVM parameter optimization based on Spark. J. Nanjing Univ. (Nat. Sci.) **52**(2) (2016). https://doi. org/10.13232/j.cnki.jnju.2016.02.016
17. Deng, J.F., Zhang, X.L.: Deep learning algorithm optimization based on combination of auto-encoders. J. Comput. Appl. **36**(3), 697–702 (2016). https://doi.org/10.1172/j.issn.1001-9081.2016.03.697
18. Guan, J.Q., Yang, B.H., Ma, S.W., Yuan, L.: Classification of motor imagery EEG based on PCA and SVM. Beijing Biomed. Eng. **29**(3), 261–265 (2010). https://doi.org/10.3969/j.jssn.1002-3208.2010.03.09
19. Modai, I., Ritsner, M., Kurs, R., Mendel, S., Ponizovsky, A.: Validation of the computerized suicide risk scale–a backpropagation neural network instrument (CSRS-BP). Eur. Psychiatry **17**(2), 75–81 (2002). https://doi.org/10.1016/s0924-9338(02)00631-4
20. Kandaswamy, K., Pugalenthi, G., Suganthan, P., Gangal, R.: SVMCRYS: an SVM approach for the prediction of protein crystallization propensity from protein sequence. Protein Peptide Lett. **17**(4), 423–430 (2010). https://doi.org/10.2174/092986610790963726
21. Dannlowski, U., Kersting, A., Lalee-Mentzel, J., Donges, U.S., Arolt, V., Suslow, T.: Subliminal affective priming in clinical depression and comorbid anxiety: a longitudinal investigation. Psychiatry Res. **143**(1), 63–75 (2006). https://doi.org/10.1016/j.psychres.2005.08.022
22. Zhang, J.D., Qin, G.H., Cui, Y.: An EEG signal system based on FastICA and SVM. J. Comput. Res. Dev. **45**(z1), 255–258 (2008)
23. Murphy, S.T., Zajonc, R.B.: Affect, cognition, and awareness: affective priming with optimal and suboptimal stimulus exposures. J Pers. Soc. Psychol. **64**(5), 723 (1993). https://doi.org/10.1037//0022-3514.64.5.723
24. Jiang, Z.Q., Yang, L.Z., Liu, Y.: Study on adults' subliminal affective priming effect. J. Liaoning Normal Univ. (Soc. Sci. Edn.) **28**(4), 48–50 (2005). https://doi.org/10.3969/j.issn.1000-1751.2005.04.015

Retraction Note to: A New Information Exposure Situation Awareness Model Based on Cubic Exponential Smoothing and Its Prediction Method

Weijin Jiang, Yirong Jiang, Jiahui Chen, Yang Wang, and Yuhui Xu

Retraction Note to:
Chapter "A New Information Exposure Situation Awareness
Model Based on Cubic Exponential Smoothing
and Its Prediction Method" in: Y. Sun et al. (Eds.):
Computer Supported Cooperative Work and Social Computing,
CCIS 1042, https://doi.org/10.1007/978-981-15-1377-0_17

The authors have retracted this article [1] because of overlap with doctoral dissertation [2]. Figures 5, 6 and 7 were taken from the dissertation without permission or attribution. Part 2.3 of the article "Evaluation of information Outburst prediction model" quotes the data from this doctoral dissertation and makes some erroneous changes. All authors agree to this retraction.

[1] Jiang, W., Jiang, Y., Chen, J., Wang, Y., Xu, Y.: A new information exposure situation awareness model based on cubic exponential smoothing and its prediction method. In: Sun, Y., Lu, T., Yu, Z., Fan, H., Gao L. (eds.) Computer Supported Cooperative Work and Social Computing. ChineseCSCW 2019. Communications in Computer and Information Science, vol. 1042. Springer, Singapore (2019). https://doi.org/10.1007/978-981-15-1377-0_17
[2] Yi, C.: Research on Mechanisms of Information Propagation And Control Strategies in Social Networks. Dissertation at Harbin University of Science and Technology (2016)

The retracted version of this chapter can be found at
https://doi.org/10.1007/978-981-15-1377-0_17

© Springer Nature Singapore Pte Ltd. 2020
Y. Sun et al. (Eds.): ChineseCSCW 2019, CCIS 1042, p. C1, 2020.
https://doi.org/10.1007/978-981-15-1377-0_61

Author Index

Printed in the United States
By Bookmasters